21st Century Chemistry

SECOND EDITION

Kimberley Waldron
Regis University

w. h. freeman
Macmillan Learning
New York

Vice President, STEM: Daryl Fox
Editorial Program Director, STEM: Brooke Suchomel
Senior Program Manager: Beth Cole
Developmental Editor: Allison Greco
Marketing Manager: Maureen Rachford
Marketing Assistant: Savannah DiMarco
Director of Content: Kristen Ford
Media Editor: Amanda Nietzel
Senior Media Project Manager: Chris Efstratiou
Director, Content Management Enhancement: Tracey Kuehn
Senior Managing Editor: Lisa Kinne
Senior Workflow Project Supervisor: Susan Wein
Senior Content Project Manager: Edward J. Dionne
Project Manager: Sumathi Kumaran, Lumina Datamatics, Inc.
Director of Design, Content Management: Diana Blume
Creative Director: Emiko-Rose Paul
Cover Design Manager: John Callahan
Cover Design: Lumina Datamatics, Inc.
Chapter Opening Art: Quade Paul, Echo Medical Media
Art Manager: Matthew McAdams
Senior Photo Editor: Cecilia Varas
Illustrators: Quade Paul and Tom Webster
Photo Researcher: Brittani Morgan, Lumina Datamatics, Inc.
Composition: Lumina Datamatics, Inc.
Printing and Binding: LSC Communications
Cover Image: Ionov Artem/Shutterstock

ISBN-13: 978-1-319-10617-1
ISBN-10: 1-319-10617-X

Library of Congress Control Number: 2018960105

Printed in the United States of America

1 2 3 4 5 6 22 21 20 19 18

W. H. Freeman and Company
One New York Plaza
Suite 4500
New York, NY 10004-1562
www.macmillanlearning.com

For Michele.

Brief Contents

Contents

Preface

Chemistry moves fast. From one week to the next, new findings can shift the path that chemistry research takes. Where the discipline of chemistry was once divided, dependably, into several distinct subdisciplines and chemists found a home in one of them, today's chemists often straddle two or more subdisciplines and frequently reach over into neighboring sciences. Since our field has become so fluid and so flexible, right now is one of the most exciting times to be a chemist or to be learning chemistry.

The first edition of this textbook was published in 2014 and, since then, the field of chemistry has continued to spread out and mix with other sciences. The field is moving in directions no one could have anticipated in 2014, and so it is important to continuously update and invigorate textbooks like this one with up-to-the-minute material, including a new type of question at the end of each chapter. These new and relevant "internet research questions" ask students to investigate their own neighborhoods and towns to see where this new content fits into their everyday lives and experiences.

There are several people who have made this second edition possible: Program Manager Beth Cole for her dedication to my vision for this book, and for her thoughtful attention to all aspects of publication; Development Editor Allison Greco for coordinating all changes made to this edition; Project Managers Ed Dionne and Sumathi Kumaran and Workflow Supervisor Susan Wein, for keeping us all on track, and for their endless patience.

I owe an enormous debt of gratitude to Emiko and Quade Paul, the artists whose beautiful art program and chapter openers continue into the second edition. The following individuals helped to continue that legacy in this revision: John Callahan, who designed the cover, and Cecilia Varas and Brittani Morgan for tracking down beautiful photos that bring the book's objectives to life.

I would also like to thank Priscilla Burrow Crocker from the University of Colorado, Denver, whose answers to the end-of-chapter questions carried from the first edition to the second. I am grateful to Priscilla, Jacob White, Jason Dunham, Stephen Tsonchev, Teri Hall, and AJ Kwolek for authoring content for the LearningCurve platform that will accompany this edition.

Most of all, I wish to thank my family. Without their patience and support, writing books would not be possible.

Print and Media Resources

Student and Instructor Support

Supplemental learning materials have been designed to engage students with the content and to provide practice with key concepts, all in a streamlined and easy-to-use package.

Instructor Supplementary Materials

Valuable teaching tools are provided with the text to support instructors.

- Images from the text in PPT and PNG files
- Lecture Slides in PPT

Student Supplementary Materials

Achieve Read & Practice for Waldron, *21st Century Chemistry*, Second Edition, provides a streamlined and affordable online experience for students. Read & Practice focuses students on the most important concepts in the course and includes:

- complete enhanced e-book with note-taking, highlighting, and search capability.
- adaptive quizzing in LearningCurve for each chapter.
- links to video content that relates content to our environment and everyday life.

☑ Put "testing to learn" into action. Based on research, LearningCurve really works: Game-like quizzing motivates students and adapts to their needs based on performance. It's the perfect tool to engage students before class and review after! Additional reporting tools and metrics help teachers to analyze what their class knows and doesn't know. Each LearningCurve question links back to the relevant sections of the e-book so students can review content directly from their text. Innovative scoring ensures that students who need more help with the material are given more practice than students who are already proficient.

A Look Inside

Chapter Openers. Energy-generating dance floors, the world's first lab-grown hamburger, graphene, disappearing gravestones, and the Fukushima-Daiichi power plant disaster: the chapter openers in this book explore every conceivable corner of chemistry. But they all have one thing in common: They draw students into the chapter, gently, and introduce them to the key topics that will be covered in the chapter in a seamless and engaging way. Each chapter opening story is tucked beneath a dramatic two-page mosaic of artwork that illustrates the opening story and selected concepts from the chapter. Each of these black-and-white montages is overlaid with watercolor, offering a vibrant prelude to the chapter.

GH1/Pavegen/WENN.com/Newscom

David Hall/USDA ARS

natureboxes. Embedded in the center of every chapter, natureboxes forge a link between nature or the environment and the material introduced in the text. These boxes are more than asides; they are short, cogent essays that ask readers to apply what they have just learned to the real world outside the pages of the text. They offer a mid-chapter respite and an opportunity for students to take fresh knowledge and apply it in a gratifying way.

The Green Beat. *The Green Beat* is a short, scientific news article, intended to be like a newpaper clipping. Placed at the end of every chapter, *The Green Beat* explores a news story related to the environment. And, for a complete understanding of each story, the student needs the essential chemistry taught to them in that chapter. *The Green Beat* feature wraps up the chapter, giving the student the chance to apply what they know to a science-based news story.

Sakis Papadopoulos/Getty Images

QUESTION 8.4 True or false? A hydrogen bond is a covalent bond.

ANSWER 8.4 False. A covalent bond is a bond within a molecule. A hydrogen bond is a type of intermolecular force.

For more practice with problems like this one, check out end-of-chapter Question 19.

FOLLOW-UP QUESTION 8.4 Do you think it is possible for a hydrogen bond to form within a molecule, as shown here?

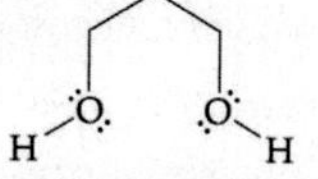

In-Chapter Practice. As students make their way through each chapter, they can pause and check their grasp of the material. The in-chapter questions, which are usually conceptual in nature, are sprinkled liberally throughout the book. Answers are provided for all questions and each question is keyed to similar questions in the end-of-chapter material. In addition, each in-chapter question is followed by a follow-up question that also assesses student understanding of the last section that they read. Answers to follow-up questions are provided at the end of every chapter.

Molecular-to-Macro Illustrations. A strong art program supports the chemistry in the book. Clear molecular art reinforces key concepts and provides a visual snapshot for the student to remember. Whenever possible, the student is shown how the molecular view of chemistry connects to the scale of everyday life—the macro scale. With art, the links between structure and function are illustrated in full color.

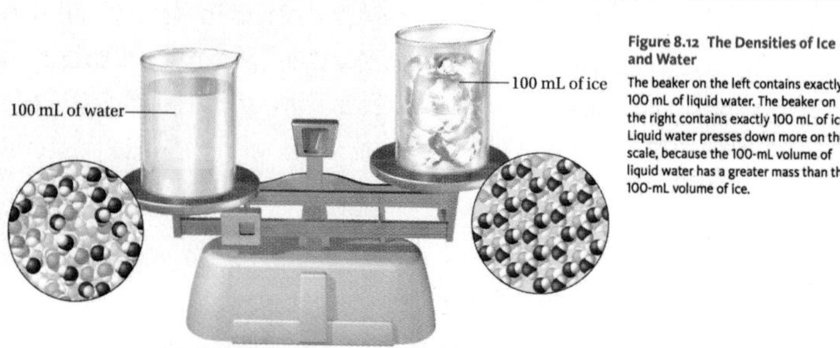

Figure 8.12 The Densities of Ice and Water

The beaker on the left contains exactly 100 mL of liquid water. The beaker on the right contains exactly 100 mL of ice. Liquid water presses down more on the scale, because the 100-mL volume of liquid water has a greater mass than the 100-mL volume of ice.

wait a minute . . .

Why don't we remove the salt from salt water to make freshwater?

We *can* remove the salt from salt water in facilities called desalination plants. Worldwide, more than 15,000 desalination plants produce more than 1.58×10^{10} gallons of freshwater each year. This may seem like a lot of water, but it's only a tiny fraction of the total freshwater used on Earth annually. If desalination works, why aren't desalination plants turning more salt water into freshwater? Because the plants are extremely expensive to build, and

Wait a Minute . . . . Every now and then, a student will pause and think, perhaps wondering about some facet of the material just introduced. We have tried to anticipate questions that naturally pop into curious minds. The Wait a Minute . . . feature pauses the text to ask that question, and then explains the answer in detail. Because these questions often detour away from the main thread of the text, they are framed in boxes and can be skipped.

pery because it is made
arbon atoms that slide across one another. Diamond is extremely
t is composed only of strong carbon–carbon linkages in a matrix
ional tetrahedrons. This important theme was first introduced in
e'll see many other examples of it throughout this book.

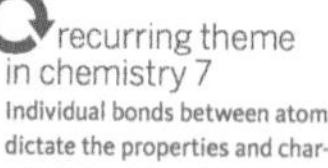

recurring theme in chemistry 7

Individual bonds between atoms dictate the properties and characteristics of the substances that contain them.

cording to Table 5.1, it takes 0.624 aJ of energy to break one single car-
. Given this, how much energy, in attojoules, is required to break 100

flashback

See *The Green Beat*, Chapter Five Edition, for more on the topic of carbon capture and sequestration (CCS).

In one day, a coal-burning pow
kilograms of sulfur dioxide from its
ide is going into the air and formin
are producing a lot of calcium sulfa

Recurring Themes. There are truisms in this book that come up over and over again in the study of chemistry. For example: "Chemical reactions are nothing more than exchanges or rearrangements of electrons." These axiomatic statements, which we call Recurring Themes, reappear in new contexts throughout the book. They are the fundamental, take-away ideas that we hope every student will remember long after they finish their course in chemistry.

Flashback. From time to time, the reader needs material from a previous chapter to understand new material. Rather than providing a simple back reference, we have created the Flashback, a margin note that succinctly explains the earlier concept and also provides the back reference for students who want to review earlier material. Each Flashback enables students to continue reading, refreshing concepts along the way. This enhances the flow of each chapter and re-introduces significant ideas in new contexts.

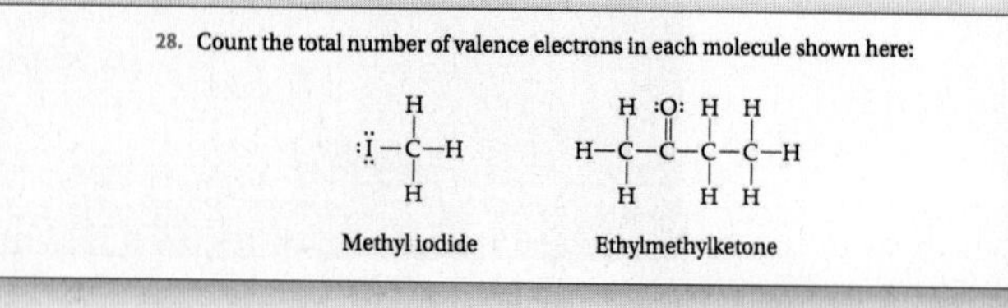

28. Count the total number of valence electrons in each molecule shown here:

More practice! There is a generous collection of questions at the end of every chapter. They are grouped according to topic, and also according to level of difficulty.

NEW to the second edition are internet research questions. These interactive, critical-thinking exercises encourage student participation and bring the book's major objectives to life.

About the Author

Trevor Brown Photography

Kimberley Waldron is a Professor of Chemistry at Regis University in Denver, Colorado, where she teaches courses at all levels. Her area of specialty is inorganic chemistry. Professor Waldron was raised in the Washington, DC, area and completed her undergraduate work at the University of Virginia, where she did undergraduate research in analytical chemistry. She worked as an analytical chemist at E.I. DuPont de Nemours before entering graduate school at Virginia Commonwealth University. She received a Ph.D. in inorganic chemistry with a specialization in biological chemistry. Professor Waldron then became a post-doctoral fellow at the California Institute of Technology in Pasadena, California, before moving to Boulder, Colorado, to work in the biotechnology industry. She joined the faculty at Regis University in 1995. Professor Waldron lives in Denver with her family, including two cats, one dog and two kids.

age fotostock/age fotostock

The Story of Chemistry

The Scientific Method: Think, Measure, Rethink

Lawsonia inermis

CHAPTER 1

This book teaches chemistry with stories because stories provide relevance and context to this sometimes complex topic. Our first story takes us to India, where a wedding is about to take place and the bride is preparing. On this day before her wedding, she is devoting all of her time not to the food or the music or the venue for the wedding, but to tattoos. She waits patiently as intricate tattoos that she has designed are applied to her hands and feet.

These tattoos (see photo on p. 2) are made with henna, also called *mehndi,* a reddish brown pigment that wears off of the skin within a few weeks. The henna is not just beautiful, it is also thought to have antiseptic properties that keep the bride healthy. It is also believed to be a powerful aphrodisiac.

Several steps are involved in making henna tattoos. First, the henna is obtained from the *Lawsonia inermis* bush by soaking its leaves in lemon juice. Then, when it is applied, the henna attaches to substances in skin. Finally, as the decorated skin is exposed to air, the henna turns darker in time for the ceremony. In each of the three steps in this henna tattoo process, chemistry is involved. A colored substance is extracted from a plant with the aid of lemon juice, the substance binds to specific structures in the skin, and then the substance changes color in air. Chemistry, chemistry, and chemistry.

As this story demonstrates, chemistry is not limited to beakers and laboratory benches. Chemistry occurs wherever substances are, whether henna is being prepared for a wedding or a fuel is being burned in an engine. Wherever something is rusting, exploding, glowing, rotting, or dissolving, some chemistry is going on. And, as we will see, an understanding of these chemical processes is useful no matter who you are or who you plan to be. Perhaps most important, though, an understanding of chemistry will make you a better-informed citizen and help you to cast thoughtful votes on topics such as the environment, public health, food, and energy.

age fotostock/age fotostock

In this first chapter, we explore the process of science and what scientists do on a day-to-day basis. Then you will learn how they disseminate their work and how others are encouraged to challenge it. Finally, you will learn to use some tools that all scientists must master and see how scientists evaluate the information that they collect.

1.1 The Scientific Method

This book will help you understand current science-related issues.

Major changes in the way people think about nature can alter the course of history. The twentieth century saw significant, world-changing advances in our understanding of nature, especially the characteristics of **matter**, the tangible material that makes up the universe and everything in it. Careful scientific work revealed the structure of the **atom**, the smallest unit of matter that cannot be broken down, except in nuclear reactions. One discovery followed quickly upon the last, until our understanding of the atom was completely new. These scientific advances were made using a method of inquiry that scientists have developed and tweaked over centuries of experimentation. (We explore this method of inquiry—the scientific method—later in this section.)

In the middle of the last century, our newfound understanding of the atom led us to the point where we could use nuclear reactions to split it. This knowledge eventually brought World War II to a dramatic end, when the United States developed the first atomic bombs and dropped them on the Japanese cities of Hiroshima and Nagasaki in the summer of 1945 (**Figure 1.1**). Since then, our understanding of the atom has allowed us to tap its energy for peaceful uses, such as the generation of electrical energy in nuclear power plants. Chapter 11, "Nukes," will explore the topic of nuclear reactions.

AFP/Getty Images

Figure 1.1 The Intensity of an Atomic Bomb

The blast from the Hiroshima bomb created enough light to make an imprint of a valve handle on the wall of the building behind it. The bomb fell on August 6, 1945.

Because nuclear power plants produce dangerous waste products, however, the use of nuclear energy for power has been, and continues to be, one of the most divisive areas of public debate. We must decide if the power we get from nuclear plants is worth the risk of producing the hazardous waste. Countries all over the world that rely on nuclear power, such as France, Hungary, South Korea, and Belgium, have had to ask and answer tough questions. Do our power plants put people at risk from potentially faulty waste receptacles? From sabotage by terrorists? From power plant accidents?

Some of these complex questions were answered for the government of Germany after an earthquake that reached 8.9 on the Richter scale was followed by a devastating tsunami on the northeast coast of Japan. The tsunami severely damaged the Fukushima power plant on March 11, 2011, and the release of radioactive substances increased the tsunami's death toll, which eventually rose to more than 15,000. In the wake of this disaster, the German government decided that, even though German nuclear power plants are not vulnerable to a tsunami, the facilities are not worth the risk they pose to the citizens of their country. By German government decree, by the year 2022, every nuclear power plant in Germany will be closed.

If you were a German citizen, how would you feel about this decision? Do you live near a nuclear power plant now? Do you benefit from the low-cost energy it provides? You may someday need to weigh in on questions like these, and that's one of the reasons it is useful to know how science works. Reading this book will help you develop the critical thinking skills you will need to form an educated opinion about questions like these. In your lifetime, there will be many issues that will force policymakers and the public to decide how far our knowledge should be allowed to take us. That is part of what this book is about: providing you with some of the fundamentals necessary to understand scientific issues.

Can you understand a newspaper article discussing a topic such as nuclear power? Do you know how to inform yourself about issues concerning the environment? Can you spot a ruse designed to influence you to buy a dubious product? Hopefully, after reading this book, your answers to these questions will be yes, and taking this course will help you learn to intelligently evaluate issues and products and come to conclusions that fit your beliefs and your way of life. And ultimately the approach used in this book, which teaches you chemistry by showing you chemistry, will make you a better voter, parent, employee, and citizen.

Scientists follow a system of peer review and reproducibility.

One question people may hear in public debate is, "Should scientists be permitted to pursue any avenue of research they choose?" Should researchers be permitted to clone any organism or create any molecule, regardless of the dangers that organism or molecule might pose? Should governments take control of the natural progression of science? What *is* the natural progression of science? How do scientists decide what is important to study, which questions should be answered, and which should be ignored?

The questions a scientist chooses to work on can depend on several things. First, as the natural momentum of scientific research rolls along, scientists often are tempted to try answering the next logical question. For example, if energy research progresses to a point where a new fuel source is discovered, a logical next step is to find out as much as possible about this fuel. A second criterion a scientist often uses for selecting a research project is how important the topic is to the well-being of the human race.

Each scientist usually addresses only a very small part of a larger problem. For example, consider a chemist who is trying to figure out why a certain substance in aerosol cans harms the environment. This chemist may decide to change the compound slightly to see if the adverse effects still occur. A second chemist working independently on the same problem may try to figure out what other, more benign substances can be substituted in the aerosol formulation. A third chemist may study how that substitute substance affects the chemistry of natural waters. As experiments are completed, each scientist submits his or her findings to a scientific journal. The journal editors then send out the findings for extensive peer review, and other scientists evaluate it (**Figure 1.2**).

Peer reviewers scrutinize the data to be sure that the experiments are designed properly and the most logical conclusion is reached. Sometimes the peer reviewers may repeat the experiment to verify that the results are reproducible in another laboratory. If the work is determined to be acceptable, it is published and becomes available to the wider scientific community.

The three scientists working on the questions relating to aerosol cans and the environment can then read one another's findings in the literature and update their own experimental designs accordingly. In this way, small, incremental gains

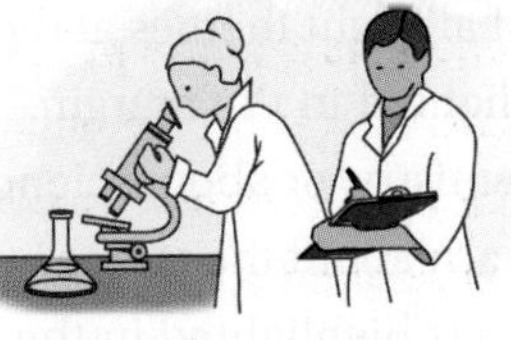
Scientists study a question.

Scientists write
about their results.
Editor receives an article
and sends it out for peer
review.
Peer reviewers read
the article and
provide feedback
to the editor.
Editor may send reviewer comments to
the scientists who may then revise and
resubmit the article for further
review. If an article does not maintain
sufficiently high scientific standards, it
may be rejected at this point.
If an article finally meets
editorial and peer standards,
it is published.

Figure 1.2 The Peer Review Process

are made toward a broader understanding of the original question. Breakthroughs happen and make way for newer breakthroughs. This whole approach to doing science—including peer review and verification of reproducibility—is how science gets done, and notable findings eventually filter out to the public.

The scientific method tests scientific hypotheses.

The general approach to scientific experimentation used worldwide is the **scientific method**. Let's look at an example of how it works. Suppose you are having coffee at a bookstore. You overhear a debate about whether a piece of toast accidentally knocked off a kitchen table lands more frequently buttered side up or buttered side down. You know that the prevailing consensus is that toast always lands buttered side down, but you are not convinced that it's true.

You decide to perform a series of exhaustive bread and toast experiments in your kitchen. You begin by formulating a **hypothesis,** which is an initial best guess about some question concerning nature.

> **Your Hypothesis:** The top side of a slice of bread or toast knocked off a kitchen table always lands facing the floor.

It's important to note that a scientific hypothesis is different from other kinds of hypotheses because it can be tested by experiment. If you ever take a class in philosophy, you may be asked to discuss some hypothesis about the meaning of life, say, or the existence of an afterlife. This type of hypothesis is not a scientific hypothesis, because it cannot be tested by experiment.

recurring theme in chemistry 1

Scientists worldwide use the scientific method as a basis for experimentation and deduction. The scientific method is based on the principles of peer review and reproducibility.

There are eight recurring themes that are woven throughout the book. (You can find a list of all eight themes at the front of the book on p. xv.) The first of these themes is highlighted in the margin. The recurring themes are fundamental concepts about chemistry, or about science in general, that will come up repeatedly as you read and learn about the topics in this book. Because they are important, each recurring theme is highlighted in the text near the discussion it accompanies. All the recurring themes for each chapter are also listed in the material at the end of the chapter.

To prepare your kitchen-laboratory, you stock up on bread and butter and then set to work. Done correctly, your experiments should convince even the most stubborn skeptic. To that end, it's important to anticipate every objection that might arise when you unveil your results. Therefore, you design an experiment to include every combination of buttered and unbuttered bread and toast, as shown in **Figure 1.3**.

Figure 1.3 Toast and the Scientific Method
This kitchen-based experiment tests the hypothesis that the top of a piece of toast always lands facing the floor.

The obvious way to begin your experiment is by pushing slices off a typical kitchen table. Because table height might be an important factor, you measure the height of yours. Having decided to start with buttered toast, you then slide a slice, buttered side up, off the table. You repeat the experiment 19 more times and record your results. To be as consistent as possible, you try to use the same force and wrist motion each time.

You're not finished yet, however. You repeat your experiment with unbuttered toast, buttered bread, and unbuttered bread, doing 20 trials each time and recording the results. With the buttered bread, you again begin with the buttered side up. With the unbuttered bread and unbuttered toast, you put a tiny spot of red ink on one side and make it the side that initially faces up in each trial. You also try the experiment with tables of different heights to see if it affects your results. After five hours of flinging toast and bread off of tabletops, you have all your data.

QUESTION 1.1 Which of these statements are scientific hypotheses? (a) People get sunburns when they are exposed to yellow light. (b) Every human being has a soul. (c) You can drive from Chicago to Philadelphia in 12 hours while obeying the posted speed limits.

ANSWER 1.1 Choices (a) and (c) are scientific hypotheses because they can be tested. Choice (b) is a hypothesis but not a scientific one, because it cannot be tested.
For more practice with problems like this one, check out end-of-chapter Question 3.

FOLLOW-UP QUESTION 1.1 In the toast experiment, why is it important to test bread that is buttered and unbuttered?

A theory is a well-substantiated and tested hypothesis.

Your data reveal the following facts:

1. With the buttered side initially facing up, both buttered toast and buttered bread land buttered side down 90% of the time from your kitchen table.
2. If the toast and bread are unbuttered, the side of the bread that is initially facing up, as indicated by the red dot, ends up facedown.
3. Table height does matter. Slices that fall from tables only slightly taller than your kitchen table land buttered/red-dot side up, and slices that fall from extremely tall tables land buttered/red side down.

Your hypothesis was right! Buttered bread lands buttered side down 90% of the time. But, you reason, the butter does not cause the bread/toast to land buttered side down. Even without butter, the side that initially faces up is the one that hits the floor. And, whether the bread is toasted or not is irrelevant. What *is* relevant is the table height.

Now you are ready to propose a **theory,** which is a detailed explanation of the experimental results. Having observed that each slice rotates as it falls, you formulate the following theory: table height determines how many rotations can be completed before the slice hits the floor. Your recommendation to the clumsy, buttered-toast-eating public is this: eat your toast at a slightly taller table. This buttered toast example may be tongue in cheek, but it demonstrates how a theory can develop from a testable hypothesis.

A theory can never be proved. Did we prove anything with our toast "experiments"? Can a scientist ever prove something by experimentation? The answer to all these questions is no. A theory provides the latest and best interpretation of experimental results, but theories are not inherently provable, because the next set of experiments may show that the theory is wrong. In other words, a theory is a hypothesis that has been found to be true in the most recent experiments. Next year, for example, someone may devise a new method of measurement that could possibly shed more light on the question and give researchers an opportunity to use a different set of experiments to test the hypothesis. Thus, we can say that evidence supports a theory, and we can say that evidence disproves a theory, but we cannot say that evidence proves a theory.

QUESTION 1.2 Describe the difference between the words *hypothesis* and *theory* in science.

ANSWER 1.2 In science, a hypothesis is a best guess or preliminary prediction of the outcome of one or more experiments. You start with a hypothesis and then test it to determine if the data you collect supports it or not. A theory explains the data that are collected, whether from one experiment or several.

For more practice with problems like this one, check out end-of-chapter Question 2.

FOLLOW-UP QUESTION 1.2 Complete the sentence with one of the choices provided. A scientific theory is (a) true forever, no matter what (b) subject to revision with further experimentation (c) an observation about nature.

Scientists create models that make experimental data easier to understand.

In chemistry, it's sometimes useful to represent atoms as balls connected by sticks. But remember that a ball-and-stick structure like this is a **model** invented by scientists to classify and organize matter. Models are representations of reality and a way of understanding nature. After reading this book, you may find yourself picturing atoms as balls connected to one another by sticks. Keep in mind, though, that these balls and sticks are models, not reality.

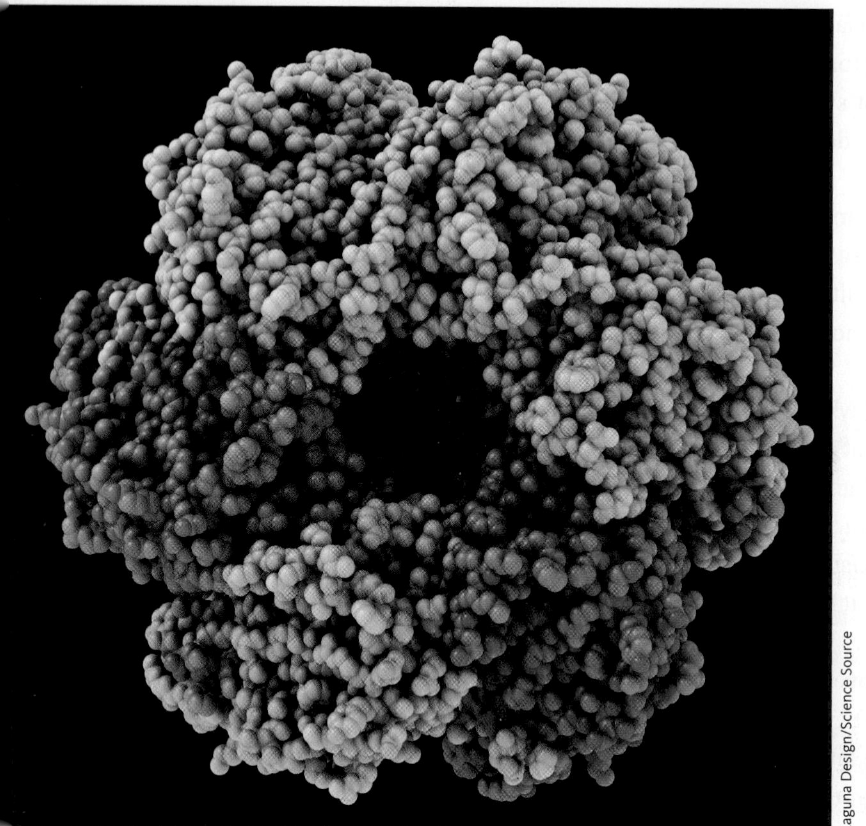
Laguna Design/Science Source

Figure 1.4 A Computer Model of a Complex Substance
This substance contains hundreds of thousands of atoms. Each ball in this computer model represents one atom.

Advances in computer technology have brought models from balls and sticks to beautiful three-dimensional graphics. With the storage space and processing speeds available today, a reasonably priced computer can display magnificent, elaborate chemical structures containing hundreds of thousands of atoms (**Figure 1.4**). With today's computing power, computational chemists often design experiments that are done completely on a computer, not in a laboratory. Every year, more and more experiments like this are performed on computers, and this capability can save months of work in the laboratory.

Are computational chemists subject to the same rules of peer review and reproducibility as chemists who work in laboratories and wear white coats? Certainly. The fundamental ideas behind the scientific method apply to every corner of the scientific community, even corners that are utterly virtual. The scientific method works because every piece of data is criticized and questioned. Findings that survive this process become the new benchmarks until the next piece of information that supersedes or clarifies what came before is determined, and new benchmarks are set. That is the way science happens.

1.2 Coming to Terms with the Very Large and the Very Small

recurring theme in chemistry 2

Things we can see with our eyes contain trillions upon trillions of atoms. Atoms are much too small to view with the naked eye.

Scientific notation allows us to work with really large and really small numbers.

A small glass of water like the one depicted in **Figure 1.5** contains about 7.9×10^{24} water molecules. (A *molecule* is a collection of atoms that are connected to one another. In Chapter 4, "Bonds," this term is defined in a formal and detailed way.

For now, all you need to know is that a water molecule includes three atoms, and it's extremely small.) Because water molecules are extremely small, we can rightly assume this glass contains an enormous number of them. But how large is the number 7.9×10^{24}, and why do we write it using a power of 10?

Figure 1.5 Molecules in a Glass of Water

When a number is exceedingly small or large, scientists use **scientific notation** for convenience. For example, to express the number 457,000 in scientific notation, we move its decimal point to the location that results in a number between 1 and 10, which in this case is 4.57. We refer to this as the *number part.* Next, we express the number of places we moved the decimal point as 10 raised to that power: 4.57×10^5. We refer to this as the *exponential part* (pronounced *ex-po-NEN-shul*).

When we start with a large number and move the decimal point of the original number to the left, as we did here, the exponent is positive. When we start with a number that is less than 1, we move the decimal point to the right to obtain a number between 1 and 10. In these cases, the exponent is negative because the number is small. For example, 0.000 004 57 becomes 4.57×10^{-6}.

Let's return to our glass of water molecules. We now know that the notation 7.9×10^{24} represents a very, very large number of water molecules. In this example, the decimal point of the original number has been moved 24 places to the left of its original location, like this:

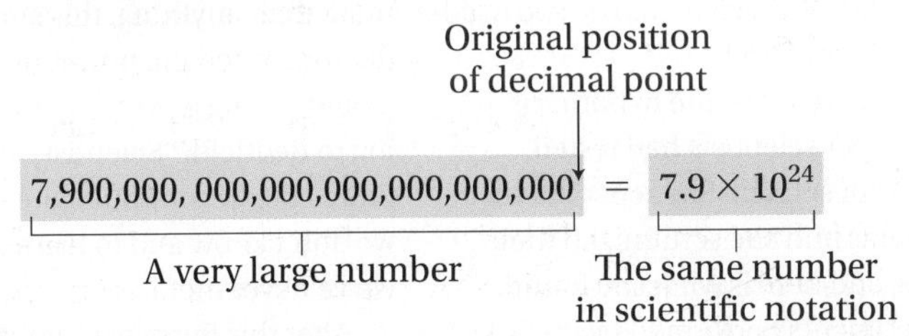

QUESTION 1.3 Label each of the following numbers in scientific notation as either "very large" or "very small.": (a) 7.70×10^{28} (b) 5.41×10^{-11} (c) 4×10^{-9} (d) 3.1×10^{12}

ANSWER 1.3 Choices (a) and (d) are very large numbers because they have positive exponents. The numbers in (b) and (c) are very small numbers because they have negative exponents.

For more practice with problems like this one, check out end-of-chapter Questions 9 and 10.

FOLLOW-UP QUESTION 1.3 For each of the following large numbers, determine what the value of the exponent is when the number is expressed in scientific notation: (a) 5,100,000 (b) 1,000,000,000,000 (c) 3000 (d) 1,234,567

natureBOX • Alien Bacteria and the Scientific Method

Rosie Redfield is a detective, of sorts. She starts with a mystery that has not been solved...yet. Then she asks a question about the mystery, a question that can be answered with facts. She collects evidence, evaluates that evidence, and draws conclusions from it. Finally, she attempts to solve the mystery by using those facts. If the mystery remains unsolved, then she asks the next logical question, until she cracks the case. Is Rosie Redfield a homicide detective or a police investigator? No, she is a scientist (**Figure 1A**). But in many ways her job is similar to a police detective's, as you are about to see.

New scientific findings often do not make their way into the public sphere. But every now and then, a scientific discovery is so astonishing, it makes headlines everywhere. That's what happened on December 2, 2010, when a press conference was called at the National Aeronautics and Space Administration (NASA) headquarters in Washington, D.C. For days, rumors of something big had been bouncing around the astronomy community; blogs predicted that there might be news of life on one of the moons of Saturn. There was a buzz of excitement as the NASA scientists took the stage.

The scientists announced that NASA researchers had discovered a bacterium—a one-celled organism—in a California lake that could survive in arsenic-laden water. Why was this obscure bacterium such a big deal? It was a big deal, NASA said, because these remarkable little organisms actually live on arsenic, and arsenic is known to be very toxic to many life-forms. In fact, NASA claimed that the bacteria could swap arsenic for phosphorus, a critical part of the scaffolding that holds together the genetic material of all known living things. (Genetic material stores an organism's hereditary information.) This was indeed big news because there never has been evidence to show that the phosphorus in our genetic material can be replaced by anything, let alone arsenic! A wave of excited electronic chatter reverberated through the scientific community. Arsenic can replace phosphorus? Really?

What made this announcement so exciting was its implication that living among us is an organism that originally may have come from a place where arsenic is used as the basis for living things. Wild speculation ensued: Had alien life-forms been living in a lake in California, and we were only now learning of their existence? If this were true, then could other, more sinister alien life-forms be living on Earth with us?

© Martin Dee

Figure 1A Rosie Redfield: Detective and Troublemaker

Enter Rosie Redfield, our scientist-detective. She wanted to know this: Is it actually possible to replace the phosphorus in these cells for something else? Suspecting that this claim was false, she asked a question—"How did NASA arrive at its results?"—and she decided to answer it. Then she did what scientists do: she used the scientific method.

She wanted to test the hypothesis that the NASA scientists had tested, starting from scratch. She repeated their experiments and she scrutinized their methods, and here is what she found: The experiments performed by the NASA scientists did not provide the evidence needed to support their conclusions. They had not followed the scientific method carefully, and their work was not reproducible. NASA had cut corners, and soon everyone knew it.

Other groups of scientists weighed in as the story unfolded. Working in their own labs and thinking hard about the problem, they posted their latest findings and insights on blogs around the clock. Every new blog post within the scientific community was a new morsel of information for everyone to digest. It was an exciting time for those involved. In fact, more than anything, this story may have demonstrated the power of social media in addressing a pressing issue. According to Redfield, "Scientists are much more able to communicate with people we don't know and to learn from people we've never met."

After this flurry of scientific detective work, the mystery was solved. People can rest easy knowing that bacteria with arsenic-based genetic material are not known to exist after all. Indeed, as far as we know, every organism on Earth does, in fact, have phosphorus, not arsenic, in its genetic material. Happily, the fundamental tenets of chemistry regarding the role of phosphorus have not been violated by alien bacteria in a California lake.

There are two rules to remember when using scientific notation.

Figure 1.6 summarizes scientific notation. The first rule is that whenever we write a number in scientific notation, the "number part" must be equal to or greater than 1, and it must be less than 10. We place the decimal point right after the first digit—which is 7 in the preceding example (7.9). Note that we do not move the decimal point one place to the left of the first digit (0.79) or one place to the right (79). We must always have a number between 1 and 10.

Here is the second rule: whenever we write a number in scientific notation, the "exponential part" is the number 10 raised to the number of positions we moved the decimal point. In our example, the power is +24, and so the number is written 7.9×10^{24}. Remember that the exponent is positive for big numbers (greater than one), and negative for small numbers (less than one). In 1.24×10^{-26}, for example, the negative exponent tells us that the decimal point has been moved to the right of its original position:

Original position of decimal point

$$0.000\ 000\ 000\ 000\ 000\ 000\ 000\ 000\ 0124 = 1.24 \times 10^{-26}$$

A very small number — The same number in scientific notation

Figure 1.6 Using Scientific Notation

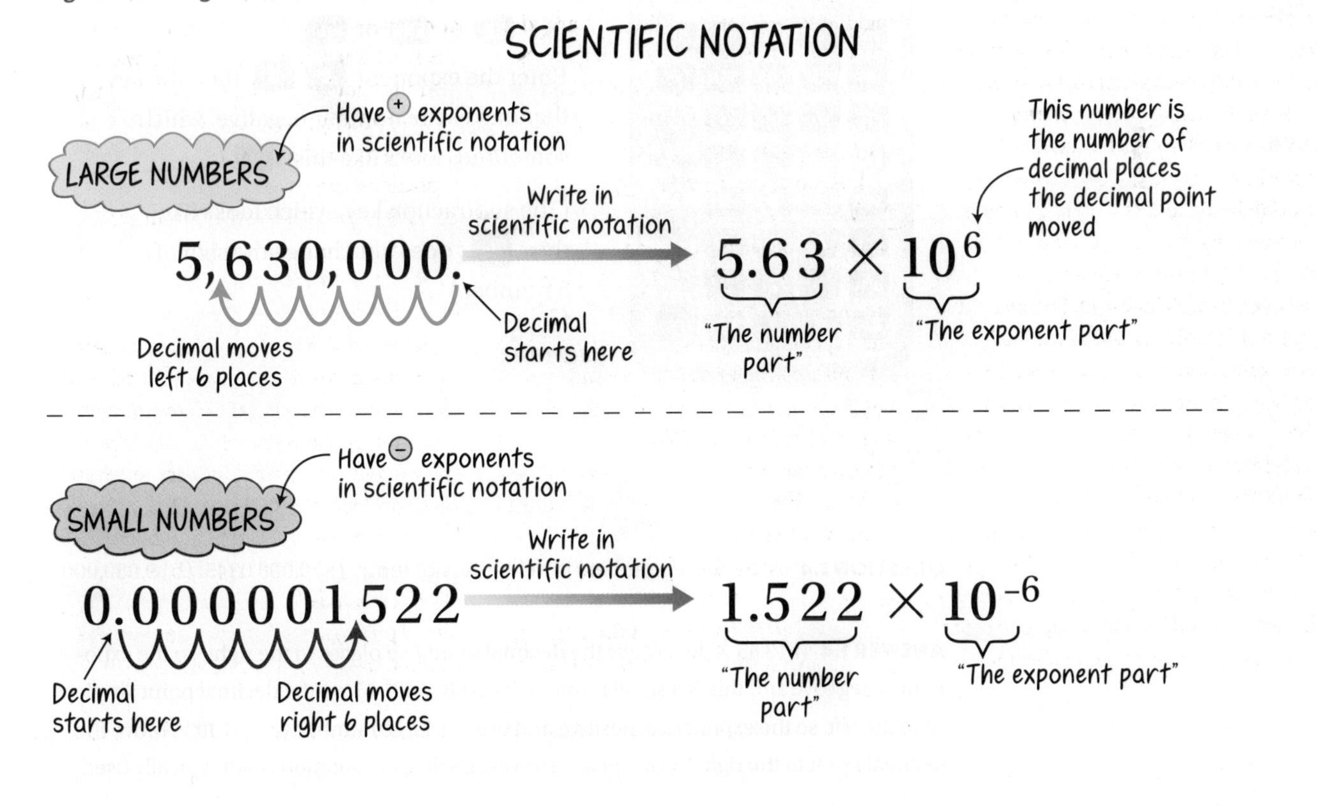

wait a minute . . .

How do I enter scientific notation into my calculator?

You will save time and keystrokes if you learn to use scientific notation with your calculator. Dozens of calculator brands are available, but most scientific calculators have a button marked as *EE* or *EXP* or $\times 10^y$ (**Figure 1.7**). To enter a number written in scientific notation, enter the decimal number, press the *EE/EXP/*$\times 10^y$ key, and then enter the exponent. For example, for 1.24×10^{-26} you enter 1.24, then EE/EXP/$\times 10^y$, then 26, then the key that makes the number negative, which sometimes looks like this: +/−. (The subtraction key looks like this: −. It does not change the sign of a number.)

Often, when students arrive at a number in a calculation that is incorrect by a factor of 10, it's because they enter the 10 part of 10^{-26} instead of just the −26 part. (Doing so multiplies the decimal number by an extra 10.) *It is not necessary to enter the number 10 when you use the EE/EXP/*$\times 10^y$ *key.*

You will sometimes see numbers in scientific notation where the first number is 1. In writing such a number, the 1 is often omitted for convenience. So, writing 1×10^{-19} is the same as writing 10^{-19}. To enter the number 10^{-19} into your calculator, however, you must include the 1: enter 1, then EE/EXP/$\times 10^y$, then −19. •

Figure 1.7 Entering a Number in Scientific Notation into a Calculator

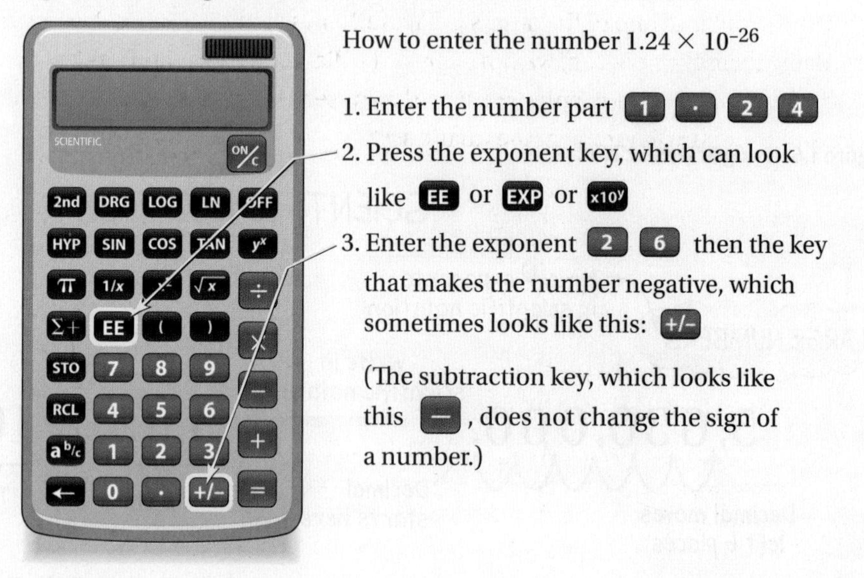

SCIENTIFIC
How to enter the number 1.24×10^{-26}
1. Enter the number part 1 . 2 4
2. Press the exponent key, which can look like EE or EXP or x10y
3. Enter the exponent 2 6 then the key that makes the number negative, which sometimes looks like this: +/-
(The subtraction key, which looks like this −, does not change the sign of a number.)

QUESTION 1.4 Write these numbers in scientific notation: (a) 0.000 0345 (b) 9,080,000 (c) 0.2

ANSWER 1.4 (a) 3.45×10^{-5}. Move the decimal point five places to the right, so the exponent is negative and this is a small number. (b) 9.08×10^{6}. Move the decimal point six places to the left, so the exponent is positive and this is a large number. (c) 2×10^{-1}. Move the decimal point to the right by one place. However, scientific notation is not typically used for numbers this close to 1, because it is more straightforward to write 0.2 than to write 2×10^{-1}. Scientific notation is usually reserved for very large and very small numbers.

For more practice with problems like this one, check out end-of-chapter Questions 11 and 15.

FOLLOW-UP QUESTION 1.4 What is incorrect about the number 88.6×10^5?

1.3 Metric Units, Conversion Factors, and Dimensional Analysis

Chemists work with very large numbers of atoms because atoms are very small.

As our discussion of scientific notation has made clear, molecules and atoms are exceedingly small. They are so small, in fact, that microscopes with the necessary resolution to view images of single atoms have become available only recently. **Figure 1.8a** is an image from such a microscope, a STEM (scanning tunneling electron microscope), at extraordinarily high resolution. In this microscopic image, images of individual atoms are visible. Each pink circle encloses a ring of six atoms. The rings are interconnected in a honeycomb pattern, just like the hexagons in an authentic bee-made honeycomb in **Figure 1.8b**. But there is a big difference between these two kinds of honeycomb: we can fit more than 10,000,000 of the hexagons from Figure 1.8a across one hexagon in Figure 1.8b. That's how small an atom is.

The diameter of one of the atoms in Figure 1.8a is about 0.000 000 000 08 meters. Using scientific notation, we can express the diameter of that atom as 8×10^{-11} m. In this section of the chapter, we introduce a second tool for managing really large and really small numbers: the metric system.

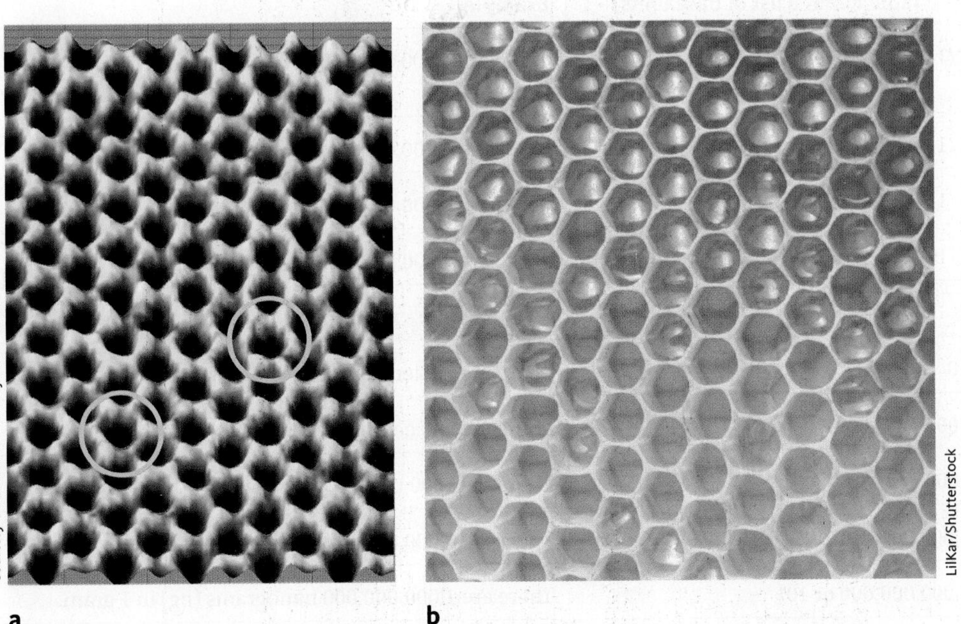

Figure 1.8 Honeycombs

(a) This honeycomb is a STEM image of atoms arranged in hexagons, circled in pink. The vertex of each hexagon is one atom. The diameter of each hexagon is about 3×10^{-10} meters. (b) A honeycomb made by bees. The diameter of one hexagon is about a half a centimeter, which is equal to 0.005 meters.

The metric system is an agreed-upon method of measurement used around the globe.

If you've ever tried to convert tablespoons to cups or ounces to pounds, then you can appreciate the beauty of a system of units based on multiples of 10. The **metric system** is such a system, and it is used to express measured quantities like time and length. Converting a length from meters to centimeters requires multiplying by 100, which is the same as moving a decimal to the right two places. Because the metric system is logical and straightforward, scientists worldwide use it.

Back in 1960, members of the scientific community came together to decide on some ways to standardize the metric system. They created an internationally agreed-upon version of the metric system, which they called the *Système International* (SI). In this book, we will use the metric units that most chemists use every day. These come mostly from the SI system, with a few exceptions. Here, this system of units is called simply "the metric system."

In the metric system, each type of measurement, such as length or volume, has a **base unit. Table 1.1** lists the base units used in this book. In the metric system, units that are multiples of 10 of each base unit are created by adding a prefix to the name of the base unit. **Table 1.2** (p. 14) lists the prefixes and the letter used to designate each one.

Table 1.1 Base Unit

Physical Quantity	Name of Base Unit	Abbreviation
Number of entities	mole	mol
Length	meter	m
Mass	gram	g
Volume	liter	L
Temperature	degrees Celsius	°C
Time	second	s
Information storage	byte	b

Table 1.2 Metric Prefixes

Metric Prefix	Symbol or Letter	Quantity in One Base Unit (see Table 1.2 for a list of base units)	Example
tera-	T	1/1,000,000,000,000 or 10^{-12}	There are 1,000,000,000,000 meters in 1 tetrameter (Tm).
giga-	G	1/1,000,000,000 or 10^{-9}	There are 1,000,000,000 bytes in 1 gigabyte (Gb).
mega-	M	1/1,000,000 or 10^{-6}	There are 1,000,000 grams in 1 megagram (Mg).
kilo-	k	1/1000 or 10^{-3}	There are 1,000 meters in 1 kilometer (km).
base unit		1	
deci-	d	10	There are 10 deciliters (dL) in 1 liter.
centi-	c	100	There are 100 centimeters (cm) in 1 meter.
milli-	m	1000 or 10^{3}	There are 1000 milliseconds (ms) in 1 second.
micro-	μ	1,000,000 or 10^{6}	There are 1,000,000 microliters (μL) in 1 liter.
nano-	n	1,000,000,000 or 10^{9}	There are 1,000,000,000 nanograms (ng) in 1 gram.
pico-	p	1,000,000,000,000 or 10^{12}	There are 1,000,000,000,000 picomoles (pmol) in 1 mole.
femto-	f	1,000,000,000,000,000 or 10^{15}	There are 1,000,000,000,000,000 femtoseconds (fs) in 1 second.

Here's an example of how to use these two tables: In this book the base unit used for length is the **meter**, and its abbreviation is m. One length unit that is a multiple of the meter is the kilometer. Table 1.2 tells us that the letter representing *kilo-* is *k,* and so the abbreviation for kilometer is km. We can combine any of the base units in Table 1.1 with any prefix in Table 1.2 to make a metric system unit, like this: femtosecond (fs); nanogram (ng); megamole (Mmol); picoliter (pL).

In Table 1.2, each prefix, when combined with a base unit, gives a larger quantity than the prefix below it. For example, a gigabyte is larger than a megabyte and a microgram is larger than a picogram. Notice, too, that the base unit—for example, the meter, liter, mole, and so on—occupies one horizontal row in the center of the table. This is a point of reference for the rest of the table. For example, 1 gigaliter contains 1,000,000,000 liters, and 1 liter contains 1000 milliliters. Thus, a gigaliter is larger than a liter, the base unit, and a milliliter is smaller than a liter.

QUESTION 1.5 Which of the following is *not* a legitimate metric unit? (a) picopound (b) femtosecond (c) nanomole (d) terayard

ANSWER 1.5 Answers (a) and (d) do not include any of the base units in Table 1.2. Therefore, these are not valid metric system units.

For more practice with problems like this one, check out end-of-chapter Question 21.

FOLLOW-UP QUESTION 1.5 Which unit in each pair represents a greater quantity? (a) 1 nanoliter or 1 milliliter (b) 1 microgram or 1 femtogram (c) 1 picosecond or 1 millisecond

Conversion factors are fractions that express the same value—in different units—on the top and bottom.

The base units in Table 1.1 express the type of quantity being measured, for example, grams for mass or liters for volume. (We discuss many of these base units in more detail as we move through this chapter.) Metric prefixes modify the base unit to indicate a quantity that is larger or smaller. For example, Table 1.1 shows that 1 meter (m) is equivalent to 1000 millimeters (mm). In other words, a millimeter is smaller than a meter, and there are 1000 of them in 1 meter. We can express this relationship by linking the units with an "equals" sign:

$$1000 \text{ mm} = 1 \text{ m}$$

In addition to writing these values next to each other separated by an equals sign, we can write them in fraction form. The value on one side of the equals sign becomes the numerator (the top), and the value on the other side of the equals sign becomes the denominator (the bottom). It does not matter which value we put on the top or bottom. Thus, we can write the following:

$$\frac{1000 \text{ mm}}{1 \text{ m}} \quad \text{OR} \quad \frac{1 \text{ m}}{1000 \text{ mm}}$$

We call these fractions **conversion factors** because they allow us to convert from one metric unit to another. Any fraction is equal to 1 when the numerator is equal to the denominator. So each conversion factor is equal to 1 because the numerator and denominator represent the equivalent quantities but with different units. Here is the conversion factor that relates meters to millimeters again:

$$1 = \frac{1000 \text{ mm}}{1 \text{ m}} = \frac{1 \text{ m}}{1000 \text{ mm}}$$

The metric system was designed to be easy to use; everything is based on multiples of 10. Even though you may be able to convert between metric units in your head, other units can be more challenging to convert. Happily, the method called **dimensional analysis** allows you to convert any unit to any other unit, no matter how complicated the units are.

Let's consider one straightforward example of dimensional analysis in which we convert from one unit of length—the meter—to another—the nanometer. In the honeycomb-like STEM image shown in Figure 1.8a, the distance from one side of one hexagon in the honeycomb pattern to the other side is about 0.000 000 000 3 meters. We can use dimensional analysis to express this distance with a less zero-laden number, by converting it to a metric unit such as nanometers.

We start by writing a conversion factor that gets us from one unit to the other. Table 1.1 tells us that 1 meter is equivalent to 1 billion nanometers, which means that:

$$\frac{1{,}000{,}000{,}000 \text{ nm}}{1 \text{ m}} \quad \text{OR} \quad \frac{1 \text{ m}}{1{,}000{,}000{,}000 \text{ nm}}$$

Because our goal is to change meters (m) to nanometers (nm), we use the conversion factor on the left. Now we multiply the measured number with the unit we wish to convert—0.000 000 000 3 m—by the conversion factor, and we get:

$$0.000\ 000\ 000\ 3 \cancel{\text{m}} \times \frac{1{,}000{,}000{,}000 \text{ nm}}{1 \cancel{\text{m}}} = 0.3 \text{ nm}$$

We can express the distance across a hexagon as 0.000 000 000 3 meters—the number we started with—or as 0.3 nm. We also could have chosen to convert our original quantity to femtometers or micrometers, for example. All are valid ways to express the same distance, and the flexibility of using different prefixes with the base unit "meter" allows us to choose how we express it. Of course, we could also elect to use our other method, scientific notation, to express the original quantity: 3×10^{-10} m. These are all valid ways to express the same length.

QUESTION 1.6 In 2009, Michael Phelps was the first man to swim 100 meters with the butterfly stroke in under 50 seconds. His time was 49,820 milliseconds. Convert this time to seconds using the following conversion factor:

$$\frac{1 \text{ s}}{1000 \text{ ms}}$$

ANSWER 1.6

$$49{,}820 \cancel{\text{ms}} \times \frac{1 \text{ s}}{1000 \cancel{\text{ms}}} = 49.82 \text{ s}$$

For more practice with problems like this one, check out end-of-chapter Question 28.

FOLLOW-UP QUESTION 1.6 At the 2012 Olympic Games in London, USA swimmer Missy Franklin swam the 200-meter backstroke in 124,000,000 μs, a new world record.

(Recall that μs is the symbol for a microsecond.) Express Missy's time in seconds using the following conversion factor:

$$\frac{1\ \text{s}}{1{,}000{,}000\ \mu\text{s}}$$

1.4 The Metric Epicurean

There are several legitimate ways to express volume and temperature.

Many years from now, kitchens in the United States and the United Kingdom may be the last places on Earth where the metric system is not the accepted and preferred system of measurement. English measuring devices such as cups, teaspoons, and tablespoons are old friends, and many cooks would balk at the idea of trading them in for metric measures. Although we may find the metric system lurking in the refrigerator—those 2-liter bottles of soda—the old standbys quart and gallon are still the most frequently seen volume units in the United States.

Figure 1.9 is a recipe for metric brownies. Everything in the recipe is expressed in metric units. At first glance, it may look more like a chemistry laboratory manual than a recipe for brownies. To bake these in our own kitchen, we need chocolate, butter, eggs, vanilla extract, flour, walnuts, and sugar, as well as the following English-to-metric conversions and a clear understanding of dimensional analysis:

$$454\ \text{g} = 16\ \text{ounces}$$

$$1\ \text{L} = 1.0567\ \text{quarts}$$

$$1\ \text{cm} = 0.394\ \text{inches}$$

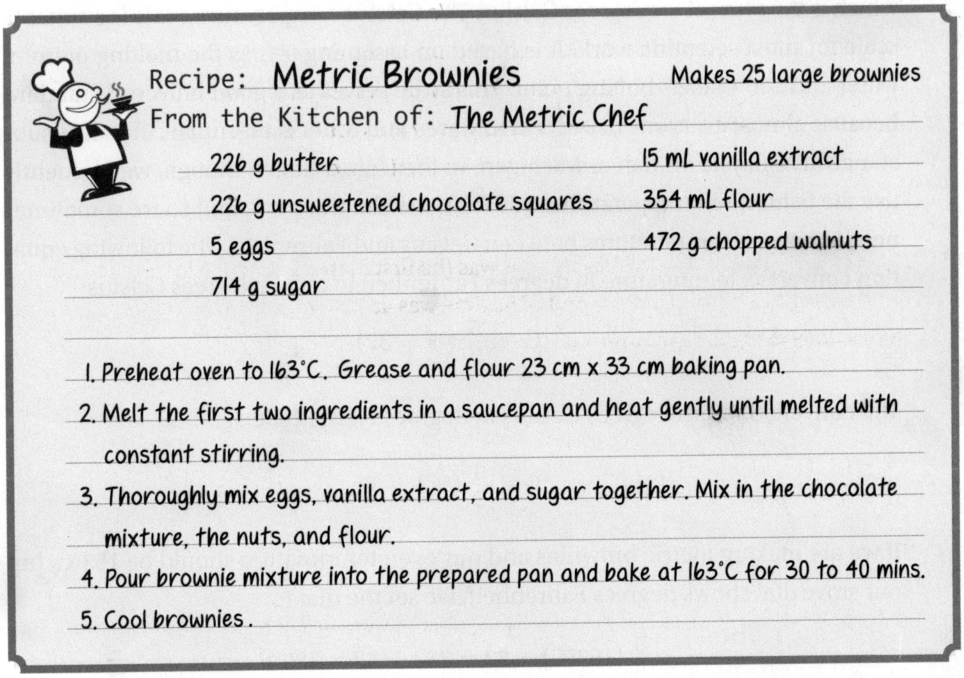

Recipe: Metric Brownies — Makes 25 large brownies
From the Kitchen of: The Metric Chef

226 g butter
226 g unsweetened chocolate squares
5 eggs
714 g sugar
15 mL vanilla extract
354 mL flour
472 g chopped walnuts

1. Preheat oven to 163°C. Grease and flour 23 cm x 33 cm baking pan.
2. Melt the first two ingredients in a saucepan and heat gently until melted with constant stirring.
3. Thoroughly mix eggs, vanilla extract, and sugar together. Mix in the chocolate mixture, the nuts, and flour.
4. Pour brownie mixture into the prepared pan and bake at 163°C for 30 to 40 mins.
5. Cool brownies.

Figure 1.9 Recipe for Metric Brownies

In this recipe, all quantities are given in metric units rather than English units.

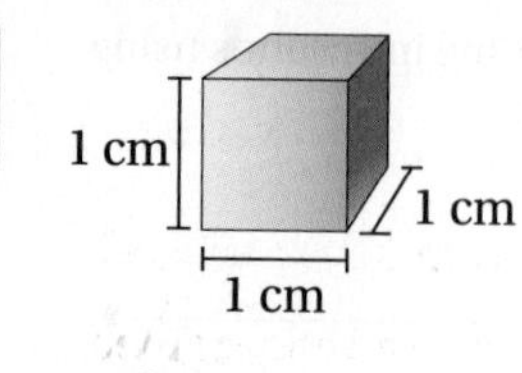

Emiko Paul

Figure 1.10 How Much Is a Milliliter?

A milliliter is approximately the same volume as the part of your pinkie fingertip that is covered by the fingernail.

When making metric brownies, we use the volume unit milliliter (mL) to measure out the sugar, flour, and walnuts. We know from Table 1.1 that the base unit for volume is the **liter**, a common metric unit. Volume is three-dimensional, and so it can also be expressed in units of length cubed. Thus, the milliliter can be defined as 1 cubic centimeter (cm^3), a volume about the size of your pinkie fingertip (only the part that's covered by the fingernail, as shown in **Figure 1.10**). An alternative abbreviation for cubic centimeter, one you might see in nonscientific writing, is cc; so 1 mL = 1 cm^3 = 1 cc. These are just different ways of expressing the same volume.

QUESTION 1.7 Referring to Table 1.1, write a conversion factor that relates the liter to each of the following units: (a) the femtoliter (b) the milliliter (c) the nanoliter

ANSWER 1.7

(a) $\frac{1{,}000{,}000{,}000{,}000{,}000 \text{ fL}}{1 \text{ L}}$ OR $\frac{1 \text{ L}}{1{,}000{,}000{,}000{,}000{,}000 \text{ fL}}$

(b) $\frac{1000 \text{ mL}}{1 \text{ L}}$ OR $\frac{1 \text{ L}}{1000 \text{ mL}}$

(c) $\frac{1{,}000{,}000{,}000 \text{ nL}}{1 \text{ L}}$ OR $\frac{1 \text{ L}}{1{,}000{,}000{,}000 \text{ nL}}$

For more practice with problems like this one, check out end-of-chapter Questions 23 and 24.

FOLLOW-UP QUESTION 1.7 Express the number of nanoliters in 1 liter using scientific notation.

In our recipe for metric brownies, the unit used for the oven temperature is °C, which is the symbol for degrees Celsius. The **Celsius temperature scale** is a suitable scale for most scientific work. It is based on assigning 0°C as the melting point of water and 100°C as its boiling point. This scale serves as a good universal standard because almost everyone has access to water, and water is essentially the same substance in Malaysia or Mali or Michigan. In the United States, though, we frequently use the Fahrenheit temperature scale for nonscientific work, and so we sometimes need to convert temperatures between Celsius and Fahrenheit. The following equation converts a temperature in degrees Fahrenheit to one in degrees Celsius:

$$°C = \frac{5}{9}(°F - 32)$$

The conversion from degrees Celsius to degrees Fahrenheit is:

$$°F = \frac{9}{5}(°C) + 32$$

If we are making metric brownies and our oven temperature should be 163°C, but our stove dial shows degrees Fahrenheit, we set the dial to:

$$\frac{9}{5}(163°C) + 32 = 293 + 32 = 325°F$$

This book uses the Celsius scale when referring to temperature measurements.

wait a minute . . .

Why do the answers to some questions have more decimal places than others?

The formal rules for deciding how many digits belong in a measured number are called *the rules for significant figures.* In this book, those rules are always followed. However, your professor may or may not require that you learn these rules and know how to apply them. Therefore, the detailed rules for significant figures are given in Appendix A. •

QUESTION 1.8 Express the temperature −152°C in degrees Fahrenheit. Is this temperature much hotter or much colder than ambient temperatures on Earth?

ANSWER 1.8

$$\frac{9}{5}(-152°C + 32) = -242°F$$

That's pretty cold! The lowest temperature recorded on Earth was −128.6°F in Vostok, Antarctica, on July 21, 1983.

For more practice with problems like this one, check out end-of-chapter Question 38.

FOLLOW-UP QUESTION 1.8 The temperature on a hot July day in Las Vegas reaches 115°F. What is this temperature in degrees Celsius?

We use the gram when we measure mass and the second when we measure time.

Let's look once more at our metric brownie recipe. The butter and chocolate amounts are expressed in grams. Recall from Table 1.1 that in this book, grams are the base unit for **mass,** the quantity of matter contained in an object. As with all other metric units, we can manipulate this one using metric prefixes. For example, if we decide to make 10 recipes of brownies, we need 2260 g of butter. We can express this mass in kilograms by using the conversion factor 0.001 kg = 1 g from Table 1.2:

$$2260 \not{g} \text{ butter} \times \frac{0.001 \text{ kg}}{1 \not{g}} = 2.26 \text{ kg butter}$$

Likewise, if we decided to make one brownie (which no one would ever want to do), we would need:

$$1 \text{ \sout{brownie}} \times \frac{226 \text{ g butter}}{25 \text{ \sout{brownies}}} = 9.0 \text{ g butter}$$

In this case, we used the fact that 25 brownies are made with 226 grams of butter to create a conversion factor.

The only unit in our recipe we have not discussed yet is the unit for time. Happily, most of the world uses a similar system for expressing time. The official metric unit is the *second,* and larger and smaller units can be made from it. Units like the millisecond (ms), microsecond (μs), and nanosecond (ns) are all legitimate ways to express time. For longer lengths of time, minutes, hours, and years are commonly used, too.

QUESTION 1.9 Before you leave for college, you get some news about your roommate, who is from France, a country that uses the metric system. He tells you in an e-mail that his weight is 69.0 kg. Given that the average man in France is 1.75 meters or 5 ft 9 in. tall, is your new roommate thin or heavy? You may use the following conversion factor in your calculation:

$$\frac{2.20 \text{ lb}}{1 \text{ kg}}$$

ANSWER 1.9 Use dimensional analysis to convert from metric units to English units:

$$69.0 \cancel{\text{kg}} \times \frac{2.20 \text{ lb}}{1 \cancel{\text{kg}}} = 152 \text{ lb}$$

Your new roommate is quite thin!

For more practice with problems like this one, check out end-of-chapter Question 37.

FOLLOW-UP QUESTION 1.9 Find the height, in meters, of a young giraffe that is 10.2 feet tall by using the following conversion factor:

$$\frac{1 \text{ meter}}{3.28 \text{ feet}}$$

wait a minute . . .

Aren't mass and weight the same thing?

No, mass and weight are *not* the same thing, although the terms are often used interchangeably in everyday conversation. Mass, as we discussed earlier in this chapter, is the amount of matter contained in something. Weight, on the other hand, is related to the force that an object feels from the pull of gravity. Mass is used by scientists because it does not change when you move an object from a place where it feels the pull of gravity (for example, on Earth) to a place where it feels less gravity (for example, on the Moon; **Figure 1.11**) or no gravity at all. So if you travel to the Moon, where the pull of gravity is about one-sixth of that on Earth, your weight drops, but your mass is still the same (unless you have an aversion to space food). •

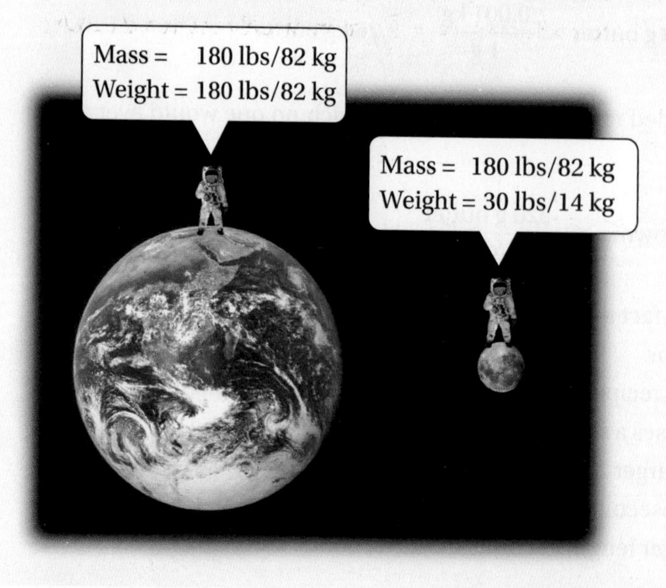

Figure 1.11 Mass and Weight: Not the Same Thing

On Earth, weight results from the pull of gravity. On the Moon, where there is much less gravity, the weight of an object—in this case an astronaut—is less than it is on Earth. However, because mass relates to the amount of matter contained in something or someone, it is the same in both places. The mass of the astronaut does not change; her weight does.

1.5 Juggling Measured Numbers

We can evaluate precision and accuracy for any repeated measurement that has a known standard.

We now know how scientific experiments are planned and how results are expressed in metric units. We know how to convert those units to more practical units or to scientific notation, if necessary. We also understand that reproducible results are valued in the scientific community, and that one scientist's data are often checked or repeated in other laboratories by other scientists. In addition to these useful skills, we need to know how to tell whether data are valuable or not. When an experiment is done well, what exactly does that mean? Certainly, it means that measurements have been made in a meticulous way, but experiments can be meticulous to different degrees.

Scientists use two complementary terms to describe the worth and quality of data. The term **precision** refers to the closeness of measured values to one another. As an example, let's consider a series of actual experiments that were performed to determine the average distance that a cookie is dunked into a cup of tea in Great Britain (where they refer to cookies as *biscuits*). In this experiment, a special cup with gradations on the inside was used to determine how much liquid tea was displaced when a cookie was dunked (**Figure 1.12**).

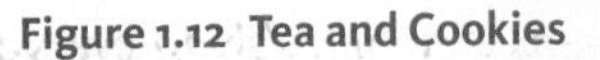

Figure 1.12 Tea and Cookies
When a cookie is dunked into tea, the level of the tea rises. This cup can be used to measure how much the level rises, in centimeters, when a cookie is dunked.

Table 1.3 includes the recorded data for successive measurements on the height of tea displaced by a dunked cookie, in centimeters, from one experiment.

Which trial is most precise? All three give the same average value, but Trial A is most precise because the measured values are closest to one another. This is evident from the range of values at the bottom of the table. As we go from A to B to C, the values become less precise. That is, they are more widely scattered around the average value.

Table 1.3 Tea Displacement

	Trial A (centimeters)	Trial B (centimeters)	Trial C (centimeters)
Measurement 1	3.4	4.0	5.0
Measurement 2	3.3	2.9	2.4
Measurement 3	3.4	2.7	3.5
Measurement 4	3.5	3.7	2.2
Measurement 5	3.2	3.5	3.7
Average	3.4	3.4	3.4
Range	3.2–3.5	2.7–4.0	2.2–5.0

What could cause three measurements of the same thing to have different levels of precision? In this case, it's possible that the three measuring cups had scales of differing fineness. It's also possible that three different people performed these measurements. Maybe the person who measured the values in Trial C needs a new pair of glasses? Maybe the person who measured the values in Trial A is a tea volume-measuring expert?

All this brings us to an important lesson in experimental design: it is important to keep the conditions as uniform as possible from one measurement to the next. The same person using the same measuring device will provide the most consistent results.

QUESTION 1.10 For the tea-and-cookie experiment, Table 1.3 indicates that each of the trials gave the same average value, in the end. If the same number results from each group of measurements, why are the three groups of measurement not equally precise?

ANSWER 1.10 Even though the same average number is obtained for each trial, the range of answers is different. For Trial A, each measurement was closer to the average value than each measurement in Trials B and C. Therefore, Trial A was more precise than either Trial B or Trial C.

For more practice with problems like this one, check out end-of-chapter Question 48.

FOLLOW-UP QUESTION 1.10 Referring again to Table 1.3, express the average measured value for Trials A, B, and C in millimeters.

Accuracy describes the closeness of a measurement to a known value.

For the preceding experiment about tea, we cannot evaluate **accuracy,** which tells us how close a number is to a known and accepted standard value. Why can't we evaluate accuracy in this case? Because there is no known standard to use for comparing the measured values. So let's look at a slightly different experiment, one that has the benefit of a known **standard,** something for which some characteristic, such as mass or length, is known very precisely.

Imagine that our standard is a 5.0000-gram sample of a metal that has been calibrated by a team of very careful scientists at the National Institutes for Standards and Technology. We plan to use this sample of metal as a standard to calibrate three balances that we have in our laboratory. **Table 1.4** lists the masses for our 5.0000-gram sample of metal from our three balances.

Because we have a known standard—5.0000 grams—we can compare the averaged measured values to it, and we can draw conclusions about the accuracy of these trials. Of our three balances, Balance III is most accurate because its average value, 5.0017 g, is closest to the standard. Balance II ranks second in accuracy, and Balance I ranks last.

Now let's evaluate the precision. The values from Balance III are tightly grouped together, and all of the values average to a number close to the standard. Thus, we can say that Balance III is both accurate and precise. Balance I is less accurate than Balance II, and its precision is lower because the Balance I values span a larger

Table 1.1 Mass Measurements from Three Balances

	Mass from Balance I (grams)	Mass from Balance II (grams)	Mass from Balance III (grams)
Measurement 1	5.4403	5.0009	4.9988
Measurement 2	5.3344	4.8992	4.9907
Measurement 3	4.6677	5.1200	5.0120
Measurement 4	4.8009	5.0130	5.0066
Measurement 5	5.2230	4.8895	5.0004
Average	5.0933	4.9845	5.0017
Range	4.6677–5.4403	4.8895–5.1200	4.9907–5.0120

range than the values for Balance II. Thus, we say that Balance I is less accurate and less precise than Balance II. Both I and II need a tune-up to reach the levels of accuracy and precision of Balance III.

In this example, the balances are high-end, state-of-the-art instruments that report to four decimal places for every measured mass. The number of decimal places we report for a given measurement depends on the measuring device we use. We can report more decimal places with a sophisticated analytical balance worth $10,000 than we can with a $10 balance from the grocery store.

QUESTION 1.11 Consider the distribution of holes on each of the targets shown here. (a) Which have high precision? (b) Which have high accuracy? (c) Which have both high accuracy and high precision?

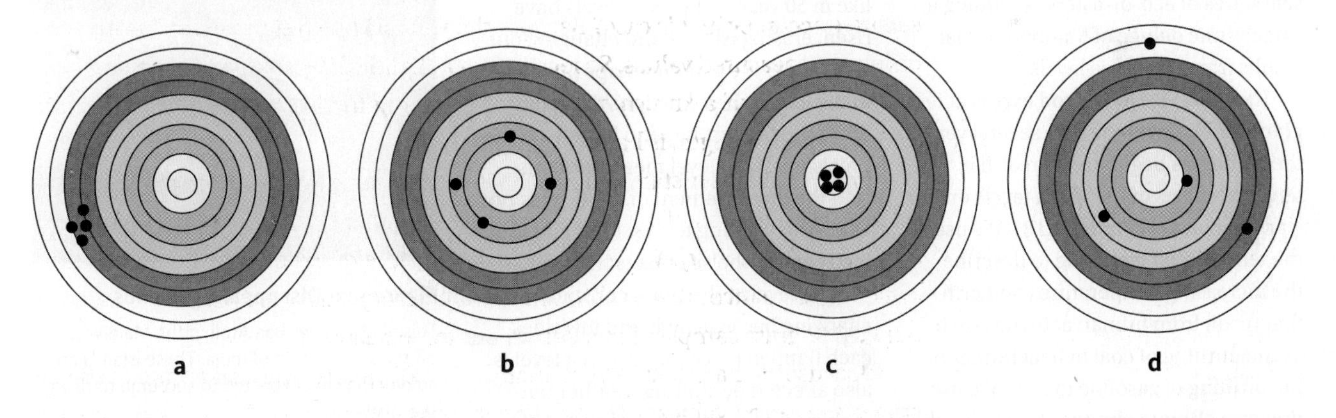

ANSWER 1.11 Targets a and c are very precise because they have holes clustered close together. Targets b and c are accurate because they have an average value near the bull's-eye. Target c is both accurate and precise.

For more practice with problems like this one, check out end-of-chapter Question 50.

FOLLOW-UP QUESTION 1.11 Is there a known standard associated with targets like those shown in Question 1.11 above?

Human beings use common sense to interpret results from calculators and scientific instruments.

When conducting experiments, scientists make every effort to be consistent and to follow some simple rules. They keep careful records, perform multiple trials in every experiment, use instruments that are properly calibrated, and agree upon internationally acceptable units of measure.

Despite these efforts, though, mistakes are made, often with disastrous results. A disastrous measuring blunder occurred when the *Mars Climate Orbiter,* sent into space in 1999, simply disappeared. The calculations made to determine how to track the orbiter were done by combining measurements from navigation teams in Colorado and California. As it turns out, one team was using English units while the other was using metric. So the lesson here is this: whether you're measuring spacecraft orbits at NASA or the parameters for dunking cookies in Great Britain, be scrupulous in your experimental design, keep track of units, and—perhaps most of all—use common sense to look critically at your own data.

THEgreenBEAT • news about the environment, chapter one edition

TOP STORY How Do We Measure Sea Level?

The 2006 movie *The Day After Tomorrow* stars Dennis Quaid as a climatologist whose warnings about the dangers of *climate change* are not heeded. The politicians and others who ignore him regret that, of course, as the effects of a warming planet play out in an impressive series of eco-disasters, including a cataclysmic deluge of Manhattan that results from rising sea levels.

The movie is Hollywood excess at its grandest but, over the past several years, the world has begun to see real-life, tangible effects of rising sea levels on our warming planet (**Figure 1.13**). We use the term **global warming** to describe the increasing temperatures on Earth that result from human activities such as the burning of coal to heat homes or the burning of gasoline in cars. We use the term **climate change** to describe the effects of global warming, such as rising sea levels and disappearing islands. We will explain these terms in much greater detail in future chapters and future *Green Beat* boxes. Chapter 12, "Energy, Power, and Climate Change," is devoted entirely to these topics.

At the time of this writing, several islands have been completely covered with water as a result of rising sea levels. An Internet search of "disappearing islands" brings up websites that track the islands that are at highest risk of disappearing and documents the most recent evacuations from inhabited islands that flood.

What will a map of the world look like in 50 years, when sea levels have risen, erasing islands and changing our coastlines? We can find maps online that show such predictions, although these predictions depend largely on the precision of sea level measurements. And the measurement of sea level is not as easy as walking to the shore and looking at where the sea meets the land.

The level of the sea is constantly changing because high and low tides each happen twice per day. Sea level is also affected by shifting weather patterns as well as seasonal changes. Thus, no single measurement of sea level is very useful; data must be collected and averaged over years. In fact, sea level measurements traditionally have been averaged over 19-year spans.

Part of the problem with tracking sea level is the precision of the measuring devices researchers use. Traditionally, a *tide gauge* has been used to measure sea level at shorelines. But even the best tide gauges are precise to only ±1 cm. That means that a tide gauge can make measurements in whole numbers of centimeters, but it cannot measure to tenths of a centimeter.

Sakis Papadopoulos/Getty Images

Figure 1.13 Disappearing Islands
This photo shows Baa Atoll in the Maldives, off the southern tip of India. These islands are among the next expected to succumb to rising sea levels.

A consortium of scientists at NASA and at CNES, the French space agency, are working on more sophisticated ways to measure sea level. One approach uses a satellite named *Jason-2,* whose orbit—and therefore its exact altitude—is tracked very precisely. Using complex measurements, the satellite tells the scientists how far it is located from the ocean's surface. With this data, the scientists are able to pinpoint the sea's level to ±1 mm, which is ten times more precise than a tide gauge. **Figure 1.14** shows the sea level readings from *Jason-2* (shown in yellow) and its two predecessors, *Jason-1* (green) and *TOPEX* (red), over a span of 20 years.

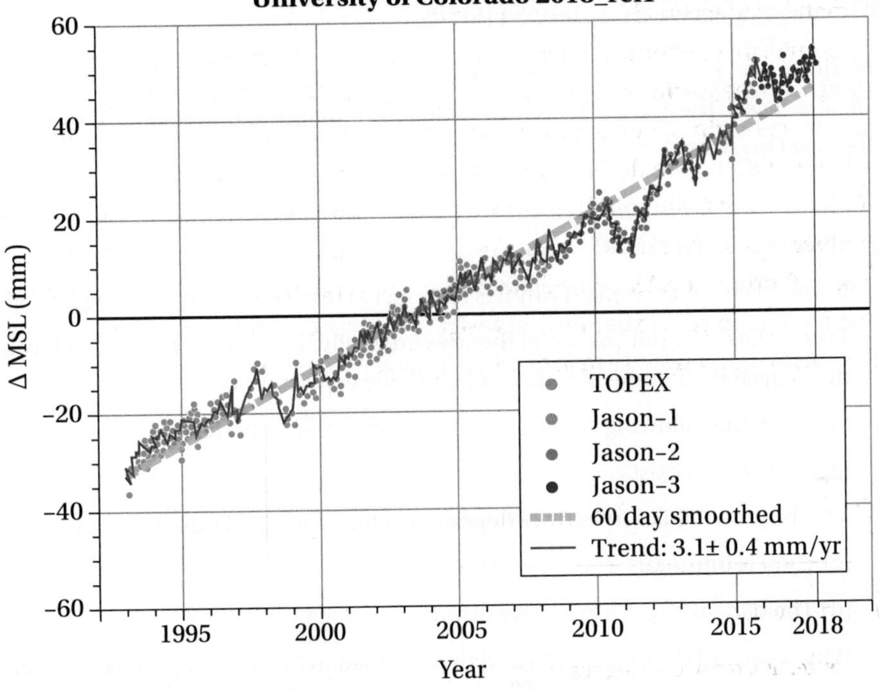

Figure 1.14 Rising Sea Level Measurements
This graph shows how sea level gradually has increased from 1992 to 2012.

QUESTION 1.12 Roughly how much deeper was the sea in 2012 compared to 2004? Express your answer in millimeters.

ANSWER 1.12 The depth of the sea is tracked on the vertical axis, which has units of millimeters. The difference between the 2012 value (about 54 millimeters) and the 2004 value (about 28 millimeters) is about 26 millimeters.

FOLLOW-UP QUESTION 1.12 Roughly how much deeper was the sea in 2008 compared to 2002? Express your answer in millimeters.

Chapter 1 Themes and Questions

recurring themes in chemistry chapter 1

recurring theme 1 Scientists worldwide use the scientific method as a basis for experimentation and deduction. The scientific method is based on the principles of peer review and reproducibility. (p. 6)

recurring theme 2 Things we can see with our eyes contain trillions upon trillions of atoms. Atoms are much too small to view with the naked eye. (p. 8)

A list of the eight recurring themes in this book can be found on p. xv.

ANSWERS TO FOLLOW-UP QUESTIONS

Follow-Up Question 1.1: because you want to know if the butter affects the results. *Follow-up Question 1.2:* (b) subject to revision with further experimentation. *Follow-up Question 1.3:* (a) 6 (b) 12 (c) 3 (d) 6. *Follow-up Question 1.4:* The number part of this expression is not between 1 and 10. The correct way to write it is 8.86×10^4. *Follow-up Question 1.5:* (a) 1 milliliter (b) 1 microgram (c) 1 millisecond.

Follow-up Question 1.6: 124.06 seconds. *Follow-up Question 1.7:* There are 10^9 nL in 1 liter. *Follow-up Question 1.8:* 46 degrees Celsius. *Follow-up Question 1.9:* 3.11 meters. *Follow-up Question 1.10:* 34 mm. *Follow-up Question 1.11:* Yes, the known standard for a target is its bull's-eye. *Follow-up Question 1.12:* about 20 mm deeper.

age fotostock/age fotostock

End-of-Chapter Questions

Questions are grouped according to topic and are color-coded according to difficulty level: **The Basic Questions***, **More Challenging Questions****, and **The Toughest Questions*****.

The Scientific Method

*1. In your own words, explain why it is incorrect to refer to a theory as being proved.

*2. True or false? Hypotheses and theories are really the same thing. Both tell you the expected outcome of a series of scientific experiments.

*3. Which of the following statements are not scientific hypotheses?

(a) Santa Claus exists.

(b) The growth of corn in Iowa depends on the amount of rain that falls.

(c) Theft is immoral.

(d) The bark of a specific tree is poisonous to mice.

*4. Why is each of the various toast trials discussed in Section 1.1 performed multiple times? In other words, why is reproducibility an important part of the scientific method?

*5. In the buttered toast experiment in Section 1.1, why was it important to put a red dot on some pieces of bread?

**6. Consider the following hypothesis: *Microwave ovens heat all foods to the same temperature in the same amount of time, regardless of the water content of the food.* Formulate a series of experiments you could perform to test this hypothesis. Describe the possible outcomes of your experiments.

**7. In your own words, describe the controversy surrounding the NASA finding about bacteria and arsenic discussed in the Chapter 1 naturebox. Why was this story such big news? How did scientists disprove the theory set forth by NASA?

***8. An old saying asserts that cold water boils more rapidly than hot water. Develop a strategy for testing this urban myth, beginning with a hypothesis and outlining a series of experiments you could perform to test it. What are the possible outcomes of your experiment?

Scientific Notation

*9. Some of the following numbers are not written correctly in scientific notation. Identify those that are incorrect, and write the correct answer for each of them.

(a) 44.5×10^4

(b) 3.5555×10^1

(c) 67.00×10^{12}

(d) 1×10^{-7}

*10. Some of the following numbers are not written correctly in scientific notation. Identify those that are incorrect, and write the correct answer for each of them.

(a) 0.5×10^8

(b) 0.0007×10^4

(c) 7.50×10^{-3}

(d) 5.1×10^{-3}

*11. Express these numbers in scientific notation.

(a) 45,000,000,000

(b) 0.000 000 000 000 444 12

(c) 13.445

(d) 130.000

*12. In your own words, explain why scientific notation is used frequently in scientific work.

**13. Roughly how many molecules of water are there in a small glass of water like the one shown in Figure 1.5? Express this value in scientific notation and as a number that is not written in scientific notation.

**14. Convert the following to numbers that are not in scientific notation:

(a) 6.5×10^4

(b) 4.45×10^{-7}

(c) 6.000×10^6

(d) 1×10^{-19}

**15. Convert the following numbers to numbers that are not in scientific notation:

(a) 5.1×10^1

(b) 1×10^{-9}

(c) 2.200×10^{18}

(d) 9.1×10^3

***16. A certain city has a population of 4,500,000 citizens. On average, each person has 5.5×10^9 nanoliters of blood. In total, how many nanoliters of blood do all the citizens of the city possess? Express this number in scientific notation.

***17. Suppose a mountain range is made up of 18 mountains, each mountain has 8800 trees on it, and each tree has 125,000 leaves on it. How many leaves are on all of the trees in the mountain range? Express your answer in scientific notation.

Fundamentals of the Metric System

*18. In your own words, explain why the metric system is the universal system of units used by scientists. What makes it easy to work with compared to some other systems of measurement?

*19. Using the information in Tables 1.1 and 1.2, give the full name for each of these abbreviations.

(a) μmol (b) fL (c) Mm (d) ds

*20. Using the information in Tables 1.1 and 1.2, give the full name for each of these abbreviations.

(a) Gb

(b) nmol

(c) Ts

(d) pm

*21. Which of the following is *not* a recognized metric system base unit?

(a) ounce

(b) gram

(c) second

(d) mile

(e) gallon

(f) minute

*22. (a) How many milliters are in 1 liter?

(b) How many nanograms are in 1 gram?

(c) How many microseconds are in 1 second? (Hint: Refer to Tables 1.1 and 1.2.)

**23. (a) How many joules are in 1 kilojoule?

(b) How many bytes are in 1 megabyte?

(c) How many meters are in 1 gigameter? (Hint: Refer to Tables 1.1 and 1.2.)

**24. (a) How many femtomoles are in 1 mole?

(b) How many milliliters are in 1 liter?

(c) How many seconds are in 1 terasecond? (Hint: Refer to Tables 1.1 and 1.2.)

**25. In each pair, indicate which unit is larger.

(a) a femtosecond or a millisecond

(b) a kiloliter or a nanoliter

(c) a megabyte or a gigabyte

***26. In each pair, indicate which quantity is larger.

(a) 1000 microseconds or 1 second

(b) a megameter or 10 million meters

(c) 100 nanoliters or 1 microliter

***27. The density of any substance is the amount of mass of the substance contained in a given volume of space.

(a) What metric unit could be used to express the numerator of a density unit?

(b) What metric unit could be used to express the denominator?

(c) If you know the density of some substance and also know the mass of a sample of that substance, what else do you know about the sample?

Converting between Units

*28. Convert the following masses into kilograms:

(a) 2,300,000 grams

(b) 5,000 milligrams

(c) 23,000 micrograms

(d) 0.155 megagrams

*29. A sumo wrestler has a mass of 140 kilograms. What is his mass in pounds? (Hint: There are 2.2 pounds in 1 kilogram.)

*30. A bottle of soda has a volume of 2.11 quarts. What is the volume of the soda in liters? (Hint: One liter is equivalent to 1.054 quarts.)

**31. There are roughly 20 drops in 1 milliliter. Roughly how many milliliters are in 1 drop?

**32. It takes the average Londoner about 5,000,000,000 ns to dunk a cookie into tea. How many seconds is this?

**33. Which mass is larger: 1.5 g, 0.000 045 kg, or 450,000 μg? Use dimensional analysis to defend your choice.

***34. The mass of a certain atom is 3.0×10^{-22} grams.

(a) What is the mass of 1 million of these atoms?

(b) Write your answer to (a) using a number that is not in scientific notation.

***35. You are calling the telephone company to discuss your bill. Your expected wait time on hold is 20,000,000,000,000,000,000 fs.

(a) Convert this time to seconds and decide if you would be willing to wait this long.

(b) What is this length of time in minutes?

(c) What is this length of time in years?

***36. If you line up atoms so that each atom is touching its two neighbors and the diameters form a straight line 1.0 nm long, how many atoms will you have? Assume that each atom has a diameter of 79 pm.

Applying the Metric System: Mass, Volume, and Temperature

*37. The Olympic marathon distance is 42,195,000,000,000,000,000 fm. Express this distance using a more reasonable metric unit. (Hint: Which metric unit is often used to measure race distances?)

*38. The temperature of the lava at the Kilauea volcano in Hawaii reaches 2120°F. Convert this temperature to degrees Celsius.

*39. A recipe for shortbread cookies tells you to set your oven at 375°F. What is this temperature in degrees Celsius?

**40. An Olympic swimming pool contains about 2.5 million liters of water. Express this number in kiloliters and milliliters.

**41. On a certain September day in Utah, the maximum and minimum temperatures differed by 30°F. Express this temperature difference in degrees Celsius.

**42. Which of the following objects has the largest mass? Which has the smallest mass?

(a) a German shepherd that weighs 34 kilograms

(b) a wheelbarrow that weighs 28,000 grams

(c) a 0.0300-megagram box of books

(d) a small pig that weighs 50,000,000,000 micrograms

***43. Imagine that you have 27,000 cm^3 of water in a big bucket. You pour the water into a plastic box. The bottom of the box is 30 centimeters wide and 30 centimeters long. How deep is the water in the box, in centimeters?

***44. A typical nerve impulse lasts 6.00 ms from start to finish. How many consecutive nerve impulses can occur in 1.00 second?

***45. When light enters the eye, a series of chemical processes occur, the first of which takes 200 fs. How many of these processes could occur consecutively during a 10-second period?

Accuracy versus Precision

***46.** In your own words, explain the difference between precision and accuracy.

***47.** Consider this statement: Joe uses his bathroom scale to weigh himself every 20 seconds over a span of 10 minutes. Over this period of time, his mass fluctuates by 7 pounds. What can you say about the precision of his bathroom scale?

****48.** A systematic error may be caused by faulty calibration of a measuring device. This results in consistently high or consistently low measured values from that device. For example, a laboratory balance is not calibrated properly and it is used to make ten consecutive measurements. Each of the ten readings comes to within 1.0 milligram of an average value, which is 0.250 grams higher than it should be. In this example, is the accuracy of the mass measurement high or low? Is the precision of the mass measurement high or low? Explain.

*****49.** Is it possible for a balance to be

(a) accurate and precise?

(b) inaccurate and imprecise?

(c) precise but not accurate?

(d) inaccurate and precise?

*****50.** Consider holes in a target, such as those shown accompanying in-chapter Question 1.11 (on p. 23). The bull's-eye of the target serves as the "true value" or "target" that you are aiming at. To be accurate, a measurement must have some standard (like the bull's-eye) against which you can compare your measured values. Cite two examples of measurements that would have a standard that could be used for comparison. What is that standard in each case?

Integrative Questions

***51.** Which quantity represents the smallest volume?

(a) 5.34×10^4 mL

(b) 1.53 kL

(c) 2,535 L

(d) 3.85×10^{12} nL

***52.** Change each of the units in the following measured numbers to its base unit (see Table 1.1). Then, express the number in scientific notation.

EXAMPLE: 5,325,000 mL = 5,325 L (the base unit) or 5.325×10^3 L

(a) 40,333 kg

(b) 0.003 706 μs

(c) 0.000 000 000 067 921 Gm

(d) 70,644 Mb

***53.** Consider these temperature data, which were collected in the laboratory by three students.

Rory	Joaquin	Michael
45.83°C	50.35°C	48.02°C
46.92°C	55.89°C	48.15°C
44.83°C	48.63°C	47.95°C
42.90°C	45.97°C	47.99°C
46.02°C	56.33°C	48.04°C

(a) Find the average temperature measured by each student in degrees Celsius.

(b) Convert the three average temperatures in degrees Celsius to degrees Fahrenheit.

(c) Which student's data is most precise?

(d) If the actual temperature measurement should have been 109.40°F, then which student's data is most accurate?

***54.** (a) Return to Figure 1.14 and determine the approximate shift in sea level, in millimeters, in the year 1998 and in the year 2008.

(b) By what amount, expressed in centimeters, did sea level rise over the decade mentioned in question 54a?

(c) If the trend in sea level rise continues to increase at the same rate shown in Figure 1.14, then what will be the sea level shift in centimeters in the year 2038?

(d) Convert your answer to question 54c to inches (1 inch = 2.54 cm).

55. INTERNET RESEARCH QUESTION

Open your Internet search engine and type in "best metric recipes." Scan the results and choose a recipe that sounds interesting. Open up the recipe, copy-and-paste it into a blank document, and look at the measured values in it. Convert each of the metric units in the recipe to its English equivalent, and replace the values in your document. For example, you may convert degrees Celsius to degrees Fahrenheit, milliliters to tablespoons, grams to ounces and centimeters to inches. If you have a kitchen, consider cooking the new recipe!

Science and Society/Superstock

Atoms

All about Atoms and What's Inside Them

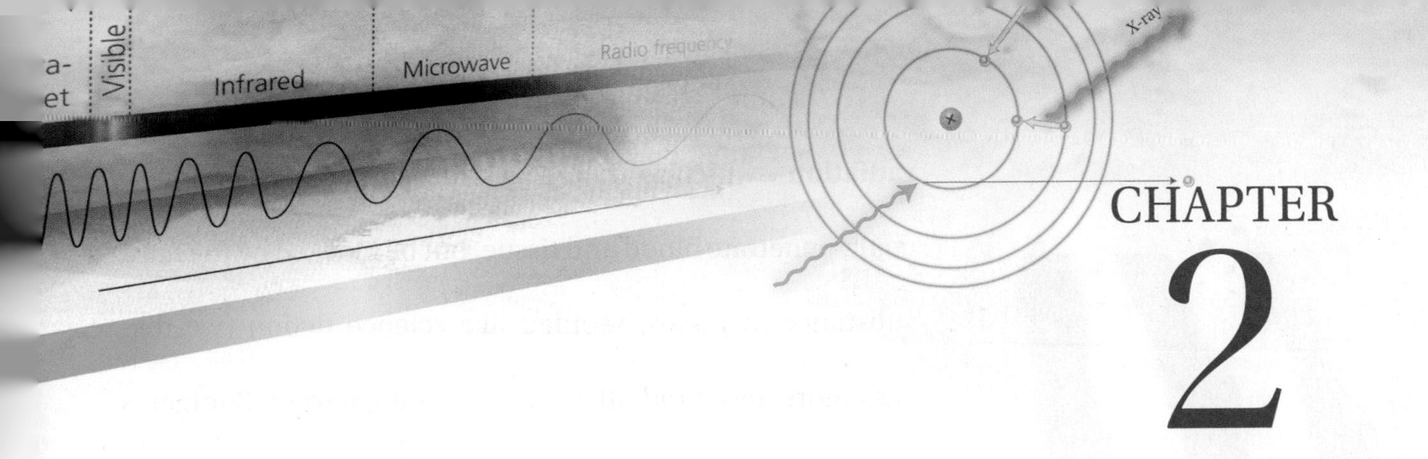

CHAPTER 2

Sometimes sacrifices must be made in the name of love, and sometimes sacrifices must be made in the name of science. Anna Bertha Röntgen made sacrifices for both on December 22, 1895, when her husband, Wilhelm, asked her to place her left hand in front of the new "rays" he had created. She held her hand in the rays for 15 minutes, and the first medical X-ray was created.

Because X-rays can pass through flesh but not through bone or metal, in her X-ray image (see photo on p. 32) Anna's wedding ring and her bones are clearly visible. In the twenty-first century, people are used to seeing X-ray images, but images of bones within flesh were unheard of at the end of the nineteenth century. Anna was, understandably, disturbed by the image of her own skeleton. Because her bones were plainly visible in the X-ray image, she believed that she would die soon after the image was taken, but she lived for another 24 years.

Science and Society/Superstock

To people at that time, the notion that a type of **radiation**—which is defined as the emission of energy—could penetrate blood and tissue, but be blocked by the hard substance in bones, seemed like science fiction (we discuss more about radiation in the coming pages). Röntgen's amazing invention made him an overnight celebrity. His X-rays were immediately useful; within a year, X-ray machines were being widely used to peek inside the human body for medical purposes.

In Chapter 2, we learn more about X-rays and other kinds of light and how light interacts with matter. But at its heart, this chapter is about the smallest units of matter—atoms—and what's inside them. Although these topics might not seem related right now, as we'll see in this chapter, there is a close link between atoms, the particles within atoms, and the behavior of light.

2.1 Atoms: The Basis for Everything

All matter is composed of atoms.

Every single thing in our physical world—everything we breathe, touch, taste, and see—is matter. And, as we learned in Chapter 1, all matter is made up of atoms, which cannot be broken down by chemical or physical means. (Atoms can be broken down by *nuclear* means; we will discuss this topic in Chapter 11, "Nukes.")

Almost all atoms contain three types of particles: positively charged particles, negatively charged particles, and neutral (uncharged) particles. What does it mean to say that something has a positive or negative charge? Matter that possesses **charge** is attracted to or repelled by other matter. Two positively charged particles of matter repel one another, as do two negatively charged particles. But positive and negatively charged particles attract each other.

recurring theme in chemistry 3
Opposite charges attract one another.

Atoms always have no *net* charge, meaning that their positively charged and negatively charged particles are in balance, and sum to zero. Thus, an atom that contains three negatively charged particles must also contain three positively charged particles in order to be charge-neutral. All three of the particles in atoms have mass and contribute to the overall mass of the atom, although some contribute much more than others, as we will soon see. We discuss each of these three types of particles, and their charges and masses, in the upcoming pages.

flashback
Recall from Chapter 1 that *mass* is the quantity of matter contained in an object.

At the time of this writing, there are 114 known different types of atoms, and scientists call each type of atom an **element**. We use the term *atom* generically, the way we would use the word *car.* We use the word *element* to describe the specific types of atoms, the way we would use *Honda* or *Ford* to describe a specific type of car. Just as we can say, "That car is a Honda or this car is a Ford," we refer to "an atom of this element" or "an atom of that element." We humans have given each element a unique name and a corresponding one-, two-, or three-letter **atomic symbol** (or, simply *symbol*). You can find a table of element names and symbols inside the front cover of this book.

Atoms are composed of three types of particles: protons, neutrons, and electrons.

All atoms share the same fundamental structure: a dense, positively charged **nucleus** (the plural is *nuclei,* pronounced *nu-klee-EYE*) surrounded by a cloudlike smear of negative charge. The tiniest type of atom is hydrogen, and its symbol is H. Hydrogen contains one positively charged particle, called a **proton,** in its nucleus. That single positively charged particle is balanced by one negatively charged particle, an **electron,** in the area around the nucleus. The nucleus of a typical atom usually contains neutral particles, called **neutrons,** too. In this respect, hydrogen is a unique exception—most atoms of hydrogen contain no neutrons, and others contain one or two neutrons. In general, the number of neutrons in a given atom is not obvious. This is something that most chemists have to look up in a book or online. In Section 2.3, we look more closely at the role that neutrons play in the atom.

The protons and neutrons in a nucleus are packed together tightly, so the nucleus takes up only a very *very* small portion of the atom. To get an idea of that scale, imagine a map of North America on which someone has drawn a circle passing through Washington, D.C., and San Francisco (**Figure 2.1**). If the diameter of this circle represents the diameter of an atom, the nucleus is about the size of a department store somewhere in the middle of Kansas. Atoms are extremely small to begin with, and their nuclei are that much smaller.

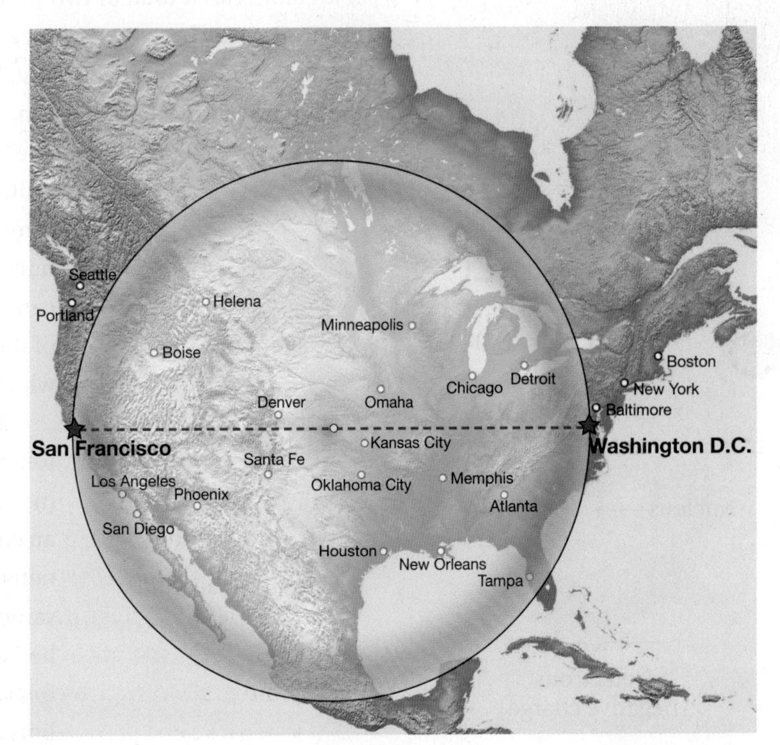

Figure 2.1 An Atom and Its Nucleus
Imagine that the circle drawn on this map of North America represents the circumference of an atom. If the diameter of the circle is equal to the distance from San Francisco to Washington, D.C., then a department store located in Kansas (shown with a dot) represents the relative size of the atom's nucleus.

Because nuclei of atoms take up a very small portion of the *space* in the atom, we might also assume that nuclei also represent a small portion of the *mass* of each atom. But that assumption is not true. Even though the particles in the nucleus—the protons and neutrons—have roughly equal masses, each of them is nearly two thousand times more massive than an electron, which occupies the space around the nucleus in the atom. So, even though electrons have mass, their masses are tiny compared to those of protons and neutrons; electrons make up an insignificant portion of the mass of an atom.

QUESTION 2.1 Which of the following are *not* particles found inside the nucleus of an atom?

ELECTRON PROTON TRITON ELEMENT NEUTRON

ANSWER 2.1 Protons and neutrons are particles found inside the nuclei of atoms. Electrons are particles located inside an atom, but outside the nucleus. Triton was the Greek god of the sea. An element is a type of atom, not a type of particle within an atom.

For more practice with problems like this one, check out end-of-chapter Question 6.

FOLLOW-UP QUESTION 2.1 Which two words in the list in Question 2.1 describe the particles in the atom that are located in the nucleus?

Scientists define each element by its atomic number—the number of protons its atoms contain.

If an atom has a total of two protons, which give it two positive charges, and two electrons, which give it two negative charges, then we know it is helium and its symbol is He. Helium also contains neutrons and, if we add up the masses of all these particles in a helium atom, we find that the mass of helium is greater than the mass of an atom of hydrogen because it contains more particles. The next element is lithium, and its symbol is Li. Lithium atoms contain three protons balanced by three electrons. The lithium nucleus also contains some neutrons (that number can vary, as we have seen), and the total mass of all of these particles makes lithium heavier than helium, which is, in turn, heavier than hydrogen.

And so it goes. Each time we add one more proton, we have a new type of atom with a different element name and a different symbol. In fact, we name an element according to the number of protons it contains. That is, all atoms of a given element contain the same number of protons. For example, every atom of helium—no matter what—always contains two protons. And, any atom with two protons is an atom of helium. The number of protons is the element's **atomic number,** and it defines that element. The number of protons, the element's symbol, and the element name all go hand in hand. **Figure 2.2** is a schematic diagram showing the three particles in the atom.

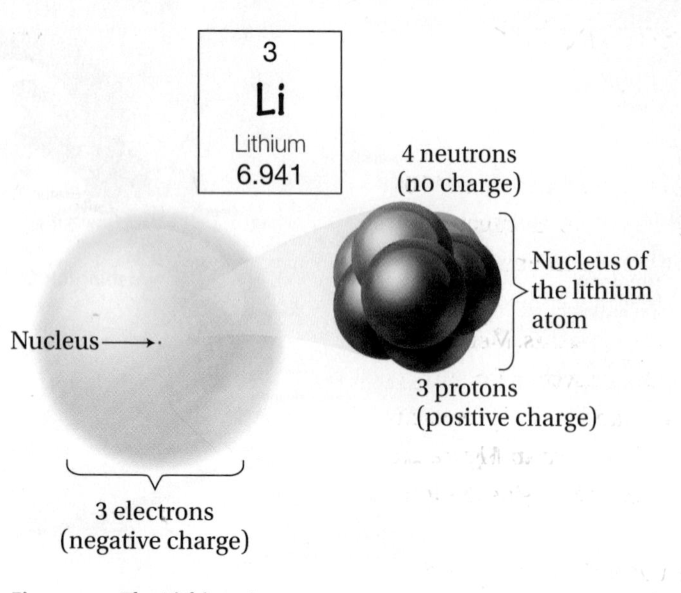

Figure 2.2 The Lithium Atom

The nuclei of all lithium atoms contain three protons. Most lithium atoms also have four neutrons in their nuclei, although some have three. All atoms of lithium have three electrons in the space around the nucleus to balance the charge of the three protons.

In all the examples we have seen so far, including hydrogen, helium, and lithium, the number of protons is always equal to the number of electrons. This will be the case for all atoms, because atoms, as mentioned earlier, have no net charge. In other words, the number of protons an atom contains in its nucleus must always be balanced exactly by the number of electrons it has around its nucleus.

QUESTION 2.2 Some elements have symbols that include the first letter(s) of the element name. Refer to the list of elements inside the front cover. Write the name and symbol for the element with each atomic number: (a) 56 (b) 52 (c) 10 (d) 87

ANSWER 2.2 (a) barium (Ba) (b) tellurium (Te) (c) neon (Ne) (d) francium (Fr)

For more practice with problems like this one, check out end-of-chapter Questions 1 and 2.

FOLLOW-UP QUESTION 2.2 What is the name of the element that contains 54 protons? How many electrons does this atom contain?

QUESTION 2.3 Some elements have symbols that have no obvious relationship to the first letter(s) of the element name. (In many cases like this, the element name and symbol have different origins.) Refer to the list of elements inside the front cover. Write the name and symbol for the element with each atomic number: (a) 19 (b) 74 (c) 82 (d) 26

ANSWER 2.3 (a) potassium (K; from the Latin word for potassium, *kalium*)
(b) tungsten (W; from the German *wolfram*) (c) lead (Pb; from the Latin *plumbum*)
(d) iron (Fe; from the Latin *ferrum*)

For more practice with problems like this one, check out end-of-chapter Question 5.

FOLLOW-UP QUESTION 2.3 The element californium (symbol Cf) was first created at the University of California, Berkeley, in 1950. Refer to the list of elements inside the front cover. How many protons and electrons does one atom of californium contain?

2.2 A Brief Introduction to the Periodic Table

The periodic table is the central organizing principle in chemistry.

Today, 114 known elements are officially recognized and named. But in the 1870s, the list of known elements was 65 elements long. Up to that time, elements were put into lists and assigned names, but they were not organized in any meaningful way. That all changed when Russian chemist Dmitri Mendeleev published his **periodic table**, which presented a new way of displaying the elements. Mendeleev, an eccentric university professor, also was known for creating a system for analyzing Russian vodka, for bringing the metric system to Russia, and for his promiscuity. A painting of Mendeleev is inset into the periodic table shown in **Figure 2.3** (p. 38). A copy of the periodic table is provided for easy reference inside the front cover of this book.

The modern periodic table, which grew out of Mendeleev's original version, displays the symbols of the elements. But, as we will see in this chapter and the next one, the periodic table is more than just a way to display the elements; it's also a grand scheme for categorizing them, and a way to understand the relationship that elements have to one another.

	1	2	3	4	5	6	7	8	9	10	11	12	13	14	15	16	17	18
1	1 H Hydrogen 1.008																	2 He Helium 4.003
2	3 Li Lithium 6.941	4 Be Beryllium 9.012											5 B Boron 10.81	6 C Carbon 12.01	7 N Nitrogen 14.01	8 O Oxygen 16.00	9 F Fluorine 19.00	10 Ne Neon 20.18
3	11 Na Sodium 22.99	12 Mg Magnesium 24.31											13 Al Aluminum 26.98	14 Si Silicon 28.09	15 P Phosphorus 30.97	16 S Sulfur 32.07	17 Cl Chlorine 35.45	18 Ar Argon 39.95
4	19 K Potassium 39.10	20 Ca Calcium 40.08	21 Sc Scandium 44.96	22 Ti Titanium 47.87	23 V Vanadium 50.94	24 Cr Chromium 52.00	25 Mn Manganese 54.94	26 Fe Iron 55.85	27 Co Cobalt 58.93	28 Ni Nickel 58.69	29 Cu Copper 63.55	30 Zn Zinc 65.39	31 Ga Gallium 69.72	32 Ge Germanium 72.64	33 As Arsenic 74.92	34 Se Selenium 78.96	35 Br Bromine 79.90	36 Kr Krypton 83.80
5	37 Rb Rubidium 85.47	38 Sr Strontium 87.62	39 Y Yttrium 88.91	40 Zr Zirconium 91.22	41 Nb Niobium 92.91	42 Mo Molybdenum 95.94	43 Tc Technetium 98	44 Ru Ruthenium 101.1	45 Rh Rhodium 102.9	46 Pd Palladium 106.4	47 Ag Silver 107.9	48 Cd Cadmium 112.4	49 In Indium 114.8	50 Sn Tin 118.7	51 Sb Antimony 121.8	52 Te Tellurium 127.6	53 I Iodine 126.9	54 Xe Xenon 131.3
6	55 Cs Cesium 132.9	56 Ba Barium 137.3	57–71 Lanthanides	72 Hf Hafnium 178.5	73 Ta Tantalum 180.9	74 W Tungsten 183.8	75 Re Rhenium 186.2	76 Os Osmium 190.2	77 Ir Iridium 192.2	78 Pt Platinum 195.1	79 Au Gold 197.0	80 Hg Mercury 200.6	81 Tl Thallium 204.4	82 Pb Lead 207.2	83 Bi Bismuth 209.0	84 Po Polonium	85 At Astatine	86 Rn Radon
7	87 Fr Francium	88 Ra Radium	89–103 Actinides	104 Rf Rutherfordium	105 Db Dubnium	106 Sg Seaborgium	107 Bh Bohrium	108 Hs Hassium	109 Mt Meitnerium	110 Ds Darmstadtium	111 Rg Roentgenium	112 Cn Copernicium	113 Nh Nihonium	114 Fl Flerovium	115 Mc Moscovium	116 Lv Livermorium	117 Ts Tennessine	118 Og Oganesson

57 La Lanthanum 138.9	58 Ce Cerium 140.1	59 Pr Praseodymium 140.9	60 Nd Neodymium 144.2	61 Pm Promethium	62 Sm Samarium 150.4	63 Eu Europium 151.9	64 Gd Gadolinium 157.3	65 Tb Terbium 158.9	66 Dy Dysprosium 162.5	67 Ho Holmium 164.9	68 Er Erbium 167.3	69 Tm Thulium 168.9	70 Yb Ytterbium 173.0	71 Lu Lutetium 175.0
89 Ac Actinium	90 Th Thorium 232.0	91 Pa Protactinium 231.0	92 U Uranium 238.0	93 Np Neptunium	94 Pu Plutonium	95 Am Americium	96 Cm Curium	97 Bk Berkelium	98 Cf Californium	99 Es Einsteinium	100 Fm Fermium	101 Md Mendelevium	102 No Nobelium	103 Lr Lawrencium

Dmitri Mendeleev

Figure 2.3 Periodic Table of the Elements

Reading the periodic table is like reading a page in a book—start at the top left and read to the right, then return to the left and read to the right again. The horizontal rows in the periodic table are referred to as **periods** (sometimes called *rows*), and they are numbered down the left side of the table. Along the top of the periodic table, we can see that each column is marked with a number starting with 1 on the far left to 18 on the far right. Each column is called either a **group** or a *family*.

The symbol for hydrogen (H) is at the top left corner of the periodic table. Above it is the number 1, its atomic number. (Don't worry yet about the number below the symbol. We will get to that soon.) Starting at the box for hydrogen, H, and moving across the page directly to the right, we see the second element, He, along with its atomic number, 2. Then, going back to the left to the beginning of the next line, we see the next element, lithium, followed by beryllium, one box to the right. Another jump to the right brings us to boron (symbol B, atomic number 5). Let's pause on the box for boron for just a moment.

Each jump we make to a new box adds 1 to the atomic number of the element. Each jump also means one more proton, because the atomic number is equal to the number of protons. It also means one more electron, because atoms must have enough electrons to balance out the protons. Thus, boron contains five protons and five electrons (and some neutrons).

Now let's go to atomic number 42. Element 42 is molybdenum (symbol Mo)—one of the trickiest element names to pronounce. This is the element that puts the "moly" in *cromoly,* the type of steel used to make mountain bike forks. Molybdenum (pronounced *muh-LIB-den-um*) contains 42 protons and 42 electrons (and some neutrons). As we move through the periodic table from left to right, all the atoms of a given element contain more protons and electrons (and generally more neutrons) than do the atoms of any element coming before it in the periodic table.

As we continue to make our way down the periodic table, we run into a problem at element 57, lanthanum (La). Here, we must take a detour to the first of two rows at the bottom of the periodic table. These elements, called the *lanthanides* (the top row) and the *actinides* (the bottom row), got stuck at the bottom because they don't fit nicely into the main part of the periodic table—if we squeezed in the lanthanides and actinides, the periodic table would have 32 columns rather than just 18.

QUESTION 2.4 The two rows at the bottom of the periodic table are the lanthanides (top row) and the actinides (bottom row). Looking at the periodic table, can you guess how the names for these two periods arose?

ANSWER 2.4 The element that begins the lanthanide elements is lanthanum (La), and the element that begins the actinides is actinium (Ac). These periods are named for the element that comes right at the beginning of each series.

For more practice with problems like this one, check out end-of-chapter Question 18.

FOLLOW-UP QUESTION 2.4 An atom of which element contains 104 protons and 104 electrons?

Scientists make and study transuranium elements in specialized laboratories.

Finally we jump back up to the main part of the periodic table to atomic number 104, rutherfordium (symbol Rf), and we are nearly at the end of the table. Things get interesting here. All of the elements after element 92, uranium (symbol U), are artificial and do not exist in nature, except in rare instances. We refer to them as the *transuranium elements.* These very last elements, those with the largest atomic numbers, are the ones that have been created recently.

Right now, as you are reading this, scientists somewhere in the world are trying to either create or verify the existence of the next element. They create new atoms in specialized laboratories, where atoms of smaller elements can be smashed together with enormous force to make a larger atom. At the time of this writing, the element with the highest atomic number to have an official name is organesson, element 118 (Og), named for the Russian physicist YURI Organessian.

In 2004, Japanese scientists reported that they had created element 113 (**Figure 2.4**). Their evidence was compelling

RIKEN Nishina Center for Accelerator-based Science

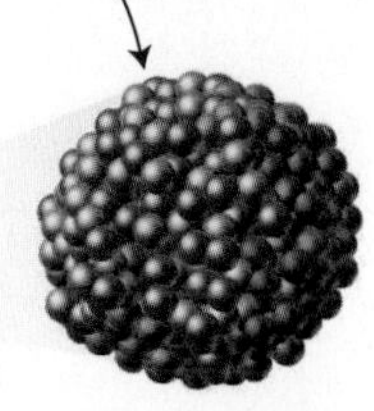
The nucleus of element 113 contains 113 protons and many neutrons.

Figure 2.4 A New Element?
This Japanese scientist was part of the Japanese team of scientists who reported the creation of element 113.

enough for element 113 to earn a temporary name, ununtrium from the International Union of Pure and Applied Physics (IUPAP), the agency that oversees the naming of new elements. In 2016, the element's discovery was confirmed and the Japanese team named the element Nihonium (Nh), which is derived from the Japanese word for Japan.

wait a minute . . .

Why aren't the elements discovered in sequence?

Good question! Look closely at period 7 of the periodic table, and you'll notice that the elements are listed continuously up through 118, but they were not discovered, confirmed, and named in sequence. Why aren't elements discovered in sequence? The answer is that some elements are naturally easier to create and observe than others. When elements 114 and 116 were created, other scientists were able to verify these observations and these elements received official names. Other elements are more elusive and take longer to be confirmed and officially named.

2.3 Why Neutrons Matter

Atoms of a given element can contain different numbers of neutrons.

Recall that atoms contain equal numbers of protons and electrons, plus some neutrons. We know that the number of protons gives an element its identity and that the electrons are the outermost particles in the atom. But what of the neutrons in the nucleus? What do they do? Why should we pay attention to them? Let's check out the neutrons in the atom and see why they matter.

The total number of neutrons plus protons in the atom's nucleus is the atom's **mass number**:

$$\text{mass number} = \text{number of protons} + \text{number of neutrons}$$

This mass number is sometimes shown either as a superscript before the symbol for an element or as a number following the element name, as shown in **Figure 2.5**.

The number of neutrons in one atom of a given element can be different from the number of neutrons in another atom of the same element. (Remember, the number of protons determines the identity of an element.) **Isotopes** are atoms that have the same number of electrons and the same number of protons, but different numbers of neutrons. For example, as we can see in Figure 2.5, both titanium-48 and titanium-50 represent the element titanium, and the two forms are isotopes of the element. The periodic table tells us that titanium has 22 protons. We know therefore that titanium-48 and titanium-50 must each contain 22 protons, because they are both titanium. The difference is in the numbers of neutrons

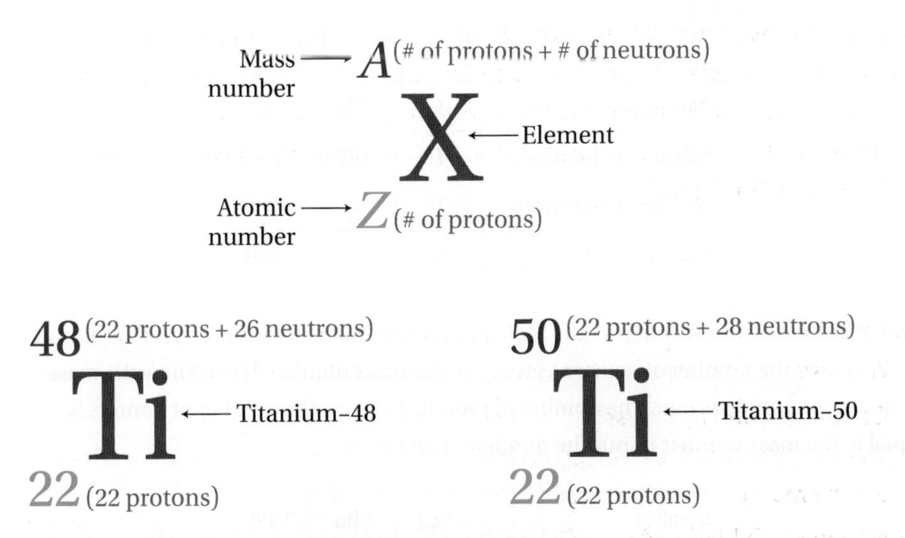

Figure 2.5 Atomic Symbols for Two Isotopes of Titanium

indicated by the two symbols. For the isotope that has a mass number of 48, the number of neutrons is:

$$\text{number of neutrons} = \text{mass number} - \text{number of protons} = 48 - 22 = 26$$

For the isotope that has a mass number of 50, the number of neutrons is:

$$\text{number of neutrons} = \text{mass number} - \text{number of protons} = 50 - 22 = 28$$

Both isotopes have 22 protons because they are both titanium. But they have different numbers of neutrons, so titanium-48 and titanium-50 are isotopes of one another.

QUESTION 2.5 Silver exists in nature as silver-107 or silver-109. (a) How many protons, neutrons, and electrons does an atom of each of these isotopes contain? (b) Of the two silver isotopes, which would you expect to have the greater mass?

ANSWER 2.5 (a) According to the periodic table, any isotope of silver must contain 47 protons. Because the isotopes are electrically neutral, the number of electrons must be equal to the number of protons. So both of these isotopes of silver contain 47 electrons. The number of neutrons is the mass number minus the number of protons: $107 - 47 = 60$ neutrons for silver-107, and $109 - 47 = 62$ neutrons for silver-109. (b) Silver-109 has two more neutrons than silver-107. Neutrons are tiny, but they do have mass. Therefore, silver-109, with 62 neutrons, is more massive than silver-107, which contains only 60 neutrons.

For more practice with problems like this one, check out end-of-chapter Questions 22 and 23.

FOLLOW-UP QUESTION 2.5 A certain isotope has a mass number of 192 and 76 protons. This is an isotope of which element? How many neutrons does it contain?

QUESTION 2.6 Complete the table (at the top of p. 42) by writing the appropriate numbers and symbols in each empty box. The first element is done for you.

Symbol	^{35}Cl		
Number of protons	17	56	82
Number of neutrons	18	82	
Number of electrons	17		
Mass number	35		208

ANSWER 2.6 The number of protons indicates the symbol of the element and the atomic number, and the number of neutrons gives you the mass number. If you know the mass number, you can determine the number of protons because the number of protons is equal to the mass number minus the number of neutrons.

Symbol	^{35}Cl	^{138}Ba	^{208}Pb
Number of protons	17	56	82
Number of neutrons	18	82	126
Number of electrons	17	56	82
Mass number	35	138	208

For more practice with problems like this one, check out end-of-chapter Questions 25 and 26.

FOLLOW-UP QUESTION 2.6 Germanium-74 is the most common isotope of this element. How many protons does this isotope of germanium contain? How many neutrons?

Samples of matter from different locations have distinctive distributions of isotopes.

When atoms interact with one another, the number of neutrons in an atom is often unimportant; atoms usually react in the same way regardless of the isotopes involved. However, in many instances, isotopes are crucial. For example, most carbon atoms are carbon-12, a few are carbon-13, and very few are carbon-14 and carbon-11. In living organisms, the ratio of carbon-12 to carbon-14 remains constant, but the concentration of carbon-14 decreases when an organism dies. (We will discuss why this happens in Chapter 11, "Nukes.")

This means that once an organism dies, the ratio of C-12 to C-14 gradually increases because C-12 does not change, but C-14 slowly disappears. Therefore, the ratio of carbon-12 to carbon-14 is an indicator of the current age of something that was once alive. This method of measurement, called *carbon dating*, was used to determine the age of Ötzi, the prehistoric Austrian "iceman" discovered by hikers in 1991. Carbon dating showed that he was buried in ice about 5300 years ago (**Figure 2.6**).

Neutrons are useful for much more than carbon dating. The pie charts in **Figure 2.7** show, for four elements, the percentages of each isotope found in a typical naturally occurring sample. As the figure illustrates, a given element can have a small number of isotopes, as is the case for chlorine (two isotopes) and for carbon (three isotopes), or there can be a large number of isotopes, as is the case for calcium and tin (six and 10 isotopes, respectively).

Figure 2.6 Ötzi
Carbon dating was used to determine the age of 5300-year-old Ötzi, who was found mummified in ice near the border between Italy and Austria in 1991. Isotopic studies on Ötzi's teeth have revealed that he lived 12 miles away from the site where he was found. We also know that his last meal was venison and cereal, he died of an arrow shot to the chest, and he had 11 tattoos.

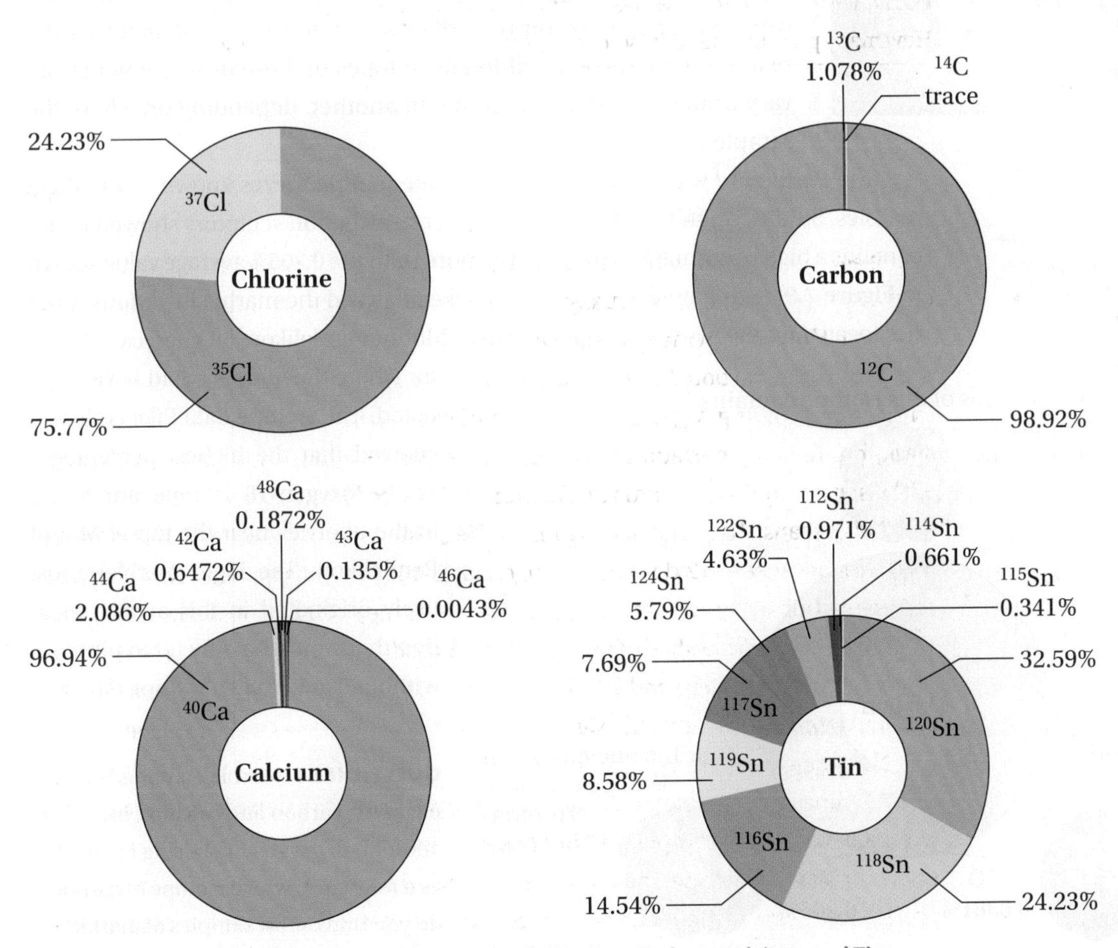

Figure 2.7 The Natural Abundances of Isotopes of Chlorine, Carbon, Calcium, and Tin

The percentages shown in Figure 2.7 are average values calculated from large numbers of samples taken from around the world. The isotope percentages in any one sample taken from a specific location may differ from those shown in the figure. A sample of dirt taken from a particular farm in Nebraska, for instance, may contain 75.75% chlorine-35 and 24.25% chlorine-37. Or, a sample of dirt collected in the Mekong Delta of Vietnam may contain 74.31% chlorine-35 and 25.69% chlorine-37. We can use such variation in isotope ratios to determine where a particular sample of dirt or piece of rock—or any other natural object—came from, as the next example illustrates.

Isotopic measurements can reveal answers to questions, such as the origin of marble in ancient monuments.

Figure 2.8 Isotopes in Ancient Greece

This photo shows a portion of the Elgin marbles, which once adorned the Parthenon, the temple to the goddess Athena built in 447 B.C. in Athens, Greece.

The Parthenon is a temple in Athens, Greece, dedicated to the goddess Athena. Construction of this most important surviving building of Classical Greece began in 447 B.C., when the Athenian Empire was at the height of its power. As a graduate student at the University of Georgia, Scott Pike wanted to answer a question that has been on the minds of art historians and archeologists for years: where did the marble used by the builders of the Greek Parthenon (**Figure 2.8**) come from? Marble is a natural substance made up of atoms of calcium (Ca), oxygen (O), and carbon (C). Pike was interested in the fact that the relative amounts of the different isotopes of these three elements can vary from one sample of marble to another, depending on where the sample was mined.

The Parthenon was decorated with intricate marble friezes known as the Elgin marbles. Studies on the marble used by ancient Parthenon sculptors showed that it contains a high percentage of oxygen-18, more than the 0.205% average value shown in **Figure 2.9**. Acting on this information, Pike analyzed the marble in various quarries located on the southern slope of nearby Mount Pendelikon, all good candidates for where the marble could have originated. The isotope data Pike collected showed that the highest percentages of the oxygen-18 isotope are found in the quarries near the top of Mount Pendelikon. The Elgin marbles most likely originated in this area, a location that would be difficult to pinpoint without using oxygen isotope data.

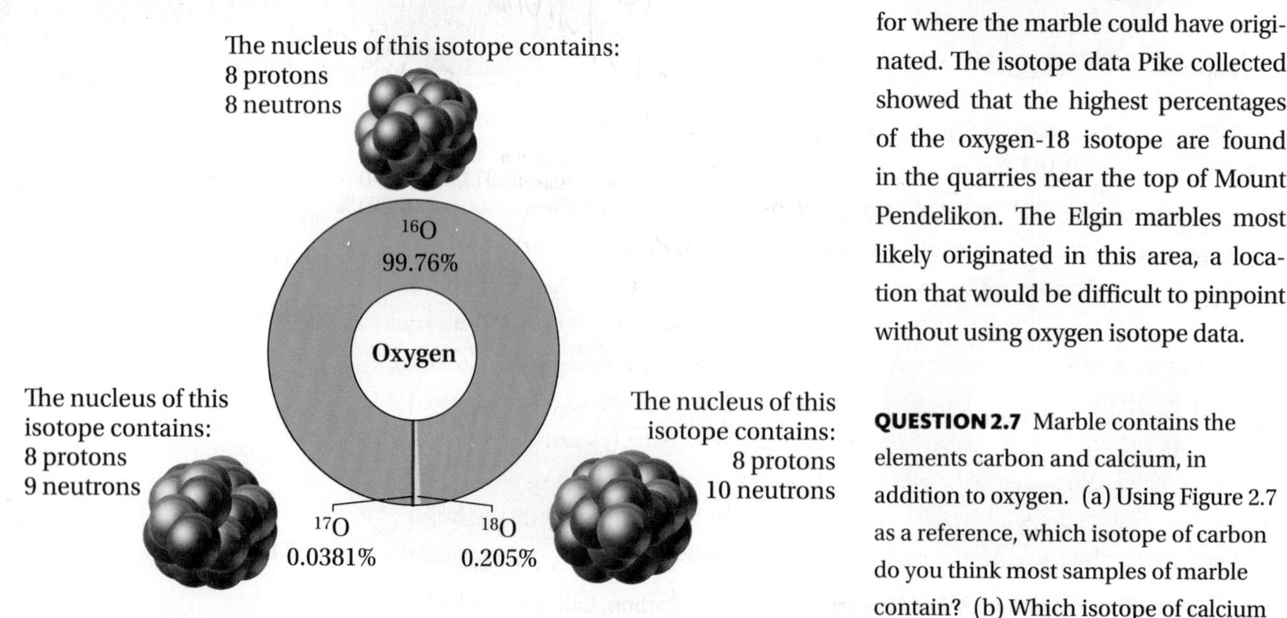

Figure 2.9 Distribution of the Isotopes of Oxygen

QUESTION 2.7 Marble contains the elements carbon and calcium, in addition to oxygen. (a) Using Figure 2.7 as a reference, which isotope of carbon do you think most samples of marble contain? (b) Which isotope of calcium do most samples of marble contain?

ANSWER 2.7 (a) carbon-12 (b) calcium-40

For more practice with problems like this one, check out end-of-chapter Question 28.

FOLLOW-UP QUESTION 2.7 Which isotope of oxygen do most samples of marble contain?

2.4 Electrons: The Most Important Particle for Chemists

Electrons are elusive.

Thus far, we have learned why protons are important: their numbers are always fixed for a given element. Protons define an element. We also know that neutrons are important because they usually are not present in equal numbers in all atoms of an element. This fact allows us to find the age of ancient skeletons or the source of Greek marble. In the world of chemistry, though, electrons are the undisputed superstars. Chemists spend a lot of time thinking about them, and we are about to see why.

Imagine this scenario: You take your camera out into the city late at night to photograph cars moving down a busy urban highway. You set up your tripod next to the interstate and snap a picture. The photograph looks something like **Figure 2.10**. Now answer this question: exactly where was each of the cars in the photo in the instant when you snapped it?

The answer to this question is tricky because the cars are moving really fast, and we cannot say that they are in one exact spot in the instant the photo was captured. Electrons are like this, too. They are difficult to pinpoint, so we often use vague terms to talk about where they are. Just as we could call the cars in the photo a "blur of lights," we could refer to electrons as "a smear of negative charge."

zodebala/E+/Getty Images

Figure 2.10 Fast-Moving Traffic at Night
Can you determine the exact location of each car in the photograph? This situation is analogous to fast-moving electrons in the atom.

Although electrons are elusive and hard to pinpoint, they do occupy different areas of space around the nucleus. In a general sense, we can think of all the electrons around the nucleus, collectively, as **electron density,** a term we use to describe the elusiveness of fast-moving electrons. But, in fact, some electrons are located closer to the nucleus than others. Some are farther away. The electrons that are farther away are more likely to be involved in **chemical reactions,** which are interactions between atoms or groups of atoms. In a chemical reaction, atoms in substances can give up electrons to other atoms, take away electrons from other atoms, or share electrons with other atoms. When these things happen, it is the outermost electrons that are easiest to pluck away from the atom, because they are held less tightly by the positive charge of the nucleus. The electrons closer to the nucleus are held more tightly, so they tend to stay put during chemical reactions. Chapter 7, "Chemical Reactions," is devoted to this topic.

recurring theme in chemistry 4

Chemical reactions are nothing more than exchanges or rearrangements of electrons.

QUESTION 2.8 How many electrons does one atom of each of these elements contain? (a) zirconium, Zr (b) fermium, Fm (c) samarium, Sm

ANSWER 2.8 The number of protons is equal to the number of electrons for atoms. (a) Zirconium atoms have 40 electrons. (b) Fermium atoms have 100 electrons. (c) Samarium atoms have 62 electrons.

For more practice with problems like this one, check out end-of-chapter Question 27.

FOLLOW-UP QUESTION 2.8 If you sum the electrons in one atom of sulfur (S), one atom of carbon (C), and one atom of scandium (Sc), what total number of electrons would you have? An atom of which element possesses this number of electrons?

Energy levels are a way to imagine the distribution of electrons around the nucleus.

Because some electrons are involved in chemical reactions and others are not, it is useful to think of individual electrons as having "locations" around the nucleus, even though we understand that those exact locations are tricky to pinpoint.

We determine the location of electrons in an atom, in part, by the amount of space around the nucleus. Each electron has one negative charge, and negative charges repel one another. So, for instance, 20 electrons cannot cram into the space immediately next to the nucleus. Instead, electrons are distributed around the nucleus in a way that puts each one as close to the nucleus as possible while allowing for the fact that electrons are negatively charged and repel one another.

If we imagine that atoms possess a series of **energy levels,** which get progressively farther from the nucleus, we can then fill these imaginary energy levels with electrons. The level closest to the nucleus is the level of lowest energy, and the energy of each successive level increases as we move away from the nucleus. We call the level closest to the nucleus the *first energy level,* the next one out from the nucleus the *second energy level,* and so forth.

The energy levels in an atom are depicted in **Figure 2.11**. In the figure, the number indicated on each level is the number of electrons that can be placed in that level. Notice that the number of electrons for each energy level gets larger and larger as we move farther from the nucleus. The first level can hold only two electrons, and the fourth level can hold 32 electrons. Keep in mind that Figure 2.11 presents a simplified model of electrons in atoms; the actual energy-level model is much more complex than the figure suggests. Each energy level is really made up of a complicated system of energy sublevels that we will not examine in this book.

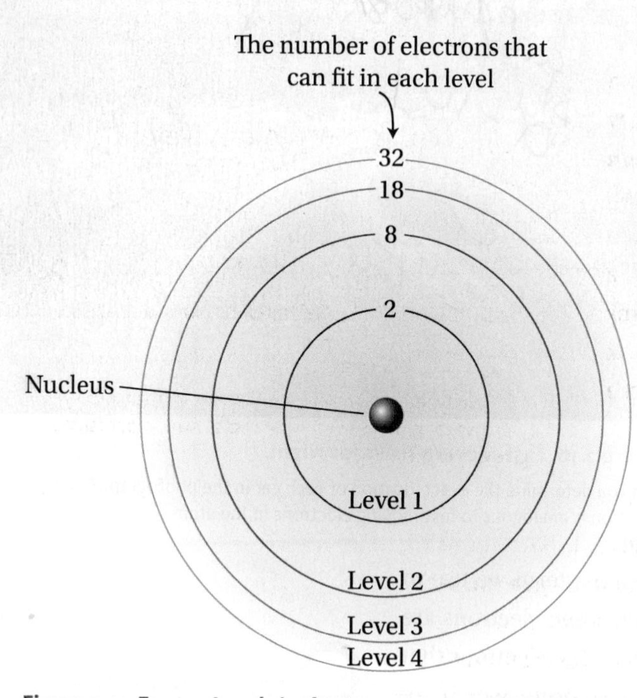

Figure 2.11 Energy Levels in the Atom

This schematic drawing indicates the number of electrons that each energy level can "hold." Atoms with just a few electrons keep those electrons close to the nucleus in the first few energy levels. In larger atoms, which have more electrons, the electrons in an atom are located at greater and greater distances from the nucleus. The assignment of electrons to locations around the nucleus is part of this model to help us understand the behavior of atoms; it does not actually show what an atom looks like.

The location of an electron with respect to the nucleus determines its role in the atom.

It's important to remember two things about this system of classifying electrons in an atom. First of all, each electron is the same kind of particle as all the other electrons in the atom. Each has one negative charge and the same mass. However, they differ from one another because of their location in the atom, and an electron's location dictates how tightly the nucleus will hold onto it. The second thing to remember when we're talking about the layered arrangement of electrons in an atom is that this is merely a model scientists have developed so we can visualize the atom more easily. Electrons do not actually stay in fixed orbits at the same distance from the nucleus, as Figure 2.11 seems to suggest, but this model can help us understand that electrons exist in different environments around the atom.

2.5 Light and Its Interaction with Atoms

Electrons can be excited to higher energy levels.

If you have ever had a window seat on an evening airplane flight, you may have noticed that street lights, like the ones used in parking lots, are typically one of two colors: bluish or yellowish. **Figure 2.12** shows a city scene in Helsinki, Finland, at night. The two distinct colors are easy to see.

These two colors of light are linked to two elements: mercury (bluish) and sodium (yellowish). The lamps that give off bluish light, called *mercury vapor lamps,* contain mercury atoms; the lamps that give off yellowish light, called *sodium vapor lamps,* contain sodium atoms. Both types of lamp have a transparent housing that is filled with atoms. When a source of electricity is applied to the atoms, their electrons become energized and move to higher energy levels. When those atoms relax and return to a lower energy level, they may give back energy in the form of light. To see how this happens, we can use a simple atom—hydrogen—as an example.

Estea/Shutterstock

Figure 2.12 Streetlights in Helsinki, Finland, at Night
Two common types of bulb are used in street lights: mercury vapor lamps that give off bluish light and sodium vapor lamps that give off yellowish light.

Let's consider exactly what happens when we energize the electrons in some hydrogen atoms with electricity. Recall that hydrogen atoms each have one electron located in energy level 1. Because this is the lowest energy level possible for that electron, we say it is in its **ground state**. When that electron is energized, the added energy allows it to move to a higher energy level. When that happens, we say the electron occupies an **excited state**. So, how do we know which energy level is the excited state? How do we know how far that electron will jump from the ground state? The answer depends on the amount of energy applied to the atom. The more energy that is applied, the higher the energy level the electron can attain.

When the electron in a hydrogen atom is energized, it can jump to another energy level, but it cannot jump halfway to another energy level. For example, a certain amount of energy is required for an electron to jump from energy level 1

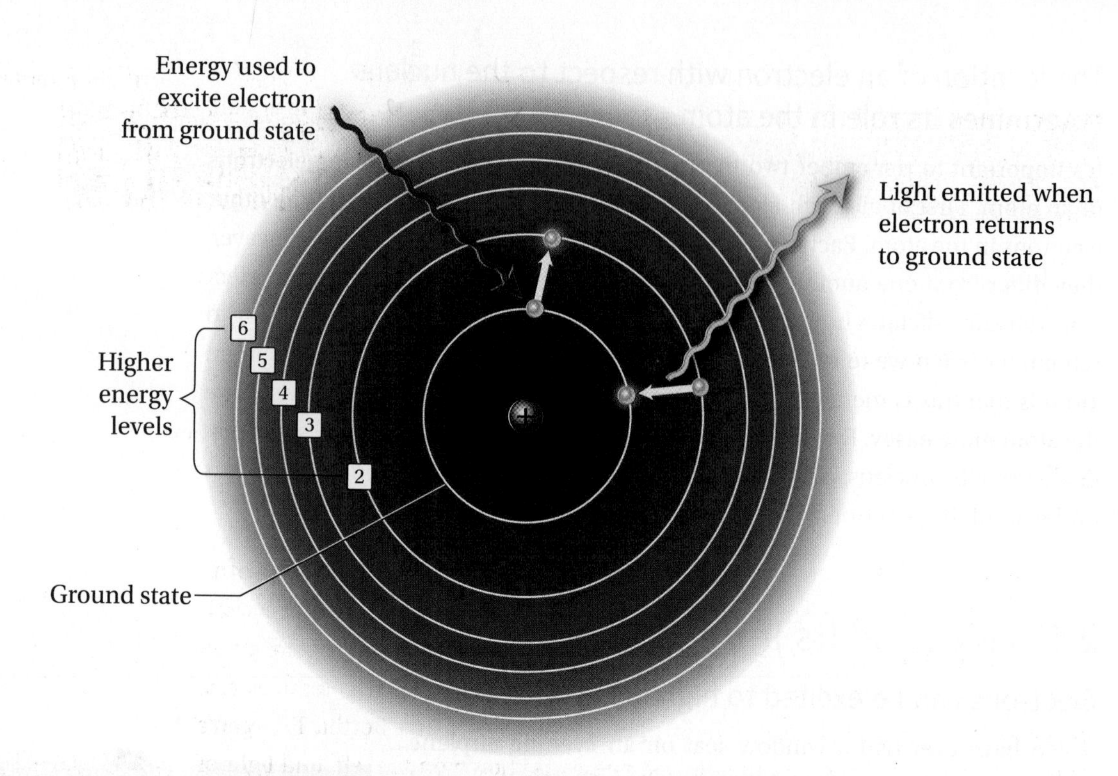

Figure 2.13 Moving between Energy Levels in a Hydrogen Atom

An electron in the ground state of a hydrogen atom can be excited by the exact amount of energy required to move it to level 2. That energy can then be emitted in the form of light when the electron returns to the ground state.

to the empty energy level 2 in a hydrogen atom. When that amount of energy is supplied to the atom, exactly that set amount of energy is used when the electron makes that jump. If too little energy is supplied, the jump to level 2 cannot happen.

Once the electron has moved to a higher energy level, it can return to the ground state. When it does, the amount of energy it used to move to the higher energy level is released, often in the form of radiant energy. Radiant energy is also known as **light**. The energy of that light is equal to the energy difference between levels 1 and 2 of a hydrogen atom, as depicted in **Figure 2.13**. The set amount of energy required to jump between these two specific energy levels is the same for every hydrogen atom.

If we excite an electron in a hydrogen atom to energy level 5 and it then jumps down to energy level 2, the light that is given off happens to look blue to the human eye. It appears blue because the energy of blue light is exactly the same as the difference in energy between levels 5 and 2 in a hydrogen atom. But what is blue light, exactly, and how is the color of light related to the energy of light? To fully understand the relationship between light and color, we must understand the nature of light first. Once we understand light, we can return to our sodium and mercury vapor lamps and see why each one has the color it has.

Light is electromagnetic radiation.

Most people think of light as the brightness we see from a light bulb or from the sun, but scientists define this term much more broadly. In a scientific context, light is a form of radiant energy. Light is also referred to as **electromagnetic radiation**.

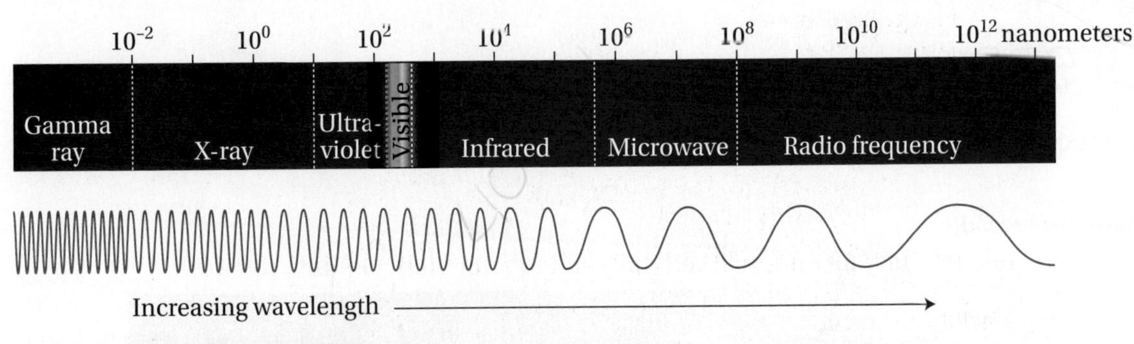

Figure 2.14 The Electromagnetic Spectrum

The various forms of light make up the electromagnetic spectrum. The highest-energy forms of light are to the left and the lowest-energy forms of light are to the right.

There are several kinds of light. The **electromagnetic spectrum** is a collection of all types of light. It is a continuum of energies, and different regions of the spectrum represent different types of light, as shown in **Figure 2.14**. These types are given different names, many of which are probably familiar to you, such as radio waves, microwaves, and X-rays—the type of light used to see Anna Bertha Röntgen's bones. In Figure 2.14, light of the highest energy is shown on the left, and light of the lowest energy is shown on the right.

The kinds of light we usually classify as dangerous are at the high-energy end of the spectrum. For example, we know that ultraviolet (UV) light can cause severe sunburn and that we should protect ourselves from X-rays by using heavy metal shields. On the right in Figure 2.14 are the more benign varieties of light, including radio waves and infrared light. Most people don't worry about the adverse effects of radio waves on their health (except when bad music is transmitted by them!). **Visible light,** the only light that human eyes are sensitive to, is located in the middle of the electromagnetic spectrum.

Infrared light is most often associated with heat, and cameras loaded with film sensitive to infrared light give us a picture of hot and cold areas in a photograph. An infrared photo of a cat's face, for example, might show the warmer areas in red and the cooler areas in blue. Which parts of the face would you expect to show up in which color—in other words, which part of the face is warmer and which part is cooler? See **Figure 2.15** to find out the answer.

Ted Kinsman/Science Source

Figure 2.15 Infrared Photograph of a Cat's Face

In this photograph, the blue region is cool and the red regions are warm.

QUESTION 2.9 Rank the following types of light from lowest energy to highest energy: microwave, radio, gamma, infrared, visible.

ANSWER 2.9 radio $<$ microwave $<$ infrared $<$ visible $<$ gamma

For more practice with problems like this one, check out end-of-chapter Question 43.

FOLLOW-UP QUESTION 2.9 Between which two types of light on the electromagnetic spectrum does the energy of the light start to become harmful to human bodies?

flashback

In Section 1.3, we discussed metric units and how to convert between them. Table 1.2 shows that a nanometer, the unit used to express the wavelength of visible light, is much, much smaller than the meter. In fact, there are 1,000,000,000—1 billion—nanometers in 1 meter. We can also say there are 1×10^9 (or, simply, 10^9) nanometers in 1 meter. The first letter of the prefix for *nano-* is *n*, so nanometer is abbreviated as nm.

The greater the energy of light, the shorter its wavelength.

Why are some kinds of light more harmful than others? A closer look at Figure 2.14 indicates that light is classified according to its **wavelength**—the distance from one peak to the next of a wave of light. The closer together the peaks in a wave of light,

natureBOX • Are Cell Phones Safe to Use?

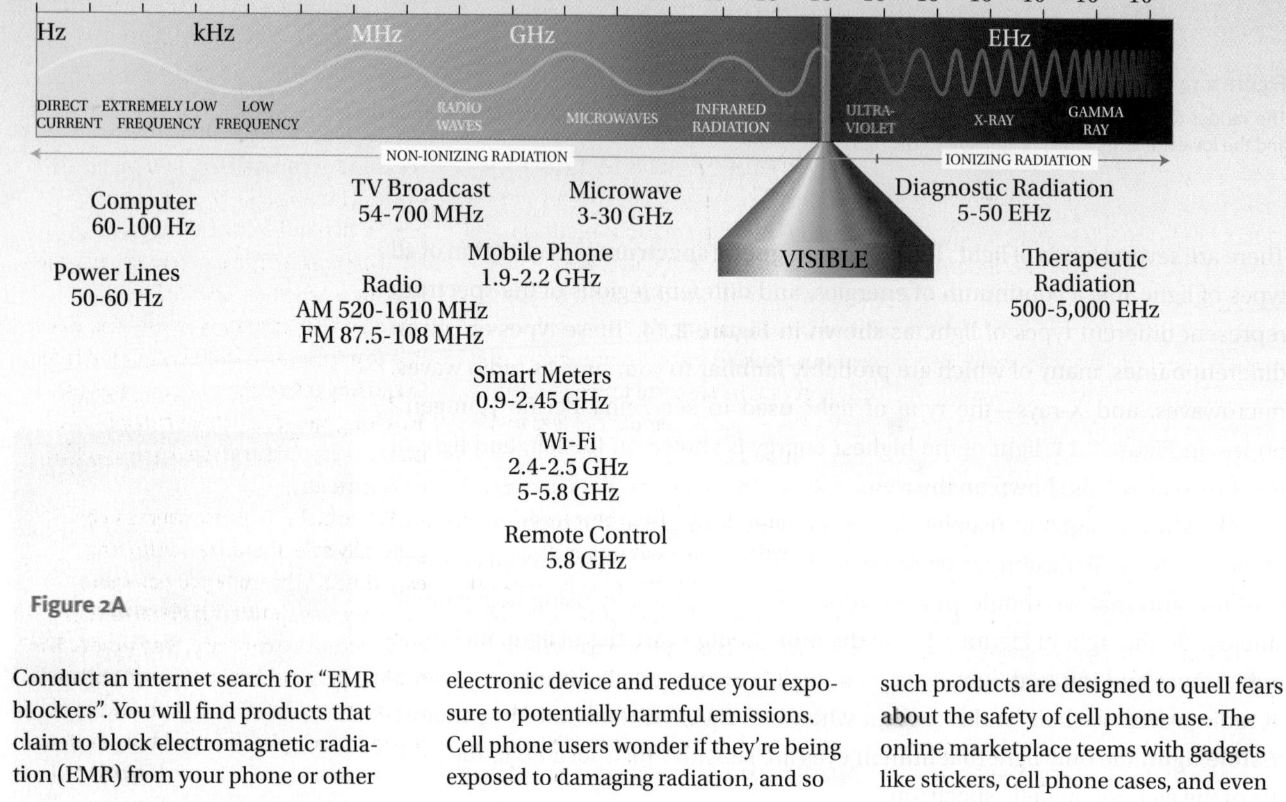

Figure 2A

Conduct an internet search for "EMR blockers". You will find products that claim to block electromagnetic radiation (EMR) from your phone or other electronic device and reduce your exposure to potentially harmful emissions. Cell phone users wonder if they're being exposed to damaging radiation, and so such products are designed to quell fears about the safety of cell phone use. The online marketplace teems with gadgets like stickers, cell phone cases, and even

the higher the energy of the light. Or, to say it another way, the shorter the wavelength, the higher the energy, as shown in **Figure 2.16**. The more energy light has, the shorter its wavelength and the more capable it is of penetrating solid matter, for example, human bodies.

The visible portion of the electromagnetic spectrum is expanded in **Figure 2.17**. Visible light spans the wavelengths of the electromagnetic spectrum from

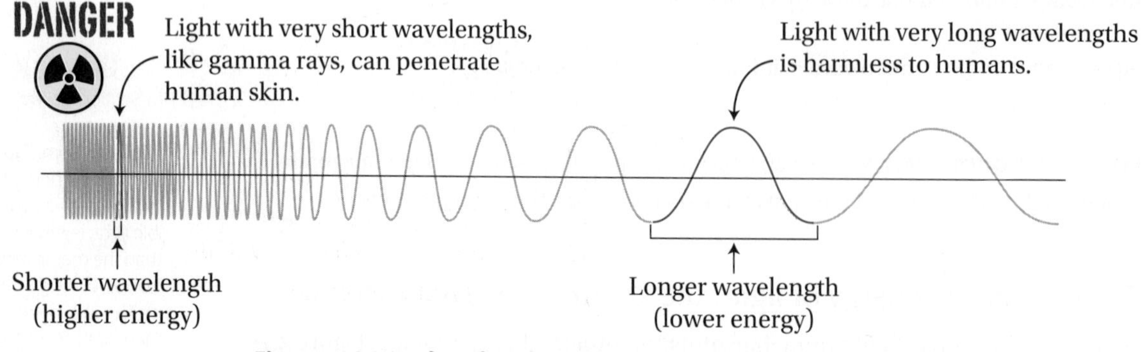

Figure 2.16 Wavelength and Energy

One wavelength is the distance between any two adjacent peaks in a wave. In the figure, shorter wavelength light is shown to the left and longer wavelength light is shown to the right.

jewelry that claim to protect you from harmful electromagnetic rays.

Let's say you invest in a new $45 EMR blocker, which often is nothing more than a small sticker that attaches to your phone case. You can test the sticker's efficacy by making a phone call. If your call goes through and you can hear the person on the other end, then the phone is still communicating with its cell phone tower, and you have not successfully blocked radiation that the phone is emitting. Many EMR-blocking devices are actually mere money-making scams.

Although many EMF blockers are scams, should we still be concerned about cell phone safety? Do cell phones really emit harmful radiation? To answer these questions, recall the electromagnetic spectrum from earlier in this chapter. **Figure 2A** features a version of the electromagnetic spectrum that shows where radiation is emitted by modern electronic devices. The highest-energy types of radiation are shown on the right side of the figure. Since these types of radiation, such as X-rays, are high in energy, they are also the most penetrating and the most dangerous. We protect ourselves from X-rays with lead aprons because high-energy forms of radiation like X-rays cause damage to our bodies. Gamma and X-ray radiation are types of "ionizing radiation" because they produce ions that lead to damage of cells and deoxyribonucleic acid (DNA), and perhaps cancer.

Cell phones and other wireless devices communicate with cell phone towers using radio waves, which appear toward the left end of the electromagnetic spectrum in the figure. Radio waves are types of nonionizing radiation and should be safe. Nonionizing types of radiation are found on the low-energy end of the electromagnetic spectrum, lower in energy than visible light. They are nonionizing because they are not energetic enough to cause significant cellular and DNA damage. We are bombarded with these lower-energy types of radiation constantly because they are widely considered to be innocuous, and they are therefore used for everything from Wi-Fi and Bluetooth devices to AM and FM radio.

Since cell phone usage has become ubiquitous, studies have been commissioned to answer this question: Radio waves are supposed to be safe, but does that mean that cell phones really are safe? Since cell phone and Wi-Fi use has blossomed globally, dozens of experiments have been performed to determine their safety. At the time of this writing, dozens of studies have been published, and meta-studies have summarized the state of the latest research performed. Currently, the general consensus is that nonionizing radiation, like radio waves, is not harmful and does not cause cancer.

However, there are some caveats to that general finding. First, children have thinner skulls and radiation of all types can penetrate their skulls more easily than radiation can penetrate adult skulls. Thus, even though there is not convincing evidence that childhood cancers are caused by cell phone use, caution is recommended. Second, a long-term study in the United States on EMR exposure in rats and mice suggests that there may be a low incidence of brain and heart tumors in rats and mice due to exposure to radiation in the radio frequency range used by cell phones. This long-range, multiyear experiment is ongoing and results are posted intermittently.

While cell phone use seems to be generally safe, there are continuing experiments that may provide more evidence to support this claim or more evidence to the contrary. Stay tuned. The National Cancer Institute posts online the latest results of all experiments related to cell phones, radiation, and cancer.

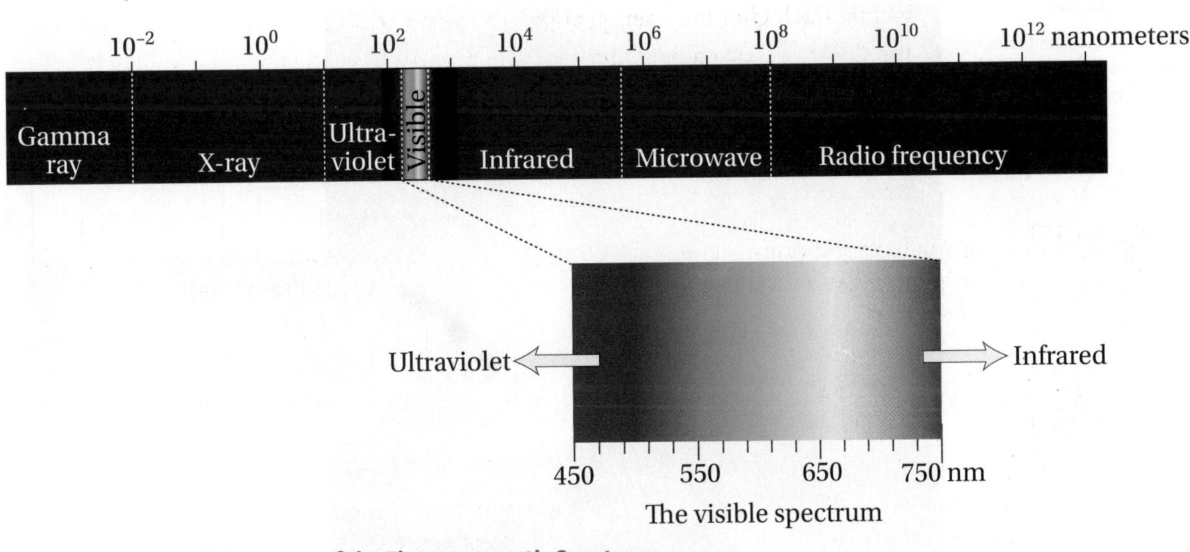

Figure 2.17 The Visible Region of the Electromagnetic Spectrum

The visible region of the electromagnetic spectrum includes wavelengths between 400 and 750 nm. The "red end" of the visible spectrum is the low-energy end. The "purple end" of the visible spectrum is the high-energy end.

400 to 750 nanometers (nm). These wavelengths of light are visible because it is the only region that can be seen by the human eye. It is not energetic enough to be harmful to humans, unless we count the effects of light pollution (see the naturebox above). The most energetic light in the visible spectrum is the light with the smallest wavelength: 400 nm. Light that has a wavelength of 400 nm is violet, and the visible spectrum continues to longer wavelengths from there: violet, blue, green, yellow, orange, and red. Past the high-energy end of the visible region, the violet end (below 400 nm), we enter the ultraviolet region. Past the low-energy end of the spectrum, the red end (above 750 nm), we enter the infrared region.

A line spectrum is a pattern of lines of light that is characteristic of a given element.

Light from the sun includes all the colors of the visible spectrum, in addition to light from other regions of the electromagnetic spectrum. We can see the entire visible portion of the electromagnetic spectrum—violet, blue, green, yellow, orange, red—when sunlight or a source of white light is shone through a prism. This assortment of colors should look familiar. The colors we see in a rainbow are those of the visible electromagnetic spectrum separated by water droplets, which act as tiny prisms in the air after a rainstorm (**Figure 2.18**).

Recall that when an atom is excited, electrons can be promoted to higher energy levels. When those electrons return to the ground state, energy is released, often in the form of light. Unlike light from the sun that gives off all the colors of the rainbow, excited atoms of one specific element give off only very specific wavelengths of light. These wavelengths correspond to the spacing between the energy levels of atoms of that given element. We refer to this series of specific wavelengths of light as a **line spectrum** (or *emission spectrum*).

To understand how a line spectrum arises, recall that specific, discrete energy levels are present in every atom and that any two energy levels in an atom are

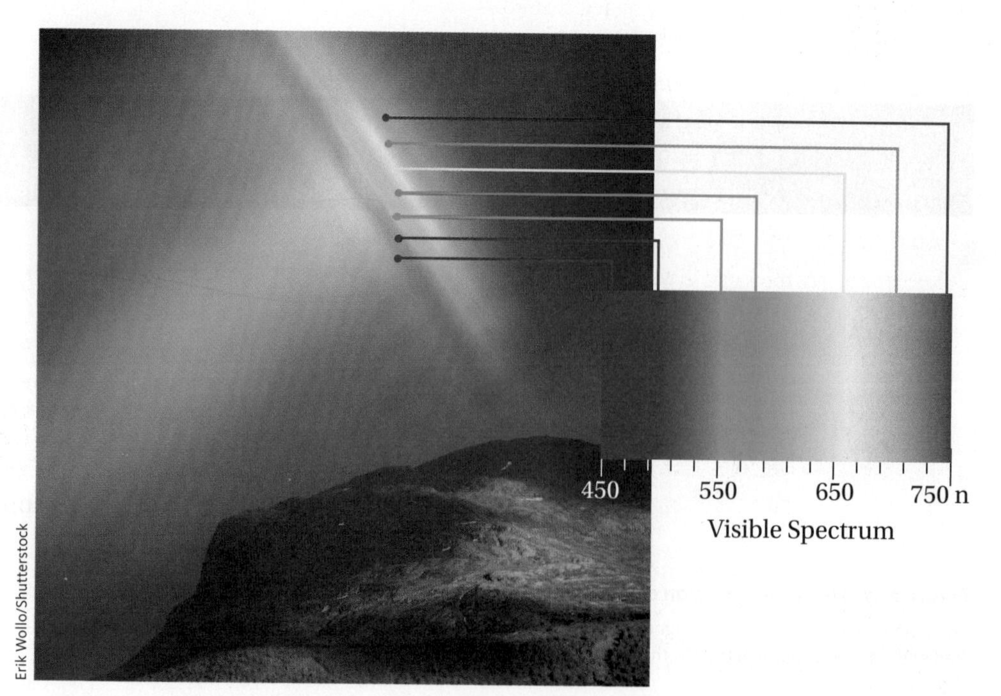

Figure 2.18 A Rainbow
The colors of the visible spectrum are separated by water droplets in the sky after it rains. The result is a rainbow.

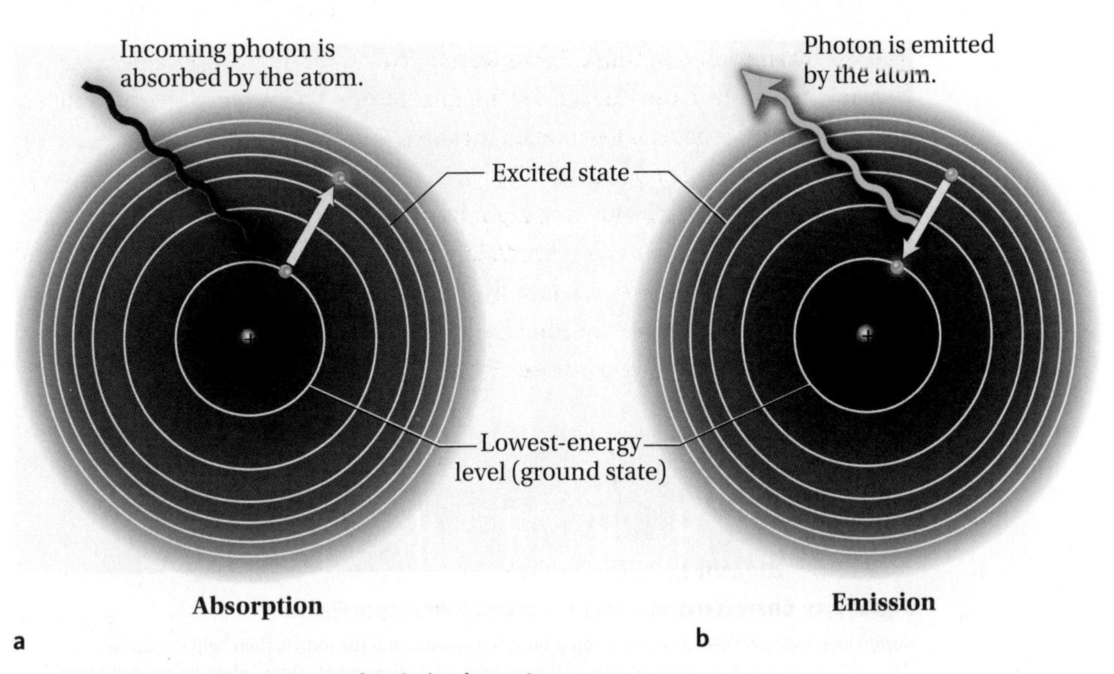

Figure 2.19 Energy Absorption and Emission by an Atom

(a) When energy is supplied, most often in the form of light or heat, an electron can absorb the energy and move from a lower energy level to a higher energy level. (b) When the electron returns to its original energy level, energy is given off (emitted), in this case in the form of light.

separated by a specific amount of energy. When a single atom is excited by a specific amount of energy, one of its electrons absorbs the energy and moves from the ground state to a higher energy level (**Figure 2.19a**). For this change to occur, the energy must be at least as great as the difference in energy between the energy level the electron leaves and the energy level the electron attains. The absorbed energy is then emitted when the electron goes back to its original state. In **Figure 2.19b,** the energy is released in the form of light. The light that results when the electron moves between a higher energy level and a lower energy level gives rise to a characteristic line spectrum for a given element.

Many elements produce a line spectrum in the visible region. For example, the excitation of sodium produces a brilliant yellow color. A line spectrum for sodium (**Figure 2.20**) is obtained when electrons in a sodium atom are excited by electricity, just as electricity excites electrons in a sodium vapor lamp in a parking lot. The relaxation of those electrons back to the ground state causes the emission of several wavelengths of light, among them visible light. For sodium, the dominant color of that visible light is yellowish, the characteristic color given off by a sodium vapor lamp.

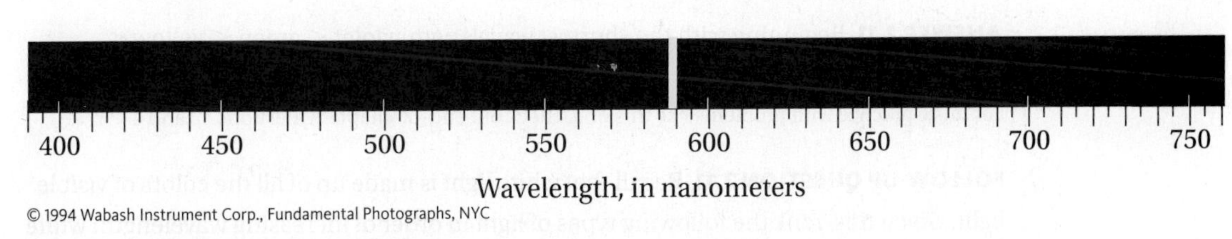

Figure 2.20 Line Spectrum of Sodium

Excited sodium atoms emit bright light at 590 nm, in the visible region of the visible spectrum. This yellow light gives sodium vapor lamps their characteristic color.

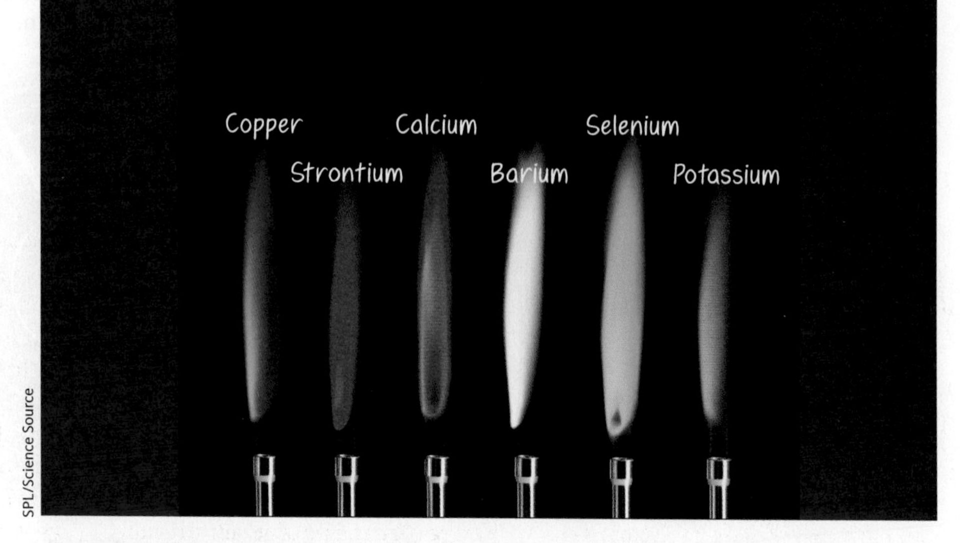

Figure 2.21 Characteristic Colors of Certain Elements in Flame
A wire loop is dipped into a solution containing an element, and the loop is then held in a flame. The color of the flame is characteristic of the element. Not all elements show bright colors, but these elements do.

Figure 2.21 shows other elements being excited, not by electricity, but by the heat of a flame. When these elements are heated in a flame, their electrons become excited and then give off brightly colored light as they return to a lower energy level. Mercury, which is not included in Figure 2.21, has a dominant blue color, the same color emitted by mercury vapor lamps. Still other elements are used to make fireworks, in which certain elements will produce flashes of light in specific colors.

QUESTION 2.10 To perform a flame test, you must dip a wire into a solution containing an unknown element and then hold the wire in a flame. The color of the flame allows you to identify the unknown element. Why can elements be distinguished in this way?

ANSWER 2.10 Each element has a unique distribution of energy levels. The colors of light that correspond to those energy differences are characteristic of that element.

For more practice with problems like this one, check out end-of-chapter Question 48.

FOLLOW-UP QUESTION 2.10 Which type of streetlamp light gives off light of lower energy, a mercury vapor lamp or a sodium vapor lamp?

QUESTION 2.11 Rank the following colors of visible light in order of increasing wavelength: yellow, orange, violet, green, red.

ANSWER 2.11 Beginning with the shortest wavelength: violet < green < yellow < orange < red.

For more practice with problems like this one, check out end-of-chapter Questions 42 and 43.

FOLLOW-UP QUESTION 2.11 Recall that white light is made up of all the colors of visible light. Given this, rank the following types of light in order of increasing wavelength: white light, ultraviolet light, infrared light.

THEgreenBEAT • news about the environment, chapter two edition

TOP STORY Have We Improved Our Light Bulbs?

Friday, September 24, 2010, was a dark day for incandescent light bulbs. On that day, in a town near Winchester, Virginia, the very last incandescent bulb factory in the United States shut its doors and chained its gates. Finally, the switch was flipped on Edison's nineteenth-century invention, which has provided us light for nearly 150 years. The factory closing was a bitter one for the more than 200 employees who lost their jobs that day. It was bittersweet for many who may feel nostalgia for the old bulbs, also known as incandescents, but won't miss the frequent trips to the light bulb box to replace them.

Incandescent bulbs typically run at about 10% efficiency; most of the energy put into them is lost as heat. In an effort to boost energy efficiency, the United States passed the Energy Independence and Security Act in 2007. The new law calls for the phaseout of any incandecent bulbs that do not meet certain energy efficiency requirements by 2014. By law, incandescents can still be sold, but they must abide by strict new standards that older-generation incandescents could not meet. A new generation of incandescents may one day be available.

Figure 2.22 A CFL Light Bulb
Consumers must wrestle with the fact that these newer light bulbs contain mercury.

Enter the CFL, the compact fluorescent light. You know these as the curly glass bulbs found nearly everywhere. For about $3.00—roughly three times the price of the old incandescent bulbs—you can purchase a CFL. But because CFLs last on average 10 times longer, you end up saving big bucks by switching to CFLs despite their higher cost per bulb.

To understand how CFLs work, recall your understanding of atomic structure. The element used in CFLs is mercury (Hg), which for decades has been used in mercury vapor lamps. Mercury lamps look bluish because the spacing between the energy levels in mercury atoms happens to result in emission of the color blue in the visible spectrum. (Mercury also emits in nonvisible regions of the electromagnetic spectrum, as discussed later in this article.)

Mercury is toxic to humans. So if the mention of mercury in the last paragraph made you wonder about the safety of CFLs, know that many other people share your concern. Because a CFL contains mercury, it should be disposed of in a way that prevents its mercury from entering the environment. Even though the amount of mercury is small, any amount of mercury means that these bulbs require special handling and special disposal (**Figure 2.22**). The old incandescents, on the other hand, were mercury-free.

The mercury issue is a concern, yes. But the benefits of the CFL have, in the minds of most consumers, made them worth the trouble that mercury may pose. With CFLs, we're spending a lot less time buying light bulbs, changing light bulbs, and paying for the inefficiency of incandescent bulbs. Besides the little problem of mercury, what's not to like about a CFL?

This question was answered recently by researchers at Stony Brook University in New York. They know a lot about atomic structure, and they wanted to know if the mercury in a CFL causes the bulb to give off harmful ultraviolet rays, in addition to visible light. In fact, as mentioned earlier, mercury *does* emit significant light in the nonvisible regions of the electromagnetic spectrum, especially the ultraviolet.

If you look closely at a CFL bulb, you may have noticed that it has a white powdery coating inside the glass. The job of that coating, called a *phosphor*, is to convert the ultraviolet light emitted by the mercury to visible light. The only light that makes it through the glass is supposed to be visible light. But the Stony Brook scientists figured out that, because the bulbs are made with thin glass tubes that are tightly coiled, the phosphor coating on the insides of those tightly curving tubes tends to crack. And when it cracks, ultraviolet light escapes.

The Stony Brook study also found that most CFLs available to consumers have cracks in the phosphor coating, and the ultraviolet light that escapes is significant and harmful. Ultraviolet light causes not only sunburn but also skin cancer, premature aging, and slow-healing wounds, among other things. If you are using CFLs in your home, it is recommended that you stay at least 2 feet away from them. Does a lampshade help to block harmful ultraviolet rays from your CFL? No, so it might be wise to apply sunscreen before you turn on your lamp! A *Scientific American* article (July 25, 2012) about the dangers of CFLs and the Stony Brook research findings states that "at close range, around a foot or so, CFL exposure is 'the equivalent of sunbathing at the equator.'"

Chapter 2 Themes and Questions

recurring themes in chemistry chapter 2

recurring theme 3 Opposite charges attract one another. (p. 34)

recurring theme 4 Chemical reactions are nothing more than exchanges or rearrangements of electrons. (p. 45)

A list of the eight recurring themes in this book can be found on p. xv.

ANSWERS TO FOLLOW-UP QUESTIONS

Follow-Up Question 2.1: PROTON and NEUTRON. *Follow-Up Question 2.2:* xenon; 54 electrons. *Follow-Up Question 2.3:* 98 of each. *Follow-Up Question 2.4:* Rutherfordium, Rf. *Follow-Up Question 2.5:* osmium; 116 neutrons. *Follow-Up Question 2.6:* This germanium isotope contains 32 protons and 42 neutrons. *Follow-Up Question 2.7:* Oxygen-16 is the most abundant isotope. *Follow-Up Question 2.8:* 43 electrons; technetium (Tc). *Follow-Up Question 2.9:* UV light is harmful; visible light is not (except in the case of light pollution). *Follow-Up Question 2.10:* a yellow sodium vapor lamp. *Follow-Up Question 2.11:* Beginning with the shortest wavelength: ultraviolet $<$ white light $<$ infrared light.

End-of-Chapter Questions

Questions are grouped according to topic and are color-coded according to difficulty level: **The Basic Questions***, **More Challenging Questions****, and **The Toughest Questions*****.

Particles in the Atom

*1. Atoms containing 33 protons are found in a murder victim's blood.

(a) These are atoms of which element?

(b) How many electrons does each of these atoms contain?

*2. What is the most massive element that has a name beginning with the letter *P*? What is the symbol for this element, and how many protons does it contain?

*3. In your own words, describe what is meant by the term *atom* and name the particles contained in an atom.

*4. A solid piece of the element indium (In) makes a high-pitched screeching noise as the structure of the metal changes when it's bent. How many protons and neutrons does the indium-115 isotope contain?

*5. The element curium (Cm) was named after Marie Curie, a pioneer in the understanding of the heavier elements on the periodic table. How many protons and electrons does one atom of curium contain?

*6. Which of the following is a particle found within the nucleus of an atom?

(a) photon
(b) meson
(c) neutron
(d) krypton
(e) radon

**7. The mass of an electron is 9.11×10^{-31} kilograms. Express this number in units of grams, and in units of femtograms. (Hint: Refer to the table of metric prefixes in Chapter 1.)

**8. Is this statement true or false? Why? Every atom has equal numbers of protons, neutrons, and electrons.

**9. Is this statement true or false? Why? The lanthanides and actinides are found at the bottom of the periodic table because they are all human-made, artificial elements.

***10. Atoms of element A have three times more protons than atoms of element B, and A and B are in the same column of the periodic table. Subtract the number of protons in B from the number in A, and the resulting answer is equal to the number of protons in atoms of a third element in the same column as A and B. Identify elements A and B.

***11. Atoms of element C have six times more protons than atoms of element D. These two elements are not in the same column of the periodic table, but their symbols and names do begin with the same letter. The sum of the protons in atoms of these two elements is 70. Identify elements C and D.

***12. Atoms of element E and F, together, have the same number of protons as an atom of element G. The names of all of these elements begin with the letter *T*, and one of them is part of the lanthanide series. What are these three elements?

***13. Protons, neutrons, and electrons have the following respective masses: 1.67×10^{-27} kilograms (kg), 1.67×10^{-27} kg, and 9.11×10^{-31} kg.

(a) How many times more massive is a proton as compared to an electron?

(b) How many electrons would have a mass equal to one neutron?

(c) Express each of these masses in units of picograms.

(Hint: Refer to the table of metric prefixes in Chapter 1.)

Elements and the Periodic Table

*14. Write the full name and symbol of each element that has the following numbers of protons:

(a) 31 protons
(b) 54 protons
(c) 102 protons
(d) 93 protons

*15. Is this statement true or false? Why? Atoms have equal numbers of protons and electrons.

**16. An unknown element has 104 protons and 104 electrons. Write the name and the symbol for this element.

**17. As we will learn in Chapter 3, "Everything," the elements in the far right column of the periodic table are called the noble gases. Which noble gas has more protons than the element gold (Au)?

**18. Which of the following is a transuranium element? (Choose all answers that apply.)

(a) dysprosium
(b) thorium
(c) silver
(d) lawrencium
(e) roentgenium

***19. The element on the periodic table with atomic number 74 is called tungsten. Its symbol, *W*, comes from wolframite, a mineral that contains it. If tungsten has 74 protons, why does it occupy the 59th box on the main part of the periodic table?

***20. In your own words, describe the terms *atom* and *element.* Why don't we use just one term or the other? When is each term used?

***21. A recently discovered element, named for a female scientist, has the same number of protons as praseodymium and tin combined. Which element is it?

Neutrons and Isotopes

*22. An isotope of a given element contains 19 protons and 22 neutrons. What is this element? What is its mass number?

*23. An isotope of a given element has a mass number of 188 and contains 112 neutrons. What is this element? What is its atomic number?

*24. According to Figure 2.7, which is the most abundant isotope for the element tin? How many protons, neutrons, and electrons does it contain?

**25. Determine the total number of electrons, protons, and neutrons in the following isotopes:

(a) ^{180}W (b) ^{54}Cr (c) ^{2}H (d) ^{104}Pd

**26. Determine the total number of electrons, protons, and neutrons in the following isotopes:

(a) thallium-205 (b) tin-119 (c) iron-56 (d) helium-3

**27. (a) If an element has 56 protons, 82 neutrons, and 56 electrons, what are its symbol and mass number?

(b) How is an atom with 56 protons, 78 neutrons, and 56 electrons related to the atom in part (a)?

**28. Figure 2.7 shows that the rarest isotope of tin is tin-115. How many neutrons does this isotope have?

**29. Is the following statement true or false? Why? All of the atoms in a sample of an element must have the same number of protons and neutrons.

***30. Imagine you are a professional scientist who has created 300 atoms of a new element. You determine that the element has 116 protons, and you decide to give it the symbol Oh, for ohgreatium. Of the 300 atoms, 24 exist as the ohgreatium-232 isotope, 266 as the ohgreatium-234 isotope, and 10 as the ohgreatium-235 isotope. How many neutrons does each isotope contain?

***31. The element radon spontaneously gives off energy and becomes a different element with a smaller mass. Write the symbol and mass number for an atom of the element that is formed from radon; the element contains 134 neutrons, 84 electrons, and is in the same group on the periodic table as oxygen.

***32. The following is a list of the most common isotopes for several elements with atomic numbers increasing in jumps of 10. The list begins with neon, which has 10 protons: neon-20, calcium-40, zinc-64, zirconium-90, tin-120, neodymium-142, ytterbium-174, mercury-202, thorium-232. Do you recognize any trend in the ratio of protons to neutrons in this series of isotopes? State the trend.

Electrons

*33. Consider five atoms of five different elements. Each of them is, in one way or another, an element that you might find around your house. They have the following respective numbers of electrons:

(a) 82 (b) 79 (c) 29 (d) 53 (e) 17

Give the name and symbol for each of these elements.

*34. Is this statement true or false? Why? Chemical reactions are nothing more than rearrangements of protons.

**35. In your own words, explain what is meant when electrons are described as elusive.

**36. Imagine an atom of calcium, atomic number 20, with energy levels around the nucleus as shown in Figure 2.11. Indicate how many electrons in this atom would occupy each energy level around the atom, beginning with energy level 1.

**37. Imagine an atom of chlorine, atomic number 17, with energy levels around the nucleus as shown in Figure 2.11. Indicate the number of electrons that would occupy each energy level around the atom, beginning with energy level 1.

**38. Is the following statement true or false? Why? In any atom with multiple electrons, the electrons are all exactly the same.

**39. An unknown atom has the same number of electrons as an atom of chlorine, an atom of nickel, and an atom of silver combined. What is the unknown atom?

***40. We discussed three particles within the atom. Of these three, only the number of electrons can change during a chemical reaction. Explain why this is so. Which electrons in an atom are most likely to take part in chemical reactions?

***41. The mass of an electron is 9.11×10^{-31} kg.

(a) Write out this number without using scientific notation.

(b) How many electrons, when combined, have a mass of 1.0 gram?

(c) Express the mass of an electron in units of grams.

Light and Line Spectra

*42. Rank the following types of light in order of increasing energy: radio waves, gamma rays, infrared light, microwaves, ultraviolet light.

*43. Rank the following types of visible light in order of decreasing energy: yellow, blue, violet, green, red, orange.

*44. Which color of visible light has a wavelength of 500 nanometers?

**45. In your own words, describe what is meant by the term *wavelength* when it is used to refer to light.

**46. What color of visible light has a wavelength of 6.2×10^{-7} meters?

**47. Is the following statement true or false? Why? When an atom absorbs energy, an electron in the atom may move from a higher energy level to a lower energy level.

**48. Is the following statement true or false? Why? A line spectrum arises when electrons in an atom of a given element are excited and then emit light as they relax to the element's ground state.

***49. Gamma rays have very high energies. The wavelength of a gamma ray is about 0.0015 nm. Express this wavelength in units of picometers.

***50. Radio waves can have very long wavelengths, up to 10^{12} nm.

(a) What does this say about the energy of a radio wave?

(b) Express the wavelength of a radio wave in units of kilometers.

(c) Which is greater, the distance between Baltimore, MD, and Washington, D.C., or the wavelength of this radio wave?

Integrative Questions

**51. Certain isotopes have properties that make them ideal for treating and diagnosing diseases. Samarium-153 is such an isotope. It is used to relieve bone pain caused by cancer treatment and is part of a larger molecule that is marketed as the drug Quadramet.

(a) How many protons are in the nucleus of a samarium atom?

(b) How many neutrons are in the nucleus of the samarium-153 isotope?

(c) The release of electromagnetic radiation is one way that the samarium-153 isotope acts as a therapeutic agent. (Chapter 11 is dedicated to substances like this which are radioactive.) If the energy of this electromagnetic radiation is 0.001 nanometers, then what type of electromagnetic radiation is it? Use Figure 2.17 to determine the answer.

***52. Tennessine (symbol Ts) is an element that was discovered in 2010 and named in 2015. Its name comes from the national laboratory in Oak Ridge, Tennessee where it was first created in collaboration with scientists in Dubna, Russia.

(a) The two known isotopes of Ts are tennessine-294 and tennessine-293. If tennessine possesses 117 protons, then how many neutrons do each of these two isotopes possess?

(b) Tennessine is the heaviest known element in the halogen group on the periodic table (column 17). Consider the names of all of the halogen elements including tennessine. What do they all have in common?

(c) Newly-created elements exist for a fraction of a second. Thus, there is no way to measure the physical properties of these new elements. However, some physical properties can be predicted based on trends from other elements on the periodic table. For example, it is estimated that if we could create enough tennessine to measure, it would be a solid element with a density of 7.2 grams per cubic centimeter. Given this fact, what would be the mass of 2.0 mL of this element (1 cubic centimeter = 1 mL)?

***53. The black-and-white version of the line spectrum for mercury is shown below. There are six lines in this image, although the two farthest to the right appear as one line.

(a) Based on the wavelengths given, use the visible spectrum in Figure 2.18 to predict what color each of these six lines would be if this spectrum were shown in color. All numbers in the figure have units of nanometers.

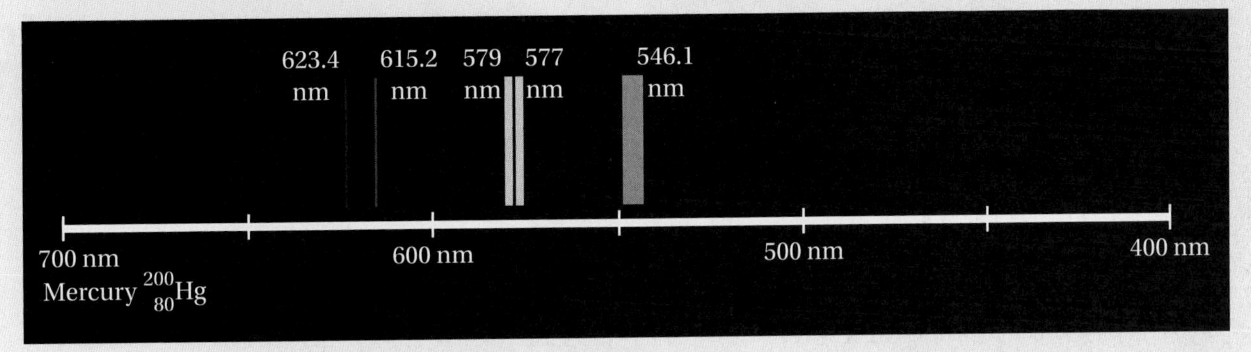

(b) What is the wavelength of the highest-energy line in the mercury line spectrum? What is the wavelength of the lowest-energy line?

(c) Mercury-200 is the isotope of mercury that was used to create this line spectrum. How many protons, neutrons, and electrons does the mercury-200 atom possess?

54. INTERNET RESEARCH QUESTION

Across the United States, there are still places where you can experience a completely dark sky at night. Go to the Internet, search for "dark sky places map," and find your location. How much light pollution exists where you live? If you live in a bright spot on the light pollution map, where and how far would you have to travel to reach a dark place? Expand your web search to answer this question: is there a city, town, or park near you that is registered as an International Dark Sky Place or International Dark Sky Park?

55. INTERNET RESEARCH QUESTION

Consider these five elements: hydrogen (H), fluorine (F), cesium (Cs), xenon (Xe), and mercury (Hg). Use the Internet to determine the number of known isotopes for each of these elements.

(a) Which of the five elements has the most isotopes?

(b) Which of the five elements has the fewest isotopes?

(c) Organesson is a new element, number 118, the element with the greatest atomic number. How many isotopes of organesson are known, and what is the mass number for each isotope?

Everything

The Ways We Organize and Classify Matter

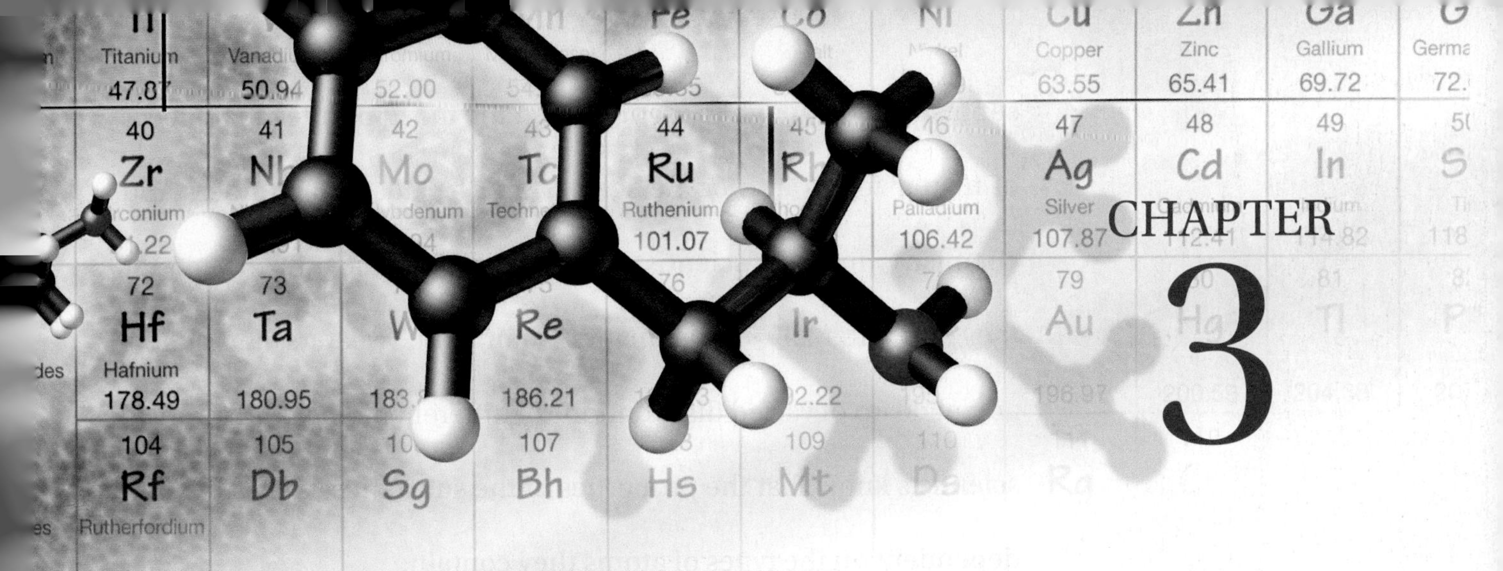

CHAPTER 3

Locard's principle is a basic tenet of forensic science, the application of science and technology to investigate crime. It says that no matter what, when two things come into contact with one another, material from one material always transfers to the other. For example, if you sit down on a black velvety sofa and you are wearing a white fuzzy sweater, you will always leave some of your sweater on that sofa. And some of the sofa will come away on your sweater. Locard's principle asserts that most criminals leave something behind at a crime scene and take some of the crime scene away with them. That "something" could be anything from a piece of hair to a skin cell to sweater fuzz to bullet residue.

Each tiny piece that a criminal leaves behind is an opportunity for the authorities to determine the criminal's identity. If you watch crime dramas on television, you know that forensic laboratories are sophisticated enough to identify even the most infinitesimal pieces of matter found at a crime scene. And, because they understand chemistry, forensic scientists can figure out what's in the matter they find. After all—like all matter—hair, skin, sweaters, and bullets are made up of varied distributions of different atoms. And forensic scientists know that the properties of the substances they study are dependent on the types of atoms they contain.

OSTILL/iStock/Getty Images

This chapter is all about different types of matter and how matter is organized. Why are some kinds of atoms suited to some jobs while others are not? In this chapter, we begin to look at everything and see what's in it. And we begin to answer some compelling questions like, Why are different kinds of atoms necessary to do all of the tasks required by nature? Why are some types of atoms found in sweaters, but not in bullets? How do we distinguish different kinds of atoms from one another and organize them in a coherent way? Read on and find out.

3.1 How Elements Are Distributed in Nature

We can divide the periodic table into metals, nonmetals, and metalloids.

In Chapter 2, we learned that all matter is composed of the elements on the periodic table. But 114 elements are a lot to think about. How can we possibly understand how they are distributed throughout everything we know? In Chapter 2, we also learned that the elements after atomic number 92 are artificial and must be synthesized in specialized laboratories. We can set that group of elements aside for now, because we don't encounter them often. (We will revisit them in more depth when we explore nuclear chemistry in Chapter 11, "Nukes.") That leaves us with the elements that have atomic numbers 1 through 92, the elements that make up the vast majority of the matter we encounter every day. This is still a significant group. How can we begin to make sense of it?

flashback ◀ Recall from Chapter 1 that all matter is made up of atoms and that we give names and symbols to specific types of atoms, which we call elements.

Figure 3.1 is a color-coded version of the periodic table. All of the pink elements are **metals**. Most of us encounter metals, such as copper, gold, silver, and nickel, every day. What do these metallic elements have in common? First of all, except for mercury (Hg), they are all shiny solids. They are malleable, which means they can be shaped into things like coins, piercings, paper clips, and forks. They are also ductile, which means they can be pulled into wires. Electrical wires are made of metal because metals conduct electricity very easily. Metals also conduct heat really well. That is why we use them to make pots and pans; they are very good at passing heat from the flame on your stove to the food you want to eat.

The blue elements in Figure 3.1 are the **nonmetals.** Nonmetals are usually solids or gases, and the only liquid nonmetal is bromine (Br). Unlike metals, the nonmetals that are solid are dull, not shiny. They are brittle, not malleable or ductile. Finally, nonmetals are not good conductors of heat or of electricity. Think for a moment about the things that we use to protect ourselves from changes in temperature: sweaters, socks, oven mitts, blankets. These things are all made of insulating materials, and those materials are composed of elements that are nonmetals. Now consider the substances that protect us from electricity. The covering on an electrical cord separates us from the electricity carried by the metal wires inside it. Those coverings are usually made of plastic, and household plastics are made of elements that are nonmetals.

Finally, there are the **metalloids**, shown in green in Figure 3.1. These elements sit between the metals and the nonmetals, and they have properties in common with both. The metalloid elements silicon, germanium, and antimony are elements commonly found in semiconductors, which are solid materials that conduct electricity under certain conditions. Semiconductors are used to make microprocessor chips, where memory is stored in your computer.

We just considered several examples of objects that are made primarily of metals, nonmetals, or metalloids. However, most natural systems are complicated mixtures of these elements. In the sections that follow, we look at two examples

	1	2	3	4	5	6	7	8	9	10	11	12	13	14	15	16	17	18
1	1 H 1.008																	2 He 4.003
2	3 Li 6.941	4 Be 9.012											5 B 10.81	6 C 12.01	7 N 14.01	8 O 16.00	9 F 19.00	10 Ne 20.18
3	11 Na 22.99	12 Mg 24.31											13 Al 26.98	14 Si 28.09	15 P 30.97	16 S 32.07	17 Cl 35.45	18 Ar 39.95
4	19 K 39.10	20 Ca 40.08	21 Sc 44.96	22 Ti 47.87	23 V 50.94	24 Cr 52.00	25 Mn 54.94	26 Fe 55.85	27 Co 58.93	28 Ni 58.69	29 Cu 63.55	30 Zn 65.39	31 Ga 69.72	32 Ge 72.64	33 As 74.92	34 Se 78.96	35 Br 79.90	36 Kr 83.80
5	37 Rb 85.47	38 Sr 87.62	39 Y 88.91	40 Zr 91.22	41 Nb 92.91	42 Mo 95.94	43 Tc 98	44 Ru 101.1	45 Rh 102.9	46 Pd 106.4	47 Ag 107.9	48 Cd 112.4	49 In 114.8	50 Sn 118.7	51 Sb 121.8	52 Te 127.6	53 I 126.9	54 Xe 131.3
6	55 Cs 132.9	56 Ba 137.3	57 La 138.9	72 Hf 178.5	73 Ta 180.9	74 W 183.8	75 Re 186.2	76 Os 190.2	77 Ir 192.2	78 Pt 195.1	79 Au 197.0	80 Hg 200.6	81 Tl 204.4	82 Pb 207.2	83 Bi 209.0	84 Po	85 At	86 Rn
7	87 Fr	88 Ra	89 Ac	104 Rf	105 Db	106 Sg	107 Bh	108 Hs	109 Mt	110 Ds	111 Rg	112 Cn	113 Nh	114 Fl	115 Mc	116 Lv	117 Ts	118 Og

Figure 3.1 The Periodic Table

The elements shown in pink on this version of the periodic table are metals, the elements shown in blue are nonmetals, and the elements shown in green are metalloids.

of complex natural systems, in turn, and try to understand how the elements on the periodic table—the metals, nonmetals, and metalloids—are distributed within each of them. Let's begin with elements in soil, and then we can compare those elements to those in another interesting natural system: the human body.

QUESTION 3.1 Identify each element as a metal, nonmetal, or metalloid.
(a) titanium (b) sulfur (c) radon (d) helium (e) silicon

ANSWER 3.1 Titanium is a metal. Sulfur, radon, and helium are nonmetals. Silicon is a metalloid.

For more practice with problems like this one, check out end-of-chapter Questions 1 and 2.

FOLLOW-UP QUESTION 3.1 Which of the following descriptors could be used to describe a nonmetal but not a metal? (Choose all that apply.) (a) brittle (b) gaseous (c) shiny (d) ductile (e) dull

The appearance of soil is partly a function of the elements it contains.

The photographs in **Figure 3.2** are images of soil in two different landscapes. The striking variations in color confirm that these soil samples are composed of different things, and that the natural components of soil vary from one location to another. Besides making different soils look different, the elements making up any given soil sample determine its capacity for supporting living plants. We know, for instance, that certain crops thrive in some areas, but not in others. If we examine the color and dampness of the soil in the desert versus that in a rain forest, we find that the former is typically pale and dry and the latter is rich, dark, and moist. These are shown on the left and right, respectively, in Figure 3.2.

Figure 3.2 Soils in Different Colors

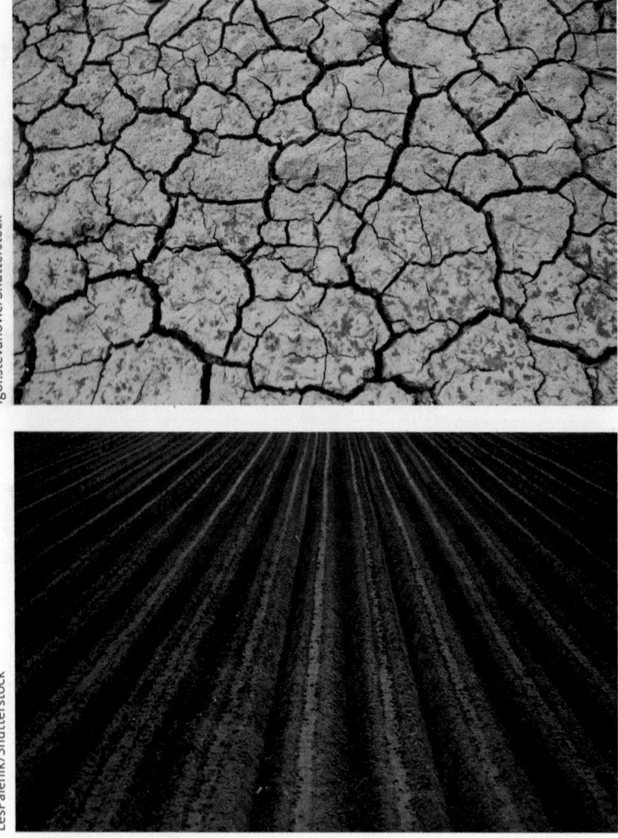

Differences in soils result from variations not only in the amount of water in the soils but also from variations in elemental composition. Red soils are typically high in the element iron (Fe). Black soils often contain higher than normal amounts of the element manganese (Mn). And soils high in calcium (Ca) tend to have a white color.

Figure 3.3 shows the periodic table of the elements, drawn with the height of each element's box proportional to that element's abundance in an average soil sample. If we look closely, we can see that soil is dominated by the elements oxygen and silicon, both nonmetals. Besides oxygen and silicon, dozens of other elements are present in soil, all of them important to the processes occurring there.

Carbon, another nonmetal, is soil's 16th most abundant element. Most of the carbon present in soil comes from decomposed animal and vegetable matter. These carbon-based substances, which we refer to as **organic,** were once part of living things that are now being recycled by the soil. We refer to substances that are not organic as **inorganic**. Soils contain metals, too, but typically in

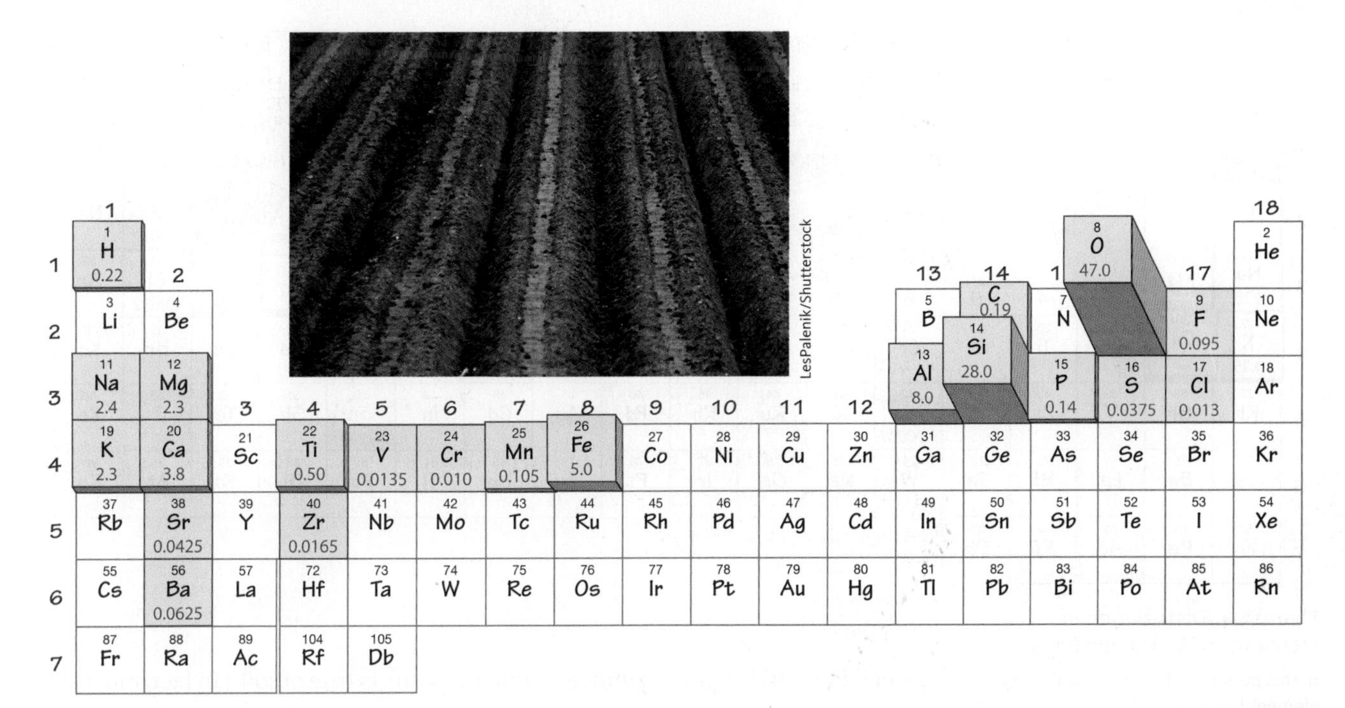

Figure 3.3 The Distribution of Elements in Soil

In this periodic table, the height of each element box is proportional to the element's percentage in a typical soil sample, which is given below each element's symbol.

smaller amounts than nonmetals. The parts of soil that contain metals, like minerals and salts, are the parts we often think of as inorganic. (We delve more deeply into the differences between these two terms in the Chapter 5, "Carbon.")

QUESTION 3.2 Rank the following elements in order of decreasing abundance in soil: oxygen, sodium, carbon, aluminum, titanium.

ANSWER 3.2 According to Figure 3.3, the order, from highest abundance to lowest, is oxygen > aluminum > sodium > titanium > carbon.

For more practice with problems like this one, check out end-of-chapter Question 7.

FOLLOW-UP QUESTION 3.2 Which of the following elements are not found in soil? (a) krypton (b) fluorine (c) scandium (d) hydrogen (e) calcium

Like soil, the human body is a mixture of metals and nonmetals.

Now let's look at the elements that make up the human body. **Figure 3.4** (p. 68) shows a periodic table that is similar to the periodic table we just saw for soil. But in this figure, the height of each column represents the percentages of that element in the human body. In comparing the two tables, what similarities can we identify? First of all, it's clear that the human body is also a mixture of both metals and nonmetals. Second, we can see that the most abundant elements in the human body are the nonmetals oxygen, carbon, and hydrogen. Soil is also made up mainly of nonmetals.

Now, notice something else: the elements in the human body are generally grouped in the top half of the periodic table. The heaviest elements on the list are molybdenum, iodine, and tin—each present in the human body in very small amounts. Because of this, we can say that the human body is made up primarily of

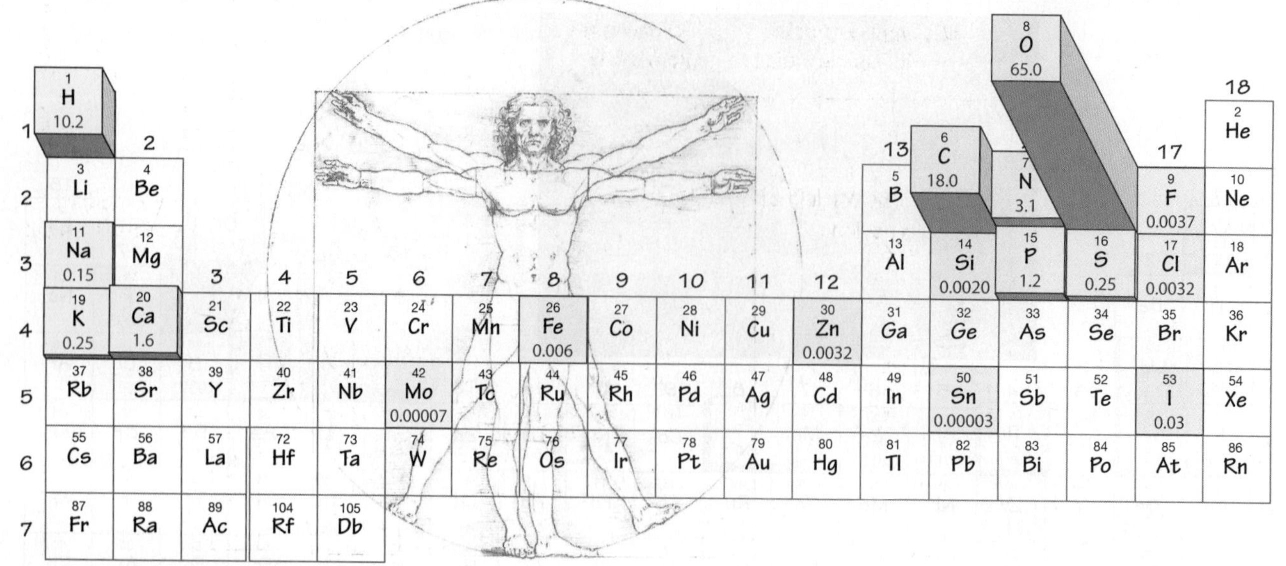

Figure 3.4 Distribution of Elements in the Human Body

In this periodic table, the height of each element box is proportional to the percentage of that element in the human body.

lighter elements. (Figure 3.3 indicates that the same is true of soil.) In fact, most living things are made up mostly of lighter elements and only trace amounts of some heavier elements.

QUESTION 3.3 According to Figures 3.3 and 3.4, which element makes up the most significant percentage of both soil and the human body?

ANSWER 3.3 Oxygen. In soil, oxygen exists primarily in the form of silicon dioxide. In the body, it is present in water and an essential part of many molecules required for life.

For more practice with problems like this one, check out end-of-chapter Question 8.

FOLLOW-UP QUESTION 3.3 Which contains a higher percentage of iron, the human body or soil?

3.2 Tour of the Periodic Table

Scientists organize the elements vertically and horizontally on the periodic table.

The periodic table, which we first looked at in Chapter 2, is the main organizing principle in chemistry. Recall that we learned to "read" the table like a book, starting at the upper left and reading to the right. Then we go back to the beginning of the next horizontal row and read to the right again, and so on. Each time we get to the end of the row on the periodic table, we complete one **period,** and when we start at the left and move across again, we begin the next period. The periods on the table are numbered 1 through 7, and these numbers are written vertically to the left of the table. These features of the periodic table are shown in **Figure 3.5**.

The elements in a given period usually do not have similar properties to one another. For example, Period 2 begins with lithium (symbol Li), which is a very reactive, shiny, silvery metal. What do we mean when we say that something is reactive? We mean that it tends to undergo chemical reactions easily. As we jump boxes to the right, we pass by beryllium (Be)—also a shiny, silvery metal, but it does

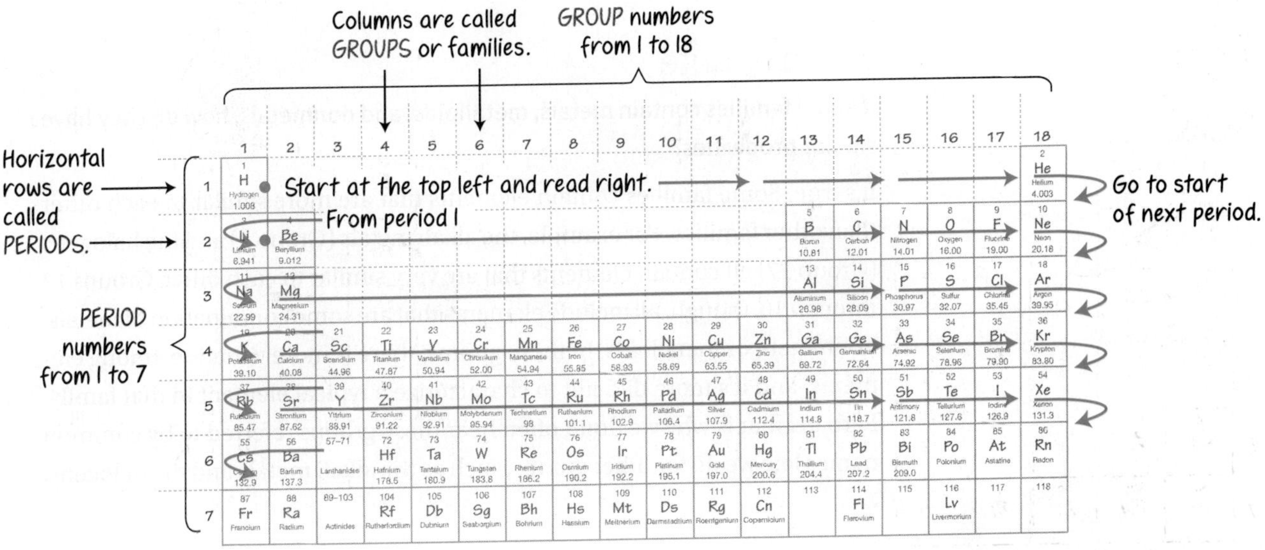

Columns are called GROUPS or families.

GROUP numbers from 1 to 18

Horizontal rows are called PERIODS.

PERIOD numbers from 1 to 7

Start at the top left and read right.

From period 1

Go to start of next period.

Figure 3.5 How the Periodic Table Is Organized

not interact with other substances as easily as lithium does. Then we jump across to boron (B), a chunky black substance. Carbon (C) is next. Carbon exists in several different forms, and one of them is a chunky, black, coal-like substance. (Carbon is such an important element that all of Chapter 5, "Carbon," is devoted to it). The next element is nitrogen (N), which makes up 78% of the air we breathe. Then comes oxygen (O), which makes up about 21% of air. The last two elements, fluorine and neon, are also found in air.

In this quick survey of Period 2, we can see that the Period 2 elements do not share properties with each other. However, if we start to examine the elements in Period 3, we notice something interesting. The properties of the Period 2 elements start to repeat themselves. The first element, sodium (Na), is a very reactive, shiny metal—just like lithium. And the next Period 3 element, magnesium (Mg), behaves similarly to the Period 2 element just above it, beryllium. Silicon (Si) shares some common properties with C; O and S also have similarities, as do F and Cl, and Ne and Ar. You know from Chapter 2 that each of these columns is either a group or a family.

recurring theme in chemistry 5

Elements in the same group on the periodic table often behave in similar ways and share properties with one another.

Notice that the periodic table inside the front cover of your book has numbers at the top of each column of elements. They are numbered from 1 (far left) to 18 (far right). It is common to use the word *group* when referring to these numbered columns. So we often say, "That's a Group 6 element," or "Group 18 elements are gases." On the other hand, the word *family* is often used when we need a more descriptive name for a column of elements. For example, the elements in the second column (from the left) are sometimes called the *alkaline earth family*. These elements share many properties. For example, they are all shiny metals, and they all interact with certain other substances in the same way. The word *family* is fitting because, just as human families share characteristics like freckles or curly hair or pug noses, elements in the same column of the periodic table often share characteristics, too.

The grouping of elements into families helps us predict the behavior and characteristics of an atom in a given group. For example, if oxygen tends to undergo chemical reactions with other substances in certain ways, then sulfur might also do the same thing, because they are in the same family. This is why the table is periodic: families on the periodic table often contain elements with similar properties, and those properties appear again and again as we read from left to right across the periodic table.

wait a minute . . .

If some families contain metals, metalloids, and nonmetals, how do they have similar properties?

It's true. Some families contain elements that are more similar to each other than other families. For example, the alkali metals (Group 1) and the halogens (Group 17) all contain elements that are very similar to each other. Groups 13 through 16, though, all include elements that are some combination of metals, nonmetals, or metalloids. In these groups, there is more variation among elements, and it's more difficult to characterize a typical element in that family. Perhaps that is why we don't often hear these groups referred to by common names like we do for other groups, such as the alkali metals and the halogens.

QUESTION 3.4 Sort the following elements into their groups on the periodic table, and indicate the group number for each one:

yttrium (Y)	germanium (Ge)	iodine (I)	bromine (Br)
scandium (Sc)	radon (Rn)	helium (He)	
carbon (C)	fluorine (F)	lead (Pb)	

ANSWER 3.4 Group 3 includes Sc and Y; Group 14 includes C, Ge, and Pb; Group 17 includes F, Br, and I; Group 18 includes He and Rn.

For more practice with problems like this one, check out end-of-chapter Questions 11 and 12.

FOLLOW-UP QUESTION 3.4 The elements in Group 17 are called the *halogens*. Which of the following elements are halogens? phosphorus, zinc, iodine, mercury, astatine

Each family of elements on the periodic table has its own family traits.

It is time to properly introduce some of the families that reside on the periodic table. **Figure 3.6** shows where several families are located. Let's begin on the left with the leftmost two columns. Groups 1 and 2 are known as the *alkali metals* and the *alkaline earth metals,* respectively.

The two leftmost columns of the periodic table are separated from the large groups of elements on the right by the *transition metals,* which include the elements in Groups 3 through 12. This is where we will find the metals that we encounter almost every day: copper, gold, iron, silver. So, why are some of the transition metals known to us, while others—like yttrium and niobium—are comparatively obscure? There are two reasons. First, the elements we know well tend to exist in larger quantities in Earth's crust—the outermost layer of the Earth—so we are able to mine them more readily. Second, many of these elements are very robust and do not react with other substances easily. Because of this, the metals used in coins—copper and gold, for example—tend to tarnish very slowly, so they last a long time.

The elements in Group 17 are the *halogens.* The halogens at the top of the family—fluorine and chlorine—are both gases. Bromine, as we know, is a liquid. Iodine is a solid. All of the halogens react readily with other substances. Unlike the halogens, the elements of Group 18, the *noble gases,* are all gases. And, unlike the halogens, the

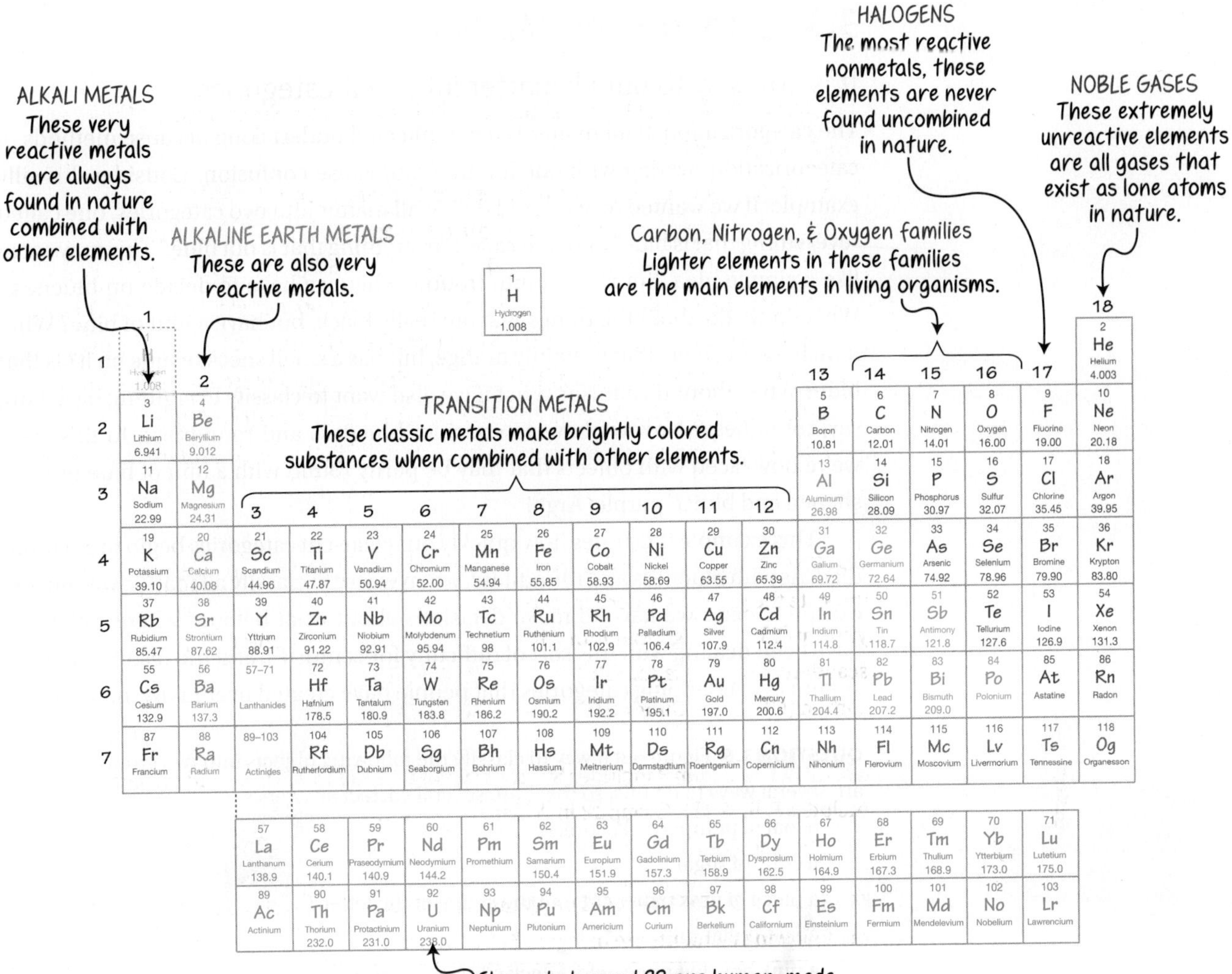

HALOGENS
The most reactive nonmetals, these elements are never found uncombined in nature.
NOBLE GASES
These extremely unreactive elements are all gases that exist as lone atoms in nature.
ALKALI METALS
These very reactive metals are always found in nature combined with other elements.
ALKALINE EARTH METALS
These are also very reactive metals.
Carbon, Nitrogen, & Oxygen families
Lighter elements in these families are the main elements in living organisms.
TRANSITION METALS
These classic metals make brightly colored substances when combined with other elements.
Elements beyond 92 are human-made and not found in nature.

Figure 3.6 A Family Portrait

This periodic table shows the characteristics of some well-known periodic families.

noble gases are very, very *un*reactive. This is where they get the name "noble": they are the snobbish elements that do not interact with atoms of other elements.

Why are the alkali metals very reactive while the noble gases don't interact with other elements at all? What makes metals conductors, and what makes nonmetals good insulators? What is it about the alkali and alkaline earth metals that makes them reactive? These are all excellent questions, but we are not ready to answer them yet. We devote all of Chapter 4, "Bonds," to exactly these types of questions. For now, let's focus on the more basic task at hand: organizing matter in general terms. We've learned a bit about which elements are most common and about how the periodic table is organized. Now we are ready to ask the larger question that we have been working toward: "How can we classify matter?"

QUESTION 3.5 Assign each of the following elements to a group on the periodic table, and identify that group with a number: (a) helium (b) chlorine (c) sodium (d) magnesium

ANSWER 3.5 (a) Group 18 (b) Group 17 (c) Group 1 (d) Group 2

For more practice with problems like this one, check out end-of-chapter Questions 13 and 14.

FOLLOW-UP QUESTION 3.5 To which family of the periodic table does each of the elements in Question 3.5 belong?

3.3 Categorizing Matter

It is not easy to put all matter into neat categories.

The categorization of all matter is a complicated undertaking because methods of categorization overlap with one another and cause confusion. Consider this silly example: If we wanted to, we could divide all matter into two categories, one called "everything that is blue" and one called "everything that is not blue." This may seem like a simple idea, but we run into trouble when we have to decide on blueness. What do we do about the things that are really black, but have a hint of blue? What if we have an object that is mainly orange, but has a small speck of blue on it? Is that blue? What about turquoise? What if we also want to classify everything based on, say, roundness? Now we have to consider blueness and roundness. In this case, we're now faced with objects that may be partly round with a hint of blue or oval-shaped and bluish purple. Argh!

This example illustrates how quickly our clear-cut categories begin to get muddled. As with blueness and roundness, when we try to classify all matter, we run into exceptions and we have to make decisions about what belongs where. Scientists have been dealing with this sort of difficulty for centuries, and the result is a messy system of overlapping categories that people have created over many, many years.

QUESTION 3.6 Figure out a way to classify the following objects into two groups. There are several ways to do this, so there are several correct answers.

a loaf of pumpernickel bread
a cube of sugar
a piece of pink paper cut into the shape of the letter *J*
a five-dollar bill
a pink box that once held a deck of playing cards

ANSWER 3.6 There are many ways to answer this question. One possible answer is this: You can categorize these items into one group that contains two-dimensional objects (pink paper and dollar bill) and one group that contains three-dimensional objects (bread, sugar, and pink box). Or, you can place them into a group of pink things and a group of not pink things, and so on.

For more practice with problems like this one, check out end-of-chapter Question 28.

FOLLOW-UP QUESTION 3.6 Suppose you have a deck of playing cards. You could classify the cards based on suit and make piles with spades, clubs, hearts, and diamonds. Think of a second way you could categorize the deck of cards.

In everyday life, pure substances are not truly pure.

Look around you. Everything we can see is made up of some combination of the 114 elements on the periodic table. For example, the clothes you are wearing right now are mostly made up of atoms of carbon, hydrogen, and oxygen. These elements are clustered in the upper right corner of the periodic table. If you are a knight and are wearing chain mail, then perhaps the clothes you are wearing right now are mainly made of metal. Familiar metals, like iron and copper, are found in the middle of the periodic table.

Even things we cannot see are made up of some combination of the 114 elements on the periodic table. In the last section of this chapter, you will learn that the air we are breathing right now is mainly nitrogen and oxygen, with some other elements mixed in. With the staggering number of substances on Earth, how can we possibly get a handle on the diversity offered by our 114 elements? In this section, we attempt to classify everything into some simple categories.

Let's begin by dividing all matter into substances that are pure and substances that are not. But how shall we define *pure*? According to the strictest definition, if a specimen of matter contains just one atom that doesn't belong, then it's not absolutely pure. In reality, we cannot be this strict with our definition of what is pure and what is not, because if we are, we quickly come to the conclusion that everything is impure and that this categorization is useless to us. So we need to be a bit more lenient.

For example, when chemists buy a chemical from a chemical company and they pay top dollar for high purity, that chemical usually has a purity of about 99.99%. In the real world, we consider this to be pretty good, and we would say that it is very pure—even though, technically, it is not absolutely pure. So, we will use the term **pure substance** loosely, and we acknowledge that pure substances may contain trace impurities. We define a pure substance, then, as a single element or compound that may have trace impurities. **Figure 3.7** shows examples of reasonably pure substances that you may see every day.

Mixtures contain more than one pure substance.

Most of the materials we encounter every day are not pure substances, even by our more relaxed definition. Instead, most substances are mixtures. We define a **mixture**

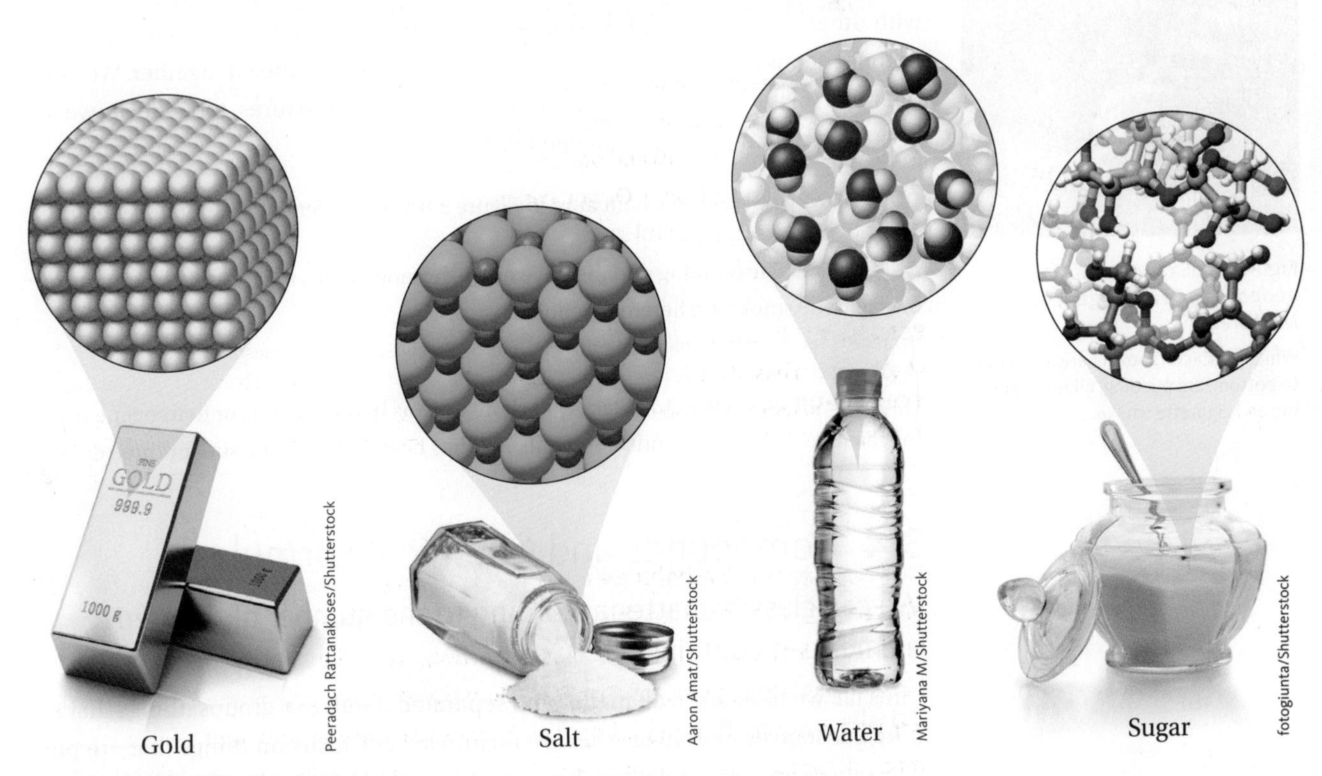

Figure 3.7 Examples of Pure Substances
These samples of gold, salt, water, and sugar are pure substances.

as two or more pure substances combined together. Most things that we see every day are mixtures: orange juice, shampoo, asphalt, smog, doorknobs, guacamole, ink.

Mixtures can be well-mixed or chunky or anywhere in between. A mixture can have a smooth texture that makes it impossible to see boundaries between the various components. We refer to a mixture like this as a *homogeneous mixture* (also called a *solution*) because it is completely uniform in appearance. For example, hot tea is a homogeneous mixture. When we put a tea bag into hot water, the tea leaves release substances into the water and a homogenous mixture results. Once we remove the tea bag, the tea has no boundaries or layers. It is thoroughly mixed.

Heterogeneous mixtures are chunky. They have boundaries where we can see when one component stops and another begins, because the composition changes from one place to another in the sample. Iced tea is a heterogeneous mixture. In a glass of iced tea, we clearly have one thing mixed into another—ice cubes and liquid tea. This mixture has boundaries—we can tell which part is tea and which part is ice. But if we leave our iced tea sitting on our porch on a hot July day, a few hours later, we may find that our heterogeneous mixture has turned into a homogenous mixture.

What about the things that require shaking before use, such as vinaigrette, nail polish, or house paint? These are also heterogeneous substances. They naturally form layers and must be shaken to be homogeneous. Some mixtures are heterogeneous regardless of how much you shake them. For example, toothpaste containing little flecks or stripes of color is a heterogeneous mixture that cannot be shaken to homogeneity. **Figure 3.8** shows examples of homogeneous and heterogeneous mixtures. The classification of each is given in Question 3.7.

We need not limit ourselves to liquids when discussing mixtures. Consider a chunk of very pure gold. This is a pure substance because, minor impurities aside, it is made up of only gold atoms. Now consider a gold ring. In jewelry, gold is mixed with other metals so that it is stronger and can be shaped into jewelry. This is a homogeneous mixture because it contains several metals mixed together. We often refer to a mixture of metals as an **alloy**. Gases may be mixtures, too. The air we are breathing right now is a homogeneous mixture.

Viktor1/Shutterstock

ifong/Shutterstock

givaga/Shutterstock

T.Dallas/Shutterstock

Figure 3.8 Examples of Homogeneous and Heterogeneous Mixtures

Which is which? Shown here, from top to bottom, are pudding, olive oil, granite, and cigarette smoke.

QUESTION 3.7 Classify each mixture in Figure 3.8 as homogeneous or heterogeneous.

ANSWER 3.7 The pudding and the olive oil are homogeneous mixtures. The granite and the cigarette smoke are heterogeneous mixtures.

For more practice with problems like this one, check out end-of-chapter Questions 21 and 22.

FOLLOW-UP QUESTION 3.7 Classify each mixture as homogeneous or heterogeneous: (a) gasoline (b) a glass window (c) milk that has been sitting in the sun for three days.

3.4 Compounds and Chemical Formulas

We can classify matter according to the number of different elements it contains.

Thus far, we have taken all matter and separated it into two groups: things that are pure substances and things that are mixtures. Let's focus on things that are pure substances and leave mixtures for now. We can further divide all pure substances into two categories: pure substances that contain just one element and pure substances that contain more than one element. If a pure substance contains only one element, we simply say, "It's an element." If it contains more than one element, we call it a **compound**. A compound is any substance made up of atoms of two or more elements.

According to this scheme, an element can be just a single atom of that element. The element neon, for example, is made up of individual neon atoms unattached to one another. An element may also consist of more than one atom of that element, with those atoms joined together in some way. For example, the element gold is a collection of many, many gold atoms all joined together. It is an element. The element fluorine exists in nature in pairs of fluorine atoms that are connected together. This is also an element because it contains only fluorine and nothing else. **Figure 3.9** shows examples of elements and compounds.

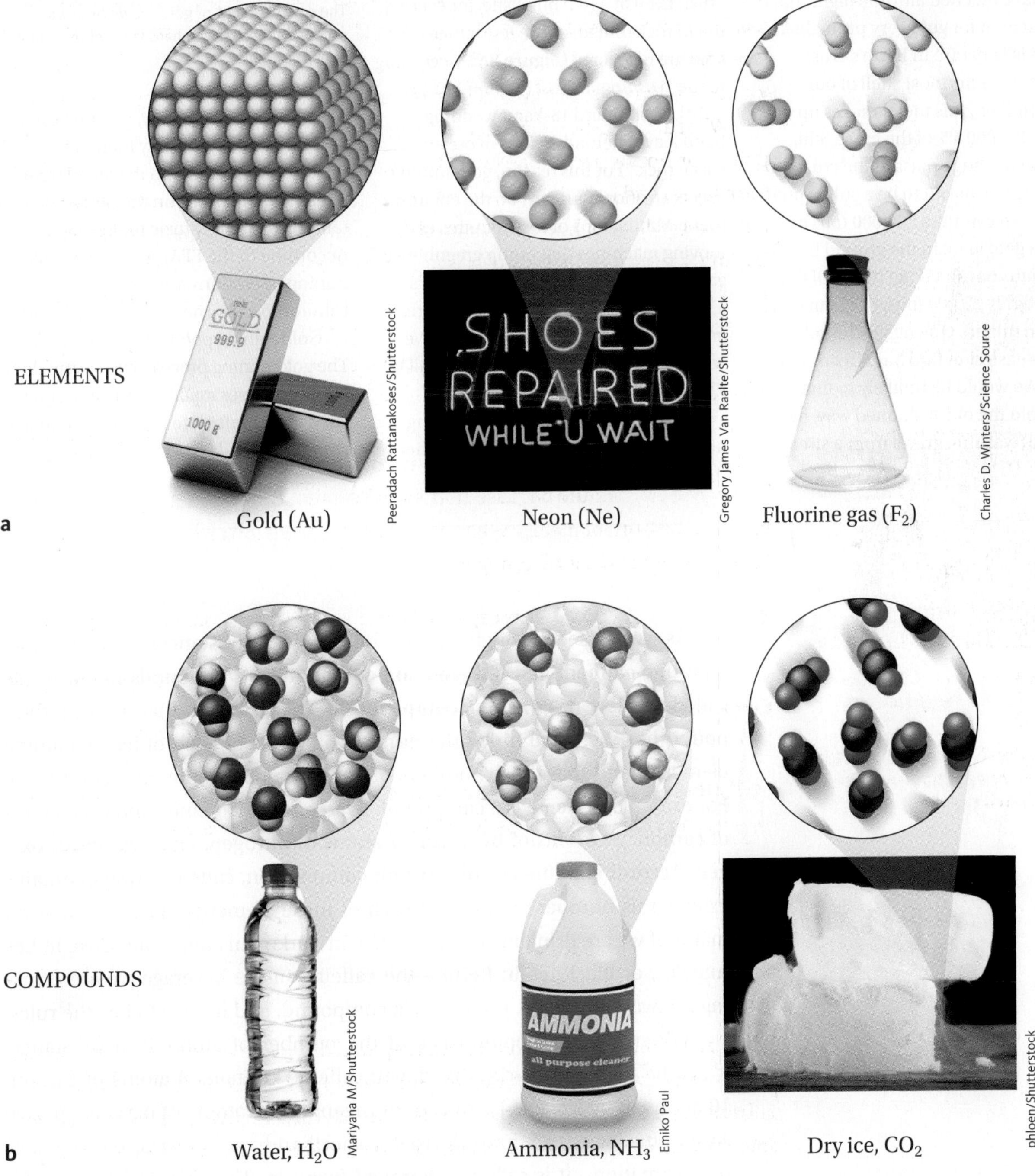

Figure 3.9 Elements and Compounds

(a) Elements contain only atoms of the same element. However, they can be individual atoms, such as neon, or atoms that are joined together, such as gold or fluorine gas. (b) Compounds contain atoms of more than one element.

natureBOX • The Problem with Gold Mining

Most matter on Earth is a mixture. This is because most elements combine with other elements to make compounds, and those compounds blend together into mixtures. So it's somewhat rare to find an element that is pure and uncombined in nature. Some metals do exist in relatively pure form on Earth—and, among those metals, gold is the ultimate find. It has been the most sought after and valued metal for centuries. In recent years, gold prices have reached all-time highs, making the search for gold very profitable.

Gold is very rare in Earth's crust, which is the outermost shell of our planet. Indeed, this metal makes up only about 0.000 000 4% of the crust. Still, thanks to the huge mass of the crust, this percentage turns out to be a substantial amount. An estimated 36,000 tons of minable gold were in the ground before we humans began taking it out. Of that total, roughly 22,000 tons, or about 60%, has been mined. Obviously, the gold that was easiest to find has already been found. We would be unlikely to find much gold the old-fashioned way, by panning (washing gravel from a stream or river in a pan to separate the precious material) or by locating veins that are easily accessible. To get to the 40% that remains, we would have to invest first in exploration and development and then in extraction and refining.

Gold mining goes on around the globe and, even compared to other mining industries, it is among the dirtiest. Why? Gold mining is especially dirty because the gold is so widely distributed in Earth's crust. Even at a rich mine site, for every ton of rock mined, only a few ounces of gold are recovered (**Figure 3A**). According to the *Washington Post* (September 21, 2010), "A standard 18-karat wedding band leaves behind 20 tons of ore and waste rock." For this reason, gold mining exacts an enormous cost on the environment. Mining tons of rock requires earth-moving machines that pump greenhouse gases into the air. (A greenhouse gas is a gas that traps heat in the atmosphere and contributes to global warming. We will discuss greenhouse gases in detail in Chapter 5, "Carbon.")

Gold mining also involves roasting, a heating step that makes the gold easier to isolate. Unfortunately, roasting causes mercury that is naturally found in the rock to be released into the air. Mercury, which is element 80 on the periodic table, is extremely toxic for humans. According to the EPA, small-scale gold mining operations are the biggest contributor to airborne mercury worldwide.

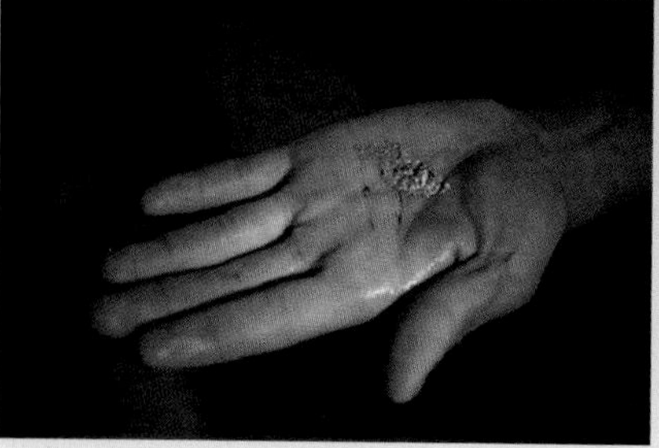

Jonas Gratzer/LightRocket/Getty Images

Figure 3A A Small Reward for an Enormous Effort
The small amount of gold shown here was produced through the hard labor of one family over one week.

Gold mining pollutes the water, too. The gold mining process, even when done legally, releases metals like arsenic, mercury, and copper into the water. Mining operations hold the runoff from their

As you'll see throughout this book, most substances contain more than one element and therefore are compounds. Compounds follow a rule called the **law of constant composition**. This law says that a given compound has a fixed and definite number of atoms of each of its constituent elements, and that those numbers of atoms stay the same, no matter what. For example, caffeine is a compound. Every unit of caffeine contains 8 atoms of carbon, 10 atoms of hydrogen, 4 atoms of nitrogen, and 2 atoms of oxygen. According to the law of constant composition, caffeine always contains exactly this number of atoms of each of these elements. *Always.* It doesn't matter if we are drinking a cup of coffee in Turkmenistan, a Diet Coke in Los Angeles, or black tea in Berlin—the caffeine in the beverage we are drinking is always the same. Caffeine is a compound, and it must follow the rules.

When we are keeping track of the number of atoms in a substance, it can be tedious to write this down: caffeine contains 8 atoms of carbon, 10 atoms of hydrogen, 4 atoms of nitrogen, and 2 atoms of oxygen. To save some time, scientists have devised a shorthand way to write out this sort of information—it is called a **chemical formula**. The chemical formula for

Peter Jordan/Alamy

Figure 3B Gold Mining in the Madre de Dios Rain Forest of Peru
This photo shows one area of the Madre de Dios rain forest that has been mined for gold. In these regions, the water is contaminated with the toxic by-products of gold mining, including mercury.

operations in small lakes called *tailings ponds,* and these ponds spread their contaminants into streams, lakes, and rivers. Fish and other organisms living in carefully balanced ecosystems are put at risk.

Where gold mining is illegal, the story becomes even uglier. For example, in eastern Peru, an area of the rain forest called Madre de Dios is being pillaged by both legal and illegal gold mining operations (**Figure 3B**). According to the Peruvian government, every day, 50 tankers of fuel make their way into this area. That fuel powers enormous earth-moving machines working around the clock to mine gold from the jungle. Because these operations are often illegal, they do not adhere to environmental regulations. In Peru, gold mining operations use mercury by the vat-full. Because it binds to gold readily, the mercury is used to pull gold from the rock. Eventually, this mercury finds its way into the local ecosystem, and local soil and water become laced with poison.

Is it possible to buy gold that has been mined in a more responsible way? Maybe. Attempts have been made to rate gold according to how it is mined, but most gold is still sold without a label. A better approach is to recycle gold. Gold mining is largely unnecessary, because tons of mined gold are circulating around the world in the form of bullion, jewelry, and coins. The best way to reduce the deleterious effects of gold mining on the environment is to keep reusing the gold we already have.

caffeine is $C_8H_{10}N_4O_2$. In a chemical formula, the element symbols are usually written alphabetically (we will see some exceptions later in the book). A subscript after the symbol tells us how many atoms of each element the compound contains. If there is only one atom of a given element in a compound, then the number 1 is left out of the formula. We can also write chemical formulas for elements. For example, neon, which exists just as single atoms, is written simply as Ne. Sulfur, which often exists in nature in rings of 8 atoms, is written as S_8. **Figure 3.10** shows the anatomy of a typical chemical formula.

We often see parentheses in a chemical formula. For example, consider $Mg(OH)_2$. In this case, the OH represents one atom of oxygen and one atom of hydrogen. The subscript 2 indicates that there are two OH units. This formula indicates that, in total, this compound has one atom of magnesium (Mg), two atoms of oxygen (O), and two atoms of hydrogen (H). The O and the H are grouped together for a reason, but we will not discuss that until we get to Chapter 10, "pH and Acid Rain," where you will learn about acids and bases. For now, just know that the subscript next to parentheses is a way to multiply everything inside the parentheses by that number.

The chemical formula of caffeine, a compound
$C_8H_{10}N_4O_2$
Element symbols are often listed alphabetically.
Subscript shows the number of atoms of that element in this compound.
Caffeine contains a total of 24 atoms.
Emiko Paul

Figure 3.10 The Anatomy of a Chemical Formula
Caffeine is a compound. Its chemical formula is shown here.

QUESTION 3.8 Which substances are elements and which are compounds? (a) silver chloride, AgCl (b) bromine, Br_2 (c) boron triiodide, BI_3 (d) molybdenum trioxide, MoO_3 (e) uranium, U (f) fluorine, F_2

ANSWER 3.8 An element contains atoms of one kind only. Choices (b), (e), and (f) are elements. Because (a), (c), and (d) all contain more than one type of atom, they are compounds.

For more practice with problems like this one, check out end-of-chapter Questions 25 and 26.

FOLLOW-UP QUESTION 3.8 Identify each substance as an element or as a compound. (a) Fe (b) PF_6 (c) H_2 (d) C_6H_6 (e) F_2

QUESTION 3.9 How many atoms of each element are in each compound? (a) $(NH_4)_2SO_4$ (b) $(C_6H_5)_3P$

ANSWER 3.9 (a) There are 2 atoms of nitrogen, 8 atoms of hydrogen, 1 atom of sulfur, and 4 atoms of oxygen (b) There are 18 atoms of carbon, 15 atoms of hydrogen, and 1 atom of phosphorus.

For more practice with problems like this one, check out end-of-chapter Question 29.

FOLLOW-UP QUESTION 3.9 Methane is the fuel that is the main component of *natural gas.* It contains only two elements: 1 atom of carbon and 4 atoms of hydrogen. What is its chemical formula?

The chemical formula for Adderall can be determined by looking at its structure.

For our final example of a chemical formula, let's turn to the most-abused prescription drug among college students. Adderall is prescribed for adults and children who have attention deficit hyperactivity disorder (ADHD). But this drug is a mixture of amphetamines, which are potent stimulants, so it has a high potential for abuse. A recent study found that more than one-third of U.S. college students admitted to trying Adderall in 2012.

The four active components of Adderall, a mixture, are all forms of amphetamine, which is shown in **Figure 3.11**. Notice that Figure 3.11 looks different from earlier images in the book. That's because we are not yet used to seeing the structures of compounds drawn with lines. The lines connecting the atoms in Figure 3.11 represent linkages between atoms; we will learn more about these linkages in Chapter 4, "Bonds."

We can write a chemical formula for the compound shown in Figure 3.11. To do this, we take stock of the number of each type of atom and then write them in the style we just introduced. In this compound, there are 9 atoms of carbon, 13 atoms of hydrogen, and 1 atom of nitrogen. To write the chemical formula for amphetamine, we alphabetize these elements and then use subscripts to indicate the number of each atom it contains: $C_9H_{13}N$. Remember that we don't need to include the number 1 after the symbol N, because the number 1 is always implied when no number is written.

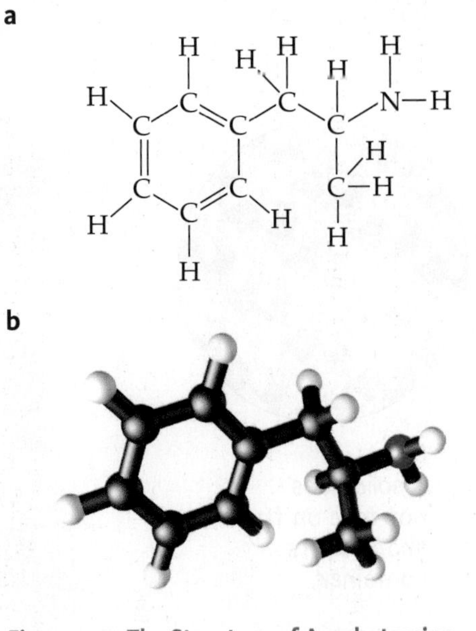

Figure 3.11 The Structure of Amphetamine
(a) This illustration shows the structure of amphetamine, the compound upon which the four active ingredients in Adderall are based, and how the atoms it contains are connected to one another. (b) Amphetamine is shown here as a model whose black spheres are carbon atoms, white spheres are hydrogen atoms, and the blue sphere is a nitrogen atom.

QUESTION 3.10 The structure of the compound caffeine is shown here. Use its structure to write its chemical formula.

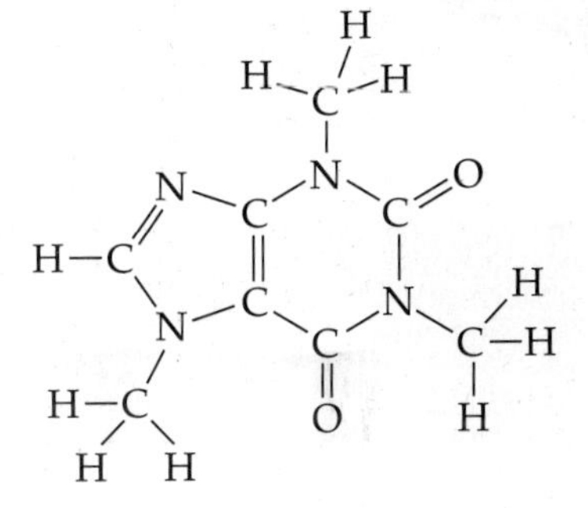

ANSWER 3.10 The chemical formula, which we saw in Figure 3.10, is $C_8H_{10}N_4O_2$.

For more problems like this one, check out end-of-chapter Questions 35 and 39.

FOLLOW-UP QUESTION 3.10 Another drug with stimulant properties is ecstasy, also known as MDMA. The *A* in MDMA stands for "amphetamine," and you may notice similarities between the way the atoms are connected in this compound (shown below) and in amphetamine, shown in Figure 3.11. Write the chemical formula for ecstasy.

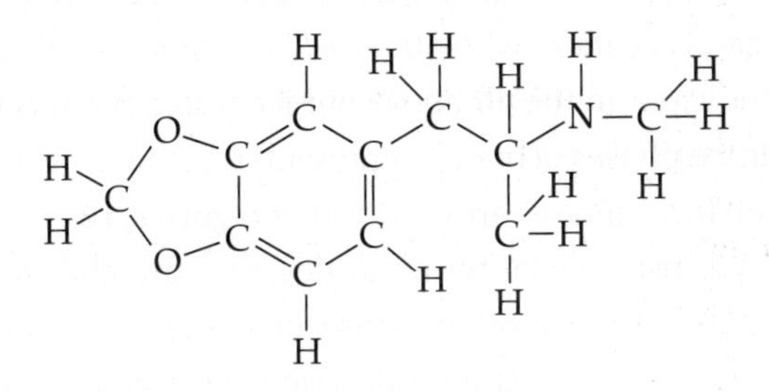

3.5 When Substances Change

The three states of matter are gas, liquid, and solid.

Let's do a thought experiment. Imagine we are standing in a kitchen. We get a glass jar with a lid, add some ice and some water, and close it tightly. Now we will let it sit for a moment while you read the next paragraph.

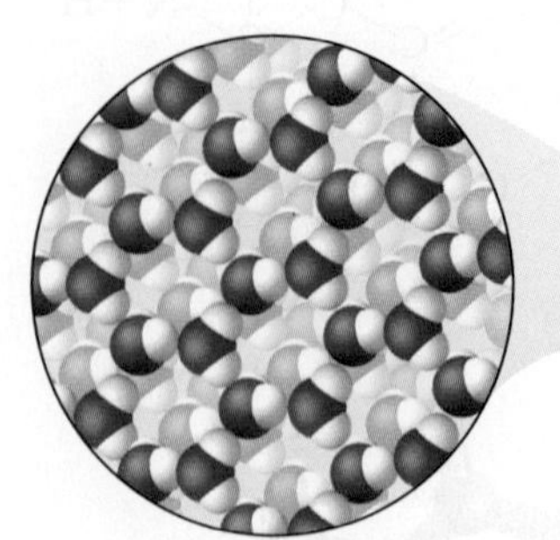

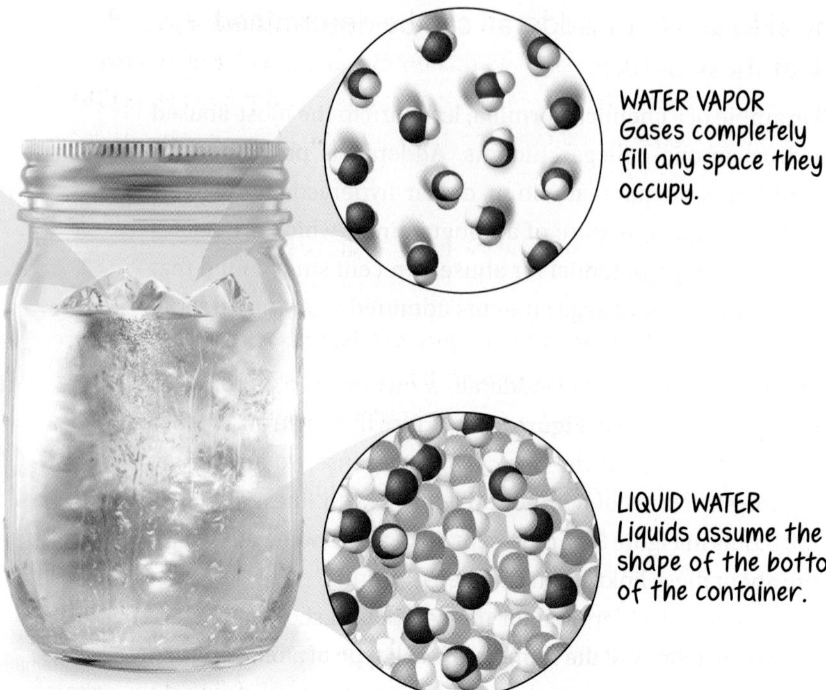

WATER VAPOR
Gases completely fill any space they occupy.
LIQUID WATER
Liquids assume the shape of the bottom of the container.
Emiko Paul

Figure 3.12 Three Phases in a Jar of Water

What phases of matter are represented in that jar? There are three **phases** of matter to choose from: solid, liquid, or gas. We can recognize a **solid** because intuitively we know that if we put it into a container, a solid does not take on the shape of that container. In other words, a solid retains its shape. A solid may be soft or it may be flexible, but it still keeps its shape. A **liquid** doesn't. If we put liquid into a container, we know that the liquid will spread out and assume the shape of the bottom of the container. The surface of the liquid will be flat and level. Gases, on the other hand, do not have surfaces. A **gas** completely fills any space, not just the bottom of a container.

Now let's return to the kitchen and check our half-full jar of icy water. What phases of water are in that jar? We know that we have liquid water and solid water, also known as ice. But what about a gas? Do we see a gas in the jar? Probably not. Water vapor—and the air it's suspended in—are both colorless, and we can't see them with our eyes. Plus, there is only a small amount of water vapor in the air above our ice water. But it *is* there. So we have all three phases of matter represented in the jar. The attributes of each of the three phases are depicted in **Figure 3.12**.

We use special terms to describe the change from one phase of matter to any other phase of matter. For example, the words *boiling* and *vaporization* describe the changes from liquid to gas. Condensation is the opposite. If we have a solid, such as ice, and we put it in a hot oven, a change from solid to liquid phase occurs. This, of course, is melting; its opposite is freezing. It's also possible to jump directly from the gas phase to the solid phase (deposition) or from the solid phase to the gas phase (sublimation). All of these terms are summarized in **Figure 3.13**. We will discuss changes of phase in more detail in Chapter 8, "Water."

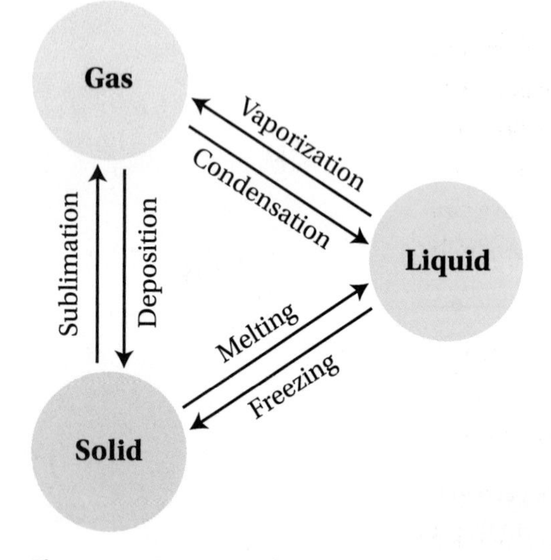

Figure 3.13 Summary of Phase Changes

QUESTION 3.11 Which term from Figure 3.13 describes each of the following phase changes? (a) Water vapor from the air makes a slippery film on a glass full of ice-cold lemonade. (b) The water level in the local reservoir drops during a dry summer. (c) The snow on your favorite ski run turns to slush on a warm spring day. (d) The water in your dog's bowl turns to ice on a cold winter night.

ANSWER 3.11 (a) condensation (b) vaporization (c) melting (d) freezing

For more practice with problems like this one, check out end-of-chapter Question 47.

FOLLOW-UP QUESTION 3.11 Which term from Figure 3.13 describes each of the following phase changes? (a) A certain solid compound changes into a gas when heated. (b) The water in a tea kettle makes the tea kettle whistle after sitting on the hot stove for 10 minutes. (c) A gas forms a solid residue on a chilled surface. (d) A pat of butter changes consistency when it is placed on top of a hot baked potato.

Many mixtures can be easily separated by physical means.

Mixtures can include different phases of matter. Ice water is an example of a solid-and-liquid mixture. Aerosols and aerogels are mixtures of liquid and gas. Alternatively, mixtures can include the same phase of matter. For example, creams and lotions are mixtures of oil and water, both liquids, forced together by mixing. (Lotions usually have less oil than creams.) Whatever their composition, mixtures can often be separated into pure substances by taking advantage of the differences in the physical properties of those pure substances. Physical properties include things like the temperature at which the substance evaporates, boils, melts, or freezes; its density; its hardness; or its color.

This idea is best explained by example. Let's consider a solution of saline, the primary component of many contact lens solutions. Saline solution is a mixture; it is a combination of table salt, or sodium chloride, (NaCl) and water, both pure substances. Because this is a mixture, it should be possible to separate the two components by physical means, but how? One way is to let the solution sit open until all the water has evaporated. What remains is NaCl. This is a separation by physical means because it takes advantage of the differences in the physical properties of the two substances: water evaporates, but sodium chloride does not.

When a substance undergoes a **physical change,** that substance does not become a different substance. One way to recognize a physical change is to ask, does the substance change into a different substance? If the answer is no, then it is probably a physical change. Phase changes are physical in nature, because they do not change a substance into something else. Consequently, changes of phase can be reversed. For example, water vapor turns from steam into water if you cool it. It returns to the liquid phase from the gas phase.

QUESTION 3.12 (a) Describe two ways that you could separate a mixture of wood shavings and iron filings. (b) When the separation is complete, what components do you have? Identify each component as element, compound, homogeneous mixture, or heterogeneous mixture.

ANSWER 3.12 (a) You could add water and allow the wood to float to the surface and the iron filings to sink. Alternatively, because iron is magnetic and wood is not, you could

use a magnet to pull the iron away from the wood. (b) When the separation is complete, you have iron filings (element) and wood shavings (heterogeneous mixture). The shavings must be a mixture because trees are complex mixtures of compounds and elements.
For more practice with problems like this one, check out end-of-chapter Question 46.

FOLLOW-UP QUESTION 3.12 One dozen glass beads are embedded in a stick of cold butter. How could you separate the beads from the butter? What physical change has taken place when you do this?

When a chemical change happens, substances become other substances.

Besides physical changes, substances can also undergo chemical changes. A **chemical change** is one that converts a substance into one or more other substances. We can recognize a chemical change because the chemical formula of the substance changes. To see what is involved in a chemical change, let's think more about the saline solution we separated into NaCl and water.

We know that these compounds—NaCl and water—are both pure substances, and pure substances cannot be broken down further by physical change. What happens, though, if we subject the water to a chemical change?

The chemical formula for water is H_2O. It is a compound, and it contains 2 atoms of hydrogen and 1 atom of oxygen. A chemical change can break apart the strong interactions between the atoms in water. For example, if a bolt of lightning hits a lake, some of the water in the lake is changed into hydrogen and oxygen. Because lightning is a powerful source of energy, it disrupts the forces that hold the oxygen atom to the two hydrogen atoms within water. Lightning hitting a lake makes a chemical change occur because it changes a pure substance, water, into something else: hydrogen and oxygen, two pure substances that are both elements. **Figure 3.14** illustrates the difference between a physical change and a chemical change.

QUESTION 3.13 Identify each of the following as a physical change or a chemical change. (a) Octane burns in a car engine. (b) Water freezes at 0°C. (c) Lake water evaporates on a summer day.

ANSWER 3.13 (a) Octane burns by combining with oxygen in the air to produce carbon monoxide, carbon dioxide, water, and other compounds; the burning of octane is a chemical change. (b) When water freezes, it forms ice, which is still H_2O; freezing is a physical change. (c) The liquid water from the lake enters the air in the form of a gas. The change from liquid water to gaseous water is another example of a physical change.
For more practice with problems like this one, check out end-of-chapter Question 45.

FOLLOW-UP QUESTION 3.13 Is a physical change or a chemical change taking place in each situation? (a) A match that is struck gives off a sulfur-like smell. (b) An opened bottle of rubbing alcohol is empty after sitting for one year. (c) A bomb explodes in a crowded shopping mall.

Figure 3.14 Chemical and Physical Changes

On a stormy day at the lake, a lightning strike can change water to hydrogen and oxygen. This is a chemical change. At the freezing point of water, 0°C, the water in the lake exists as both liquid water and ice. When ice becomes water or water becomes ice, it is a physical change because water does not change to another substance.

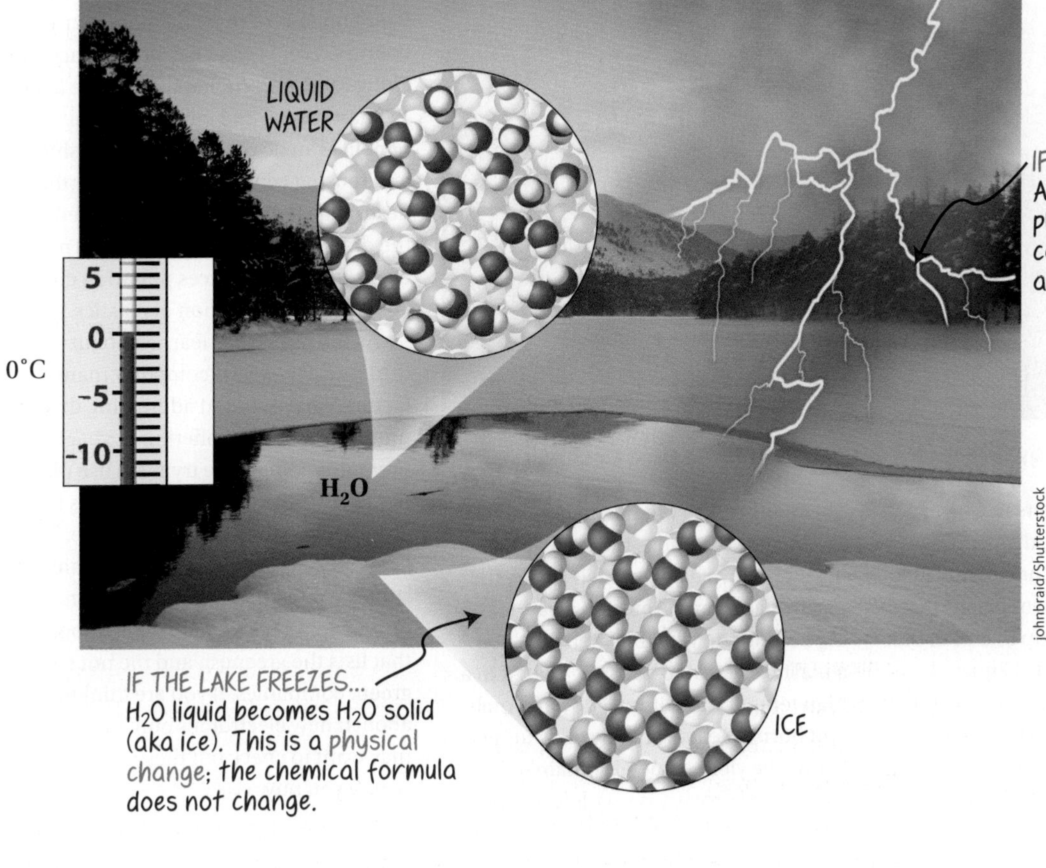

LIQUID WATER
5
0
-5
-10
0°C
H_2O
IF LIGHTNING STRIKES...
A chemical change takes place and some H_2O is converted to hydrogen and oxygen.
IF THE LAKE FREEZES...
H_2O liquid becomes H_2O solid (aka ice). This is a physical change; the chemical formula does not change.
ICE
johnbraid/Shutterstock

THEgreenBEAT • news about the environment, chapter three edition

TOP STORY E-Waste

On June 14, 2008, a group of Greenpeace volunteers quietly arrived at the harbor in Hong Kong. They approached the docks where a cargo vessel, which had just traveled from the docks in Oakland, California, was preparing to unload its goods. The Greenpeace group boarded the vessel and put a banner on the side of the containers reading, "Toxic waste not welcomed here" (**Figure 3.15**). They demanded that the containers be detained and inspected. What was inside the containers? In fact, the containers held used electronics such as cell phones, computers, and monitors. Lots of them. And they were headed to the Shantou Province in China, a popular destination for used electronics.

So why did Greenpeace care about a shipment of broken electronics headed to China? And why should you care about what happens to your old cell phones and old computer monitors? Because surprisingly, these shipments represent a major—and rapidly growing—source of pollution.

Used electronics are pollutants because they contain toxic substances, and those substances end up in the environment if they are not disposed of correctly. The United States ships its old electronics, also called **e-waste,** to other countries rather than recycling or refurbishing them at home. Our country does so for the same reasons it ships jobs overseas: labor is cheaper. Cheap labor means cheap disposal costs. Plus, in many of these countries with lower labor costs, there are fewer restrictions on how and where toxins are dumped. According to the UN website, "Labor is cheapest in countries with the lowest health and safety standards."

E-waste includes a wide range of elements. Several that are toxic to humans and present in large amounts include lead (Pb), beryllium (Be), cadmium (Cd), and mercury (Hg). **Figure 3.16** shows a periodic table with the elements commonly found in electronics highlighted. These elements are some of the biggest threats to people who handle used electronics and to the environment where they are discarded.

AP Images/Kin Cheung

Figure 3.15 Sign Posted by Greenpeace on Containers Full of Used Electronics

Figure 3.17 shows workers in Guiyu, a town in southern China, dismantling e-waste. In this town alone, there are more than 5,000 family workshops that dismantle e-waste, and up to 80% of that waste comes from outside China.

The amount of pollution that comes from used electronics is determined largely by what happens to them. In China, many electronics are broken down by melting them in backyard incinerators. Because the metal parts in such a device do not melt easily, the plastic parts can be burned away, leaving behind the valuable metals. When a plastic electronic device is heated to a high temperature, its plastic parts melt and burn, releasing highly toxic fumes. Sadly, the yield of metals obtained by this type of process is very low, while the release of toxins into the air and water is very high.

It is estimated that millions of tons of e-waste is produced around the world each year. The United States contributes 3 million tons of that, the most of any country. This is what led Greenpeace to Hong Kong harbor that day in June 2008.

Many social and environmental benefits can be realized by refurbishing and reusing computer parts rather than destroying them. Refurbishment reduces the drain on raw materials used to make electronics and reduces pollution created by the incineration of plastics.

As the public has learned about the issue of e-waste, computer manufacturers have started addressing this problem. Some are offering buyback programs. Others are trying to use plastics in their computers that are less hazardous when they become e-waste. The Greenpeace website conducts an annual survey of electronics manufacturers and publishes a *Greener Electronics* report that lists the greenest, and the not so green, companies. If you are thinking about a new electronics purchase, you may want to check out these rankings before you buy.

	1	2	3	4	5	6	7	8	9	10	11	12	13	14	15	16	17	18
1	1 H 1.008																	2 He 4.003
2	3 Li 6.941	4 Be* 9.012											5 B 10.81	6 C 12.01	7 N 14.01	8 O 16.00	9 F 19.00	10 Ne 20.18
3	11 Na 22.99	12 Mg 24.31											13 Al 26.98	14 Si 28.09	15 P 30.97	16 S 32.07	17 Cl 35.45	18 Ar 39.95
4	19 K 39.10	20 Ca 40.08	21 Sc 44.96	22 Ti 47.87	23 V 50.94	24 Cr 52.00	25 Mn 54.94	26 Fe 55.85	27 Co 58.93	28 Ni 58.69	29 Cu 63.55	30 Zn 65.41	31 Ga 65.39	32 Ge 72.64	33 As 74.92	34 Se 78.96	35 Br 79.90	36 Kr 83.80
5	37 Rb 85.47	38 Sr 87.62	39 Y 88.91	40 Zr 91.22	41 Nb 92.91	42 Mo 95.94	43 Tc 98	44 Ru 101.1	45 Rh 102.9	46 Pd 106.4	47 Ag 107.9	48 Cd* 112.4	49 In 114.8	50 Sn 118.7	51 Sb 121.8	52 Te 127.6	53 I 126.9	54 Xe 131.3
6	55 Cs 132.9	56 Ba 137.3	57–71	72 Hf 178.5	73 Ta 180.9	74 W 183.8	75 Re 186.2	76 Os 190.2	77 Ir 192.2	78 Pt 195.1	79 Au 197.0	80 Hg* 200.6	81 Tl 204.4	82 Pb* 207.2	83 Bi 209.0	84 Po	85 At	86 Rn
7	87 Fr	88 Ra	89–103	104 Rf	105 Db	106 Sg	107 Bh	108 Hs	109 Mt	110 Ds	111 Rg	112 Cn	113 Nh Nihonium	114 Fl Flerovium	115 Mc Moscovium	116 Lv Livermorium	117 Ts Tennessine	118 Og Organesson

57 La 138.9	58 Ce 140.1	59 Pr 140.9	60 Nd 144.2	61 Pm	62 Sm 150.4	63 Eu 151.9	64 Gd 157.3	65 Tb 158.9	66 Dy 162.5	67 Ho 164.9	68 Er 167.3	69 Tm 168.9	70 Yb 173.0	71 Lu 175.0
89 Ac	90 Th 232.0	91 Pa 231.0	92 U 238.0	93 Np	94 Pu	95 * Am	96 Cm	97 Bk	98 Cf	99 Es	100 Fm	101 Md	102 No	103 Lr

Figure 3.16 Periodic Table Showing Elements Found in Electronic Devices

Elements found in large amounts in electronic devices are shown in dark purple, moderate amounts are in medium purple, and trace amounts are shown in light purple. Elements marked with asterisks are highly toxic to humans.

Figure 3.17 Chinese Workers Dismantling Electronics

QUESTION 3.14 What do the elements that are toxic to humans, as shown in Figure 3.16, have in common?

ANSWER 3.14 They are all metals.

For more practice with problems like this one, check out end-of-chapter Question 11.

FOLLOW-UP QUESTION 3.14 Rank the following toxic elements found in electronic devices according to their abundance in e-waste: americium, lead, cadmium.

Chapter 3 Themes and Questions

recurring themes in chemistry chapter 3

recurring theme 5 Elements in the same group on the periodic table often behave in similar ways and share properties with one another. (p. 69)

A list of the eight recurring themes in this book can be found on p. xv.

ANSWERS TO FOLLOW-UP QUESTIONS

Follow-Up Question 3.1: Nonmetals can be described as brittle (a), gaseous (b), or dull (e). *Follow-Up Question 3.2:* Krypton (Kr) and scandium (Sc) are not found in soil. *Follow-Up Question 3.3:* Soil contains more Fe. *Follow-Up Question 3.4:* Iodine (I) and astatine (At) are halogens. *Follow-Up Question 3.5:* (a) noble gases (b) halogens (c) alkali metals (d) alkaline earth metals. *Follow-Up Question 3.6:* You could separate according to color (two colors: red and black). *Follow-Up Question 3.7:* (a) homogeneous (b) could be homogeneous or heterogeneous (c) probably heterogeneous. *Follow-Up Question 3.8:* Substances (a), (c), and (e) are elements. The others are compounds. *Follow-Up Question 3.9:* CH_4. *Follow-Up Question 3.10:* $C_{11}H_{15}NO_2$. *Follow-Up Question 3.11:* (a) sublimation (b) vaporization (c) deposition (d) melting. *Follow-Up Question 3.12:* Melt the butter and pour it off of the beads; the butter melted. *Follow-Up Question 3.13:* (a) chemical change (b) physical change (c) chemical change. *Follow-Up Question 3.14:* lead > cadmium > americium

End-of-Chapter Questions

Questions are grouped according to topic and are color-coded according to difficulty level: **The Basic Questions***, **More Challenging Questions****, and **The Toughest Questions*****.

The Distribution of Elements in Nature

***1.** Which of the following is a metal?

(a) bromine, Br (b) oxygen, O (c) titanium, Ti

(d) cerium, Ce (e) erbium, Er

***2.** Which of the following is a nonmetal?

(a) selenium, Se (b) sodium, Na (c) protactinium, Pa

(d) molybdenum, Mo (e) neon, Ne

***3.** Which of the following elements is typically found in organic compounds?

(a) nobelium, No (b) uranium, U (c) rhodium, Rh

(d) carbon, C (e) potassium, K

****4.** Which of the following descriptors can be used for metals?

(a) brittle (b) shiny (c) good conductor of heat

(d) poor conductor of electricity (e) malleable

****5.** Which of the following descriptors can be used for nonmetals?

(a) dull (b) always a solid (c) brittle

(d) ductile (e) good conductor of electricity

****6.** Comment on the following statement: Most things in nature are composed entirely of either metal or nonmetal, but not mixtures of both.

****7.** Refer to Figure 3.3 and rank the following elements in order of increasing abundance in soil: manganese, sulfur, chlorine, potassium.

****8.** Refer to Figure 3.4 and rank the following elements in order of decreasing abundance in the human body: nitrogen, zinc, phosphorus, calcium.

*****9.** If a person has a mass of 85 kilograms and the value, in dollars, of iron is 0.13 dollars per kilogram, what is the mass of iron in that person's body worth, in dollars? (Refer to Figure 3.4 for data.)

*****10.** The mass of a sample of soil is 2370 grams. What is the mass, in grams, of the aluminum in that sample of soil? (Refer to Figure 3.3 for data.)

The Periodic Table

***11.** Which of the following elements would you expect to share common properties with sodium?

(a) lithium (b) zinc (c) arsenic (d) rubidium (e) potassium

***12.** Which of the following elements would you expect to share common properties with chlorine?

(a) astatine (b) krypton (c) iodine (d) tellurium (e) francium

***13.** Are there more periods or more groups on the periodic table?

****14.** Is the following statement true or false? Why? As you move across Period 3 on the periodic table, you encounter elements with similar physical properties.

****15.** In your own words, explain what the terms *period, group,* and *family* mean with respect to the periodic table.

****16.** Is the following statement true or false? Why? The periodic table is periodic because, as we move from low to high atomic number, we see a repeating pattern of similar properties for elements that fall below one another.

****17.** Which of the following gaseous elements do not make up a significant part of the air we breathe?

(a) nitrogen (b) fluorine (c) xenon (d) oxygen (e) helium

*****18.** Is the following statement true or false? Why? Elements in one family must be all metals or all nonmetals. There are no families that have both.

*****19.** Identify the following elements by name:

(a) a noble gas with more protons than osmium

(b) the halogen with the fewest number of electrons

(c) an alkaline earth metal found in electronic devices

*****20.** Identify the following elements by name:

(a) a reactive metal with fewer than four protons

(b) a nonmetal that begins with the 16th letter of the alphabet

(c) a metal whose name and symbol do not share the same first letter

Pure Substances, Mixtures, Compounds, and Elements

***21.** Classify each item as a homogeneous mixture, a heterogeneous mixture, or a pure substance:

(a) glucose, $C_6H_{12}O_6$ (b) filtered apple juice

(c) unfiltered apple juice (d) chocolate chip ice cream

***22.** Classify each item as a homogeneous mixture, a heterogeneous mixture, or a pure substance:

(a) orange juice with pulp (b) milk (c) bottled water

(d) Plexiglas (e) iced tea

***23.** Classify each item as a homogeneous mixture, a heterogeneous mixture, or a pure substance:

(a) gravel (b) granite (c) glass (d) plywood (e) white glue

***24.** Isoflurane, $C_3H_2OF_5Cl$, is an inhaled anesthetic used in hospital operating rooms. Which term(s) describes isoflurane: mixture, pure substance, element, compound?

***25.** Which of the following are elements and which are compounds?

(a) BF_3 (b) H_2 (c) S_8 (d) C_6H_{12}

*26. Which of the following are elements and which are compounds?

(a) C (b) CH_4 (c) Br_2 (d) $C_{10}H_{19}O_2$

**27. Is the following statement true or false? Why? An element must consist of individual atoms that are separate from one another.

**28. Speculate on how you could separate the pure substances in the following mixtures from one another:

(a) a bowl containing water, small pieces of copper pipe, and table salt

(b) a bucket that contains water, cooking oil, and pieces of marble

**29. A homogeneous mixture contains only aspirin, $C_9H_8O_4$, and vitamin C, $C_6H_8O_6$.

(a) Is aspirin a compound or an element?

(b) Is aspirin a pure substance or a mixture?

(c) Is vitamin C a compound or an element?

(d) Is vitamin C a pure substance or a mixture?

**30. Using your own words and full sentences, explain the meaning of the law of constant composition. Give one example of a compound, and explain what it means for that compound to follow this law.

***31. The element phosphorus exists in several elemental forms in nature. Go to the Internet or another resource and research the most common form of phosphorus in nature. What are its common name and its chemical formula?

***32. Caffeine is a compound that has the chemical formula $C_8H_{10}N_4O_2$. Samples of caffeine are collected from sources in Quebec, Morocco, Turkey, Norway, and Chile. The samples from Turkey and Chile both contain five hydrogen atoms for every oxygen atom. In the other three samples, the hydrogen/oxygen ratio varies widely from this value. What can you conclude about these three samples?

***33. In the laboratory, you come across an unlabeled bottle containing some chemical substance. Because it was purchased at a chemical company, you suspect that the bottle contains a pure substance and not a mixture. Is it possible that this substance is absolutely pure with no impurities? Why or why not?

***34. The element carbon exists in nature in a form called *buckminsterfullerene*. Go to Chapter 5, "Carbon," of this book and find the three-dimensional structure of this compound. Write its chemical formula. How many carbon atoms does it contain?

***35. In your own words and using complete sentences, define the following terms and describe when they are used: *element, compound, mixture, pure substance*. Next, pair each word with every other word, and explain if each pair of words can be used to describe the same substance. For example, can a substance be both an element and a pure substance?

Chemical Formulas

*36. List the total number of atoms and the number of atoms of each element contained in these substances:

(a) benzene, C_6H_6 (b) bromine, Br_2 (c) nitrous oxide, N_2O

*37. List the total number of atoms and the number of atoms of each element contained in these substances:

(a) pentane, C_5H_{12} (b) hydrogen, H_2 (c) hydrogen sulfide, H_2S

*38. Classify each substance as a compound or an element.

(a) CH_4 (b) S_8 (c) NO_2 (d) Cu (e) P_4

**39. The drug levonorgestrel is the hormone in the "morning-after pill," used to prevent implantation of a fertilized egg in a human uterus. The chemical formula for levonorgestrel is $C_{21}H_{28}O_2$. How many atoms are there in levonorgestrel?

**40. A mystery element exists in units of four atoms. For each group of four atoms of that element, there are 60 protons. What is the chemical formula for this element?

**41. A compound is composed only of the elements carbon, hydrogen, and oxygen. The number of carbon atoms is 10, and the number of carbon atoms is equal to the number of hydrogen atoms *plus* five times the number of oxygen atoms. What is the total number of atoms in the chemical formula for this compound?

***42. A mystery compound is made up of carbon atoms, oxygen atoms, nitrogen atoms, and hydrogen atoms. It contains 35 atoms in total. If the number of carbon atoms is twice the number of oxygen atoms, the number of hydrogen atoms is 8 more than the number of carbon atoms, and the number of nitrogen atoms is two, what is the chemical formula for this compound?

***43. A mystery compound is made up of the elements carbon, hydrogen, bromine, and nitrogen. There are 16 atoms in total, and two of the elements are present as one atom only. The number of carbon atoms is two more than the number of hydrogen atoms and eight times the number of bromine atoms. What is the chemical formula for this compound?

***44. Adderall XR is a prescription medication that is used in the treatment of attention deficit disorders. It is a mixture of four different types of amphetamines, which are compounds that stimulate the central nervous system. Plain amphetamine has the chemical formula $C_9H_{13}N$.

(a) How many atoms of carbon, hydrogen, and nitrogen does amphetamine contain?

(b) How many total atoms does amphetamine contain?

Chemical and Physical Change

*45. Identify each change as chemical or physical.

(a) gasoline burning in a car engine (b) snow melting

(c) alcohol boiling (d) dynamite exploding

*46. You have a mixture of oil and vinegar. Describe one way that you could separate these from one another.

**47. Which of the following only represents a change of phase of a substance?

(a) Water levels in the lake drop during a drought.

(b) You bake a cake and leave it in the oven too long, and it burns.

(c) A candle melts in the sun.

(d) You spill gasoline on the ground when you are filling your gas tank, and the gasoline quickly disappears.

**48. It is often possible to separate a mixture of a solid and liquid by decantation. What does it mean to decant a liquid? You may need to refer to your dictionary to find the answer.

**49. In your own words, explain why, when a substance undergoes a phase change, it is a physical and not a chemical change.

**50. (a) Describe two ways that a mixture of water and sand (sand is primarily silicon dioxide, SiO_2) can be separated into its two components.

(b) Is the SiO_2 present in sand an element or a compound?

(c) Can a physical or chemical change (or both) be used to break down the water and silicon dioxide further?

Integrative Questions

*51. The structure of nicotine, the highly addictive compound found in cigarettes, is shown to the left.

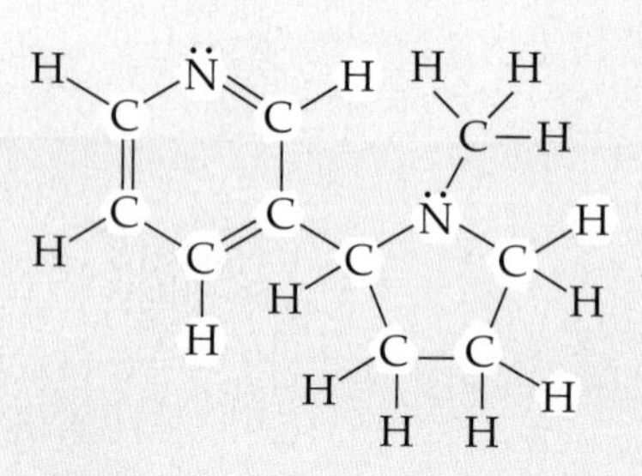

(a) Write the chemical formula for nicotine.

(b) Refer to Figure 3.13. Use one of the six phase changes shown to describe each of the changes that could occur for nicotine.

i. When nicotine is cooled from room temperature down to –80°C, it solidifies.

ii. When nicotine is cooled from 300°C down to room temperature, it changes from a gas to a liquid.

iii. If you take solid nicotine and heat it to a very high temperature under certain conditions, it can change directly from a solid to a gas.

(c) Based on the work you did in question 51b, in which phase does nicotine exist at room temperature?

*52. Carbon dioxide is a gas found naturally in our atmosphere. The natural levels of carbon dioxide in the air are increased by human activities such as driving cars.

(a) Dry ice is the solid form of carbon dioxide. It turns directly into a gas at room temperature. What is the word for this process?

(b) How many atoms does one molecule of carbon dioxide possess?

(c) Are the elements found in carbon dioxide metals, metalloids, or nonmetals?

**53. Insulin is a biological molecule that is involved in the regulation of glucose levels in the human body. Its chemical formula is $C_{256}H_{387}N_{65}O_{79}S_6$.

(a) The periodic table is composed of metals, nonmetals and metalloids. Into which of these categories do all of the atoms in insulin fall?

(b) How many total atoms are found in insulin?

(c) What percentage of the atoms in insulin are carbon atoms? Hydrogen atoms? Nitrogen atoms? Oxygen atoms? Sulfur atoms? Add up the percentages you just determined. The value you obtain should be very close to 100%.

**54. Return to the Greenbeat, "E-Waste," to find answers to the following questions.

(a) Which five elements are the most toxic elements found in e-waste?

(b) Are these elements metals, metalloids, or nonmetals?

(c) Would you consider e-waste to be a mixture or a pure substance?

(d) Referring to Figure 3.6, determine which of the toxic elements you identified in question 54(a) are transition metals. Which are alkaline earth metals?

55. INTERNET RESEARCH QUESTION

Coffee and tea contain caffeine, but so do other foods like chocolate. Let's investigate the caffeine content of several foods. Use your Internet search engine to determine the content of these foods: dark chocolate, white chocolate, milk chocolate, coffee, decaffeinated coffee, black tea, green tea, white tea, herbal tea, diet cola, and root beer.

We can only compare amounts of caffeine if the amounts are expressed with the same units for each food and drink. Most values are expressed in milligrams of caffeine in some number of ounces of the food or beverage. Record this value and then convert the amount to milligrams of caffeine per one ounce of the food or beverage.

EXAMPLE: A latte contains 95 mg of caffeine in 16 ounces of drink. Divide these (95 mg/16 oz.) to find that it contains ~5.9 milligrams of caffeine per one ounce of drink.

(a) Take a look at your data. Which item has the most caffeine per ounce? Which has the least? Do any of these values surprise you?

(b) Consider your data on the four types of tea. Which type has the most caffeine? Which has the least? Which type might be best to drink before bedtime? Which type might be the best to drink in the morning in place of coffee?

age fotostock/age fotostock

Bonds

An Introduction to the Forces within Substances

B C N Ne

CHAPTER

4

In the 2002 film Die Another Day, *James Bond faces off with villain* Gustav Graves for control of a solar focusing device called Icarus. This device is so powerful, it can incinerate anything, and Graves must be stopped before he uses it to take control of planet Earth.

Icarus may seem like science fiction but, in fact, such devices—called *solar furnaces* in the real world—do exist. The biggest, most powerful solar furnace (see photo on p. 94) is located in the sunny little French town of Odeillo on the border of Spain and France. At this facility, 10,000 mirrors focus the sunlight that hits them, and this concentrated sunlight is indeed powerful. In fact, no known substance can withstand the temperatures in this solar furnace, which can reach as high as 3800°C.

age fotostock/age fotostock

Diamond, which has a melting point of approximately 3500°C, is one of the toughest substances known, and even diamond melts in this solar furnace.

Iron has a much lower melting point, 1535°C, so it doesn't stand a chance in the solar furnace. But pots and pans, many of them made primarily of iron, do not melt on our stovetops, because a stove doesn't get hot enough. A stovetop is hot enough to burn food, but table salt can withstand even the hottest stove. Clearly, all of these substances differ in some fundamental way, but how?

This chapter is all about the forces that hold substances together. We'll explore this topic throughout the chapter, because these forces help us account for the properties of matter. For example, the forces in substances are directly related to the fact that food can burn on a stove, but salt and the pan holding the food do not. These substances—food, the salt, and a metal pot—are made up of different combinations of atoms, and those atoms are held together by different kinds of forces, which are disrupted easily in some foods and not nearly as easily in salt and the metal pot. Why does this happen? To find out, we have to return to one of the fundamental particles in the atom and the most important particle for chemists: the electron.

age fotostock/age fotostock

4.1 The Octet Rule

The noble gases are especially stable.

Throughout this chapter, we will see how atoms of the same element interact with one another as well as how atoms of different elements interact with one another. But some elements do not interact with anything: they are the **noble gases** (Group 18). All the noble gases—from helium (He) down to radon (Rn)—exist as lone atoms. This group is the last column on the far right of the periodic table, and its location on the periodic table is the key to its unique properties, as we're about to see.

Let's begin with this basic assumption: atoms interact with one another as a means of gaining something we call *stability*. Because noble gases already possess this thing called stability, they do not need to interact with other atoms to attain it. So what makes an atom stable? Stability is related to an atom's number of electrons, which is related, in turn, to an atom's location on the periodic table. So, we can learn something from the noble gases by thinking about their position on the periodic table.

flashback
Recall from Section 2.4, you know that electrons are negatively charged particles located outside the nucleus of an atom. Electrons in an atom can be assigned to energy levels.

When we are reading the periodic table from left to right, like a book, we read across a period until we get to the last column. As we learned in Chapter 3, each new box that we pass as we go along represents an element with one additional proton. And, because atoms have a net charge of zero, that element also has one additional electron. Thus, each element, in turn, has one more proton and electron than the element before it. Recall from Section 2.4 that electrons occupy energy levels in an atom. Figure 2.11 is repeated here as a reminder (**Figure 4.1**). As we're about to see, this distribution of electrons tells us about the stability of an atom of that element.

flashback
Recall from Section 2.1 that protons are positively charged particles located in the nucleus of an atom. They define an element.

Consider the first noble gas: helium. Helium has two protons and two electrons. Those two electrons fill up the first energy level.

Now let's look at energy level 2, which can hold eight electrons. If we consider Period 2, beginning with lithium (Li), we see that this period includes eight boxes. And, accordingly, eight electrons fill up energy level 2. Each element in Period 2, as we move our way across the table, has one additional electron in its second energy level. The period ends with neon (Ne), a noble gas, which has eight electrons in its second energy level. And this pattern is repeated. Each period on the periodic table (p. 96) represents a new energy level, as shown in **Figure 4.2**. Within a period, each successive element from left to right has one more electron than the element before it. The outer level of the last element in each period is completely filled, and this pattern characterizes the noble gases, which all appear in the last column of the table.

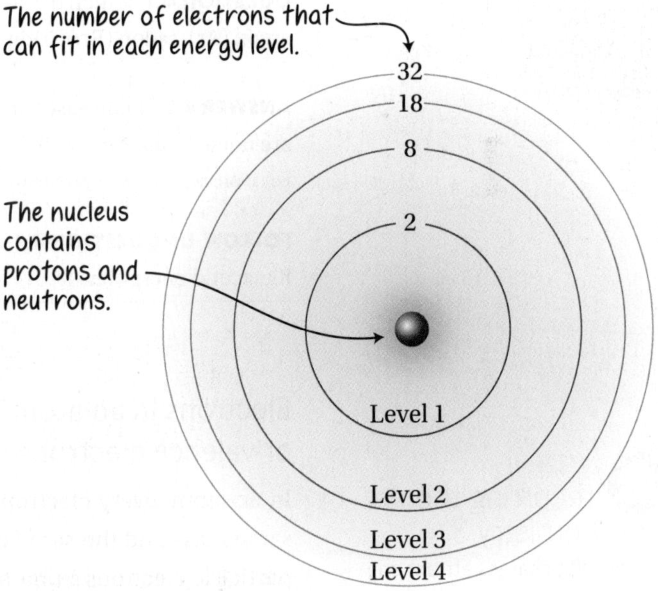

Figure 4.1 Energy Levels in the Atom
This diagram indicates the number of electrons that each energy level in an atom can "hold." Atoms with few electrons keep those electrons close to the nucleus in the first few energy levels. In larger atoms, the outermost electrons in the atom are located at greater and greater distances from the nucleus. The assignment of electrons to locations around the nucleus is a model to help us understand the behavior of atoms.

Now it's clear why that rightmost column of the periodic table, Group 18, has special significance: each noble gas has a number of electrons that represents

The number of elements that span period I is equivalent to the number of electrons in the first energy level of the atom because each successive element adds one electron.

The noble gases lie in the rightmost column of the periodic table and always are the last element in a given period.

Two elements span period I, and each element has one more electron than the one before it.

Proceed to period 2

Proceed to period 3

Eight elements span period 2. Reading left to right, each successive element has one more electron than the one before it.

Figure 4.2 Relationship between Energy Levels and the Periodic Table

a period filled completely from left to right. This "ideal" number of electrons gives the noble gases special stability. For these elements, the number of electrons is just right; they do not become more stable by gaining additional electrons or by losing some electrons. This also explains the behavior of noble gases. Each noble gas has an ideal number of electrons, so noble gases do not need to undergo chemical reactions to obtain a more optimal number of electrons.

QUESTION 4.1 Which of these elements are not noble gases? xenon (Xe), yttrium (Y), argon (Ar), radon (Rn), iridium (Ir), helium (He), californium (Cf)

ANSWER 4.1 Noble gases are all found in Group 18 of the periodic table. These elements are noble gases: Xe, Ar, Rn, He. These elements are not noble gases: Y, Ir, Cf.

For more practice with problems like this one, check out end-of-chapter Question 2.

FOLLOW-UP QUESTION 4.1 According to Figure 4.2, which element has four electrons in its second energy level?

Electrons in an atom are either core electrons or valence electrons.

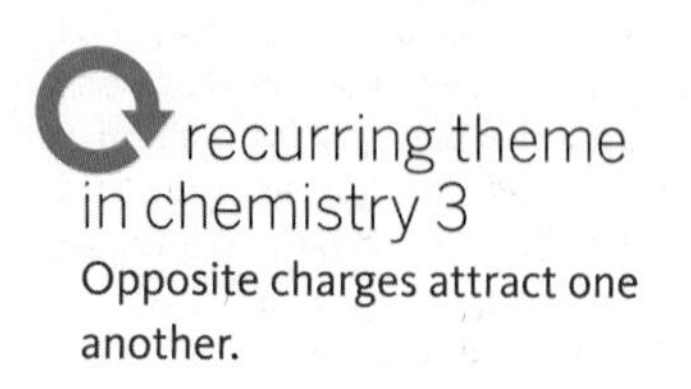

Opposite charges attract one another.

In an atom, every electron is the same as every other electron. They all have the same mass and the same charge, −1. But, even though they are exactly the same particles, electrons in an atom differ based on their location with respect to the nucleus. We can imagine that an atom is like an onion, and energy levels of the atom are like the layers of that onion. Period 1, which is like the innermost layer of the onion, includes hydrogen and helium. These Period 1 elements have tightly held electrons that are right next to the nucleus, which is positively charged. These electrons fill the innermost layer, which holds only two electrons.

The next period, Period 2, begins with lithium and continues through neon. These Period 2 elements have a new layer of electrons that is outside the inner layer of two electrons. Thus, for neon, there are 2 electrons in the inner layer and 8 electrons in the outer layer, for a total of 10 electrons. The outermost electrons in an atom are the **valence electrons**. All of the remaining electrons—those that are between the valence electrons and the nucleus—are **core electrons**. Larger elements can have several layers of core electrons.

We can determine the distribution of core and valence electrons for any atom by looking at the last row of electrons it contains. The number of electrons in that last row is the number of valence electrons in that atom. All of the other electrons in the atom are, by default, core electrons.

For example, **Figure 4.3** shows how we assign electrons in an atom of aluminum, which is overlaid onto an onion. Aluminum is in Period 3 of the periodic table. That means the valence electrons in aluminum are in the third energy level. Counting left to right, Na then Mg then Al, we can figure out that the outermost layer of aluminum has three electrons. Thus, aluminum contains three valence electrons.

We know that aluminum has a total of 13 electrons. Because three of those electrons are valence electrons, the remainder—10 electrons—are core electrons. This makes sense because—counting left to right and moving from top to bottom—the first period of elements has two elements in it. These represent the two electrons of aluminum that are closest to the nucleus. The next period has eight elements, representing eight electrons; these are also core electrons.

The outermost electrons in an atom do all of its chemistry. These valence electrons are the furthest electrons from the nucleus and feel less of its pull. Thus, the valence electrons are involved when atoms undergo chemical reactions. The core electrons are held more tightly by the protons in the nucleus, and they are not easily removed. Consequently, they do not take part in chemical reactions.

recurring theme in chemistry 4

Chemical reactions are nothing more than exchanges or rearrangements of electrons.

When there are more and more layers of electrons in an atom, the valence electrons get farther and farther from the nucleus. The farther from the nucleus a valence electron is, and the more layers of core electrons that shield it from the pull

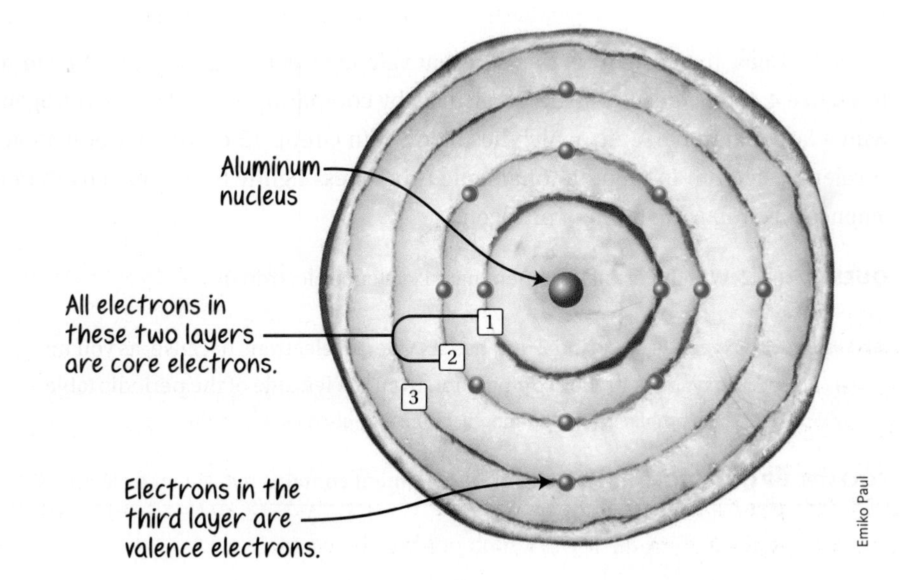
Aluminum nucleus
All electrons in these two layers are core electrons.
1
2
3
Electrons in the third layer are valence electrons.
Emiko Paul

Figure 4.3 An Atom of Aluminum Depicted as an Onion

An atom is like an onion because its electrons can be grouped into layers. If the center of the onion represents the nucleus of an atom of aluminum, then the first layer of the onion represents energy level 1, the second layer is energy level 2, and so on. For aluminum atoms, the third layer holds the valence electrons, while the first two layers hold core electrons. Keep in mind that this is only a model. Electrons do not exist in defined orbits around the nucleus.

Figure 4.4 How Does a Small Atom Differ from a Large Atom?

A small atom like nitrogen has fewer core electrons to shield valence electrons from the pull of the nucleus (top). A larger atom, like arsenic, has more core electrons (bottom).

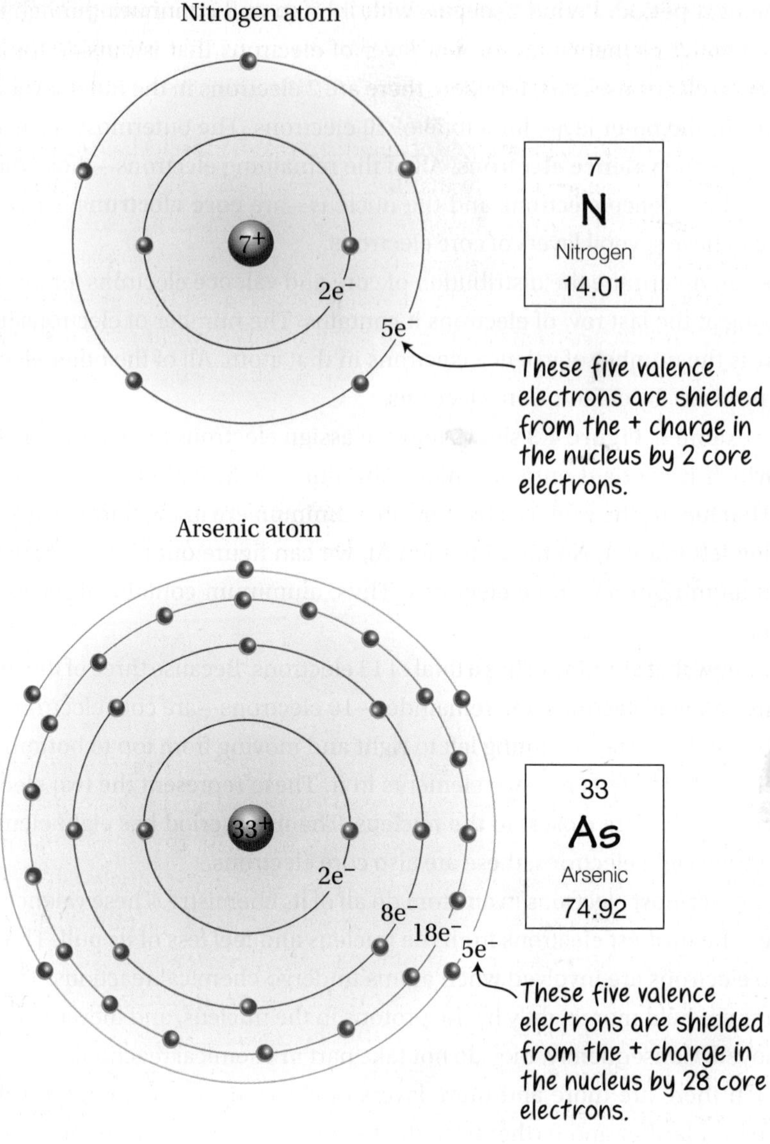

Nitrogen atom
7+
2e⁻
5e⁻
7
N
Nitrogen
14.01
These five valence electrons are shielded from the + charge in the nucleus by 2 core electrons.
Arsenic atom
33+
2e⁻
8e⁻
18e⁻
5e⁻
33
As
Arsenic
74.92
These five valence electrons are shielded from the + charge in the nucleus by 28 core electrons.

of the nucleus, the easier it is to pluck that valence electron away from the atom. In **Figure 4.4,** this idea is illustrated visually by comparing a small atom, nitrogen, with a larger atom, arsenic, which lies below it in Group 15 on the periodic table. A valence electron in an atom of arsenic is held less tightly to the nucleus than a valence electron in an atom of nitrogen.

QUESTION 4.2 Which Period 2 element has six valence electrons?

ANSWER 4.2 Oxygen is in Period 2 and has six valence electrons because, as you move across Period 2, oxygen is in the sixth position from the left side of the periodic table. For more practice with problems like this one, check out end-of-chapter Question 11.

FOLLOW-UP QUESTION 4.2 Which Period 3 element contains two valence electrons?

QUESTION 4.3 (a) How many electrons do atoms of each of the following elements possess? lithium, sodium, potassium, rubidium (b) How many of the electrons in each of the elements listed in (a) are valence electrons and how many are core electrons?

ANSWER 4.3 (a) The total number of electrons for an atom of each element is equal to the total number of protons, which is the atomic number. Therefore, the numbers of electrons are lithium (3), sodium (11), potassium (19), rubidium (37). (b) The number of valence electrons for each atom is the number of boxes counted from left to right in the row that element occupies. An atom of each of these elements possesses one valence electron, because each element occupies the first box in its row. The number of core electrons equals the total minus the valence electrons. Lithium has 2 core electrons, sodium has 10, potassium has 18, and rubidium has 36.

For more practice with problems like this one, check out end-of-chapter Questions 12 and 13.

FOLLOW-UP QUESTION 4.3 How many total electrons do atoms of each of the following elements possess? boron, chlorine. Of these totals, how many are valence electrons and how many are core electrons?

Atoms with eight valence electrons have special stability and obey the octet rule.

Clearly, the position occupied by the noble gases on the periodic table is linked to their unique stability. In fact, it's virtually impossible to take an electron away from a neon atom, for example, because all 10 electrons—even the valence electrons—are held so tightly by the nucleus. Nor will a neon atom accept any more electrons.

Table 4.1 shows how the electrons in each of the noble gases are assigned as either core or valence electrons. According to Table 4.1, what do all of the noble gas atoms have in common? They all have eight valence electrons, and we say that these atoms obey the **octet rule** because they all have eight electrons in their outermost layers. Another way to say that these atoms obey the octet rule is to say that they possess a **noble gas configuration** of electrons. As we will see later in this chapter, non-noble gas atoms use some tricks to achieve a noble gas configuration of electrons. And when they do, they achieve some extra stability, too.

Table 4.1 Core and Valence Electrons in the Noble Gas Atoms

Element	Total Number of Electrons	Number of Valence Electrons	Number of Core Electrons
Neon (Ne)	10	8	2
Argon (Ar)	18	8	10
Krypton (Kr)	36	8	28
Xenon (Xe)	54	8	46
Radon (Rn)	86	8	78

The total number of electrons is equal to the number of protons in the nucleus, given by the atomic number. This is the number shown above the element symbol on the periodic table. The number of valence electrons also is found on the periodic table: It is the number of electrons in the outermost layer of electrons in the atom. All electrons that are not valence electrons are core electrons.

wait a minute . . .

Why does krypton (Kr) have an octet when there are 18 boxes across Period 4?

When we get to Period 4, a strange thing happens. After K and Ca, there is a whole new block of elements beginning with scandium (Sc). As we learned in Section 3.2, these elements, which span Groups 3 through 12, are known collectively as the *transition metals.*

When we count valence electrons in an element in Period 4 or below it, we skip the transition metals and do not include them in our valence electron count. So, for krypton, we start with K at the beginning of Period 4 and start counting from left to right. After two boxes, we jump across the transition metals over to element 31, gallium (Ga), which we count as 3. If we continue counting across, we reach 8 when we get to krypton. Using this counting method, krypton obeys the octet rule.

You may have noticed that one noble gas is missing from Table 4.1: helium (He). Helium occupies a special spot on the periodic table. Right at the top of the table with hydrogen (H), helium has only two electrons. Obviously, it cannot obey the octet rule, but helium is still a noble gas and is especially stable. Because of its unique circumstances, we give this petite element its own rule, the **duet rule**. According to the duet rule, elements at the very top of the periodic table can obtain special stability with two electrons rather than eight.

QUESTION 4.4 True or false? Helium atoms have special stability even though they cannot achieve an octet of electrons.

ANSWER 4.4 True. Helium atoms follow the duet rule, not the octet rule.

For more practice with problems like this one, check out end-of-chapter Question 1.

FOLLOW-UP QUESTION 4.4 True or false? Atoms that achieve an octet of electrons are said to have a noble gas configuration of electrons.

4.2 An Introduction to Bonding

Atoms achieve stability by gaining or losing electrons.

Any element that is not a noble gas can become more stable by gaining or losing some electrons to achieve a noble gas configuration of electrons. Consider sulfur (S), a Group 16 element that is located two groups to the left of the noble gases. By accepting two electrons, a sulfur atom (with 16 electrons) obtains an octet of electrons like that of argon, the nearest noble gas (**Figure 4.5**).

Likewise, magnesium (Mg), a Group 2 element, becomes more stable by losing two electrons in order to have an electron configuration like that of neon, the noble gas closest to magnesium on the periodic table. This is easier to visualize with a cylindrical periodic table in which Group 1 and Group 18 are aligned next to each other (see Figure 4.5). When an atom gains or loses electrons, those electrons are

Figure 4.5 A Cylindrical Periodic Table
Sulfur can gain two electrons to obtain a noble gas configuration of electrons. Magnesium can lose two electrons to obtain a noble gas configuration of electrons.

always valence electrons because they are farthest from the nucleus. In the following sections, we discuss two ways that atoms gain electrons and one way that atoms lose electrons.

We have raised some important questions: How do atoms gain and lose electrons, and how do atoms interact with other atoms? Sections 4.3 through 4.5 of this chapter answer these questions. But first we must discuss how chemists keep track of electrons.

Lewis dot diagrams are a way to keep track of electrons.

Chemists are constantly thinking about valence electrons, because they are the electrons involved in chemical reactions. It's possible to count valence electrons using the periodic table, but chemists have devised a simpler, more visual way to imagine them. To see how this is done, let's consider some Period 2 elements, beginning with its noble gas, neon. Neon has an octet of electrons, and we can place each electron in one of eight imaginary vacancies on the four sides of the element symbol, like this:

$$:\ddot{\underset{\cdot\cdot}{\text{Ne}}}:$$

This illustration is called a **Lewis dot diagram**. It is named for Gilbert Lewis, a renowned American chemist who first used this dot notation to represent atoms and their electrons.

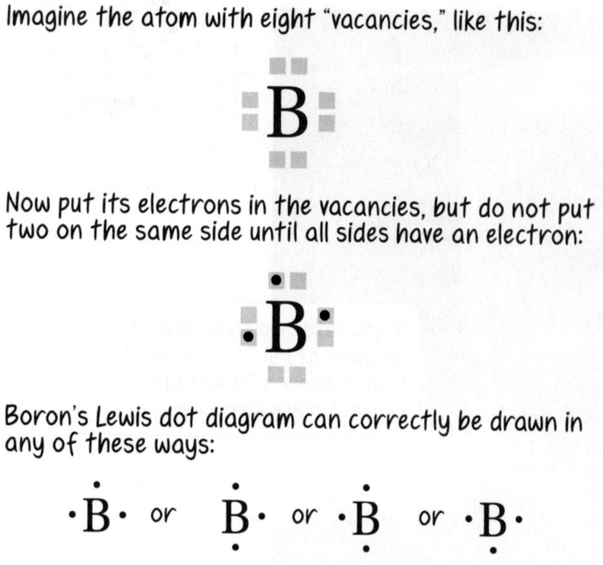

Imagine the atom with eight "vacancies," like this:

B

Now put its electrons in the vacancies, but do not put two on the same side until all sides have an electron:

B

Boron's Lewis dot diagram can correctly be drawn in any of these ways:

·B· or B· or ·B or ·B·

Figure 4.6 How to Draw Lewis Dot Diagrams

Consider another element: boron. Boron has three valence electrons, so we imagine the B surrounded by eight vacancies, of which only three are filled. When filling the vacancies around atoms, we do not put two electrons on the same side until each side has at least one electron. It doesn't matter which three sides we choose. There are several correct ways to draw boron's Lewis dot diagram, as shown in **Figure 4.6**.

Consider another example. Nitrogen has five valence electrons, so we pair the fifth one with one of the other four to make a pair. That is, one side has a pair of electrons, which we refer to as a **lone pair** of electrons, or sometimes a *nonbonding pair.* The other three sides have one electron each, for a total of five. The next element in Period 2 is oxygen. When we draw the Lewis dot diagram for oxygen, we pair its sixth valence electron with one of the other single electrons (it doesn't matter which one). When we follow this pattern, fluorine has three lone pairs and one single electron, and the noble gas neon has four lone pairs. Here is the whole Period 2 series of Lewis dot diagrams:

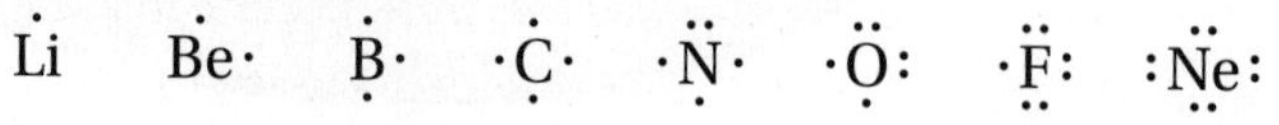

Li   Be·   B·   ·C·   ·N·   ·O: ·F: :Ne:

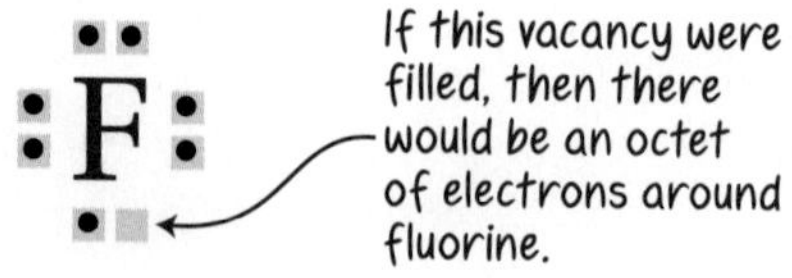

F

If this vacancy were filled, then there would be an octet of electrons around fluorine.

Figure 4.7 Lewis Dot Diagram for an Atom of Fluorine

Lewis dot diagrams allow us to see where atoms have vacancies. In the example in **Figure 4.7,** fluorine is shown with its seven valence electrons. Fluorine has one vacancy, and it is one electron away from having an octet of electrons, just like a noble gas.

QUESTION 4.5 Draw a Lewis dot diagram for a sulfur atom.

ANSWER 4.5 The Lewis dot diagram for a sulfur atom will look identical to that shown for oxygen above, with the letter "O" replaced with the letter "S."

For more practice with problems like this one, check out end-of-chapter Questions 9 and 10.

FOLLOW-UP QUESTION 4.5 Draw a Lewis dot diagram for potassium.

4.3 Ionic Bonds

Atoms gain stability by taking or giving away electrons.

We just saw that an atom of fluorine has a vacancy, which is evident in its Lewis dot diagram. Filling that vacancy gives fluorine a noble gas configuration of electrons. Fluorine can obtain that octet of electrons by undergoing a chemical reaction. The reaction results in the formation of a **chemical bond,** an attractive force that links atoms together. We discuss two ways that fluorine can take part in chemical bonding. We discuss the first way, called *ionic bonding,* here. We discuss the second way, *covalent bonding,* in Section 4.4.

An atom of fluorine can fill its single vacancy by taking an electron from another atom. To see how this happens, let's first look at the source of that electron. In this example, the source of that electron is a magnesium atom. Magnesium (Mg) is a Group 2 element that possesses 12 electrons, two of which are valence electrons.

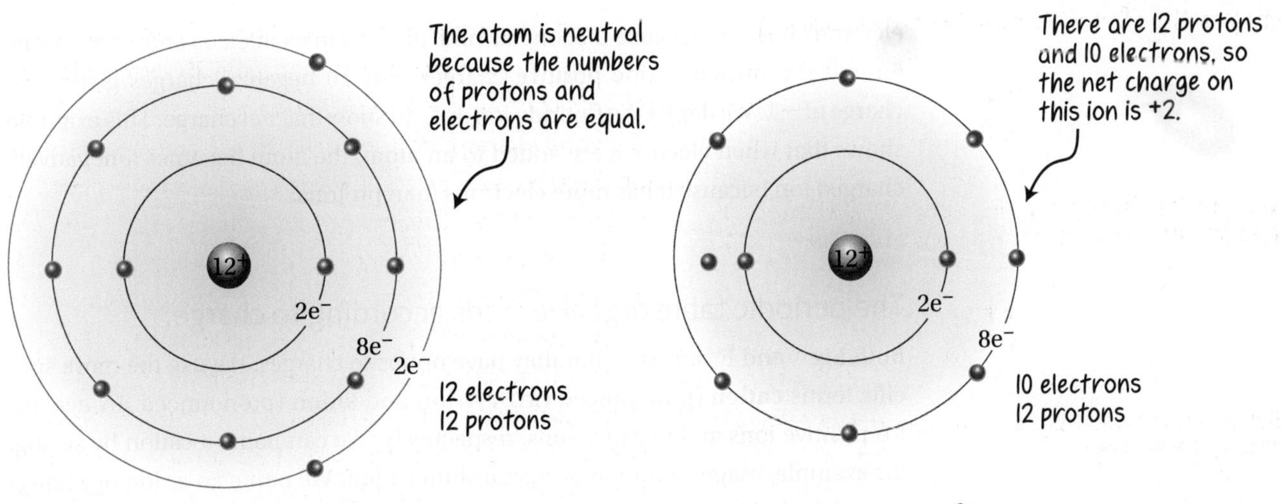

The atom is neutral because the numbers of protons and electrons are equal.
12+
2e⁻
8e⁻
2e⁻
12 electrons
12 protons
There are 12 protons and 10 electrons, so the net charge on this ion is +2.
12+
2e⁻
8e⁻
10 electrons
12 protons

A magnesium atom, Mg

A magnesium ion, Mg^{2+}

Figure 4.8 An Example of a Positive Ion

Each magnesium atom has two electrons more than the nearest noble gas, which is neon (Ne, with 10 electrons).

A magnesium atom can lose its two outermost electrons to form an **ion,** an electrically charged atom (**Figure 4.8**). What is the resulting charge on magnesium when it loses its two valence electrons? It is +2, because the ion has 12 protons in its nucleus and 10 electrons around the nucleus. Twelve positive charges plus 10 negative charges give a net charge of +2. We depict the magnesium ion as Mg^{2+} to show that net charge. This example shows that when electrons are removed from an atom, the atom becomes a positively charged ion because it has more protons than electrons.

Let's reconsider fluorine and its electron vacancy. We know that fluorine, in Group 17, is one column away from the column of noble gases, in Group 18. Therefore, a fluorine atom requires one more electron to fill its single vacancy and achieve a full complement of electrons like the nearest noble gas, neon (Ne, with 10 electrons). Fluorine (F) can add an extra electron to make an ion, as shown in **Figure 4.9**. What is the resulting charge on fluorine when it gains its one valence

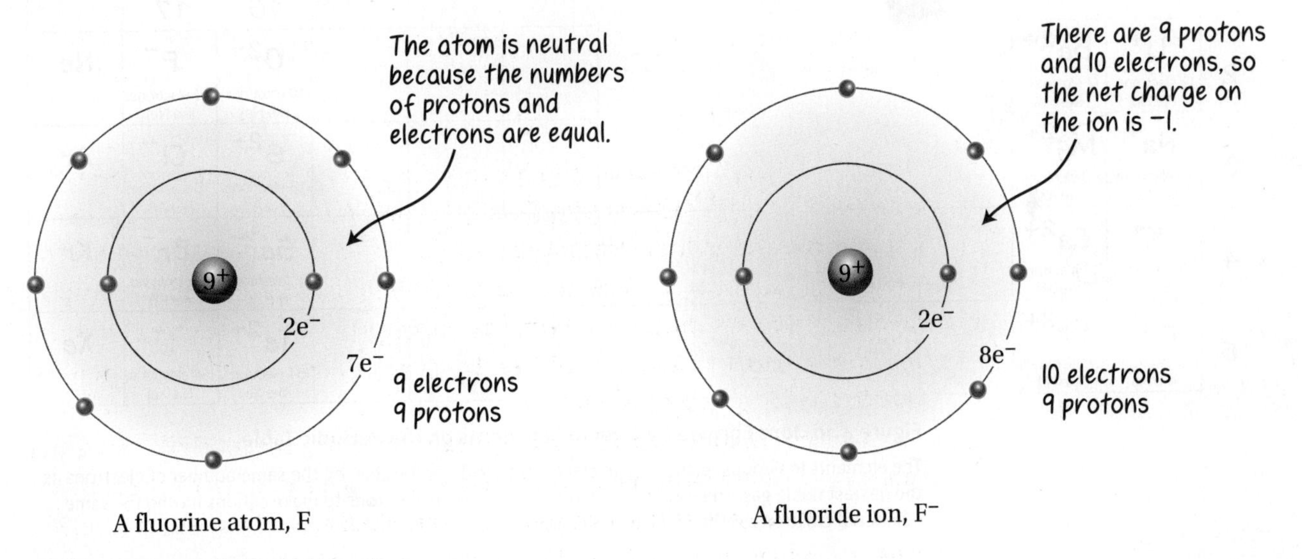

The atom is neutral because the numbers of protons and electrons are equal.
9+
2e⁻
7e⁻
9 electrons
9 protons
There are 9 protons and 10 electrons, so the net charge on the ion is −1.
9+
2e⁻
8e⁻
10 electrons
9 protons

A fluorine atom, F

A fluoride ion, F^-

Figure 4.9 An Example of a Negative Ion

electron? It is −1, because the ion has nine protons in its nucleus and 10 electrons around the nucleus. Nine positive charges plus 10 negative charges give a net charge of −1. We depict the fluoride ion as F^- to show that net charge. This example shows that when electrons are added to an atom, the atom becomes a negatively charged ion because it has more electrons than protons.

The periodic table organizes ions according to charge.

Both Mg^{2+} and F^- are ions, but they have opposite charges. We use the more specific terms **cation** (pronounced *CAT-eye-un*) and **anion** (pronounced *AN-eye-un*) for positive ions and negative ions, respectively. We can name a cation by saying, for example, magnesium ion or magnesium cation. We name an anion by changing the end of an element's name to *-ide*. Thus, a fluorine atom gains an electron to make a fluoride anion, F^-. Examples of other anion names are chloride, oxide, sulfide, and bromide.

Elements located in groups just before and just after the noble gases always form ions that have an octet or a duet of electrons, which is the same number of electrons as the nearest noble gas (**Figure 4.10**). For example, all elements in Group 16 require two electrons to attain an octet of electrons, because they are all two jumps away from the nearest noble gas. Thus, oxygen forms the anion O^{2-} and sulfur forms the anion S^{2-}. The anions O^{3-} and S^{3-} do not exist, because these anions would not have an octet of electrons. Likewise, all elements in Group 2 shed two electrons to achieve an octet of electrons. For example, calcium and magnesium form the cations Ca^{2+} and Mg^{2+}, respectively, but they never form Ca^+ or Mg^{3+}.

Figure 4.10 is a valuable reference. It summarizes the ions that form for selected families on the periodic table. Refer to it when you assign a positive or negative charge to an ion to check that the charge is correct.

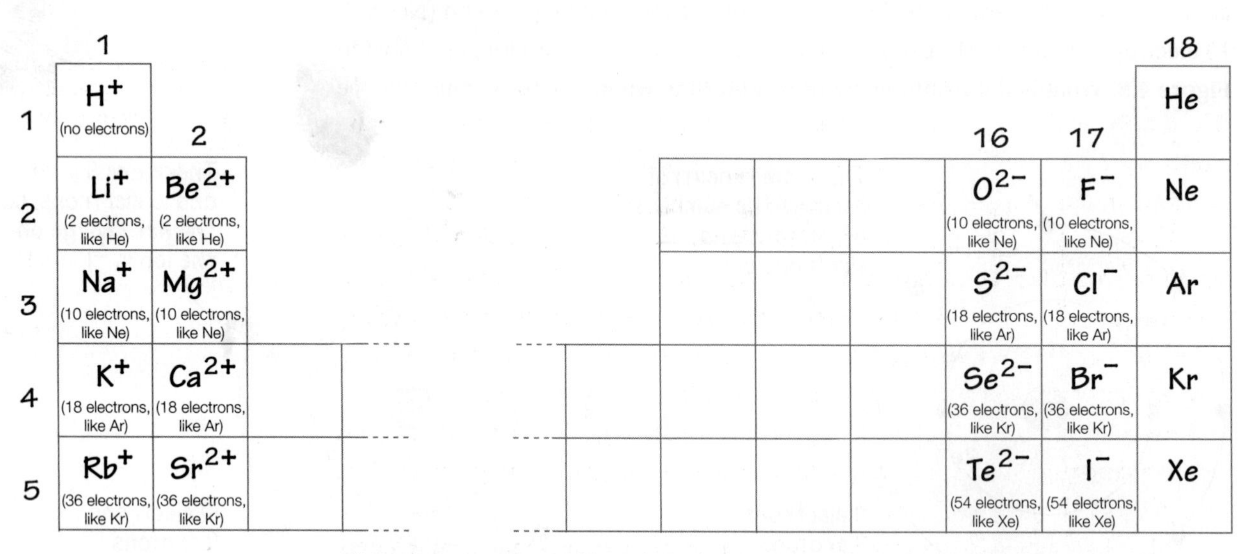

Figure 4.10 Ions Formed by Selected Elements on the Periodic Table

The elements in Groups 16 and 17 gain electrons to make anions having the same number of electrons as the nearest noble gas. The elements in Groups 1 and 2 lose electrons to make cations having the same number of electrons as the nearest noble gas.

wait a minute . . .

Why are ions shown only for selected groups on the periodic table in Figure 4.10?

Each of the elements in these four groups form only one ion, ever, and that is the ion listed in each box in Figure 4.10. For example, bromine atoms form only one ion, the bromide ion (Br^-). You will never encounter Br^+ or Br^{2-}. The two groups that flank the noble gases always behave this way. However, the groups on the periodic table that are farther away from the noble gases are not as straightforward. For instance, many of the transition metals, which are located in Groups 3 through 12, each form several different ions. Consider an atom of manganese, element number 25. Manganese can lose electrons to attain charges anywhere from +2 through +7!

QUESTION 4.6 Does each element form an anion or a cation? (a) magnesium, Mg (b) iodine, I (c) potassium, K (d) sulfur, S

ANSWER 4.6 (a) Magnesium loses two electrons to become a cation with a +2 charge. (b) Iodine gains one electron to become an anion with a −1 charge. (c) Potassium loses one electron to become a cation with a +1 charge. (d) Sulfur gains two electrons to become an anion with a −2 charge.

For more practice with problems like this one, check out end-of-chapter Question 19.

FOLLOW-UP QUESTION 4.6 Which ion does each of the following elements form? oxygen, rubidium, beryllium, bromine

Cations and anions combine in a way that balances their charges.

Ions do not exist in isolation; they pair with oppositely charged ions. An **ionic bond** is formed when ions of opposite charge come together to make an **ionic compound,** also known as a **salt**. (Chapter 9, "Salts and Aqueous Solutions," is devoted to salts. We'll explore ionic bonds in more depth there.)

flashback
Recall from Section 3.4 that a compound is any substance made up of two or more elements.

Table salt is the most famous salt. Its formal name is *sodium chloride,* and its chemical formula is NaCl. (We write NaCl, not Na–Cl.) In NaCl, both sodium and chlorine are in their ionic forms. Each ion carries one electrical charge: a charge of +1 on each sodium cation, Na^+, and a charge of −1 on each chloride anion, Cl^-. We know this must be the case, because chlorine always forms a −1 ion and sodium always forms a +1 ion, giving both ions noble gas configurations. These two ions always pair one to one, and the positive and negative charges balance so that the net charge in NaCl is zero.

What happens when we combine ions that do not have the same charge? In this case, we balance positive and negative charges by combining multiple anions or cations in a way that makes the charges equal. For example, as we know, magnesium fluoride contains magnesium cations, Mg^{2+}, and fluoride anions, F^-.

To write the chemical formula for this salt, we include two fluoride ions for each magnesium ion for a total of two positive and two negative charges, like this: MgF_2. The subscript 2 after the fluoride ion tells us that there are two fluoride anions for every magnesium cation. The absence of any subscript after the magnesium ion tells us that there is just one magnesium ion. Thus, the chemical formula tells us that there is one magnesium cation for every two fluoride anions, which is a total of three ions.

wait a minute . . .

If ions have charges, why aren't the charges written in the formula of a salt?

In general, although ions have charges, we do not write the charge on each ion in a salt. We write NaCl but not Na^+Cl^-. Likewise, we write CaF_2 but not $Ca^{2+}(F^-)_2$. Each of the cations and anions in these salts could have only one possible charge, as we know from Figure 4.10. Writing in the charge would be redundant, and it would clutter the chemical formula for the salt. Thus, doing so is incorrect.

QUESTION 4.7 Which salt forms from the elements calcium and chlorine?

ANSWER 4.7 $CaCl_2$. Calcium forms the cation Ca^{2+}, and chlorine forms the anion Cl^-. Because salts must have zero net charge, one ion of calcium combines with two chloride ions. The resulting salt is calcium chloride.

For more practice with problems like this one, check out end-of-chapter Question 20.

FOLLOW-UP QUESTION 4.7 What salt forms from the elements magnesium and oxygen?

QUESTION 4.8 Identify and count the cations and anions in these ionic compounds: (a) K_3N (b) K_2O (c) CaF_2 (d) $BaCl_2$

ANSWER 4.8 (a) three K^+ cations and one N^{3-} anion (b) two K^+ cations and one O^{2-} anion (c) one Ca^{2+} cation and two F^- anions (d) one Ba^{2+} cation and two Cl^- anions

For more practice with problems like this one, check out end-of-chapter Question 21.

FOLLOW-UP QUESTION 4.8 List the names and chemical formulas for the salts formed from each pair of elements: (a) lithium and iodine (b) potassium and sulfur

---

Most salts exist as crystals that have a repeating pattern of cations and anions.

Most salts exist as solids that have a three-dimensional, repeating pattern of ions. This lattice of ions makes up a **crystal,** and we say that salts exist as *crystalline solids.* In a solid crystal of a salt, the ions alternate in a regular, geometric pattern as **Figure 4.11** shows. This crystal has alternating cations and anions, and it resembles

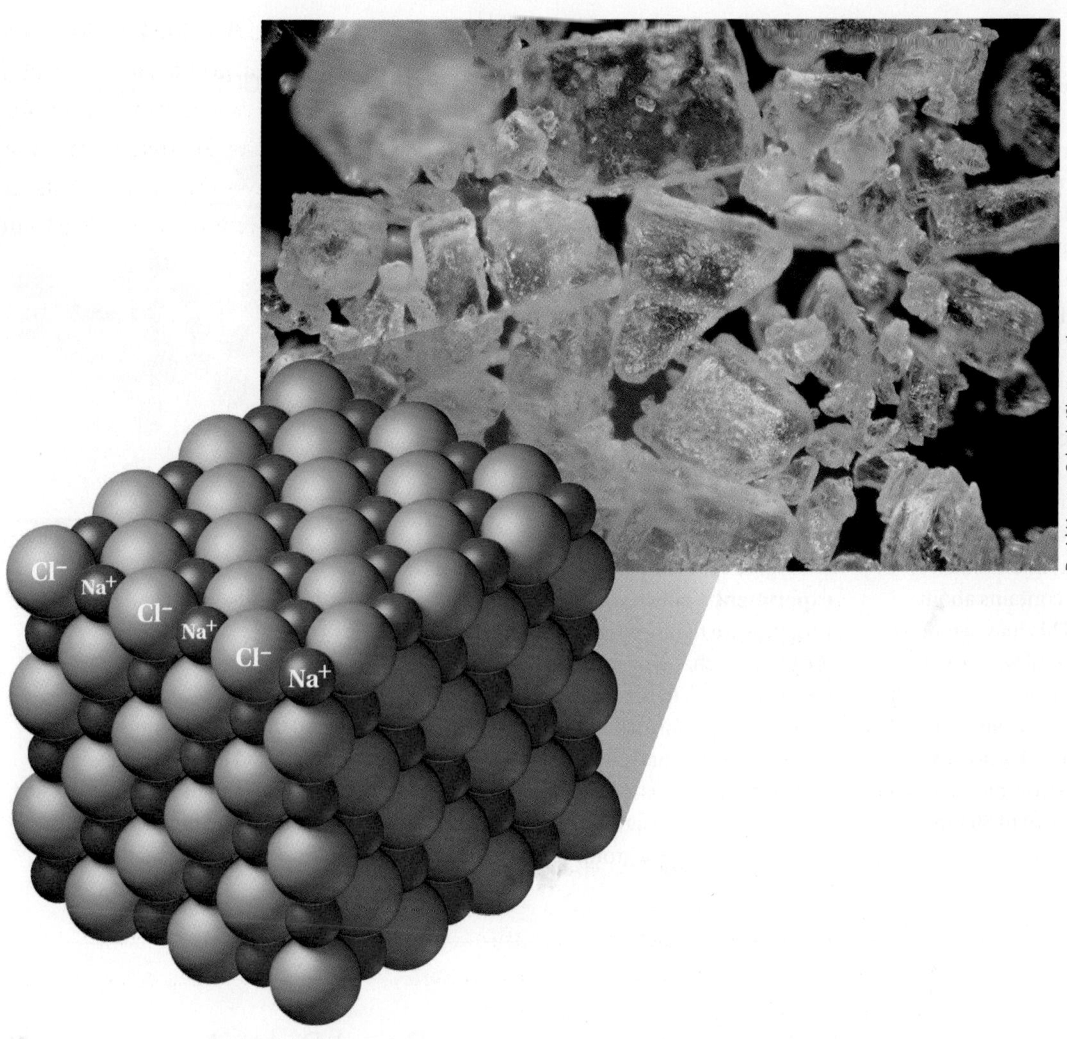

Figure 4.11 The Ionic Compound Sodium Chloride
This schematic drawing is of a portion of a sodium chloride crystal and shows the regular array of sodium ions and chloride ions. The background image is of magnified sodium chloride crystals.

a box of fruit packed with alternating lemons and limes, with layers stacked on top of each other.

A typical crystal of a salt contains billions and billions of alternating ions, and accounting for every ion is impractical (and impossible). So, when we write the name or chemical formula of a salt, we do not attempt to indicate all of the anions and cations in the crystal. Instead, we look for the simplest unit in the crystal that is repeated again and again. For example, for the pattern cation-anion-cation-anion-cation-anion-cation-anion, the repeating motif is cation-anion repeated four times, and this is the unit that we write in the chemical formula. Or, for the pattern cation-anion-anion-cation-anion-anion-cation-anion-anion, the repeating motif is cation-anion-anion. We use the term **formula unit** to indicate this smallest repeating motif in a salt.

One formula unit for sodium chloride is simply NaCl. Thus, that is also the chemical formula for this salt. This formula tells us that, in a crystal of NaCl, one Na^+ ion alternates in a pattern with one Cl^- ion. Here is another example: for CaF_2, one Ca^{2+} stacks with two F^- ions in an alternating pattern.

natureBOX • Should We Use Silver Nanoparticles to Reduce Body Odor?

Anyone who has spent time at the gym knows this: exercise makes some people very smelly. Fungi and bacteria are one cause of these offensive odors. Luckily, science has created a solution to this problem: silver nanoparticles. Silver nanoparticles are known to have antifungal and antibacterial properties, meaning they can kill fungi and bacteria. For this reason, hundreds of clothing products now are infused with silver nanoparticles.

A silver nanoparticle is a very, very small piece of silver that is roughly 1 to 100 nanometers (nm) in diameter. A typical nanoparticle contains about 10^{19} atoms of silver. This may seem like a lot of silver atoms until we consider that, if a dime were made of pure silver, it would contain two thousand times that number of silver atoms. **Figure 4A** is a microscopic view of silver nanoparticles that range from about 10 to 50 nm in diameter.

How do silver nanoparticles kill bacteria and fungi? This is a complex question that scientists have been struggling to answer. Over the years, many experiments have been conducted in which silver nanoparticles were added to cultures of bacteria or fungi to evaluate their toxicity. Until recently, though, one experiment had not yet been done: no one had looked at how the oxygen in air affected the toxicity of nanoparticles. To unravel the role of oxygen in toxicity, the researchers performed an experiment in the absence of oxygen. This is an example of a **control experiment**—one in which one or more variables are tightly controlled so that the effect of changing one variable can be determined. In this case, the control experiment performed in the absence of oxygen reveals what effect, if any, oxygen had on the previous experiments.

To see if the presence or absence of air made a difference, a group of scientists performed the experiments without air. They learned that when no air was present, the silver nanoparticles did not kill the bacteria and fungi. This piece of information told them that the air was

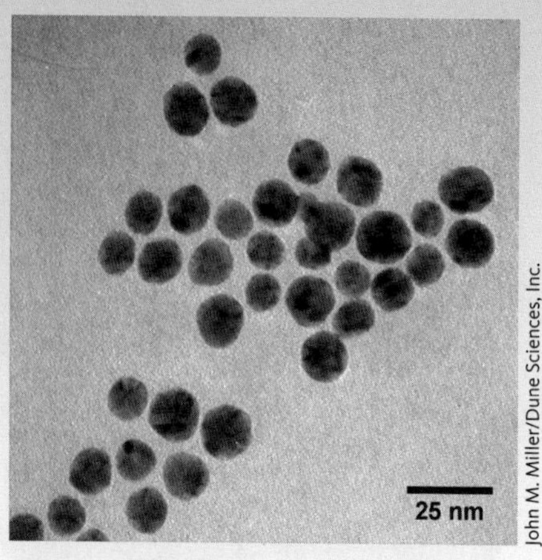

Figure 4A A Micrograph of Silver Nanoparticles

4.4 Covalent Bonds and Bond Polarity

Atoms gain stability when they make molecules by sharing electrons in covalent bonds.

Let's return to our fluorine atom with one vacancy:

$$:\ddot{\text{F}}:$$

We know that a fluorine atom can fill that vacancy by becoming an anion, fluoride (F^-). A fluorine atom can fill that vacancy in another way: by sharing electrons with another atom, like this:

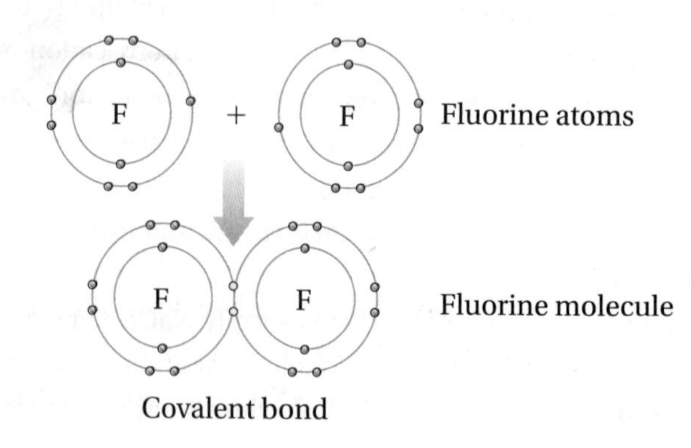

somehow making the silver nanoparticles toxic to bacteria and fungi. The researchers postulated that the oxygen in air allowed the silver in the nanoparticles to become a cation by losing an electron. That electron is designated with the symbol e^- in the chemical equation shown here:

$$Ag\,(s) \xrightarrow{\text{AIR}} Ag^+\,(aq) + e^-$$

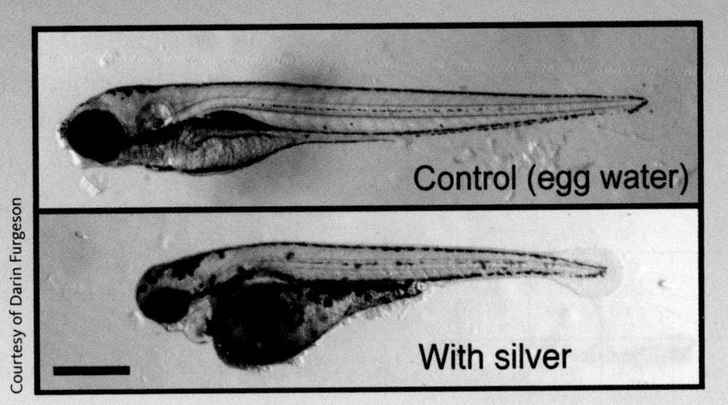

Figure 4B Dr. Darin Furgeson's Mutated Fish The top photo shows the control fish. The bottom photo shows a fish that has been exposed to silver nanoparticles.

The scientists concluded that, because silver ions can move through a watery medium, the silver ions formed in the presence of air can travel to bacteria or fungi and interact with them. And, because the silver ions are toxic to these organisms, the organism is killed when it interacts with the silver ions. So the nanoparticles themselves are not toxic to bacteria and fungi, but silver ions that form when the nanoparticles are exposed to air *are* toxic.

Now that we know how silver nanoparticles work, we can ask the next obvious question: do they actually reduce smell when embedded in athletic clothing? The answer is yes—and no. There is evidence that silver nanoparticles do reduce the growth of bacteria and fungi. But scientific studies show that the silver nanoparticles wash away after a few launderings. So expensive silver-nanoparticle-embedded socks might keep feet smelling fine for a while, but only as long as they are never washed!

Silver nanoparticles in clothing aren't just a rip-off. The use of silver in clothing has become an issue for environmental groups, whose members fear that silver is being introduced into natural waters at alarming rates. Fish researcher Darin Furgeson has shown that silver nanoparticles cause malformations in the eyes, tails, and hearts of zebrafish embryos (**Figure 4B**). Many of the embryos exposed to silver nanoparticles died. So when people wash their fancy silver-embedded socks, they are polluting natural waters, and organisms that live there, with silver. Silver in socks is an innovative invention, but it is not the environmentally responsible way to reduce body odors at the gym.

The two electrons between the two fluorine atoms represent a chemical bond between those atoms. When pairs of electrons are shared by two atoms, we call that type of bond a **covalent bond**. Now the fluorine atoms are no longer separated atoms; they are connected by a chemical bond. A **molecule** is a grouping of atoms connected by covalent bonds. If sharing electrons between atoms gives each of those atoms an octet of electrons, like that of the nearest noble gas, a covalent bond is likely to form.

In a fluorine molecule, each fluorine atom shares one of its valence electrons with its partner, giving each atom an octet of valence electrons. The individual fluorine atoms, F, become the fluorine molecule, F_2. We refer to F_2 as a **diatomic molecule**, because it contains two (*di-*) of the same atoms (*-atomic*). We depict the single covalent bond between the atoms of fluorine with a line to indicate that two electrons are being shared:

$$:\ddot{\underset{..}{F}}-\ddot{\underset{..}{F}}:$$

Because elements in the same family of the periodic table have the same number of valence electrons, they generally have similar properties. It makes sense, then, that other elements in Group 17—the halogens—exist in nature as diatomic

flashback
Recall from Chapter 3 that the elements in Group 17 are the halogens.

The elements shaded in yellow all make diatomic molecules because they can share electrons with one another.

1	13	14	15	16	17	18
1 H Hydrogen 1.008						2 He Helium 4.003
	5 B Boron 10.81	6 C Carbon 12.01	7 N Nitrogen 14.01	8 O Oxygen 16.00	9 F Fluorine 19.00	10 Ne Neon 20.18
	13 Al Aluminum 26.98	14 Si Silicon 28.09	15 P Phosphorus 30.97	16 S Sulfur 32.07	17 Cl Chlorine 35.45	18 Ar Argon 39.95
		32 Ge Germanium 72.64	33 As Arsenic 74.92	34 Se Selenium 78.96	35 Br Bromine 79.90	36 Kr Krypton 83.80
		50 Sn Tin 118.7	51 Sb Antimony 121.8	52 Te Tellurium 127.6	53 I Iodine 126.9	54 Xe Xenon 131.3

Figure 4.12 Elements That Exist as Diatomic Molecules

The elements Br_2, I_2, N_2, Cl_2, H_2, O_2, and F_2 exist as diatomic molecules in nature. Some people use this mnemonic device to remember these elements: BrINClHOF (pronounced *BRINK-el-hof*).

molecules, just like fluorine does (**Figure 4.12**). Thus, the following molecules are stable and do exist: F_2, Cl_2, Br_2, and I_2.

Now we can see why some atoms exist as diatomic molecules: they are more stable as molecules than as individual atoms because, by being part of a molecule, each atom achieves an octet of electrons. In nature, hydrogen, nitrogen, and oxygen also exist as the diatomic molecules H_2, N_2, and O_2, respectively.

QUESTION 4.9 Draw a Lewis dot diagram for an iodine molecule.

ANSWER 4.9 instructions for drawing I_2 in art manuscript.:

:I:

For more practice problems like this one, check out end-of-chapter Question 30.

FOLLOW-UP QUESTION 4.9 Hydrogen atoms can gain a duet of electrons by making the molecule H_2. Draw a Lewis dot diagram for this molecule.

Covalent bonds can form between atoms of different elements.

Covalent bonds can form between atoms of the same element, as we have just seen with F_2, or they can form between atoms of different elements. As a result, an enormous assortment of covalent bonds exist in nature, each forming in a way that usually gives each participating atom an octet of electrons. Chapter 5, "Carbon," is devoted to covalent bonding, and there you will learn more about the octet rule and why the previous sentence includes the word *usually*.

Covalent bonds are not all the same. The nature of a given covalent bond depends on the elements involved. In a fluorine molecule, the bond forms thanks to equal sharing of electrons between atoms. This is a **nonpolar** covalent bond, because there is no lopsidedness to the electron sharing. Each fluorine atom pulls on the shared electrons with exactly the same force, and the sharing is equal. Some covalent bonds, however, are **polar**. That is, there is unequal sharing of the electrons in the bond, and this unequal sharing makes one side of the molecule electron rich and the other side electron poor.

Water is a classic example of a molecule containing polar covalent bonds. Each water molecule contains two covalent bonds, one between each hydrogen atom and the central oxygen atom. In **Figure 4.13a**, each arrow points along a polar covalent bond in the water molecule. The "plus end" of the arrow is closest to the atom that pulls less electron density toward itself, and the head of the arrow is closest to the atom that pulls more electron density toward itself. In the water molecule, the oxygen in each O−H bond pulls more electron density from the covalent bond toward itself, and as a result this bond is very polar. In **Figure 4.13b**, we see that most of the electron density in the molecule resides on the oxygen atom, and less resides on the hydrogen atom. The shape of the molecule is easily seen when it is drawn as a space-filling model (**Figure 4.13c**). Chapter 8, "Water," is devoted

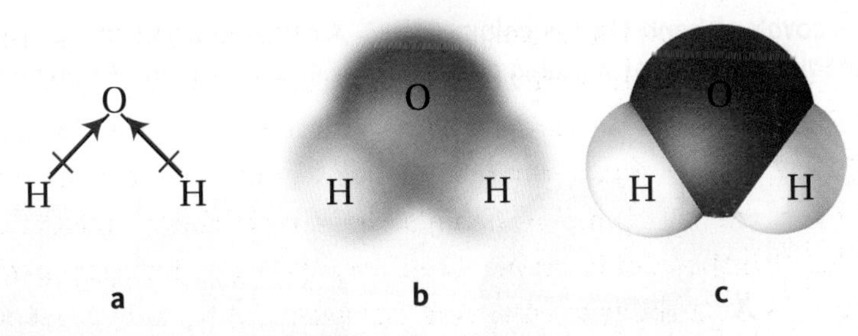

Figure 4.13 The Bonds in Water Are Polar: Three Views

(a) The arrow points to the atom in a polar bond where more electron density resides. (b) In the water molecule, more electron density resides on the oxygen atom (dark red) and less on the hydrogen atoms (light yellow). (c) This space-filling model of water emphasizes the molecule's overall shape.

entirely to this unique molecule. As we will see, the polarity of the bonds in water molecules profoundly affects the properties of water, including what dissolves in it.

Fluorine molecules, which are completely nonpolar, and water molecules, which are quite polar, are examples of the range of polarities in covalent bonds. There are bonds with every degree of intermediate polarity between them.

recurring theme in chemistry 7

Individual bonds between atoms dictate the properties and characteristics of the substances that contain them.

QUESTION 4.10 Which of the following covalent bonds would you expect to be completely nonpolar? (a) iodine and iodine (b) hydrogen and oxygen

ANSWER 4.10 The iodine–iodine bond is nonpolar, so choice (a) is correct. Whenever you have a bond between two atoms that are the same, you should always expect the electrons in the covalent bond to be shared equally between them. Therefore, this pair of atoms makes a nonpolar covalent bond. The oxygen–hydrogen bonds in water molecules are polar covalent bonds, because the electrons are not shared equally.

For more practice with problems like this one, check out end-of-chapter Question 42.

FOLLOW-UP QUESTION 4.10 Oxygen atoms bond to make oxygen molecules, O_2. Is the bond in an oxygen molecule polar or nonpolar?

Scientists depict salts and molecules differently because they have different types of bonds.

We now know that covalent bonds can be completely nonpolar—as with a fluorine molecule, F_2—or they can have polarity, as with a water molecule. The bonds in water molecules are polar due to lopsidedness in the distribution of electron density in the bond. Still, though, the electrons in a covalent bond are always shared, even when that sharing may not be equal.

Salts take the notion of polarity to the extreme. Rather than being lopsided in their electron density, like polar covalent bonds are, salts are completely polarized. The ionic bonds in salts are the most polar bonds imaginable. The electrons are not simply concentrated near one or the other atom in a shared bond. Rather, they are transferred fully—completely lost or gained. In the continuum of bond polarities, ionic bonds fall at the most polar end. The three examples of polarity that we have discussed are shown side by side in **Figure 4.14**.

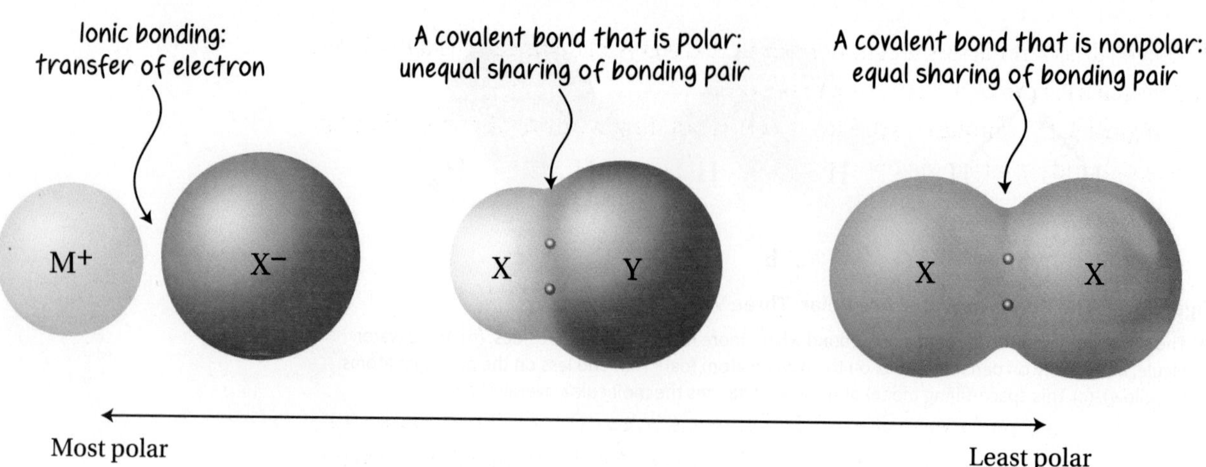

Figure 4.14 A Side-by-Side Comparison of Bonds with Different Polarities

Whenever electrons are shared in a covalent bond, we depict this with a line between atoms. But there is no sharing of electrons in ionic bonds; these are not covalent interactions. Therefore, we do not draw lines between ions in the chemical formula for a salt. In other words, we *do* write F–F to depict a molecule by drawing a line to represent a covalent bond, but we *do not* use a line in the formula for an ionic compound such as NaCl.

In addition, molecules, in which atoms are connected to one another by covalent bonds, are different from salts because molecules usually exist as discrete entities and not in extended crystalline arrays of alternating ions. Water, for example, is composed of individual molecules, each containing exactly two hydrogen atoms and exactly one oxygen atom. The atoms are linked by covalent bonds, so we use lines to link the atoms together. And the molecular formula, H_2O, tells us exactly how many atoms are present in each molecule. A formula for a salt such as NaCl,

Figure 4.15 An Overview of Bonding

HOW ATOMS ACCEPT ELECTRONS

1. Atoms can *accept electrons* by sharing another atom's electrons. This results in the formation of a bond between these two atoms called a **covalent bond**.

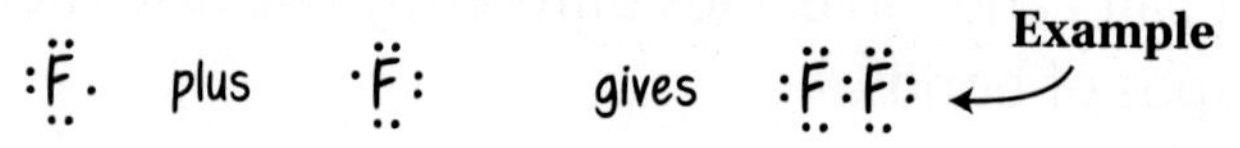

   :F· plus ·F: gives :F:F: ← **Example**

2. Atoms can *accept electrons* by getting electrons from another atom. This does not involve sharing. These atoms are participants in ionic bonding.

   :S: plus 2 electrons gives :S:$^{2-}$ ← **Example**

   This is now an anion. It has an electron configuration like the noble gas argon, Ar.

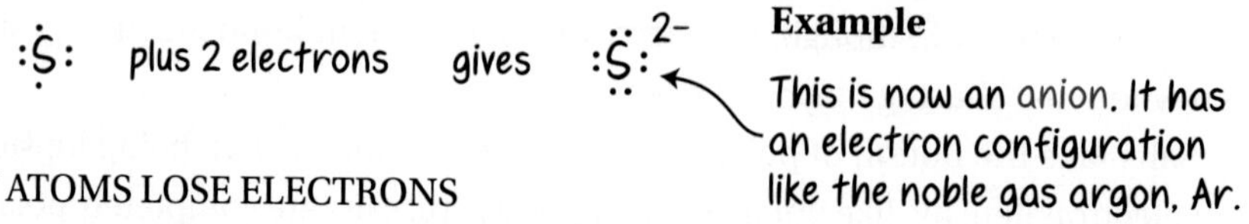

HOW ATOMS LOSE ELECTRONS

1. Atoms can *lose* electrons by giving one or more electrons away to another atom. This does not involve sharing. These atoms are also participants in ionic bonding.

   Li· minus 1 electron gives Li^+ ← **Example**

   This is now a cation. It has an electron configuration like the noble gas helium, He.

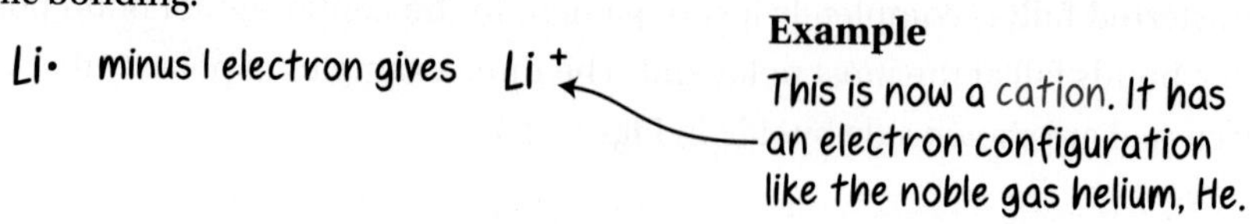

however, does not tell us the number of ions that a crystal of the salt contains. What it does tell us is the ratio of one ion to the other(s).

Figure 4.15 summarizes the two types of bonding we have discussed thus far, covalent bonding and ionic bonding.

4.5 Bonding in Metals

For substances that are composed only of metals, metallic bonds hold atoms together.

We've discussed two types of bond so far: covalent bonds and ionic bonds. Now we'll consider one more type: bonds within a pure chunk of metal. Let's turn our attention to a pure, unadulterated chunk of magnesium, and to the bonds that exist within it. Magnesium is located in Group 2 of the periodic table, so we know that by losing two electrons, a Mg atom can form the ion Mg^{2+} and achieve an electron configuration like that of neon. This is true of all the metals in Group 2. They all form cations with a +2 charge.

This situation is different from ionic bonding, however, because magnesium is the only element in solid magnesium. There is no atom to accept the electrons lost by Mg and thus form an anion. When there is no anion around to balance the charge on a cation, the electrons produced from the ionization of the atoms form a sea of negative charge that bathes the cations, in this case Mg^{2+} ions, in electrons. These mobile electrons surround and neutralize the positive charges of the magnesium cations, which stay in a fixed lattice, as shown in **Figure 4.16**.

Metals are malleable and conduct electricity.

The **metallic bonds** formed within a sea of electrons are similar to covalent bonds in that electrons are shared. But metallic bonds are different because the electrons are not just shared between two atoms. Instead, the sea of electrons moves through the whole lattice of cations in the solid chunk of metal. As a result, solid metals are malleable (easy to shape or bend) and can be formed into things like nickels and paper clips and nose rings. Plus, because electrons in a solid metal are mobile, these materials readily conduct heat and electricity.

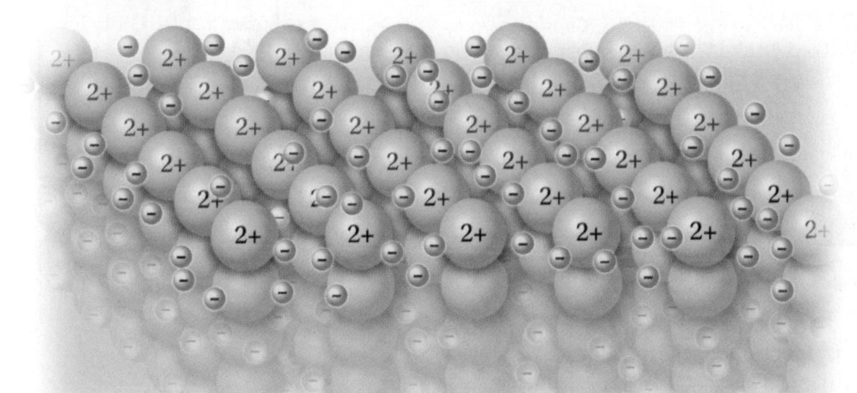

Figure 4.16 Bonding in Metals Represented as a Sea of Electrons
In metallic bonding, electrons are free to move through the whole lattice of cations.

Now that we've learned about the three major types of bonds between atoms, it is appropriate to ask, how can we predict which of the three types of bonds exists in a given substance? To answer this question, we must return to our now familiar organizer for all the elements: the periodic table.

4.6 Determining the Type of Bond between Two Atoms

Electronegativity is the tendency of an atom to draw electrons toward itself.

We have discussed three types of bonds. For a given pair of atoms, how do we know which type of bond connects them? When choosing among metallic, ionic, or covalent bonds, it's easiest to spot the metallic bonds. In general, when the atoms forming the bond are both metal atoms in a chunk of metal, it's metallic bonding. It's that straightforward.

The more difficult part is deciding when a bond is covalent versus ionic. We can use the concept of **electronegativity,** the tendency of an atom to draw electrons toward itself, to figure out what type of bond forms between any two atoms on the periodic table. **Figure 4.17** is a periodic table that includes electronegativity values for every element. The electronegativity values range from 0 to 4, and the higher the number, the more that atom pulls electrons to itself. For any two atoms that come together to make a chemical bond, the more similar their electronegativities, the less polar the bond. Why is this true? Because when two elements have similar electronegativities, they pull electrons to themselves to roughly the same extent. Therefore, the distribution of electrons across the bond and the two atoms is roughly equal.

												1 H 2.1					2 He
3 Li 1.0	4 Be 1.5											5 B 2.0	6 C 2.5	7 N 3.0	8 O 3.5	9 F 4.0	10 Ne
11 Na 0.9	12 Mg 1.2											13 Al 1.5	14 Si 1.8	15 P 2.1	16 S 2.5	17 Cl 3.0	18 Ar
19 K 0.8	20 Ca 1.0	21 Sc 1.3	22 Ti 1.5	23 V 1.6	24 Cr 1.6	25 Mn 1.5	26 Fe 1.8	27 Co 1.9	28 Ni 1.9	29 Cu 1.9	30 Zn 1.6	31 Ga 1.6	32 Ge 1.8	33 As 2.0	34 Se 2.4	35 Br 2.8	36 Kr
37 Rb 0.8	38 Sr 1.0	39 Y 1.2	40 Zr 1.4	41 Nb 1.6	42 Mo 1.8	43 Tc 1.9	44 Ru 2.2	45 Rh 2.2	46 Pd 2.2	47 Ag 1.9	48 Cd 1.7	49 In 1.7	50 Sn 1.8	51 Sb 1.9	52 Te 2.1	53 I 2.5	54 Xe
55 Cs 0.7	56 Ba 0.9	57 La 1.0	72 Hf 1.3	73 Ta 1.5	74 W 1.7	75 Re 1.9	76 Os 2.2	77 Ir 2.2	78 Pt 2.2	79 Au 2.4	80 Hg 1.9	81 Tl 1.8	82 Pb 1.9	83 Bi 1.9	84 Po 2.0	85 At 2.2	86 Rn
87 Fr 0.7	88 Ra 0.9	89 Ac 1.1	90 Th 1.3	91 Pa 1.4	92 U 1.4	107 Bh	108 Hs	109 Mt	110 Ds	111 Rg	112 Cn	113 Uut	114 Fl	115 Uup	116 Lv	117 Uus	118 UUo

Figure 4.17 Electronegativity Values

Fluorine, in the top right corner of the periodic table, has the largest electronegativity value of all the elements. Cesium and francium, in the lower left corner, have the smallest value. Electronegativity values change gradually across the periodic table between these two extremes.

For example, when a bond forms between two atoms of the same element, the bond between them is completely nonpolar because they both pull the electrons to themselves equally. This bond is covalent because electrons are shared, and in this case the sharing is exactly equal. This is one extreme. Now consider what happens when two atoms in a chemical bond have very different electronegativities. In that case, one of the atoms in the bond pulls electrons to itself much more than the other atom does, and the bond is very polar. It is so polar that one atom has taken one or more electrons and become an anion; the other atom has lost one or more electrons and become a cation. This is the other extreme.

The proximity of elements on the periodic table is a clue to the type of bond that will form between them.

In Figure 4.17, we can see that the electronegativities vary from high values in the upper right corner to low values in the lower left corner. The values change gradually across the table, and no large jumps in electronegativity values occur between adjacent elements. Thus, adjacent elements have similar electronegativities, and elements that are far apart on the periodic table have very different electronegativities. In general, we can conclude that elements very close to each other on the periodic table will make less polar bonds with one another, and atoms of elements that are far away from each other will make more polar bonds.

Let's consider some real examples. Consider a bond between potassium and chlorine. Because these atoms are very far apart on the periodic table, they have markedly different tendencies to pull electrons to themselves, and the bond between them will be extremely polar. Chlorine tends to pull electrons to itself, and potassium does not. Because this difference is extreme, the type of bond that forms between them is extreme: chlorine takes an electron to become a chloride ion, Cl^-, and potassium gives up an electron to become a potassium ion, K^+. Together, these two ions form an ionic bond and, in this case, the salt is potassium chloride. Each of these ions forms by gaining or losing electrons; there is no sharing of electrons. In general, when a nonmetal and a metal come together to make a bond, that bond is usually ionic.

We now know that two metal atoms form a metallic bond and that a nonmetal and a metal form an ionic bond. But what happens when two nonmetals form a bond? Figure 4.17 indicates that nonmetals have pretty high electronegativities. And because the nonmetals are close to one another, they tend to have rather small differences in electronegativities. Therefore, these bonds between nonmetals tend to be slightly polar or nonpolar.

If we look back to Section 4.4, we will see that all of the covalent bonds we considered were bonds between two nonmetal atoms. Covalent bonds have electrons distributed across both atoms in the bond. That distribution can be equal across both atoms, as in the case of a fluoride–fluoride (F–F) bond. Or the bond can be lopsided, as it is for the bond between H and O atoms in water molecules. Covalent bonds come in a range of polarities from completely nonpolar to fairly polar, but none of them is as extremely polar as ionic bonds are.

These ideas are summarized in **Table 4.2**.

Table 4.2 The Three Types of Bonds Discussed in Chapter 4

Bond Forms between a ___ and a ___		Type of Bond That Forms	Difference in Electronegativity between the Atoms in the Bond	Example
Metal	metal	metallic bond	smaller	two atoms of iron in a chunk of steel
Metal	nonmetal	ionic bond	larger	sodium chloride, NaCl
Nonmetal	nonmetal	covalent bond	smaller	a Cl−Cl bond

QUESTION 4.11 Refer to the periodic table to predict whether the bond between each pair of atoms is ionic or covalent. (a) K bonding with Cl (b) C bonding with Cl

ANSWER 4.11 (a) Potassium is a metal and chlorine is a nonmetal, so they form an ionic bond. (b) Carbon and chlorine are both nonmetals, so they form a covalent bond.
For more practice with problems like this one, check out end-of-chapter Question 45.

FOLLOW-UP QUESTION 4.11 Which pair of elements forms the most polar bond? (a) chlorine and oxygen (b) fluorine and fluorine (c) lithium and bromine (d) copper and copper

THEgreenBEAT • news about the environment, chapter four edition

TOP STORY Collaborative Consumption

Now we turn from bonds between atoms to bonds between humans. Imagine a block in a typical suburban U.S. neighborhood. On this particular block there are 10 houses, and each home on the block has three bedrooms, a lawn, a driveway, and two cars in the garage. Now, if we peek into each garage, we'll find that each garage has many of the same things in it. For example, every one of these 10 houses has a ladder. Besides that ladder, these 10 homes have scores of other things that are used only now and then: lawn mowers, ice cream makers, drills, snowblowers, rug shampooers, weed whackers, and on and on. People in the United States use an estimated 10% of the items in their garages once per month, on average, and we use the rest of the things less often than that. Because so much of our stuff is not used as much as it could be, a movement is afoot to make better use of what we have.

Collaborative consumption is an economy based on sharing and on forming new bonds between people. For example, OpenShed is a website that neighbors can join that allows everyone who takes part to share the items in their neighbors's sheds. So, if you need a ladder on Tuesday, you can go to your neighbor Ben and use his ladder. Next Saturday, Ben arranges to borrow your pasta maker, which hasn't been used in two years (**Figure 4.18**). This model of collaborative consumption is like a

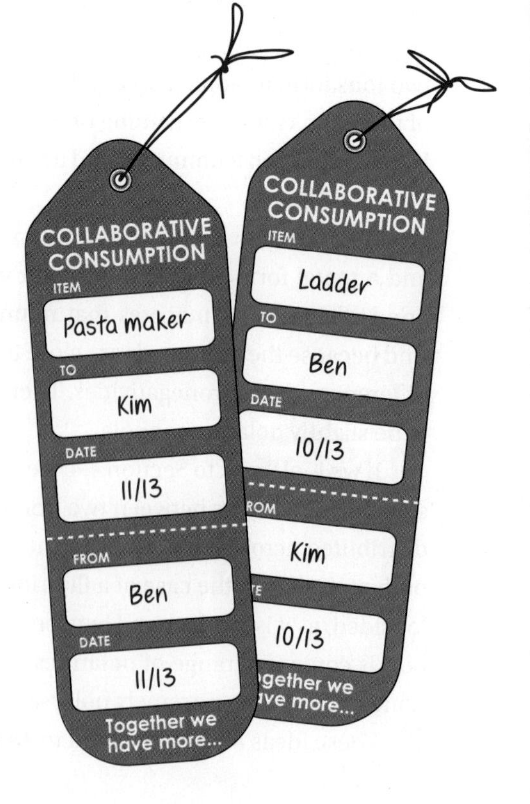

Figure 4.18 Tags Keep Track of Shared Household Items

These tags are used to track who has what and when. Like a library lending system, this type of collaborative consumption maximizes the use of household items that sit in kitchens and garages all over the United States.

covalent bond in which electrons are not concentrated on one atom or another, but shared across atoms.

But why are we talking about collaborative consumption in a news article about the environment? Because by manufacturing less stuff, we can save energy—and the environment wins. But collaborative consumption helps the environment in more obvious ways. For example, in the past six months alone, more than 30 million rides have been shared by people who use the website www.ridester.com. And it's becoming more common to see people renting a car only for the time they need it rather than leaving a car unused for part of the time. The nonprofit eGo Carshare, for example, rents out cars at a rate of $4 per hour, including insurance and gasoline. And it's estimated that every car that is shared takes 5 to 20 cars out of circulation.

Collaborative consumption also means getting groups of people together to buy something in bulk. For example, One Block Off the Grid (http://1bog.org) is a way to form a group with other people who want to invest in solar power. The experts at the website give advice on how to buy solar panels and which are best, and then group purchasing power allows a group to get a cheaper rate.

By participating in a sharing economy, each of us can reduce our impact on the environment. If everyone saves just a small amount of energy, but millions of people do it, then the result is significant.

Collaborative consumption can be more than sharing. It can be swapping, lending, borrowing, renting, or donating. You can share office space, lend textbooks, donate toys, borrow bikes, trade art, share gardens, swap electronics, reuse wedding dresses, and lend parking spots. You can participate in time banking, in which you contribute your time and expertise for a certain number of hours and then use someone else's expertise later. So, if you're an expert at teaching Swahili, you can teach a plumber to speak Swahili, and the plumber, in turn, fixes your toilet.

The idea of collaborative consumption has been driven partly by the weak economies in many countries, including the United States. When people cannot afford to buy a new car or a new ladder, they use the steadily growing online infrastructure of collaborative consumption to get access to what they need. It's not surprising that millennials—people from 18 to 34 years old—are investing most in collaborative efforts (**Figure 4.19**).

A modern, pared-down, sharing economy means that we have less stuff and make better use of the stuff we have. It also means that we can move from hyperconsumption to reasonable and efficient consumption. Finally, it means that we are learning to make new connections—yes, *bonds*—with people we may not otherwise meet.

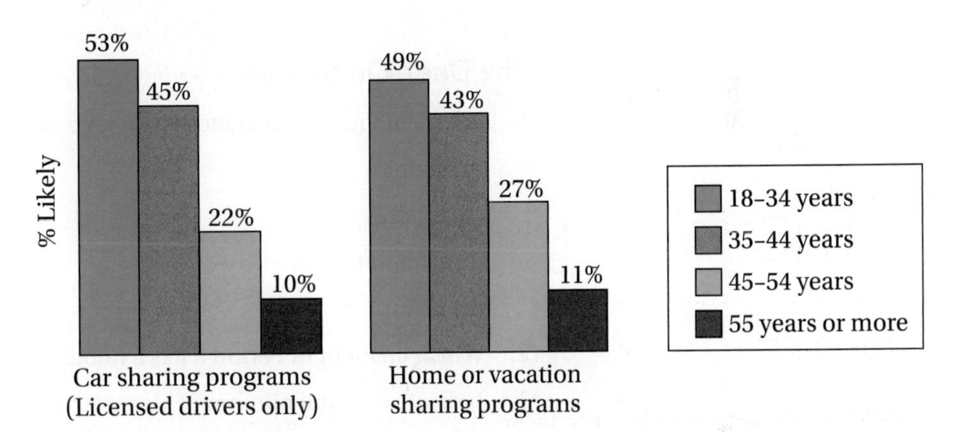

Figure 4.19 Millennials Embrace Sharing Economy
This study showed that Americans aged 18 to 34 years are more likely than older adults to be involved in car sharing or home sharing.

Chapter 4 Themes and Questions

recurring themes in chemistry chapter 4

recurring theme 3 Opposite charges attract one another. (p. 96)

recurring theme 4 Chemical reactions are nothing more than exchanges or rearrangements of electrons. (p. 97)

recurring theme 7 Individual bonds between atoms dictate the properties and characteristics of the substances that contain them. (p. 111)

A list of the eight recurring themes in this book can be found on p. xv.

ANSWERS TO FOLLOW-UP QUESTIONS

Follow-Up Question 4.1: carbon (C). *Follow-Up Question 4.2:* magnesium (Mg). *Follow-Up Question 4.3:* Boron has a total of 5 electrons: 3 valence electrons and 2 core electrons. Chlorine has a total of 17 electrons: 7 valence and 10 core. *Follow-Up Question 4.4:* true. *Follow-Up Question 4.5:* The Lewis dot diagram for a potassium atom looks similar to the diagram shown for lithium on p. 102 but the letter "K" replaces the letters "Li." In your diagram, there should be a single dot beside a letter "K." *Follow-Up Question 4.6:* O^{2-}, Rb^{+}, Be^{2+}, Br^{-}. *Follow-Up Question 4.7:* magnesium oxide (MgO). *Follow-Up Question 4.8:* (a) LiI, lithium iodide (b) K_2S, potassium sulfide. *Follow-Up Question 4.9:* H—H. *Follow-Up Question 4.10:* nonpolar. *Follow-Up Question 4.11:* (c) lithium and bromine

End-of-Chapter Questions

Questions are grouped according to topic and are color-coded according to difficulty level: **The Basic Questions***, **More Challenging Questions****, and **The Toughest Questions*****.

The Duet/Octet Rule

***1.** Which of the following atoms has an octet of electrons?

(a) sulfur

(b) radon

(c) selenium

(d) neon

***2.** Which element in Period 5 has a noble gas configuration of electrons?

***3.** Indicate the total number of electrons as well as the number of core and valence electrons for the following:

(a) sulfur

(b) fluorine

(c) lithium

(d) aluminum

***4.** Indicate the total number of electrons and the number of core electrons and valence electrons for the following:

(a) beryllium

(b) calcium

(c) helium

(d) phosphorus

****5.** Which atoms or ions contain a duet of electrons?

(a) H^{+} (b) He (c) Li^{+} (d) H^{-} (e) Be^{2+}

****6.** In your own words, explain why an atom or ion that obeys the duet rule or the octet rule is said to have a noble gas configuration of electrons.

***7. The element bromine exists in nature as a diatomic molecule. Draw a Lewis dot diagram for an oxygen molecule in a way that gives each bromine atom access to an octet of electrons.

***8. The element nitrogen exists in nature as a diatomic molecule. Draw a Lewis dot diagram for a nitrogen molecule in a way that gives each nitrogen atom access to an octet of electrons. (Hint: It may be necessary to connect the nitrogen atoms with more than one bond. Each pair of electrons shared between atoms represents one bond and is shown with one line.)

Electron Bookkeeping

*9. Draw Lewis dot diagrams for the following atoms: argon, sulfur, sodium, calcium, nitrogen.

*10. Draw Lewis dot diagrams for the following atoms: bromine, krypton, phosphorus, rubidium, carbon.

*11. How many core and valence electrons do each of the following atoms have?

(a) phosphorus

(b) lithium

(c) helium

(d) fluorine

**12. An atom has 10 core electrons and 4 valence electrons. Which element is it?

**13. How many core and valence electrons do each of the following ions have? Draw the symbol for each.

(a) sulfide ion

(b) magnesium ion

(c) potassium ion

(d) chloride ion

***14. Is it possible for an atom to have zero valence electrons? Justify your answer.

***15. Take all of the core electrons from these atoms—Cl, F, Na—add them together, and multiply them by three. If this is the number of protons in an atom, which element is it?

Ionic Bonding and Salts

*16. Calcium forms calcium oxide, CaO, as well as calcium bromide, $CaBr_2$.

(a) Identify the cation and anions in these formulas.

(b) Explain why the ions in calcium oxide have a one-to-one ratio, but the ions in calcium bromide have a one-to-two ratio.

*17. Draw a Lewis dot diagram for ions of the following elements: astatine, oxygen, magnesium, lithium.

*18. Draw Lewis dot diagrams for ions of the following elements: sulfur, fluorine, calcium, potassium.

*19. Which of the following ions is not likely to exist?

(a) O^{2-} (b) Br^- (c) Mg^+ (d) Na^+

**20. Write a chemical formula for the salt that forms from the ions of sodium and sulfur.

**21. Write the chemical formula for the ionic compound that forms between the following elements:

(a) barium and oxygen

(b) rubidium and nitrogen

(c) magnesium and chlorine

**22. Students are sometimes confused by the differences among the terms *salt, crystal,* and *formula unit.* Pretend you are explaining these terms to a friend (or find a friend who is willing to listen). In what contexts are these terms used, and how are they similar or different from one another?

**23. Find the repeating motifs in each example.

(a) ABBBABBBABBBABBBABBB

(b) CDCDCDCDCDCDCD

(c) EEFEEFEEFEEFEEF

(d) GHHGHHGHHGHHGHH

***24. Some anions and cations have more than one atom. When a chemical formula contains more than one of these *polyatomic ions,* the atomic symbols are enclosed in parentheses and the number of ions is indicated as a subscript outside the parentheses. Parentheses are not needed when one polyatomic ion is present. We will look closely at several examples of polyatomic ions in later chapters. For now, let's make salts from them by applying the rule we already know: salts must have a balance of positive and negative charges. Write the chemical formula for the salt that forms from each of these ion pairs.

(a) Na^+ and PO_4^{3-}

(b) NH_4^+ and Se^{2-}

(c) Mg^{2+} and CO_3^{2-}

***25. An anion is formed when a neutral atom takes on additional electrons. For example, sulfur forms S^{2-}, chlorine forms Cl^-, and tellurium forms Te^{2-}.

(a) Explain why each of these elements has a specific number of negative charges in its anionic form.

(b) Why do you think Ne^- or Ar^{2-} would be unlikely to form?

Covalent Bonds

*26. In which situation is a covalent bond most likely to form?

(a) when a metal bonds to another metal

(b) when a nonmetal bonds to a metal

(c) when a nonmetal bonds to a nonmetal

*27. True or false? When drawing a covalent bond, it is proper to use a line to connect atoms.

*28. In your own words, describe the difference between a bond that is nonpolar covalent and polar covalent. Cite one example of each type of bond.

*29. Consider two atoms of iodine. Describe two ways that each of these atoms—either together or separately—can achieve a noble gas configuration of electrons.

**30. Draw a Lewis dot diagram for a molecule of chlorine, Cl_2.

**31. Draw a Lewis dot diagram for a molecule of ammonia, NH_3. This molecule has nitrogen in the center with three hydrogen atoms around it.

**32. In your own words, explain the duet rule. Why is it necessary to have a duet rule?

**33. How many vacancies does each of the following atoms have in its Lewis dot diagram?

(a) argon

(b) nitrogen

(c) carbon

(d) phosphorus

**34. How many vacancies does each of the following atoms have in its Lewis dot diagram?

(a) chlorine

(b) sulfur

(c) oxygen

(d) neon

Bonding in Metals

*35. Identify each element as a metal or a nonmetal.

(a) xenon

(b) copper

(c) manganese

(d) carbon

*36. Which of the following is not used to describe metals?

(a) malleable

(b) good heat conductor

(c) poor electrical conductor

(d) ductile

**37. Based on what you know about metallic bonding, explain why metals are exceptionally good conductors of heat and electricity.

**38. A certain piece of copper pipe is composed of (nearly) pure copper.

(a) Describe the bonding in this piece of metal.

(b) How is it possible that there are no anions in the metal?

**39. Write symbols for the ions of each metal.

(a) sodium

(b) radium

(c) lithium

(d) barium

***40. The temperature at which a metal melts is directly related to the strength of the bonds between atoms in the metal. Based on what you know about the sea of electrons model of metallic bonding, and the way that atoms form ions, which group of metals do you think has the higher melting point—the Group 1 elements or the Group 2 elements? Why?

***41. A transition metal has 42 protons and 37 electrons. Write a symbol for this ion.

Electronegativity and Bond Polarity

*42. (a) Give an example of two atoms that form a nonpolar bond with each other.

(b) Is it possible for two atoms of different elements to form a nonpolar bond? How?

*43. True or false? Atoms with the highest electronegativity tend to be located in the lower right corner of the periodic table.

*44. In your own words, describe what is meant when the term *polarity* is used to describe a covalent bond.

**45. Referring to the periodic table and the electronegativity values given in Figure 4.17, rank the bonds between these pairs of atoms, from most polar to least polar. For each pair, indicate whether you expect the bond to be covalent or ionic.

(a) Cs bonding with Cl

(b) Cl bonding with Cl

(c) Mg bonding with O

(d) O bonding with Pb

**46. Consider the electronegativity values in Figure 4.17. Why is it unlikely that two nonmetals will pair together in an ionic bond?

**47. Using the electronegativity values given in Figure 4.17, rank the bonds between these pairs of atoms, from most polar to least polar. For each pair, indicate whether you expect the bond to be covalent or ionic.

(a) K bonding with F

(b) O bonding with N

(c) C bonding with Co

***48. Water molecules are polar, as we learned in Section 4.4. A similar molecule, H_2S (hydrogen sulfide), also exists. Do you think that water or H_2S has more polar bonds? Justify your answer.

***49. Carbon dioxide molecules contain one atom of carbon attached to two atoms of oxygen, one on each side. The molecule is straight; the three atoms form a line. Carbon dioxide molecules contain polar bonds; each carbon–oxygen bond is polar. However, the molecule as a whole is nonpolar. Think about the shape of the molecule, and explain why this molecule is nonpolar when it contains polar bonds.

***50. In your own words, explain why diatomic molecules that contain the two atoms of the same element always form a nonpolar bond.

Integrative Questions

*51. Think about the types of bonding discussed in this chapter: ionic bonding, covalent bonding, and metallic bonding. Describe each type of bonding. Which involves sharing electrons? Which does not involve sharing electrons?

*52. Organic compounds tend to include elements that are nonmetals, such as hydrogen. For example, drug molecules like acetaminophen ($C_8H_9NO_2$), valium ($C_{16}H_{13}ClN_2O$), and benzocaine ($C_9H_{11}NO_2$) are all organic compounds. Look at the elements present in these compounds and speculate about what type of bond—ionic, covalent, metallic—holds the atoms in these molecules together. Provide a brief explanation for your choice.

*53. Over the first four chapters, we have encountered three particles within the atom: the proton, the neutron, and the electron. List the location and the charge of each, and then answer this question: Why is the electron the most important particle in the atom for chemists? That is, why does the electron play a central role in chemistry?

***54. Return to Figure 4.10 and look again at the elements in columns 16 and 17 of the periodic table. Those in column 16 tend to make anions with a charge of -2, while those in column 17 tend to make anions with a charge of -1.

(a) Given these facts, speculate on the ion called *nitride ion* that would form from an atom of nitrogen and the ion called *phosphide ion* that would form from an atom of phosphorus . What charge would each have?

(b) Write the chemical formula for the salt that would be made from the ions of potassium and nitrogen. What do you think this salt would be named?

(c). Write the chemical formula for the salt that would be made from the ions of magnesium and phosphorus. What do you think this salt would be named?

55. INTERNET RESEARCH QUESTION

The temperature at which a pure substance melts is an indication of the strength of the bonds within it. The stronger the bonds are within a substance, the higher the temperature required to melt it. In this chapter, we discuss the bonds that hold metal atoms together in a sample of pure metal. Consider some of the metals from the first two columns of the periodic table: lithium (Li), sodium (Na), potassium (K), beryllium (Be), magnesium (Mg), and calcium (Ca). Look up and record the melting points in degrees Celsius of these metals. and compare them. What trend in melting points, if any, do you see among the metals that tend to make $+1$ cations (those in column 1) versus the metals that tend to make $+2$ cations (those in column 2)? Can you suggest a reason for this trend?

Andre Geim & Kostya Novoselov/Science Source

Carbon

Elemental Carbon, Organic Molecules, and Carbon Footprints

aladin66/iStock/Getty Images

CHAPTER 5

Billboard charts keep track of the top 40 most popular songs, and The New York Times best-seller list follows the hottest books from week to week. If such a list existed for discoveries in science, it would have been dominated by one thing over the past several years: graphene. Graphene is the name given to the single sheets of carbon atoms that, when stacked together, make up graphite. The figure shown on p. 126 is an image of sheets of graphene obtained with a high-resolution microscope. We'll learn a lot about graphite in this chapter, but for now all you need to know is that graphite is a substance made up of flat sheet upon flat sheet of carbon atoms.

Why so much hoopla over slices of graphite, the substance we've been using in our pencils for decades? It's because, until recently, everyone thought that one single layer—graphene—was impossible to isolate. Physicists and chemists tried for decades to tease apart individual layers from a chunk of graphite, until there was a collective throwing up of hands.

But then, in 2004, a group of scientists in Britain decided to try an old-school approach. Instead of using fancy, state-of-the-art scientific instruments or chemical reactions to pull away a single layer from a piece of graphite, they used a tool that you can find on almost any desk or in any kitchen drawer: Scotch tape. In the most inelegant, unsophisticated way imaginable, they stuck some graphite between some Scotch tape and pulled, and voilà—graphene!

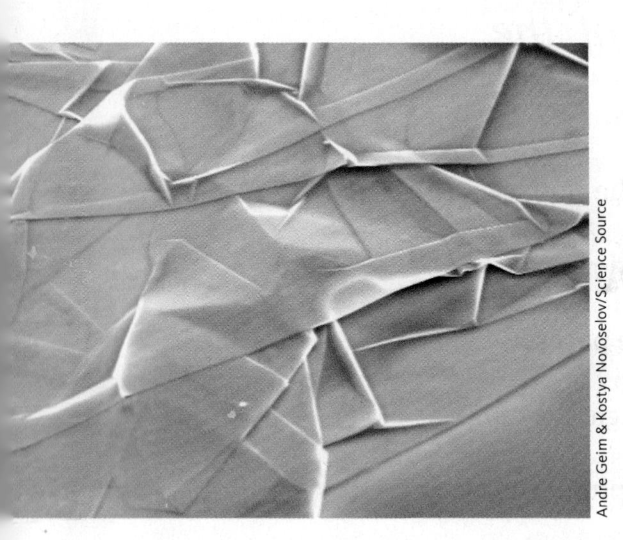
Andre Geim & Kostya Novoselov/Science Source

Graphene's single layer of honeycombed carbon atoms gives it strange and amazing properties. Graphene conducts current with unheard of speed, and it's the strongest material known. According to scientists who work with it, if you could stretch a layer of graphene over a space no larger than a coffee cup, that small piece of graphene would be strong enough to hold the weight of an 18-wheeler!

Right now graphene is not useful for much, because we don't know how to make it in large enough pieces. But this is changing fast. Research groups around the world are trying to make bigger pieces and trying to better understand its unique properties. A very intriguing thing about graphene is that it's made up of only one element and nothing else. So the chemistry of graphene is really the chemistry of the element carbon, the subject of this chapter.

5.1 Why Is Carbon Special?

Carbon has special properties because of its small size.

Carbon has always been a flashy element, one of the most notorious on the periodic table. It forms the framework of all organic molecules, including the molecules that make up living things. This chapter is about carbon, and it is also about the molecules that carbon makes. Our story begins with one simple question: Why is carbon special?

We learned in Chapter 2 that the periodic table is an element organizer. Stepping through the periodic table from left to right, moving from top to bottom, each new step represents a new element with a new atomic number and an additional proton. The numbers of electrons and neutrons increase, too. So, in general, as we move across and down the periodic table, the elements become more and more massive because they contain more and more particles.

Trends within the periodic table, such as this mass trend, are typically reliable, but they aren't always manifested in a regular, incremental way. **Figure 5.1** plots atomic size versus atomic number for elements in the carbon family (Group 14 on the periodic table). As you can see, carbon is smaller than you would expect based on the size trend for silicon, germanium, tin, and lead. The same is true for all the elements in that top row of the periodic table, Period 2, which includes carbon: their size is much smaller than you would expect by looking at other elements in the same group.

What causes this anomaly? Atoms of an element are extra-small when they have only a few electrons and those electrons are held close to the nucleus, which is the case for atoms across the very top of the periodic table, in Period 2. For atoms farther down in a group, there are additional electron layers farther from the nucleus, and these outer electrons feel less pull from the positive charge of the nucleus. Thus, those atoms are larger. We refer to this phenomenon as the **uniqueness principle**. The uniqueness principle states that the Period 2 elements are uniquely small because they have so few electrons, and those electrons are so close to the nucleus.

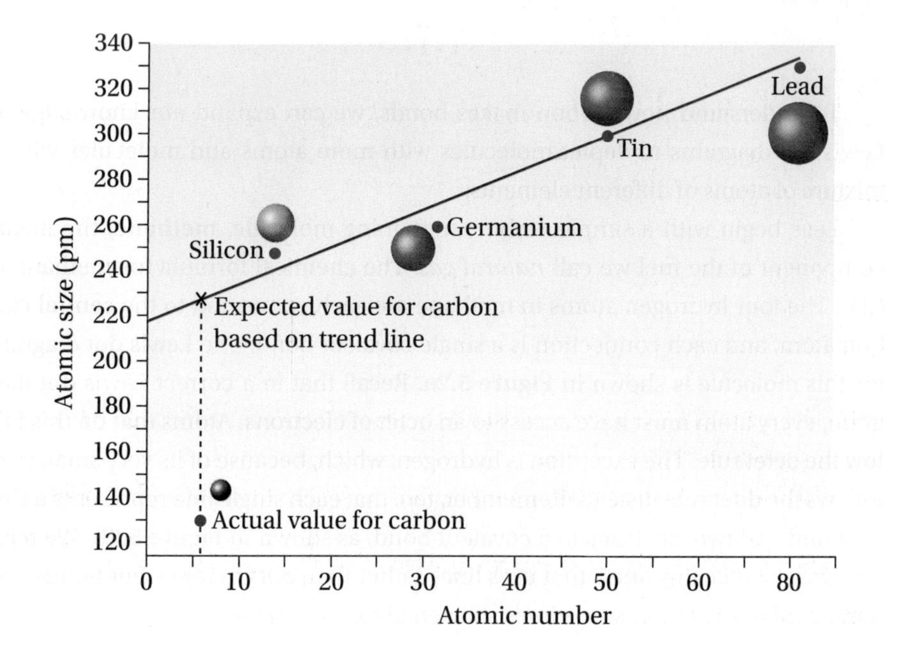

Figure 5.1 Atomic Size Trend
This graph shows the diameter of atoms, in picometers (pm), of five carbon-family elements (Group 14) as a function of atomic number. The trend line is determined from the diameters of the elements silicon, germanium, tin, and lead.

QUESTION 5.1 Refer to the periodic table inside the front cover of this book to determine which element in each pair has smaller atoms. (a) gallium (Ga) or boron (B) (b) calcium (Ca) or magnesium (Mg) (c) xenon (Xe) or neon (Ne)

ANSWER 5.1 (a) A boron atom is smaller because it is higher on the periodic table than gallium. (b) A magnesium atom is smaller because it is higher on the periodic table than calcium. (c) A neon atom is smaller because it is higher on the periodic table than xenon.

For more practice with problems like this one, check out end-of-chapter Questions 1 and 2.

FOLLOW-UP QUESTION 5.1 Rank the following atoms in order of increasing atomic diameter: argon, helium, radon, krypton, neon.

A carbon atom can form four bonds.

recurring theme in chemistry 6

The structure of carbon-containing substances is predictable because each carbon atom almost always makes four bonds to other atoms.

Because it is so tiny, a carbon atom is able to approach other atoms closely, sharing electrons in a covalent bond. When a carbon atom bonds to other carbon atoms, the atoms are especially close, and the bond is especially strong. Carbon atoms can link together to form long chains, and those chains may be branched or may form rings. In total, each carbon atom can make four bonds to other atoms.

In Section 4.2, we introduced Lewis dot diagrams, which are also referred to as *molecular structures* or simply *structures.* The Lewis dot diagram for carbon shows us that carbon atoms make four bonds because they have four electrons and four vacancies, as shown here:

$$\cdot\dot{\underset{\cdot}{\mathrm{C}}}\cdot$$

We can also use a Lewis dot diagram to show how valence electrons are distributed around atoms involved in covalent bonds. Recall that the examples we covered in Chapter 4 were simple diatomic molecules, such as I_2, the iodine molecule:

$$:\ddot{\underset{\cdot\cdot}{\mathrm{I}}}:\ddot{\underset{\cdot\cdot}{\mathrm{I}}}:$$

To understand how carbon makes bonds, we can expand our knowledge of Lewis dot diagrams to depict molecules with more atoms and molecules with a mixture of atoms of different elements.

Let's begin with a simple carbon-containing molecule, **methane,** the major component of the fuel we call *natural gas.* The chemical formula for methane is CH_4. The four hydrogen atoms in methane are each connected to the central carbon atom, and each connection is a single covalent bond. The Lewis dot diagram for this molecule is shown in **Figure 5.2a.** Recall that in a correct Lewis dot diagram, every atom must have access to an octet of electrons. Atoms that do this follow the octet rule. The exception is hydrogen, which, because of its very small size, follows the duet rule instead. Remember, too, that each single line represents a single bond and two electrons in a covalent bond, as shown in **Figure 5.2b**. We refer to a drawing like this one—that uses lines rather than dots to represent bonds—as a **molecular structure,** sometimes shortened to just *structure.*

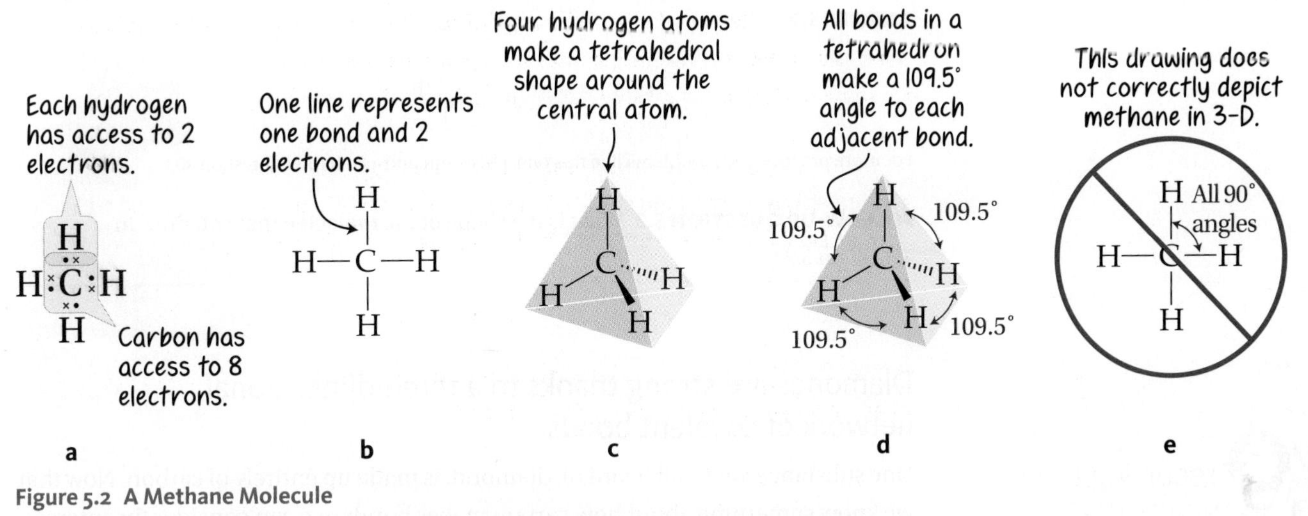

Figure 5.2 A Methane Molecule
Methane can be correctly depicted in several different ways, as shown here: (a) a Lewis dot diagram of methane; (b) the two-dimensional structure of methane with pairs of electrons in a bond represented by a single line; (c) a three-dimensional drawing of methane with its tetrahedral shape indicated by wedges and dashes; (d) the same three-dimensional drawing with angles between bonds depicted; and (e) a three-dimensional drawing of methane that is incorrect because it indicates that the bond angles are 90 degrees.

We draw molecules in a way that shows they are three-dimensional.

Notice that the methane molecule in **Figure 5.2c** is drawn in a way that makes it appear three-dimensional (or 3-D for short). Of its four bonds, two are drawn in the plane of the paper. Another is drawn as a wedge, and it appears to be coming out of the page toward us. The fourth is drawn with dashes, meant to show that hydrogen is beneath the plane of the paper. **Figure 5.2d** is the methane molecule again, but this time the angles between each hydrogen atom are indicated. As the figure shows, each of the H–C–H angles in this molecular structure has an identical value: exactly 109.5 degrees. This shape and these angles are common in organic molecules, and the shape has a special name: **tetrahedron**. As we'll see throughout this text, the tetrahedron is the most recognizable structural motif in carbon-containing molecules. And, whenever carbon makes bonds to four atoms, the shape is almost always tetrahedron-like.

Suppose that we drew the 3-D molecular structure flat, as shown in **Figure 5.2e**. Without the wedges and dashes to indicate three-dimensionality, the H–C–H angles now become 90-degree right angles. This shape is incorrect and is not observed for carbon because the 90-degree angle puts the hydrogen atoms too close together. In the tetrahedron, the hydrogen atoms can spread farther from each other, and each hydrogen atom occupies more space around the central carbon atom. We encounter the tetrahedron whenever we consider organic molecules.

QUESTION 5.2 Using Figure 5.2c as a guide, draw a 3-D sketch of the molecule CF_4. Connect the four fluorine atoms to carbon, which is in the center of the molecule, as it is in methane. Use lines to indicate bonds, and do not worry about placing electrons around atoms.

ANSWER 5.2

For more practice with problems like this one, check out end-of-chapter Question 10.

FOLLOW-UP QUESTION 5.2 What is the shape of the molecule that you drew in Question 5.2?

Diamonds are strong thanks to a three-dimensional network of covalent bonds.

> **recurring theme in chemistry 2**
> Things we can see with our eyes contain trillions and trillions of atoms. Atoms are much too small to view with the naked eye.

One substance we've all heard of, diamond, is made up entirely of carbon. Now that we know something about how carbon makes bonds, we can consider the interesting molecular structure of diamond. Diamond's structure, depicted in **Figure 5.3,** is an extraordinary example of covalent bonding. Notice that, except for those atoms at the perimeter of the drawing, each carbon atom forms a covalent bond with four other carbon atoms. This drawing shows only a small part of the whole. In a real diamond, billions of carbon atoms extend out from this molecular structure in all directions.

A solid, like diamond, that has an extended system of repeating covalent bonds is a **network solid**. Network solids are generally very robust because, as the pattern of covalent bonding repeats, the strength of the overall structure increases. This three-dimensional network of covalent bonds makes diamond the hardest natural substance known.

The extraordinary toughness of diamond makes it unlikely to undergo any type of chemical reaction. All of its carbon atoms are satisfied, in the sense that they all have a full octet of electrons. To understand why this is so, consider one carbon atom. As **Figure 5.4** shows, a single carbon atom has four valence electrons and needs access to four more to achieve the ideal: an octet like that of the nearest noble gas, neon.

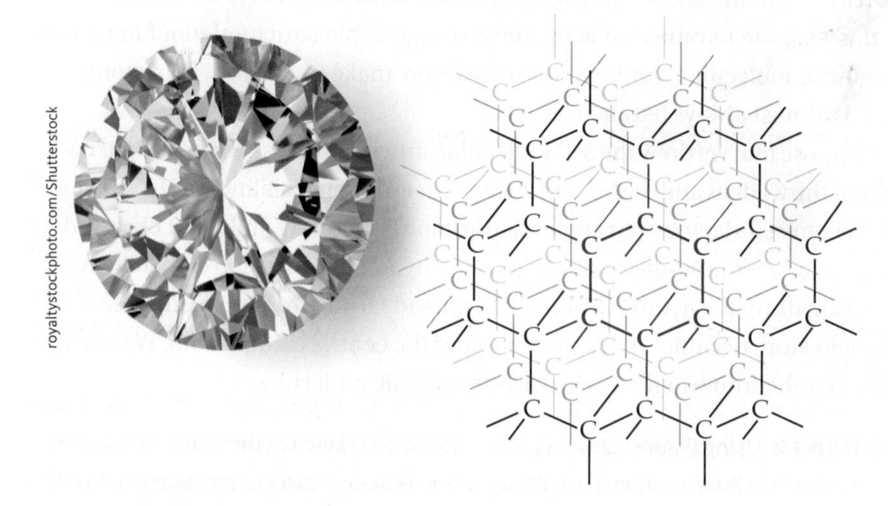

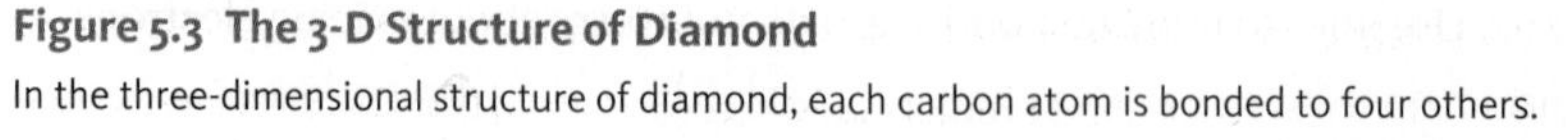

Figure 5.3 The 3-D Structure of Diamond
In the three-dimensional structure of diamond, each carbon atom is bonded to four others.

The molecular structure of methane showed us that carbon makes a tetrahedral shape when it makes bonds to four other atoms. Diamond is an extreme example of that idea. Each carbon atom in a diamond makes a tetrahedral shape but, rather than making bonds to hydrogen, every carbon atom makes four bonds to other carbon atoms. Of the eight valence electrons around each carbon, four are from that carbon and four come from the atoms on the other end of each covalent bond.

A single diamond is made up of billions and billions of carbon atoms, and an octet of electrons is achieved by every one of those carbon atoms. The strong covalent bonds made from carbon atom to carbon atom extend throughout the network of atoms, and this continuous network of bonds makes diamond inert like a noble gas. Within diamond, each atom has an octet of valence electrons and thus has no reason to undergo a chemical reaction. Recall from Chapter 4 that a chemical reaction occurs when the circumstances of the atom or molecule could become closer to achieving an octet or duet by the reaction. In the case of diamond, its ideal arrangement of electrons makes it resistant to reactive chemical compounds and explains why, at least in theory, a diamond is forever.

recurring theme in chemistry 4
Chemical reactions are nothing more than exchanges or rearrangements of electrons.

QUESTION 5.3 What are the angles between adjacent bonds within the structure of diamond?

ANSWER 5.3 Diamond is made up of many tetrahedrons (or *tetrahedra*), each one a carbon atom attached to four other carbon atoms. The angles within one tetrahedron are 109.5 degrees. Therefore, the angles between bonds in diamond also have this value.
For more practice with problems like this one, check out end-of-chapter Question 5.

FOLLOW-UP QUESTION 5.3 Silicon, the element just below carbon in Group 14, also can make four bonds to obtain an electron configuration like its nearest noble gas. Silicon will have the same number of electrons as which noble gas if it makes four bonds?

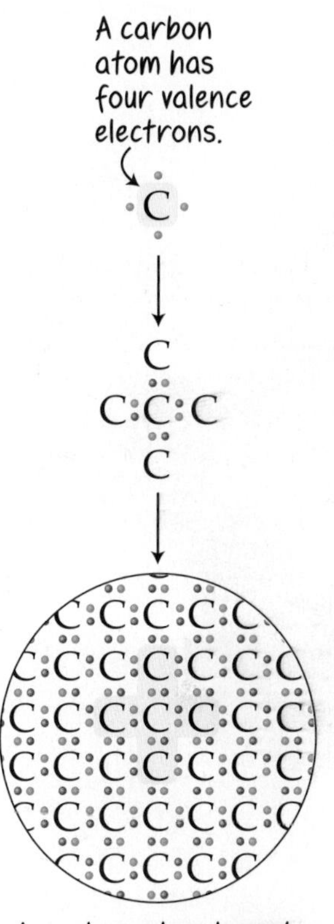
A carbon atom has four valence electrons.
A carbon atom bound to four other carbon atoms can share four electrons and achieve a stable octet. Recall that, in diamond, the structure is three-dimensional.

Figure 5.4 Covalent Bonds between Carbon Atoms in Diamond

Carbon atoms have four valence electrons. To have an octet of electrons, a carbon atom may share one electron with each of four other carbon atoms. In this rendering, each covalent bond appears as a blue dot and a purple dot. The blue dot represents one valence electron from one of the atoms making the bond, and the purple dot represents one valence electron from the other atom in the bond.

5.2 Graphite, Graphene, Buckyballs, and the Multiple Bond

A carbon atom can form multiple bonds to other atoms.

In the previous section, we considered two examples—methane and diamond—in which carbon atoms make bonds to four other atoms. And, although carbon always makes four bonds, it does not necessarily always form bonds to four other atoms. How can carbon make four bonds, but not be attached to four other atoms? The answer is that carbon, and many other atoms, form not only single bonds (in which two electrons are shared by two atoms), but multiple bonds as well. In a **multiple bond,** two atoms share more than two electrons.

Recall from Section 5.1 that, according to the uniqueness principle of atomic size, Period 2 elements are very small. Thus, they can get very close to one another. This is, in part, what makes multiple bonding between carbon atoms possible. The multiple bond known as a double bond is a pair of bonds, each containing two electrons, for a total of four shared electrons. We represent a double bond with two

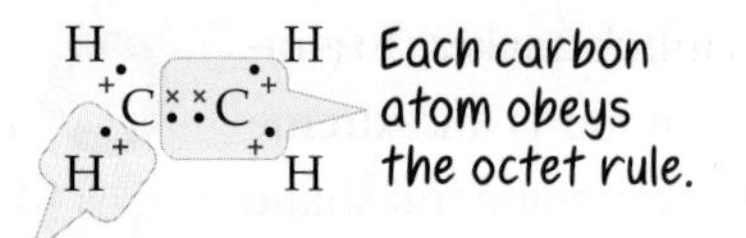

a The structure of ethylene

$H{-}C{\equiv}C{-}H$

b The structure of acetylene

Figure 5.5 Multiple Bonds in Ethylene and Acetylene
For each molecule, a molecular structure and Lewis dot diagram are shown.

lines between atoms. Ethylene is an example of a simple molecule that contains a double bond. **Figure 5.5a** shows the molecular structure and Lewis dot diagram of ethylene. Note that each carbon atom in the ethylene molecule is able to access eight electrons. Each carbon atom in ethylene obeys the octet rule, and each hydrogen atom follows the duet rule.

The multiple bond known as a triple bond is three bonds, each containing two electrons, for a total of six shared electrons. We represent a triple bond with three lines between atoms. Acetylene, a gas used in welding, is a simple molecule that contains a triple bond. **Figure 5.5b** shows acetylene's molecular structure and Lewis dot diagram. Each carbon atom in acetylene is able to access eight electrons and obeys the octet rule, and each hydrogen atom follows the duet rule.

Figure 5.6 illustrates some typical carbon-containing molecules featuring single and multiple bonds. Notice that in every molecule in Figure 5.6, every carbon atom forms a total of four bonds. Recall from Chapter 3 that a chemical formula tells us how many atoms of each element are present in a substance. We can deduce the chemical formula for each molecule in Figure 5.6 by counting the number of each type of atom. Try this for each molecule shown. The answers are given in the figure caption.

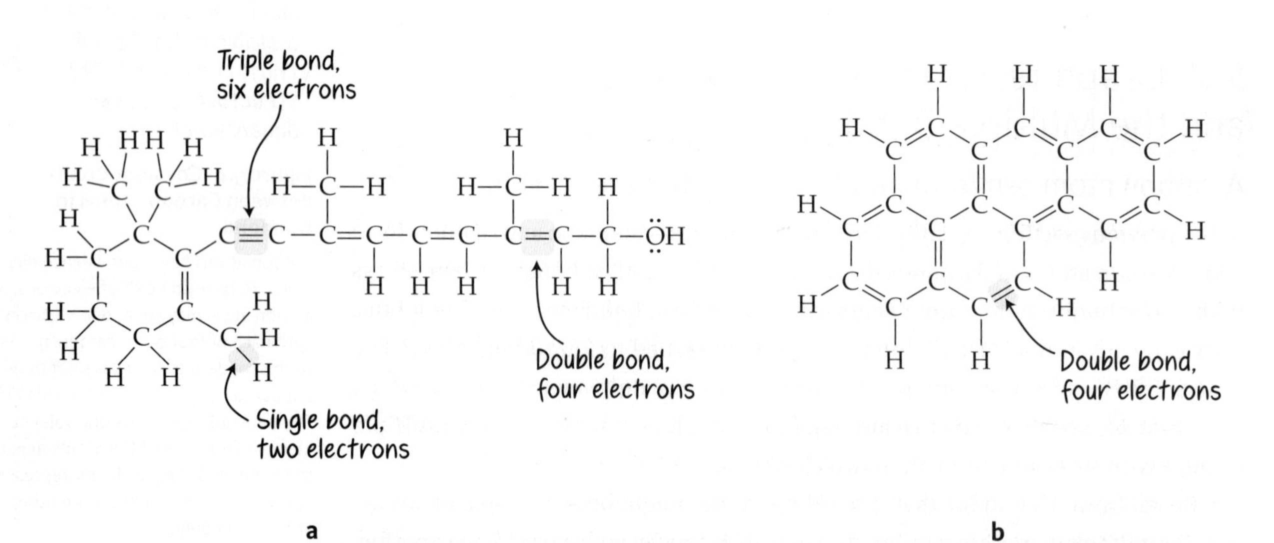

Figure 5.6 Multiple Bonds in Carbon Compounds
Notice that each carbon atom always forms four bonds in organic compounds; it does not matter if the bond is single, double, or triple. Each carbon atom in these molecules obeys the octet rule, and each hydrogen atom obeys the duet rule. (a) The molecule shown here has the chemical formula $C_{20}H_{28}O$. (b) The chemical formula for the molecule shown here is $C_{20}H_{12}$.

QUESTION 5.4 Linalool is a molecule with a pleasant lavender scent that is used in perfumes. Here is the molecular structure for linalool:

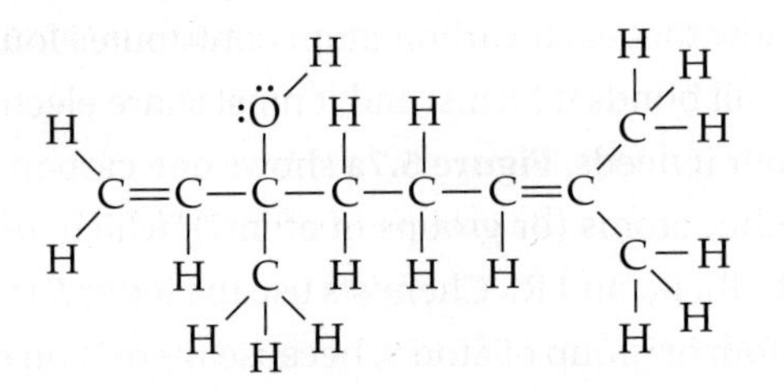

(a) How many double bonds between carbon atoms are in linalool? (b) How many triple bonds between carbon atoms are in linalool? (c) What is the chemical formula for this molecule?

ANSWER 5.4 (a) 2 (b) 0 (c) $C_{10}H_{18}O$

For more practice with problems like this one, check out end-of-chapter Questions 11 and 15.

FOLLOW-UP QUESTION 5.4 The molecular structure of a molecule similar to vitamin A is shown here. (a) How many single bonds between carbon atoms are in this molecule? (b) How many double bonds between carbon atoms? (c) How many triple bonds between carbon atoms?

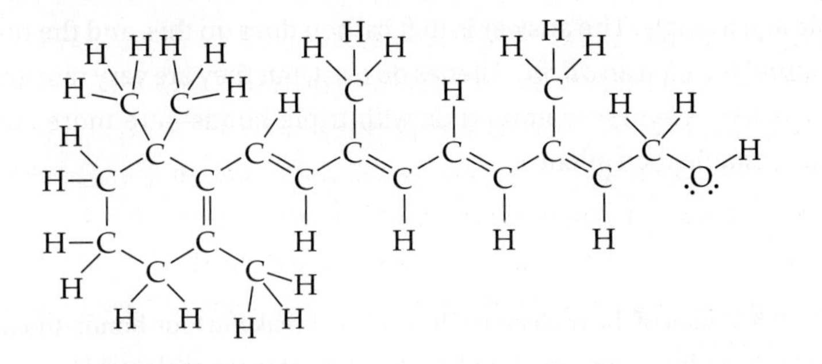

---

wait a minute . . .

Why are organic molecules sometimes drawn flat when tetrahedral carbons are three-dimensional?

We sometimes draw organic molecules so that they look flat when they are actually three-dimensional. For example, in Question 5.4, no 3-D perspective is included in those molecular structures, even though some of the carbon atoms could be drawn with wedges and dashes. Why do we sometimes use a 3-D perspective to draw a molecule and then draw exactly the same molecule so that it looks flat as a pancake? The answer is this: It's tedious to draw wedges and dashes to indicate three-dimensionality, and, sometimes, the 3-D perspective is not always the point of the drawing.

For example, Question 5.4 asks you to identify multiple bonds, not the shape of the molecule. So, depending on the context in which organic molecules are presented, sometimes we draw them in a way that suggests three-dimensionality, and sometimes we draw them as flat structures.

Carbon atoms form four bonds in various ways.

Let's pause for a moment to review the differences between single, double, and triple bonds. We know that each carbon atom contributes four electrons, in total, to all bonds it forms, and it must share electrons to get the other four it needs. **Figure 5.7a** shows one carbon atom bound to four other atoms (or groups of atoms), which are given the symbols R_1, R_2, R_3, and R_4. Chemists use the letter *R* to represent a generic atom or group of atoms, because there is no element that has the symbol R. **Figure 5.7b** shows one carbon bound to three other atoms (R_1, R_2, and R_3) and **Figure 5.7c** shows one carbon bound to two other atoms (R_1 and R_2). Note that to accommodate the four-bonds-to-carbon rule, the molecules shown in Figures 5.7b and 5.7c have double and triple bonds.

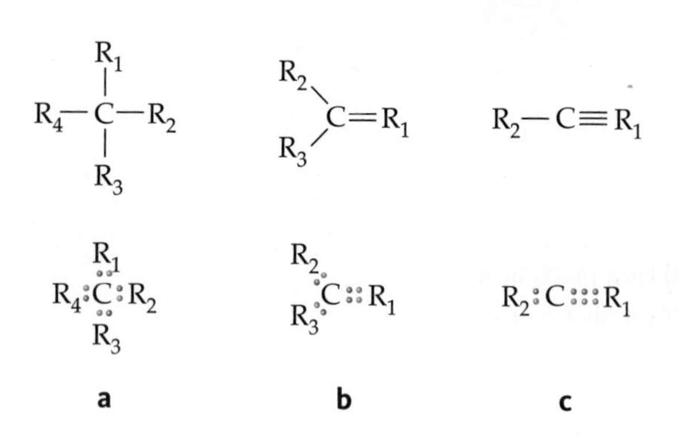

Figure 5.7 Bonding Patterns in Generic Carbon-Containing Molecules

Carbon can bond to (a) four other groups by making single bonds to each, (b) three other groups by making two single bonds and one double bond, and (c) two other groups by making one single bond and one triple bond. The bottom drawings show the Lewis dot diagram for each type of pattern. Remember that each line in a chemical bond is equivalent to two electrons.

wait a minute . . .

Can't carbon make two double bonds rather than one single plus one triple?

Readers may wonder why carbon cannot make two double bonds instead of a triple and a single. The answer is that carbon does do this, and the resulting compound is called an *allene.* Allenes do exist, but they are very rare and usually unstable. Alkynes—compounds with triple bonds—are more common and more stable than allenes.

QUESTION 5.5 Which of the molecules shown here break the four-bonds-to-carbon rule or the duet rule for hydrogen and therefore do not exist in the real world?

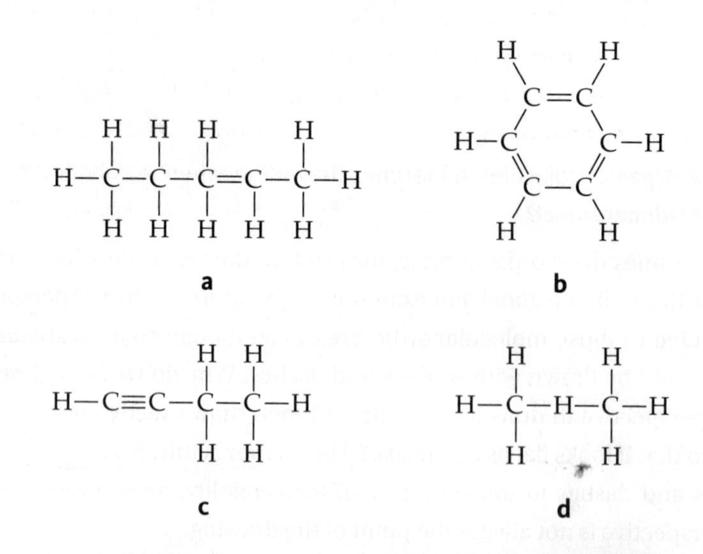

ANSWER 5.5 All of the atoms in (b) and (c) obey the octet or duet rules. These molecules, therefore, could exist in the real world. One carbon in (a) makes five bonds and breaks the octet rule. One hydrogen in (d) makes two bonds and breaks the duet rule. Molecules (a) and (d) do not exist; they are imaginary.

For more practice with problems like this one, check out end-of-chapter Questions 25 and 27.

FOLLOW-UP QUESTION 5.5 Which of the molecules shown here break the four-bonds-to-carbon rule?

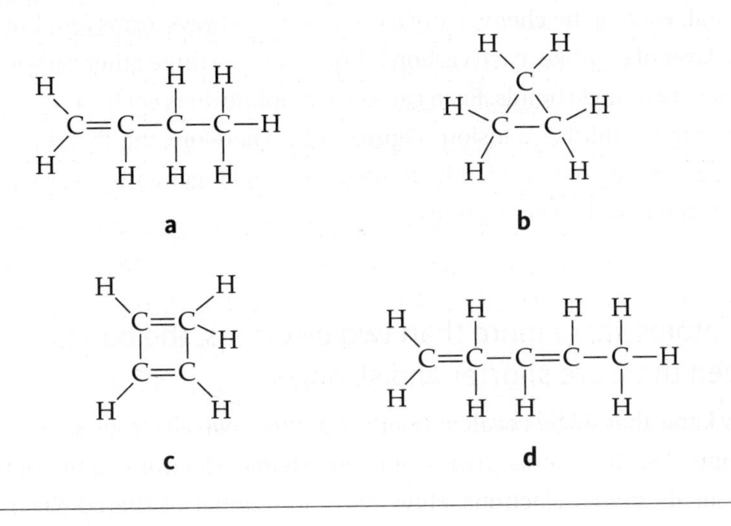
a
b
c
d

Carbon's allotropes feature examples of single bonds and multiple bonds.

Some elements on the periodic table exist in different forms in nature. For example, phosphorus is found in two common forms: red phosphorus and white phosphorus. We refer to different forms of a pure element as **allotropes**. For allotropes of an element, the atoms are the same, but they are arranged differently. We've just met the most famous allotrope of carbon: diamond. Now let's meet the rest.

We discussed graphite and graphene in the introduction to this chapter. Because graphite and graphene exist in nature and are forms of pure carbon, they are allotropes of carbon. Graphite is very different from diamond. It is a greasy black solid, so slippery that it makes an excellent lubricant. If, like diamond, graphite is pure carbon, how can it have such a radically different look and feel? The answer lies in the covalent bonds that hold graphite together.

Figure 5.8a shows a schematic drawing of graphite. What is the most striking difference between diamond's structure (see Figure 5.3) and that of graphite? The most noticeable difference is that all the bonds in graphite are in the same plane (the plane of the page). In contrast, the bonds in diamond are in many planes.

Figure 5.8 The Molecular Structure of Graphite

(a) All the carbon atoms in this sheet of graphite are in the same plane. Each carbon atom forms a total of four bonds: two single and one double. An isolated sheet of graphite, like the one shown here, is known as graphene. (b) This figure illustrates how graphite is made up of multiple sheets of carbon—graphene—stacked on one another. The individual sheets are held together by very weak forces.

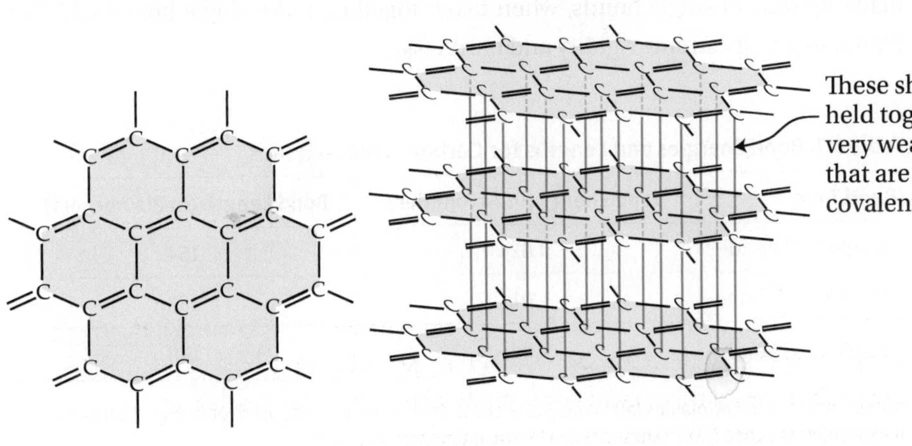
These sheets are held together by very weak forces that are not covalent bonds.
a
b

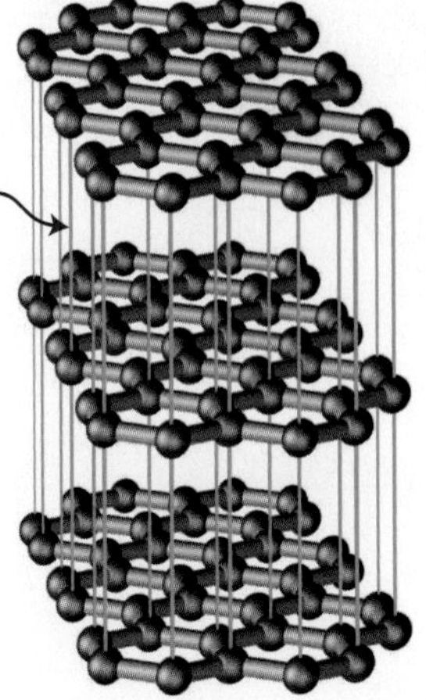

In other words, graphite's molecular structure is a stack of flat, two-dimensional sheets—graphene—and diamond's molecular structure is three-dimensional. This is a natural result of the chemistry of carbon, which always forms four bonds.

In a layer of graphite, each carbon is bound to only three other carbons via one double and two single bonds. Each carbon in graphite does *not* have a carbon atom to bond to in the third dimension (**Figure 5.8b**). Therefore, the sheets of graphene within graphite are not covalently bonded, but held together by very weak forces that allow them to slide across one another easily.

When atoms share more than two electrons, the bonds between them are shorter and stronger.

We now know that single covalent bonds arise from two electrons shared between two atoms. Double bonds arise from four shared electrons, and triple bonds arise from six shared electrons. How does the number of shared electrons, and therefore the number of bonds, affect the nature of the bond? Are multiple bonds stronger or weaker than single bonds? Are double bonds shorter or longer than triple bonds?

The strength of a bond, referred to as the **bond energy,** is measured in energy units. The metric unit for energy is the joule (J). **Table 5.1** shows the bond energies for single, double, and triple bonds between carbon atoms, and the distance between atoms in each bond, referred to as **bond length**.

What do the data in Table 5.1 tell us about the relative strengths of single and double bonds? Clearly, a double bond is stronger, which means that the amount of energy needed to break a double bond is greater than the amount needed to break a single bond. What happens to the distance between carbon atoms as we go from a single bond to a double bond? The data tell us that the bond shortens. These trends continue when we consider double versus triple bonds: triple bonds are stronger and shorter. Thus, the more bonds there are between two atoms, the shorter the bond length. This is true for all sorts of chemical bonds, not just those between two carbon atoms.

What does this tell us about the properties of graphite compared to diamond? Both structures are very robust, but in different ways. Diamond has an extremely tough three-dimensional structure: it is tough in every direction. Even though it is made up only of single bonds, when taken together, these single bonds within a diamond give it extreme rigidity and hardness.

Table 5.1 Bond Energies and Lengths for Carbon–Carbon Bonds

Bond Type	Bond Energy (in Attojoules)	Bond Length (in Picometers)
Single C–C bond	0.624	154
Double C=C bond	1.01	133
Triple C≡C bond	1.39	120

Notes: There are 10^{18} attojoules (aJ) in one joule; these are the metric units used for energy. There are 10^{12} picometers in one meter; these are the metric units used for length.

Graphite, on the other hand, is super-tough within each layer of carbon atoms because each layer contains single bonds as well as stronger double bonds, as shown in Figure 5.8a. It makes sense that, because graphene is a single layer of graphite, a sheet of graphene is extra-strong, too. But, because the carbon atoms in a single layer of graphite do not have bonds available to attach to carbon atoms in other layers, graphite is a stack of slippery sheets that are very loosely held together. This is what makes graphite such a great lubricant: the flat sheets within graphite can slide easily across one another. The tube of lubricant shown in **Figure 5.9** lists graphite as its first ingredient.

This brings us to an important lesson in chemistry, and perhaps the most important recurring theme in this book: the nature of the bonds between atoms dictates the properties and characteristics of the substances that contain them. Graphite is slippery because it is made up of sheets of carbon atoms that slide across one another. Diamond is extremely tough because it is composed only of strong carbon–carbon linkages in a matrix of three-dimensional tetrahedrons. This important theme was first introduced in Chapter 4, and we'll see many other examples of it throughout this book.

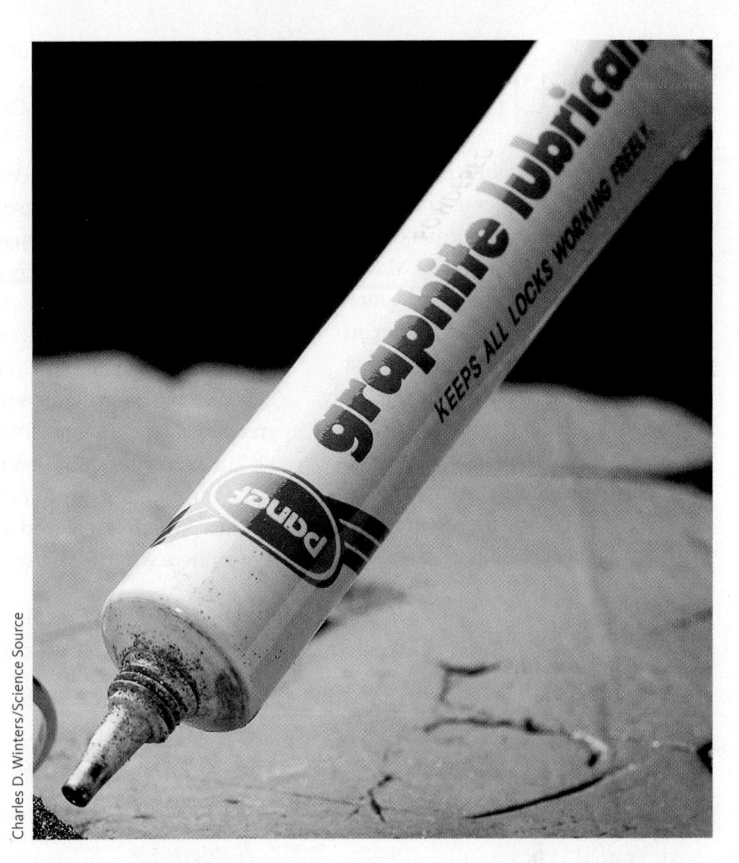

Charles D. Winters/Science Source

Figure 5.9 Slippery Stuff
Lubricants often contain graphite because of its slipperiness.

recurring theme in chemistry 7

Individual bonds between atoms dictate the properties and characteristics of the substances that contain them.

QUESTION 5.6 According to Table 5.1, it takes 0.624 aJ of energy to break one single carbon–carbon bond. Given this, how much energy, in attojoules, is required to break 100 single carbon–carbon bonds?

ANSWER 5.6 One hundred times the amount of energy required to break one bond is the amount of energy required to break 100 bonds: 62.4 aJ.

For more practice with problems like this one, check out end-of-chapter Question 11.

FOLLOW-UP QUESTION 5.6 Express the length of a typical double C=C bond in meters. Recall from Table 1.2 that there are 10^{12} picometers in one meter.

The buckyball is an allotrope of carbon.

To understand the last allotrope of carbon we will discuss, we have to rewind to 1985. That year, a paradigm shift in chemistry occurred when three scientists discovered a new form of natural carbon: C_{60}. The scientists realized that this new form of carbon was extraordinarily stable and highly symmetrical, but they did not know how the 60 carbon atoms were arranged in the molecular structure. It occurred to the chemists working with C_{60} that the number 60 is found in some well-known

natureBOX • Assessing Carbon Footprints

Over the years, food labels have become more and more complex as consumers have demanded more information about what they put into their bodies. Modern grocery shoppers are used to considering several pieces of information about an item, including its price, its nutritional value, and where it came from. But lately, a different kind of information has started appearing on grocery shelves: information about a product's carbon footprint.

The **carbon footprint** of a product is the amount of greenhouse gas emission associated with the production of that item. **Greenhouse gases** are gases that trap heat in Earth's atmosphere and contribute to climate change and global warming (see *The Green Beat,* Chapter One Edition). Greenhouse gases are so called because they trap the sun's energy in Earth's atmosphere in the same way that a greenhouse traps energy from the sun.

The most notorious greenhouse gas is the carbon dioxide molecule, CO_2. This gas, which is released into the atmosphere when fossil fuels are burned, contributes most to heating of our atmosphere because it's released in the largest quantities in the atmosphere. Fossil fuels are carbon-based natural fuels such as coal or gas that were formed from the remains of living organisms; we humans burn them to power our cars and other forms of transportation, to fuel our factories, and to heat our homes and offices.

It makes sense that anything that's directly powered by fossil fuels such as gasoline, methane, or coal has a carbon footprint. But how can a product such as

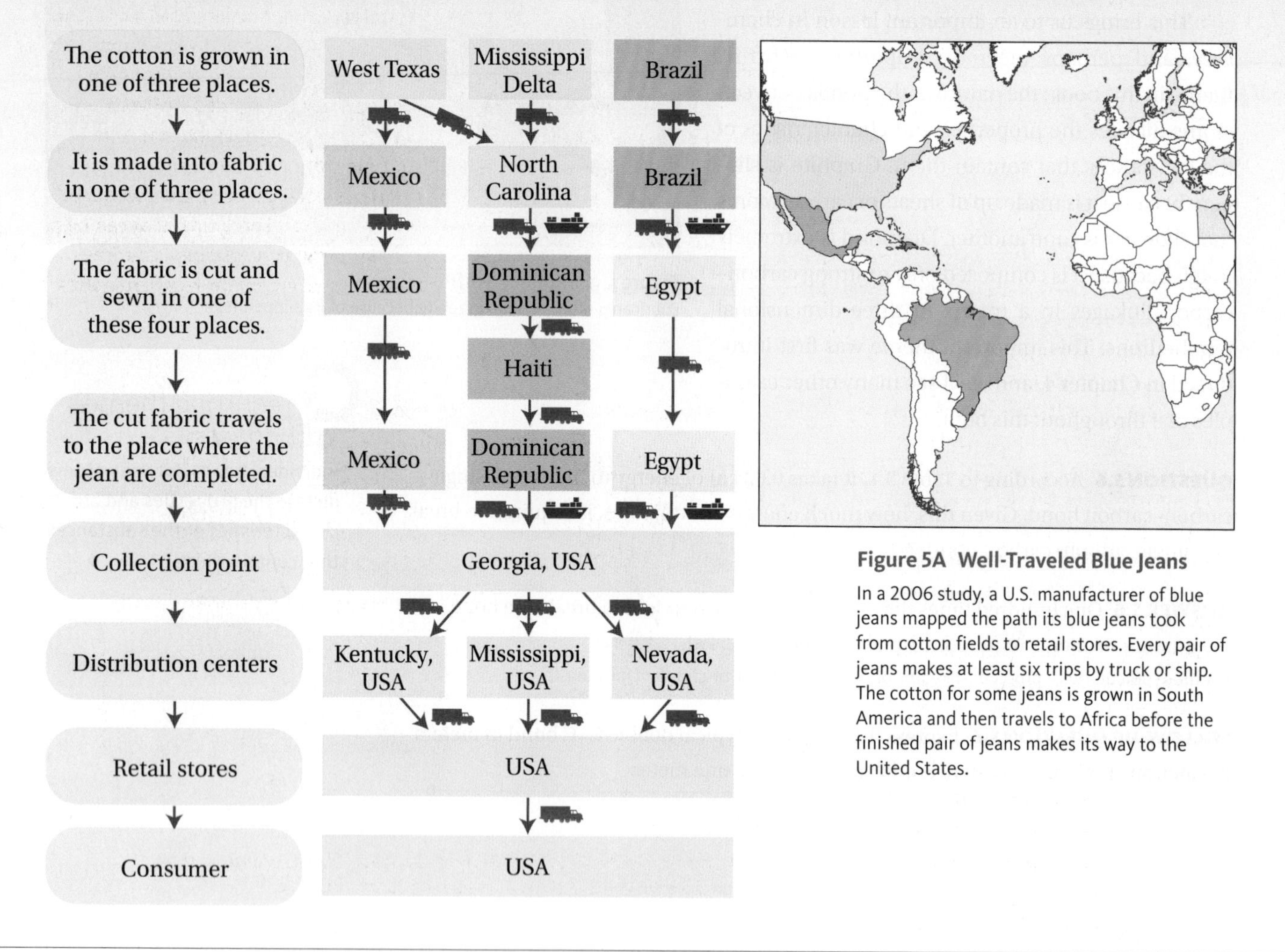

Figure 5A Well-Traveled Blue Jeans

In a 2006 study, a U.S. manufacturer of blue jeans mapped the path its blue jeans took from cotton fields to retail stores. Every pair of jeans makes at least six trips by truck or ship. The cotton for some jeans is grown in South America and then travels to Africa before the finished pair of jeans makes its way to the United States.

flashback

Recall from Chapter 3 that the subscript in a chemical formula refers to how many atoms of that element the compound contains. For example, C_{60} contains 60 carbon atoms.

manufactured products, such as soccer balls (**Figure 5.10a** on p. 140). The familiar shape of a soccer ball is formed from pentagonal rings surrounded by hexagonal rings to make what's referred to as a closed-cage arrangement.

This motif is found both in nature and in manufactured products because it's the only way to arrange 60 vertices in a closed symmetrical structure. (A vertex—the plural is *vertices*—is a meeting point of two lines that form an angle.) Knowing

a loaf of bread have a carbon footprint? Even though the loaf of bread doesn't produce greenhouse gases directly, at some point in that loaf's production and distribution, greenhouse gases inevitably are released.

Perhaps the loaf of bread was made from wheat that was harvested using a gasoline-powered harvester. And, at the bakery, the bread made from that wheat might have been baked in a methane-fired oven. Then the bread could have been put into a plastic bag made in a factory powered by coal. Maybe the bread then took a trip from the factory to a distribution center, and then to a grocery store in a diesel-powered truck. All of these steps use carbon-based fuels and contribute to the bread's carbon footprint.

Estimating the footprint of a product can be tricky. For example, if we want to account for every possible contribution to the carbon footprint of our loaf of bread, we need to widen our scope even further. We can take into account the gas that the harvester operator burned on her way to work. Or we can figure out how much heat was used to maintain an ideal loaf-storage temperature in the distribution center. As you can imagine, the tallying of a carbon footprint can go on and on.

Figure 5A illustrates the manufacturing steps used by a U.S. maker of blue jeans. Starting with this diagram, we can unearth myriad sources of greenhouse gas emissions if we consider every step within each part of the process.

To get a handle on the problem of tallying carbon footprints, a British organization known as the Carbon Trust has taken the lead on calculating carbon footprints for more than 27,000 consumer products ranging from bread to lumber to tires. The organization has developed a series of ground rules that place boundaries on the size of a carbon footprint calculation. When the Carbon Trust performs an official carbon footprint calculation for a product, that number is displayed on its packaging along with the words "working with the Carbon Trust." If the company selling the product demonstrates that it is actively working to reduce the footprint of the product, then the logo reads "reducing with the Carbon Trust" (**Figure 5B**).

Some British consumers look for the Carbon Trust's black foot logo because they want to buy products from companies that care about the environment. And the very process of obtaining a carbon footprint for an item can lead to greener practices, as the following story demonstrates.

The first product to display the Carbon Trust logo was a bag of UK-made potato chips (the British call them crisps). The company that makes them, Walkers, decided to have the carbon footprint of its crisps tallied. By doing this, they learned that the farmers who harvested their potatoes were storing them in buildings that were humidified to increase the mass of the potatoes. Because humid potatoes are heavier than unhumidified potatoes, the farmers made more money because Walkers paid the farmers according to the total mass of the potatoes.

The humidification step turned out to be a significant factor in the carbon footprint of the crisps. Fossil-fuel-based energy was used to humidify the buildings, and the heavier potatoes required more fuel when transported. Upon learning this, Walkers began to pay for the potatoes according to their dry weight, and the farmers stopped humidifying the potatoes. This small change reduced the carbon footprint of the crisps and saved the company money.

Carbon footprints apply to more than just consumer products. For example, the country of Brazil has enlisted the Carbon Trust to help determine its government's carbon footprint. You can even tally your own footprint. An Internet search for "footprint calculator" brings up several programs that can help you figure out your personal contribution to global greenhouse gas emissions.

Courtesy of The Carbon Trust

a

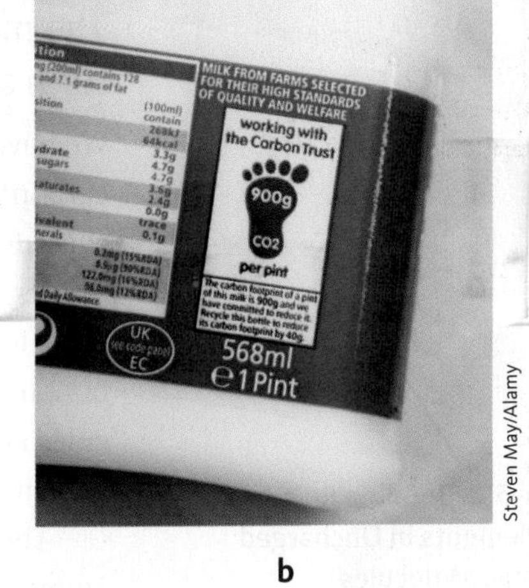

Steven May/Alamy

b

Figure 5B The Carbon Trust Logo

(a) A logo like this is placed on a product whose carbon footprint has been evaluated by the Carbon Trust. (b) This British product sports a Carbon Trust logo.

this structure, the scientists suggested it for their carbon allotrope (**Figure 5.10b**). The structure was confirmed, and the three men took home the 1996 Nobel Prize in Chemistry for their work. They named C_{60} buckminsterfullerene after Buckminster Fuller, the architect who patented the geodesic dome, which looks similar to it (**Figure 5.10c**). This molecule is now affectionately known as the **buckyball**.

Figure 5.10 Buckminsterfullerene Large and Small

(a) A soccer ball's structure is made up of alternating pentagons and hexagons. (b) The geodesic dome, designed by Buckminster Fuller, has a similar structure to a soccer ball. (c) The molecular structure of buckminsterfullerene—better known as the buckyball—is like that of the soccer ball and geodesic dome. Each carbon atom occupies one vertex of the molecule's shape.

Since the discovery of the buckyball, this carbon framework has been shaped into different structures. For example, the *nanotube* is a tube-shaped molecule with the same type of carbon framework that makes up the buckyball.

5.3 Making Sense of Organic Molecules

The most common elements in organic molecules each form a predictable number of bonds.

We have discussed two things that make drawing organic molecules easier: First, carbon always makes four bonds, because it has four valence electrons and must share four more electrons with other atoms to form an octet. Second, hydrogen always makes one bond, because it has one valence electron and needs only one more to form a duet. But organic molecules may contain elements other than carbon and hydrogen. The most common "other" elements found in organic molecules are nitrogen, oxygen, sulfur, and the halogens. These elements are typically present in small numbers in organic molecules.

Table 5.2 Bonding Rules for Elements in Uncharged Organic Molecules

Element	Number of Bonds It Always Forms
H	1
O	2
S	2
N	3
C	4

The term **hydrocarbon** refers to organic molecules that contain only carbon atoms and hydrogen atoms. In most organic molecules, the hydrocarbon framework forms the skeleton of the molecule, and atoms of these other elements—nitrogen, oxygen, sulfur, and the halogens, which we call **heteroatoms**—can be found within or appended to the hydrocarbon skeleton. The heteroatoms are like ornaments on a holiday tree (the hydrocarbon). Like carbon and hydrogen, these heteroatomic elements tend to make the same number of bonds when they appear in organic molecules, as **Table 5.2** shows.

For example, oxygen and nitrogen, when present in uncharged organic molecules, always make two and three bonds, respectively. A look at the Lewis dot diagrams for these atoms shows you why:

$$:\dot{\underset{\cdot}{O}}: \quad \cdot\ddot{\underset{\cdot}{N}}\cdot$$

Oxygen needs to share two electrons to obtain a total of eight electrons. Nitrogen needs to share three.

QUESTION 5.7 Consider the molecular structures of the following organic molecules, all of which contain the elements carbon, oxygen (or sulfur), nitrogen, and hydrogen. Identify molecules that violate the octet or duet rules.

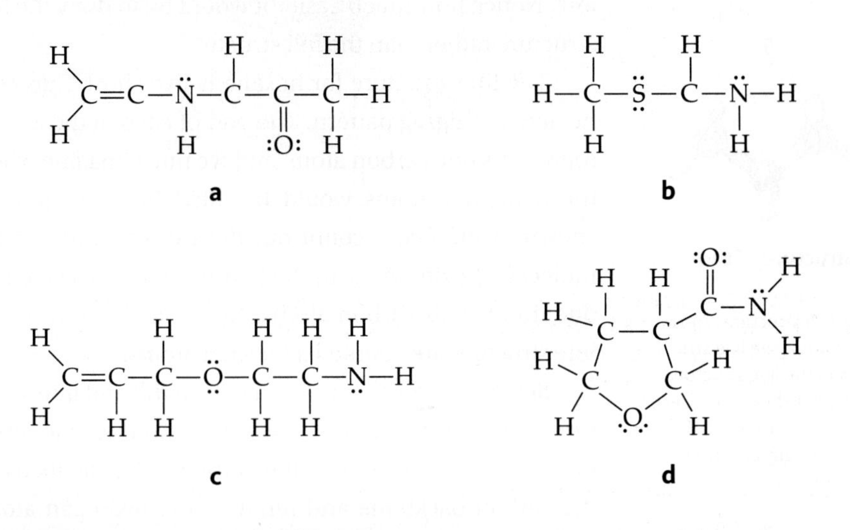

ANSWER 5.7 Molecules (a) and (b) disobey the octet rule, because each has at least one carbon that makes only three bonds. Therefore, these molecules cannot exist. The molecules in (c) and (d) are okay. Each carbon makes 4 bonds, each hydrogen makes 1 bond, each nitrogen makes 3 bonds, and each oxygen makes 2 bonds. Nonbonding pairs are shown on the oxygen and nitrogen atoms, as required.

For more practice with problems like this one, check out end-of-chapter Question 38.

FOLLOW-UP QUESTION 5.7 Write the chemical formula for each of the correct molecular structures shown in Question 5.7.

Chemists use line structures to draw organic molecules.

Some chemists spend a lot of time drawing molecules on paper or with special molecule-drawing computer programs. For example, if a chemist wants to draw the molecule hexane, C_6H_{14}, she can connect six carbon atoms together and then fill in the 14 hydrogen atoms around the carbon backbone. It's clear where to put the hydrogen atoms, because each carbon atom must always make a total of four bonds. She would end up with something like this:

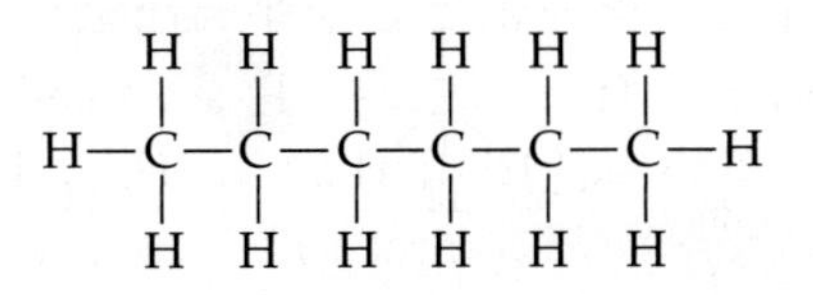

Drawing hexane is pretty straightforward. But when molecules get bigger, it can be tedious to keep writing the letter *C* over and over again, and then filling in all of the hydrogen atoms. For this reason, chemists use a shorthand notation called a **line structure** to depict molecules. In this book, we don't teach you how to draw line structures, but we do show you what they are and what they mean. If you read about a topic related to chemistry—for example, the insert for a prescription drug

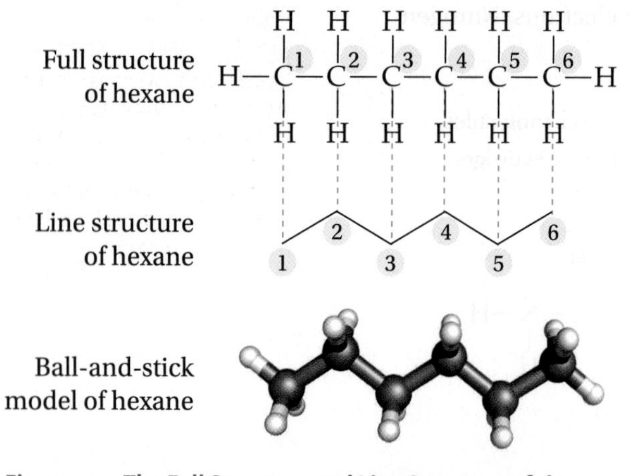

Figure 5.11 The Full Structure and Line Structure of the Hexane Molecule

The top structure is the full structure of hexane. Full structures show every bond and every atom in the molecule. Below it is the line structure for the same molecule. The end of each line in the line structure represents one carbon atom; hydrogen atoms are not shown at all, even though they are present in the molecule. A ball-and-stick model of this molecule (bottom) shows how the backbone makes a zigzag pattern through the molecule.

bottle—then you may encounter line structures, and it's useful to recognize them.

In **Figure 5.11,** we have redrawn the **full structure** of the hexane molecule—the structure that shows every atom and every bond. Below it is the line structure for hexane. Notice how much easier it would be to draw the line structure rather than the full structure.

The line structure for hexane is five short lines connected in a zigzag pattern. The end of each of those lines represents one carbon atom, and we must imagine where the hydrogen atoms would be filled in. The numbers shown in the figure count out the carbon atoms in the molecule. Beginning with the leftmost end, we can count one through six carbon atoms. Note that the ends of the line structure are counted as carbon atoms.

Seeing a large organic molecule in full and line structure forms demonstrates how much simpler line structures are in some cases. A line structure lets us focus on the carbon backbone and removes the hydrogen atoms that clutter a full structure. Beta-carotene is a large organic molecule that is bright orange in color. As its name implies, it is found in carrots. It also gives pumpkins and sweet potatoes their orange color. The full and line structures of beta-carotene are shown together in **Figure 5.12**. The backbone of the beta-carotene molecule is highlighted in yellow in the full structure.

Many, many organic molecules contain a special kind of ring—called a benzene ring—that's drawn in a distinctive way. The benzene ring is a ring of six carbon atoms. The bonds between these atoms alternate around the ring, first a single bond, then a double bond; then single, double, single, double. The full and line structures of a benzene ring are shown in **Figures 5.13a** and **5.13b,** respectively. The lengths of all six bonds in the ring of benzene are actually between a double and single bond, more like a bond and a half. Because the bonds are all the same length,

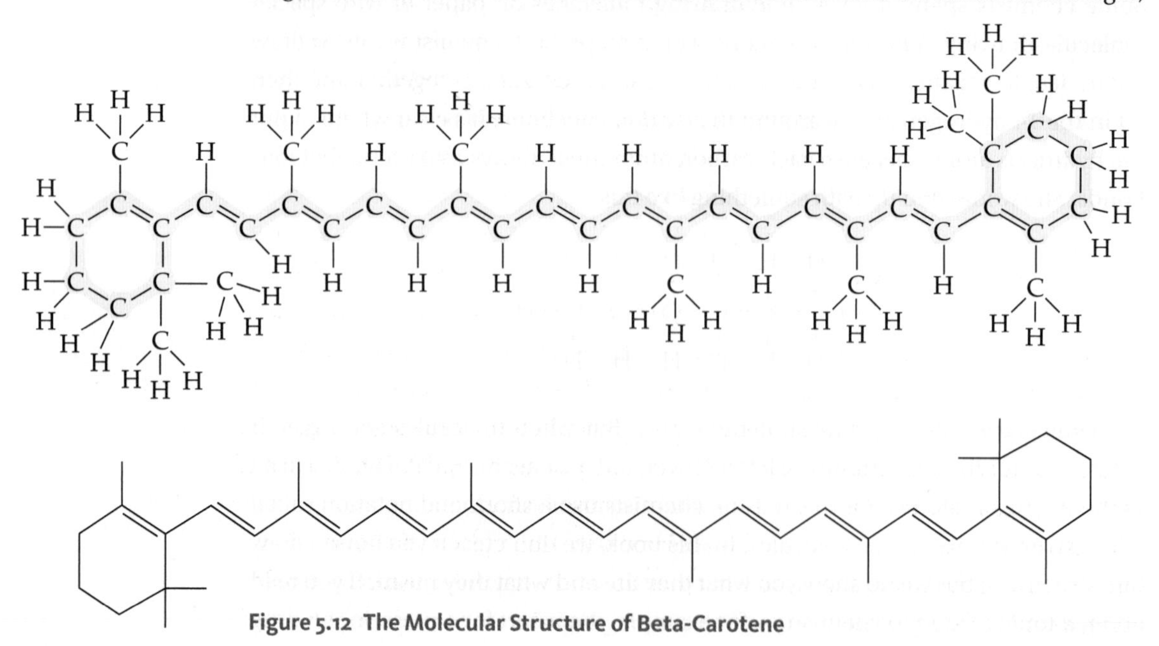

Figure 5.12 The Molecular Structure of Beta-Carotene

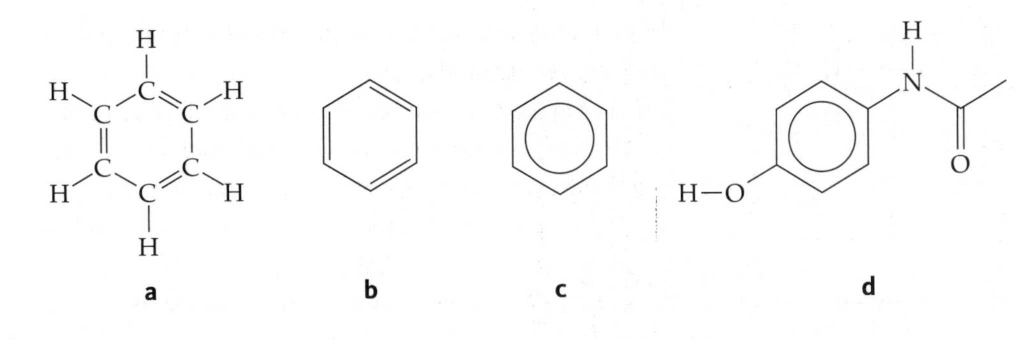

Figure 5.13 Benzene Rings

(a) The full structure of a benzene ring is shown here. In this figure we also show: (b) one way of drawing a line structure of a benzene ring; (c) another way of drawing the line structure of a benzene ring; and (d) the line structure of Tylenol.

the line structure can also be drawn as a hexagon with a circle inside, as shown in **Figure 5.13c**. The pain relief drug acetaminophen (also known as Tylenol) contains a benzene ring and is shown in **Figure 5.13d**.

QUESTION 5.8 How many carbon atoms does the following organic molecule contain?

ANSWER 5.8 It has four carbon atoms.

For more practice with problems like this one, check out end-of-chapter Questions 40 and 42.

FOLLOW-UP QUESTION 5.8 How many carbon atoms does the following organic molecule contain?

5.4 Selected Organic Functional Groups

There is a remarkable diversity of organic molecules.

Organic molecules are incredibly diverse. There are literally millions of known organic molecules, and tens of thousands of different organic molecules exist in the human body alone. Why is there such incredible diversity among them? It's because carbon atoms can make chains, and those chains can be branched or can make rings of various sizes, as we are about to see. To demonstrate how diverse organic molecules are, let's focus on hydrocarbon molecules for a moment.

Figure 5.14 on p. 144 shows all of the ways you can connect six carbon atoms using only single bonds. The figure shows that there are 15 ways to arrange six carbon atoms without a ring or with only a single ring. There are more than 100 ways to arrange 10 carbon atoms connected by single bonds! This number mushrooms into the hundreds and thousands as you increase the number of carbon atoms.

To this point, we have limited ourselves to hydrocarbons that contain only single bonds. If we start to consider molecules with double and triple bonds, then our already staggering number of possible molecules gets even bigger. The number of shapes and sizes of organic molecules is indeed impressive.

Hydrocarbon molecules are the molecules found in fossil fuels, which we introduced in the naturebox in this chapter (fossil fuels are discussed in detail in Chapter 12, "Energy, Power, and Climate Change"). These molecules are the stuff

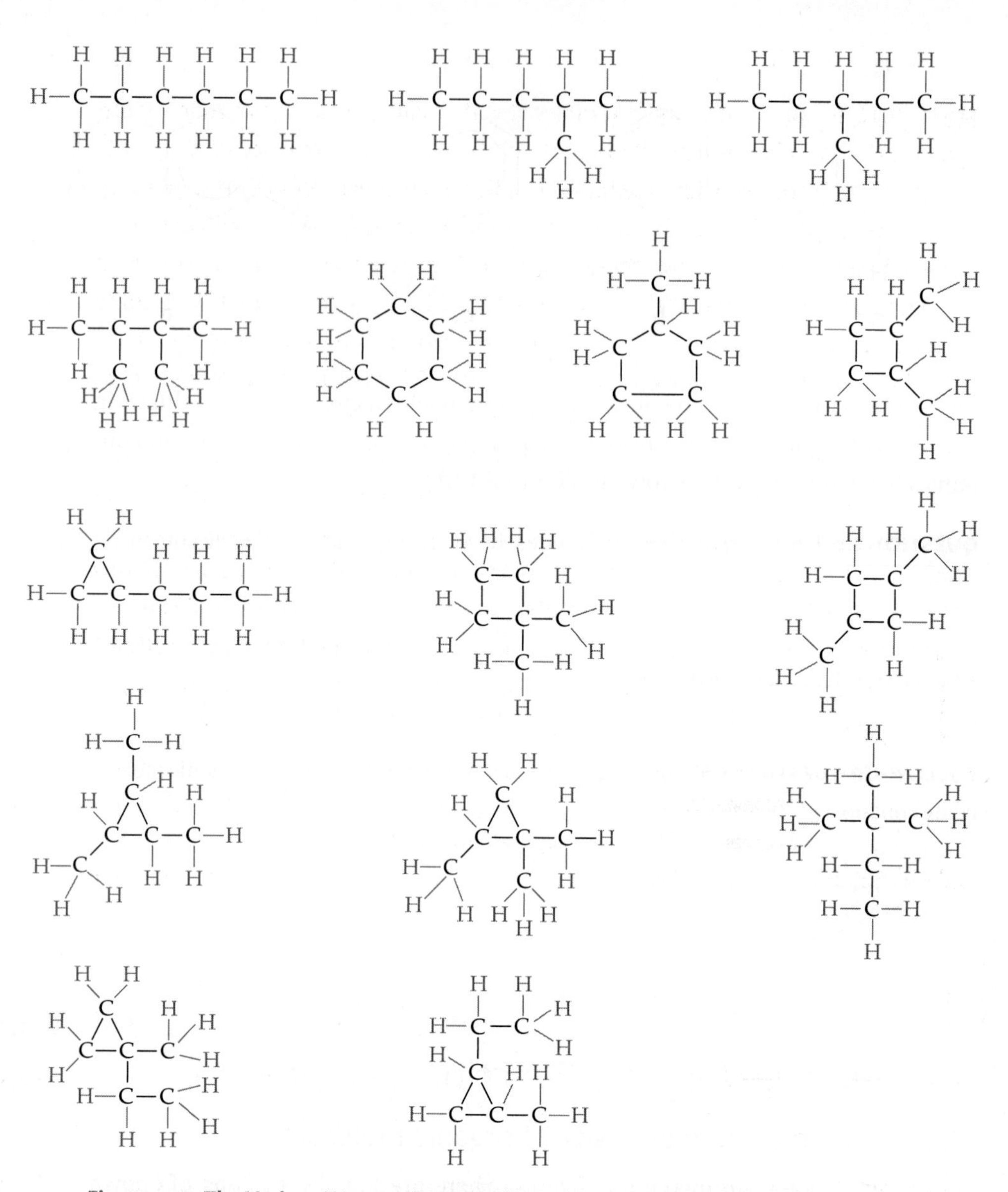

Figure 5.14 The Various Ways to Arrange Six Carbon Atoms to Make Molecules with Only Single Bonds

of fuels such as gasoline, diesel, kerosene, and coal. But as we already know, many organic molecules are not just hydrocarbons, they also can contain other elements such as oxygen, sulfur, nitrogen, and the halogens. These heteroatoms add complexity and functionality to organic molecules, as we will see.

Heteroatoms appear in organic molecules in particular groupings that are found again and again in nature. We refer to these common groupings of heteroatoms as **functional groups,** and this section of the chapter includes several examples that demonstrate what we mean by the term *functional group*. There are too many functional groups to cover all of them, so we have selected just a few important ones as examples. Because they are interesting, we have chosen to illustrate functional groups in molecules that have particularly pungent or obnoxious odors, such as the molecules that give a characteristic smell to bad breath, cooked cabbage, and dead bodies.

Sulfides contain a sulfur atom in a hydrocarbon framework.

Figure 5.15 shows several organic molecules that can exist in the human mouth and contribute to bad breath. Bad breath is caused by the presence of anaerobic bacteria in our mouths. These bacteria must live in an anaerobic environment—an environment that is devoid of oxygen. In our mouths, this environment can be created by the buildup of plaque. When a layer of plaque reaches a thickness of about 1 to 2 millimeters, it's possible for anaerobic bacteria to grow within it. Anaerobic bacteria produce offensive-smelling molecules as waste products, and these generate the odors that we associate with bad breath. This is why dentists urge us to keep our mouths plaque-free and to use antibacterial mouthwash.

The first smelly molecule we will consider is dimethylsulfide. Its molecular structure is shown in Figure 5.15a. The functional group in this molecule is simply the sulfur atom, and it is known as a **sulfide**. If asked, the one word most chemists would use to describe sulfides might be *stinky* or *smelly*. Sulfides represent, hands down, the stinkiest functional group in all of organic chemistry. They contribute to the bad smell of human sweat, garlic, skunk spray, urine, and "skunky-smelling" beer. The dimethylsulfide molecule is one of the culprits in the smell of bad breath as well as the odor of cooked cabbage.

The strong odors of sulfur-containing compounds can be useful. Back in 1937, a natural gas leak caused the New London School in Texas to explode, killing nearly 300 students and teachers. This accident still stands as the deadliest school accident in U.S. history. Afterward, new laws were introduced that required odorant

Dimethyl sulfide is associated with the odor of boiled cabbage.

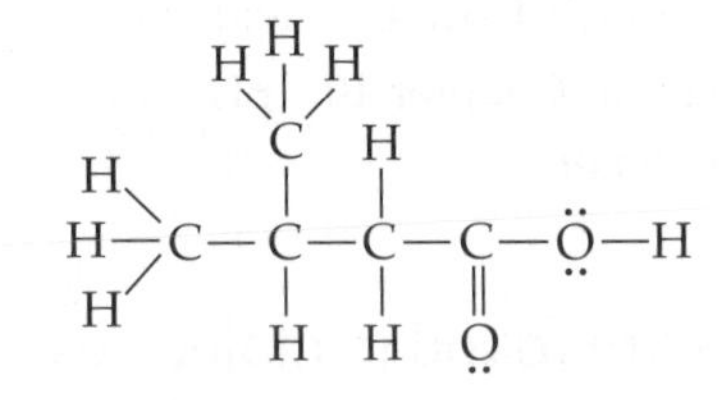

Isovaleric acid is associated with the odor of feet at the end of an active day.

Skatole is associated with the odor of cooked beets.

a

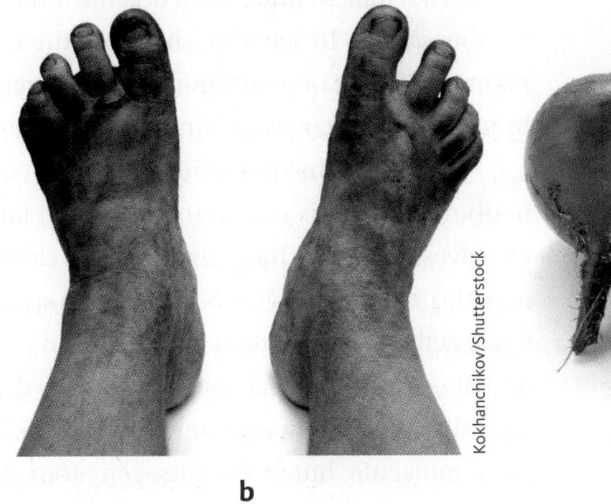

b

c

Figure 5.15 Bad-Breath Molecules

Boiled cabbage (a), smelly feet (b), and cooked beets (c) all have distinctive odors that arise from the same molecules that contribute to bad breath.

molecules (molecules with a strong odor) to be added to natural gas, which is naturally odorless, so people are able to smell a leak. What did the natural gas industry choose as their odorant molecules? They chose sulfur-containing compounds, because humans can smell them even in very small quantities. The mixture of sulfur-containing molecules used to odorize natural gas today includes dimethlysulfide, the sulfide in bad breath.

Carboxylic acids are an important class of organic molecules with many varied uses.

The molecule shown in Figure 5.15b is isovaleric acid. The functional group in this molecule appears at the right end of the molecule. This specific grouping of one carbon atom, two oxygen atoms, and one hydrogen atom is a functional group known as a **carboxylic acid**. Carboxylic acids are an incredibly useful class of organic molecules. They can combine with other molecules to make new, more complex molecules. For example, carboxylic acids are widely used as building blocks in the synthesis of plastics. They are also found in the human body, where they are used to make complex biological molecules.

Carboxylic acids with one to five carbon atoms are powerful chemicals. They can be used as disinfectants and in cleaning products. These small organic molecules have strikingly pungent smells. For example, formic acid, a one-carbon carboxylic acid found in ant venom, has a powerful, penetrating odor. The two-carbon carboxylic acid is known as acetic acid, the molecule that gives vinegar its sour smell. Butanoic acid has four carbon atoms, and it's responsible for the unpleasant smell of rancid butter. Isovaleric acid contributes to bad breath, and it's associated with smelly feet. In Chapter 10, "pH and Acid Rain," we discuss the topic of acids and acidity in detail.

Amines are found in molecules that affect the brain.

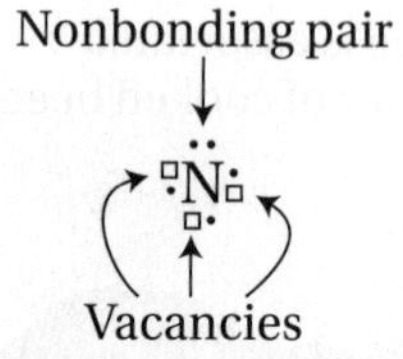

The last functional group we will consider is the amine. The **amine** functional group is a single nitrogen atom in a hydrocarbon framework. Recall from the discussion in Section 5.3 that, in uncharged organic molecules, nitrogen atoms make three bonds to other atoms. To see why, check out the Lewis dot diagram for a nitrogen atom in **Figure 5.16**. The nitrogen atom has three vacancies and, when it fills those vacancies by making three covalent bonds, the nitrogen atom obtains an octet of electrons.

Nonbonding pair

N

Vacancies

Figure 5.16 The Lewis Dot Diagram for Nitrogen

Notice, too, that the nitrogen atom has one nonbonding pair of electrons. This nonbonding pair is important because it takes up space just like a bond does, and this gives the amine functional group a characteristic shape. **Figure 5.17** shows two organic molecules. Figure 5.17a shows a space-filling drawing of a hydrocarbon molecule with a straight chain of five carbon atoms. (The space-filling drawing depicts each atom as a large ball and emphasizes the shape of the molecule's surface rather than the bonds between atoms.) Figure 5.17b shows a space-filling model of the same molecule, but with a nitrogen atom substituted for the middle carbon atom. You can see from these drawings that the surface contour of a molecule depends on the atoms it contains; the shapes of the two molecules are subtly different.

In living systems, molecules often interact with one another when their surfaces match one another, allowing them to join. Thus, a molecule that can join with

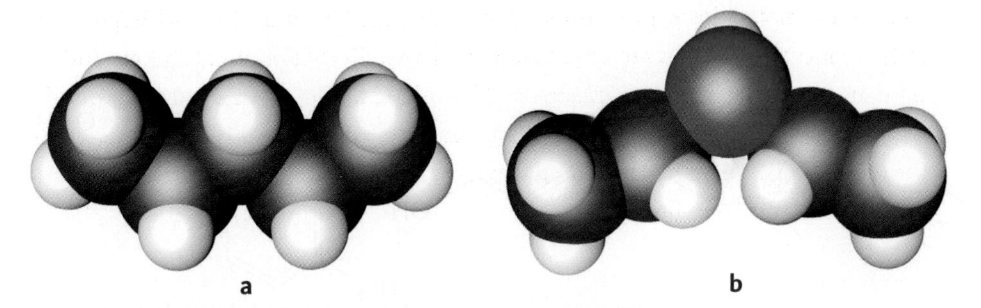

Figure 5.17 The Atoms in a Molecule Dictate the Shape of Its Surface

the molecule in Figure 5.17b may not bind to the molecule in Figure 5.17a, because their surfaces are slightly different.

The framework of any organic molecule is made up of its carbon chain. That carbon chain is often made up of carbon atoms attached to four things. Those carbon atoms each create a tetrahedron-shaped contour. When an amine is inserted into a carbon chain, however, the amine creates a new feature on the surface. That is, the amine group punctuates the hydrocarbon chain with a new and different shape. That shape makes amines the most common functional group found in drug molecules. Why? Because the contour of an amine molecule makes it recognizable to biomolecules in the human body.

> **flashback**
> Recall from Chapter 4 that atoms can obtain an octet of electrons when they share electrons with other atoms through covalent bonding.

Many drugs, both legal and illegal, that affect the brain contain the amine functional group. This list includes amphetamines (such as those found in Adderall), heroin, nicotine, cocaine and crack, LSD, oxycodone, morphine, codeine, novocaine, Valium, and diphenhydramine, the active ingredient in Benadryl. The brain has many ways to recognize amines because many of the natural molecules in the brain, such as adrenaline and dopamine, are also amines. This is why so many drug molecules are amines; they can mimic or replace the brain's natural amines.

Some amines, such as the one shown in Figure 5.15c, have unpleasant odors. Skatole is a molecule that contributes to bad breath, and it's also the molecule that gives cooked beets their characteristic smell. The structures of two other malodorous amines, putrescine and cadaverine, are shown in **Figure 5.18**. As their names imply, these amines give characteristic smells to rotting meat and cadavers, respectively.

After reading this section, you might have the impression that all organic molecules are stinky. But in fact some sulfides, for example, smell pretty good. Perfumes and scented oils, such as oil of banana and oil of wintergreen, are great-smelling

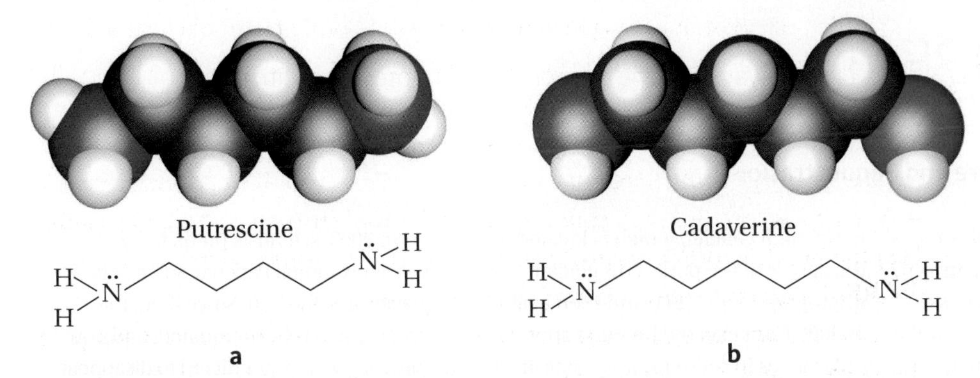

Figure 5.18 Malodorous Molecules

The odors of putrescine and cadaverine give them their evocative names. Each molecule contains two amine functional groups.

Figure 5.19
A Sampling of Great-Smelling Molecules

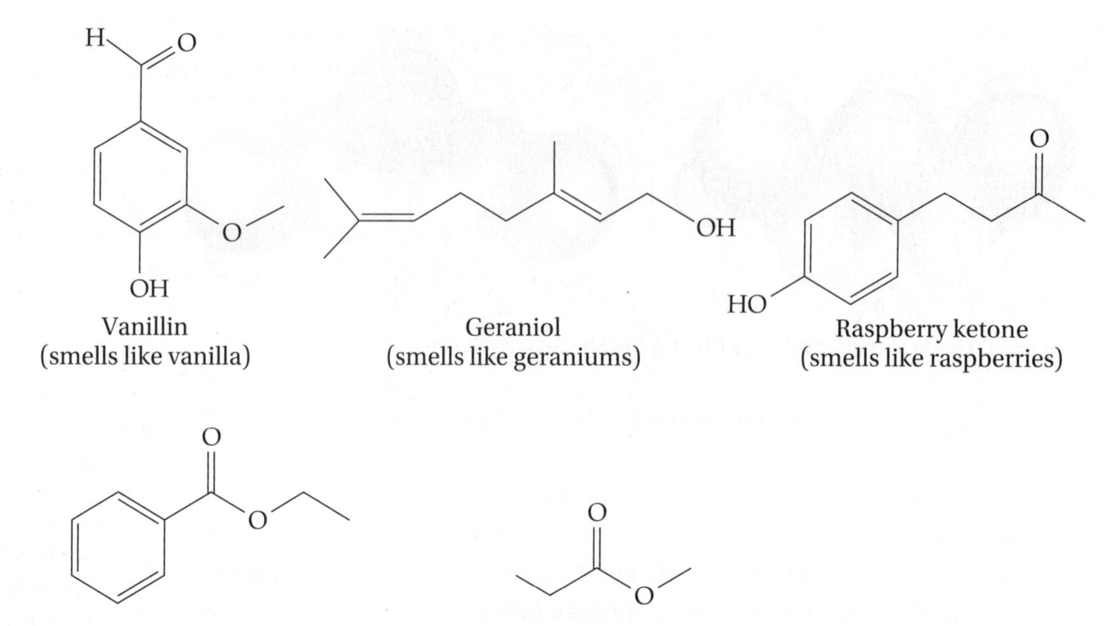

organic molecules that contain other functional groups not covered here. **Figure 5.19** shows several pleasant-smelling organic molecules containing a variety of functional groups.

QUESTION 5.9 Which functional group discussed in the chapter is found in the following organic molecule?

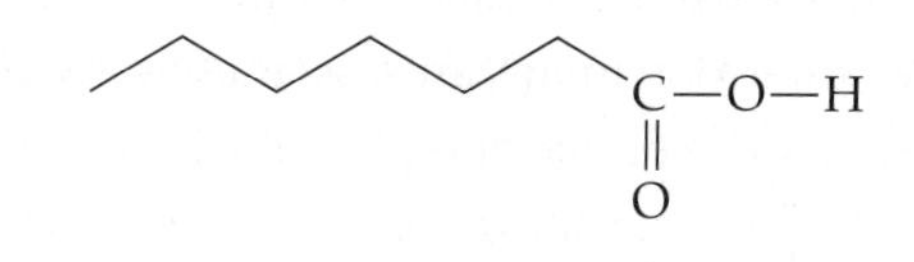

ANSWER 5.9 It contains a carboxylic acid functional group.

For more practice with problems like this one, check out end-of-chapter Questions 45 and 46.

FOLLOW-UP QUESTION 5.9 Which of the following is not the name of a functional group mentioned in this chapter? (a) amine (b) anthracene (c) heptane (d) carboxylic acid

THEgreenBEAT • news about the environment, chapter five edition

TOP STORY Carbon Capture and Sequestration

Glacier National Park, in the northern Rocky Mountains, has been home to many of the most impressive glaciers in the United States. In the year 1850, the park was graced by 150 of these magnificent ice formations; photographs of these glaciers have been taken over the years. **Figure 5.20** shows photographs of Grinnell Glacier, which is located at the center of the park. This glacier was photographed in 1940 and then again in 2006 from exactly the same spot. The differences in these photographs are surprising and unambiguous: the glacier is melting. In fact, of the 150 glaciers in the park in 1850, only 26 remain.

In 2003, scientists predicted that Glacier National Park would be entirely glacier-free by 2030. Since then, the prediction has been updated, and now the last glacier is predicted to disappear around 2020. When that happens, the park will need a new name: Glacier National Park may well be renamed "National Park."

unknown photographer, circa 1940, GNP archives
a
Grinnell Glacier from Overlook, 2006.
Karen Holzer photo, USGS
b

Unknown Photographer/Courtesy of GNP Archives/USGS Repeat Photo Project

Karen Holzer/USGS Repeat Photo Project

Figure 5.20 Grinnell Glacier in 1940 and 2006

The glacier was photographed in 1940 (a) and in 2006 (b) from the same spot.

Why are the glaciers melting? The vast majority of scientists who work in the area of climate change would say that glaciers are melting because of accelerated global warming, which is due largely to the gases that are produced when humans burn fossil fuels such as gasoline. (Recall from Chapter 1 that global warming is the increase in temperature of the Earth's atmosphere caused by heat-trapping gases released through human activities. Climate change refers to the measureable effects of global warming on Earth's climate.)

As we discussed in the naturebox in this chapter, of the gases that are released when fuels are burned, the main culprit is carbon dioxide gas. The chemical formula for carbon dioxide is CO_2, and its molecular structure looks like this:

$$:\ddot{O}::C::\ddot{O}:$$

It's hard to refute the evidence: the nine warmest 12-month periods ever recorded in the United States have all occurred since the year 2000 (**Table 5.3**). In fact, 2012 was the hottest year in U.S. history—during that summer, 63% of the United States was officially in a drought state. And it's not just happening here: Russia suffered a similar heat wave in 2010, and other countries have experienced the same trend. It appears that the planet is getting warmer quite quickly. Chapter 12, "Energy, Power, and Climate Change," is devoted to a discussion of global warming, climate change, and greenhouse gases.

Theories abound on how we can most effectively slow down the warming of the planet (some of them are topics of other *Green Beats* in other chapters). One that's being pursued by several governments of industrialized nations is carbon capture and sequestration (CCS). About 10% of the gas coming out of a coal-fired power plant is carbon dioxide. Rather than releasing it into the atmosphere, CCS theory posits that the CO_2 can be captured and sequestered (that is, stored) somewhere. But where?

Table 5.3 The 10 Warmest Consecutive 12-Month Periods in the Contiguous United States through 2013

Rank	Consecutive 12-Month Period	Temperature Departure (from 20th Century Average)
Warmest	August 2011–July 2012*	+3.37°F
2nd Warmest	July 2011–June 2012*	+3.31°F
3rd Warmest	October 2011–September 2012*	+3.27°F
4th Warmest	January 2012–December 2012*	+3.26°F
	June 2011–May 2012*	+3.26°F
	September 2011–August 2012*	+3.26°F
7th Warmest	December 2011–November 2012*	+3.18°F
8th Warmest	November 2011–October 2012*	+3.17°F
9th Warmest	February 2012–January 2013*	+2.91°F
10th Warmest	May 2011–April 2012*	+2.90°F
11th Warmest	April 2011–March 2012*	+2.68°F
	November 1999–October 2000	+2.68°F

Source: National Oceanic and Atmospheric Administration's National Climatic Data Center.

*Preliminary data

(continued)

Where can we put unwanted carbon dioxide? One idea that has attracted significant funding in recent years is to sequester carbon dioxide underground. For example, it could be stored in depleted underground cavities that formerly contained oil, or in underground deposits of salt or coal. Thus, a coal-fired power plant could pump its carbon dioxide into the ground, and that carbon dioxide would not contribute to warming global temperatures. In theory, that is.

In practice, the sequestration of carbon in underground reservoirs has not gained much momentum, for several reasons. First, it's expensive and it takes energy to pump a gas into the Earth. Second, people are concerned about possible risks from the leakage of carbon dioxide and, on top of that, the potentially deleterious effects of forcing gas into the Earth. It's possible that large-scale underground sequestration of carbon dioxide could trigger an earthquake.

There have been some creative alternatives to pumping carbon dioxide underground. Green Fuel Technologies, a business collaboration between Harvard University and MIT, experimented with using gas-eating algae to sequester carbon dioxide. The organization shut its doors in 2009 after a few frustrating years of business. Luminant, a company in Texas, is successfully making carbon dioxide waste into baking soda and other carbon-containing products that can be recycled and sold to the public. A pioneering Canadian engineer named David Keith has created large, windmill-sized, air-capturing towers that pull in air and remove the carbon dioxide. One thing is clear: carbon dioxide waste is a monumental problem that needs a creative solution that has to work on a grand scale.

Chapter 5 Themes and Questions

recurring themes in chemistry chapter 5

recurring theme 6 The structure of carbon-containing substances is predictable because each carbon atom almost always makes four bonds to other atoms. (p. 128)

recurring theme 2 Things we can see with our eyes contain trillions and trillions of atoms. Atoms are much too small to view with the naked eye. (p. 130)

recurring theme 4 Chemical reactions are nothing more than exchanges or rearrangements of electrons. (p. 131)

recurring theme 7 Individual bonds between atoms dictate the properties and characteristics of the substances that contain them. (p. 137)

A list of the eight recurring themes in this book can be found on p. xv.

ANSWERS TO FOLLOW-UP QUESTIONS

Follow-Up Question 5.1: From smallest to largest: $He < Ne < Ar < Kr < Rn$. *Follow-Up Question 5.2:* tetrahedron. *Follow-Up Question 5.3:* argon. *Follow-Up Question 5.4:* (a) 15 (b) 5 (c) 0. *Follow-Up Question 5.5:* (a) and (b) are fine, but (c) and (d) each have one instance of five bonds to carbon. *Follow-Up Question 5.6:* 1.33×10^{-10} meters is the same as 133 picometers. *Follow-Up Question 5.7:* (c) $C_5H_{11}NO$ (d) $C_5H_9NO_2$. *Follow-Up Question 5.8:* 12 carbon atoms. *Follow-Up Question 5.9:* (b) and (c) are not functional groups.

End-of-Chapter Questions

Questions are grouped according to topic and are color-coded according to difficulty level: **The Basic Questions***, **More Challenging Questions****, and **The Toughest Questions*****.

Carbon and the Periodic Table

***1.** (a) Which is smaller, an atom of iodine or an atom of fluorine?

(b) Which is larger, an atom of tungsten or an atom of chromium?

(c) Which is smaller, an atom of calcium or an atom of beryllium?

***2.** Rank the following atoms in order of increasing atomic diameter: bromine, iodine, chlorine, fluorine, astatine.

****3.** The elements of Group 16 are the chalcogens. What trend do you expect to see in the sizes of the atoms in this family? Do you expect oxygen to have properties representative of the other elements in this family? Why or why not?

****4.** The halogens, found in Group 17 of the periodic table, are common in organic molecules. In general, how many bonds does a halogen make to other atoms?

****5.** On the vertical axis of Figure 5.1, atomic diameter is expressed in picometers. Atomic size is also sometimes expressed in angstroms ($1 \text{Å} = 10^{-10}$ m) and nanometers. Express the atomic diameter of carbon in these units.

*****6.** The elements of Group 18 of the periodic table are *never* found in organic molecules. In your own words, briefly explain why this is the case.

*****7.** The radius and atomic number of each of the noble gases are listed here. Based on the trend for carbon shown in Figure 5.1, can you guess what the radius of helium is?

Element	Atomic Number	Radius (pm)
Helium	2	?
Neon	10	71
Argon	18	98
Krypton	36	112
Xenon	54	131
Radon	86	140

Bonds in Organic Molecules

***8.** Using only the elements carbon and hydrogen, sketch a Lewis dot diagram for any molecule that has only single bonds. Write its chemical formula.

***9.** Using only the elements carbon and hydrogen, sketch a Lewis dot diagram for any molecule that has one double bond and all the rest are single bonds. Write its chemical formula.

***10.** In your own words, describe the three-dimensional shape of a methane molecule.

****11.** Refer to Table 5.1 to answer these questions:

(a) Which is/are the weakest carbon–carbon bond(s) in this molecule?

(b) Which is/are the strongest carbon–carbon bond(s)?

```
    H           H
    |           |
H—C—C≡C—C—H
    |           |
    H           H
```

****12.** The length of the carbon–carbon bonds in the molecule known as benzene is 139 pm. Express this value in the following units:

(a) nanometers (b) micrometers

****13.** Bond energies are most often expressed in either kilojoules or kilocalories (1 kcal = 4.184 kJ). How many kilojoules of energy are contained in a piece of cherry pie that contains 750 kilocalories?

****14.** Consider the molecule shown here.

(a) What is the chemical formula of this molecule?

(b) How many atoms does it contain?

```
    H   H   H   :O:
    |   |   |   ||
H—C—N—C—C—O—H
    |   ..  |       ..
    H       H
```

***15. A molecule contains five carbon atoms and 10 hydrogen atoms. It has only single bonds, and the carbon atoms are connected sequentially with no branches in the chain. Sketch the Lewis dot diagram of this molecule.

Carbon Allotropes

*16. Which allotrope of carbon is the toughest natural substance known?

*17. In your own words, describe the structure of graphite. What is graphene?

*18. True or false? Some elements on the periodic table possess allotropes and some do not.

**19. On average, would you expect the bonds between the carbon atoms within a sheet of graphite to be shorter or longer than those in diamond? Why?

**20. Name the three allotropes of carbon discussed in this chapter. Briefly describe the physical appearance of each allotrope and comment on how its appearance is related to its structure.

**21. Like graphite, buckminsterfullerene is known to have lubricating properties. From an understanding of the molecular structures of these two molecules, speculate on why buckyballs are able to act as lubricants.

**22. Is the way the carbon atoms are connected in graphite (Figure 5.8a) the same as the way they are connected in a buckyball (Figure 5.10c)?

***23. The diagram shown here features a piece of graphite at the top. Identify the various carbon-based substances (a, b, c, and d) shown in the figure.

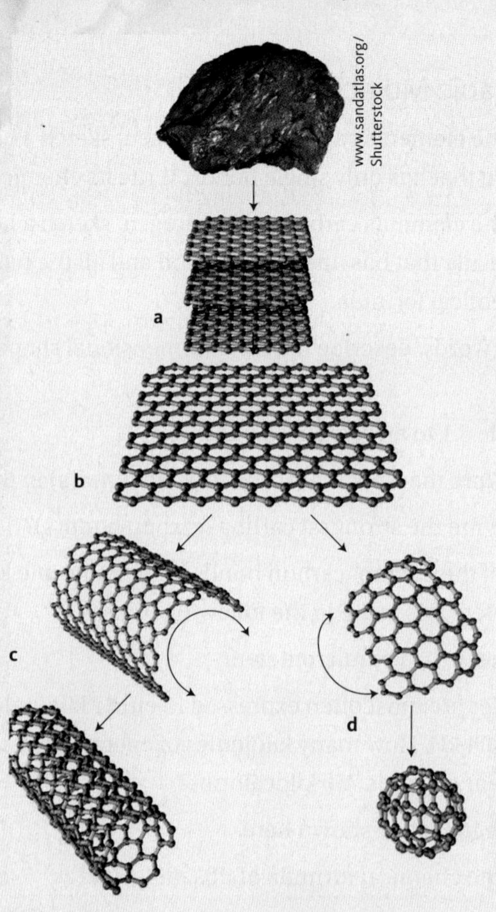

www.sandatlas.org/
Shutterstock

Lewis Dot Diagrams, the Octet Rule, and the Duet Rule

***24.** Which group of elements has a tendency to make two covalent bonds to other atoms? Why?

***25.** Which Period 4 element has a tendency to make one covalent bond to other atoms?

***26.** Draw the Lewis dot diagrams for the following atoms:

(a) F (b) I (c) He (d) S

***27.** In your own words, explain the octet rule and the duet rule. When are these terms used? To which type of atom do they apply?

****28.** Count the total number of valence electrons in each molecule shown here:

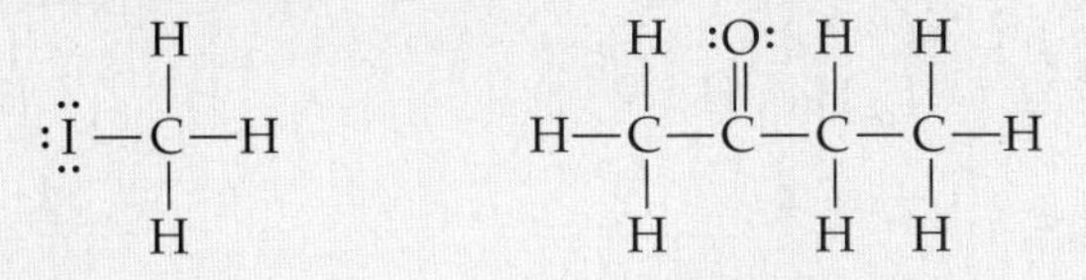

****29.** (a) Draw the Lewis dot diagram for phosphorus.

(b) How many covalent bonds does phosphorus usually form?

(c) What other nonmetal forms the same number of bonds in organic molecules?

****30.** In which of the following molecules do all of the atoms follow the duet rule?

(a) F_2 (b) Cl_2 (c) C_2 (d) H_2

*****31.** In nature, nitrogen gas exists as a simple diatomic molecule, N_2, in which the nitrogen atoms are connected by a triple bond. Sketch the Lewis dot diagram for this molecule. Be sure each nitrogen atom has an octet of electrons.

*****32.** For each of these chemical formulas, draw one molecular structure that gives each atom its correct number of bonds and an octet or duet of electrons. (Hint: In each case, link the carbon atoms in a chain with single, double, or triple bonds, as needed. Attach the remaining atoms to them.) There are several ways to draw each molecule.

(a) $C_5H_{10}Cl_2$ (b) $C_6H_{10}O$ (c) $C_6H_{15}N$

Drawing Organic Molecules

***33.** Which of the molecules shown here obey the octet and duet rules?

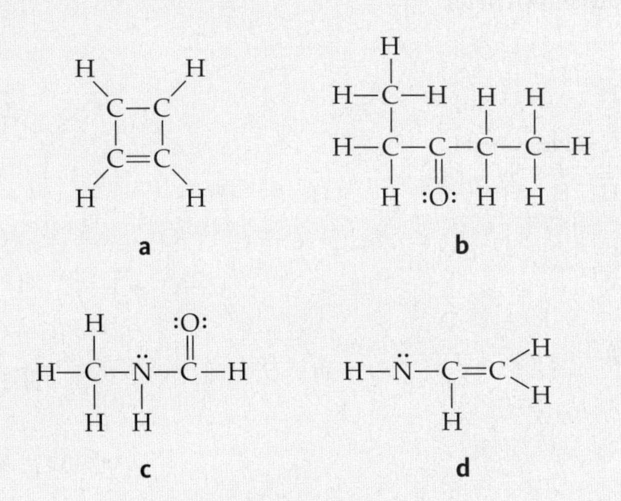

***34.** Which of the molecules shown here obey the octet and duet rules?

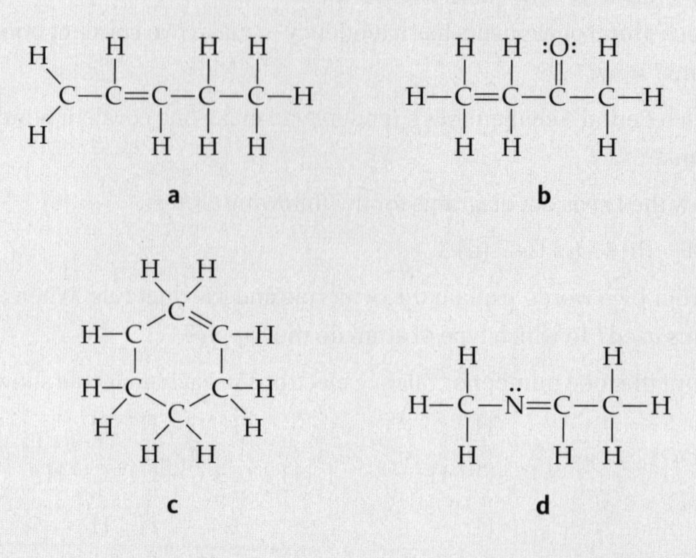

****35.** Cinnamaldehyde and anethole are both sweet-smelling organic compounds used in flavorings and perfumes. Their structures are shown here. For each compound, write the chemical formula.

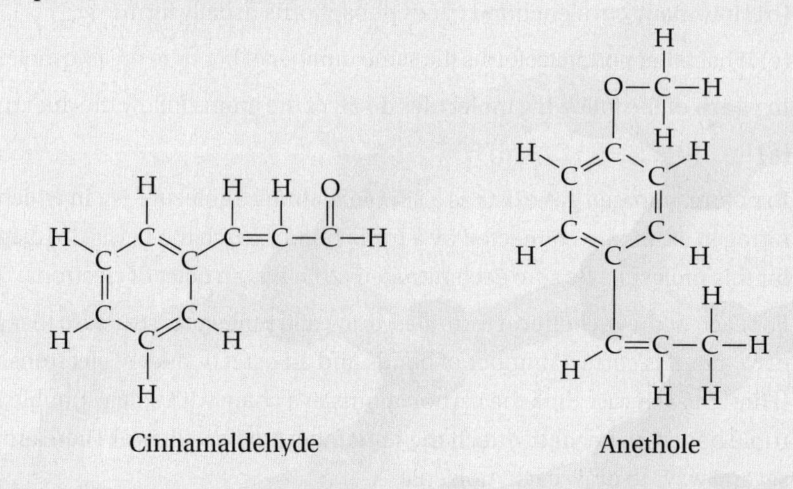

Cinnamaldehyde   Anethole

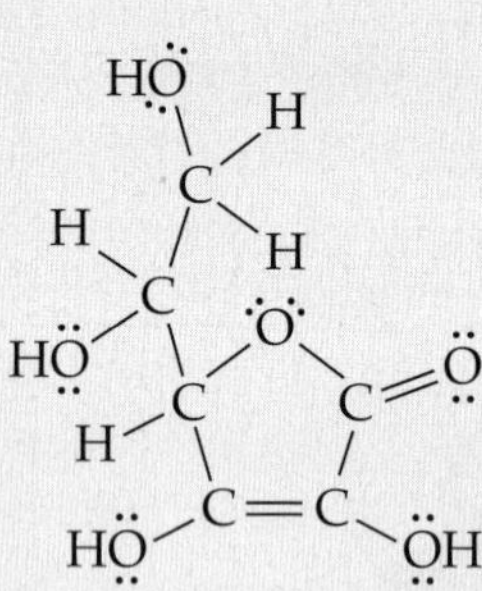

Ascorbic acid (vitamin C)

****36.** Ascorbic acid, more commonly known as vitamin C, is shown here. It is important for the maintenance of connective tissues in the human body. Write its chemical formula.

*****37.** Write chemical formulas for fluoxetine (marketed as Prozac) and cocaine, which are shown here.

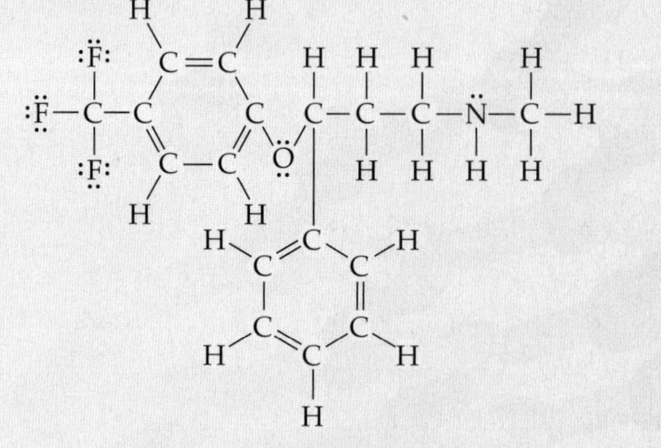

Prozac

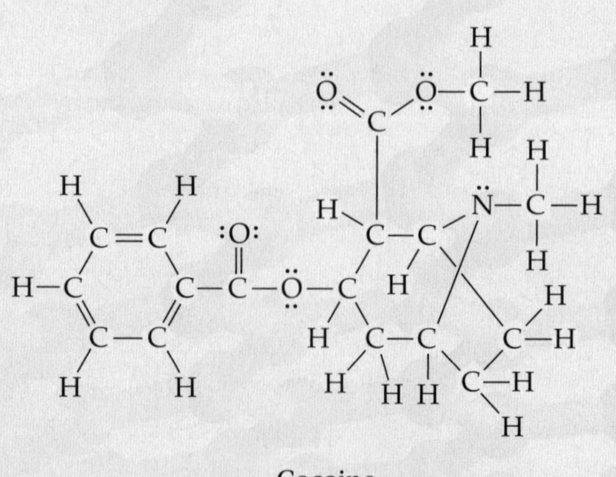

Cocaine

***38. Identify the mistakes in the two molecular structures shown here.

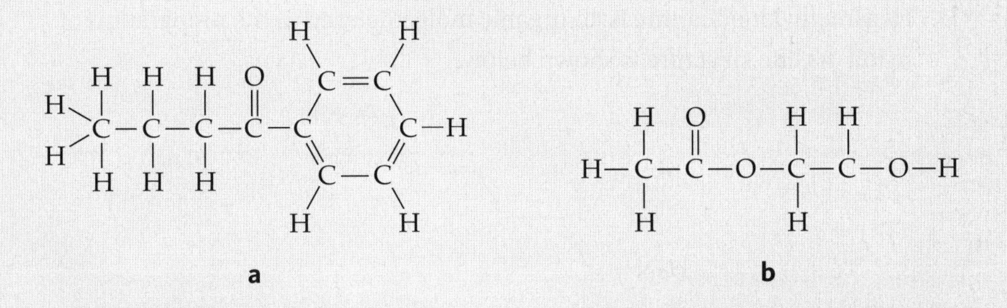

Line Structures of Organic Molecules

*39. In your own words, explain why line structures rather than full structures are often used to depict organic molecules.

*40. How many carbon atoms are there in the line structure shown here?

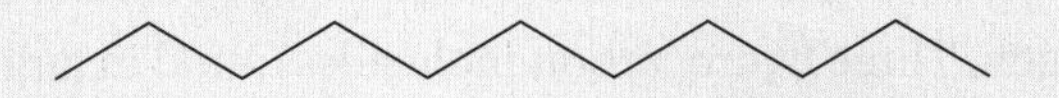

**41. Sketch a benzene ring as a full structure and as a line structure.

**42. How many carbon atoms are there in the line structure shown here?

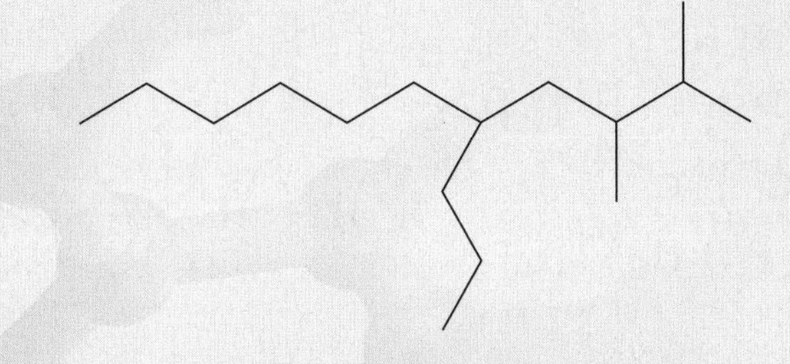

**43. In the molecular structure shown here, what does the ring inside the hexagon represent?

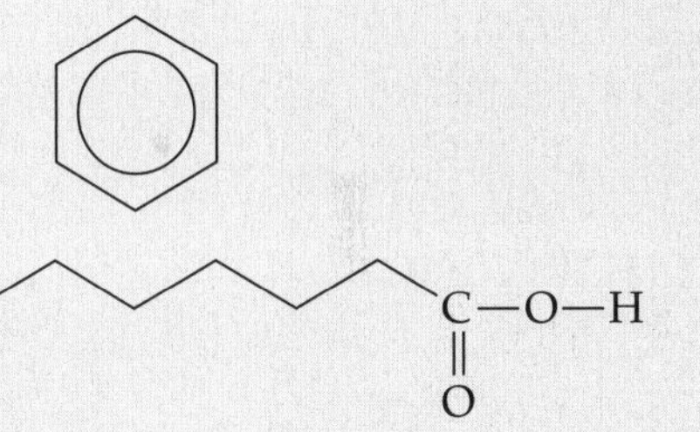

***44. Write a chemical formula for the organic molecule shown here. Which functional group does it contain? (Hint: Draw its full structure first!)

Functional Groups in Organic Molecules

*45. Which functional group discussed in the chapter can be found in the organic molecule shown here?

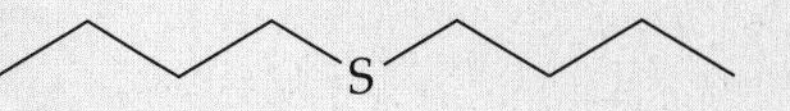

*46. Which functional group discussed in the chapter can be found in the organic molecule shown here?

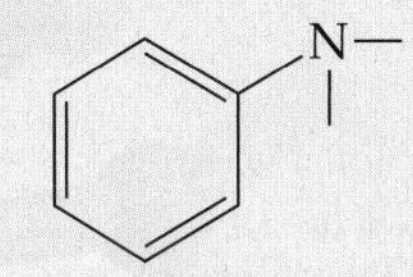

*47. In your own words, explain what is meant by the term *functional group.* In general, are organic molecules composed mostly of a hydrocarbon framework or of functional groups?

**48. Which amine, putrescine or cadaverine, do you think has the greatest mass per molecule? Both molecules are shown in Figure 5.18.

**49. In your own words, explain how an amine functional group changes the contour of a molecule's surface.

***50. Return to Figure 5.19 and write chemical formulas for all of the sweet-smelling molecules in that figure.

Integrative Questions

****51.** Hexamethylenediamine is an organic molecule used in the preparation of nylon. Its line structure is shown below.

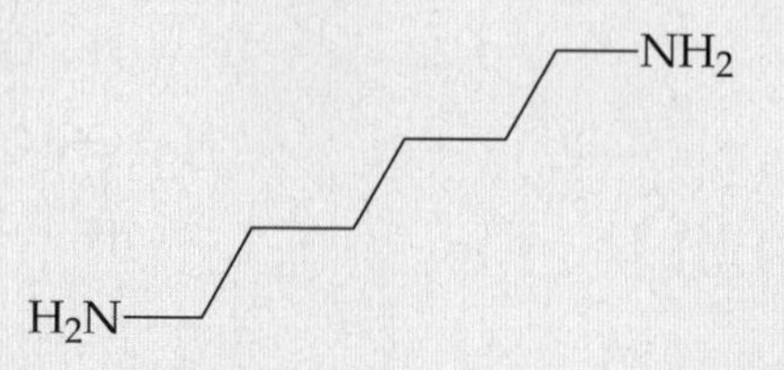

(a) What organic functional group does this molecule contain?

(b) Based on your reading from this chapter, can you predict what unpleasant characteristic this molecule might have?

(c) Draw out the full structure of this molecule. What is its chemical formula?

****52.** Gasoline is a complex mixture that contains mostly hydrocarbon molecules, which are organic molecules that include only the elements carbon and hydrogen. Three hydrocarbon molecules found in large quantities in gasoline—isooctane, butane, and 3-methyltoluene—are shown below from left to right, respectively.

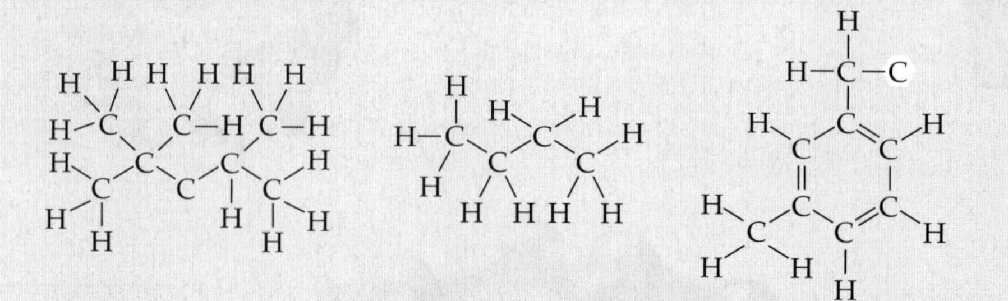

(a) Draw a line structure for each of these molecules.

(b) Table 5.2 tells you how many bonds are always formed by various elements found in organic molecules. Do the atoms shown in these three molecules obey these rules?

(c) Write a chemical formula for each of these molecules.

(d) Do these molecules contain any heteroatoms? If so, what are they?

*****53.** Diclofenac is used to treat arthritis. It is an anti-inflammation drug that reduces joint stiffness and pain. Its line structure is shown left.

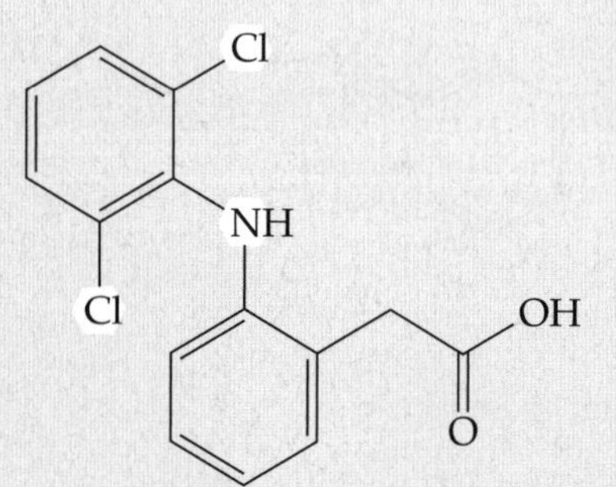

(a) Draw a full structure for diclofenac from the line structure.

(b) Write a chemical formula for diclofenac.

(c) Table 5.2 tells you how many bonds are always formed by various elements found in organic molecules. Do the atoms shown in diclofenac obey these rules? (Note that chlorine does not appear in Table 5.2. All halogens, including chlorine, make one bond in uncharged organic molecules.)

(d) Diclofenac contains one C=O double bond and one C—O single bond. Which one do you expect to be shorter and stronger? Why?

***54. Nifedipine is a drug used to relax blood vessels and reduce the effects of heart disease. Its line structure is shown below.

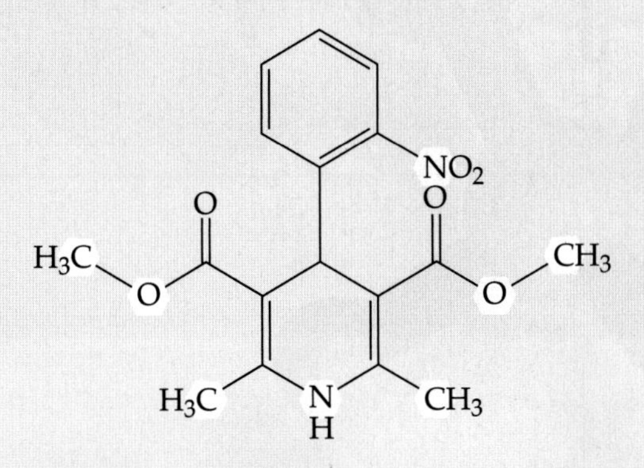

(a) Draw a full structure for nifedipine from the line structure.

(b) Write a chemical formula for nifedipine.

(c) Table 5.2 tells you how many bonds are always formed by various elements found in organic molecules. Do the atoms shown in nifedipine obey these rules?

(d) How many carbon—carbon single bonds does nifedipine contain? How many carbon—carbon double bonds does nifedipine contain? Which of the two types of carbon—carbon bonds do you expect to be shorter and stronger? Why?

55. INTERNET RESEARCH QUESTION

Carbon capture facilities are being built worldwide. Go to your Internet search engine and type in "carbon capture facility map". Try to answer these questions about the facility nearest to your location.

(a) What type of plant is it, and how does that type of plant operate?

(b) MTPA is the number of metric tons of carbon dioxide collected per year. Can you locate the MTPA for the location nearest you? This value typically varies from 0.3 to 5.0 MTPA. Where does the facility nearest you fall in this range?

(c) Perform a follow-up search using the name of the specific facility nearest you and summarize the work being done there. Is carbon dioxide stored at that facility, or is it transferred to other locations via pipeline? If it is stored, where is it stored?

Air

A Study of the Gases in Our Atmosphere

CHAPTER

6

March 20, 1995, promised to be a typical spring day in Tokyo as people left their homes to go to work. What they did not know was that members of the Aum Shinrikyo terrorist cult were en route to various points along the Tokyo subway system. At the height of the morning rush hour, members of this group punctured containers of sarin, a deadly nerve agent that has no color or odor. The gas spread through the air in several crowded subway cars.

When sarin changes from a liquid to a gas, which it does very readily, the gas can be absorbed through the skin or the lungs and enter a person's bloodstream. Sarin quickly disrupts the normal functioning of the nervous system, causing paralysis, convulsions, loss of vision, and violent headaches. In doses greater than 0.5 milligrams, exposure is usually fatal. During the Tokyo subway attack, many of the people who were exposed to near-fatal levels of sarin escaped death but suffered permanent neurological damage. Those exposed to smaller quantities of sarin recovered completely. All told, the attack resulted in the death of 12 people; more than 5000 others were injured. The fate of any one person riding the Tokyo subway on that horrifying day depended on the behavior of the gases around that person.

AP Images/Ju wen/Imaginechina

Our bodies are surrounded by gases at all times. Clean, dry air is colorless and odorless and contains about 21% oxygen, 78% nitrogen, and small amounts of other benign gaseous atoms and molecules (**Table 6.1**). In the same way that we would not eat contaminated food or drink contaminated water, we don't want to see or smell the air that we breathe. Gases we can see or smell may be cause for alarm. Gases we can see may be laden with undesirable specks of airborne pollution (see the photo of people cycling through polluted air in Tiananmen Square in Beijing, China). Gases we can smell contain molecules that trigger the olfactory nerve. This nerve transmits odor information, which may include signals of possible contamination, from the nose to the brain.

Table 6.1 Composition of Clean Air at Sea Level

Atoms or Molecules	Percentage (%)
Nitrogen molecules, N_2	78.1
Oxygen molecules, O_2	20.9
Argon atoms, Ar	0.934
Carbon dioxide molecules, CO_2	0.036
Trace gases	<0.03

In this chapter we explore gases, such as air, the gas that constantly surrounds us. To learn about air, though, we must first understand how gases behave. We'll do this by learning some of the rules we expect gases to follow.

6.1 The Nature of Gases

Of the three phases of matter, gases are the simplest.

Of the three phases of matter, gases may be the simplest. They are simple because, unlike liquids and solids, they tend to be completely homogeneous: except under special circumstances, gases mix thoroughly with one another. For example, although all the gases listed in Table 6.1 are present in clean air, clean air is completely homogeneous because it's the nature of all gases to blend together.

> **flashback**
> Recall from Section 3.5 that there are three phases of matter: gas, liquid, and solid.

Gases are also simple because individual atoms or molecules within a gas typically do not interact with one another. Sure, they collide with one another now and then, but for the purposes of our discussion in this chapter, we can assume that the atoms or molecules of a gas do not change when they collide. In other words, they do not undergo any chemical reactions. This means it's safe to assume that all of the atoms or molecules in a gas act the same way, no matter what elements they contain. For this reason, we can think of any of the atoms or molecules of a gas as a generic "particle of a gas." These gas particles behave like tiny billiard balls that collide and bounce.

Although they are homogeneous, and particles of a gas do not interact with one another, gases are high maintenance. **Figure 6.1** highlights the differences between the three phases of matter. One important difference between liquids, solids, and

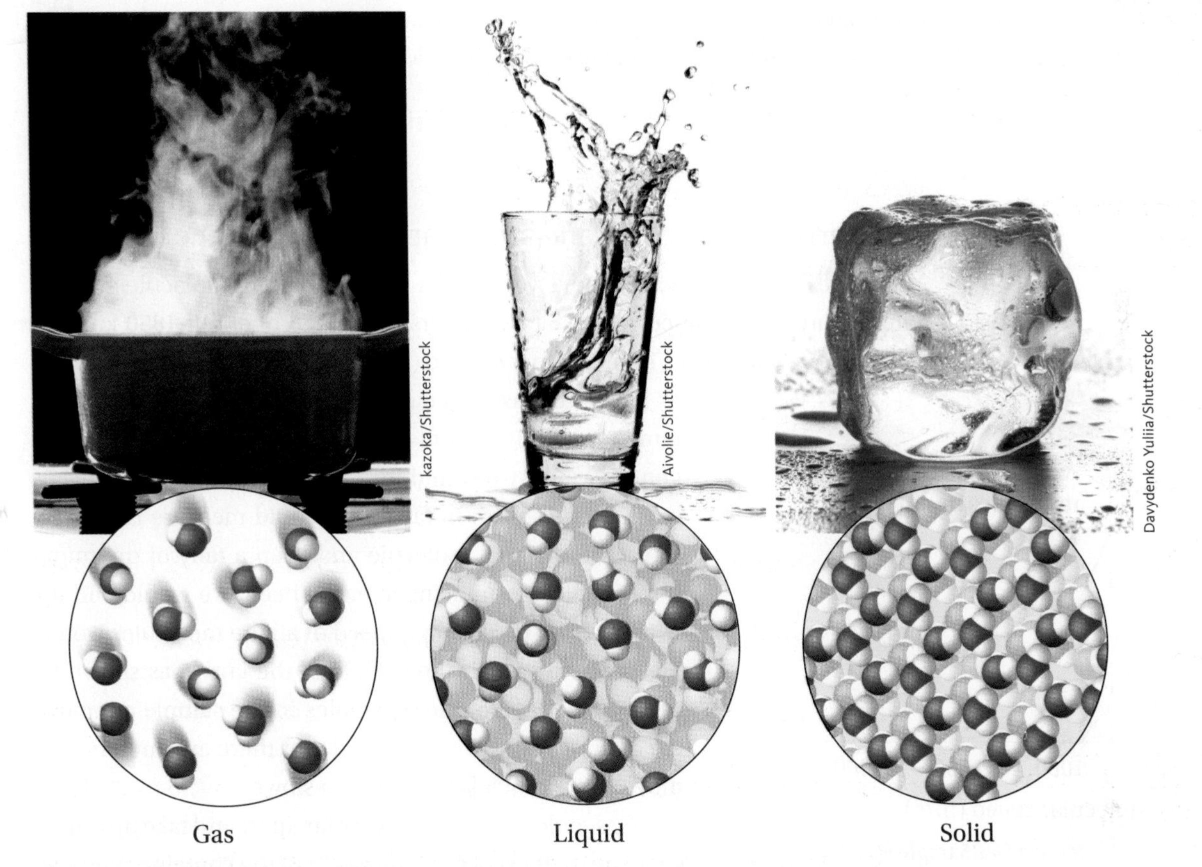

Figure 6.1 The Three Phases of Matter

(a) In a gas, the particles are far apart from one another and fill all the available space in the container holding them. (b) In a liquid, molecules are close together but move relative to one another. The liquid conforms to the shape of the container but does not fill the entire container. (c) In a solid, molecules are tightly packed in a regular array and move very little relative to each other. The solid retains its own shape regardless of the shape of the container holding it.

gases is how they must be stored. We can leave a sample of most solids and liquids sitting out in the open and not worry about losing any of it (except small amounts that are lost when the liquid evaporates or a solid sublimes—that is, turns from a solid directly into a gas). That's because, within solids and liquids, there are forces that hold together individual units. With gases, though, the forces of attraction are not strong enough to keep the individual gas molecules together. That is why we store most gases in airtight containers to prevent leakage.

Think about gas-filled things we see or use every day: a tire, a pipe carrying natural gas into a building, a neon sign, an unopened can of soda. In all these cases, the container must be leakproof. If a leak develops, the product or device ceases to do its job or ceases to be useful. Sometimes this is no big deal, as with a burst soda can. But when the gas in the container has the potential for harm, as in the case of highly flammable natural gas, escaping gas can lead to an explosion.

QUESTION 6.1 Imagine you have a glass containing ice cubes and some water. If the glass is tightly covered with a lid, how many phases of water exist in the glass?

ANSWER 6.1 All three phases. Liquid water is mixed with ice (solid water), and water vapor (gaseous water) exists in the space above the liquid surface.

For more practice with problems like this one, check out end-of-chapter Question 3.

FOLLOW-UP QUESTION 6.1 Identify the phase of matter described in each example. (a) the phase that conforms to the shape of its container and has a surface (b) the phase that is always homogenous (c) the phase that is least likely to disappear from an open container if left undisturbed.

Gas particles move fast and are far apart from one another.

All gas atoms and molecules move very fast. A typical nitrogen molecule at room temperature, for instance, moves at the supersonic speed of about 1850 kilometers per hour (km/h) or 514 meters per second (about 1150 miles per hour), nearly as fast as the fastest jet airplanes. If we could point a radar gun at the individual molecules in a sample of pure nitrogen, however, we would find that they are not all moving at the same speed. In fact, if we could measure the speed of each individual molecule and keep a tally of the number of molecules moving at each speed, we would end up with a curve. The average speed of all the molecules would be somewhere near the middle of the curve, as shown in **Figure 6.2**. Most of the molecules in the sample are moving at or near the average speed, but there are outliers that move either much faster or much slower.

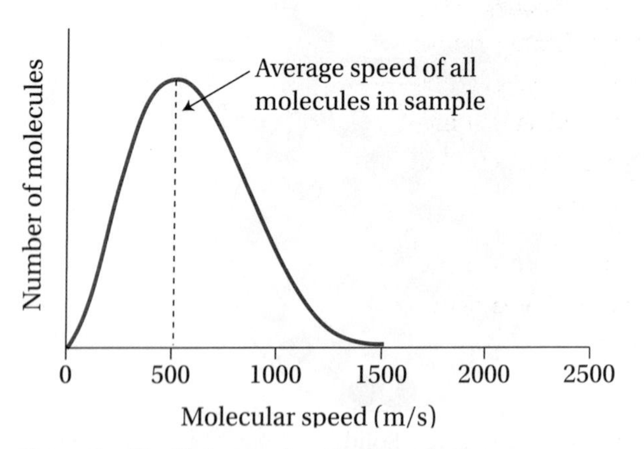

Figure 6.2 The Molecules in a Gas Sample Move at Various Speeds

This graph shows the speed of all the molecules in a sample of nitrogen gas at 27°C. The speeds are distributed in a curve because some molecules move at speeds lower than the average speed of all the molecules in the sample while other molecules move at speeds higher than the average speed.

Gas molecules are also very far apart and take up only a small percentage of the total space of the container holding them. Even so, it's inevitable for them to come into contact with one another occasionally. The collisions that occur in a gas are assumed to be completely elastic: The molecules simply bounce off each other. We can think of the molecules in a gas as tiny, high-speed billiard balls that are in continuous, random motion through space.

The random motion of gas molecules leads to some interesting effects. For instance, although it's easy to see your brother enter the house after cleaning fish in the backyard, it takes a while before you smell him, because the smelly gas particles he picked up from the fish do not travel directly from his body to your nose. Rather, they take a zigzag path across the room, all the while colliding with other gas particles as well as walls and tables and lamps (**Figure 6.3**).

double_p/iStock/Getty Images

Figure 6.3 A Real Stinker
The molecules associated with strong odors eventually make their way to your nose.

Fast-moving gas particles mix together quickly and completely.

Gas molecules move into every space available and completely fill any container they occupy. Imagine we introduce a new molecule to a population of gas molecules speeding around in a container. The new molecule swiftly becomes integrated and assimilated into the existing mixture of fast-moving molecules. This process, known as **diffusion,** is analogous to a busy highway where cars entering the highway quickly reach the same speed as the other moving cars. Once a new car enters the highway, it becomes part of the traffic flow, where all the cars are moving at slightly different individual speeds but at a similar average speed.

The assimilation of one gas into another has important practical consequences because the air we breathe naturally mixes with other gases quickly and completely. As described at the start of this chapter, the canisters of sarin used in the Tokyo subway released poisonous gas into the air. Just like cars merging into fast-moving traffic, the deadly molecules became mixed in with the rest of the molecules in the air breathed by thousands of commuters. Evidence gathered after the attack showed that people closest to the canisters had the most severe symptoms of nerve gas poisoning. This finding is reasonable because, as the poison gas molecules moved farther from a canister, they became farther apart from one another as they diffused into the air and became diluted. A passenger farther from a canister therefore inhaled air containing fewer sarin molecules, and a passenger closer to the canister inhaled more. The closer passenger would reach the fatal limit of sarin with fewer breaths.

QUESTION 6.2 Which of these statements about gases is untrue? (a) Gas particles move fast. (b) In a sample of a gas, all gas particles move at the same speed. (c) Gas particles mix completely with each other. (d) Gas particles bounce off each other and off container walls.

ANSWER 6.2 (b) In a sample of gas, there is a distribution of speeds for gas particles, as shown in Figure 6.2.

For more practice with problems like this one, check out end-of-chapter Question 2.

FOLLOW-UP QUESTION 6.2 True or false? When gases diffuse, gas molecules do not collide with walls or with other gas particles.

6.2 Pressure

The pressure that a gas exerts is related to the collisions the gas particles make with their container.

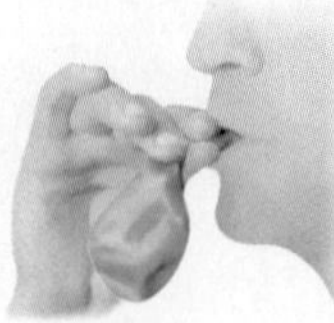

Zero breaths

Two breaths

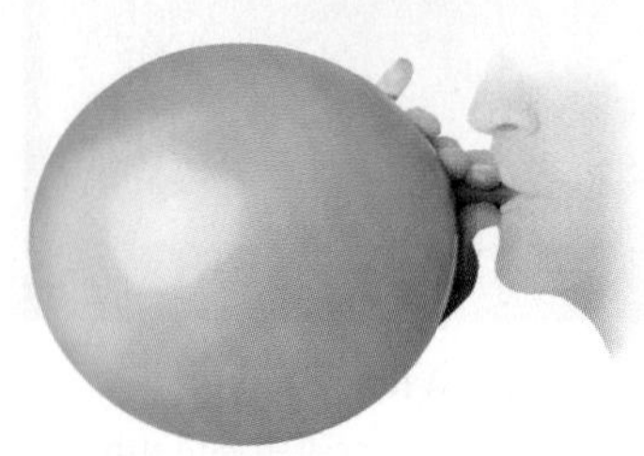

Four breaths

Figure 6.4 Three Balloons

Consider a simple gas-holding device: a balloon. Suppose that we begin with three identical, uninflated balloons. If we take the first uninflated balloon and tie off the end of it, the balloon contains only the air that was in the room before we tied it. Next, we take the second uninflated balloon and blow two deep breaths into it and tie it off. Finally, we take the third uninflated balloon and blow into it until we think it might pop. Imagine that this process takes four breaths. Our three balloons look very different now: one is flaccid, one is partly filled, and the third is completely filled, as shown in **Figure 6.4**.

What makes these three balloons different? They are different because they all contain different amounts of gas. (Remember, we can ignore the fact that the gas we blew into the balloon is a different mixture of gases than the gases in the room. We consider all gas particles to be equal.) The first balloon contains the fewest gas particles, the third balloon contains the most gas particles, and the second balloon contains some intermediate number of particles. So how do gas particles cause a balloon to inflate?

Inside the second balloon, which is now partially inflated, the gas particles that we forced in with the strength of our lungs are moving very fast, as gas particles tend to do. But these gas particles are not free to roam anywhere; they encounter the balloon's inside surfaces constantly. And every time a particle of gas hits the balloon's inside surface, it pushes it outward very slightly and then bounces away to the inside of the balloon. Quite soon, each particle is colliding with the surface again, then again, and again. Billions and billions of particles inside the balloon are behaving like this, and they're all pushing against the inside surfaces of the balloon over and over again. This is what inflates the second balloon and gives it some shape.

The third balloon is a more extreme example of this phenomenon. It contains even more particles, and all of those particles, together, are pushing even harder on the inside surfaces of the balloon. Thus, the third balloon is fully inflated, as illustrated in **Figure 6.5**. Eventually, if we blow enough gas particles into a balloon, the force of the particles pushing on the interior of the balloon is too much. The balloon material is not strong enough to withstand the force, and it pops!

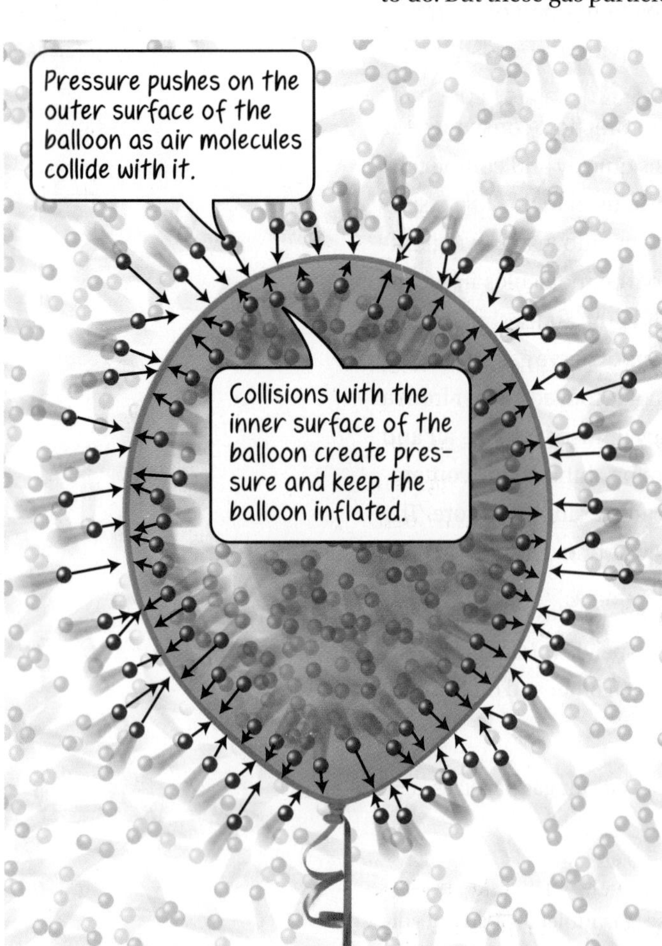
Pressure pushes on the outer surface of the balloon as air molecules collide with it.
Collisions with the inner surface of the balloon create pressure and keep the balloon inflated.

Figure 6.5 Why a Balloon Inflates

This balloon is inflated because the gas particles within it push against the inside walls whenever particles collide with them. The pressure inside the balloon pushes against the pressure of the atmosphere on the outside of the balloon.

Figure 6.6 An Uninflated Tire versus an Inflated Tire

(a) A flat tire contains fewer gas particles than a full tire. As a result, less pressure is pushing against the inside tire walls. (b) A fully inflated tire contains enough gas particles to press on the inside tire walls and keep the tire inflated.

Pressure is a force applied to a surface.

Now we can see why a balloon that contains more gas particles inflates more. The more gas particles, the more collisions on its inside surfaces, the more inflated the balloon must be. We refer to the constant pushing by gas particles that's taking place on the interior surface of the balloon as pressure. **Pressure** is a force, such as the force from gas particles, applied to a surface, such as the inside surface of a balloon.

The term *pressure* is not used exclusively for gas particles that are pushing on a balloon's inside surface. We can use it for a hand pressing on the surface of a door, or a foot stepping on an accelerator pedal, or a thumb pushing onto a fingerprint card. For this reason, we encounter units of pressure like pounds per square inch, also known as psi. This pressure unit, which we might see on a tire gauge, indicates the number of pounds of force being pressed into an area—in this case, each square inch of the surface. **Figure 6.6** shows how the pressure of a tire changes as it is filled with air.

Atmospheric pressure changes with altitude.

Imagine you live at sea level—say, in San Francisco. After deciding to take a skiing vacation in the Colorado Rockies, you pack your bags and hop on a plane. When you arrive in Colorado and unpack your bags, you notice that your plastic shampoo bottle is bigger than it was when you packed it back home at sea level (**Figure 6.7**). You carefully open it, and a gentle whish of air comes out. On the return trip, you arrive in San Francisco to find that your shampoo bottle is now smaller than it was when you packed it in Colorado. What's going on?

Although we cannot see them, atoms and molecules in the air are bouncing around and ricocheting off us and off everything around us. Our bodies provide a surface for bouncing gas molecules that is no different from the interior surface of a balloon or a tire. Every surface exposed to air is constantly bombarded by the gas molecules making up the air. However, surfaces exposed to air at different altitudes are bombarded with different numbers of molecules.

To understand why, we must know two things: (1) gravity pulls everything downward toward Earth's surface; and (2) atoms and molecules are not exempt from the pull of gravity. The atoms and molecules in air are pulled downward just like everything else. So we can imagine that, at sea level, the molecules in air are relatively close together because a given volume contains more of them than that same volume does in, say, the mountains of Colorado. Put another way, air is denser at sea level than at high altitude, and its density changes through all points in between (**Figure 6.8**).

We can solve the mystery of the distorted shampoo bottle by considering the density of air at the two altitudes. The denser air in San Francisco means that, relative to Colorado, more collisions are occurring between air molecules and the outside of the bottle. This means that the pressure on the outside of the bottle is greater in San Francisco than in Colorado. When you pack your shampoo in San Francisco, the density of air is the same on the inside and outside of the bottle, and so the bottle maintains its normal shape. When you carry that sealed bottle to the mountains, though, the pressure on the inside of the bottle (the sea level pressure) is now greater than

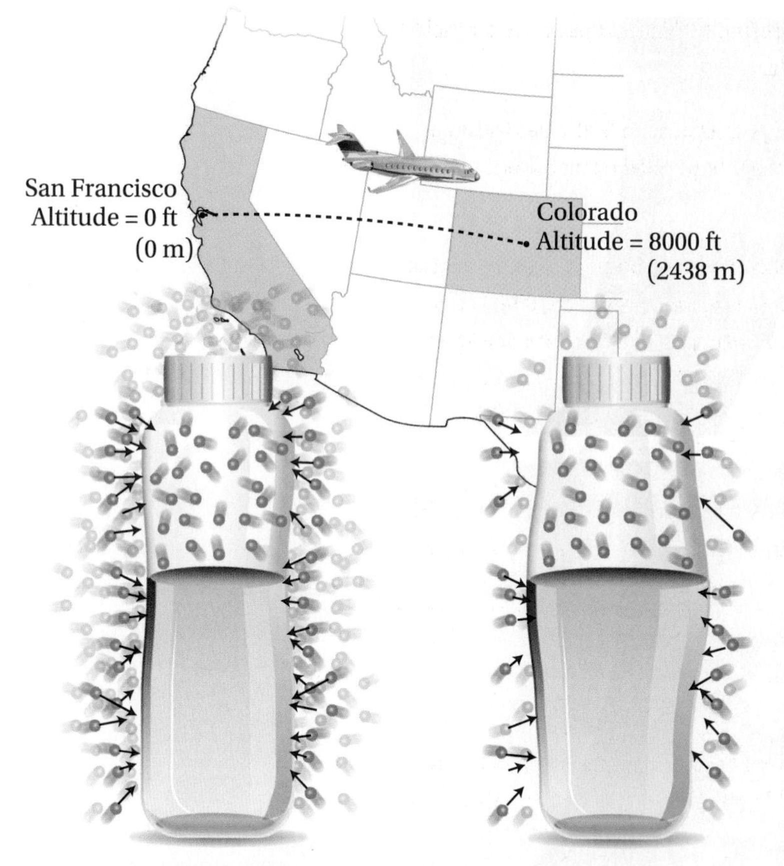

Figure 6.7 I Left My Air Pressure in San Francisco

When the cap is put on a plastic shampoo bottle at sea level, the air pressure inside the bottle is the same as the air pressure outside. Later, at an altitude where the air pressure is lower than at sea level, the air trapped inside the bottle exerts enough pressure on the inside walls to distort the shape of the container, which is flexible.

Figure 6.8 Into Thin Air

An imaginary column of air starting at Earth's surface (sea level) and extending into the atmosphere has fewer and fewer air molecules and atoms per volume as the distance from sea level increases. In other words, the air density decreases with increasing altitude.

the pressure on the outside (the high-altitude pressure). Molecules in the air inside the bottle push outward, and the bottle expands.

QUESTION 6.3 Imagine you arrive in Colorado and your shampoo bottle has leaked all over the new ski clothes in your suitcase. Describe what happened using the concepts discussed in this section of the chapter.

ANSWER 6.3 When you get to Colorado, the pressure inside the bottle is higher than the pressure outside. The inside pressure pushes against the shampoo, forcing it up against the covered hole in the bottle cap. The pressure pushing on the cap from the outside of the bottle is lower than the pressure pushing from the inside of the bottle. Thanks to this difference in pressure, the cap is pushed outward and pops off.

For more practice with problems like this one, check out end-of-chapter Question 15.

FOLLOW-UP QUESTION 6.3 On average, are the particles of gas in the air closer together (denser) or farther apart (less dense) on top of Mount Whitney, the tallest mountain in California, than they are at the beach in San Diego?

A gas particle's mean free path is the distance it travels between collisions.

One way to imagine the density of gas is to think about the motion of a handful of gas particles. We know that they travel at very high speeds and randomly careen from one place to another, colliding occasionally with other gas particles and with the walls of their container. It follows that those collisions are more frequent in a high-density gas than in a low-density gas. We can quantify this difference in terms of the **mean free path** of a molecule, defined as the average distance the molecule travels between collisions.

For example, on average, the mean free path of the atoms and molecules in air at sea level is about 60 nanometers (nm). This means that, on average, a particle in air collides with something else every 60 nm that it travels. At high altitude, though, we would expect the mean free path to be greater because at high altitude there are fewer molecules in a given volume. And this is exactly what we find: the mean free path of air particles on the summit of Mount Everest is about 180 nm, three times longer than the mean free path at sea level.

At the top of Everest, molecules in the air are about three times farther apart than at sea level. Thus, a single breath of air on the summit of Everest, represented by the clear box on the left of **Figure 6.9,** supplies the body

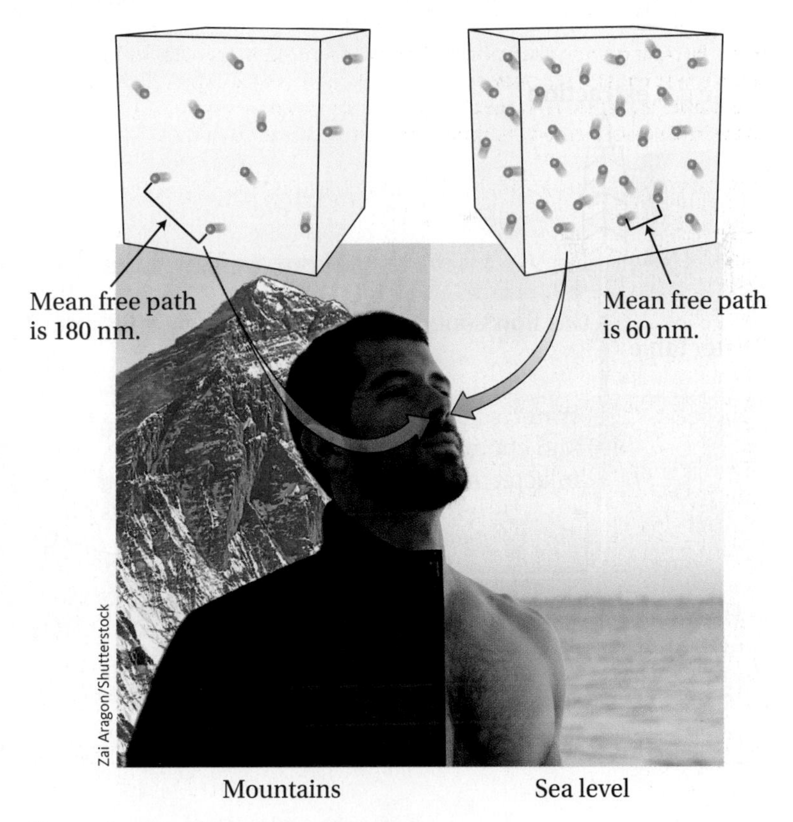

Figure 6.9 Breathing at Two Altitudes
Consider a person in the mountains (left) and at sea level (right). The clear boxes represent one big breath of air at each location, and the particles inside show the mean free path at these two elevations. Gas particles in air are farther apart in the mountains (left) compared to sea level (right).

natureBOX • Is Natural Gas the Ideal Energy Resource?

In 2005, the United States was poised to make a dramatic shift in its energy use. Many people favored ending military and political conflicts stemming from our foreign oil dependence, and natural gas entered the arena as the new, cleaner energy source.

Natural gas is a mixture of gases, but it is predominantly composed of methane. Methane is a molecule with one carbon atom surrounded by four hydrogen atoms, and it is a fossil fuel because it comes from organic matter decay in the earth. It is considered relatively clean because it is a gas and is not mixed with undesirable substances like mercury and sulfur, which make coal and crude oil "dirty."

Since 2005, natural gas use has mushroomed in the United States. In the areas that have the most abundant underground natural gas—Pennsylvania, Ohio, and West Virginia—the natural gas withdrawal has increased from about 2 billion cubic feet per day in 2009 to a whopping 24 billion cubic feet per day in October 2017.

Natural gas is most often extracted using *hydraulic fracturing*, also known as *fracking*. As the figure illustrates, a well is drilled down through the water table and into *shale*, a dense rock whose fissures hold natural gas. The drill then turns and drills horizontally into the shale. When a watery mixture is pumped into the shale at very high pressure, the rock fractures. Natural gas is sucked out and up, where it is collected at the well head.

If fracking produces cleaner natural gas and reduces our foreign oil reliance, then why are some people against it? The clearest reason is because natural gas, which is mostly methane, produces carbon dioxide when burned. Carbon dioxide produced by burning fossil fuels causes global warming, and this temperature increase is rapidly changing our climate. Furthermore, methane, which leaks into the air unburned as part of fracking, contributes to global warming four times more than does carbon dioxide. Thus, while natural gas may decrease our foreign oil dependence, it does not solve the problem of global warming (see Chapter 12).

Additionally, fracking fractures rock using fluids, which may contain toxins like benzene, a potent carcinogen. These toxins have begun appearing in the water of residents living near drilling sites. Since fracking releases methane into the surrounding rock and into the water table, there are reports of "flammable water" coming from kitchen faucets. The water is flammable because it can be lit with a match.

Even more dramatic than flaming faucets is the increase in earthquakes in areas where fracking is common. For example, before fracking began in Oklahoma, the state saw about 1.5 earthquakes that measured 3.0 or more on the Richter scale, per year on average. Since fracking wells have been operating, this number has risen to more than 500 per year. Texas, Arkansas, and Colorado have seen a similar increase.

Given these facts, is fracking the ideal solution to our energy needs? Perhaps it is a reasonable first step that will free us from imported oil. However, as a fossil fuel, it is not renewable or sustainable. In Chapter 12, we will explore energy alternatives that *are* renewable and sustainable.

Shale gas extraction

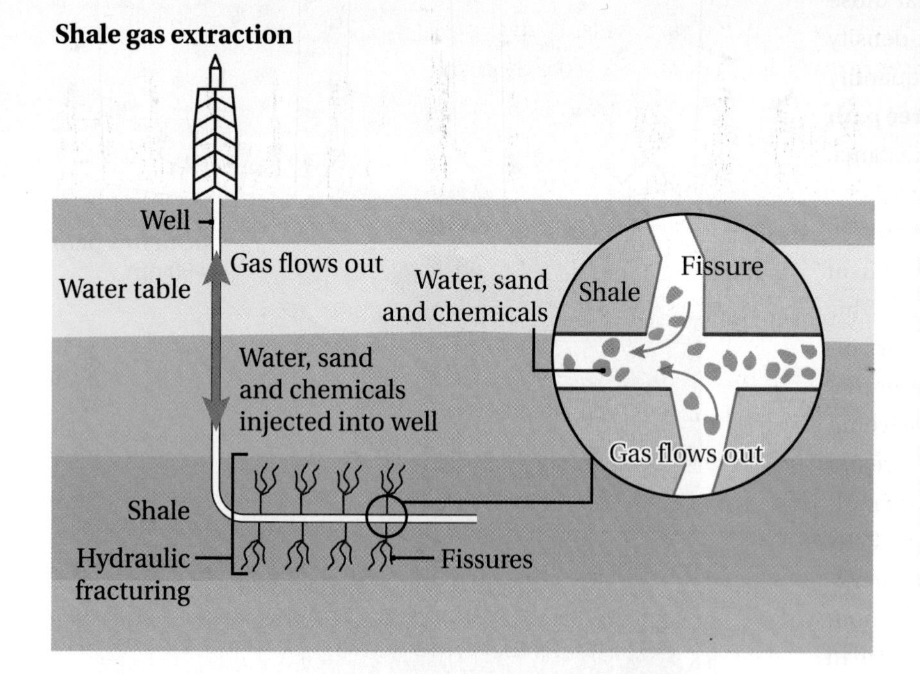

with fewer particles of air than the same-sized breath at sea level, represented by the clear box on the right in Figure 6.9. Because about one in five air particles are oxygen molecules, this also means there are fewer molecules of oxygen on Everest as compared to a lower elevation. This is why most climbers who scale Everest rely on supplemental oxygen to get them to the top.

wait a minute . . .

Is the percentage of oxygen in air lower at high altitude than at sea level?

No. The composition of air is roughly the same on a mountaintop and at sea level; in both places, air contains about 21% oxygen. However, at high altitude there are fewer gas molecules per volume of air. One breath of air at 18,000 feet (about 5500 meters) will contain about half the number of oxygen molecules as the same breath taken at sea level. So, even though the composition of air does not change between sea level and a mountaintop, the density of the air *does* change and this is why we breathe less oxygen at high altitude.

Atmospheric pressure is the pressure exerted on us by air in the environment.

Although we cannot see it and we really do not feel it, we have evidence that air is always colliding with our bodies. As we have just seen, however, the pressure pushing on us varies from place to place. It also changes with the weather. A weather system may bring high pressure (when the pressure of air molecules on us is greater than normal) or low pressure (when the pressure of air molecules is lower than normal). The pressure exerted on us by the air in the environment, called **atmospheric pressure,** fluctuates constantly. As a part of weather prediction, meteorologists keep track of pressure fluctuations in the atmosphere. When the pressure changes, the weather changes.

How is atmospheric pressure measured? **Figure 6.10** (on p. 170) shows a rudimentary, yet accurate, version of a **barometer**—a device for measuring atmospheric pressure. A glass column sits in a pool of mercury that is open to the air and therefore subject to the force of air molecules in its environment. Because pressure is defined as the force exerted over a given surface area, the higher the pressure on the surface of the mercury in the open dish in Figure 6.10, the farther up the tube the mercury moves. The height of the mercury column in the tube is therefore proportional to the atmospheric pressure. A common unit for expressing pressure is **millimeters of mercury (mm Hg)**. This is a measurement of the distance, in millimeters, that the mercury in a barometer tube travels as a result of atmospheric pressure.

Now let's consider another unit of pressure commonly used in scientific measurements. The **atmosphere (atm)** is a pressure equal to 760 millimeters of mercury. It's a convenient unit of measure because at sea level the atmospheric pressure is typically about one atmosphere, more or less.

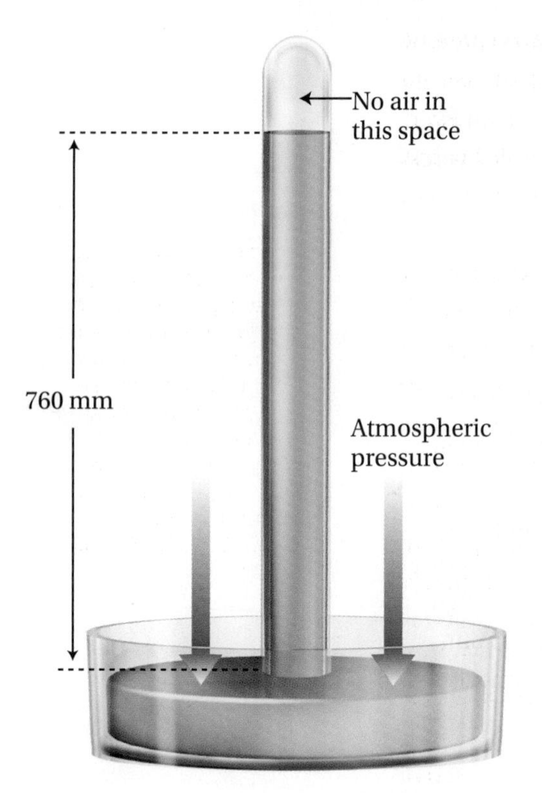

Figure 6.10 A Simple Barometer

In a barometer, atmospheric pressure pushes on the surface of the mercury liquid in the bowl, forcing the mercury up into the evacuated tube. The height of the mercury column is one way to express the magnitude of the atmospheric pressure.

QUESTION 6.4 Which of these statements about pressure is untrue? (a) Air pressure does not vary with altitude. (b) Atmospheric pressure is measured with a barometer. (c) Atmospheric pressure can affect weather. (d) Pressure is a force applied to a surface.

ANSWER 6.4 (a) Air pressure *does* vary with altitude, as our shampoo bottle example demonstrates.

For more practice with problems like this one, check out end-of-chapter Question 21.

FOLLOW-UP QUESTION 6.4 Which of the following is *not* a unit of pressure discussed in this chapter? (a) millimeters of mercury (b) joules (c) psi (d) atmospheres

6.3 Variables That Affect Gases: Moles, Temperature, Volume, and Pressure

The mole allows us to count very small things, such as atoms and molecules.

Consider an imaginary box measuring 28.19 centimeters in each of its three dimensions: length, width, and height. This is approximately one foot on each side. If we do the math, we find that the volume of the box is 22.4 liters. It's also an unusual box, because it can be hermetically sealed—when it's shut, nothing can go in and nothing can come out. So, it's a perfect place to keep a gas. Now, imagine that the temperature in the box is 0°C and pressure in the box is 1.00 atmosphere. We refer to this pair of conditions as *standard temperature and pressure,* and we abbreviate it **STP**. In this case, we can say that our box is at STP. For simplicity, let's impose one final constraint on our box: the box contains a gas, and it can be any gas at all. The gas could exist in the form of molecules or just atoms, but we will think of the gas—whatever it is—as individual particles flying around inside the box.

Suppose we would like to know how many particles of gas are in the box. As it happens, under these special conditions, we know the answer to this question: there are 602,000,000,000,000,000,000,000 particles in the box. This may seem like an enormous number of particles, and it is, but it's a typical number of particles for a container like this to hold. Under reasonable conditions, in a box of a size that would fit on your car seat, this is roughly the number of particles that are present. Now, what if we have two identical boxes rather than just one box? In that case, we have twice as many particles of gas: 1,204,000,000,000,000,000,000,000 particles in the two boxes. Imagine that we have one box again, but the new box is one half the volume of the original box. How many particles does it contain? Half as many particles: 301,000,000,000,000,000,000,000.

It's pretty straightforward to see how the number of particles changes as we change the size of the box or the number of boxes, but it can get pretty tedious to write numbers like the blue ones in the last paragraph. Luckily, we have two ways to get around this problem. First, we can use scientific notation and say that our original box contains 6.02×10^{23} particles. That is much easier! But there's an even quicker way to express this number. Rather than saying, "This box contains 6.02×10^{23} particles of gas," we can simply say, "This box contains one mole of gas."

If we have one **mole** of anything, then we have 6.02×10^{23} of those things. Now we see why our original box is special: under those exact conditions, that specific box holds *exactly* one mole of gas. For this reason, we say that the volume of that box—22.4 liters (L)—is the **molar volume** of a gas. In other words, this is the volume that holds exactly one mole of gas at STP. **Figure 6.11** shows an everyday item that has a volume of about 22.4 L.

A mole is a counting unit that makes it easy to talk about the number of particles in a quantity of a substance. It's like the word *dozen,* which is an alternate way of talking about 12 things, be they eggs or hats or snow shovels. Likewise, we can use the mole to count anything. We can have a mole of doorknobs or coffee beans or helium atoms or ballet slippers. Whatever it is, a mole is 6.02×10^{23} of that thing. The number 6.02×10^{23} is sometimes called **Avogadro's number** after Amedeo Carlo Avogadro, a nineteenth-century Italian scientist.

Although the mole can be used to count anything, it makes sense that we typically use the word *mole* when we're talking about atoms or molecules or other very tiny things. We won't often need to talk about a mole of doorknobs because, if we set out to cover Earth with a layer of doorknobs, we would need nearly 10 million Earths to distribute one whole mole of doorknobs. So, even though we *could* use moles to count doorknobs, the mole is most useful when we use it to count really tiny things, such as atoms and molecules.

flashback

Recall from Section 1.2 that scientific notation is a way to express large or small numbers using the number 10 raised to an exponent. In the number 6.02×10^{23}, the exponent is 23. We can express this number in non-exponential form by starting at the decimal point and moving it to the right 23 times. The result is 602,000,000,000,000,000,000,000.

recurring theme in chemistry 2

Things we can see with our eyes contain trillions upon trillions of atoms. Atoms are much too small to view with the naked eye.

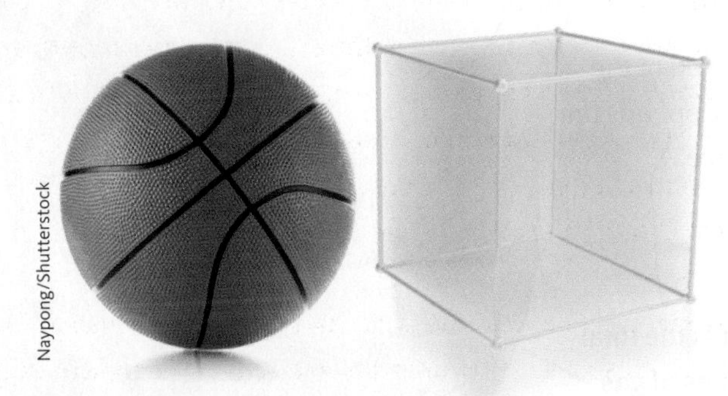

Figure 6.11 Everyday Molar Volume

Both the items shown here have a volume of about 22.4 L. At STP, these volumes contain about one mole of gas particles.

QUESTION 6.5 Which of the following objects would likely not be counted using moles? (a) carbon atoms (b) sugar cubes (c) molecules of methane (d) electrons

ANSWER 6.5 (b) Sugar cubes are not tiny, but the rest of the items listed are. Therefore, we would probably not count sugar cubes in mole units.

For more practice with problems like this one, check out end-of-chapter Question 31.

FOLLOW-UP QUESTION 6.5 A sealed box at STP has a volume of 2240 liters which is equal to 100 times the molar volume of a gas. How many gas particles does the box contain?

Four variables dictate the behavior of a gas.

When we defined molar volume, we were careful to specify the exact conditions of the box that contained our gas. One of the conditions that we specified was temperature, because it's one of the variables that must be controlled if we want to understand how a gas is behaving. A **variable,** such as temperature, is a condition that can be changed.

How does varying temperature affect the behavior of a gas? Temperature (T) changes the speed at which gas particles move through space, and this in turn affects how quickly a given gas particle reaches the next wall it is going to collide with. This change in speed, therefore, affects the pressure of the gas. Because the pressure can change as the temperature is varied, pressure (P) is also a variable.

So far, we have seen that T and P can be altered for a sample of gas, and that changing one of these variables causes a change in the other variable. What else might change for a sample of gas? Or, put another way, what are other examples of variables? As we saw from our balloon example in Section 6.1, the balloon got bigger when we blew more gas into it. This tells us that the volume (V) the gas occupies is another variable.

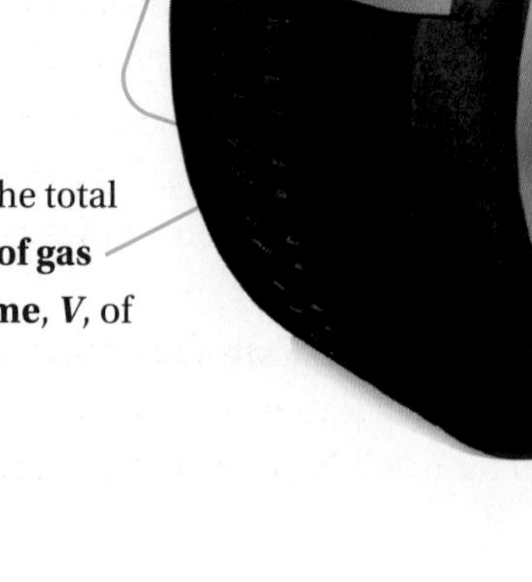
Pressure, *P*, depends on the frequency of collisions with the walls of the container.
The average speed of the gas molecules depends on the **temperature, *T*.**
The quantity ***n*** is the total number of **moles of gas** found in the **volume, *V*,** of the tire.

Figure 6.12 The Four Variables That Affect Gas Behavior

Pressure, temperature, volume, and number of moles all play a part in how a gas behaves.

We have just one more variable to consider. Remember that we changed the balloon's size by changing the number of gas particles in it: the more breaths we blew in, the more particles of gas were in the balloon. More gas particles mean more collisions with the inside walls of the balloon. More collisions expand the balloon. This effect tells us that the number of particles is a fourth variable that can be changed. Because the mole is the way we count particles, we use moles to keep track of particles, and we use the variable n to represent this number.

We now have a list of four variables that can change for any sample of a gas: temperature (T), pressure (P), volume (V), and the amount of gas (n), which we count in mole units. When we change any one of those variables, one or more of the other variables must also change, as we'll see in the next section. These four variables are summarized in **Figure 6.12**.

6.4 The Gas Laws: An Introduction

Pressure and volume are inversely proportional to one another.

In Section 6.3, we learned that each of our four gas variables can change as the other variables are changed. Now we will choose some specific pairs of variables, in turn, and create statements about how they change in relation to one another. We call these statements **gas laws,** but they are really nothing more than formal statements of commonsense observations about the way gases behave. You will read about several gas laws in this chapter, but there's no need to memorize them (unless your instructor asks you to do so). Your intuition about the fundamental behavior of gases is enough to remind you about each of them, as you will see.

Let's revisit Figure 6.12, which summarizes the four variables that we can change when we are considering a gas: T, P, V, and n. If we determine these four values, then we know the condition of our gas. In this section, for each gas law, we consider an experiment in which we change one variable and see what happens to a second variable, leaving the remaining two fixed.

For our first gas law, we return to Figure 6.11 and the 22.4-liter box at STP that contains exactly one mole of gas particles. If we sit on the box and make it half the size, but keep the box sealed and at the same temperature, what happens? Of our four variables, what has changed and what has stayed the same? In this case V has changed and T has stayed the same. The variable n has also stayed the same, because we did not allow any gas to escape. But what about P? What happened to it?

When we made the box smaller, how did the environment of the gas inside it change? In the smaller box, each gas particle has a smaller distance to travel to reach a wall. That means there are more collisions with the inside surface of the box, which means the pressure of the gas must have increased. We have just deduced that when we keep T and n the same and make V smaller, we make P bigger. We can express this using an if-then statement with arrows, or we can express it mathematically as shown in **Figure 6.13**.

If $V\downarrow$ then $P\uparrow$

or

$$P \propto \frac{1}{V}$$

Figure 6.13 The Relationship between Volume and Pressure

This figure illustrates the relationship between two variables that dictate the behavior of gases. We can use an if-then statement and up/down arrows (top) or a mathematical relationship (below) to show a relationship between variables.

The symbol $\propto$ between the P and the $1/V$ on the bottom of Figure 6.13 is a *proportionality symbol.* It relates the thing to its left to the thing to its right. For example, imagine that that the volume, V, is big and then gets smaller, as illustrated

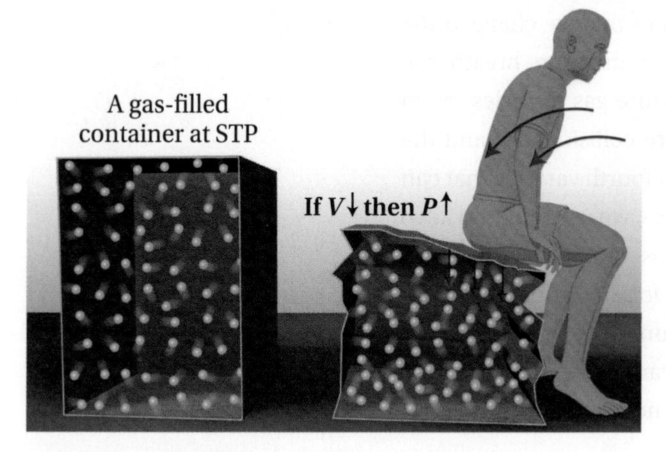

Figure 6.14 Boyle's Law

Reducing the volume of a gas-filled container, like a box, increases the pressure of the gas. Notice that the box contains equal numbers of gas particles before and after it is compressed. The compressed box contains gas at a higher pressure because the particles collide with the walls of the box more often. As volume goes down, pressure goes up. In other words, pressure is inversely proportional to volume.

in **Figure 6.14**. According to our proportionality relationship, when *V* gets smaller, *P* gets bigger. And because *V* and *P* vary in opposite directions, we say that *V* and *P* are *inversely proportional*: as one goes up, the other goes down.

This principle about the behavior of gases is known as **Boyle's law,** after Robert Boyle, a seventeenth-century British scientist. It's not necessary to memorize Boyle's law, because the relationship between volume and pressure is already second nature to us. We already know, for example, that an inflated balloon compresses when we sit on it. Compression means the balloon volume decreases, and we may worry that the balloon will burst because we sense that its pressure has increased due to the decrease in volume.

QUESTION 6.6 If you take the sealed box in Figure 6.14 and somehow make it bigger, what happens to the pressure of the gas inside the box?

ANSWER 6.6 The pressure inside the box decreases. Why? Because you have increased the volume of the box. When the volume of the box increases, the surface area also increases and there are fewer collisions per area inside the box.

For more practice with problems like this one, check out end-of-chapter Question 40.

FOLLOW-UP QUESTION 6.6 True or false? Pressure inside a container depends not only on the number of collisions in the box, but on the number of collisions per area of the inside of the box.

If we change the number of moles of a gas, the volume of the gas changes.

Our second gas law is named for Avogadro, the same man whose number defines one mole. **Avogadro's law** explains how a change in the numbers of atoms and molecules—given by the variable *n*—affects the volume of a gas. To understand this law, let's imagine that we have a cylinder capped with an airtight piston. The cylinder contains a small amount of gas and also has an inlet valve through which we can add gas in much the same way that air is added to a tire (**Figure 6.15**). The piston is in a position some distance up from the bottom of the cylinder because the gas molecules are colliding with the piston. These collisions keep the piston from falling to the bottom.

If we add more gas to the cylinder, more gas molecules are hitting the interior surface of the cylinder, and the piston is forced upward. As the piston ascends, the volume of the cylinder increases. When the pressure inside the cylinder equals the pressure of the atmosphere pushing down on it, the piston stops rising. Thus, in this experiment, the variables *T* and *P* are held constant and the variables *V* and *n* are changing. We can describe this relationship with these words: As the number of moles of gas increases, so does the volume. We can write this relationship using symbols and a proportionality symbol ($\propto$):

$$V \propto n$$

where n represents the number of moles of gas, and V represents the volume of the gas. Recall that the proportionality sign tells us that as one quantity goes up, so does the other. Likewise, when one quantity goes down, the other does as well.

As we all know from personal experience, Avogadro was right in saying that the volume of a gas increases when more gas is added to a container. For example, we know that when we blow up a balloon, the balloon gets bigger (**Figure 6.16a**). We know that when we push air from our lungs into our cheeks while keeping our mouth shut, our cheeks puff out (**Figure 6.16b**). We know that when we inhale air into our lungs, our chest cavity increases in size (**Figure 6.16c**). It's second nature to us: as the number of moles of gas increases, volume also increases.

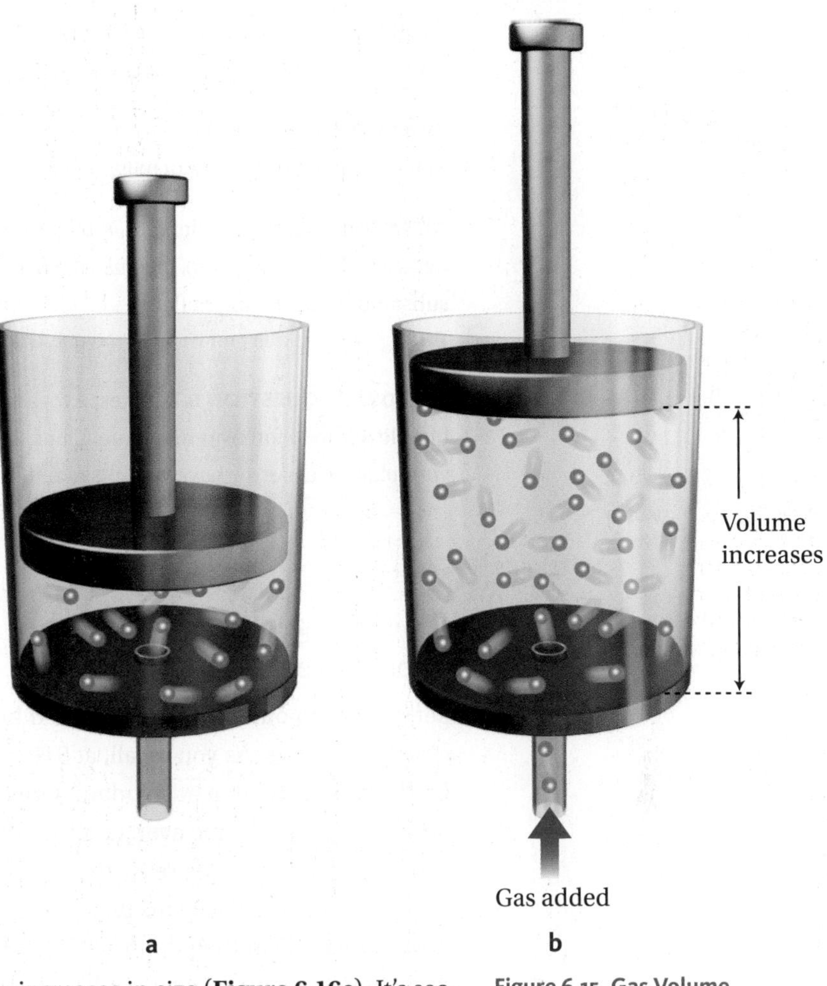

Figure 6.15 Gas Volume and Amount of Gas

(a) In a cylinder fitted with a piston, the pressure of the gas molecules determines the height of the piston and therefore the volume of the cylinder. (b) When more gas molecules are added to the cylinder, the piston rises and the volume increases.

Notice this common thread for the gas laws: they all involve a change in only two of the four variables. In Figure 6.15, the number of moles of gas is increased by forcing gas molecules into the cylinder. Volume is the only other thing permitted to change, and the experiment is designed to tell us what happens to the volume (second variable) as the value of n (first variable) increases. The other two variables

Figure 6.16 Avogadro's Law in Action

In any flexible container, gas volume is proportional to the number of moles of gas present. (a) When you blow up a balloon, you are increasing the number of moles of gas (air) it contains. The volume of the (flexible) balloon increases. (b) Your cheek volume increases when the number of moles of gas (air) in your (flexible) mouth increases. (c) When you inhale a larger-than-usual amount of air, the volume of your (flexible) lungs increases your chest size noticeably.

depicted in Figure 6.15, pressure and temperature, are not permitted to change, and we say that they are *held constant.*

QUESTION 6.7 A bicycle tire runs over a tack and deflates. How do the four variables that we usually keep track of in gas experiments change when this occurs?

ANSWER 6.7 As the tire deflates, it releases moles of pressurized gas. Thus, n and P decrease. V decreases, too, because the tire gets smaller, but T will not change substantially as the tire deflates.

For more practice with problems like this one, check out end-of-chapter Question 39.

FOLLOW-UP QUESTION 6.7 To test Avogadro's law for yourself, you decide to use a bicycle pump to force air into a metal box that has a gas inlet valve. What is wrong with your plan? What error have you mistakenly introduced into your experiment?

If we change the temperature of a gas, the volume or pressure changes.

Let's return to our rigid, tightly sealed box from Section 6.3 that has a volume of exactly 22.4 liters. As you recall, at STP our box contained exactly one mole of gas particles. Imagine that at standard temperature, 0°C, the gas molecules inside the box are moving at some average speed. If we increase the temperature in the box, what happens? As we increase the temperature, the average speed of the gas molecules increases. This means that each molecule collides with the inside surface of the box more often than it did at 0°C. Because the box is rigid and cannot expand, the volume remains constant. What does change is the pressure, because the number of collisions on the inside surface of the box has increased. We can see that as the temperature increases, the pressure increases, too. Pressure and temperature are proportional to one another:

$$P \propto T$$

We can describe this event in words: *If the volume of a gas and the number of moles of gas are fixed, an increase in gas temperature results in an increase in gas pressure.* This relationship is sometimes called **Amontons' law** after the scientist who first studied it.

Let's change things a bit. Imagine now that the box is flexible, like a cube-shaped balloon. In this case, when the temperature increases, and the number of collisions increases, the box gets bigger. And, because we are allowing the box to get bigger, the pressure is held constant. P and n stay the same while V and T change. We can say that as the temperature increases, the volume also increases. The opposite must also be true: if we lower the temperature, the volume of the cube will decrease. We can write this as a proportion:

$$V \propto T$$

We can also state this in words: *When the temperature of a gas is increased, the volume also increases.* This relationship is sometimes called **Charles's law** after the scientist who first studied it. These two laws, in which temperature is changing, are illustrated in **Figure 6.17**.

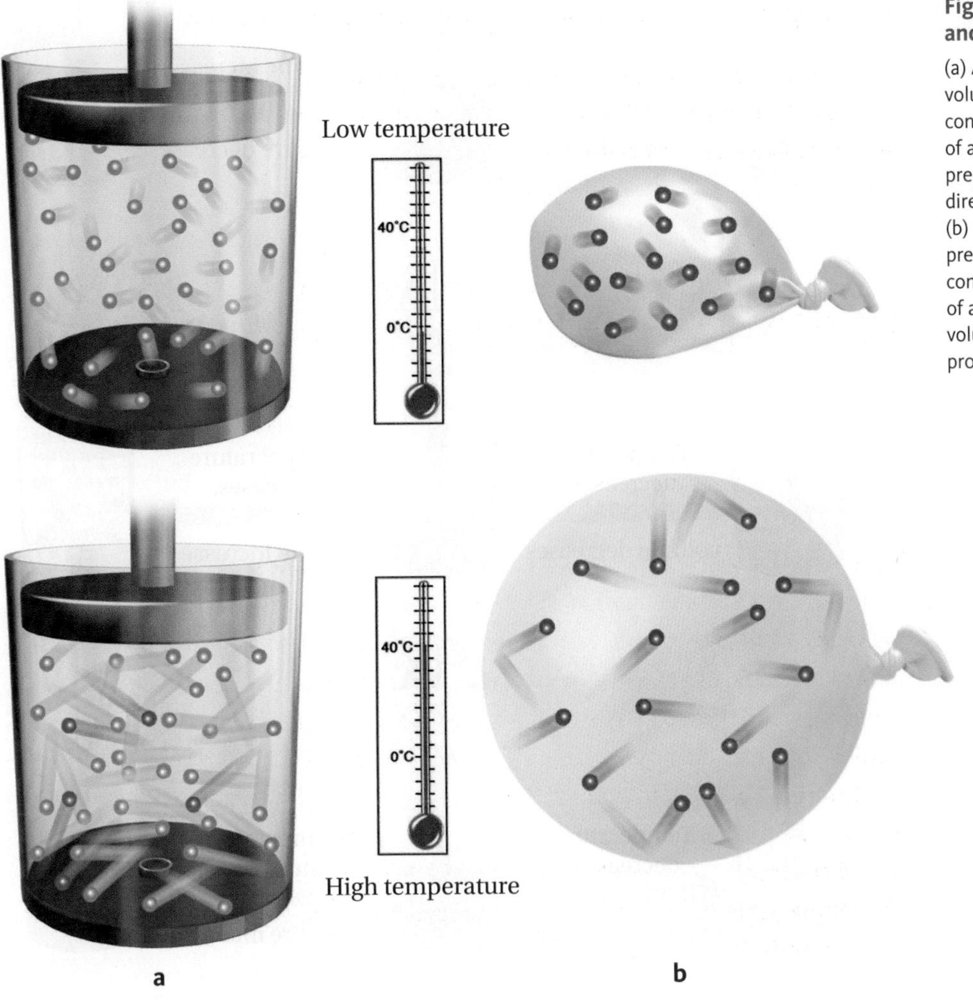

Figure 6.17 Charles's Law and Amontons' Law

(a) Amontons' law tells us that with volume V and number of moles n held constant, increasing the temperature of a gas-filled container increases the pressure of the gas. Pressure is directly proportional to temperature. (b) Charles's law tells us that with pressure P and number of moles n held constant, increasing the temperature of a gas-filled container increases the volume of the gas. Volume is directly proportional to temperature.

QUESTION 6.8 What happens to the mean free path of the molecules in a sample of gas when the volume of the gas is decreased and the temperature is held constant?

ANSWER 6.8 When the volume decreases as the temperature stays unchanged, a molecule collides with the walls of the container and with other molecules more frequently. The mean free path gets shorter.

For more practice with problems like this one, check out end-of-chapter Question 47.

FOLLOW-UP QUESTION 6.8 You decide to buy your cousin a balloon for her birthday, which is in February. Leaving the warm store, you walk out into subzero temperatures. Which one of the four gas variables does not change significantly when you walk outside?

The four gas laws are summarized in **Figure 6.18** (on p. 178).

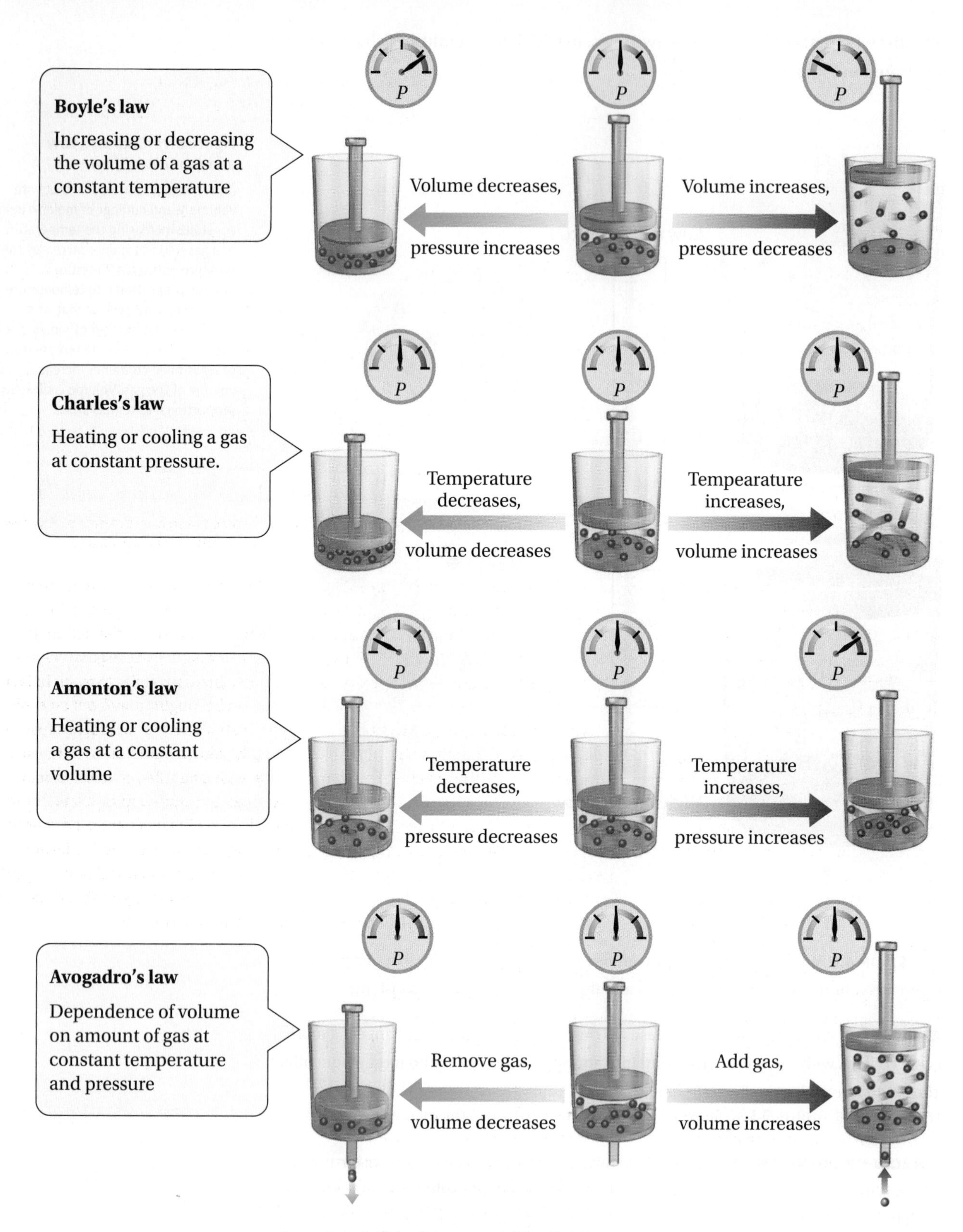
Boyle's law
Increasing or decreasing the volume of a gas at a constant temperature
P
Volume decreases,
pressure increases
Volume increases,
pressure decreases
Charles's law
Heating or cooling a gas at constant pressure.
Temperature decreases,
volume decreases
Tempearature increases,
volume increases
Amonton's law
Heating or cooling a gas at a constant volume
Temperature decreases,
pressure decreases
Temperature increases,
pressure increases
Avogadro's law
Dependence of volume on amount of gas at constant temperature and pressure
Remove gas,
volume decreases
Add gas,
volume increases

Figure 6.18 A Visual Summary of the Gas Laws

TOP STORY Bees as Gas Detectives

Bees are devoted busybodies. Because they are small—a typical bee weighs about half a gram—they can fly in and out of tiny nooks and crannies, all the while interacting with molecules in the air. They can take those molecules back to the hive along with the pollen, the dusty particles from certain plants, they collect. The exact number of particles a bee encounters on a flight depends on gas-law variables such as temperature and pressure. But because the mean free path of the particles in air is very short, on the order of nanometers, any bee encounters particles constantly as it flies.

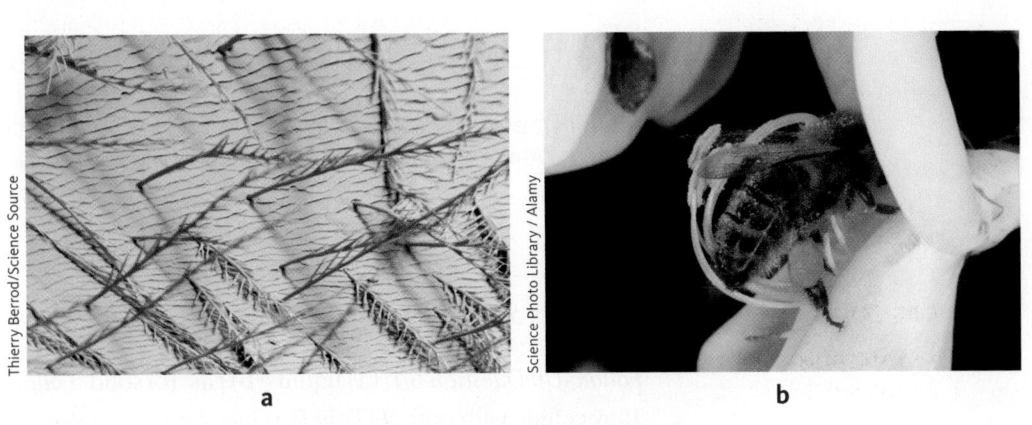

Figure 6.19 How Bees Bring the Local Environment Back to the Hive

(a) A microscopic view focuses on barbed bee hairs, which catch particles in air and pollen from flowers. (b) This bee is deep inside a flower and coated with pollen, which it takes back to the hive along with other particulates.

The hairs on a bee are specially designed to attract pollen, hold on to it, and transport it back to the hive. Dust particles in the air naturally stick to those bee hairs as they fly, and those particles can contain molecules of any substances found in the air. In fact, bees have been described as "flying dust mops." And, because bees are breathing while they are flying, their bodies take in the surrounding air and the water contained in that air. They bring those air and water samples back to the hive in their bodies. In this way, bees serve as a means of sampling water, gases, particulates, and plant matter in the area around the hive they call home. **Figure 6.19a** shows a micrograph of jagged bumblebee hair, and **Figure 6.19b** shows a bee happily headfirst in a yellow flower.

When bees return to the hive, they beat their wings furiously to cool off the temperature inside the hive. This beating motion releases the molecules that were stuck to the bees' bodies and hair. The molecules then become concentrated in the hive space. If a bee takes nectar or pollen from a plant containing a foreign molecule of some sort, or if the bee flies through air contaminated with some foreign molecule, those molecules eventually will end up in the hive, sometimes in the honey and at other times in the bees themselves.

Scientists like bee expert Jerry Bromenshenk of the University of Montana are trying to exploit the natural talent that bees have for taking home samples of the air they fly through. Bromenshenk and his research group have taught bees to seek out specific non-nectar molecules, and he says this task is easier than training a bloodhound to follow a scent.

How can this work possibly be relevant to humans? One application is in the detection of land mines, which sometimes contain the explosive molecule trinitrotoluene (TNT). After being trained to sniff out TNT molecules (**Figure 6.20**), bees are amazingly adept at locating buried land mines. Bees can be trained to fly to spots where they smell the explosive compound TNT. The bees swarm in the areas highest in TNT molecules, which are the places where the land mines are located. In countries where land mines are prevalent, such as Cambodia, Somalia, and Angola, mine removal is an urgent priority. With luck, the coming years will see mine-tracking bees working in these locations.

So why use bees rather than dogs, which are known to be great mine-sniffers? The answer is that although both bees and dogs have great senses of smell, bees are easier to train. In fact, they can be taught to sniff out an explosive in about two days. And, because bees fly solo and don't need to be on a leash with a handler, concerns about a human or a dog tripping a land mine are allayed when bees are used instead. Bees are also being trained to locate dead bodies and chemical or biological warfare agents, other jobs that dogs have traditionally performed.

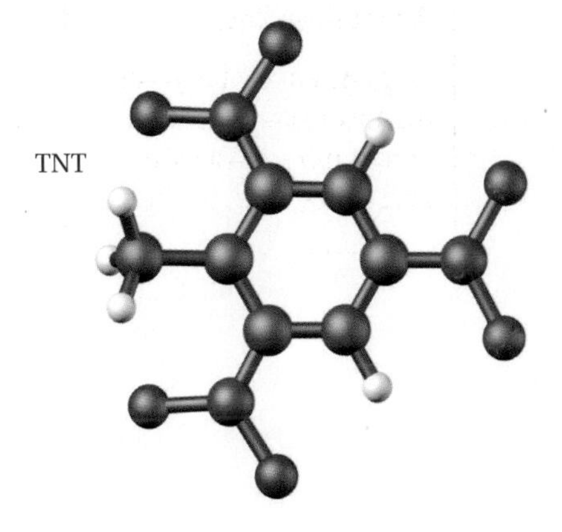

Figure 6.20 The Hunt for TNT

TNT is an organic molecule with the chemical formula $C_7H_5N_3O_6$. (Black balls = carbon atoms; red = oxygen atoms; white = hydrogen atoms; blue = nitrogen atoms.)

Chapter 6 Themes and Questions

recurring themes in chemistry chapter 6

recurring theme 2 Things we can see with our eyes contain trillions upon trillions of atoms. Atoms are much too small to view with the naked eye. (p. 171)

A list of the eight recurring themes in this book can be found on p. xv.

ANSWERS TO FOLLOW-UP QUESTIONS

Follow-Up Question 6.1: (a) liquid (b) gas (c) solid. *Follow-Up Question 6.2:* False. They collide with both. *Follow-Up Question 6.3:* They are less dense on Mount Whitney and denser at the beach. *Follow-Up Question 6.4:* (b) joules. *Follow-Up Question 6.5:* One hundred times Avogadro's number, which is the same as adding two to the exponent: 6.02×10^{25} gas particles. *Follow-Up Question 6.6:* This is true. *Follow-Up Question 6.7:* To test Avogadro's law, the box must be flexible so that volume (V) can change. *Follow-Up Question 6.8:* The variables P, V, and T will decrease. The variable n will stay the same.

End-of-Chapter Questions

Questions are grouped according to topic and are color-coded according to difficulty level: **The Basic Questions***, **More Challenging Questions****, and **The Toughest Questions*****.

The Nature of Gases

*1. An odor brought into a room by a very smelly person takes a while to travel from that person to your nose. Imagine this scenario on a hot day and a cold day. Does the travel time differ on those two days? On which day does the scent travel faster? Why?

*2. Imagine you have a 2-L sealed flask containing 1 mole of xenon atoms. You measure the average speed of these atoms to be 700 km/h. Is every atom of xenon in the sample moving at this speed? Explain.

*3. Is the following statement true or false? All gases mix together completely regardless of the types of atoms or molecules they contain. This is true because the atoms or molecules in a gas are all very far apart from one another and interact only rarely during collisions.

**4. Your roommate is becoming more and more malodorous every day. You decide to adjust your thermostat to try to slow down the movement of her "fragrance" through your apartment. Will you turn on the heat or the air conditioning? Explain your answer.

**34. For each pair of gas variables, describe an experiment that would allow only the following two variables to change:

(a) P and V (b) n and P

**35. For each pair of gas variables, describe an experiment that would allow only the following two variables to change:

(a) n and V (b) V and T

***36. What is the approximate volume of a balloon at STP that contains 9.00×10^{23} atoms of helium gas?

The Gas Laws and Gas-Law Variables

*37. In the imaginary lakeside town of Goolikobruquik, the price of potatoes always goes up when the temperature of the lake water goes down. When the water temperature goes up, the price of cherries inevitably goes up. For each pair of variables, indicate whether they are directly proportional or inversely proportional to each other.

*38. In your own words, explain what is meant by the terms *inversely proportional* and *directly proportional*. Give one example of each relationship from this chapter.

*39. Which of these pairs of gas variables are inversely proportional to one another?

(a) P and V (b) n and V (c) V and T (d) T and P

*40. A rigid container has a piston that can move up and down freely. If the temperature of the gas in the cylinder is increased, what other gas variable will change?

*41. A box is sealed and rigid. If you add gas to the box through a valve, which variables are changing? Which variables do not change?

*42. On a hot summer day, a car tire bursts when the temperature inside it increases. Which of the gas laws does this scenario describe?

**43. A kid pumps up his bike tire, and it becomes very firm. If the temperature is assumed to stay constant, which of the four gas variables change when the tire is pumped up? Which do not change?

**44. A balloon is heated by a candle flame held at a distance to prevent the balloon from melting. It pops. If you assume that the balloon stays the same size until just before it pops, which variables change when the balloon is heated? Which do not change?

**45. A window breaks in an airplane traveling at 32,000 feet. Does the pressure within the plane increase or decrease when this happens? Explain.

**46. A container fitted with a movable piston is used to demonstrate Avogadro's law. As gas is introduced into the cylinder, the piston moves upward and the volume increases. Why does the pressure of the gas remain constant during this experiment?

**47. For each of these situations, indicate which gas particles have the longer mean free path.

(a) 2 billion gas particles in a 1-L container or 4 billion of gas particles in the same container

(b) 4 billion gas particles in a 4-L container or 4 billion gas particles in a 2-L container

**48. A rigid, sealed box contains two molar volumes of gas at STP. What is the size of the box? If you want to increase the pressure in the box, which gas variable could you change?

***49. A box at 0 degrees Celsius has a volume of 11.2 liters and contains one-half mole of gas particles. The gas particles are a 50–50 mixture of nitrogen gas (N_2) and oxygen gas (O_2). What is the pressure in the box, expressed in millimeters of mercury?

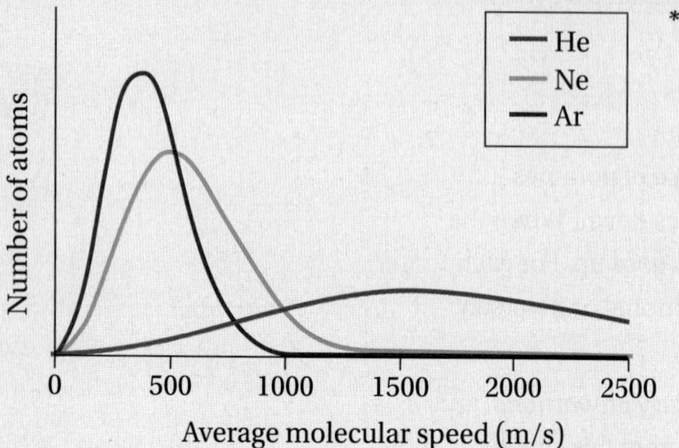

***50. This graph shows the average molecular speeds of three gases at 27°C.

(a) Use the periodic table to rank these gases in order of increasing atomic mass.

(b) What correlation do you observe between molecular speed and molar mass?

(c) Estimate the average speed of a neon atom at 27°C.

*51. In this chapter, we discuss the idea of gas pressure and see what can happen to a shampoo bottle that is taken from a place with high atmospheric pressure to a place with low atmospheric pressure. But while the place where we live is at a specific atmospheric pressure (although that may vary a bit with changing weather), there are things in your environment that may be at higher or lower pressure. Imagine the following objects and predict if the pressure in that object will be greater than or less than the atmospheric pressure.

(a) a new can of Fix-a-Flat on a garage shelf

(b) a bottle of vitamins bottled and shipped from a location at very high altitude

(c) a piece of fish vacuum-sealed in a plastic bag

(d) a can of whipped cream

*52. The tire pressure on your bike or your car's tires will vary with changes in temperature throughout the year. Would you expect your tire pressure to be higher or lower in winter versus summer? Tire pressure can also change as the tire is used. Would you expect tire pressure to increase or decrease as you travel on the tire?

**53. If you have a gas stove that uses natural gas, then you probably have a pipe leading into your home that delivers it. These pipes, which typically are 20 to 42 inches in diameter, are built to withstand pressures greater than what is needed. The pressure of natural gas in municipal pipelines is typically in the range of 200 psi to 1,500 psi. "Psi" is a nonmetric unit for pressure that stands for "pounds per square inch." It can be converted to pascal (P), the metric unit for pressure, using this conversion:

$$1 \text{ psi} = 6{,}894 \text{ P}$$

If the pressure of natural gas in a municipal pipeline is four million pascals, what is the pressure expressed in psi?

***54. Figure 6.2 in Section 6.1 shows the average speed of a collection of nitrogen molecules at a given temperature. That graph shows that at 27°C, the typical nitrogen molecule moves at a speed of about 500 meters per second (m/s). The graph below shows the same speed of nitrogen molecules at the same temperature but also shows several other atoms and molecules as well. Examine this graph and try to answer this question: Why are some atoms or molecules moving faster than others? Here is a hint: consider the masses of each atom or molecule. Is there a relationship between the mass and the speed? Does this trend make sense to you?

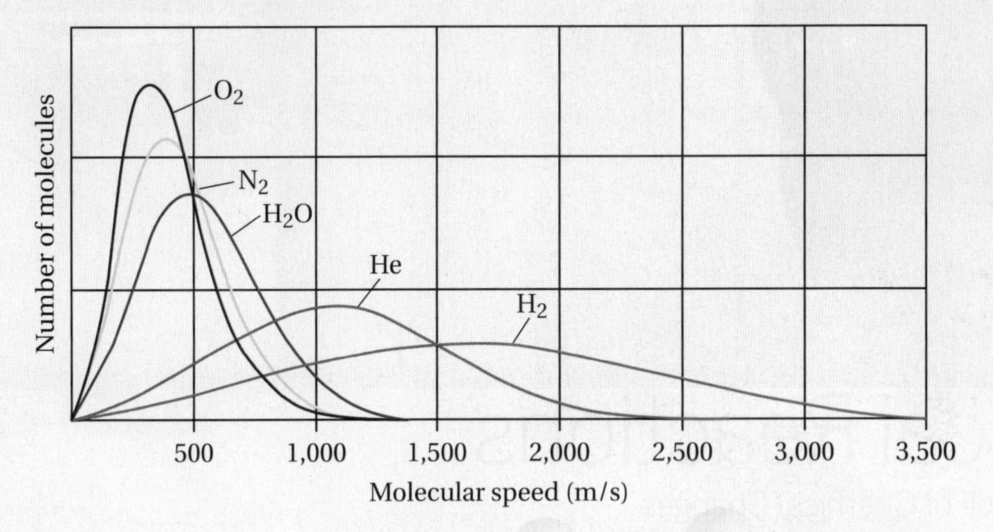

55. INTERNET RESEARCH QUESTION

Return to the Nature Box in this chapter entitled, "Is Natural Gas the Ideal Energy Resource?" In this essay, the dramatic rise of hydraulic fracturing in Pennsylvania, Ohio, and West Virginia is discussed. Earthquakes due to hydraulic fracturing in Oklahoma, Arkansas, and Colorado are also mentioned. These states all have underground deposits of natural gas that are being tapped by hydraulic fracturing, but there are many other states where hydraulic fracturing is used, or could be used, to produce natural gas.

Is hydraulic fracturing being used in your area? To find out, go to the website of the U.S. Energy Information Administration. Under the Sources and Uses tab, select Natural Gas. Now select the U.S. Natural Gas Infrastructure Map. Under "Find address" enter your zip code or full street address.

Answer these questions about natural gas drilling and use in your area.

(a) Are there power plants in your area that run on natural gas?

(b) Do you live over an area with shale gas, colored in tan on the map?

(c) Do you see blue dots in your area on the map? These dots represent gas well locations.

(d) If you don't see any of these features, zoom out away from your local area until you do. What do you see? How far away are the nearest gas power plants, the nearest shale deposits, or the nearest gas wells?

GREENLAWN PRODUCTIONS/Album/Superstock

Charles D. Winters/Science Source

Chemical Reactions

How We Keep Track of Chemical Changes

CHAPTER 7

Chemical reactions sometimes change the course of history, or, in one case, the history of women's fashion. The development of the substance we know as nylon caused a revolution. As soon as it became available in the early 1940s, nylon was found in everything from toothbrushes to fishing line and, of course, ladies' hosiery. In fact, 64 million pairs of nylon stockings were sold in 1940, the first year they were available. Over the next several decades, nylon stockings evolved into the modern pantyhose we know today. The first person to conceive of the idea for pantyhose, and successfully patent it, was Julie Newmar, the woman who also played Catwoman in the Batman television series in the 1960s. Her image is shown on the facing page.

Nylon was one of the first engineered materials. When we say nylon is engineered, we mean that it does not come from a natural source. Instead, nylon is made in a simple chemical reaction. **Figure 7.1** shows how nylon can be made by using a beaker and two chemicals. The two chemicals are layered in a beaker, where they undergo a chemical reaction to form nylon. Then the nylon rope is pulled from the beaker and wrapped around something, in this case a circular piece of plastic.

GREENLAWN PRODUCTIONS/Album/Superstock

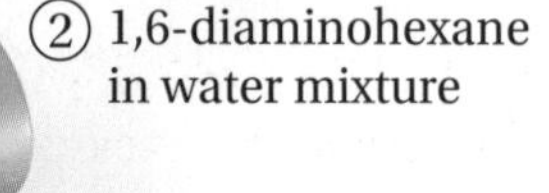

The two chemicals that are put into the beaker each contain a different molecule, and those two molecules have an affinity for one another. When they interact, a chemical reaction takes place, making a new, larger molecule that we call nylon. And because there are formal ways of keeping track of chemical reactions like this one, we can write the reaction in the form of an equation. Then we can use that equation when we want to perform the reaction in the lab.

How exactly is this done? That is, how do we keep track of chemical reactions? How do we know how much of each substance to mix together? These are the central questions we'll answer in this chapter. We begin with another chemical reaction called the thermite reaction, which is performed with a rusty nail, a piece of aluminum foil, and a hammer.

Charles D. Winters/Science Source

Figure 7.1 A Chemical Reaction Demonstrated: The Synthesis of Nylon

Top: Two chemicals used to make nylon are mixed together in a beaker. Bottom.: The nylon that forms in the beaker is pulled out of the beaker and wrapped around a circular piece of plastic.

7.1 Sparks! The Thermite Reaction

Balanced chemical equations represent chemical reactions.

Warning: The reaction we are about to consider can be dangerous! With that forewarning, let's consider what happens when we take a rusty nail, wrap it with a single layer of aluminum foil, and crack it with a hammer. What do we get? Sparks! This chemical reaction, known as the *thermite reaction,* is easy to execute, but how do we write it down? We begin with the chemical formulas for each pure substance involved, and then we arrange them into a chemical equation. A **chemical equation** is a notation chemists use to represent a chemical reaction. It consists of chemical formulas separated by an arrow. The chemical equation for the thermite reaction is shown in **Figure 7.2.**

flashback Recall from Section 2.4 that a chemical reaction is an interaction between atoms or groups of atoms. In a chemical reaction, atoms in substances can give up electrons to other atoms, take away electrons from other atoms, or share electrons with other atoms.

The starting materials in any chemical reaction are **reactants** (they *react*). The substances present after the reaction has been carried out are **products** (they are *produced*). By convention, we place all reactants to the left of the arrow in a chemical equation and we place all products to the right of the arrow.

recurring theme in chemistry 4 Chemical reactions are nothing more than exchanges or rearrangements of electrons.

In most chemical reactions, chemical bonds in the reactants break and new bonds form as the products are created. All the atoms present before a chemical reaction begins are still there after the reaction is finished. What is different after the reaction takes place is the way these atoms are connected to one another. That is, during the reaction, the atoms in the reactants have been rearranged to become products.

The left side of the chemical equation is the *before* side, and it tells you what substances are going to undergo the reaction. The chemical formula for the rust on the nail used in the thermite reaction is Fe_2O_3. Rust, or Fe_2O_3, is also referred to as ferric oxide. Recall from the discussion of compounds in Section 3.4 that rust is a compound because it contains more than one element. Rust is also a solid, and we indicate that with the letter *s* in parentheses after its chemical formula. The chemical formula for the element aluminum is Al. It is the second reactant in the chemical equation in Figure 7.2, and it is also a solid.

flashback Recall from Section 3.4 that in a chemical formula, the element symbols are usually written alphabetically (but there are some exceptions), and a subscript after the symbol tells us how many atoms of each element the compound contains. If there is only one atom of a given element in a compound, then the number 1 is left out of the formula.

There are two products of the thermite reaction. Both appear to the right of the arrow in the chemical equation. The first one, Fe (*s*), is the element iron, which is

Figure 7.2 The Thermite Reaction

A chemical equation shows the reactants on the left and the products on the right, separated by an arrow. The chemical formulas for each of the reactants and products are shown, along with the state of each. In this chemical equation, all reactants and products are solids, indicated with (*s*). Iron oxide is also known as rust.

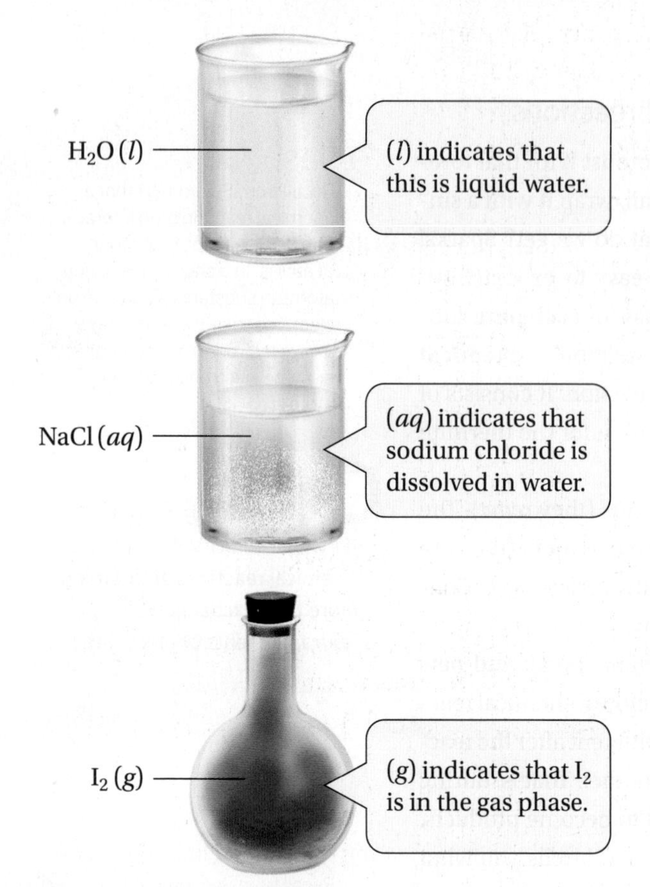

H_2O (*l*)
(*l*) indicates that this is liquid water.
NaCl (*aq*)
(*aq*) indicates that sodium chloride is dissolved in water.
I_2 (*g*)
(*g*) indicates that I_2 is in the gas phase.

Figure 7.3 Symbols That Indicate Phase or State in a Chemical Equation

These symbols for the states of matter, as well as (*s*) for the solid phase (which is not shown here), may be noted in chemical equations. For each state, an example of a substance is given. These examples show the states for water (liquid), sodium chloride in water (aqueous solution), also known as saline, and iodine (gas).

a solid. The second product is aluminum oxide, Al_2O_3, and it too is a solid.

All of the reactants and products in the thermite reaction are solids, but three other *states* (see nearby Wait a Minute) are commonly indicated in chemical equations. These are summarized in **Figure 7.3**. We indicate a pure liquid reactant or product with the symbol *l*, and we indicate a gaseous reactant or product with the symbol *g*. The final symbol, *aq*, stands for "aqueous," and we use it to indicate substances dissolved in water. We'll see many examples of aqueous substances in Chapter 8, "Water," which is devoted to water and things dissolved in it. Note that sometimes the symbols that indicate state are *not* shown in a chemical equation. We commonly use them when the state of a reactant or product is important for some reason.

QUESTION 7.1 Indicate the state of each of the reactants and products in the following chemical equation:

$$Al(OH)_3\,(s) + HCl\,(aq) \rightarrow AlCl_3\,(aq) + H_2O\,(l)$$

ANSWER 7.1 The states are solid, aqueous, aqueous, and liquid, respectively.

For more practice with problems like this one, check out end-of-chapter Questions 2 and 3.

FOLLOW-UP QUESTION 7.1 Indicate the state of each of the reactants and products in the following chemical equation:

$$CaF_2\,(s) + H_2SO_4\,(l) \rightarrow CaSO_4\,(s) + 2HF\,(g)$$

wait a minute . . .

Do the terms *phase* and *state* mean the same thing?

There are three phases of matter: gas, liquid, and solid. However, when writing chemical equations, we also want to be able to identify substances that are dissolved in water, known as aqueous solutions. Because aqueous solutions are not a phase of matter, we use an alternate term, *state*, to describe them. Thus, the symbols (*g*), (*l*), (*s*), and (*aq*) indicate the four states seen in chemical equations: gas, liquid, solid, and aqueous, respectively. In this context, the terms *state* and *phase* are sometimes used interchangeably.

The total number of each type of atom must be the same on both sides of a chemical equation.

The chemical equation for the thermite reaction is shown again in **Figure 7.4**. We know that two reactants and two products are involved in the thermite reaction. Notice that in the chemical equation there is a number, which we call a coefficient, in front of the aluminum that is reacting. There is also a coefficient in front of the iron that is produced. These **coefficients**—which in this case are both 2—indicate the numbers of atoms (or molecules or formula units) present before and after the reaction. Where there is no coefficient, the coefficient is understood to be the number 1. In the thermite reaction, two atoms of Al react with one formula unit of Fe_2O_3 to produce two atoms of Fe and one formula unit of Al_2O_3.

This equation for the thermite reaction brings us to a fundamental tenet of chemistry: in any chemical reaction, matter is neither made nor destroyed. This law, which is always obeyed, *no matter what,* is the **law of conservation of mass**. This law requires equal numbers of each atom on the two sides of any correctly written chemical equation because we can't create new atoms and we can't destroy atoms in a chemical reaction.

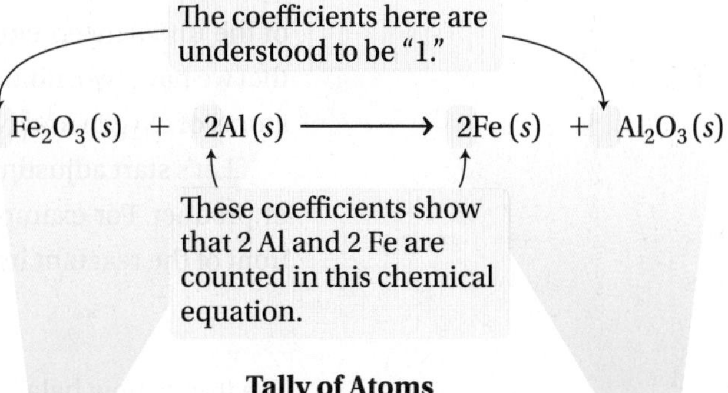

The coefficients here are understood to be "1."
$Fe_2O_3(s) + 2Al(s) \longrightarrow 2Fe(s) + Al_2O_3(s)$
These coefficients show that 2 Al and 2 Fe are counted in this chemical equation.
Tally of Atoms
Fe = 2
Al = 2
O = 3
Because these numbers are equal, this is a balanced chemical equation.
Fe = 2
Al = 2
O = 3

Figure 7.4 The Tally of Atoms in a Chemical Equation

Is the law of conservation of mass obeyed for the thermite reaction? Yes. In **Figure 7.4**, we tally each type of atom on each side of the chemical equation. There are two aluminum atoms on the left and two on the right, and there are two atoms of iron on each side of the chemical equation, and so on. We say that an equation like this one is a *balanced chemical equation* because it has equal numbers of each type of atom on each side. If an equation is unbalanced, the atoms are not equal on both sides, and that equation is not valid.

We add coefficients to reactants and/or products to balance chemical equations.

Now let's do some balancing with a different chemical equation. We've just encountered the compound ferric oxide, Fe_2O_3, otherwise known as rust. Rust can form from the reaction of iron with oxygen. This happens, for example, when a car (the car's iron contributes the iron) sits outside (the oxygen in air contributes the oxygen) for nearly 60 years, like the old rust bucket shown in **Figure 7.5**.

flashback
Recall from Section 4.3 that when we talk about salts, such as aluminum oxide and ferric oxide (rust), we use the term *formula unit* to keep track of them.

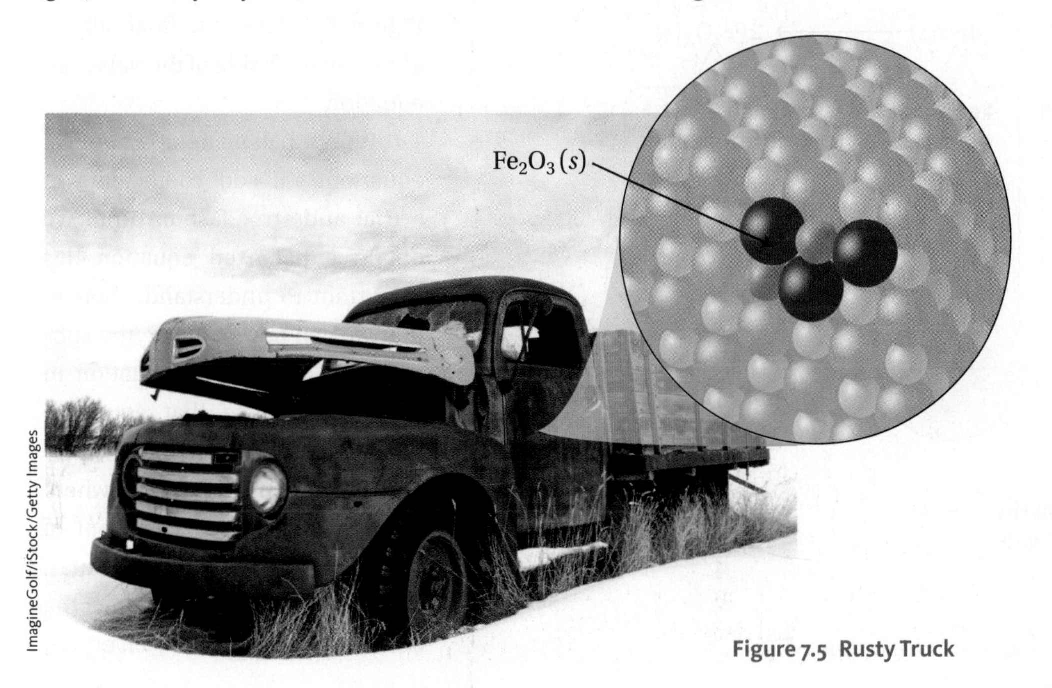

$Fe_2O_3(s)$

ImagineGolf/iStock/Getty Images

Figure 7.5 Rusty Truck

Consider the *unbalanced* chemical equation for the formation of rust from iron and oxygen gas. An unbalanced equation, such as this one, is missing coefficients or it has incorrect coefficients:

$$O_2\,(g) + Fe\,(s) \longrightarrow Fe_2O_3\,(s)$$

We balance this reaction by first looking at how many of each atom is on each side of the unbalanced equation. On the left, we can tell from the chemical formulas that we have two atoms of oxygen and one atom of iron. On the right, we have three atoms of oxygen and two atoms of iron.

Let's start adjusting the equation by adding coefficients in front of any reactant or product. For example, we could balance the iron by putting the coefficient 2 in front of the reactant iron, like this:

$$O_2\,(g) + 2Fe\,(s) \longrightarrow Fe_2O_3\,(s)$$

The iron is now balanced (there are two iron atoms on each side), but the oxygen remains unbalanced. The oxygen is tricky, because we cannot simply put the number 2 in front of the O_2 molecule to balance it. That would give us four oxygen atoms on the left and three on the right. Instead, we can try putting the number 3 in front of the reactant oxygen molecule, like this:

$$3O_2\,(g) + 2Fe\,(s) \longrightarrow Fe_2O_3\,(s)$$

This does not balance the equation, but it does give us six oxygen atoms on the left side.

Because there are three oxygen atoms on the right, let's add a coefficient on the right to balance the oxygen atoms, like this:

$$3O_2\,(g) + 2Fe\,(s) \longrightarrow 2Fe_2O_3\,(s)$$

Now there are six oxygen atoms on each side. But, unfortunately, the iron is no longer balanced: There are two iron atoms on the left and four iron atoms on the right. To fix this, we can change the coefficient on the reactant iron atom from 2 to 4 and—voilà—it's balanced! **Figure 7.6** shows the final tally of atoms on each side of the balanced equation.

When balancing a chemical equation, we add coefficients in a trial-and-error fashion until we obtain a balanced equation. It's important to understand, though, that we can *never* change the subscripts in a chemical equation in order to balance that equation. Look again at the chemical equation shown in Figure 7.6, where we added coefficients in front of each of the chemical formulas. Notice that we did not change the chemical formulas themselves. For

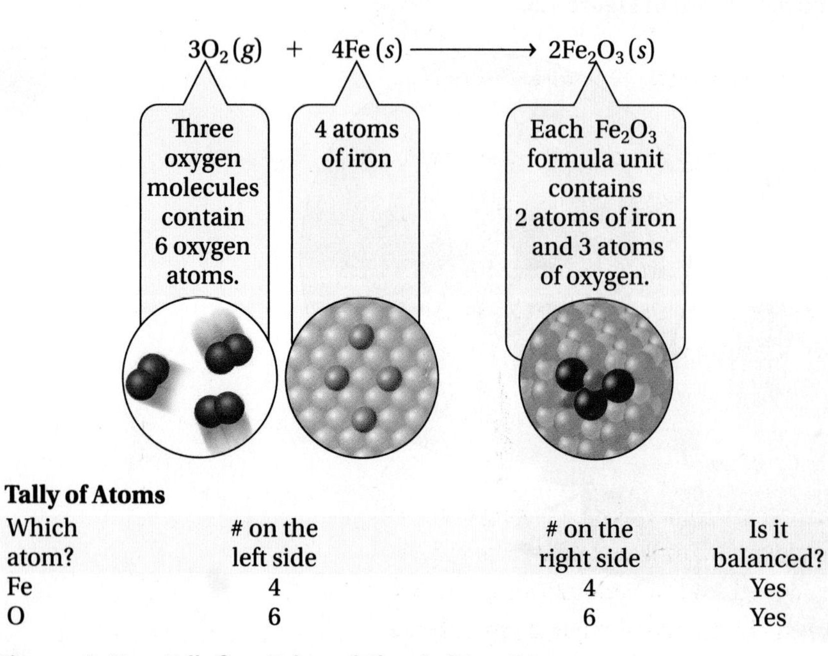

$3O_2\,(g) + 4Fe\,(s) \longrightarrow 2Fe_2O_3\,(s)$
Three oxygen molecules contain 6 oxygen atoms.
4 atoms of iron
Each Fe_2O_3 formula unit contains 2 atoms of iron and 3 atoms of oxygen.

Tally of Atoms

Which atom?	# on the left side	# on the right side	Is it balanced?
Fe	4	4	Yes
O	6	6	Yes

Figure 7.6 Atom Tally for a Balanced Chemical Equation

example, rust is always Fe_2O_3, and we cannot change that formula when balancing a chemical equation.

QUESTION 7.2 The chemical equation shown here is not balanced. Add coefficients in front of any reactant(s) or product(s) to balance the equation.

$$Li\ (s) + F_2\ (g) \longrightarrow LiF\ (s)$$

ANSWER 7.2 The correct answer leaves the fluorine molecule with a coefficient of 1.

$$2Li\ (s) + F_2\ (g) \longrightarrow 2LiF\ (s)$$

For more practice with problems like this one, check out end-of-chapter Questions 10 and 11.

FOLLOW-UP QUESTION 7.2 The chemical equation shown here is not balanced. Add coefficients in front of any reactant(s) or product(s) to balance the equation.

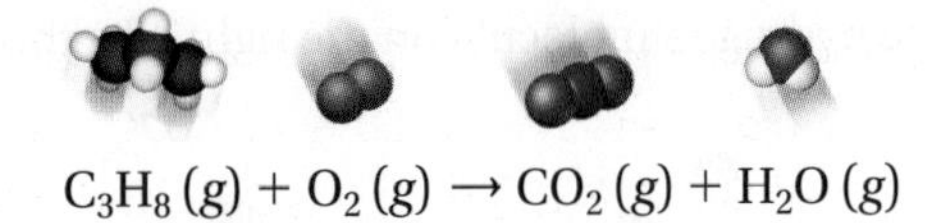

$$C_3H_8\ (g) + O_2\ (g) \longrightarrow CO_2\ (g) + H_2O\ (g)$$

---

wait a minute . . .

Aren't there different combinations of coefficients that can balance any given chemical equation?

Yes, there are different combinations of coefficients that you can use to balance the same chemical equation. For example, when we balanced the equation for that formation of rust, we ended up with this:

$$3O_2\ (g) + 4Fe\ (s) \longrightarrow 2Fe_2O_3\ (s) \quad \text{(Let's call this Rust One.)}$$

But we could also have arrived at this answer:

$$6O_2\ (g) + 8Fe\ (s) \longrightarrow 4Fe_2O_3\ (s) \quad \text{(Rust Two)}$$

Both equations are balanced. So, which one is correct? The answer is that, technically, they are both correct because they both obey the law of conservation of mass: Equal numbers of each atom are on each side of the chemical equation. However, if you divide all the coefficients in Rust Two by 2, you get Rust One. Even though Rust Two is balanced, Rust One is preferred because it's simpler. So, when you are balancing a chemical equation, pay attention to the final coefficients. If they are all divisible by the same number, then it's correct to write the simpler version of the balanced chemical equation.

Balancing equations is a trial-and-error process.

We just balanced the equation for the formation of rust. It was challenging because we had to go through three versions of the equation until we had it balanced. As this process illustrates, balancing chemical equations is a trial-and-error process. We add coefficients and remove them. We might double or triple them or erase them until we get all of the atoms properly balanced.

Let's work through one more example. The unbalanced equation shown here represents the reaction of water, H_2O, to form molecules of hydrogen, H_2, and oxygen, O_2.

$$H_2O\ (l) \rightarrow H_2\ (g) + O_2\ (g)$$

A quick survey of this equation shows us that one of the elements, hydrogen, is already balanced; each side has two atoms of hydrogen. The oxygen, however, is not. One atom of oxygen is on the left side and two are on the right side. To get started, let's put the coefficient 2 in front of the water molecule, like this:

$$2H_2O\ (l) \rightarrow H_2\ (g) + O_2\ (g)$$

Now the oxygen is balanced. However, by adding the coefficient in front of the water molecule, we have unbalanced the hydrogen: now we have four atoms of hydrogen on the left and only two on the right. To adjust this element, we can add the coefficient 2 to the hydrogen molecule on the right, like this:

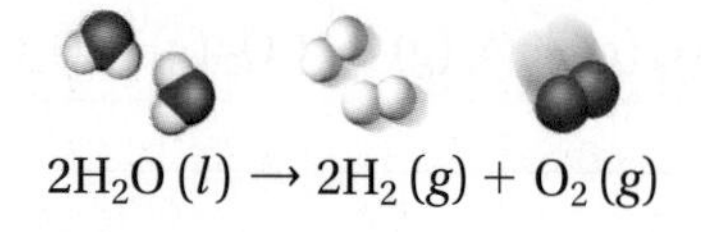

$$2H_2O\ (l) \rightarrow 2H_2\ (g) + O_2\ (g)$$

Now the equation is completely balanced.

Take note of how this example unfolded. We balanced one element, only to find that another element was no longer balanced. Then we adjusted that element. Sometimes this process happens over and over again. Balancing a chemical equation is like solving a puzzle. And when we think we've solved the puzzle, we always should check ourselves by tallying the left and right sides of the equation.

QUESTION 7.3 Balance the following chemical equation by putting coefficients in front of reactants and/or products:

$$NaN_3\ (s) \rightarrow Na\ (s) + N_2\ (g)$$

ANSWER 7.3 In this chemical equation, sodium is balanced initially, but nitrogen is not. You can use trial and error to balance it. Here is one approach: To balance nitrogen, try putting the coefficient 3 in front of N_2 and the coefficient 2 in front of NaN_3. Now sodium is no longer balanced. But if you add a coefficient in front of Na, you get the following balanced result:

$$2NaN_3\ (s) \rightarrow 2Na\ (s) + 3N_2\ (g)$$

For more practice with problems like this one, check out end-of-chapter Questions 9, 13, and 14.

FOLLOW-UP QUESTION 7.3 Balance the following chemical equation:

$$C\ (s) + O_2\ (g) \rightarrow CO\ (g)$$

QUESTION 7.4 Balance the following chemical equation by putting coefficients in front of reactants and/or products:

$$Al\ (s) + O_2\ (g) \rightarrow Al_2O_3\ (s)$$

ANSWER 7.4 In this chemical equation, nothing is balanced initially, so you need to use trial and error to balance both elements. Here is one approach: to balance the oxygen, add the coefficient 2 in front of Al_2O_3 and the coefficient 3 in front of O_2. Now put the correct coefficient in front of Al to get the following balanced equation:

$$4Al\ (s) + 3O_2\ (g) \longrightarrow 2Al_2O_3\ (s)$$

For more practice with problems like this one, check out end-of-chapter Questions 12, 15, and 16.

FOLLOW-UP QUESTION 7.4 Balance the following chemical equation:

$$KO_2\ (s) + CO_2\ (g) \longrightarrow K_2CO_3\ (s) + O_2\ (g)$$

7.2 Atomic Accounting

We can view a chemical equation from different perspectives.

Let's consider a balanced chemical equation from a few different perspectives. The first perspective is the **atomic scale**. We return to our water molecule reaction for an example:

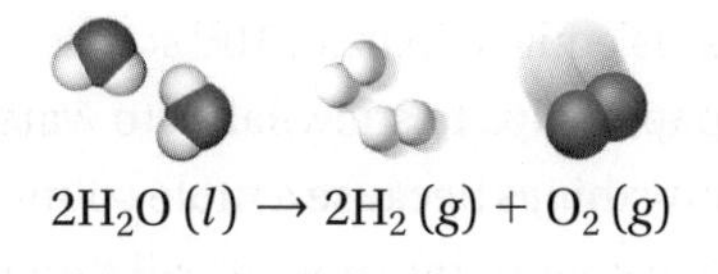

$$2H_2O\ (l) \longrightarrow 2H_2\ (g) + O_2\ (g)$$

What does it mean to view this chemical reaction on an atomic scale? It entails imagining ourselves to be as small as the atoms or molecules in the reaction. It also requires us to think of the reactants as structures moving through space and perhaps colliding and forming the new structures that are the products, as depicted in **Figure 7.7**.

On the atomic scale, this equation paints a picture of two molecules of water undergoing a chemical reaction and producing two molecules of hydrogen gas and one molecule of oxygen. The atomic perspective is useful when we want to imagine the reactants somehow rearranging and the atoms switching partners so that the products can form. For example, we might imagine the rearrangement taking place as illustrated in Figure 7.7. Does this really happen when two water molecules collide with one another? That's a complicated question, because the details of what interacts with what and how bonds change places can only be determined experimentally, and we will not describe those experiments here. But when we're thinking on the atomic scale, this is how we can imagine a reaction might happen.

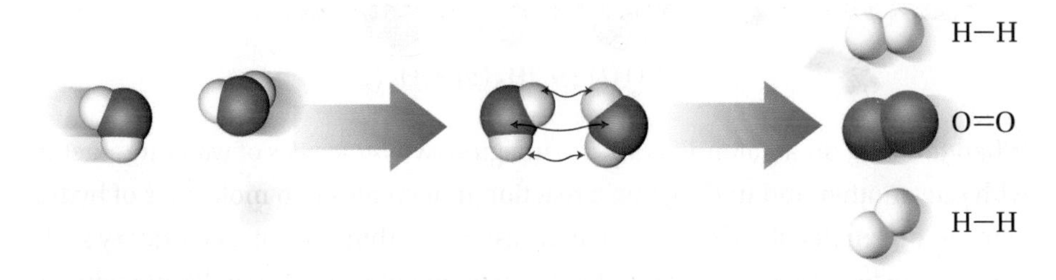

Figure 7.7 Imagining Water
We can visualize how water is converted to hydrogen and oxygen by imagining the collision of two molecules and the formation of products as the atoms change bonding partners.

On the atomic scale, we can imagine our two water molecules floating around, bumping into each other and possibly undergoing a chemical reaction. But two water molecules is not a number that we could work with in a laboratory. If we can see a substance with our naked eyes, then many, many atoms and/or molecules of that substance are present—far more than two. So, it is helpful to have a different, more practical scale. If we scale up this reaction to everyday, laboratory-sized amounts, we can make the chemical equations more useful, as we are about to see.

Our second perspective for a chemical reaction is the **laboratory scale**. This view gives us a practical, real-life perspective on the reaction. Many, many, many molecules of water are involved in the chemical reactions we can perform in a laboratory. Roughly how many water molecules are we talking about? If we could count the number of water molecules in a 6-ounce glass of water, we would come up with a number like this: 6×10^{24} molecules—a huge number! Happily, we have a better way to keep track of the huge numbers that always arise when we are dealing with atoms and molecules on a laboratory scale.

recurring theme in chemistry 2

Things we can see with our eyes contain trillions upon trillions of atoms. Atoms are much too small to view with the naked eye.

The mole is a counting device that helps us to think on a macroscopic scale.

In Chapter 6, we introduced the mole. Here in Chapter 7 we introduce the mole again, but with a different purpose. The mole is a counting device that we use to keep track of huge numbers of atoms and molecules. Recall from Section 6.3 that one mole of "somethings" is equal to 6.02×10^{23} somethings. We can use the mole to count anything from paper clips to snowflakes to water molecules, but we typically use it only for very tiny things because a mole counts an enormous number of somethings. We use the word *mole* for super-small things like formula units, molecules, and atoms.

flashback

We first discussed the very, very small sizes of atoms in Section 1.2. Figure 1.6 illustrates the number of water molecules in a glass of water.

flashback

Recall from Table 1.1 that the mole is the metric unit for the quantity of a substance. It is abbreviated *mol*.

For example, we know from Section 4.3 that salts are counted in formula units, so we might say that we have one mole of sodium chloride formula units. We could also have one mole of water molecules—because water exists as molecules. Or, because some elements, like iron, exist only as single atoms, we could also have one mole of iron atoms.

In the real world, substances don't exist only in one-mole quantities. We can also have fractions of a mole or multiples of a mole. So, for example, we may have half a mole of potassium chloride formula units, or 10 moles of carbon dioxide molecules. Or, because aluminum can exist as individual atoms, we could have 650 moles of aluminum atoms.

We can also use the mole to handle the huge numbers of atoms or molecules or formula units that undergo chemical reactions. Let's return to our water example:

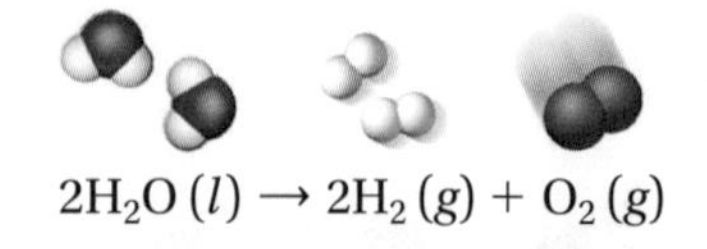

$$2H_2O\,(l) \rightarrow 2H_2\,(g) + O_2\,(g)$$

When thinking on an atomic scale, we imagine two *molecules* of water interacting with one another and undergoing a reaction that creates two molecules of hydrogen and one molecule of oxygen. In contrast, when thinking on a laboratory scale, we imagine two *moles* of water molecules reacting to form two moles of hydrogen molecules and one mole of oxygen molecules. The beauty of a balanced chemical reaction is that the ratio of reactant and products to one another is the same regardless of the scale.

Molar mass is a way to convert mole units to mass units.

If we walk into a laboratory to try a chemical reaction, we want to be thinking on the laboratory scale. We think in terms of moles of substances, because moles will give us amounts that we can see or measure or analyze. But, if we want to start with two moles of water, how do we measure out that exact amount?

To make use of moles, we need a way to convert the unit moles to something we can really use, like a volume we can measure in a graduated cylinder or a mass we can weigh on a balance. Volume is a topic we'll get into in Chapter 8, "Water." We will consider mass here.

Let's flip to the front of the book and check out the periodic table again. Recall that the number *above* the element symbol is the atomic number. This tells us the number of protons an atom of that element contains. The number *below* the element symbol is the **molar mass** of that element. As the name implies, molar mass is the mass, in grams, of one mole of atoms of an element. For example, the molar mass of radon, element 86, is 222 grams per mole. That means that one mole of radon atoms has a mass of 222 grams. Molar mass is written in several ways, as we demonstrate in **Figure 7.8**. These are all ways to say the same thing: the mass of one mole of radon atoms is 222 grams. In this book, we write molar masses like this: 222 g/mol.

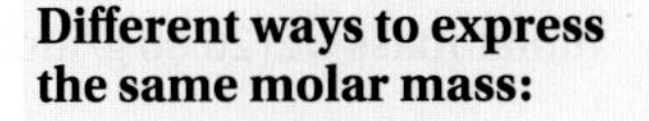

Different ways to express the same molar mass:

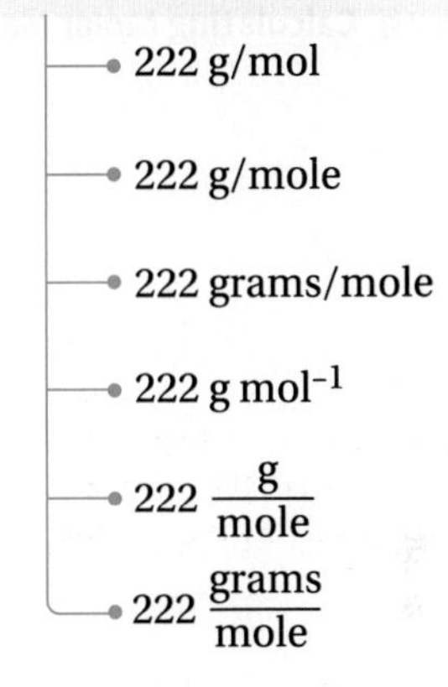

Figure 7.8 Ways to Express Molar Mass

Let's apply this concept. If we have half a mole of radon atoms, what is its mass? It weighs half of 222 grams, or 111 grams. Two moles of radon atoms have a mass that is two times 222 grams, or 444 grams. One-tenth of a mole of radon atoms has a mass that is one-tenth of 222 grams, or 22.2 grams, and so on. We are not compelled to work only in whole numbers of moles. Nanomoles, millimoles, and micromoles are all legitimate metric units. These are all laboratory-scale quantities because when we are working with moles or fractions of moles or multiples of moles, we are working with quantities of substances that we can manipulate and measure.

QUESTION 7.5 What is the mass of one-quarter mole of oxygen atoms?

ANSWER 7.5 According to the periodic table, one mole of oxygen atoms has a mass of 16.00 grams. So, one-quarter mole has one-quarter of that quantity, or 4.000 grams.
For more practice with problems like this one, check out end-of-chapter Questions 22 and 23.

FOLLOW-UP QUESTION 7.5 A balloon filled with helium atoms contains 6.02×10^{23} atoms of helium. What mass of helium is contained in the balloon?

We can calculate molar mass for any element or compound.

We just saw that, because scales measure mass and not moles, we must have a way to convert moles to mass. We can look up molar masses of various elements on the periodic table. For example, iron (Fe) has a molar mass of 55.847 g/mol, and oxygen (O) has a molar mass of 15.9994 g/mol. But what about rust, also known as ferric oxide or Fe_2O_3? How do we figure out the molar mass of a compound? It's as straightforward as adding up the molar masses of everything in the chemical formula. In the case of rust, the formula tells us that each formula unit includes two

a **Calculating the molar mass of Fe_2O_3**

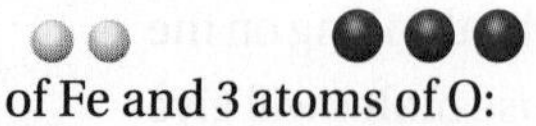

Fe_2O_3 contains 2 atoms of Fe and 3 atoms of O:

molar mass = 2(55.85 g/mol) + 3(16.00 g/mol) = 159.7 g/mol

b **Calculating the molar mass of Al_2O_3**

Al_2O_3 contains 2 atoms of Al and 3 atoms of O:

molar mass = 2(26.98 g/mol) + 3(16.00 g/mol) = 101.96 g/mol

Figure 7.9 Calculating Molar Mass for Compounds

atoms of iron and three atoms of oxygen. So we multiply the molar mass of iron times two, and we multiply the molar mass of oxygen by three and add them together, as shown in **Figure 7.9a.** The grand total is 159.7 grams per mole. Thus, the molar mass of ferric oxide is 159.7 g/mol.

Let's try one more. Aluminum oxide, Al_2O_3, includes two atoms of aluminum and three atoms of oxygen. The molar mass of aluminum is 26.9815 g/mol. Adding it all up, we get a grand total of 101.96 g/mol. The math is worked in **Figure 7.9b.**

QUESTION 7.6 Urea is a compound that is used as a fertilizer. Its chemical formula is CH_4N_2O. What is the molar mass of urea?

ANSWER 7.6 To find the molar mass, add up the molar masses of each of the elements in urea, taking into account the number of atoms of each: (1)(12.01 g/mol) + (4)(1.008 g/mol) + (2)(14.01 g/mol) + (1)(16.00 g/mol) = 60.06 g/mol.

For more practice with problems like this one, check out end-of-chapter Questions 29 and 31.

FOLLOW-UP QUESTION 7.6 The chemical name for baking soda is sodium bicarbonate, $NaHCO_3$. What is the molar mass of baking soda?

7.3 Stoichiometry

Stoichiometry allows us to use chemical equations to perform chemical reactions.

Stoichiometry (pronounced *stoy-kee-OM-uh-tree*) is a complicated word for a straightforward process. It is a way to use a balanced chemical equation to figure out how much of each reactant reacts and how much product forms.

Imagine this fictional situation: A mad scientist is working late in the lab and has a splitting headache (perhaps from balancing too many chemical equations). What the scientist really needs is two aspirin tablets, which are equal to about 650 milligrams of aspirin. She has no aspirin on hand, but she knows that the chemical name for aspirin is acetylsalicylic acid, and she happens to know how to make it. As luck would have it, she has on hand the ingredients she needs, so she decide to whip some up.

The chemical equation for the preparation of aspirin is shown in **Figure 7.10a.** Rather than write the long, chemical names for the two reactants, we call them simply Reactant A and Reactant B. The products are aspirin and acetic acid, which is the substance that is diluted to make vinegar. Whenever we encounter a new chemical equation, the first thing we should always do is check to see if it is balanced. **Figure 7.10b** is a tally of each of the three atoms involved in this chemical reaction: carbon, hydrogen, and oxygen. The chemical equation is balanced.

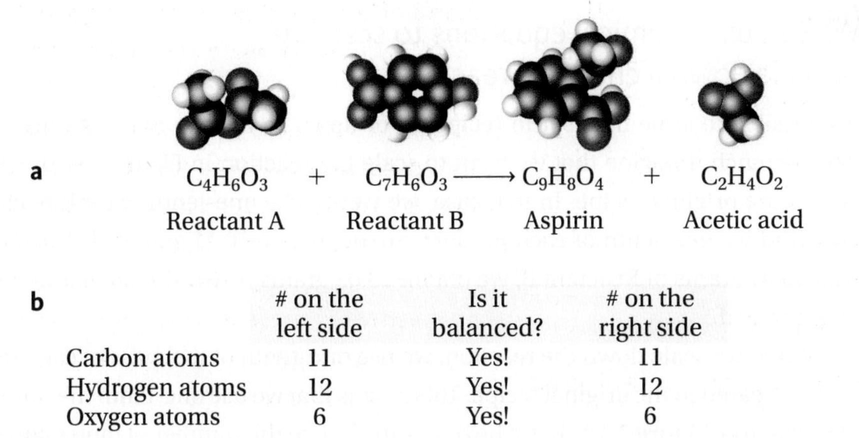

a

$$C_4H_6O_3 + C_7H_6O_3 \longrightarrow C_9H_8O_4 + C_2H_4O_2$$

Reactant A   Reactant B   Aspirin   Acetic acid

b

	# on the left side	Is it balanced?	# on the right side
Carbon atoms	11	Yes!	11
Hydrogen atoms	12	Yes!	12
Oxygen atoms	6	Yes!	6

Figure 7.10 Making Aspirin

The reaction for the synthesis of aspirin is shown here in a simplified version (a) with the tally for the balanced reaction (b).

We have a balanced chemical equation, and we have the reactants in the lab. But how do we know how much of each reactant we need? Thinking on an atomic scale, we know that one molecule of each reactant produces one molecule of each product. If we scale this chemical reaction up to the laboratory scale, we can say that mixing one mole of each reactant will give us one mole of each product. To convert this equation into a recipe we can follow in the lab, we need to know the quantity of each reactant needed. To get that information, we have to use the molar mass of each reactant. Try this on your own: What are the molar masses of Reactant A and Reactant B? See **Figure 7.11** for the answer, along with the molar masses of the two products.

Now we have something quite useful. If we mix 102.1 grams of Reactant A with 138.1 grams of Reactant B, we get 180.2 grams of aspirin and 60.1 grams of acetic acid. Is that enough aspirin to conquer the scientist's headache? Yes. The typical adult dose is only 650 milligrams of aspirin. We'll have enough for 275 headaches! In fact, we will make enough aspirin to fill a large bottle of aspirin tablets. (Remember that this example is fictitious; we *never* consume substances that we make ourselves in the chemistry lab.)

$$C_4H_6O_3 + C_7H_6O_3 \longrightarrow C_9H_8O_4 + C_2H_4O_2$$

$C_4H_6O_3$	$C_7H_6O_3$	$C_9H_8O_4$	$C_2H_4O_2$
Reactant A has a molar mass of 102.09 g/mol.	Reactant B has a molar mass of 138.12 g/mol.	Aspirin has a molar mass of 180.16 g/mol.	Acetic acid has a molar mass of 60.05 g/mol.

Figure 7.11 Molar Masses in Aspirin Reaction

Reactant A and Reactant B combine to make aspirin and acetic acid. The molar masses of these reactants and products are shown.

We can use chemical equations to scale up or scale down a chemical reaction.

It's possible to scale the aspirin recipe up or down so that we make enough, but not too much. Imagine that we want to scale the reaction in Figure 7.11 down to 10% of the original recipe. In that case, we would use one-tenth of each reactant and produce one-tenth of each product. So, by mixing 10.21 grams of Reactant A with 13.81 grams of Reactant B, we obtain 18.02 grams of aspirin and 6.005 grams of acetic acid.

When we scale down the reaction, we use one-tenth of the mass of each reactant compared to the original recipe. This means that we use one-tenth the number of moles, too. **Figure 7.12** shows how we can change the number of moles and still have a balanced chemical equation.

100 100 moles of $C_4H_6O_3$ + 100 moles of $C_7H_6O_3 \longrightarrow$ 100 moles of $C_9H_8O_4$ + 100 moles of $C_2H_4O_2$

5 5 moles of $C_4H_6O_3$ + 5 moles of $C_7H_6O_3 \longrightarrow$ 5 moles of $C_9H_8O_4$ + 5 moles of $C_2H_4O_2$

1 1 mole of $C_4H_6O_3$ + 1 mole of $C_7H_6O_3 \longrightarrow$ 1 mole of $C_9H_8O_4$ + 1 mole of $C_2H_4O_2$

0.1 0.1 mole of $C_4H_6O_3$ + 0.1 mole of $C_7H_6O_3 \longrightarrow$ 0.1 mole of $C_9H_8O_4$ + 0.1 mole of $C_2H_4O_2$

Figure 7.12 Scaling the Aspirin Reaction Up and Down

As we can see in Figure 7.12, we can scale down a chemical reaction, but we can also scale up a chemical reaction.

QUESTION 7.7 Nitrous oxide, N_2O, is also known as laughing gas. It is the anesthetic sometimes administered for dental work, and it's also the gas used as a propellant for whipped cream. Consider the following chemical equation for the reaction of nitrous oxide with oxygen gas:

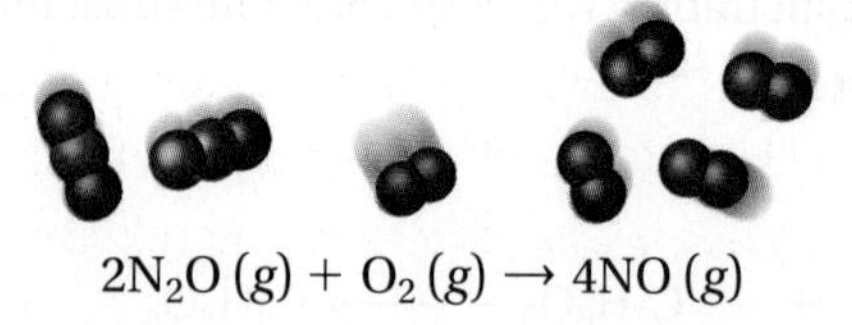

$$2N_2O\,(g) + O_2\,(g) \rightarrow 4NO\,(g)$$

(a) Is the equation balanced? (b) If 10 moles of N_2O react with 5 moles of oxygen gas, how many moles of the product form?

ANSWER 7.7 (a) The reaction is balanced. (b) The reaction is scaled up five times from what is written. Thus, $5 \times 4 = 20$ moles of the product form.

For more practice with problems like this one, check out end-of-chapter Questions 39 and 40.

FOLLOW-UP QUESTION 7.7 Returning to the equation shown in Question 7.7, consider this question: If 2 moles of NO form, how many moles of N_2O reacted?

Balanced chemical equations obey the law of conservation of mass.

We will consider one more example of stoichiometry. *Butane* gas has the chemical formula C_4H_{10}. It is the fuel found in disposable lighters. When you turn the metal wheel in a lighter, a spark is made and oxygen molecules from the air ignite the gas. This chemical reaction, called **combustion,** occurs whenever a fuel, such as butane, burns in the presence of oxygen to produce carbon dioxide (CO_2) and water (H_2O). We will see other examples of combustion reactions in this chapter and throughout this book but, for now, let's write a balanced chemical equation for the combustion of butane:

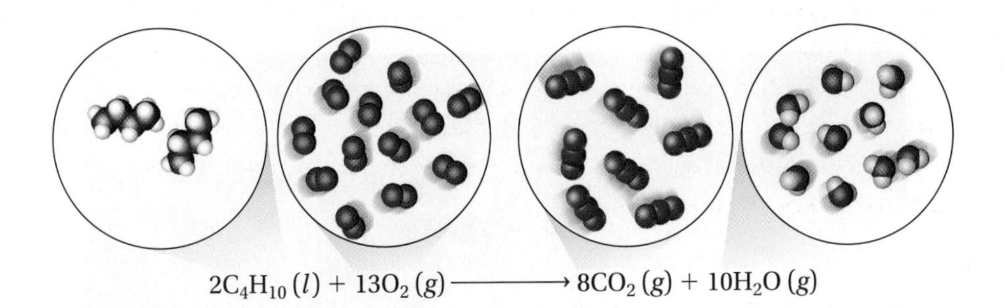

$$2C_4H_{10}\,(l) + 13O_2\,(g) \longrightarrow 8CO_2\,(g) + 10H_2O\,(g)$$

This balanced chemical equation tells us that 2 moles of butane react with 13 moles of oxygen gas to produce 8 moles of carbon dioxide and 10 moles of water. **Figure 7.13** shows the molar masses of each of the reactants and products in the balanced combustion equation. The "recipe" we are using—the balanced chemical equation—calls for 2 moles of butane, so we need 2 moles × 58.12 grams/mol = 116.24 grams. This mass of butane reacts with 13 moles of oxygen molecules. Using the molar mass, we see that 13 moles of oxygen have a mass of 416.00 grams, and so on.

Figure 7.13 shows this calculation for every reactant and product. Now let's consider this question: is the law of conservation of mass obeyed by this chemical equation? We know that the atoms on each side are balanced. But is mass conserved? Have we gained or lost mass during this chemical process?

In Figure 7.13, the total masses of all reactants and products are tallied, and we can see that no mass is gained or lost during this chemical reaction. This must be true because our balanced chemical reaction has equal numbers of each type of atom on each side. Thus, mass is conserved and the law of conservation of mass is obeyed.

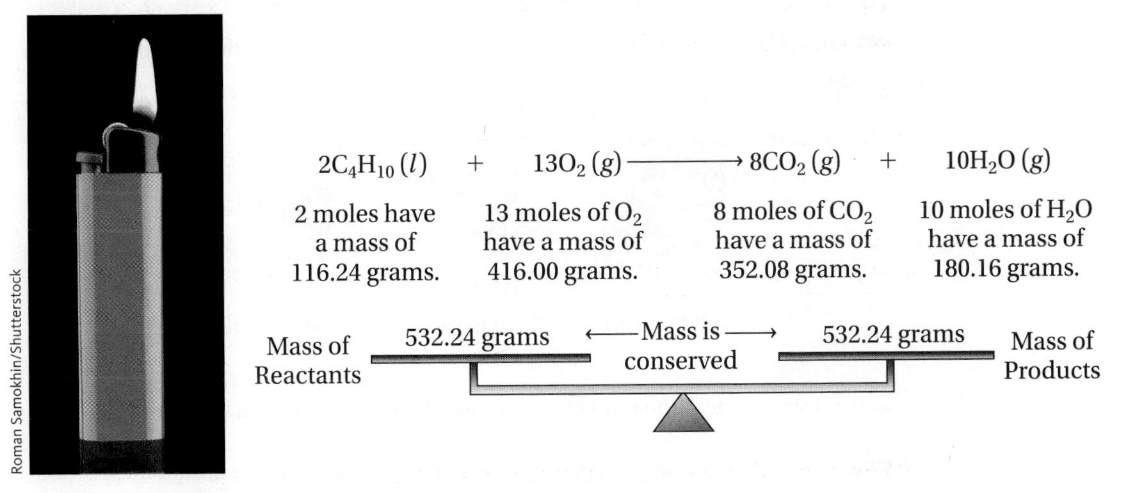

$2C_4H_{10}\,(l)$ + $13O_2\,(g)$ → $8CO_2\,(g)$ + $10H_2O\,(g)$
2 moles have a mass of 116.24 grams.
13 moles of O_2 have a mass of 416.00 grams.
8 moles of CO_2 have a mass of 352.08 grams.
10 moles of H_2O have a mass of 180.16 grams.
Mass of Reactants
532.24 grams
Mass is conserved
532.24 grams
Mass of Products
Roman Samokhin/Shutterstock

Figure 7.13 Stoichiometry in a Butane Lighter

natureBOX • Two Ozone Holes?

As our climate continues to change, we are getting used to bad news. The oceans are rising. One by one, small islands are disappearing (for more on this topic, see *The Green Beat,* Chapter One Edition). The summer of 2012 was the hottest on record. These stories are no longer shocking news. But in 2011, when the news broke about a second ozone hole, the scientific community was dumbstruck. How could that happen? To answer that question, we have to know something about ozone, and why it's in our atmosphere in the first place.

We live in the part of the atmosphere, the *troposphere,* that is closest to Earth's surface. Above that is the *stratosphere* (**Figure 7A**). The stratosphere contains an abundance of ozone, O_3, and its job is to protect the Earth from dangerous radiation that comes from the sun. We can think of the ozone layer as Earth's layer of sunscreen.

In a given calendar year, the ozone layer gets thinner as winter becomes spring. And, back in the 1970s, scientists noticed that the ozone layer had gotten very thin. In fact, it had formed a distinct hole that was centered over Antarctica. Once they identified it, scientists started keeping an eye on that hole and thinking about why it was there. In the 1980s, three scientists figured out that a pair of ozone-destroying chemical reactions were responsible for the loss of ozone in the stratosphere. **Figure 7B** shows the balanced chemical equations for the reactions that destroy ozone.

In the first chemical equation, an ozone molecule reacts with an atom of chlorine. The Cl–O that is formed takes part in a second chemical reaction that

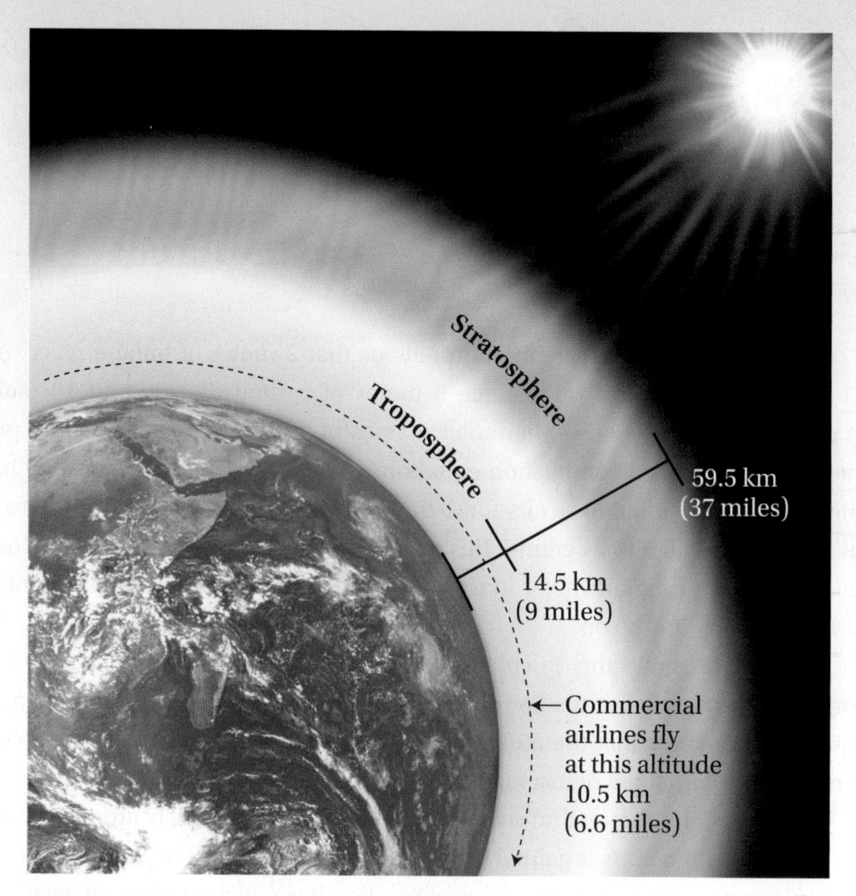

Figure 7A The Stratosphere and the Troposphere

QUESTION 7.8 The following balanced chemical equation includes the molar masses of reactants and products:

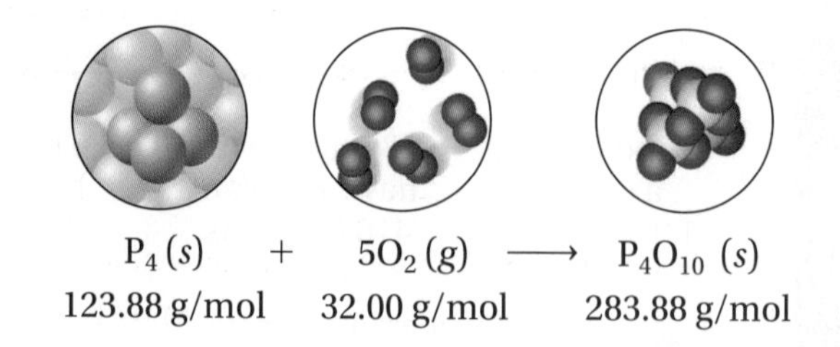

$$\underset{123.88\ \text{g/mol}}{P_4\,(s)} + \underset{32.00\ \text{g/mol}}{5O_2\,(g)} \longrightarrow \underset{283.88\ \text{g/mol}}{P_4O_{10}\,(s)}$$

Show that this chemical equation obeys the law of conservation of mass.

ANSWER 7.8 To determine if the law of conservation of mass is obeyed, sum the masses of all reactants and all products and compare them. The reactants include 1 mole of P_4,

re-forms a new chlorine atom, which can then take part in the first reaction again. The net result is loss of ozone, O_3, and formation of oxygen molecules, O_2.

These reactions take place due to the presence of chlorine atoms in the stratosphere, something that wasn't observed until the arrival, in the twentieth century, of *chlorofluorocarbons,* also known as CFC molecules. Down here on Earth's surface, we used CFCs in aerosol sprays and in refrigerators and freezers, until scientists figured out that they're the reason for chlorine atoms in the stratosphere. And those chlorine atoms are the reason the ozone layer developed a big hole in the area over Antarctica. An example of a CFC is $CFCl_3$, a molecule used to make Freon.

In 1987, a global ban on CFCs was put into place. The levels of CFCs in the stratosphere increased until about the year 2000, when they started to drop off thanks to the ban on the release of new CFCs. With CFCs out of the picture, everyone was expecting a gradual decrease in the ozone hole over Antarctica. But in 2011, the unexpected happened. Observers reported that the hole over the Antarctic was larger than usual, and an entirely new hole had opened up—this time over the Arctic.

Scientist Gloria Manney (**Figure 7C**) of the Jet Propulsion Lab, now at NorthWest Research Associates, led an international team of scientists in an investigation of this bizarre finding. Here is what they determined: The stratosphere still contains huge numbers of chlorine atoms due to CFC use in the twentieth century before the ban. This chlorine is still destroying ozone, and it destroys ozone even faster when the temperature drops below a certain chilly point. In the winter of 2011, the stratosphere over the Arctic got that cold. Those chilly chlorine atoms depleted enormous amounts of ozone, and so a second hole formed.

The loss of ozone was felt most in Europe, which received an unexpected dose of ultraviolet radiation because of the new Arctic ozone hole. Just this one episode of ozone hole formation is thought to have significantly reduced the growth of some crops in Europe, especially wheat.

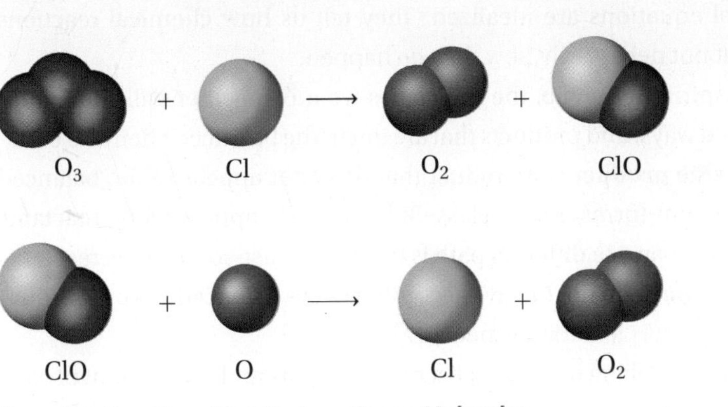

Figure 7B Reactions That Destroy Ozone Molecules in the Stratosphere

These two reactions occur in the stratosphere and account for the ozone loss occurring there.

Figure 7C Scientist Gloria Manney

which has a mass of 123.88 grams. They also include 5 moles of oxygen, which have a mass of 5 mol × 32.00 g/mol = 160.0 grams. The sum of both reactants is 123.88 grams + 160.0 grams = 283.88 grams. One mole of product forms, with a mass of 283.88 grams. This number matches the masses of the reactants, so the law of conservation of mass is obeyed.

For more practice with problems like this one, check out end-of-chapter Question 37.

FOLLOW-UP QUESTION 7.8 Perform the same analysis on the following equation that you did in Question 7.8:

$$\underset{24.31\ \text{g/mol}}{2\ \text{Mg}} + \underset{32.00\ \text{g/mol}}{\text{O}_2} \longrightarrow \underset{40.31\ \text{g/mol}}{2\ \text{MgO}}$$

What are the total masses of reactants and products for this chemical reaction?

7.4 Chemical Reactions in the Real World

Real chemical reactions are usually not as straightforward as their equations imply.

So far, we have introduced a few chemical reactions: we've made sparks by hitting a rusty nail with a hammer (the thermite reaction), we've seen how an old pickup got rusty, and we've made aspirin. For each of these examples, we wrote a balanced chemical equation showing that each obeyed the law of conservation of mass. And, if we had to, we could figure out how much of the reactants we would need to perform each reaction. We also know how to use molar mass to figure out the recipe to follow in the laboratory, and, if desired, we can scale it up or scale it down. And finally, we know how to figure out how much product we will obtain when the reaction is finished. The ability to easily scale up and scale down makes chemical reactions seem very straightforward.

But they're not straightforward, really. Chemical reactions are messy, and very few of them work in an ideal way. For example, if we were to make aspirin using the quantities given in Figure 7.11, we should obtain 180.2 grams of aspirin. But if we actually performed the reaction, we would inevitably get less than that. Why? Because chemical equations are idealized. They tell us how chemical reactions *could* happen, but not necessarily how they *do* happen.

If we make aspirin in the lab, the molecules we mix together will sometimes react in unexpected ways, and products that are not in the balanced chemical equation will form. A **side product** is a product that does not appear in the balanced chemical equation, but forms nonetheless. Side products appear when reactants interact in a different way or a different path is taken. Because some of the reactants in the synthesis of aspirin do not behave according to the balanced chemical equation, the yield of aspirin is less than expected.

A recipe used for cooking includes a list of ingredients and the amount of each substance, and it states the expected yield (e.g., "this recipe makes four servings of brownies"). This is the sort of information we get from a balanced chemical equation. But a good recipe also gives us instructions we can use in the kitchen: the oven temperature, how to mix ingredients, what kind of pan to use, how long to cook it, and so on. It's no different for chemistry: When we perform a chemical reaction in the lab, we need more than just the chemical equation. We need a recipe that tells us the details: the best temperature to use and the best flask size, how long to heat it, when to filter or decant or cool or shake. It's not realistic to perform a chemical reaction based solely on the balanced chemical reaction, any more than we can make brownies when we have only a list of brownie ingredients. You need someone's recipe. Most of us rely on the hard work of others—whether they are chefs or chemists—to be successful in the kitchen or in the lab.

A reaction energy diagram illustrates the progress of a chemical reaction.

If we type the phrase "chemical reaction" into the YouTube search box, we're treated to videos of hundreds of chemical reactions. And often the most video-worthy reactions are explosions, because they are exciting and unpredictable. Chemists

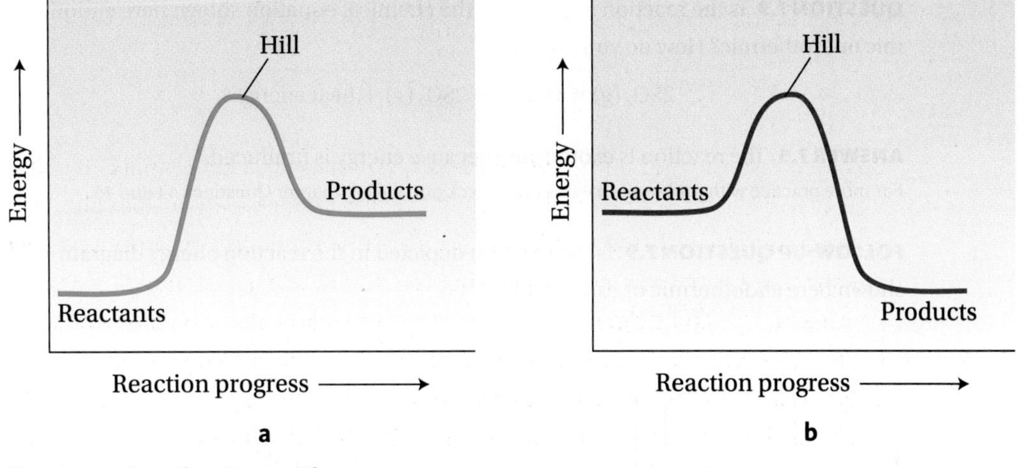

Figure 7.14 Reaction Energy Diagrams

The reaction energy diagram on the left (a) represents an endothermic reaction because energy must be put into the reaction; the reaction energy diagram on the right (b) represents an exothermic reaction that produces heat.

track the progress of chemical reactions in less dramatic fashion by using **reaction energy diagrams,** which tell us how the energy of the reaction changes as it progresses from reactants to products.

The reaction energy diagram in **Figure 7.14a** records energy on the vertical axis and reaction progress on the horizontal axis. The reaction begins at the flat area on the left, and the reactants are poised at the energy indicated. The reaction concludes at the plateau on the right, which marks the energy of the products. During the course of the reaction, the reactants must scale an energy hill to get to the products. We will discuss that hill shortly.

In the reaction energy diagram in Figure 7.14a, the energy of the products is higher than the energy of the reactants. This is indicative of an **endothermic** reaction, one that absorbs heat energy. Because an endothermic chemical reaction requires heat energy as a reactant, we can even add the words *heat energy* on the left side of the balanced equation, although it is technically not a formal part of the balanced reaction:

heat energy + reactants ⟶ products (endothermic reactions absorb heat)

Now consider the opposite situation, depicted in **Figure 7.14b**. In this case, the reactants are higher in energy than the products. This **exothermic** reaction *produces* heat:

reactants ⟶ products + *heat energy* (exothermic reactions produce heat)

Why are some reactions exothermic while others are endothermic? To answer this question, we must think about the bonds that hold together the reactant molecules taking part in a reaction. In general, a molecule that contains relatively weak bonds is more likely to undergo a reaction, because those bonds are easily broken. Explosives are a classic example of highly exothermic reactions. They are exothermic because the reacting substances—the explosives—tend to have weak bonds. The products of explosive reactions tend to have strong bonds. The reaction energy diagram for an explosive reaction looks like Figure 7.14b: reactants are up high, products are down low, energy is released, and . . . boom!

recurring theme in chemistry 7

Individual bonds between atoms dictate the properties and characteristics of the substances that contain them.

QUESTION 7.9 Is the reaction depicted by the chemical equation shown here endothermic or exothermic? How do you know?

$$2SO_2\,(g) + O_2\,(g) \longrightarrow 2SO_3\,(g) + \text{heat energy}$$

ANSWER 7.9 The reaction is exothermic because energy is produced.

For more practice with problems like this one, check out end-of-chapter Questions 44 and 45.

FOLLOW-UP QUESTION 7.9 Is the reaction depicted in the reaction energy diagram shown here endothermic or exothermic?

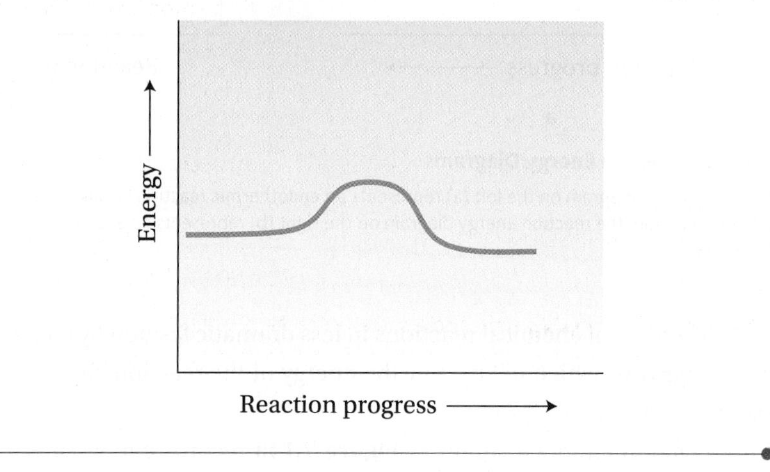

Reactants must scale an energy hill to become products.

We now know that some reactions require energy and some produce energy, and we can use a reaction energy diagram to visualize the before-and-after energy tally. Now we have one final question to consider: how fast do chemical reactions go? The speed of a reaction is related to the size of the energy hill that the reactants must scale to get to the products. That hill is the **activation energy** of a reaction. How is the size of the activation energy related to how fast a reaction goes? To answer this question, we turn to the track, where a high jumper is trying to go up and over the bar.

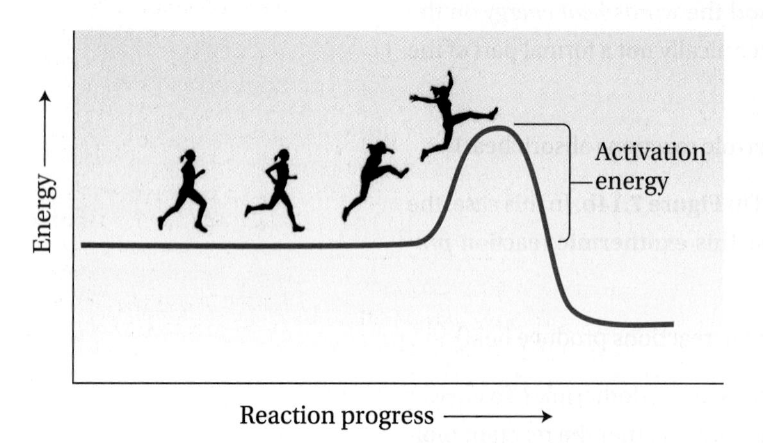

Figure 7.15 Activation Energy Can Be Compared to a High Jump

When the bar is set low, the high jumper quickly goes over it, perhaps on the first or second try. But as the bar goes up, the high jumper must try again and again to get over it. Maybe she tries a dozen times and finally makes it. Or, maybe the bar is so high that it takes her weeks of practicing and trying to finally make it over the top. Activation energy works the same way. The reactants in a chemical equation might have a small hill to scale, and so they scale it quickly, without requiring much energy. But, as the hill gets bigger, the reaction takes longer because fewer reactants have sufficient energy to get to the top of the hill and over to products (**Figure 7.15**).

Catalysts make chemical reactions go faster.

Consider a chemical reaction that we discussed earlier in this chapter: rusting, a very slow reaction between iron and oxygen. Here is the balanced chemical equation again:

$$4Fe\ (s) + 3O_2\ (g) \longrightarrow 2Fe_2O_3\ (s)$$

We know that it usually takes quite a long time for iron to rust. But if you live near or have traveled to a coast, you may have noticed that cars are rustier near the ocean. The salt in the sea air makes cars rust faster. Returning to our high jumping analogy, this means that, when salt is present, the reaction has a lower bar to get over. In chemistry lingo, we say that the activation energy is lower for the reaction that produces rust when salt is present.

The reaction energy diagrams for rusting with salt and without salt look something like the illustrations in **Figure 7.16**. Note that the starting and ending points for the reaction are the same. What is different? The size of the hill is different. The hill is smaller in the presence of salt. Because salt lowers the activation energy of the chemical reaction, but is not a reactant or product, we refer to it as a **catalyst**.

But this reaction can go even faster. If you go to your local sporting goods store, you will find an array of heat packs that can be used as sock or glove warmers for winter sports or as relief for sore muscles. Many of these products are based on the rusting reaction. But if you have cold toes, you want this reaction to happen faster than the time it takes for a car to rust.

If we urgently want this chemical reaction to proceed, we can speed it up by adding a catalyst to the heat pack. This type of heat pack contains pieces of iron. When you're ready to use it, you open it up to oxygen in the air. If we look at the list of

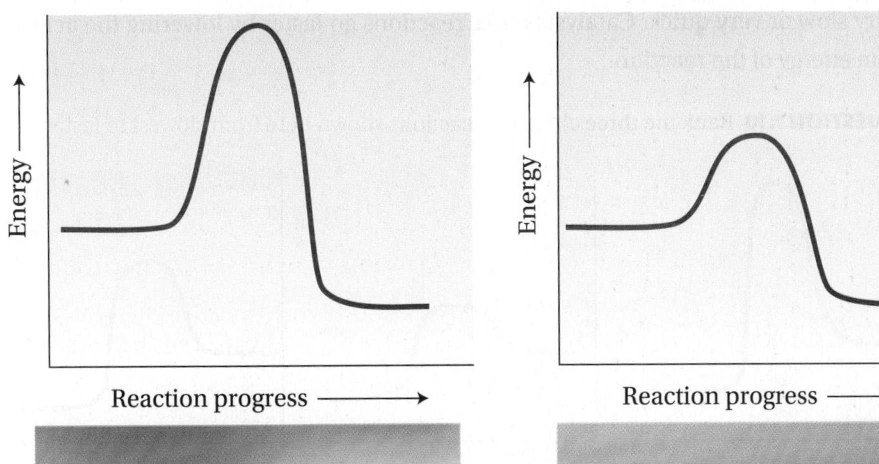

a

b

Figure 7.16 Activation Energy with and without a Catalyst

The graphs in (a) and (b) each show a reaction energy diagram for rusting. However, the reaction in (a), which occurs inland where the air is not salty, has a higher activation energy. For (b), which occurs by the ocean where the air is salty, salt acts as a catalyst and lowers the size of the activation energy hill, and thus rusting is faster under these conditions.

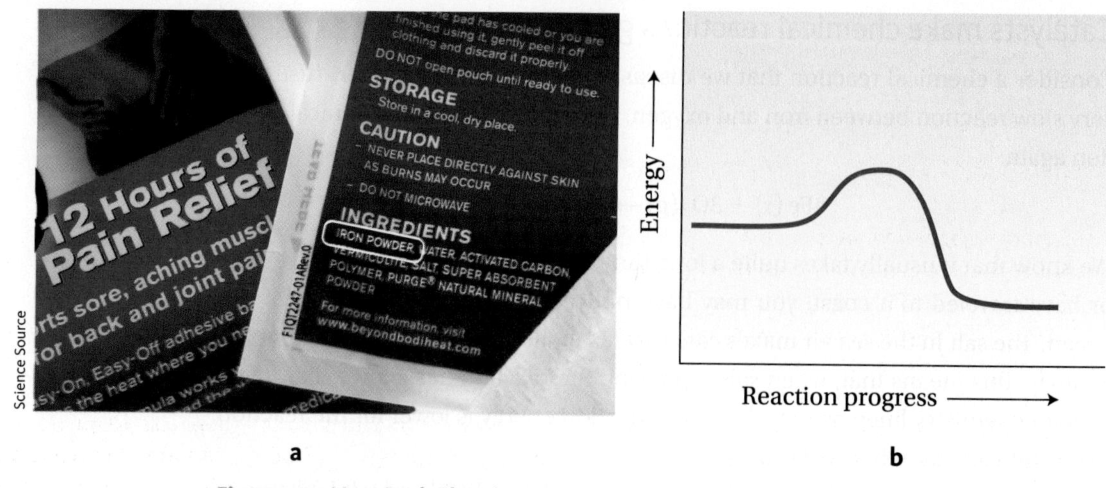

Figure 7.17 Heat Pack Chemistry

The ingredients in an iron-based heat pack are shown in (a); (b) is a reaction energy diagram for the iron-based heat pack.

ingredients for these heat packs (**Figure 7.17a**), we can see that they contain a lot of salt, because salt is a proven catalyst for this reaction (**Figure 7.17b**). In its presence, the iron in the pack rusts really fast. It's fast enough to produce heat in seconds.

What do you think the reaction energy diagram for the heat-pack-based rusting reaction looks like? Because it's the same rusting reaction, the starting and finishing energies look the same as they do for a rusting car. But now the hill is even smaller, as shown in Figure 7.17b. It is so small that reactants can easily make it over, and the reaction is fast. It is fast because there is a catalyst at work: the large amount of salt added to the heat pack lowers the activation energy and makes the hill smaller. Thus, it's possible for the same reaction—rusting, in this case—to be very slow or very quick. Catalysts make reactions go faster by lowering the activation energy of the reaction.

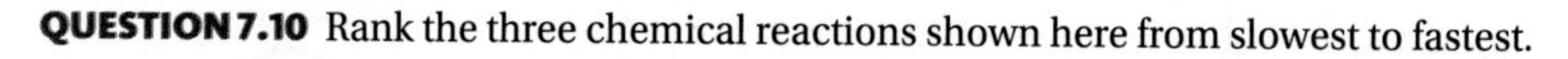

QUESTION 7.10 Rank the three chemical reactions shown here from slowest to fastest.

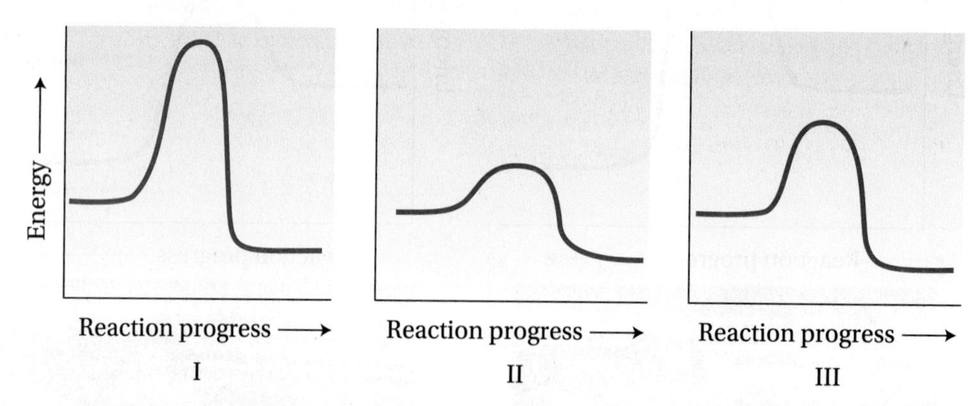

ANSWER 7.10 In all three cases, the reactions begin and end at roughly the same energies. We can compare the activation energies, which are all different. The reaction with the largest hill will have the slowest rate. Thus, from slowest to fastest: I, then III, then II.

For more practice with problems like this one, check out end-of-chapter Question 49.

FOLLOW-UP QUESTION 7.10 For each reaction shown in Question 7.10, indicate whether it is endothermic or exothermic.

THEgreenBEAT • news about the environment, chapter seven edition

TOP STORY Cow Flatulence and Global Warming

This story is about a combustion reaction, one that involves a fuel we all know well: **methane**. Methane powers our gas stoves and also happens to be a major component of flatulence. We will return to this combustion reaction and the subject of flatulence shortly.

Methane is a member of a small group of substances called greenhouse gases. Recall from Chapter 5 that a greenhouse gas is a substance that absorbs heat in the atmosphere and prevents that heat from escaping into space. This is a natural phenomenon that has always kept our atmosphere warm and hospitable. However, because industrialized nations have dumped large amounts of greenhouse gases into the environment, the atmosphere now has much more of these gases, and their levels are increasing at an alarming rate.

The most common greenhouse gases are compounds and, as their name indicates, gases. **Figure 7.18** shows the four most common gases in this group and the quantities of each produced in one year. Although methane makes up only 10% of greenhouse gas emissions, its effect on the environment is significant. There is less methane in the air, but it absorbs more heat per molecule than carbon dioxide does, so that a smaller amount has a measurable effect on the temperature of our atmosphere.

So where is all of the methane coming from, and how can we reduce the amount that's released? As it turns out, cows are one of the biggest sources of methane gas. When a cow eats, food is digested in one of its four stomachs, not in the intestines as it is in humans. Cows regurgitate partially digested food, called *cud,* and when they do, methane gas escapes in the form of a burp. And, yes, some escapes from the other end in the form of cow flatulence. How much greenhouse gas could a cow produce? Can that even be measured?

The answer is yes. Some interesting methods have been used to accurately measure the amount of methane gas a cow releases. The Argentine cow in **Figure 7.19** carries a balloon that gets filled with the gases she releases—at least from the tail end. This is one way that we have obtained reliable data on the question of how much methane comes from cows.

It's estimated that, on average, one cow produces about 200 grams of methane (from both ends) in one day, or 73,000 grams in one year. If there are about 96 million cows in the United States, we can multiply these figures to find the total methane emissions from U.S. cows: about 7,000,000,000,000 grams, or 7×10^{12} grams of methane per year.

Cows have been getting a bad rap from calculations like this. In fact, in New Zealand, the country's government considered a bill that would put a "pollution tax" on every cow, which would have to be paid by the cow's owner. This idea caused quite a controversy in a country where cows are everywhere—there are over two cows for every person in New Zealand. In protest, cow-owning New Zealanders started sending parcels to government officials. What did the parcels contain? Cow manure!

When some farmers learned that their cows were producing methane, a light bulb came on. After all, methane is *natural gas.* It's a fuel. Methane is useful! Here's an unbalanced chemical equation showing the reaction of methane with oxygen, the energy-producing, exothermic reaction in which methane is a fuel:

$$CH_4\,(g) + O_2\,(g) \longrightarrow CO_2\,(g) + H_2O\,(g)\ \text{(plus energy!)}$$

Some resourceful farmers have begun collecting the gases produced by their livestock and using it to power their farms.

Note that, depending on how this reaction is performed, carbon monoxide can be formed in addition to carbon

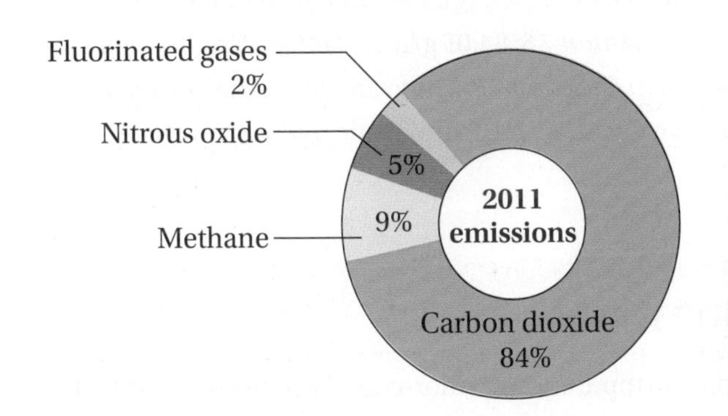

Figure 7.18 Percentage of Emissions of the Major Greenhouse Gases

Figure 7.19 Measuring Cow Flatulence

(continued)

THEgreenBEAT *continued*

dioxide. Both CO and CO_2 are themselves greenhouse gases and contribute to global warming, which is a shortcoming of this plan to rid the world of cow-related pollution. However, when methane from cows is combusted and used to power a working farm, at least some useful work has been done with that gas, which is better than releasing the methane directly into the air.

In fact, if we could somehow harness the 7,000,000,000,000 (or 7×10^{12}) grams of methane that cows produce in the United States every year, we could provide enough natural gas to power more than 150,000 average American households for one year—that's the entire population of a city the size of Pomona, California, or Savannah, Georgia.

QUESTION 7.11 Balance the chemical equation for the reaction of methane with oxygen.

ANSWER 7.11

$$CH_4\,(g) + 2O_2\,(g) \longrightarrow CO_2\,(g) + 2H_2O\,(g)$$

Chapter 7 Themes and Questions

recurring themes in chemistry chapter 7

recurring theme 4 Chemical reactions are nothing more than exchanges or rearrangements of electrons. (p. 189)

recurring theme 2 Things we can see with our eyes contain trillions upon trillions of atoms. Atoms are much too small to view with the naked eye. (p. 196)

recurring theme 7 Individual bonds between atoms dictate the properties and characteristics of the substances that contain them. (p. 205)

A list of the eight recurring themes in this book can be found on p. xv.

ANSWERS TO FOLLOW-UP QUESTIONS

Follow-Up Question 7.1: The states are solid, liquid, solid, and gas, respectively. *Follow-Up Question 7.2:* The coefficients will be 1, 5, 3, and 4, respectively. *Follow-Up Question 7.3:* The coefficients will be 2, 1, and 2, respectively. *Follow-Up Question 7.4:* $4KO_2\,(s) + 2CO_2\,(g) \longrightarrow 2K_2CO_3\,(s) + 3O_2\,(g)$. *Follow-Up Question 7.5:* 4.00 grams of helium. *Follow-Up Question 7.6:* 84.01 g/mol. *Follow-Up Question 7.7:* One mole of N_2O reacted. *Follow-Up Question 7.8:* There are 80.62 grams each of reactants and products. *Follow-Up Question 7.9:* Exothermic. *Follow-Up Question 7.10:* All are endothermic.

End-of-Chapter Questions

Questions are grouped according to topic and are color-coded according to difficulty level: **The Basic Questions***, **More Challenging Questions****, and **The Toughest Questions*****.

The Nature of Chemical Reactions

*1. In your own words, explain the difference between a chemical equation and a chemical reaction.

*2. Identify the reactants and products in the following chemical equation, and indicate the state in which each substance exists.

$$CaO\ (s) + CO_2\ (g) \rightarrow CaCO_3\ (s)$$

*3. Identify the reactants and products in the following chemical equation, and indicate the state in which each substance exists.

$$2H_2O_2\ (l) \rightarrow 2H_2O\ (l) + O_2\ (g)$$

*4. The letter following the name of a reactant or product in a chemical equation indicates the state of that substance. Must states always be indicated for each reactant and product in a chemical equation?

**5. Adenosine triphosphate (ATP) is the main energy currency in human tissues. Its chemical formula is $C_{10}H_{16}N_5O_{13}P_3$. How many atoms does one molecule of ATP contain?

**6. In your own words, explain the law of conservation of mass.

***7. True or false? The law of conservation of mass dictates that any balanced chemical equation must have the same number of reactants and products.

***8. The medical drug RU-486, formally known as *mifepristone,* is the hormone in the morning-after pill, used to prevent implantation of a fertilized egg in a human uterus. The chemical formula for mifepristone is $C_{29}H_{33}NO_2$.

(a) How many atoms are there in 1 molecule of mifepristone?

(b) How many nitrogen atoms are there in 5 molecules of mifepristone?

(c) How many carbon atoms are there in 2 molecules of mifepristone?

Balancing Chemical Equations

*9. The chemical equation shown here is balanced except for the value of the coefficient *x*. What is the value of *x*?

$$2NO + O_2 \rightarrow xNO_2$$

*10. The chemical equation shown here is balanced except for the value of the coefficient *y*. What is the value of *y*?

$$UO_2\ (s) + yHF\ (g) \rightarrow UF_4\ (s) + 2H_2O\ (g)$$

*11. The chemical equation shown here is balanced except for the value of the coefficient *z*. What is the value of *z*?

$$zO_2\ (g) + C_6H_{12}O_6\ (s) \rightarrow 6CO_2\ (g) + 6H_2O\ (l)$$

*12. Use coefficients to balance this equation.

$$C_7H_{16} + O_2 \rightarrow CO_2 + H_2O$$

*13. Consider this chemical equation representing the reaction between rust, Fe_2O_3, and acid.

$$Fe_2O_3 + 4HCl \longrightarrow 2FeCl_3 + 3H_2O$$

One of the coefficients in this equation is incorrect, and thus the equation is not balanced.

(a) Which coefficient is it?

(b) What should this number be changed to?

(c) After you make this coefficient change, show that the equation is balanced by counting atoms.

*14. In nature, zinc is sometimes found in the form of zinc sulfide, ZnS. One step in converting this compound to pure zinc metal is to convert it first to zinc oxide, ZnO, which happens in the following chemical reaction:

$$2ZnS\ (s) + \_\_O_2\ (g) \longrightarrow 2ZnO\ (s) + 2SO_2\ (g)$$

(a) What coefficient must be added in front of the oxygen molecule to balance this equation?

(b) What is the state of each reactant and product?

(c) If you begin the reaction with 1 unit of zinc sulfide, how many units of zinc oxide will be produced?

*15. Balance the following chemical equation by putting coefficients in front of reactants and/or products:

$$KClO_3\ (s) \longrightarrow KClO_4\ (s) + KCl\ (s)$$

*16. (a) Balance the following chemical equation by putting coefficients in front of reactants and/or products:

$$H_2S\ (g) + O_2\ (g) \longrightarrow S_8\ (s) + H_2O\ (l)$$

(b) Name two of the compounds in this equation.

*17. The reaction shown here takes place when hydrogen sulfide gas, H_2S, is released during a volcanic eruption:

$$2H_2S + 3O_2 \longrightarrow 2SO_2 + 2H_2O$$

(a) Is this equation balanced?

(b) How many atoms of each element make up the reactants?

(c) How many atoms of each element make up the products?

(d) If you begin the reaction with 6 molecules of hydrogen sulfide and 9 molecules of oxygen, how many molecules of water are produced?

Moles and the Laboratory Scale

*18. (a) Explain why the mole is a convenient way to count things like atoms and molecules.

(b) Why is Avogadro's number so large?

*19. In your own words, explain the difference between the atomic perspective and the laboratory perspective. When is it most useful to use each perspective?

*20. If you know exactly how many atoms or molecules comprise 1 mole, why do you need to express quantities of substances in moles? Why not simply use numbers of atoms or molecules?

*21. For each of these statements, explain whether the perspective taken is at the atomic scale or the laboratory scale:

(a) When a water molecule interacts with an atom of cesium metal, a chemical reaction occurs.

(b) To prepare aspirin, you may begin with several milligrams of each reactant.

*22. Moles are used to count things that are very small. In chemistry, which of the following are small things that are typically counted in units of moles?

(a) atoms (b) liters (c) molecules (d) formula units (e) grams

**23. Imagine that a chunk of pure metal contains 5 moles of atoms of that metal. How many atoms of the metal are in the chunk?

**24. The chemical formula of baking soda is $NaHCO_3$. If you have 3 moles of baking soda, how many moles of atoms (of all types) do you have?

**25. The chemical formula for acetic acid, which is diluted to make vinegar, is $C_2H_4O_2$. If you have 6 moles of acetic acid, how many moles of hydrogen atoms do you have?

**26. The chemical formula for magnesium chloride is $MgCl_2$. If a sample of this compound contains 4 moles of chlorine atoms, how many moles of magnesium chloride are present?

***27. A certain diamond contains 1.81×10^{24} atoms of carbon. How many moles of carbon does it contain?

***28. Your local dentist just ordered a fresh tank of laughing gas, N_2O, for his patients. The tank holds about 32.4 kilograms of gas. How many moles of laughing gas are in the tank?

Molar Mass

*29. The mineral galena has the chemical formula PbS. What is the mass of 1 mole of galena?

*30. Carbon dioxide, CO_2, is the greenhouse gas that poses the greatest threat to our planet. What is the molar mass of the CO_2 molecule?

**31. This photograph shows the mineral malachite, $Cu_2(CO_3)(OH)_2$, mixed with the mineral azurite, $Cu_3(CO_3)_2(OH)_2$. Determine the molar mass of each of these two substances.

Zelenskaya/Shutterstock

**32. *t*-Butyl mercaptan is a smelly molecule used to add odor to natural gas, which is naturally odorless. The structure of the *t*-Butyl mercaptan molecule is shown here:

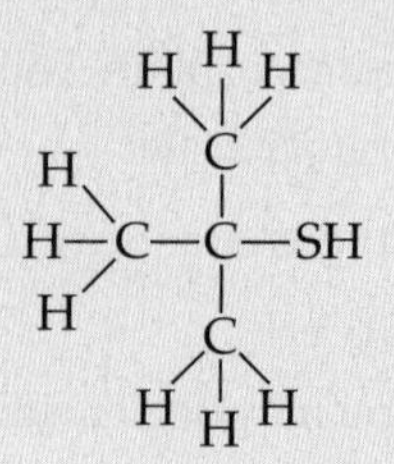

What is the molar mass of this molecule?

***33. The active ingredient in the popular cholesterol-lowering drug Lipitor has the molecular formula $C_{33}H_{35}FN_2O_5$.

(a) What is its molar mass?

(b) What is the mass, in grams, of 1 molecule of Lipitor?

***34. Because of its prismatic shape, the mineral scapolite is named for the Greek word for *shaft*. Its chemical formula is quite complex: $Na_4(AlSi_3O_8)_3Ca_4(Al_2Si_2O_8)_3(SO_4)$.

(a) Determine the number of each type of atom in scapolite.

(b) What is the molar mass of the mineral?

***35. The gemstone aquamarine is composed chiefly of the mineral beryl, $Be_3Al_2Si_6O_{18}$. The brilliant blue color of aquamarine comes from iron impurities.

(a) From the chemical formula, can you guess where beryl gets its name?

(b) Find the molar mass of this mineral.

***36. Sildenafil is an enormous drug molecule with this chemical formula: $C_{22}H_{30}N_6O_4S$. Its brand name is Viagra, and it is used to treat erectile dysfunction. What is the mass of 5 moles of sildenafil?

Stoichiometry

*37. True or false? A balanced chemical reaction can be used to determine the amounts of substances needed in a chemical reaction.

*38. True or false? A balanced chemical equation can be used to consider a chemical reaction either at the atomic scale or at the laboratory scale.

*39. Consider the following balanced chemical reaction:

$$2H_2S + 3O_2 \rightarrow 2SO_2 + 2H_2O$$

How many moles of oxygen are required to react with 6 moles of hydrogen sulfide?

*40. Consider the following balanced chemical equation in which oxygen gas reacts with glucose to produce carbon dioxide and water:

$$6O_2\,(g) + C_6H_{12}O_6\,(s) \rightarrow 6CO_2\,(g) + 6H_2O\,(l)$$

If 4 moles of glucose react with 24 moles of oxygen, how many moles of carbon dioxide are formed?

**41. In your own words, explain what information might be needed—in addition to a balanced chemical equation—to run a chemical reaction in the laboratory.

**42. Consider the following unbalanced chemical equation:

$$H_2S + 4O_2 \longrightarrow S_8 + H_2O$$

(a) Balance the equation.

(b) If 9 moles of H_2S reacts with 4 moles of oxygen gas, O_2, which of these two reactants will be left over when the reaction is complete?

(c) If 5 moles of S_8 are formed from the reaction shown, how many moles of oxygen molecules were consumed?

***43. Consider the following unbalanced chemical equation:

$$NH_3 + O_2 \longrightarrow H_2O + NO$$

(a) Balance the equation.

(b) If 68.12 grams of ammonia, NH_3, reacts with 160 grams of oxygen gas, O_2, how many moles of NO will form?

The Progress of Chemical Reactions

*44. Consider the following balanced chemical equation:

$$2H_2O_2\,(l) \longrightarrow 2H_2O\,(l) + O_2\,(g) + \text{heat energy}$$

Is this reaction endothermic or exothermic? How do you know?

**45. Consider the following balanced chemical equation:

$$\text{heat energy} + Ba(OH)_2 + 2NH_4NO_3 \longrightarrow Ba(NO_3)_2 + 2NH_3 + 2H_2O$$

Is this reaction endothermic or exothermic? How do you know?

**46. In your own words, explain what catalysts are and how they work.

**47. Consider the reaction energy diagram shown below. Is this reaction endothermic or exothermic? How do you know?

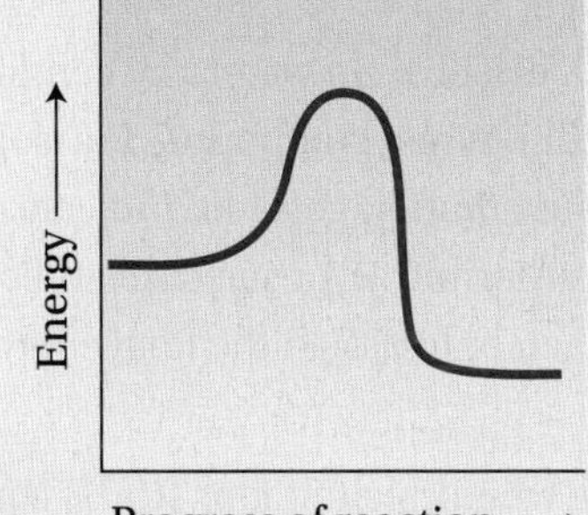

GREENLAWN PRODUCTIONS/Album/Superstock

**48. Consider the reaction energy diagram shown below. Is this reaction endothermic or exothermic? How do you know?

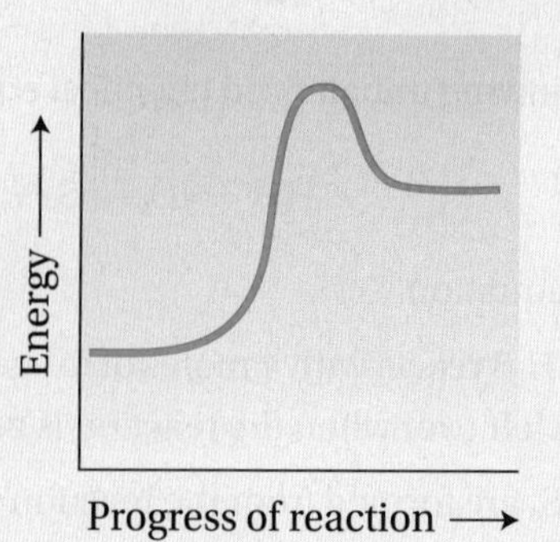

**49. Using your understanding of reaction energy diagrams, evaluate the two reaction energy diagrams shown below.

(a) Which diagram shows the faster chemical reaction?

(b) For each diagram, indicate if the reaction is endothermic or exothermic.

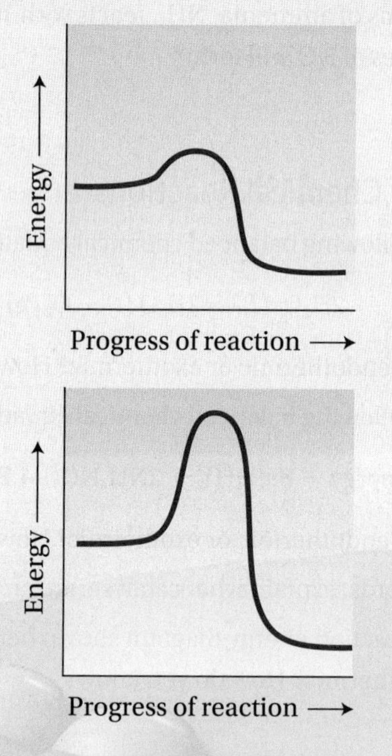

***50. When compound X dissolves in water, the container holding this aqueous solution gets cold. When compound Y dissolves in water, the container holding this aqueous solution gets hot. Indicate whether each process is endothermic or exothermic. For each case, indicate if the starting energy would be greater than or less than the final energy as shown on a reaction energy diagram.

*51. Seashells are made when calcium ions (Ca^{2+}) combine with carbonate ions (CO_3^{2-}) to make the salt calcium carbonate. When shells accumulate and are buried, they pack together to create limestone. Under pressure, limestone can become marble.

(a) Write a balanced chemical equation for the formation of calcium carbonate from its ions, referring back to Section 4.3, if necessary, to review how salts are made.

(b) Determine the molar mass of calcium carbonate.

****52.** The cisplatin molecule, shown below, has been used as an anti-cancer drug since the 1980's.

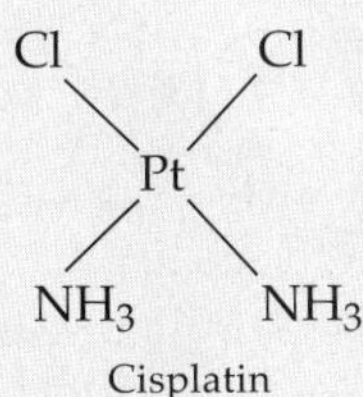

Cisplatin

(a) What is the molar mass of cisplatin?

(b) When the cisplatin molecule enters the cell, two water molecules substitute for two chlorine atoms, as shown in this chemical equation:

$$[Pt(NH_3)_2Cl_2] + 2H_2O \rightarrow [Pt(NH_3)_2(H_2O)_2]^{2+} + 2Cl^-$$

If 16 cisplatin molecules undergo this reaction, how many chloride ions are produced?

****53.** "Pharaoh's Serpent" is a chemical reaction that looks like a gnarled, snake-like creature growing out of a pile of white power. A search for this name in your Internet browser will bring up videos of this impressive reaction. The nearly-balanced chemical equation for this reaction is shown below.

(a) Determine the value of *x*.

(b) Determine the molar mass of the mercury-containing reactant, mercury thiocyanate.

$$2Hg(SCN)_2 + x\,O_2 + C_3N_4 \rightarrow CO_2 + 4SO_2 + 2Hg + N_2 + 3(CN)_2$$

****54.** In Section 7.3, we learned that in a combustion reaction, a fuel of some sort—usually a molecule that contains carbon, hydrogen, and maybe oxygen—reacts with oxygen, O_2, to make carbon dioxide and water. Ethyl alcohol, C_2H_6O, is often found as an additive in gasoline mixtures. Write a balanced chemical equation for the combustion of ethyl alcohol.

55. INTERNET RESEARCH QUESTION

Hydrogen cars operate with fuel cells, a type of battery which uses hydrogen gas, H_2, and oxygen gas, O_2, to make water.

(a) Write a balanced chemical equation that shows this process.

(b) One limitation of hydrogen-fueled cars is the lack of fuel stations that sell hydrogen gas. Open up your search engine and type in "hydrogen fueling station location near me." How far is the nearest fuel station to your current location? How many stations are located in your state? How does the number in your state compare to the number in other states in the United States?

Victor Polyakov/Shutterstock

PhotoStock-Israel/Alamy Stock Photo

Water

Why Water Is Critical for Human Beings and the Planet

PhotoStock-Israel/Alamy Stock Photo

CHAPTER 8

Virtually every culture and religion includes a ritual involving water. The Japanese Shinto religion, which is steeped in an appreciation of nature, has a water ritual for almost every aspect of human life—staving off illness, improving eyesight, finding a new job, living a long life, and having a healthy baby. In all these rituals, water cleanses and purifies. In Judaism, orthodox brides were traditionally required to go to the mikveh, or ritual bath, before marriage (see nearby illustration). And, over the past decade, this practice has seen a revival in American synagogues, where mikvehs are being installed to keep up with this resurfacing wedding trend.

The special significance of water carries over from culture and religion to the realm of chemistry. From a molecular point of view, water is unique, unlike any other substance on Earth. In this chapter, we discuss what makes water so extraordinary and why we cannot live without it.

We already know quite a bit about water. Recall from Figure 4.13 that water molecules are bent. As we'll see in this chapter, this bent shape gives water some very unusual characteristics and is responsible for some of its physical properties. We also discuss how water molecules form a special kind of bond with one another, a bond that is strong and allows water to mix readily with many other substances. And we learn what water does as it changes from one phase of matter to another—solid phase to liquid phase to gas phase and back again. To understand these transformations, we explore how water molecules shake and quiver and how these motions affect the way molecules interact with one another.

PhotoStock-Israel/Alamy Stock Photo

8.1 Water Footprints

A water footprint tallies the total water use by a person, a business, a country, or the planet.

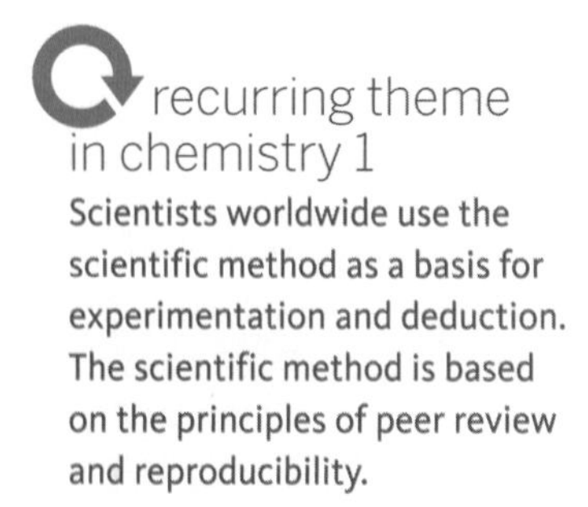

recurring theme in chemistry 1

Scientists worldwide use the scientific method as a basis for experimentation and deduction. The scientific method is based on the principles of peer review and reproducibility.

It's a familiar mantra: Drink more water! Many of us think that we should be drinking eight 8-ounce glasses of water per day. But recently, a study on water consumption asked the question: Where did that mandate come from? The answer is very interesting. The researchers found that much of the lore about drinking more water has come from businesses that tout the benefits of water to the public. Bottled water companies, for example, invest in drink-more-water advertising.

Does the medical community support the practice of drinking 64 ounces of water every day? The answer is no. To date, no peer-reviewed studies have linked

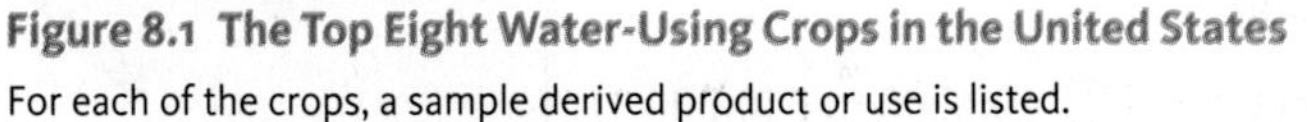

Figure 8.1 The Top Eight Water-Using Crops in the United States
For each of the crops, a sample derived product or use is listed.

high levels of water consumption with overall health. Thus, it seems that the 64-ounce rule is nothing more than an urban myth.

Recently, a movement is afoot to make people aware not of the water they drink, but of the water they use in one day. This **water footprint** can be estimated for a person, an object, a corporation, a country, or the planet. And, as it turns out, the water we drink makes up only a very small percentage of our water footprint: less than one-tenth of 1%.

If the water we drink contributes so little to our water footprint, what does contribute? Much of our water use comes from lawn watering, showering, and flushing the toilet; but we also must account for the water use associated with the food we eat and the products we use. For example, if we eat a piece of pork, we must consider not only the water content in the pork but also the water content of the food eaten by the pig. We also must consider the water used to irrigate the land that was used to grow the pig's food and the water used to clean its pen.

As this example illustrates, agriculture accounts for a significant chunk of water usage. Irrigation of crops accounts for 70% of the water used in the United States, and this usage contributes to the personal water footprint of anyone who uses or consumes products made from irrigated crops. **Figure 8.1** shows the eight U.S. crops that require the most water.

The average water footprint for a U.S. citizen is 2,842 cubic meters per year, which is equivalent to about 750,000 gallons per year or more than 2000 gallons of water per day. For context, consider that the volume used by one person in the United States in one year is slightly greater than the volume of an Olympic-sized swimming pool. Now compare this to the global value, which is about 3.8 cubic meters or about 1000 gallons per day. Thus, on average, a person in the United

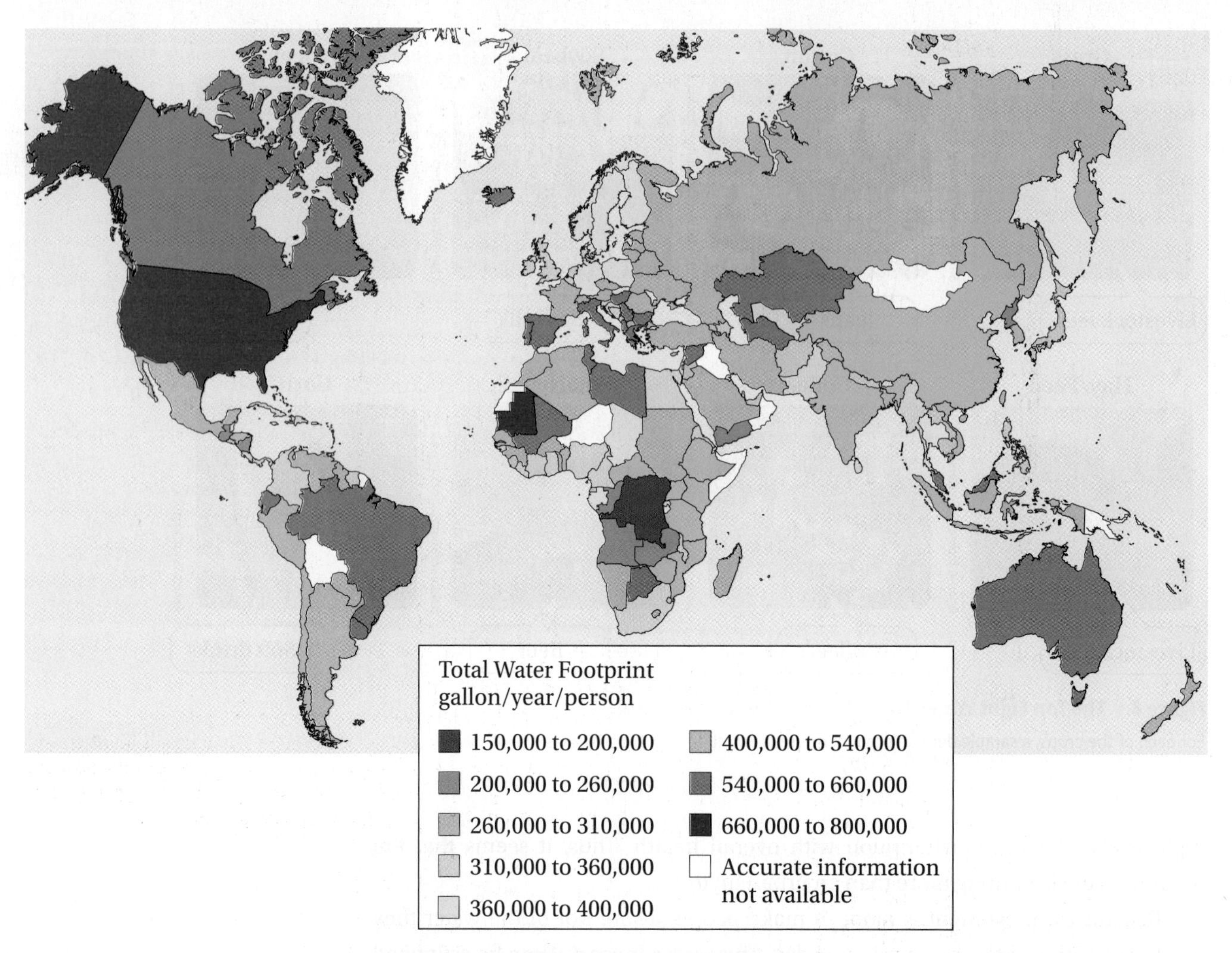

Figure 8.2 Water Footprints around the World

The average water footprint for each country is indicated on this map. The colors correspond to annual water usage per person in units of gallons per year.

States has a water footprint that is more than twice the global average. A map of global water footprints can be found in **Figure 8.2.**

QUESTION 8.1 True or false? By drinking less water every day, we can significantly reduce our water footprints.

ANSWER 8.1 False. Drinking water makes up only a tiny fraction of each person's water footprint.

For more practice with problems like this one, check out end-of-chapter Questions 1 and 6.

FOLLOW-UP QUESTION 8.1 Which of the following food products has the largest water footprint: lettuce, grapes, coffee, beans, or rice?

Fresh water is only a small percentage of all the water on Earth.

In total, about 1.23×10^{21} liters of water are present on Earth, and that volume covers about 70% of its surface. If there is so much water, then why worry so much about our water footprints? Don't we have plenty of water available on Earth?

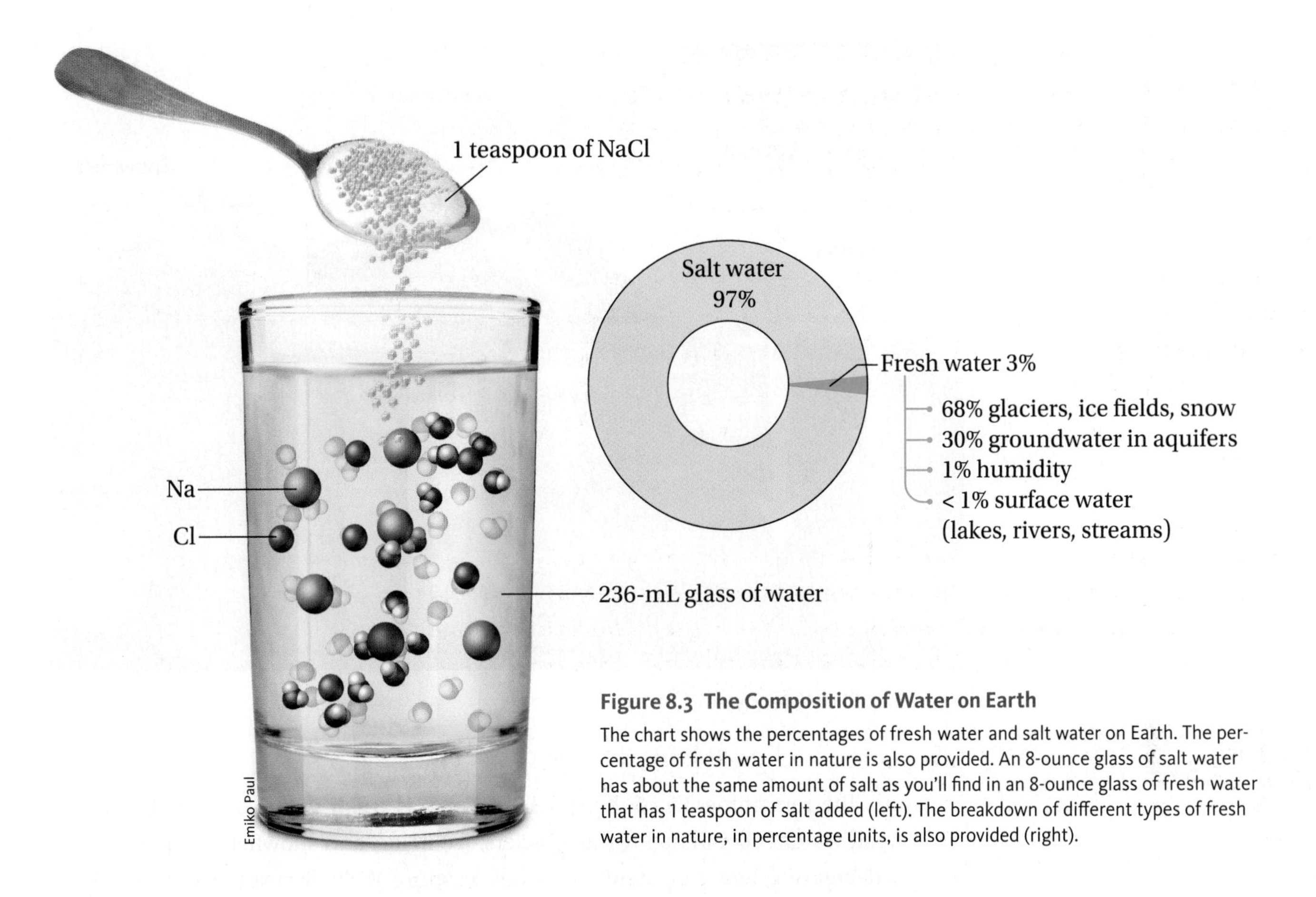

Figure 8.3 The Composition of Water on Earth

The chart shows the percentages of fresh water and salt water on Earth. The percentage of fresh water in nature is also provided. An 8-ounce glass of salt water has about the same amount of salt as you'll find in an 8-ounce glass of fresh water that has 1 teaspoon of salt added (left). The breakdown of different types of fresh water in nature, in percentage units, is also provided (right).

The answer is yes, we do have a lot of water on Earth. But 97% of that water is salt water, and salt water is not **potable**. That is, it's not safe for human consumption. Salt water contains 3.5% salt by mass, so an 8-ounce glass of salt water contains roughly one teaspoon of table salt, sodium chloride, as shown in **Figure 8.3**. Nobody wants to drink that.

wait a minute . . .

Why don't we remove the salt from salt water to make fresh water?

We *can* remove the salt from salt water in facilities called desalination plants. Worldwide, more than 15,000 desalination plants produce more than 5.98×10^4 liters of fresh water each year. This may seem like a lot of water, but it's only a tiny fraction of the total fresh water used on Earth annually.

If desalination works, why aren't desalination plants turning more salt water into fresh water? Because the plants are extremely expensive to build, and they use tremendous amounts of energy. The top three users of desalination plants are Saudi Arabia, the United States, and the United Arab Emirates—all wealthy countries. If a country has access to salt water and has limited freshwater resources, a desalination plant is an option only if the country can afford to build it and provide power to it. For this reason, desalination has not contributed significantly to the global demand for fresh water.

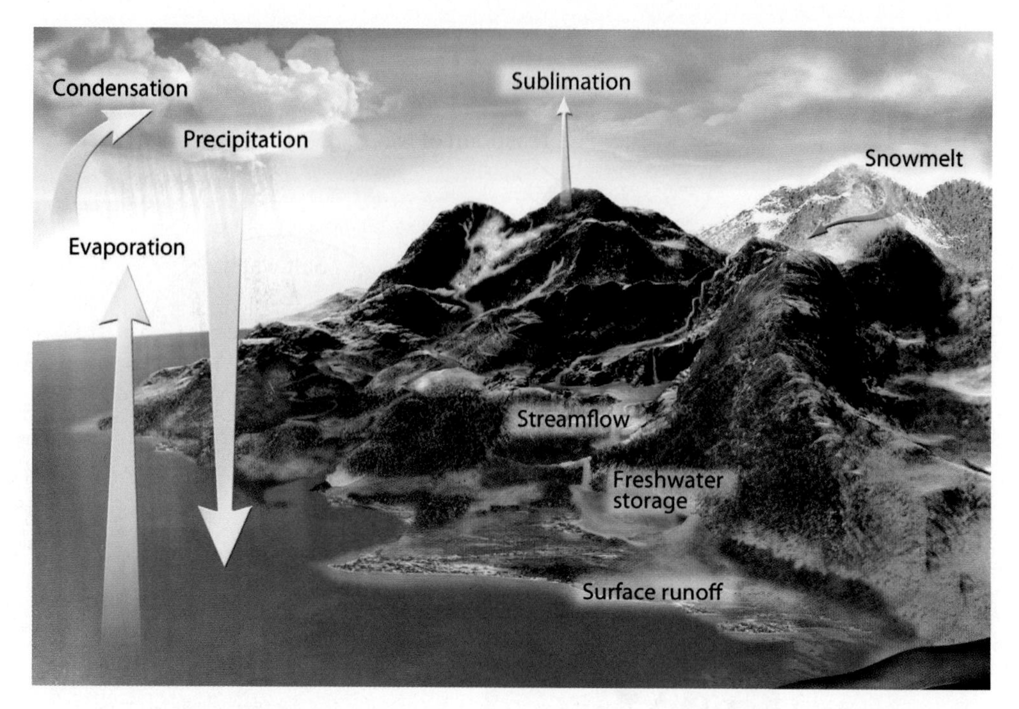

Figure 8.4 The Water Cycle

The remaining 3% of the water on Earth is **fresh water**. About 68% of this fresh water is contained within glaciers, ice fields, and snowfields. (That amount is decreasing, however, thanks to global warming. We'll discuss this more in Chapter 12, "Energy, Power, and Climate Change.") About 30% of our fresh water is underground. This *groundwater* exists in aquifers, natural underground spaces that can be tapped with wells that bring water to the surface. Another 1% of fresh water is in the atmosphere in the form of humidity. That leaves a mere 0.3% for fresh water on Earth's surface, which includes rivers, lakes, and streams. Much of this fresh water eventually makes its way to the ocean, where it mixes with salt water.

The **water cycle** moves water among these water reservoirs. Within this cycle, shown in **Figure 8.4** water can exist in any of its phases: gas, liquid, or solid. Liquid water on Earth's surface, including water in the oceans, lakes, and rivers, can undergo **evaporation**, a change of phase from liquid water to become **water vapor**. As water vapor moves upward into cooler regions of the atmosphere, it condenses. **Condensation** is the phase change from gas back to liquid. When water in the atmosphere returns to Earth in the form of snow, sleet, hail, or rain, this is called **precipitation**. (We spend more time discussing the phases of water in Section 8.3.)

Ice caps, glaciers, and snowfields are reservoirs that hold water in the solid phase, ice. When liquid water freezes, ice results. When ice melts, it becomes liquid water. It is also possible for ice to become water vapor directly, through the process called **sublimation**.

QUESTION 8.2 Using Figure 8.4 as a reference, name two examples of phase changes that are part of the water cycle.

ANSWER 8.2 Here are examples: ice and snow melt into water; ice sublimes to become water vapor; oceans and lakes evaporate to make water vapor; water freezes to make ice. For more practice with problems like this one, check out end-of-chapter Questions 3 and 5.

FOLLOW-UP QUESTION 8.2 True or false? Rivers, streams, and groundwater are all examples of fresh water.

Much of Earth's freshwater supply is polluted.

The North Brooklyn Boat Club, located in New York City, is a thriving recreational organization with more than 200 members. Every weekend, they take their boats into Newtown Creek, a branch of the East River that separates Queens from Brooklyn. One of these boats, a graceful 28-foot flat-bottom rowboat, was handmade by the club's members. The boat and its rowers, shown in **Figure 8.5,** present an idyllic image of humans enjoying themselves, their boat gliding easily through the heart of New York City. But this particular river is not known for its recreational activities. It is a Superfund site, and it's one of the most polluted rivers on the East Coast of the United States.

Robert Stolarik/The New York Times/Redux

Figure 8.5 Members of the North Brooklyn Boat Club Enjoying a Weekend Ride

The U.S. Environmental Protection Agency (EPA) defines a **Superfund site** as "an uncontrolled or abandoned place where hazardous waste is located, possibly affecting local ecosystems or people." Only the most polluted places make this list, which includes more than a thousand locations around the country. Many of these sites are lakes and rivers, like Newtown Creek. When a site receives Superfund status, this label sets into motion a series of EPA-led actions that should result in clean-up of the site over time. The Superfund division of the EPA researches the site's history and identifies the parties responsible for the contamination. Those parties are then held financially and legally responsible for cleaning up the site in collaboration with the local community.

How did Newtown Creek become so polluted? The pollution comes primarily from two sources. First, beginning in the nineteenth century, the City of New York dumped raw sewage into the creek for many years. Thus, the city is named as one party in the cleanup of Newtown Creek. Second, the shores of the creek are home to many oil refineries that have dumped thousands of gallons of oil into it over the years. Evidence of spilled oil is obvious. According to an article in the *New York Times* profiling the Boat Club, "The soil [on the shores of the creek] resembles black mayonnaise." Consequently, local oil companies are also responsible for cleaning up Newtown Creek. Despite all this, the North Brooklyn Boat Club is undeterred. Optimists every one, the club members do not let black mayonnaise and toxic water keep them from enjoying a vigorous weekend excursion.

Examples like this make it easy to see why we must pay attention to our water footprints. We do have a tremendous volume of water on Earth, but only a very small fraction of it is fresh water. And sadly, a significant amount of that water is not usable because we have allowed it to become polluted over time. Clean water is a rare and valuable resource that we must protect, just as we strive to protect our other natural resources such as wildlife, woodlands, and wetlands.

In the United States, drinking water is protected by the Safe Drinking Water Act.

The water we drink in the United States is carefully monitored. The quality of that tap water is governed by the **Safe Drinking Water Act,** passed by Congress in 1974. Some of the first environmental protests in the United States prompted the passage of this act. The act was updated in 1996.

The water we drink can come from surface waters, like lakes and reservoirs, as well as underground aquifers. The water from those sources must be treated to make it drinkable, even if the water source is very clean. Water treatment plants are designed to take unclean water and remove toxins, debris, and microorganisms. Obviously, it's more challenging to purify water from Newtown Creek compared to water, for example, from a pristine rural lake. But, with little variation, most water treatment plants use a similar process to purify their water.

Figure 8.6 depicts a typical water treatment plant. Water from a natural source is pumped into the plant, where large pieces of debris are filtered out with a screen. Next, a **flocculant** is added. A flocculant is a compound that becomes fluffy and cloudlike when mixed with water as part of the purification process. After mixing, the flocculants are allowed to settle to the bottom of the tank, and they take *particulates*—small particles—in the water with them. This step removes smaller

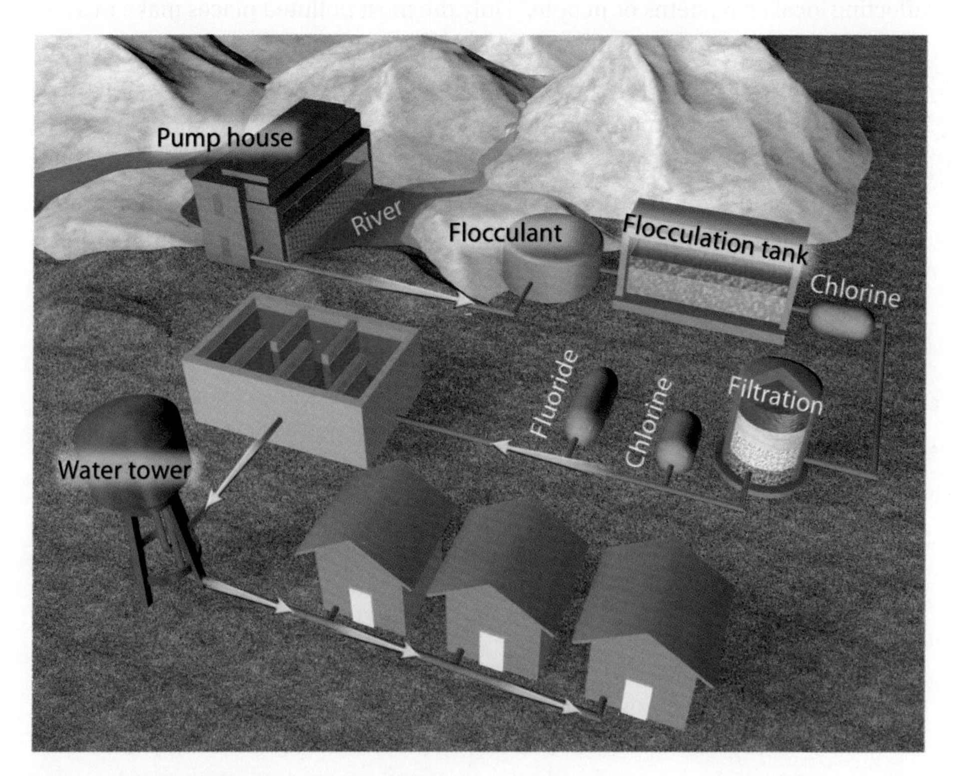

Figure 8.6 A Typical Water Treatment Plant

debris and particles that made it through the screen. The water is then disinfected with chlorine molecules, Cl_2, to kill the viruses and bacteria that naturally grow in water reservoirs. This disinfection step was introduced in the early 1900s and has all but eliminated the spread of certain diseases such as cholera and dysentery.

After being disinfected, the water passes through layers of sand and gravel to remove any remaining particles. More chlorine may be added for disinfection, and fluoride ions (F^-) also go into the water. (Read this chapter's *Green Beat* for a discussion about the debate over adding lithium to our drinking water.) The purified water then flows out to homes and businesses.

The lack of clean water is a serious problem worldwide. However, new methods are available to purify water in areas that have no municipal water treatment. For example, through a nonprofit program, Proctor & Gamble (P&G) has developed a sachet of powdered ingredients that can be added to unclean water to purify it. This product has been used extensively in Myanmar to provide potable water to rural communities. **Figure 8.7** shows Chelsea and Bill Clinton using P&G sachets to purify a local water sample.

AP Images

Figure 8.7 Purification of Water with P&G Sachets
P&G's nonprofit work in Myanmar is funded, in part, by the Clinton Foundation.

QUESTION 8.3 Which step in the water treatment process kills viruses and bacteria?

ANSWER 8.3 The addition of chlorine molecules serves this purpose.

For more practice with problems like this one, check out end-of-chapter Question 4.

FOLLOW-UP QUESTION 8.3 True or false? Water treatment plants are needed only in places where water supplies contain hazardous waste.

8.2 The Nature of Liquid Water

Intermolecular forces are forces between molecules.

Water is as essential to our planet as blood is to the human body. Earth and its inhabitants cannot live without water, a truly extraordinary liquid.

Water boils at a high temperature compared to other molecules of the same size. It also can dissolve many, many substances. (In Chapter 9, "Salts and Aqueous Solutions," we'll investigate substances dissolved in water.)

Why is water so unusual in comparison to many other molecules? To understand the answer to this question, we have to think about the interactions *between* molecules, which we call **intermolecular forces**. Intermolecular forces are different from the covalent bonds we discussed in Chapter 4, which are bonds *within* a molecule. For example, within a water molecule, covalent bonds attach the oxygen to each of the hydrogen atoms. In contrast, intermolecular forces hold one water molecule to its neighboring water molecules. This distinction, illustrated in **Figure 8.8,**

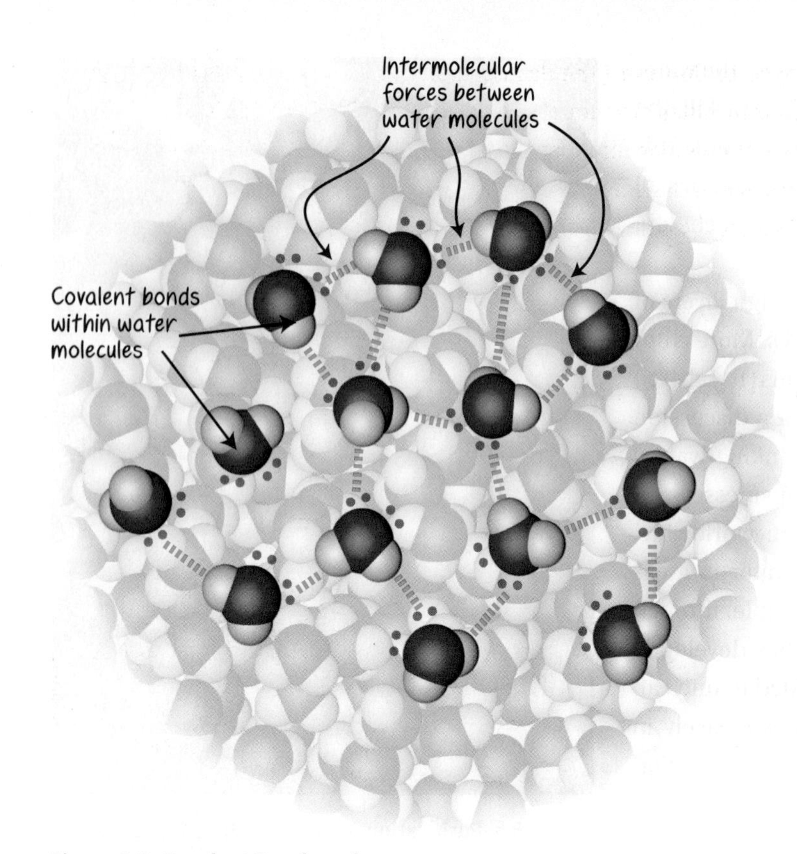

Figure 8.8 Covalent Bonds and Intermolecular Forces

is important because the interactions among water molecules break much more easily than the bonds within them. In general, covalent bonds are anywhere from about 50 to 150 times stronger than intermolecular forces.

There are many types of intermolecular forces, but two of them are most relevant for water: hydrogen bonds and dipole–dipole interactions. We'll discuss those two types here and consider the others in Chapter 9.

Hydrogen bonds are the strongest type of intermolecular force.

Within the covalent bond between oxygen and hydrogen in water, the electrons are not shared equally between the two atoms. As we discussed in Chapter 4, the electrons spend more time near the oxygen atom and less time near the hydrogen. The first type of intermolecular force we discuss in this chapter relies on the lopsided bonds between O and H in water. Those bonds are, in fact, extremely lopsided. They are so lopsided that the hydrogen's single electron is pulled from the area around the hydrogen atom toward the oxygen atom. This leaves that hydrogen atom almost bare, its single proton with its single partial positive charge exposed. That nearly naked proton is part of what gives water its unique properties.

recurring theme in chemistry 3

Opposite charges attract one another.

In Chapter 5, we saw that the oxygen atom in water has two nonbonding electron pairs, as shown in **Figure 8.9a**. Because there is a buildup of negative charge (electron density) on the oxygen atom, its nonbonding electron pair is attracted to and can interact with the exposed hydrogen atom on another water molecule, as illustrated in **Figure 8.9b**. This type of interaction is a **hydrogen bond**. Hydrogen

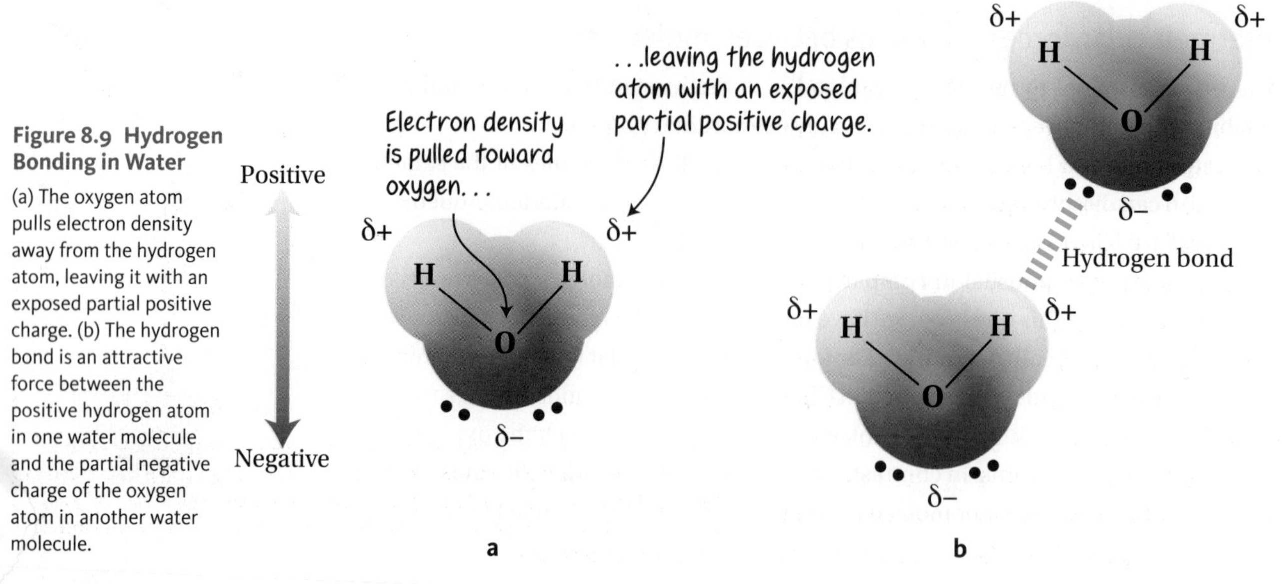

Figure 8.9 Hydrogen Bonding in Water

(a) The oxygen atom pulls electron density away from the hydrogen atom, leaving it with an exposed partial positive charge. (b) The hydrogen bond is an attractive force between the positive hydrogen atom in one water molecule and the partial negative charge of the oxygen atom in another water molecule.

bonds can also exist between hydrogen and nitrogen or fluorine. (The dashed line in Figure 8.8 represents a hydrogen bond between molecules of water.)

A hydrogen bond is a very strong type of intermolecular force, but it is still weaker than the covalent bonds that exist within the water molecule. As we'll see in Section 8.3, boiling water requires a great deal of energy because hydrogen bonds must be broken before water boils.

recurring theme in chemistry 7

Individual bonds between atoms dictate the properties and characteristics of the substances that contain them.

QUESTION 8.4 True or false? A hydrogen bond is a covalent bond.

ANSWER 8.4 False. A covalent bond is a bond within a molecule. A hydrogen bond is a type of intermolecular force.

For more practice with problems like this one, check out end-of-chapter Question 19.

FOLLOW-UP QUESTION 8.4 Do you think it is possible for a hydrogen bond to form within a molecule, as shown here?

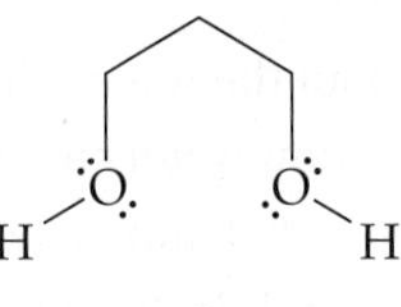

Water molecules are polar and experience dipole–dipole interactions.

Because each of the O–H bonds in water is polar, and the molecule is bent, the whole molecule is polar. The buildup of electron density around the oxygen means that part of the molecule is more negative than the rest of the molecule (see Figure 8.9a). We depict this unevenness of electrical charge, called a **dipole,** by drawing an arrow through the water molecule. The arrow points from the positive end to the negative end of the molecule, as shown in **Figure 8.10a**. That is, the head of the arrow points to the more negative end of the molecule, and the plus sign (+) points to the more positive end of the molecule.

The second of the two types of intermolecular force we'll consider in this chapter is a **dipole–dipole interaction**. When molecules of water mix with one another, the dipoles in each molecule affect how the molecules move and interact. The positive end of the dipole of one water molecule is attracted to the negative end of another water molecule's dipole, as shown schematically in **Figure 8.10b**. This means that even though liquid water contains constantly moving and shifting water molecules, part of the way the molecules move and shift is dictated by their dipoles. These dipole–dipole interactions are about 50 times weaker than the covalent bonds

Figure 8.10 Dipole within a Bent Water Molecule

(a) The dipole in a water molecule results from an excess of negative charge on the oxygen atom. The head of the arrow represents the negative end of the dipole, and the other end of the arrow represents the positive end. (b) Water's dipoles naturally align when water molecules are together so that the negative end of one dipole is close to the positive end of another.

recurring theme in chemistry 2

Things we can see with our eyes contain trillions upon trillions of atoms. Atoms are much too small to view with the naked eye.

within a water molecule, but a glass of water contains billions and billions and billions of water molecules, and all those weak interactions add up.

QUESTION 8.5 True or false? The bonds between atoms within a molecule are usually weaker than the forces between molecules.

ANSWER 8.5 False. The bonds between the atoms within a molecule are strong covalent bonds. They are stronger than the strongest type of force that exists between molecules.
For more practice with problems like this one, check out end-of-chapter Question 20.

FOLLOW-UP QUESTION 8.5 Which bond would you expect to be more polar: a bond between two oxygen atoms or a bond between an oxygen atom and a hydrogen atom?

Ice is less dense than water.

Now that we know something about the interactions that hold together the molecules in liquid water, let's consider how water changes to the solid phase—freezes—as the temperature drops. In Figure 8.8, notice how some of the water molecules make more than one hydrogen bond. In liquid water, this network of hydrogen bonds is arranged willy-nilly. Hydrogen bonds are elusive and short lived in liquid water. Hydrogen bonding partners change quickly, like dancers in a square dance.

However, in solid water—ice—the hydrogen bonds are less fickle. Rather than constantly changing partners, they stay put. In ice, hydrogen bonds hold water molecules in a rigid hexagonal pattern (**Figure 8.11a**). If we extend this structure into three dimensions, we get the arrangement you see in **Figure 8.11b**. Ice has a three-dimensional crystal lattice that's based on this repeating hexagonal motif.

One result of this open, hexagonal lattice is that ice contains a lot of empty space. In fact, 100 molecules of water in ice take up more space than 100 molecules

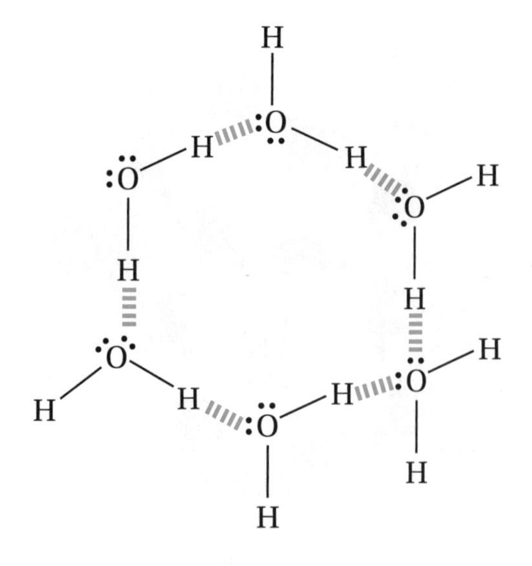

a

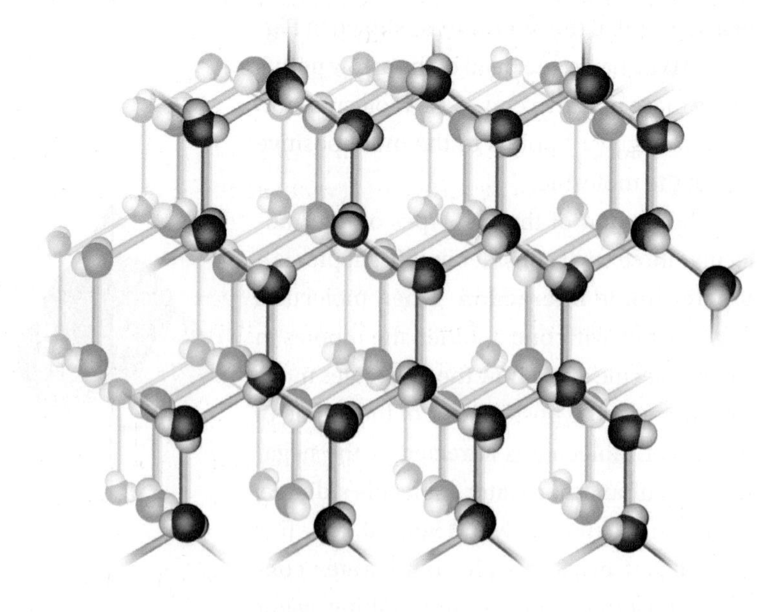

b

Figure 8.11 The Structure of Ice

(a) A hexagonal shape results when six water molecules form hydrogen bonds to one another and form a circle. (b) The H_2O molecules in ice form a hexagonal lattice held together by hydrogen bonds. When a huge number of hexagons have formed in the lattice, the result is a rigid structure that makes ice solid. The lattice structure also creates empty spaces within the ice crystal.

Figure 8.12 The Densities of Ice and Water

The beaker on the left contains exactly 100 mL of liquid water. The beaker on the right contains exactly 100 mL of ice. Liquid water presses down more on the scale, because the 100-mL volume of liquid water has a greater mass than the 100-mL volume of ice.

of liquid water. Or, to put it another way, more liquid H_2O molecules than solid H_2O molecules can fit into a certain volume of space. The mass of ice or water (or any other substance), when squeezed into a given volume, is known as the **density** of that substance. We express density as a mass unit divided by a volume unit. So, for example, we can express density as milligrams per milliliter (mg/mL) or kilograms per liter (kg/L) or grams per milliliter (g/mL).

Because of all the empty spaces within ice, liquid water has a greater density (1.00 g/mL) than ice (0.92 g/mL). In other words, if we measure the mass of 100.0 mL of water and then we measure the mass of the same volume of ice, we will find that the ice has slightly less mass (**Figure 8.12**). This variation in mass—0.08 g/mL of water—is significant, and the world would be a much different place if not for this difference.

If you have ever shaken a bottle of salad dressing, you've probably noticed that substances with lower densities (like oil) float on substances with higher densities (like vinegar). Similarly, in winter, the water in a pond or river freezes at the surface first, and because ice is less dense than water, that surface ice does not sink to the bottom. The layer of ice acts as an insulator for the water below, allowing aquatic organisms to survive through the winter. If ice sank in water, bodies of water would freeze solid and the world would be less habitable for the myriad forms of life that call Earth home.

QUESTION 8.6 Three liquid substances (A, B, and C), do not mix easily with one another. They have densities of 1.4 g/mL, 0.79 g/mL, and 1.1 g/mL, respectively. If you put 5.0 mL of substance A, 3.0 mL of substance B, and 8.0 mL of substance C into a beaker together, which one would you expect to be on top? Which one would be on the bottom?

ANSWER 8.6 The substance with the lowest density should be on the top, regardless of its volume. So, B will be on top. A has the highest density and will be on the bottom. C will be in the middle.

For more practice with problems like this one, check out end-of-chapter Questions 10 and 13.

FOLLOW-UP QUESTION 8.6 You have 300 grams of substances D and E. Substance D fills a graduated cylinder to the 150.0-mL point. Substance E fills a graduated cylinder to the 100.0-mL point. What can you say about the relative densities of substances D and E?

natureBOX • Songbirds and Hydrogen Atoms

Consider a sample of water that has been purified by the most advanced methods known, in the fanciest laboratory, by the most respected scientists. If every trace of impurity has been removed, we might then assume that we have a sample of identical, pristine water molecules. But our assumption would be wrong.

Recall from Chapter 2 that atoms can have differing numbers of neutrons in their nuclei and that those different versions are called isotopes of that atom. Hydrogen has three common isotopes, and by far the most abundant in nature is hydrogen-1, which has an abundance of 99.98%. Thus, in our super-pure water sample, 99.98% of the water molecules contain hydrogen-1, but a few water molecules here and there contain rarer isotopes of hydrogen.

As we're about to see, the existence of these rare isotopes in water can be useful. This story is about hydrogen-2 (also called deuterium) and hydrogen-3 (also called tritium). Deuterium and tritium make up just 0.02% of the isotopic abundance of hydrogen. And that exact amount varies slightly from one location to another. Thus, water samples in a given location have a distinctive ratio of one isotope to another.

The ornithologist Richard Holmes of Dartmouth University and his student Dustin Rubenstein (**Figure 8A**) used the natural distribution of hydrogen isotopes in water molecules to track migrating birds. Their work is important because scientists who study breeding birds often lose track of them during the winter migration, and a better understanding of migratory and breeding habits can help preserve a species.

Courtesy of Dustin Rubenstein

Figure 8A Researching the Black-Throated Warbler
Dartmouth ornithologist Richard Holmes and his student Dustin Rubenstein (shown) used the distribution of hydrogen isotopes in water to track black-throated warbler migration.

The traditional way to monitor migration is to tag the birds and try to recapture them and record their movement from place to place. But this method can have success rates as low as 2%. Besides, the process of tagging a bird is traumatic for the bird and can disrupt its natural pattern of migration.

The Dartmouth biologists reasoned that it may be possible to use the distribution of isotopes in water to follow the migration patterns of the black-throated warbler, an insect-eating songbird that travels between the Caribbean and the eastern United States. But what does isotope distribution have to do with birds? Birds eat insects, insects eat plants, and plants take up water from soil that has a distinctive isotope ratio. Thus, when a bird spends time in one place, it takes up water from that location. It should be possible to analyze the water in these birds and, from that analysis, determine where the birds have been and therefore identify their migration path.

The biologists did just that. They used measurements of deuterium

8.3 Changing Phases: Water and Ice

Freezing and melting occur at the same temperature.

Let's take a closer look at phase changes within the water cycle. Phase changes occur in pairs. For example, snow and ice can melt to become liquid water, and liquid water can freeze to make snow and ice. Pairs of phase changes like melting and freezing can go back and forth. Water easily cycles between these phases, because the water molecules remain intact and are not changed by these processes. When phase changes happen, what does change are the intermolecular forces between water molecules. In liquid water, for example, hydrogen bonds and dipole–dipole interactions hold molecules loosely together in a disorganized jumble. When water becomes ice, those bonds shift and become more regular, more orderly, and better organized (see Figure 8.11). The reverse happens when the ice melts.

We define the **freezing point** as the temperature at which a liquid becomes a solid. The **melting point** of a solid is the temperature at which the solid becomes a liquid. These two terms—freezing and melting—describe the same temperature, but that temperature is approached from opposite directions. This concept is illustrated in **Figure 8.13**.

to track the migration patterns of two groups of black-throated warblers by analyzing hydrogen/deuterium ratios in the water contained in the birds's feathers. Don't worry—this project did not involve a group of scientists chasing birds and pulling out their feathers. After mating, the birds naturally lose some of their feathers, so the discarded feathers were collected and analyzed.

By using their measurements of hydrogen isotopes in the bird feathers, Holmes and Rubenstein accurately tracked the migration patterns of the two groups of warblers across the United States and the Caribbean (**Figure 8B**). With this information, ornithologists can track these birds in the future and assess the risks, such as those from environmental pollutants, that may threaten the black-throated warbler population.

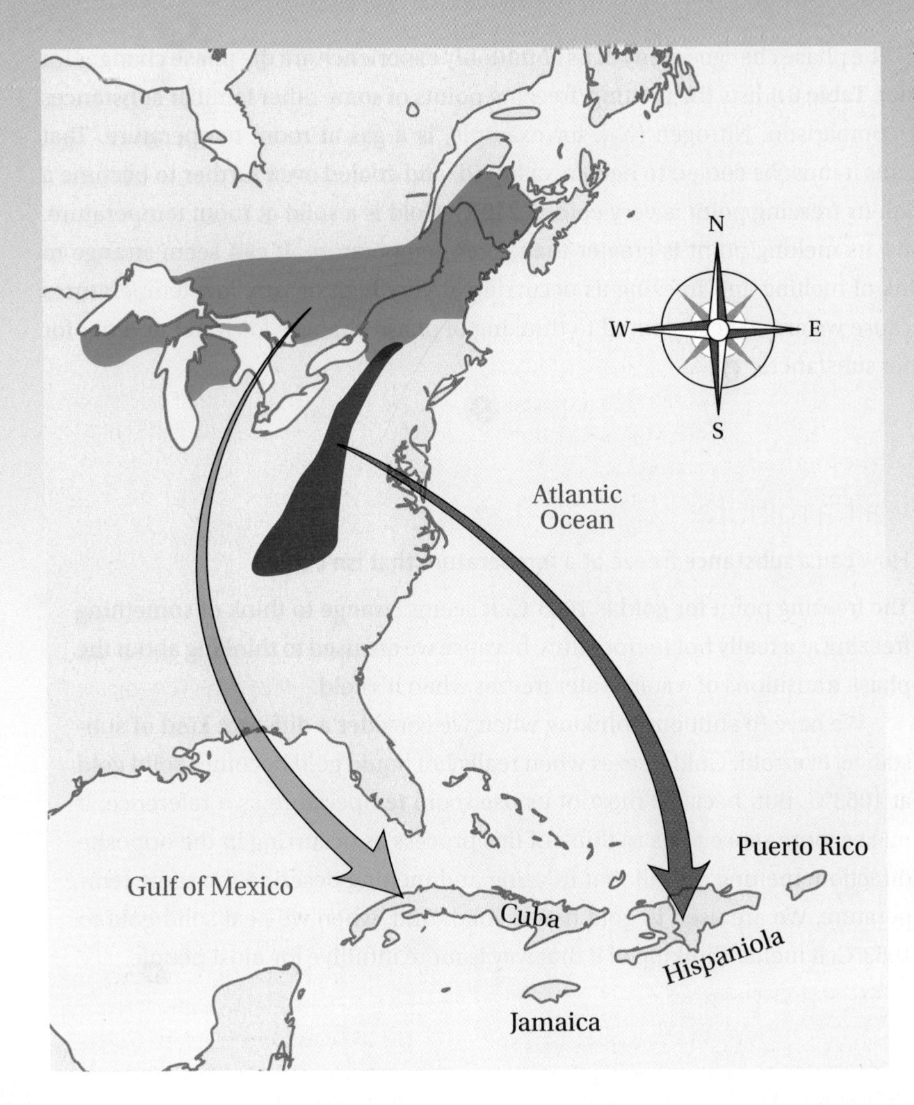

Figure 8B South for the Winter

Isotope analysis of the water contained in the feathers of molting birds helps Dartmouth University researchers identify the birds' migration destinations. Black-throated warblers that summer in the northern United States and southern Canada (green) spend the winter on islands in the western part of the Caribbean Sea. Warblers that summer in the southern United States (red) spend the cold months on islands in the eastern Caribbean.

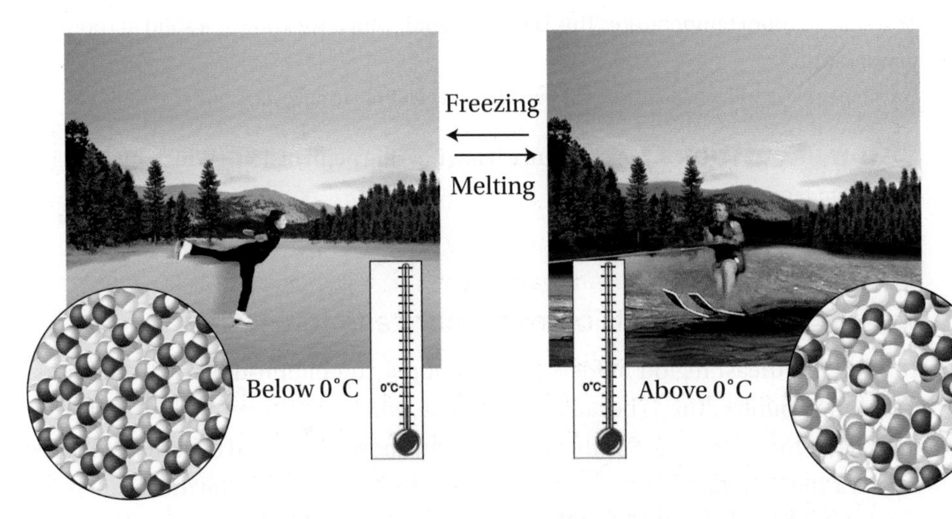

Figure 8.13 Melting and Freezing

The surface of this lake freezes when the temperature drops below 0°C. When the temperature rises again and reaches the 0°C mark, the lake's surface melts. Freezing and melting happen at the same temperature, but that temperature is approached from different directions in each case.

Table 8.1 Melting and Freezing Points for Familiar Substances

Substance	Melting/ Freezing Point (°C)
water, H_2O	0
nitrogen, N_2	−210
mercury, Hg	−39
gold, Au	1063

The phase changes many of us commonly experience are the phase changes for water. **Table 8.1** lists the melting/freezing points of some other familiar substances for comparison. Nitrogen (N_2), for example, is a gas at room temperature. That means it must be cooled to become a liquid, and cooled even further to become a solid. Its freezing point is very cold, −210°C. Gold is a solid at room temperature. Thus, its melting point is greater than room temperature. It can seem strange to think of melting and freezing as occurring at very high or very low temperatures because we are so accustomed to thinking of phase changes for water and not for other substances.

wait a minute . . .

How can a substance freeze at a temperature that isn't cold?

The freezing point for gold is 1063°C. It seems strange to think of something freezing at a really hot temperature because we are used to thinking about the phase transitions of water. Water freezes when it's cold.

We have to shift our thinking when we consider a different kind of substance, like gold. Gold freezes when really hot liquid gold becomes solid gold at 1063°C. But, because most of us use room temperature as a reference, it makes more sense to us to think of this process as occurring in the opposite direction: melting. Recall that freezing and melting describe the same temperature. We are used to gold that is solid. And, when we heat solid gold to 1063°C, it melts. Thinking of it that way is more intuitive for most people.

QUESTION 8.7 Which of the substances in Table 8.1 are likely to be a solid at room temperature—about 25°C?

ANSWER 8.7 Any substance that has a melting point above room temperature should be a solid at room temperature. This is true for gold, which we know is a solid at room temperature.

For more practice with problems like this one, check out end-of-chapter Questions 41 and 44.

FOLLOW-UP QUESTION 8.7 A substance has a freezing point of 86°C. Which phase is it in at room temperature?

Freezing of water contributes to hurricane-force winds.

The U.S. Southeast region holds a near-monopoly on billion-dollar natural disasters in this country. This is because a hurricane, which is the type of natural disaster most likely to occur in the Southeast, causes the most costly kind of damage. **Figure 8.14** is a 2005 satellite photograph of Hurricane Katrina, one of the most expensive natural disasters in U.S. history. While cruising through the Gulf of Mexico, Katrina reached the level of a Category 5 storm and then dropped to a Category 3 storm before making landfall. Nevertheless, Katrina caused $125 billion in damage and

GOES 12 Satellite, NASA, NOAA

Figure 8.14 Hurricane Katrina

submerged 80% of the city of New Orleans. The third deadliest hurricane to hit U.S. soil, Katrina took more than 1500 lives.

As a hurricane brews, ice crystals form at the top of the brewing storm, where the temperature is lowest. These ice crystals escalate the hurricane's intensity, as we're about to see. Hurricanes are much more than just heavy rain. They produce strong winds that begin to swirl, producing the familiar hurricane eddy seen in satellite photographs (Figure 8.14). How does the formation of ice crystals contribute to the strong winds that characterize a hurricane? To answer this, we need to consider the phase changes that are the freezing of water and the melting of ice. One requires an input of energy, and the other produces heat, but which is which? If we want to melt a block of ice, we must add energy in the form of heat:

$$H_2O\ (s) + \text{heat} \rightarrow H_2O\ (l)$$

To freeze water, this process must go in reverse:

$$H_2O\ (l) \rightarrow H_2O\ (s) + \text{heat}$$

At the top of a mounting hurricane, the second reaction takes place as ice crystals form and produce enormous amounts of heat. This heat causes the air to expand, and the expansion of the air, in turn, causes winds to rise. Thus, ice crystal formation is largely responsible for the devastating winds that accompany hurricanes.

While a substance changes phase, its temperature remains constant.

Imagine you are an amateur chemist living in a bitterly cold and inhospitable climate. You have a glass of water at room temperature (about 25°C) and a fairly accurate thermometer. You record the water temperature in your 25°C kitchen and then take the glass and thermometer outside, where the temperature is −25°C. You carefully monitor the water's temperature as it drops to the freezing point, 0°C. Then a peculiar thing happens. The temperature stops dropping at the freezing point and stays exactly at 0°C as ice crystals begin to form. Once the water has completely solidified into ice and the thermometer is held rigidly in place, the temperature once again begins to drop. What's going on?

The phenomenon you just imagined happens in the reverse direction, too. If you take your glass of ice back inside, the temperature begins to rise, but it ceases rising when the melting point (0°C) is reached. Until all the ice has become water, the temperature stays constant at 0°C. Then, once the ice is completely melted, the temperature begins to rise again until it reaches room temperature.

What we are describing is related to the way molecules absorb and release energy. Recall from Section 8.2 that the water molecules in ice are held in their hexagonal lattice by rigid hydrogen bonds (see Figure 8.11b). As the temperature rises toward the melting point, 0°C, the added heat energy makes the molecules vibrate more and more vigorously. The more the temperature goes up, the more they shiver and shake. At the melting point, however, the available heat energy is used not to make the molecules vibrate more vigorously, but instead to break the hydrogen bonds between molecules. The energy that goes into the water at its melting point is used to disrupt the rigid hydrogen bonds that gave the ice its structure. In other words, all the energy absorbed is used to change the ice to liquid water, and there is no heat energy available to raise the water's temperature. Once the ice is completely liquefied, however, the energy that is put into the water again goes toward increasing the movement within it.

QUESTION 8.8 When ice crystals form within a young hurricane, is heat released or is heat absorbed?

ANSWER 8.8 Heat is released, and this release causes winds to rise. This is why hurricanes have powerful winds.

For more questions like this one, check out end-of-chapter Questions 31 and 34.

FOLLOW-UP QUESTION 8.8 True or false? During a change of phase, the temperature of a substance continues to rise until the phase change is complete.

Heating curves illustrate how phases change.

Figure 8.15 shows the process of water changing phases. We call this a **heating curve**. At the lower left corner, where the temperature scale begins at −25°C, the temperature rises as heat is added to the ice. As the temperature of the ice increases, the molecules within it vibrate more and more, as described earlier. Then, when the melting point is reached, the line becomes flat. At this point, heat continues to be added, but the temperature does not increase because the added energy is used to break the rigid bonds of ice to make liquid water. When melting is complete, the temperature begins to rise again because the added heat goes toward increasing the motion and vibration of the molecules of water.

When water reaches its boiling point, the same phenomenon occurs again as it starts to change from liquid water to water vapor: the temperature stays constant until every liquid water molecule has become water vapor. Once all molecules are

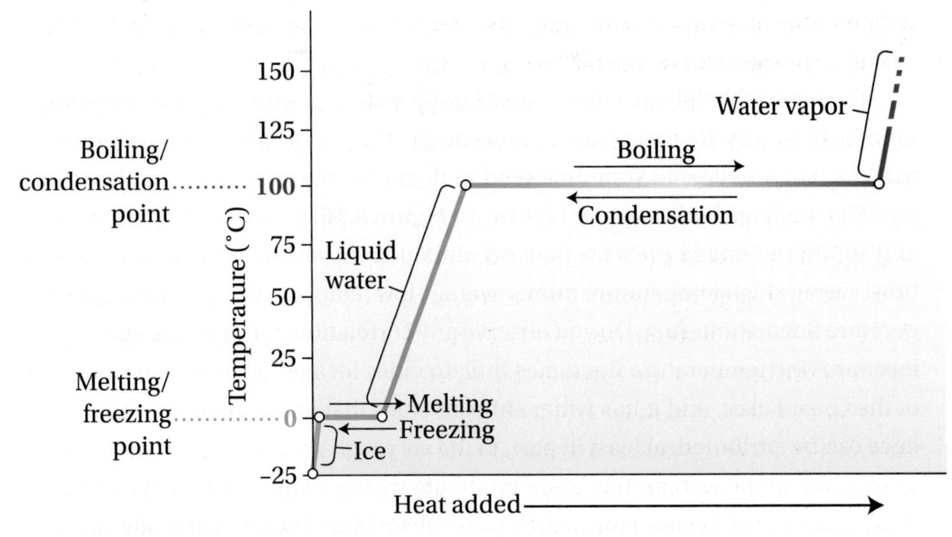

Figure 8.15 The Three Phases of Water

This heating curve shows the phase changes of water as heat is added and the water goes from ice to liquid water to water vapor.

in the gas phase, the temperature once again starts to rise, and the water reaches the gas phase shown at upper right in the figure.

Notice in Figure 8.15 that we can consider each horizontal line from two perspectives. The horizontal line at 0°C represents melting if heat is being added (which means moving left to right along the horizontal axis) and freezing if heat is being removed (moving right to left along the horizontal axis). Likewise, the horizontal line at 100°C represents boiling (phase change from liquid to gas) if heat is being added (moving left to right) and condensation (phase change from gas to liquid) if heat is being removed (moving right to left). (Recall that we discussed the term *condensation* in Section 8.1.)

QUESTION 8.9 True or false? Covalent bonds within water molecules do not break when ice melts.

ANSWER 8.9 True. When any phase change occurs, intermolecular forces are disrupted, but covalent bonds do not break.

For more practice with problems like this one, check out end-of-chapter Questions 20 and 23.

FOLLOW-UP QUESTION 8.9 What is a key difference between the behavior of a collection of water vapor molecules at 110°C and the same molecules at 210°C?

Water has a high specific heat.

If we were to compare the heating curve for water (see Figure 8.15) with the heating curve for another substance, we would notice one key difference: the green line for water is not as steep when compared to the same line for most other substances. This tells us that if a certain amount of heat is added to a certain amount of water, the temperature will rise by a relatively small value. If the same amount of heat were added to most other substances, the temperature would rise by a larger amount. The lack of steepness of that line reflects the high **specific heat** of water. The specific heat of a substance is the amount of heat energy required to raise the temperature of 1 gram of a substance by 1 degree Celsius. Water's high specific heat tells us that water can hold a lot of heat energy.

The relatively high specific heat of liquid water has important consequences for our daily lives. If you live near an ocean or a large lake, you may have noticed that the temperatures in your area tend to fluctuate much less than in places that don't have a large body of water nearby. In **Figure 8.16,** a map of the United States and southern Canada presents data on rainfall and average temperature fluctuation (average high temperature minus average low temperature). Examine the temperature fluctuations first. Do you observe any correlation between fluctuation and location? The temperature fluctuates little in cities located on an ocean or on one of the Great Lakes, and it has wider shifts in cities that are landlocked. This difference can be attributed, at least in part, to the very high specific heat of liquid water. Cities close to the water enjoy more moderate weather and maintain a more constant temperature because the nearby water absorbs and holds heat easily, and this ensures that temperatures vary less.

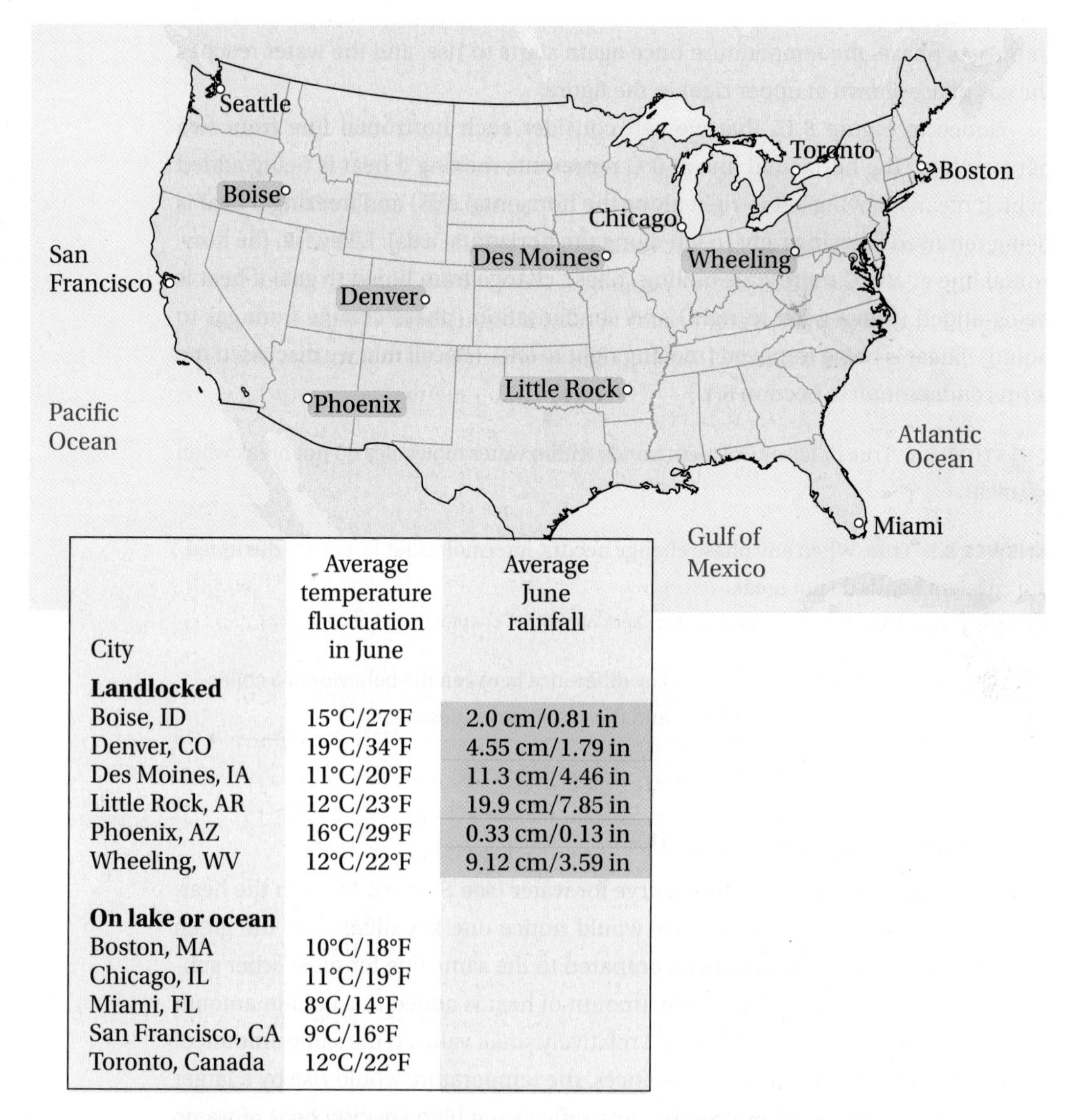

City	Average temperature fluctuation in June	Average June rainfall
Landlocked		
Boise, ID	15°C/27°F	2.0 cm/0.81 in
Denver, CO	19°C/34°F	4.55 cm/1.79 in
Des Moines, IA	11°C/20°F	11.3 cm/4.46 in
Little Rock, AR	12°C/23°F	19.9 cm/7.85 in
Phoenix, AZ	16°C/29°F	0.33 cm/0.13 in
Wheeling, WV	12°C/22°F	9.12 cm/3.59 in
On lake or ocean		
Boston, MA	10°C/18°F	
Chicago, IL	11°C/19°F	
Miami, FL	8°C/14°F	
San Francisco, CA	9°C/16°F	
Toronto, Canada	12°C/22°F	

Figure 8.16 Average Temperature Fluctuations and June Rainfall in Landlocked versus Waterfront Cities

dieKleinert/Alamy Stock Photo

Figure 8.17 A Passive Solar Home
In this passive solar home, large tubes of water absorb heat from the sun during the day and release it at night.

The temperature fluctuations for landlocked cities in Figure 8.16 vary from 12°C to 19°C. Why do some landlocked cities have much larger fluctuations than others? The answer can be found by looking at the average rainfall data. For cities with rainfall data shaded pink, the average June rainfall is quite high. This means that in June, in those cities, the air contains a great deal of humidity. Humidity is a complex topic and many variables are at play. Despite these variables, though, it is clear that humidity is one factor that helps maintain the temperatures in humid and rainy areas in the same way a nearby body of water maintains temperature. The cities shaded green in the average June rainfall column have very low rainfall, and so they experience more drastic temperature fluctuations. Thus, we can expect cities in dry areas, such as the desert Southwest, to have wider extremes of temperature than landlocked cities located in humid areas.

The extraordinarily high specific heat of water also plays a role in keeping humans comfortable. Our bodies are mostly water, and so we are able to maintain a stable body temperature because the water in our bodies can absorb heat and buffer us against changes in temperature. The same principle is at work in **passive solar** homes, which can provide a comfortable living environment by using only sunlight as an energy source (**Figure 8.17**). In some passive solar homes, large tubes of water are placed in a sunny location, where they can absorb large amounts of energy when the sun is up. When the sun sets, the water in the tubes releases that heat back into the house and provides warmth.

QUESTION 8.10 Imagine you live in a place that has no large bodies of water nearby. Despite this, the temperature there fluctuates very little in a 24-hour period. How is this possible when you know that in general temperature fluctuates more widely in cities that are landlocked?

ANSWER 8.10 You may live in a very humid or rainy place. Water in the air plays a role in reducing temperature fluctuations that occur over a 24-hour period.

For more practice with problems like this one, check out end-of-chapter Questions 42 and 45.

FOLLOW-UP QUESTION 8.10 True or false? Human body temperature varies widely, for example, when a person has a fever.

8.4 Changing Phases: Water and Water Vapor

The boiling point of water depends on altitude.

flashback

Recall from Section 6.2 that a given volume of air at high altitude contains less oxygen than is found in the same volume of air at sea level.

If you have ever been on a camping trip at high altitude, you may have noticed the effect of air pressure on water. Suppose you have stopped to camp at 5000 meters (about 16,400 ft) during a climb up Denali in south-central Alaska (**Figure 8.18**). As you unpack your stove and try to light it, you may notice you are out of breath and more tired than usual. This is your body telling you it wants more oxygen and to slow down. Once the stove is lit, though, your pot of water is boiling in minutes, and you feel a sense of pride and satisfaction. You mix up some hot chocolate and sit down to enjoy the view. To your disappointment, what you expected to be *hot* chocolate is instead only *warm* chocolate because the boiling point of water at this altitude is 83°C. You're quite sure the water was boiling, and surely boiling water is the same temperature everywhere, right? The answer is no.

To understand how water can boil at different temperatures, we must first think about how boiling occurs. A pot of water that's becoming hotter and hotter contains water molecules that are vibrating and wiggling. The hotter the water gets, the more energetic the molecules become. As the temperature goes up, more and more of the water molecules get so animated that they have enough energy to escape from the liquid and become water vapor.

You know from everyday life that molecules at the surface of a liquid can evaporate and become water vapor. But the same thing happens in the water below the surface: Some water molecules that are not on the surface become energetic enough to become a gas as the water heats up. When these few molecules change

Figure 8.18 Not-So-Hot Chocolate

If you are ever fortunate enough to climb Denali in Alaska and reach a height of 5000 meters (about 16,400 ft), boiling water will not be as hot as what you are used to at much lower elevations.

to water vapor for an instant, a tiny bubble forms. Then the pressure of the water above squashes the bubble, and it returns to the liquid phase. But as the temperature continues to increase, more and more of those below-surface water molecules simultaneously leave the liquid phase and enter the gas phase, and larger, stronger bubbles form. The pressure exerted by the gas molecules in these bubbles gets higher and higher as more water molecules take part in their formation.

Bubble formation is not the whole story, though. To understand boiling completely, we must also take the air outside the pot into account because it exerts a constant pressure on the water's surface. When water is heated, eventually a temperature is reached at which the pressure inside a submerged bubble equals the atmospheric pressure pushing down from above. At this temperature, called the **boiling point,** the pressure inside the bubble is equal to the atmospheric pressure, and the bubble stays intact and floats to the surface. This is why water that's about to boil contains small bubbles that appear and disappear. When full boiling begins, the bubbles rise and break the surface. On the camp stove in **Figure 8.19,** water is shown before it boils (left) and at the boiling point (right).

flashback

Recall from Chapter 6 that atmospheric pressure is lower at high altitude and higher at sea level.

So how is all of this related to the warm chocolate we made up on Denali? It's related because the atmospheric pressure dictates when boiling happens. If the atmospheric pressure is low, as it is on Denali, then the submerged bubbles in the water only need to reach that pressure for boiling to commence. At sea level, where the atmospheric pressure is higher, the submerged bubbles have to reach a much higher pressure before they're able to stay intact and make it to the surface. So, boiling at sea level requires a higher temperature than boiling on Denali. In fact, for roughly every 300 meters in altitude gain, the boiling point of water decreases by 1°C. And, because water boils at a lower temperature on Denali, our chocolate is not piping hot. It's just warm.

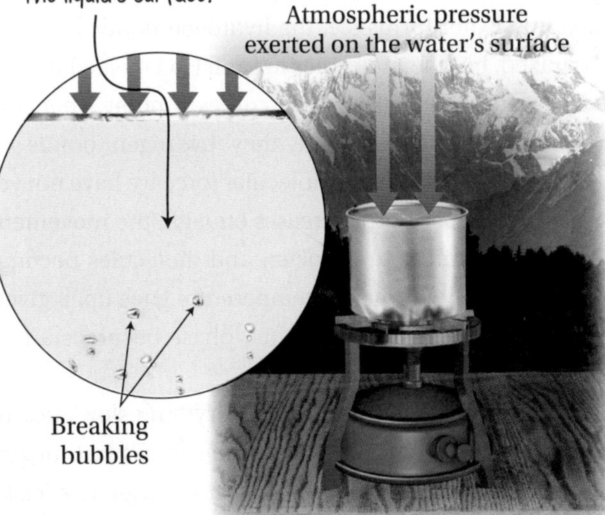

Bubbles do not persist, because the pressure inside is less than the pressure on the liquid's surface.
Atmospheric pressure exerted on the water's surface
Breaking bubbles
Below the boiling point

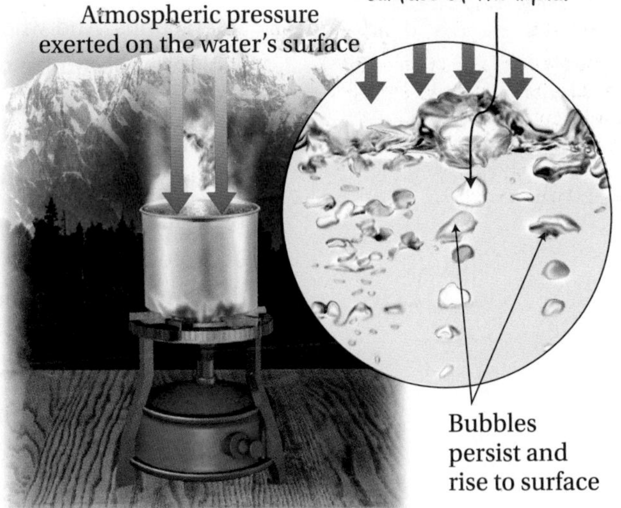

At the boiling point, the pressure in the bubble equals the pressure on the surface of the liquid.
Atmospheric pressure exerted on the water's surface
Bubbles persist and rise to surface
At the boiling point

Figure 8.19 How Water Boils: A Camp Stove Example

We already know that boiling is the phase change from liquid to gas and that condensation is the shift from gas to liquid. Although boiling is the formal word for the transition from the liquid phase to the gas phase, we use other, everyday words to describe this process. For example, we use the word "evaporation" to explain the slow loss of water molecules from a lake on a warm day. The end result of both processes is that a liquid has become a gas.

QUESTION 8.11 Would you expect the boiling point of liquids other than water to be lower at high altitude than at sea level?

ANSWER 8.11 Yes. All liquids become gases when heated to the boiling point. Furthermore, all liquids experience lower atmospheric pressures at high altitudes. Therefore, you should expect the boiling point of any liquid to be lower at high altitude than at sea level.

For more practice with problems like this one, check out end-of-chapter Question 35.

FOLLOW-UP QUESTION 8.11 Which of the following temperatures could be the boiling point of water at a pressure slightly greater than that at sea level? 100°C, 90°C, 200°C, 104°C

Summary: During phase changes, intermolecular interactions are made and broken.

Water molecules are still water molecules when they move from one phase to another. They do not change composition: each one is still made up of two hydrogen atoms and one oxygen atom, covalently bound together in the now-familiar bent shape. The only thing that changes as water moves from gas phase to liquid phase to solid phase is the amount of interaction among the molecules.

For example, H_2O molecules in the gas phase do not interact much with one another, because they're very far apart. These gas molecules may enter the liquid phase, where they become close neighbors with other water molecules and interact with them via a disorderly network of hydrogen bonds as well as dipole–dipole interactions. And when the liquid water forms ice, the hydrogen bonds between molecules become more organized, but the molecules are still H_2O molecules.

When energy is added to water molecules, there comes a point when the interactions holding those molecules together—be they hydrogen bonds or dipole–dipole interactions, or another type of intermolecular force we have not yet met—are overwhelmed. When the temperature increases enough, the movement becomes too intense, intermolecular forces are broken, and molecules become disconnected from one another. This is why, as the temperature goes up, a given substance reaches certain points where a more restricted phase becomes a less restricted one. Solid becomes liquid. Liquid becomes gas.

In the opposite direction, as the temperature drops, everything slows down. Molecules spend more time with one another. Intermolecular forces last longer. Again, substances reach points where phases shift. Gas molecules begin to stick to one another and liquefy. As the temperature continues to fall, neighboring molecules in the liquid phase spend more time with one another. Intermolecular forces take hold and lock molecules into place. A solid forms.

THEgreenBEAT • news about the environment, chapter eight edition

TOP STORY Should We Be Spiking Our Drinking Water?

In 1901, freshly educated Frederick McKay, DDS, arrived in Colorado Springs, Colorado, looking to start a dental practice. What he found was a city full of people whose teeth were stained brown (**Figure 8.20**). Even the kids had unsightly brown chompers! McKay was determined to figure out what caused this unfortunate discoloration, and he did. But he learned that it wasn't such a curse after all. Over the years he spent studying the problem, his data showed that the people with brown teeth also had very few cavities. As it turns out, the local water supply contained high levels of fluoride. The fluoride was causing the discoloration, but it was also increasing the strength of the townspeople's tooth structure.

Years later, in 1945, the entire city of Grand Rapids, Michigan, was the site of the first public fluoridation experiment. Fluoride was added to the public water supply in amounts small enough to avoid discoloring the teeth. And, lo and behold, the incidence of cavities in Grand Rapids decreased by 60% over the first year. After that compelling experiment, fluoride was added to water supplies all over the United States, and it's still being added today.

A similar story is unfolding now, in the twenty-first century, but it involves a different substance.

If you drive north out of New York City toward Syracuse for about four hours, you'll reach Cherry Valley, New York. The town is known for its fresh water spring, which has been supplying the village with "Happy Water" for two centuries. It turns out that the spring is a natural source of the element lithium, which also happens to be the most commonly prescribed drug for some serious psychiatric disorders. So how is the element lithium—the third lightest on the periodic table—related to happiness?

Dr. Gerhard Schrauzer is especially qualified to answer this question, because he was the first person to publish a paper linking lithium in water to the behavior of people who drink it. His study spanned 10 years and focused on 27 different counties in Texas, which he divided into three groups according to the levels of lithium that exist naturally in their public water supplies. The data are shown in **Figure 8.21**.

According to the data, the counties with the highest levels of lithium in the water showed significantly reduced crime rates. This included homicide, rape, and even suicide. Because of data like this, Schrauzer wants to add low levels of lithium to all public water supplies, in much the same way that we add fluoride to reduce tooth decay.

The idea of spiking public drinking water with a mind-altering drug is, to say the least, controversial. After all, lithium is a powerful drug when used for psychiatric diseases, such as bipolar disorder, and side effects of high doses are severe. Further, because the word *lithium* is associated with mental illness, people bristle at the thought of being required to imbibe it. In fact, one researcher who wrote an article that defended the idea of adding small amounts to drinking water received death threats and scores of angry letters. Lithium in water is an off-the-wall idea that has not quite made its way into the public sphere, but it might soon be on its way.

Crime rate per 100,000 people, based on a 10-year study of 27 Texas counties

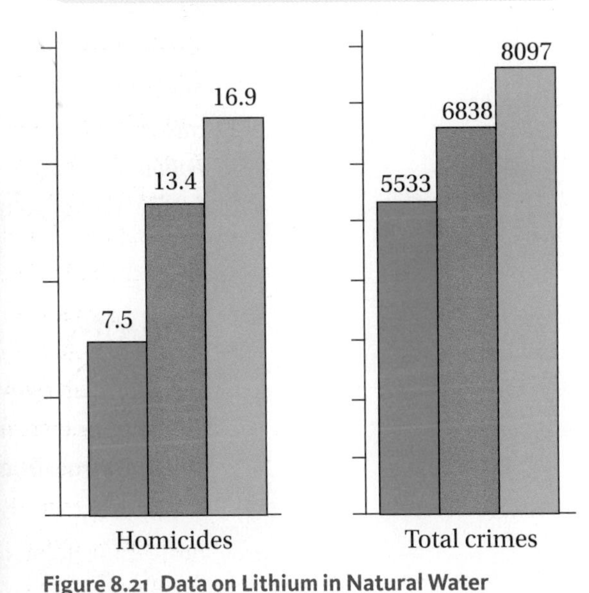

Figure 8.21 Data on Lithium in Natural Water

The data shown as green bars comes from counties with the highest levels of lithium in the public water supply. Orange and blue bars represent counties with lithium present at low and very low levels, respectively.

Dr. P. Marazzi/Science Source

Figure 8.20 Brown-Stained Teeth

Chapter 8 Themes and Questions

recurring themes in chemistry chapter 8

recurring theme 1 Scientists worldwide use the scientific method as a basis for experimentation and deduction. The scientific method is based on the principles of peer review and reproducibility. (p. 220)

recurring theme 3 Opposite charges attract one another. (p. 228)

recurring theme 7 Individual bonds between atoms dictate the properties and character (p. 229)

recurring theme 2 Things we can see with our eyes contain trillions and trillions of atoms. Atoms are much too small to see with the naked eye. (p. 230)

A list of the eight recurring themes in this book can be found on p. xv.

ANSWERS TO FOLLOW-UP QUESTIONS

Follow-Up Question 8.1: Coffee. *Follow-Up Question 8.2:* True. *Follow-Up Question 8.3:* False. In the United States, water treatment plants treat all natural water that becomes tap water, no matter how clean or dirty. *Follow-Up Question 8.4:* Yes. This molecule happens to contain both groups required for hydrogen bonding, in this case H–O and O. *Follow-Up Question 8.5:* A bond between two of the same atoms is always nonpolar, because each atom pulls the same on the electrons in the bond. Thus, the O–H bond is more polar. *Follow-Up Question 8.6:* Substance E is denser. *Follow-Up Question 8.7:* It is a solid at room temperature, and you must heat it to its freezing/melting point to melt it. *Follow-Up Question 8.8:* False. The temperature remains constant during a phase change. *Follow-Up Question 8.9:* The warmer gas molecules are more energetic and move faster. *Follow-Up Question 8.10:* False. Human body temperature does increase with a fever, but that increase is very small. Human bodies stay within a narrow temperature range because we are made mostly of water, which has a high specific heat. *Follow-Up Question 8.11:* The temperature is slightly greater than the temperature for the boiling point at sea level, which is 100°C. So, 104°C is a reasonable answer.

End-of-Chapter Questions

Questions are grouped according to topic and are color-coded according to difficulty level: **The Basic Questions***, **More Challenging Questions****, and **The Toughest Questions*****.

Victor Polyakov/Shutterstock

Water Footprints and the Water Cycle

*1. List three things that contribute to your individual water footprint.

*2. In your own words, explain why water footprints are important for modern societies. Answer this question in your response: If we have tremendous amounts of water on Earth, why do we need to be concerned about water usage?

*3. List an example of each of the three phases of water (gas, liquid, or solid) in the water cycle.

**4. Which of these materials are not removed during the water treatment process?

(a) debris

(b) viruses

(c) deuterium

(d) bacteria

**5. What percentage of total natural water is fresh water? What percentage of fresh water is groundwater?

**6. According to Figure 8.2, which country in South America has the smallest water footprint? Which has the largest?

**7. The water cycle in Figure 8.4 shows several examples of phase changes. What is the name of the phase change that involves a solid converting directly into a gas?

***8. According to this chapter, what is the mass of salt water on Earth? If salt water is 3.5% salt by mass, what is the mass of salt in all the salt water on Earth?

Liquid Water and Ice

*9. Which is heavier, 100 milliliters of water or 100 milliliters of ice? Explain your answer.

*10. Which is denser, liquid water or ice? Explain your answer.

*11. Describe how songbird migration routes can be tracked by using the percentage of deuterium in water.

**12. The process $H_2O\ (l) \rightarrow H_2O\ (s)$ + heat occurs in your freezer. How does your freezer facilitate this reaction?

**13. In your own words, explain how the structure of ice makes it less dense than water.

**14. Your friend is a graduate student in chemistry. She plans to study the levels of tritium in bumblebee wings to see if they vary from bee to bee. Why would you caution her to work on something else?

***15. Deuterium and tritium, which are discussed in this chapter's naturebox, are sometimes represented by the symbols D and T, respectively, instead of the standard H for hydrogen. Rank these forms of water in order of their abundance in a typical sample of water, from most abundant to least abundant: D_2O, HTO, HDO, H_2O, T_2O.

***16. A 0.35-mL sample of water is taken from a songbird wing. This volume of water contains 0.020 mol of water molecules. If one in every 10,000 hydrogen atoms in the sample is a deuterium atom, roughly how many deuterium atoms does the sample contain?

Intermolecular Forces

*17. List the two types of intermolecular forces mentioned in this chapter. Briefly describe each type, and state which is stronger.

*18. Which of these bonds would you expect to be more polar: a bond between two nitrogen atoms or a bond between a hydrogen atom and a nitrogen atom?

*19. Sketch a hydrogen bond between two water molecules.

*20. Which is stronger, a covalent bond between two atoms in a molecule or a dipole–dipole force between two molecules? Explain your answer.

**21. Correct the following statement: Individual atoms interact with one another through dipole–dipole forces.

**22. True or false? A dipole in a molecule always falls along a bond between two atoms.

**23. True or false? When water changes from water to water vapor, hydrogen bonds break but dipole–dipole interactions are not affected.

**24. Acetone, shown here, contains six hydrogen atoms and does not form hydrogen bonds with other acetone molecules. How is it possible for a molecule like acetone to contain hydrogen atoms but not be capable of forming a hydrogen bond?

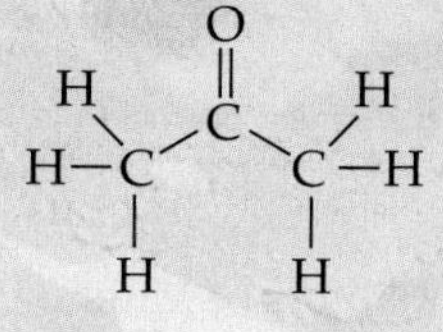

**25. While painting your nails, you uncap a bottle containing about 250 mL of nail polish remover, which is made from acetone. You also happen to be drinking from a bottle containing about the same volume of water. You run to answer the phone and forget about both of them. The next day, one of the containers is empty but the other is only half empty. Which container is empty? Why did one liquid evaporate completely and the other not? (Hint: You may want to answer Question 24 before this one.)

***26. The two compounds shown below have the same chemical formula.

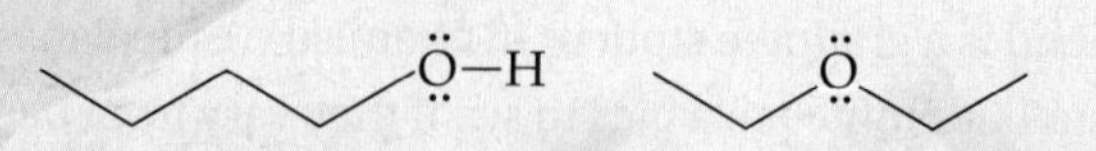

(a) What is that chemical formula?

(b) Which compound would you expect to have the highest boiling point? Explain your choice.

PhotoStock-Israel/Alamy Stock Photo

***27. Sketch 11 water molecules interacting with one another via hydrogen bonds.

***28. The ammonia molecule is shown here.

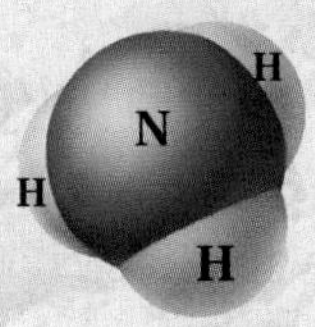

Each H–N bond is polar, because more electron density resides on the nitrogen atom than on the hydrogen atoms. Given this structure, can you predict where the dipole for the molecule will lie?

***29. The structures of the CO and CO_2 molecules are shown below.

Each one of these molecules is made up only of bonds between carbon and oxygen—and yet, only one is polar. Explain this fact.

PhotoStock-Israel/Alamy Stock Photo

Weather and Altitude

*30. Is the atmospheric pressure greater, on average, in Colorado or in Florida?

*31. Does the melting of ice require heat or produce heat?

*32. In your own words, explain why the ice on a lake freezes only on the surface and not on the lake bottom.

*33. Which phase transition is described by the term *evaporation*?

**34. Describe the way strong winds arise during a hurricane.

**35. The directions for cooking oatmeal at high altitude suggest doubling the regular cooking time. Why?

**36. When planning a trip to a landlocked city in central Africa, you are advised to take clothes for various temperatures because areas far from water tend to experience large swings in temperature. If you knew the average humidity at your destination, how would that help you decide what to pack?

**37. Without consulting your book, write down the names of three phase transitions mentioned in this chapter.

***38. In this chapter, we learned that the phase transition from liquid to gas is affected by changes in altitude. Consider this question: Is the phase transition from solid to liquid affected to the same extent? Why or why not?

***39. You and a friend who is a chef are planning to climb Annapurna, one of the tallest peaks in the Himalayas. He is planning an elaborate menu for the climb, including many foods that require heating. What advice can you give him about heating food at high altitudes?

Phase Transitions and Heating Curves

*40. True or false? The temperature of a substance stays constant while it is changing phase.

*41. A certain substance has a boiling point of −150°C. What phase is it in at room temperature?

*42. What is the temperature of boiling water in a pot sitting on a stove at sea level?

*43. Which phase transition is not included on the heating curve shown in Figure 8.15?

*44. A certain substance has a melting point of 200°C. What phase is it in at room temperature?

**45. A certain substance has a boiling point of 30°C. What phase is it in at 300°C?

**46. In your own words, explain why heating curves have regions that are flat and regions that are sloped. What do these regions signify?

**47. When ice is heated, its temperature increases until the melting point (0°C) is reached. At that point, the temperature remains at 0°C until all the ice is melted and is in the liquid phase. Additional heat then raises the temperature of the water. Why does this phenomenon occur?

**48. Cooks can save time when boiling a large quantity of water by dividing the water into two pots and then recombining the water once boiling commences. Why is this method faster than boiling all the water in a single pot?

***49. Suppose a diagram like the one shown for water in Figure 8.15 is available for substance Z. The line in the diagram for substance Z in the liquid phase is one-half as steep as the same line in the water diagram. After seeing both diagrams, a friend decides to empty the water from the tubes in her passive solar home and fill them with substance Z. Based on what you know about substance Z, is this a good idea?

***50. Cooks sometimes use pressure cookers to speed up the cooking process. A pressure cooker increases the pressure on whatever you put in it. How does this make the food cook faster?

*51. In Section 8.2, we learned that water has a density of 1.0 g/mL, whereas ice has a density of 0.92 g/mL. If one liter of water has a mass of one kilogram, what is the mass of one liter of ice?

**52. A typical bathroom sink faucet releases about two gallons of water every minute. Let's say that it takes you 2.5 full minutes to brush your teeth. That means that you would use about five gallons of water every time you brush your teeth if you leave the water running while you are brushing. Imagine that, of the 2.5 minutes it takes to brush your teeth, 0.5 minute is spent rinsing with water. The rest of the time is spent brushing. If this is true, what volume of water would you save by turning off the water when you are brushing? What percentage of the five gallons would be saved by doing this?

***53. The adult human body is composed of 50–65% water. Newborn babies, however, are about 78% water. If a baby is born weighing 8 pounds 3 ounces, what is the mass of water, in grams, in the baby? Here are some useful conversion factors: 16 ounces = 1 pound; 0.002205 pounds = 1 gram.

***54. The structure of ice is shown in Figure 8.11. When salt is added to water and the mixture is cooled, the presence of dissolved ions makes it difficult for the open lattice structure shown in the figure to form. Given this fact, why would the freezing point of water that's salty be lower than that of pure water? Why do you think it is helpful to add salt to icy roads? Speculate on whether this salt treatment would work at extremely cold temperatures.

54. INTERNET RESEARCH QUESTION

Return to Figure 8.16 and the discussion of temperature fluctuation as a function of proximity to large bodies of water and as a function of relative humidity. Now, choose three cities that are not listed in Figure 8.16. Choose one city on the ocean and two cities that are landlocked, one in a wet climate and one in a dry climate. Go to the Internet and research the weather in those cities in June. What are the average maximum and minimum temperatures in degrees Fahrenheit? Convert each of those six numbers (one pair of numbers for each of three cities) to Celsius. Next, subtract each minimum from each maximum to arrive at the average fluctuation in temperature in degrees Fahrenheit and in degrees Celsius.

Consider your new data. What do you think? Does the general trend follow the trend we saw in Figure 8.16? If so, what are examples of this? If not, how does your data differ? Describe your overall findings.

55. INTERNET RESEARCH QUESTION

In Section 8.1, we discussed the water footprints of several types of food, from wheat to coffee. Think of one type of food that was not discussed in that section and think about what the mass of one serving size would be. That number might be found on a can or box that holds the food. If you cannot figure out the mass of your food, conduct an Internet search to figure it out, if necessary. For example, you could ask your search engine, "What is the mass of one green bell pepper?" Once you know the mass of your food, convert that mass to kilograms (You can use conversions in this book, or find them online).

Enter "water footprint food search" into your search engine. This should bring up various places to find the water footprint of your food. Go to one of these websites and search for your food of choice. The result you get is likely to be given in liters per kilogram, so multiply this value by the mass of your food in kilograms to find the number of liters. This is the water footprint of your chosen food.

Danny Smythe/Shutterstock

Salts and Aqueous Solutions

The Nature of Salts and How They Interact with Water

Courtesy of James Bull, Castellina in Chianti, Italy

NaCl

CHAPTER 9

Go to a restaurant almost anywhere in the world, and you'll probably find salt on your table. For centuries, we humans have been shaking table salt, NaCl, onto our food. This gesture has become the subject of metaphors that underscore the importance of salt in our lives. A person who is the "salt of the earth," for instance, is one of a small number of people sprinkled throughout society who set an example of excellence for others.

The history of salt and the history of the human race are intertwined. For millennia, people have been making salt, consuming salt, trading salt, and dying for salt. In ancient Rome, soldiers were paid in salt, their "salarium argentum." From this term, we now have the English word salary. The ancient Roman road, Via Salaria, owes its name to the Latin word for salt. Some historians consider the trade in salt, which took place along this road, to have been the origin of the settlement of Rome. A street by this name still exists in the heart of the Italian capital (see nearby photo). In ancient Greece, salt was traded for slaves. A lazy slave, it was said, was "not worth his salt." Obviously, any chemistry book worth its salt, then, must devote some of its pages to this important substance.

In this chapter, we'll spend some time discussing the familiar salt that comes from our saltshaker. We also look at the family of compounds called salts, one member of which is our old friend sodium chloride. Then we turn to salts dissolved in liquids, especially water. We talk about what happens when salts dissolve and become ions, how ions travel from one place to another, and why these ions are critical for life. Along the way, we explore how the concentration of ions in a solution can have interesting consequences for the environment, for human blood, for football scores, and even for Egyptian mummies.

Courtesy of James Bull, Castellina in Chianti, Italy

9.1 Review: The Nature of Salts

Salts are ionic solids.

Chemists define the term *salt* more broadly than laypersons do. When the average person thinks about salt, what usually comes to mind is table salt, which is more formally known as sodium chloride or NaCl. Recall from Chapter 4 that for a chemist, a salt is any ionic solid, be it sodium chloride, magnesium bromide, potassium sulfate, or calcium iodide.

In Section 4.3, we discussed salts and learned how cations and anions form, and how they combine to make ionic compounds. Here we briefly review these ideas. In that section, we discussed examples of salts that form when a cation from the left side of the periodic table combines with an anion from the right side of the table. Figure 4.5 shows all the ions that are formed from the elements on the far left and right sides of the periodic table. That figure is reproduced here as **Figure 9.1**. Within each group shown on this cylindrical periodic table, ions of the same charge form because elements in the same group share this characteristic. For example, all the halogens (Group 17) make anions with a charge of −1.

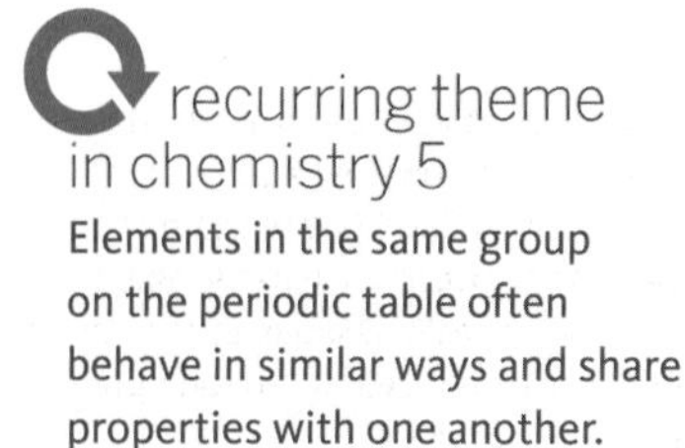
recurring theme in chemistry 5
Elements in the same group on the periodic table often behave in similar ways and share properties with one another.

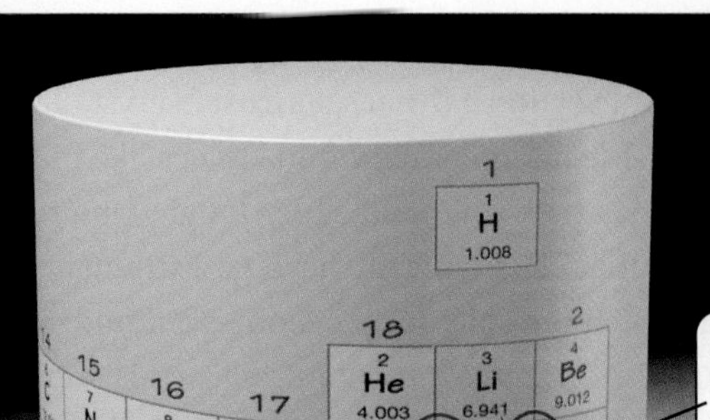

Figure 9.1 A Periodic Table Wrapped into a Cylinder

If we take a periodic table and curve it so that Groups 1 and 18 are next to each other, then the elements are numbered, in sequence, as we move across each period, around and around. Looking at the periodic table in this way makes it easier to see which noble gas is closest to any given element. (This figure also appears as Figure 4.5.)

Consider this example that reviews how ions join to form a salt: A potassium atom gives up one electron in order to have the same number of valence electrons as argon, the nearest noble gas. In giving up the electron, potassium can form a salt with, say, a fluorine atom, which accepts one electron in order to have the same number of valence electrons as neon. These two ions, K^+ and F^-, come together in an ionic bond—an attraction between positive and negative charges, as shown in **Figure 9.2**. The result is potassium fluoride, KF.

flashback

Recall from Section 4.1 that an atom can obtain an octet of electrons by losing or gaining electrons. According to the octet rule, this process results in an ion with the same number of electrons as the nearest noble gas.

recurring theme in chemistry 3

Opposite charges attract one another.

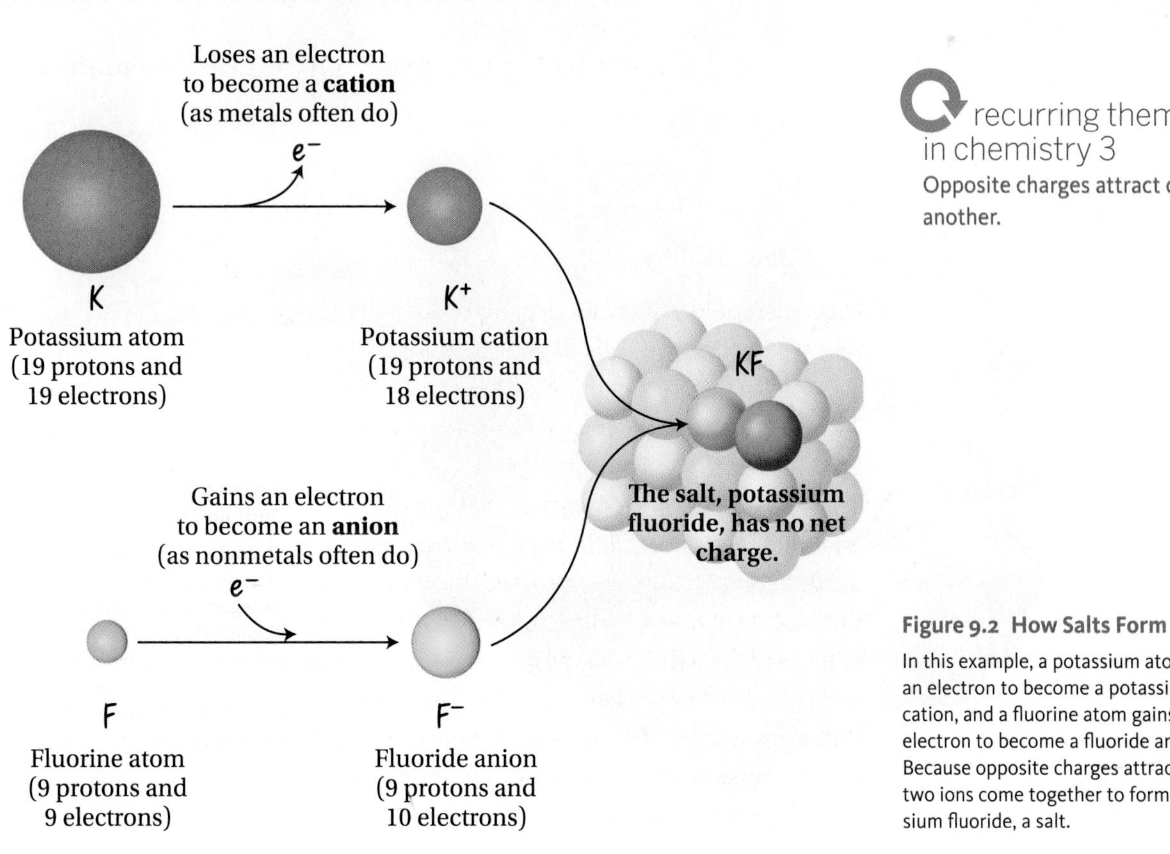

Figure 9.2 How Salts Form

In this example, a potassium atom loses an electron to become a potassium cation, and a fluorine atom gains an electron to become a fluoride anion. Because opposite charges attract, these two ions come together to form potassium fluoride, a salt.

Salts pack tightly into a crystalline lattice.

flashback

We discussed transitions between phases, including solid to gas and solid to liquid, in Chapter 8. Recall from that discussion that a substance reaches its melting point when the rigid structure of a solid melts to create a liquid.

Recall also from Section 4.3 that salts exist in crystalline lattices, like the one shown in **Figure 9.3**. The interactions within lattices like this are so strong that the resulting salt crystal is rock solid and super-tough. One result of this ultra-strong ionic interaction is that salts typically have an extremely high melting point, the temperature at which the solid form becomes a liquid. Sodium chloride, for instance, melts at 801°C. At extreme temperatures like this, most substances that are not salts have either melted to become liquids or vaporized to become gases. For example, ice melts at 0°C, and water boils at 100°C. If sodium chloride liquefies at a super-hot 801°C, at what temperature does sodium chloride become a gas? The answer: sodium chloride is a gas at or above 1413°C. The vast majority of salts pack into robust, crystalline lattices and so, like sodium chloride, they melt only at very high temperatures.

recurring theme in chemistry 7

Individual bonds between atoms dictate the properties and characteristics of the substances that contain them.

QUESTION 9.1 How many electrons are there in the following atoms and ions? Write a symbol for each ion, and remember that a symbol for an ion includes its charge. (a) a potassium atom (b) a chlorine atom (c) a calcium ion (d) a fluoride anion (e) a potassium ion (f) an iodine atom (g) a chloride anion (h) a sulfide anion.

Ted Kinsman/Science Source

Figure 9.3 The Ionic Compound Sodium Chloride

This schematic drawing of a portion of a sodium chloride crystal shows its regular array of sodium ions and chloride ions (inset). This image shows sodium chloride crystals magnified several hundred times. This figure is a reproduction of Figure 4.11 (background).

ANSWER 9.1 (a) K, 19 (b) Cl, 17 (c) Ca^{2+}, 18 (d) F^-, 10 (e) K^+, 18 (f) I, 53 (g) Cl^-, 18 (h) S^{2-}, 18

For more practice with problems like this one, check out end-of-chapter Question 1.

FOLLOW-UP QUESTION 9.1 Many atoms obtain the same number of electrons as the nearest noble gas by forming ions. For each ion, indicate which noble gas has the same number of electrons as the ion (a) Cl^- (b) S^{2-} (c) Na^+ (d) Br^-.

9.2 Polyatomic Ions

Egyptian mummies can help us understand the behavior of salts.

Being an early Egyptian was not easy. The Egyptians believed that your body as well as your spirit would continue to persist in the afterlife, but only if very specific rules were followed. When a person died, it was especially important to preserve the body so that it would be intact in the afterlife. This preservation was accomplished through mummification, a ritual that has been studied for centuries (**Figure 9.4**).

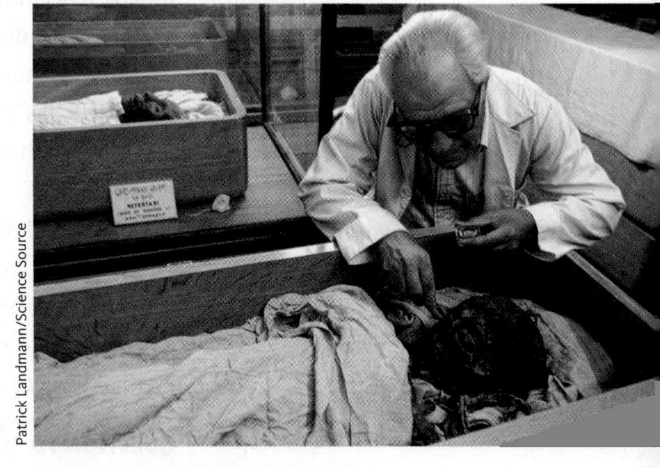

Patrick Landmann/Science Source

Figure 9.4 A Mummy Is Examined

There are still conflicting ideas about exactly how bodies were mummified. For example, Egyptologists are fairly certain that natron, a mixture of salts, was used in mummification, but scholars still debate the role of natron. Was it used to pickle the body, in much the same way NaCl pickles a cucumber? Was it used as a drying agent? Examination of the remnants of early Egyptian civilization supports the second theory.

In 1994, scientists at the University of Maryland School of Medicine decided to try mummification for themselves. They reasoned that a modern mummy would be a reasonable replica of an ancient specimen. And as a bonus, because their modern mummy would be a modern-day person whose lifetime medical history was known, they could use it as a standard for comparison to other mummies.

First, the Maryland scientists went in search of someone to mummify. Their criteria were strict: the person must have died of natural causes, have filled out a donor card, and never have had surgery. The body of a man from Baltimore who had donated his body to science fit the bill.

The researchers worried about every detail, because they wanted to mimic the Egyptian mummification ritual as closely as possible. They planned carefully for everything from the vessels that held the embalming fluids, which were handmade and painted with authentic hieroglyphics, to the embalming table, which was a replica of ancient tables found in Egypt. They even went to Egypt to procure spices and oils from a local source. They knew that the natron, in particular, would be one of the most important parts of the mummification process, so they were careful to obtain this salt mixture from local sources.

Given the important but mysterious role of natron in this process, we should probably know more about it. So let's put our team of scientists, and their mummy, on hold until we learn a bit more about the chemistry of natron. But don't worry, we'll return to the topic of mummification at the end of Section 9.3.

Table 9.1 Common Polyatomic Ions

Cation	
NH_4^+	Ammonium
Anions	
$C_2H_3O_2^-$	Acetate
$Cr_2O_7^{2-}$	Dichromate
NO_3^-	Nitrate
SO_4^{2-}	Sulfate
OH^-	Hydroxide
CN^-	Cyanide
PO_4^{3-}	Phosphate
HPO_4^{2-}	Hydrogen phosphate
$H_2PO_4^-$	Dihydrogen phosphate
CO_3^{2-}	Carbonate
HCO_3^-	Hydrogen carbonate (bicarbonate)
ClO_4^-	Perchlorate
MnO_4^-	Permanganate
CrO_4^{2-}	Chromate
O_2^{2-}	Peroxide

Polyatomic ions include multiple atoms and one or more charges.

Natron is composed of four salts: sodium carbonate, Na_2CO_3; sodium bicarbonate, $NaHCO_3$; sodium chloride, NaCl; and sodium sulfate, Na_2SO_4. From this list, we have discussed only sodium chloride so far. The simplest ions, such as sodium ions and chloride ions, are single atoms that have ionized to form a cation or anion. We call them **monatomic ions** because they each contain only one atom. A **simple salt,** such as sodium chloride, is composed of one or more monatomic cations paired with one or more monatomic anions.

The other three salts in the natron recipe contain ions that are each made up of not one atom, but several. For example, sodium carbonate contains sodium cations and carbonate anions. We have discussed the sodium ion in previous chapters, so it is familiar; but the carbonate ion, CO_3^{2-}, may be less familiar. We call ions of this type **polyatomic ions** because they contain more than one atom.

The chemical formulas for several common polyatomic ions are shown in **Table 9.1**. It's important to note a few things about this collection of ions. First, notice that they can possess various charges and different numbers of atoms and elements. For example, dihydrogen phosphate ($H_2PO_4^-$) contains three elements, a total of seven atoms, and has a −1 charge. Peroxide (O_2^{2-}) contains one element, two atoms, and has a −2 charge. Finally, most common polyatomic ions are anions, so we commonly see a salt that features a monatomic cation combined with a polyatomic ion. Three of the salts found in natron fall into this category: Na_2CO_3, $NaHCO_3$, and Na_2SO_4.

QUESTION 9.2 Which of the following are polyatomic ions and which are monatomic ions? (a) Cl^- (b) NO_3^- (c) I_3^- (d) O^{2-}

ANSWER (a) and (d) are monatomic ions because they contain only one atom. (b) and (c) are polyatomic ions because each of them contains more than one atom.

For more questions like this one, check out end-of-chapter Question 10.

FOLLOW-UP QUESTION 9.2 Which of the following is a simple salt? Which are salts that contain a polyatomic ion? (a) NaI (b) NaOH (c) $BaPO_4$ (d) K_2CO_3

Salts are electrically neutral.

flashback

Remember from Chapter 4 that the smallest repeating unit in a salt is a *formula unit.* Salt crystals contain a repeating lattice of cations and anions, and the formula unit is the motif that repeats over and over. For table salt, that motif is Na^+ then Cl^-, repeated again and again. Thus, a formula unit of sodium chloride includes one sodium ion and one chloride ion.

Simple salts have no net charge. Their charges add up to zero, so they are electrically neutral. For example, magnesium chloride contains Mg^{2+} ions and Cl^- ions. (In the name of a salt, the cation name is followed by the anion name.) To form an electrically neutral salt from these ions, there must be two chloride ions for each magnesium ion: $MgCl_2$. The same rule applies to salts containing polyatomic ions. For example, magnesium carbonate, which contains magnesium ion, Mg^{2+}, and carbonate ion, CO_3^{2-}, has zero net charge. Therefore, there must be one magnesium ion for every carbonate ion: $MgCO_3$.

We enclose polyatomic ions in parentheses when we add subscripts to them, so that the total number of atoms is clear. For example, the formula for ammonium sulfide is $(NH_4)_2S$. The subscript 2 in its formula tells us that we need to multiply the

Kim Waldron

Figure 9.5 Salts That Contain Polyatomic Ions

Each of these salts contains a polyatomic ion listed in Table 9.1: $K_2Cr_2O_7$ (potassium dichromate, red), $CuSO_4$ [copper(II) sulfate, blue].

number of atoms within the parentheses by two. We put parentheses around the ammonium ion to indicate that the subscript 2 multiplies all nitrogen and hydrogen atoms by two. In other words, one formula unit of this salt contains 2 atoms of nitrogen, 8 atoms of hydrogen, and 1 atom of sulfur.

Here's a second example: Sodium phosphate, which contains Na^+ ions and polyatomic PO_4^{3-} ions, has the formula Na_3PO_4. Because sodium ions include only one atom, Na is not enclosed in parentheses, and we understand that there are three sodium ions in this formula. Likewise, there is only one phosphate PO_4^{3-} ion, so we don't enclose it in parentheses. **Figure 9.5** shows more examples of cations and anions pairing up to form salts.

QUESTION 9.3 Refer to Table 9.1 to name the polyatomic ion in each of the following salts. (a) $CaCrO_4$ (b) Li_2CO_3 (c) $Mg(CN)_2$

ANSWER 9.3 (a) chromate ion (b) carbonate ion (c) cyanide ion

For more practice with problems like this one, check out end-of-chapter Question 12.

FOLLOW-UP QUESTION 9.3 Write a chemical formula and name for the salt that each pair of ions forms. (a) K^+ and CrO_4^{2-} (b) K^+ and PO_4^{3-} (c) Mg^{2+} and SO_4^{2-}

9.3 The Hydration of Ions

Water molecules are polar and contain a dipole.

The natron used in mummification removes the water that is present in human body tissue. We refer to salts that absorb water, such as natron, as *desiccants.* What properties of salts allow them to attract water? To answer this question, we must go back to the properties of water molecules. Recall these facts about water from Chapter 8:

1. The water molecule is bent.
2. Because oxygen is more electronegative than hydrogen, in any bond between oxygen and hydrogen, the electrons are pulled toward the oxygen atom and away from the hydrogen atom.
3. Because the oxygen atom is partially negative and the hydrogen atoms are partially positive, the water molecule is polar. We can imagine a *dipole* running through it, as shown here.

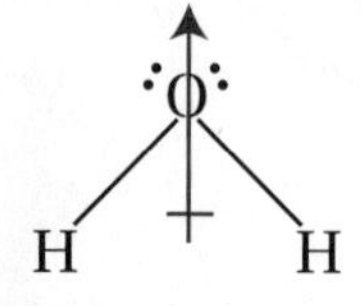

The water molecule's dipole can be thought of as an imaginary arrow running through the oxygen atom. The head of the arrow points to the more negative end of the molecule, and the plus sign (+) points to the more positive end of the molecule. Remember that these are partial charges, so they are not quite full charges like the ones associated with ions.

QUESTION 9.4 The oxygen atom in a water molecule is attached to two hydrogen atoms, and each of those bonds possesses a lopsidedness of electron density. That is, in each oxygen–hydrogen bond, there is greater electron density on the oxygen atom. Why don't those two bonds cancel out one another to create a nonpolar molecule?

ANSWER 9.4 The water molecule is bent, so the two bonds in water do not point in opposite directions. If they did, they would cancel out one another. Because the molecule is bent, it has an overall lopsidedness that we call a dipole.

For more practice with problems like this one, check out end-of-chapter Question 22.

FOLLOW-UP QUESTION 9.4 Consider a molecule that contains just two atoms of the same element—for example X_2, where X is some element. Is it possible for this molecule to contain a dipole?

Cations and anions are hydrated by water molecules.

Thinking about the partial charges on the water molecule can help us figure out how water interacts with a cation or an anion. For example, a sodium cation (Na^+) is attracted to a water molecule because the ion, which is positively charged, is attracted to the negative end of the water molecule. Thus, water molecules pack in around the sodium ion as closely as possible. We say the sodium ion is a *hydrated ion,* an ion that is surrounded by water. This process is called **hydration**.

Hydration happens with anions as well as cations, because the positive end of a water molecule is attracted to the negative charge on an anion. **Figure 9.6** shows space-filling models of these interactions. Because water possesses a dipole, hydration is an example of an *ion–dipole interaction.* The dipole is the water molecule, and the ion is the cation or anion being hydrated.

In Section 8.2, we introduced examples of intermolecular forces. An **ion–dipole interaction** is an example of an intermolecular force, because it is a force

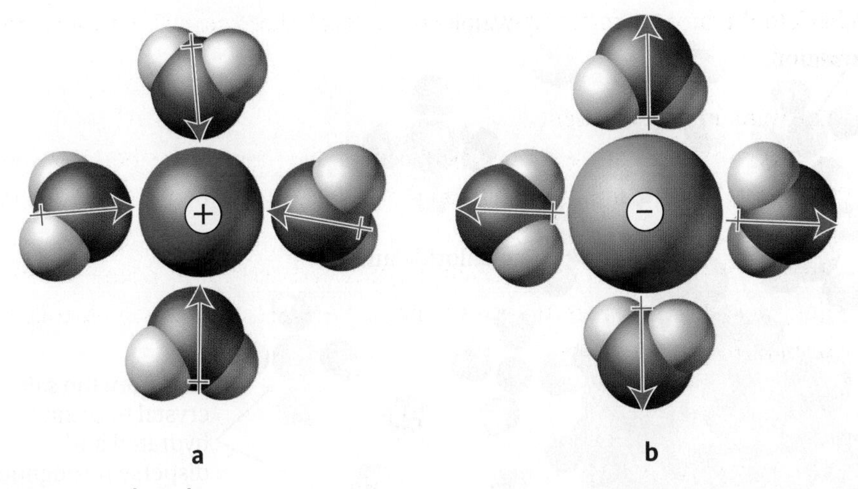

Figure 9.6 Hydrated Ions

(a) Cations are attracted to the negative end of water's dipole. (b) Anions are attracted to the positive end of water's dipole.

between an ion and a polar molecule such as water. The force holding the hydrogen and oxygen atoms together within the water molecule, though, are *intramolecular forces* because they involve the sharing of electrons in a covalent bond. For this reason, we draw lines connecting the oxygen atom to the two hydrogen atoms in the water molecule, but we do not draw a line between an ion and a water molecule.

flashback

As we learned in Section 8.2, the term *intermolecular force* describes the forces between an ion and an ion—hydrogen bonds—and interactions between molecules that have dipoles. *Intramolecular force* refers only to a covalent bond within a molecule.

QUESTION 9.5 In Figure 9.6, two different kinds of interaction are taking place at once. (a) Identify a covalent bond in this drawing. (b) Identify an ion–dipole interaction in this drawing.

ANSWER 9.5 (a) The covalent interaction is happening in the bonds between the hydrogen and oxygen atoms within the water molecules in Figures 9.6a and 9.6b. (b) The ion–dipole interaction is the interaction between the cation (the ion) and the negative end of the water molecule (the dipole) in Figure 9.6a or between the anion (the ion) and the positive end of the water molecule (the dipole) in Figure 9.6b.

For more practice with problems like this one, check out end-of-chapter Questions 19 and 20.

FOLLOW-UP QUESTION 9.5 Which of these substances can be hydrated by molecules of water? (a) an atom of neon (b) a calcium ion (c) a sulfur ion (d) a molecule of carbon dioxide

Most salts easily dissolve in water because salts and water are both polar substances.

Because water has a partial charge located on the oxygen atom and on the hydrogen atoms, it has a dipole. And, as we learned in Section 4.4, the molecule is polar. Ions are even more polar than water molecules, because they have full charges. When water molecules and ions are combined, they mix well because they are both polar. We use the term **aqueous solution** to describe a uniform mixture of water molecules and another substance, such as a salt.

Usually, when salt and water are combined, the salt dissolves in the water and does not remain in the solid phase. As each ion in the salt crystal becomes hydrated, the ions are plucked away and dispersed throughout the water. In this

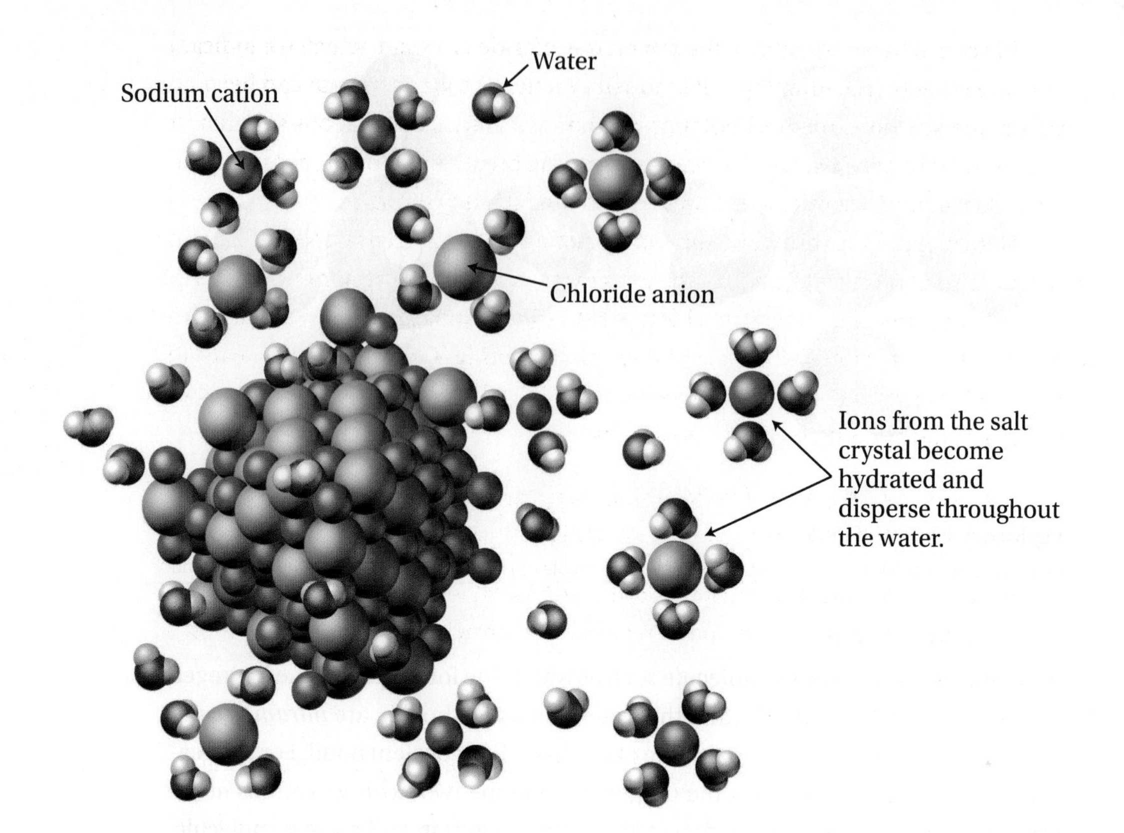
Water
Sodium cation
Chloride anion
Ions from the salt crystal become hydrated and disperse throughout the water.

Figure 9.7 How Salts Dissolve in Water

The solid crystalline lattice of the salt falls apart as each ion is plucked away from the lattice and becomes hydrated. A solution forms.

way, they **dissolve** to eventually make a clear solution, as shown in **Figure 9.7**. In this aqueous solution, the substance that's present in the greater amount in a solution is called the **solvent** (in this case, water). The **solute** (in this case, the salt) is the substance present in the smaller quantity in a solution.

Whenever we are dealing with substances in solution, we can apply this rule of thumb: like dissolves like. It means that substances with similar polarities tend to dissolve easily in one another. So if something is ionic, like a salt, it will dissolve in a solvent that is polar, like water. In the same way, nonpolar substances tend to dissolve readily only in other nonpolar substances. Thus, a nonpolar substance, like grease, easily dissolves in another nonpolar substance, like gasoline.

The hydration of ions happens so readily that most salts dissolve easily in water. Substances that carry an electrical charge, whether it's a partial charge (a dipole) or a full charge (an ion), interact with one another through attraction between positive and negative charges. If you drop a crystal of table salt into a glass of water, the salt quickly disappears as it dissolves in the water and becomes part of the solution. We can write this hydration process in the form of a chemical equation. Here are several examples:

$$\mathrm{NaCl}\,(s) \xrightarrow{\mathrm{H_2O}} \mathrm{Na^+}\,(aq) + \mathrm{Cl^-}\,(aq)$$

$$\mathrm{KCl}\,(s) \xrightarrow{\mathrm{H_2O}} \mathrm{K^+}\,(aq) + \mathrm{Cl^-}\,(aq)$$

$$\mathrm{K_2SO_4}\,(s) \xrightarrow{\mathrm{H_2O}} 2\mathrm{K^+}\,(aq) + \mathrm{SO_4}^{2-}\,(aq)$$

$$\mathrm{CaF_2}\,(s) \xrightarrow{\mathrm{H_2O}} \mathrm{Ca^{2+}}\,(aq) + 2\mathrm{F^-}\,(aq)$$

In each of these equations, the salt on the left side is a solid, which we indicate with the notation (*s*). After the salt is added to water, the ions separate and become part of the solution. The (*aq*) notation on the right stands for aqueous solution. It means the salts are dissolved in water. When a salt breaks up into its ions to become part of a solution, we refer to this event as **dissociation** of the salt.

Notice that, in cases where the formula of a salt contains multiples of an ion, the coefficient on the right side of the ion's formula indicates the number of ions in the salt. It means that potassium chloride (KCl) dissociates to make one potassium ion (K^+) and one chloride ion (Cl^-), but calcium fluoride (CaF_2) dissociates to make one calcium ion (Ca^{2+}) and two fluoride ions ($2F^-$), because each formula unit of calcium fluoride contains two fluoride ions.

flashback
Recall from Chapter 7 that the coefficient is the number located in front of a reactant or product in a balanced chemical equation. The coefficient indicates the number of units of that substance in the equation.

QUESTION 9.6 Identify each of the following as an ion, a nonpolar molecule, or a molecule with a dipole: (a) Cl^- (b) I_2 (c) water (d) Mg^{2+}

ANSWER 9.6 Chloride (a) and magnesium (d) are ions. Water (c) is a polar molecule, and therefore possesses a dipole. Iodine molecules (b) are nonpolar and do not possess a dipole.

For more practice with problems like this one, check out end-of-chapter Question 9.

FOLLOW-UP QUESTION 9.6 True or false? A polar substance does not mix easily with a nonpolar substance.

Alert readers are probably wondering if there's a limit to the amount of salt we can dissolve in water. After all, if we add a tablespoon of water to a bucket of sodium chloride, common sense tells us the salt does not dissolve. There simply isn't enough water in a tablespoon to surround and hydrate that many ions.

Because water molecules hydrate ions, most salts dissolve readily in water. In chemistry, we say that such salts are soluble in water or that their **solubility** in water is high. But there is always a limit to the solubility of any solid. If we have a lot of sodium chloride, but only a small amount of water, we run out of water before the process shown in Figure 9.7 is complete. The available water hydrates ions from the NaCl crystal one by one. Eventually, if enough NaCl is present, all the water molecules are used up in the process of hydration. When this happens, we end up with a pile of solid NaCl in our solution.

In **Figure 9.8,** sodium chloride has been added to a volume of water until the water has absorbed as many ions as it can. We describe this as a **saturated solution**—no more NaCl can dissolve in it.

Richard Megna/Fundamental Photographs

Figure 9.8 A Saturated Solution

A dynamic equilibrium exists in a saturated salt solution.

Once the solubility limit of a salt has been reached and we have a pile of salt sitting at the bottom of a solution, it may seem like nothing is happening in the solution. But this is not the case. In our saturated NaCl solution, for example, we have a pile of salt sitting on the bottom of a solution saturated with hydrated sodium ions and chloride ions moving around in the solution. Whenever an ion bumps into the solid sodium chloride crystal, the ion can shed its hydrating water molecules and return to the crystalline lattice. We refer to this return to the solid state as **precipitation**, a term that describes things that are falling, such as rain. In this case, the ions are falling out of solution and reentering the solid phase.

As ions return to the solid phase, other ions can be plucked away from the crystal by water molecules. Thus, these reactions are going in both directions at once, as indicated by a double arrow:

$$\underset{\text{Solid phase}}{\text{Solid NaCl crystalline lattice}} \rightleftharpoons \underset{\text{Liquid phase}}{\text{Na}^+ \text{ ions and Cl}^- \text{ ions in solution}}$$

When a saturated solution has been sitting long enough, there is no *net* change in the size of the salt pile, and we say the system is at **equilibrium**. And we describe this equilibrium as *dynamic*, because the ions are still moving back and forth between the aqueous and solid phases. For a solution in dynamic equilibrium, the rate at which ions leave the solid and go into solution equals the rate at which ions return to the solid lattice:

$$\text{Rate at which ions in a salt dissolve in liquid} = \text{Rate at which ions precipitate and form solid}$$

This process is illustrated in **Figure 9.9a**.

Equilibria (the plural of *equilibrium*) exist in many other situations. The surface of a liquid can be at equilibrium with gaseous molecules in the air above the liquid. Molecules of water in the solid phase (ice cubes) are at equilibrium with molecules of water in the liquid phase at the freezing point of water, as shown in **Figure 9.9b**.

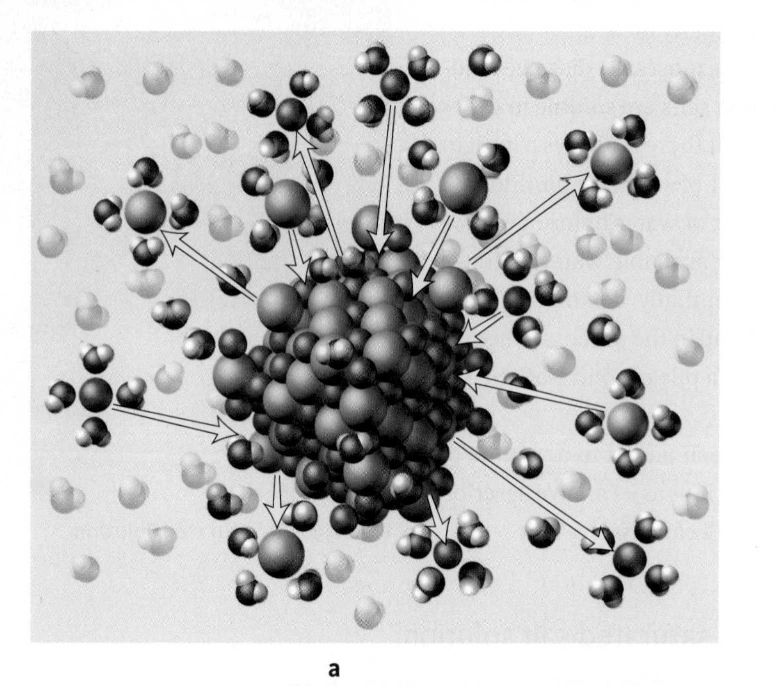

a

b

buradaki/Shutterstock

Figure 9.9 Systems at Equilibrium

(a) In a saturated solution, ions of the crystalline lattice are at equilibrium with hydrated ions in solution. This figure looks like Figure 9.7—but in this case, the solution is saturated and a dynamic equilibrium exists. (b) In a glass of water, the solid ice is at equilibrium with the water around it.

Now that we understand the concept of equilibrium, we can better appreciate the mummification project introduced earlier in the chapter. The process used to mummify a body is designed to remove water, because where there is water, there are likely to be bacteria. Bacteria decompose human flesh, and no one wants to be partially decomposed in the afterlife.

When natron is packed into the body cavity, the surrounding tissue, which is primarily water, comes into contact with the natron. Thus, the ions that compose the solid salts in the natron are surrounded by molecules of water. Because there is much more salt than water, all the water gets absorbed by the salt, and equilibrium is never reached. There is simply too much salt. Consequently, the tissues are completely dehydrated. Most bacteria do not grow in tissues that are dehydrated, so the mummified body is preserved and does not spoil.

The ability of salts to suck water from whatever they touch makes them useful as preservatives for things other than mummies. Meat is cured, for example, by coating it with a heavy layer of sodium chloride. This process removes water from the meat and thereby prevents spoilage by bacterial growth. (There are exceptions; see the *Green Beat* at the end of this chapter.)

The same properties of salts that make them useful can also make them harmful, however. Scientists at the University of Missouri–Rolla have been studying why some of Egypt's antiquities are crumbling. Many sandstone structures, such as the Temple of Luxor (**Figure 9.10**), are deteriorating at an alarming rate. The researchers have determined that a salty residue that results from Egypt's dry climate becomes wedged in the nooks and crannies of the sandstone monuments. The salts in the residue absorb water from the air and swell, pushing on the sandstone until finally the monument crumbles.

mary416/Shutterstock

Figure 9.10 The Damage Salt Can Do
The Temple of Luxor, as well as other historic monuments in Egypt, is threatened by the presence of salt.

QUESTION 9.7 True or false? In mummifying a body, it doesn't matter how much salt is used to treat the body.

ANSWER 9.7 False. You need enough salt around the body to soak up all the water in the body's cells. Therefore, the amount of salt does matter. Too little salt will not provide the desired antibacterial effect.

For more practice with problems like this one, check out end-of-chapter Questions 19 and 21.

FOLLOW-UP QUESTION 9.7 True or false? The curing of meat is based on the premise that removing water from the meat prevents bacterial growth.

9.4 Concentration and Electrolytes

Gatorade is an electrolyte solution.

The average human sweats out approximately 1.5 L of water every day, but a person exercising vigorously can lose up to 16 L of water in a day—that's about 20% of her or his body mass. The liquid lost by sweating must be replaced by drinking either water or a reasonable substitute. Sweat is not pure water, however. It contains salts because body fluids are salty.

When salts dissociate in water, they produce hydrated ions. The presence of ions means the solution can conduct an electric current. Consider **Figure 9.11**. It shows that although pure water does not conduct a current that will light up a light bulb, a solution containing salt does. Because a salt dissolved in water yields a solution capable of conducting electricity, both salts and the ions that form them are frequently referred to as **electrolytes**. This term is used in reference to ions dissolved in body fluids, and so we often hear about "maintaining electrolyte balance" in discussions about human health.

The dissolved salts in the body are critical for maintaining blood pressure, neural function, and healthy cells. Therefore, an athlete losing 16 L of sweat is also losing valuable electrolytes that must somehow be replaced.

Because sports drinks have a chemical makeup similar to that of sweat, they replace the electrolytes and water lost during physical exertion. Of course, even though these drinks are sweat impersonators, manufacturers wisely leave the word *sweat* out of drink names (no one wants to drink a product called Sweaty Juice). Instead, we have Gatorade®, the granddaddy of sports drinks.

Imagine it's the year 1964, and you are the coach of the struggling University of Florida Gators football team. Your players are dropping like flies during practice in the sizzling Florida heat. You try the usual dehydration remedy—salt tablets—but they do not help. Then you try orange juice, but it causes stomach cramps. Finally

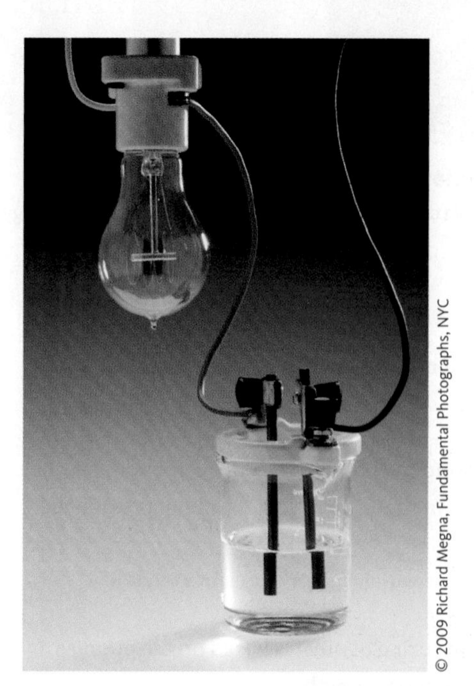

Figure 9.11 Ions Conduct Electricity

The beaker on the left contains pure water. The lightbulb is dark because pure water cannot conduct electricity. The beaker on the right contains a solution of some electrolyte dissolved in water. The bulb is lit because the ions in the solution conduct electricity.

you go to the team doctor, Dana Shires, for help. He turns to a University of Florida medical researcher by the name of Robert Cade, and together they work out a solution (literally).

What Cade and Shires realized was that salt tablets replace the salts lost by sweat, but not the water. Without water, salt tablets just make dehydration worse. Cade and Shires also found that orange juice causes stomach cramping because it contains about 10% sugar. Large amounts of sugar in the stomach keep water from passing through the stomach wall and into the body's cells, where it is needed.

Armed with this knowledge, Cade and Shires formulated a sweat-like concoction. First they determined that 1 L of sweat contains about 100 mg each of sodium (Na^+) and potassium ions (K^+) and about 25 mg of magnesium ions (Mg^{2+}). (These are balanced by an appropriate number of anions—in this case, chloride ions, Cl^-). Finally, they tried mixing these salts in water, but the solution was undrinkable. They then added about 6% sugar—enough to mask the bad taste, but not enough to cause stomach cramps. **Figure 9.12** shows an athlete replacing electrolytes lost during a vigorous workout.

auremar/Shutterstock

Figure 9.12 An Athlete Replacing Electrolytes

The drink was a godsend for the Florida Gators. No more stomach cramps. No more dehydration. In 1966, they went to the Orange Bowl. The score was Georgia Tech, 12, Gators, 27. The *Jacksonville Journal* described the impact Gatorade had on the team this way:

> The Gators went 7–4 that year [1965], then 9–2 in '66. Time after time, the Gators outplayed their opponent late in the game, even under conditions that sportscaster Keith Jackson said "would make a salamander sweat."

Molarity indicates how much solute is dissolved in a given volume of solvent.

Twenty-first-century Gatorade is not the same drink it was in 1965. Nowadays, no Mg^{2+} ions are used, but sodium chloride and potassium chloride are still added in the concentrations given in **Table 9.2** on p. 267. To understand the units in this table, we need a way to express how much of something is dissolved in something else.

If we want to understand how much of a given substance is present in, for example, a sports drink, we express that amount in units of concentration. **Concentration** is the amount of one substance, the solute, dissolved in another substance, the solvent, in a solution. A solute can be anything—from table sugar to caffeine to chocolate syrup—that dissolves in or mixes with the solvent.

One of the most common ways to express concentration is **molarity,** which is the number of moles of solute per liter of solution:

$$\text{Molarity} = \frac{\text{moles of solute}}{\text{liters of solution}}$$

flashback

We introduced the mole in Chapter 6 and again in Chapter 7. One mole of a substance contains 6.02×10^{23} units of that substance. Molar mass is the mass of one mole of a pure substance.

Let's consider a straightforward example. Suppose we are given a one-liter flask filled with 1 liter of an aqueous sodium chloride solution. That solution contains 0.052 mol of sodium chloride, so the molarity of the solution is

$$\frac{0.052 \text{ mol NaCl}}{1.00 \text{ L solution}} = 0.052 \text{ mol/L or } 0.052 \text{ M}$$

We can express molarity by saying, "The molar concentration of that solution is 0.052 mole per liter," or "The molar concentration of that solution is 0.052 M." The capital letter *M* is a unit of concentration that means moles per liter, and we can modify it with prefixes in the same way that we modify metric units. So, just as we modify gram to become nanogram or milligram, we can modify molar to nanomolar or millimolar.

Now suppose we want to prepare this solution for ourselves in our one-liter flask. We'll need some water, the flask, some table salt, a balance, a periodic table, a calculator, and a spoon. How much table salt goes into the flask if we want one liter of a 0.052-M solution of sodium chloride, aka saline? First, we turn to the periodic table to find the molar mass of sodium chloride: 58.5 grams per mole. Starting with the molarity, we cancel moles and find grams of table salt, like this:

$$\text{Molar concentration} = \frac{0.052 \text{ mol NaCl}}{1.0 \text{ L solution}} \times \frac{58.5 \text{ g NaCl}}{1 \text{ mol NaCl}} = \frac{3.0 \text{ g NaCl}}{1 \text{ L solution}}$$

In the next step, we need to weigh out 3.0 grams of table salt and put it into our one-liter flask. Next, we add some water to dissolve the salt. Now we add water up to the 1-liter line, stir, and we have a 0.052-M saline solution.

Note that the concentration of a something in something else does not depend on the total amount we have, as shown in **Figure 9.13**. For example, it doesn't matter whether it's in a swimming pool or in a beaker, a solution of NaCl with a concentration of 0.052 M stays the same: every liter of solution in the pool and every liter in the bottle contains 0.052 mole of salt.

Figure 9.13 Different Amount, but Same Solution

This saltwater swimming pool has a molar concentration of salt equal to about 0.052 M. If you take a beaker and fill it with water from this pool, the molarity of the water in the beaker is also 0.052 M. Molarity does not depend on the volume of the sample.

Table 9.2 Concentration of Electrolytes and Sugar in Some Liquids

Electrolyte	Human Sweat	Modern Gatorade	Orange Juice	Coconut Milk	Carrot Juice
Na^+	20 mM	20 mM	0.43 mM	5.7 mM	3.9 mM
Mg^{2+}	8.0 mM	0	4.5 mM	19 mM	5.8 mM
K^+	10 mM	6.6 mM	100 mM	116 mM	150 mM
Cl^-	38 mM	27 mM	110 mM	160 mM	170 mM
Sugar (mass percent)	0%	6%	10%	3%	9%

wait a minute . . .

Can we use molarity for substances, such as chocolate syrup, that are already a mixture of substances?

No. We can use molarity only when we have a way to figure out the molar mass of the solute. Therefore, the use of molar concentration is limited to situations where the solute is a pure substance, like table sugar (sucrose) or salt (sodium chloride). It's also possible to have several solutes that are each pure substances, each with their own molarities.

In cases like chocolate syrup dissolved in milk, it's more common to use a concentration unit such as **mass percent**. Mass percent is the mass of the chocolate syrup divided by the mass of the milk and chocolate syrup together, times 100%. When we figure out percent by mass, the composition of the syrup is not important—only its mass is, and that mass is easily measured. In the next section, we look at other examples.

QUESTION 9.8 If you arrive at the hospital with severe dehydration or decreased blood volume, you may be given an intravenous saline solution known as NS, or normal saline. NS typically contains 0.0154 mole of sodium chloride per 100.0 mL of solution. What is the molar concentration of NaCl in this solution?

ANSWER 9.8

$$\frac{0.0154 \text{ mol NaCl}}{100.0 \text{ mL solution}} \times \frac{1000 \text{ mL}}{1.00 \text{ L}} = \frac{0.154 \text{ mol NaCl}}{1 \text{ L solution}} = 0.154 \text{ M NaCl}$$

For more practice with problems like this one, check out end-of-chapter Questions 41 and 42.

FOLLOW-UP QUESTION 9.8 A certain pond has a volume of 56,500 L. It contains 452 moles of magnesium chloride. What is the molar concentration of magnesium chloride in the pond water?

There are several ways to express concentration.

We know that Gatorade was designed to be a substitute for sweat, but how much do the two fluids really resemble each other? To find the answer, we need to know the concentrations of various solutes in each fluid. Table 9.2 compares the ion

natureBOX • Using Ionic Liquids to Solve Modern Environmental Problems

The notion that ionic compounds are always salts and that salts are always solids at room temperature has been debunked by the development of a new class of salts: ionic liquids. What are ionic liquids? We know that regular salts like sodium chloride (table salt) are always solids, partly because they pack neatly and tightly into a crystalline lattice supported by strong ionic bonds. Conversely, the ions in ionic liquids do not pack tightly into a crystalline lattice like they do in a traditional ionic substance like sodium chloride.

Sodium ions and chloride ions are similar in size and they stack together easily like a crate stacked full of alternating lemons and limes. In contrast, **Figure 9A** shows the cation and the anion in an ionic liquid. The ring in the middle of the cation structure is flat, and two arms are attached to it. This cation is much bigger than the sodium ion in table salt, and the anion in Figure 9A is a bulky sphere. The ionic liquid is like a crate filled with lampshades and coat hangers; there is no way to stack them neatly and closely, resulting in much weaker ionic bonds in the ionic liquid than those in sodium chloride.

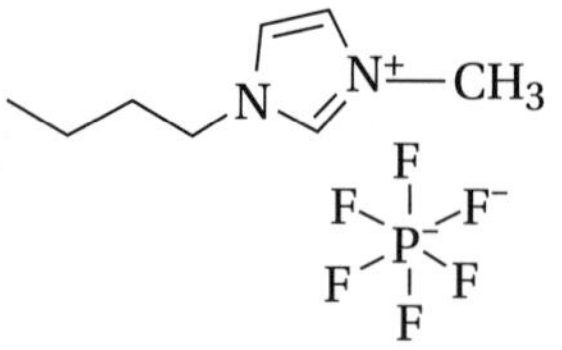

Figure 9A The Cation and Anion in an Ionic Liquid

The differences between the shapes of the ions in the ionic liquid and those in NaCl cause striking variations in melting points. Sodium chloride's melting point is 801°C, so it is a solid below and a liquid above that temperature; it's a solid at room temperature, which is approximately 25°C. However, the ionic liquid's melting point is 0°C, so it is a liquid at room temperature, and it would need to cool below 0°C to be a solid. **Figure 9B** shows the difference between the two.

Figure 9B A container holding an ionic liquid.

Generally, salts don't damage the environment because they are mostly biodegradable and benign. Additionally, ionic liquids solve many environmental problems because salts that are liquids at room temperature can be very useful: Ionic liquids can be reused and recycled; they readily dissolve carbon dioxide and can be used in carbon capture and sequestration; they do not evaporate readily and can replace the water used in new types of batteries; they can separate similar compounds and can separate plastics in waste recycling; they can decontaminate water polluted with heavy metals; they exist in the liquid phase over many temperatures; and, they are being developed as heat-storing compounds for solar energy systems.

There are many ways to make alternative structures to the cation and anion of the ionic liquid in Figure 9A, allowing these versatile compounds to be fine-tuned to specific applications. The potential for ionic liquids is just beginning to be tapped.

---

concentrations in some drinks, including Gatorade, and in human sweat. For these substances, the concentration of ions is pretty small, and so the concentrations are given in millimoles per liter (mM). "Millimolar" is a concentration unit that tells you the number of millimoles of some substance per liter of solution. (Recall that there are 1000 millimoles in one mole.)

$$\text{mM represents } \frac{\text{millimoles of solute}}{\text{liters of solution}}$$

Let's look at Table 9.2 more closely. We can see that, except for the magnesium ions, Gatorade and human sweat have pretty similar ionic compositions. But suppose an athlete would prefer to drink something else, such as orange juice, carrot juice, or coconut milk? Would any of those liquids have a composition similar to that of human sweat? The answer is not really, because all of them contain too much potassium ion, too much chloride ion, and not enough sodium ions.

In Table 9.2, the concentration of sugar is expressed in units of mass percent. Recall from the "Wait a Minute" earlier in this section that mass percent is the mass of solute divided by the mass of the solution, multiplied by 100%:

$$\text{mass percent} = \frac{\text{mass of solute}}{\text{mass of solution}} \times 100\%$$

If you have 100 grams of carrot juice, 9% of that mass, or 9 grams, is sugar:

$$9\% = \frac{\text{9 grams sugar}}{\text{100 grams of carrot juice}} \times 100\%$$

QUESTION 9.9 If you have a glass that contains 300 grams of coconut milk, how many grams of sugar are in your glass?

ANSWER 9.9 According to Table 9.2, coconut milk is 3% sugar. That means that every 100 grams of juice contains 3 grams of sugar. So 300 grams contain three times this amount: 3×3 grams = 9 grams of sugar.

For more practice with problems like this one, check out end-of-chapter Questions 35 and 36.

FOLLOW-UP QUESTION 9.9 Pomegranate juice is one of the sweetest juices. An 8-ounce serving, which is equal to 227 grams, contains 34 grams of sugar. What is the mass percent of sugar in pomegranate juice?

9.5 Osmosis and Concentration Gradients

Living cells use semipermeable membranes to control the flow of substances through the organism.

In the pomology research group, part of the Department of Plant Sciences at the University of California at Davis, researchers are interested in only one thing: tomatoes. A tomato plant has strict growing requirements. It needs plenty of sunshine, plenty of fertilizer, plenty of water, and most of all, the right soil.

About now, you might be wondering what tomatoes have to do with the topic of this chapter—salt and aqueous solutions. Tomato plants are especially finicky when it comes to the concentration of salts in the soil: in fact, too much salt means no tomatoes. The reason is that salts in the soil hold onto water (as we know from Section 9.3), so the water stays in the soil rather than flowing into the plants. When tomatoes are planted in extremely salty soil, water is actually sucked out of the plants, which become dehydrated and die. The health of the tomato depends on the balance of water and ions in its cells.

We've seen several examples in this chapter of water and ions moving through aqueous solutions. For example, ions move across fish skin to maintain electrolyte balance for fish out of water. We also know that ions move into and out of solution when a dynamic equilibrium exists. When someone is mummified, water moves out of the person's cells and into the salt that is applied to the body. In this section, we consider how salts and water behave when one substance is allowed to move and the other is not.

We can think of a tomato—or any multicellular organism, for that matter—as being constructed of a series of barriers. Every cell in that organism has a cell

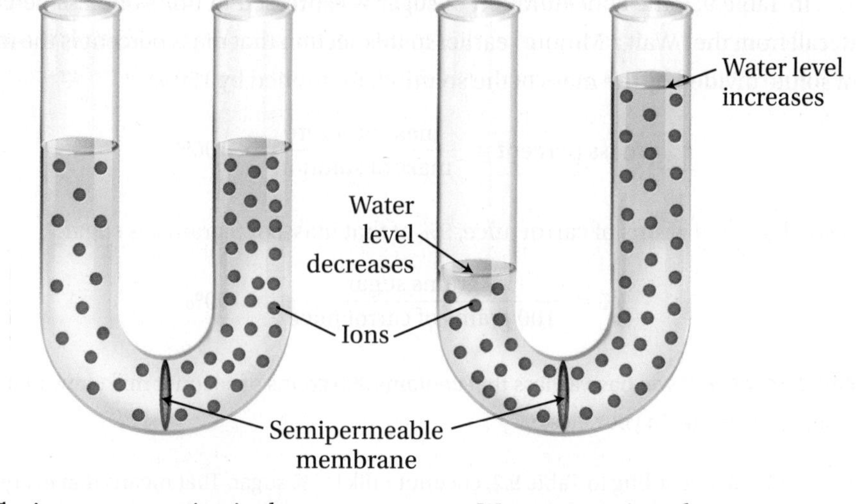
Water level
increases
Water
level
decreases
Ions
Semipermeable
membrane
The ion concentration in the solution in the right arm is higher than the ion concentration in the solution in the left arm.
Water moves into the more concentrated solution until the ion concentrations are equal.
a
b

Figure 9.14 Osmosis across a Semipermeable Membrane

(a) The solution in the right arm of the tube has a higher concentration of ions (represented by the orange dots) than the solution in the left arm. Note that the water levels are equal. (b) Due to the difference in ion concentrations, water molecules go from the left arm into the right arm, moving through the semipermeable membrane. After equilibrium is reached, the ion concentrations are equal but the water levels are unequal.

membrane that acts as a barrier for substances. Certain substances can cross those barriers and others cannot.

We model this process with a U-shaped tube containing two solutions separated by such a barrier. Suppose this system represents the membrane surrounding the cell. Then, for example, one side of the barrier represents the inside of a cell and the other side of the barrier represents the outside of the cell, as shown in **Figure 9.14**. This barrier is a **semipermeable membrane** that limits the passage of certain molecules and ions from one side to the other. Membranes can be manufactured, like the one in Figure 9.14, or they can be natural, as in the case of cell membranes.

Our U-shaped tube shows us how semipermeable membranes work in cells like those in tomatoes or in mummies. Hydrated ions are surrounded by water molecules, and this makes them very bulky. In fact, hydrated ions are too large to pass through this semipermeable membrane. But water molecules traveling solo—without ions—can pass through the membrane. The movement of those water molecules is governed by a principle that we encountered in Section 9.3: equilibrium.

Initially, conditions in the tube are unbalanced. The left arm contains water plus only a few ions, and the right arm contains water plus many ions (Figure 9.14a). But what does the system look like after a while? Over time, it moves away from the initial, unbalanced condition and toward equilibrium, which is reached when the ion concentration is the same in the two arms (Figure 9.14b). Because only water molecules can cross the membrane, equilibrium is attained by water molecules moving across the membrane from left to right. When they do that, they dilute the solution in the right arm until the ion concentration there is the same as in the left arm.

Water molecules move across a semipermeable membrane from a region of low ion concentration to a region of higher ion concentration by a process called **osmosis**. Osmosis moves the concentrations of substances on the two sides of the membrane toward equilibrium.

QUESTION 9.10 True or false? A tube like the one shown in Figure 9.14, equipped with a membrane that only water can pass through, is filled on both sides with equal amounts of the same salt solution. For this system, no water molecules pass through the membrane, because both sides have exactly the same concentration of ions.

ANSWER 9.10 This statement is not true. The membrane is permeable to water, and thus water is always moving back and forth across it. Because the two sides are in equilibrium, however, equal numbers of water molecules move across the membrane in each direction, so there is no *net* movement of water.

For more practice with problems like this one, check out end-of-chapter Questions 48 and 49.

FOLLOW-UP QUESTION 9.10 A tube like the one shown in Figure 9.14 is filled with a concentrated salt solution on the left side and a less concentrated solution of the same salt on the right side. Which way does water flow to create an equilibrium condition in the system?

Concentration gradients for various ions are maintained across cell membranes.

Osmosis attempts to equalize the concentrations of ions on the inside and outside of a cell. **Figure 9.15a** shows a red blood cell under various osmotic conditions: initially, the concentration of ions inside is greater than the concentration of ions outside. In this case, water travels from outside to inside, and the cell swells. It's possible for a cell to burst under these conditions.

Figure 9.15b shows the opposite situation. The initial concentration of ions inside is lower than the concentration of ions outside. In this case, water travels from inside to outside, and a shrunken, dehydrated cell is left behind. For comparison, **Figure 9.15c** shows a cell with equal ion concentrations on each side of the cell membrane.

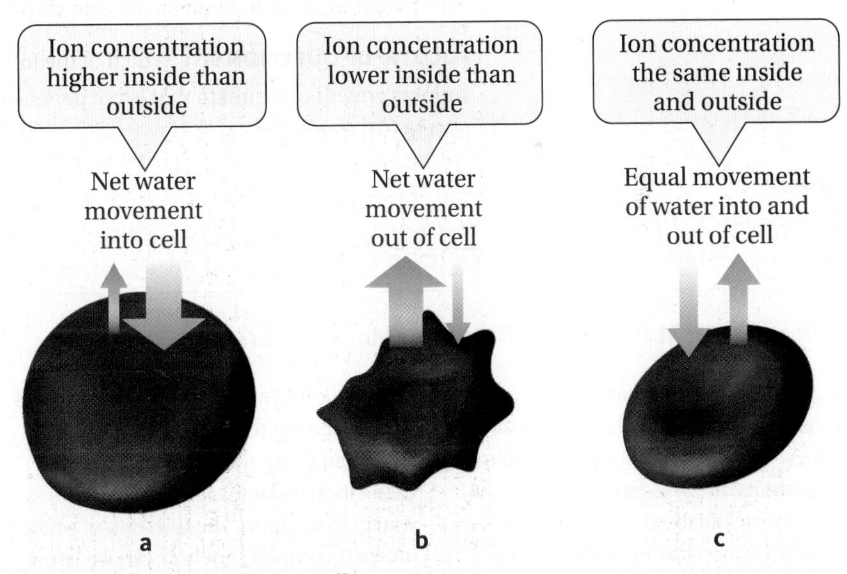

Figure 9.15 Osmosis in Living Cells
(a) A red blood cell that has a higher ion concentration inside than outside swells as water moves into the cell. (b) A cell that has a lower ion concentration inside than outside shrivels as water leaves the cell. (c) A cell in which the ion concentration inside is the same as the ion concentration outside maintains its original size.

Figure 9.15b provides the key to why a tomato plant is unhappy in salty soil. The cells in a typical tomato plant are adapted to do best when the total salt concentration in the soil is about 4 millimolar (mM). At this concentration, the cells are plump and healthy. When the soil has a higher salt concentration, the cells shrink because water exits the cells (just as it does in Figure 9.15b) and goes into the soil. Studies on how high-salt soils affect crop yields estimate that because of high salt levels, about 40% of the world's irrigated land is unusable for tomato production.

In salty soil, ions are concentrated in the leaves.
Salt content in fruit remains low.
Zilu8/Shutterstock

Figure 9.16 Tomatoes for Everyone?

Genetically engineered tomatoes can grow in salty soil because they accumulate salt in their leaves rather than in the fruit.

The pomologists (tomato scientists) at UC Davis may have a solution to this problem, however. They have genetically engineered a tomato plant that can live in soil where the total salt concentration is up to 200 mM—fifty times the normal concentration (**Figure 9.16**). How did they do it? They inserted a gene from a cabbage plant into some ordinary tomato plants. That gene tells cells to produce miniature ion pumps, which pump ions against the natural gradient. The pumps insert themselves into cell membranes and transport certain ions into the leaves of the plant, but not the fruit.

The result is a tomato plant that pumps excess ions against the gradient and into the leaves in the plant. The fruit stays the same and, according to the pomologists, tastes great. They hope this technology will be applied to other plants and that newly developed salt-tolerant crops may increase the world's food supply. Chapter 14, "Food," explains the genetic modification of food crops in detail.

QUESTION 9.11 A tomato that was genetically modified by the pomologists at UC Davis is planted in soil with a salt concentration of 0.18 M. Will it grow?

ANSWER 9.11 Recall from Chapter 1 that 1000 mM is the same as 1 M. If the tolerance of the engineered tomato is 200 mM, that is the same as 0.2 M, which is greater than 0.18 M. Thus, the tomato can grow.

For more practice with problems like this one, check out end-of-chapter Questions 37 and 41.

FOLLOW-UP QUESTION 9.11 Which of the following soil salt concentrations can support growth of a tomato that is not genetically modified? (a) 0.4 M (b) 3 mM (c) 4 M (d) 0.004 mM

THEgreenBEAT • news about the environment, chapter nine edition

TOP STORY Is It Possible to Drink Too Much Water?

In 2007, KDND radio station in Sacramento, California held an on-air contest to win a free Wii game console if contestants drank as much water as possible during the three-hour show without urinating. Contestant Jennifer Strange drank six liters of water, but only took second place! During the contest, Jennifer mentioned that she didn't feel well, and after the show ended, she went home and died. Her family won a wrongful death lawsuit against KDND, which was found to be negligent by encouraging contestants to drink too much water.

We often hear the mantra, "Drink at least eight glasses of water per day!" However, we don't hear warnings about drinking too much water, or *water intoxication.* We can guess from Jennifer's story that two liters of water per hour is too much, but the contest's winner drank more than that and didn't die of water intoxication. The question of water intake is complex, but how can we know how much water is too much? Why is too much water dangerous? And why do water requirements vary among people?

The answers lie in this chapter's discussion about salts and their concentrations in our bodies. Recall that one liter of Gatorade contains about 100 mmoles of sodium ions, Na^+, in addition to other ions. Conversely, one liter of human body fluid contains between 135 and 145 mmoles of sodium ions, equating to between 7.9 and 8.5 grams. The body requires this narrow concentration range of sodium for healthy electrolyte balance, which is tightly regulated.

When excess water is consumed, sodium concentration drops, and that level can dip below the lower limit of the healthy concentration range. When this happens, cells begin to swell due to osmosis, which occurs when water crosses a cell membrane in order to equalize ion concentrations on either side. For the cells in many bodily tissues, swelling, or *edema,* is often insignificant because the cells can grow larger and then the swelling can harmlessly go down. In the brain, however, there is no room for swelling, so a pressure increase can cause seizures, coma, or death. This problem is directly related to the concentration of sodium, so the medical term for water intoxication is *hyponatremia*, or "deficient sodium."

The Institute of Medicine of the U.S. National Academy of Sciences, the authority on water consumption, recommends 91 ounces per day, or about eleven 8-ounce glasses of water. This is likely good advice for most healthy adults, but not in every situation. **Figure 9.17** shows how the human body takes up and releases water. This mechanism works naturally to maintain the body's correct electrolyte balance. However, when the body feels stress during endurance events, the natural water balance shown in the figure can be disrupted by hormonal signals that tell the body to release less water. For this reason, many athletes experience hyponatremia because they drink copious amounts of water to stave off dehydration, but their hormones tell the body not to release water. This causes the sodium concentration in their body fluids to drop and hyponatremia to set in.

How can we avoid drinking too much water? Since every person is different and since water intake also depends on the climate in which a person lives and many other factors, experts agree that there is no universal volume of water for each person. According to the experts, the best guide is simply this: if you are thirsty, drink.

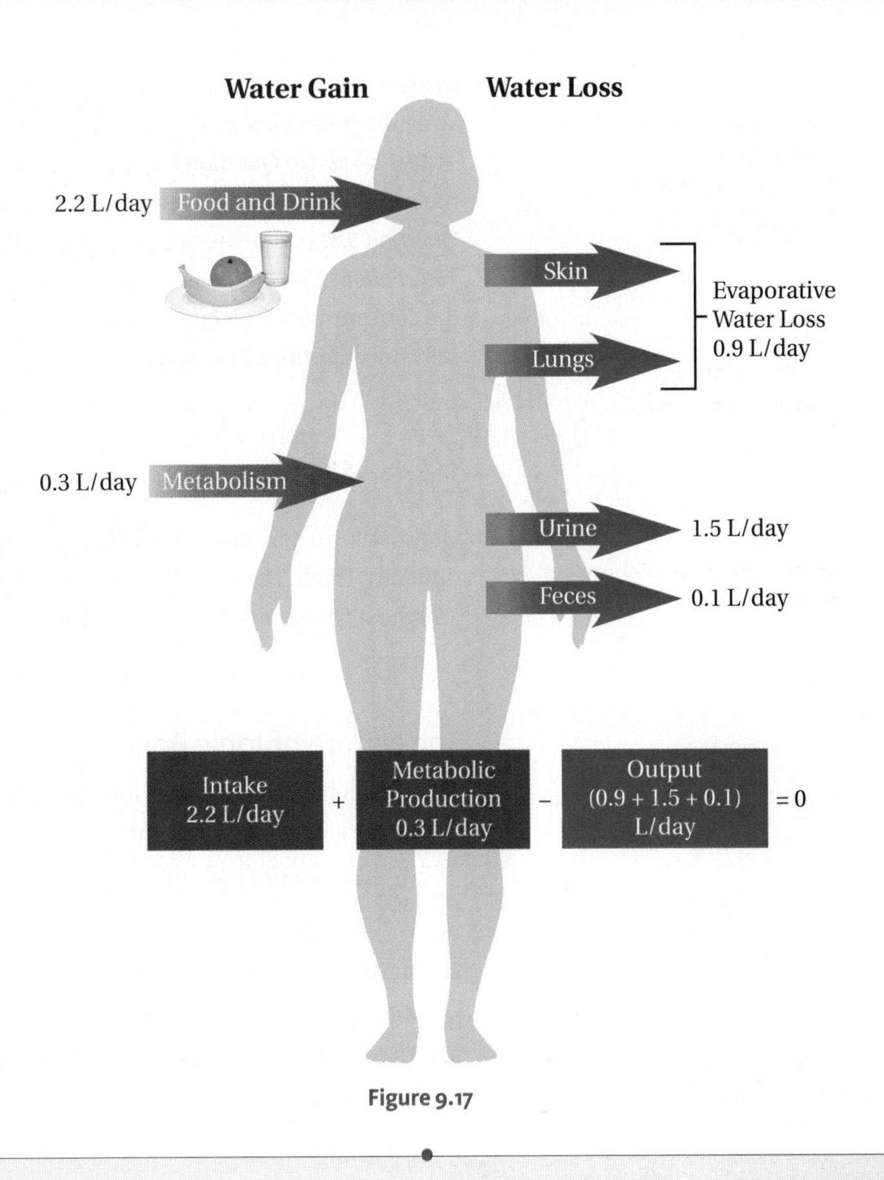

Figure 9.17

Chapter 9 Themes and Questions

recurring themes in chemistry chapter 9

recurring theme 5 Elements in the same group on the periodic table often behave in similar ways and share properties with one another. (p. 252)

recurring theme 3 Opposite charges attract one another. (p. 253)

recurring theme 6 Individual bonds between atoms dictate the properties and characteristics of the substances that contain them. (p. 254)

A list of the eight recurring themes in this book can be found on p. xv.

ANSWERS TO FOLLOW-UP QUESTIONS

Follow-Up Question 9.1: (a) argon (b) argon (c) neon (d) krypton. *Follow-Up Question 9.2:* Only (a) is a simple salt. The others each contain a polyatomic ion. *Follow-Up Question 9.3:* (a) K_2CrO_4, potassium chromate (b) K_3PO_4, potassium phosphate (c) $MgSO_4$, magnesium sulfate. *Follow-Up Question 9.4:* No. A molecule made up of two identical atoms will not possess a dipole. *Follow-Up Question 9.5:* Water hydrates ions, so (b) and (c) are correct choices. *Follow-Up Question 9.6:* True. *Follow-Up Question 9.7:* True. *Follow-Up Question 9.8:* Divide the number of moles by the number of liters to get 0.00800 mole per liter, or 0.00800 M. *Follow-Up Question 9.9:* Divide the mass of sugar by the mass of juice and multiply by 100 to get 15%. *Follow-Up Question 9.10:* Water will flow from the right to the left to equalize the concentrations on both sides. *Follow-Up Question 9.11:* Choices (b) and (d) are salt concentrations in which the unmodified tomato can grow.

End-of-Chapter Questions

Questions are grouped according to topic and are color-coded according to difficulty level: **The Basic Questions***, **More Challenging Questions****, and **The Toughest Questions*****.

The Nature of Ionic Bonds

***1.** For each of the following ions, write the symbol for the ion and indicate which noble gas has the same number of electrons as the ion:

(a) magnesium ion (b) lithium ion

(c) bromide ion (d) oxide ion

***2.** In your own words, explain what the following statement means: Crystalline ionic solids resemble neatly packed crates of fruit.

***3.** Use Figure 9.1 to write chemical formulas for these simple salts.

(a) sodium bromide (b) lithium sulfide

(c) magnesium chloride (d) calcium oxide

****4.** Use Figure 9.1 to write chemical formulas for these simple salts.

(a) potassium nitride (b) rubidium fluoride

(c) barium sulfide (d) lithium bromide

****5.** The melting points of salts are higher than those for most other substances. How does the bonding in ionic solids make their melting points so high?

****6.** The outmoded way to treat dehydration called for ingestion of salt tablets to replace electrolytes lost in sweat. Why is this an ineffective treatment? Speculate on what happens when a salt tablet is swallowed.

*****7.** Recall that the strength of the ionic bonds within salts is dictated by two things: the higher the charge on the ions, the greater the bond strength; and the smaller the ions, the greater the bond strength. Given these facts, rank the salts magnesium oxide, barium sulfide, and rubidium iodide in order of increasing bond strength.

Polyatomic Ions

***8.** In your own words, explain what a polyatomic ion is. What do polyatomic ions have in common?

***9.** Identify each of the following as a molecule or a salt. For each salt, indicate whether it is a simple salt.

(a) $C_4H_{10}O$ (b) NaBr (c) NaCN (d) KI (e) K_2CO_3

***10.** Identify each of the following as a molecule or a salt. For each salt, indicate whether it is a simple salt.

(a) Li_2CO_3 (b) CO_2 (c) NH_3 (d) KCN (e) Rb_2SO_4

****11.** The salts $Ca(NO_3)_2$ and KNO_3 both contain the nitrate polyatomic ion. Explain why one formula includes parentheses and the other does not.

****12.** Identify the polyatomic ion in each of these salts.

(a) $Mg(HCO_3)_2$ (b) KOH (c) $LiMnO_4$ (d) $NaClO_4$

****13.** Write the chemical formula for the following compounds:

(a) potassium phosphate (b) potassium permanganate

(c) calcium hydrogen carbonate (d) ammonium chloride

(e) barium hydroxide (f) rubidium perchlorate

*****14.** Write the chemical formula and calculate the molar mass of the following compounds. For each, also write a chemical equation showing its complete dissociation into ions.

(a) sodium chromate

(b) potassium cyanide

(c) lithium dihydrogen phosphate

(d) magnesium carbonate

(e) ammonium peroxide

(f) calcium hydroxide

*****15.** Write the names of these salts and calculate their molar masses:

(a) $Ca(NO_3)_2$ (b) $CaSO_4$ (c) Na_3PO_4 (d) $Ba(ClO_4)_2$

*****16.** A saline solution has a mass of 300.0 grams. If we dissolve 0.87 mole of sodium chloride into the solution, what is the percent by mass of salt in this solution?

The Polarity of Water and Ion–Dipole Interactions

***17.** Which of the following pairs interact via ion–dipole interactions?

(a) a potassium ion and hydrogen sulfide

(b) two water molecules

(c) an ammonia molecule and a calcium ion

(d) a calcium ion and a fluoride ion

***18.** Sketch the interaction between the following:

(a) a calcium ion and surrounding water molecules

(b) a carbonate ion and surrounding water molecules

***19.** On a humid day, you may find that the salt in a shaker is clumpy and difficult to shake out. What is the reason for this?

****20.** In your own words, explain why ions become surrounded by water. What interactions are at work in an aqueous salt solution?

**21. In your own words, explain why the Egyptians used natron, a mixture of salts, for mummification.

**22. Hydrogen sulfide, H_2S, is structurally analogous to water. Sketch the electron-pair geometry and the molecular geometry for hydrogen sulfide. Sketch an arrow that shows the positive and negative ends of the molecule's dipole.

***23. Review Chapter 8, "Water," and explain why water is an unusual substance. What unique type of bond does it make with itself? Sketch a group of three water molecules held together by this type of interaction.

Electrolyte Solutions

*24. In your own words, explain what an electrolyte solution is. List one ion that takes part in electrolyte balance in the human body.

*25. Explain the reasoning behind including certain salts and sugar in sports drinks. How do manufacturers of sports drinks decide how much sugar to include in their products?

*26. Rank the following salt solutions in order of how much light they would produce in the light bulb experiment depicted in Figure 9.11: a 3.0-molar solution of sodium chloride, a 400-mM solution of sodium chloride, and a 1.0-molar solution of sodium chloride.

*27. State whether each of the following electrolytes is more concentrated, less concentrated, or at the same concentration relative to human sweat:

(a) magnesium ions in coconut milk (b) chloride ions in Gatorade

(c) sodium ions in Gatorade (d) sugar in orange juice

(e) magnesium ions in carrot juice

**28. As Figure 9.11 illustrates, electrolytes dissolved in water produce ions that can conduct electricity. Indicate whether these statements are true or false, and explain each of your answers.

(a) If a beaker contains a solution of polyatomic ions, the bulb will not light up.

(b) The lower the electrolyte concentration in the beaker, the more weakly the bulb glows.

**29. Write chemical equations that show the dissociation of each of these salts.

(a) $MgCl_2$ (b) LiI (c) Na_3N (d) KOH

**30. Write chemical equations that show the dissociation of each of these salts.

(a) NH_4Cl (b) KCN (c) Na_2O_2 (d) $CaCO_3$

**31. Refer to Table 9.2 to answer these questions.

(a) Many people consider tomato juice to be a healthy drink, but would it make a good electrolyte-replacement drink? According to the data shown in the following table, which would be a better electrolyte-replacement drink, the tomato juice containing salt or the no-salt juice? Why?

(b) How do the tomato juice numbers in the following table compare with those for Gatorade in Table 9.2?

(c) Which is the best choice for electrolyte replacement: orange juice, coconut milk, or tomato juice?

Ingredient	Tomato juice, no salt	Tomato juice with salt
Sugar (mass percent)	4.2%	4.2%
Magnesium ions	4.5 mM	4.5 mM
Potassium ions	56 mM	56 mM
Sodium ions	4.4 mM	160 mM

*32. A 591-mL bottle of one brand of fortified water contains 30.0 mg of potassium ions.

(a) What is the molarity of the potassium ions in the bottle?

(b) How many potassium ions are contained in this bottle?

***33. When 87.5 g of magnesium chloride is added to a 5.0-L container and the container is then filled with water, every formula unit of the salt dissociates to make one magnesium ion and two chloride ions. Given this, what are the molar concentrations of Mg^{2+} and Cl^-, assuming all the salt dissolves?

Concentration Units

*34. What is the molarity of a salt solution made by dissolving 4.00 moles of salt in 2.0 liters of water?

*35. If you weigh out 6.50 grams of a salt, put it in a beaker, and fill the beaker with water until the total mass of the solution is 3500 grams, what is the percent by mass of salt in this solution?

*36. A certain solution is 5.0% by mass of potassium iodide, KI. What mass of KI is contained in 2000 grams of this solution?

*37. Rank the following solutions in order of increasing concentration: 500 Mm, 50 mM, 5 M, 0.5 mM, and 50 M.

*38. Which of the following are not valid ways to express concentration? Which are valid?

(a) moles per liter (b) meters per hour (c) grams per second

(d) percent by mass (e) moles per kilogram (f) meters per second

**39. If you weigh out 65.00 grams of silver nitrate, $AgNO_3$, and add enough water to make 3.0 liters of solution, what is the molarity of the solution?

**40. You can find 4% by mass solutions of hydrogen peroxide at your local pharmacy. If you buy a bottle of this solution, and its contents weigh 250 grams, how many grams of hydrogen peroxide does it contain?

**41. Express the following concentrations in millimolar units:

(a) 0.00650 M (b) 0.044 M (c) 0.887 M (d) 4,000,000 nM

**42. The genetically altered tomato plants designed by pomologists at the University of California can live in a 200-mM sodium chloride solution. How many moles of this salt would you have to add to a 1.0-L flask to make a 200-mM solution of sodium chloride in water?

**43. A saline solution is 6.60% by mass. If you have a large container of this solution, and the container contains 2.00 kilograms of sodium chloride, what is the mass of the solution in your container?

***44. A concentration of 1 ppm is equivalent to a concentration of 1 mg/L. The concentration of sodium ions in tomato juice is 0.160 M. What is this concentration of sodium ions in tomato juice expressed in parts per million? Assume that the density of the juice is 1.00 g/mL.

***45. The concentration of magnesium ion in human blood plasma is 5.5 mM.

(a) If you have 4.0 pints of blood, how many moles of magnesium ions does the sample contain?

(b) If magnesium were the only cation in blood, what molar concentration of chloride ions would be required to balance the charge on the cations? (Hint: 1 pint = 0.5 quarts, and 1.0567 quarts = 1.00 L.)

Osmosis and Ion Mobility

*46. How is it possible for a salt-tolerant tomato plant to thrive in high-salt soils? How is the salt in the soil taken up by the genetically modified plant in a way that leaves the tomatoes edible?

*47. In your own words, define these terms: *osmosis, concentration gradient,* and *semipermeable membrane.*

**48. A U-shaped device like the one shown in Figure 9.14 is made with a membrane that does not allow ions or water to pass through. Will the system work the way the U-shaped tube in the figure works? Why or why not?

**49. Imagine that an alternative semipermeable membrane allows only ions to pass through and not water. Would the system in Figure 9.14 equilibrate to the same state with this alternate semipermeable membrane compared to the traditional membrane discussed in the text, which lets only water pass?

***50. Consider a tube like the one shown in Figure 9.14. There is an aqueous solution of glucose in each arm, and the membrane is impermeable to glucose but permeable to water. Indicate whether these sentences are true or false, and explain each of your answers.

(a) If the water in the left arm is at the same level as the water in the right arm, the system must be at equilibrium.

(b) At equilibrium, the glucose concentration in the left arm is equal to the glucose concentration in the right arm.

*51. In Figure 9.5, we see examples of two salts that are brightly colored. Often, such compounds contain a transition metal, an element in columns 3 through 12 of the periodic table. We know that elements in the first column of the periodic table always make +1 ions and elements in the second column always make +2 ions, but transition metals can have more than one and often have several, different possible charges. For this reason, when naming salts containing transition metals, the charge on the metal is often indicated in parentheses using Roman numerals. For example, the blue compound in Figure 9.5 is copper (II) sulfate, a salt with a Cu^{2+} cation and an SO_4^{2-} anion. The chemical formula shows that one cation and one anion are necessary for

Courtesy of James Bull, Castellina in Chianti, Italy

the charges to balance, like this: $CuSO_4$. Write the chemical formula for these other brightly-colored salts that contain transition metals.

(a) chromium (II) chloride (bright green)

(b) lead (II) iodide (vibrant yellow)

(c) cobalt (II) chloride (royal blue)

**52. A recipe for chocolate chip cookies calls for two teaspoons of baking soda, also known as sodium bicarbonate. This quantity is equivalent to about 20 grams.

(a) Write a chemical formula for sodium bicarbonate and find its molar mass.

(b) If the total mass of all ingredients for a batch of cookies is 750 grams, what is the mass percent of baking soda in the recipe?

**53. Consider the following label for a container of yogurt.

(a) If the total mass of one serving of this yogurt is 170 grams, what is the mass percent of carbohydrate?

(b) What is the mass percent of protein?

(c) What is the mass percent of cholesterol (1000 mg = 1 g)?

(d) What is the mass percent of sodium?

(e) If we assume that water makes up all of the mass not accounted for on this label, what is the mass percent of water?

Nutrition Facts

Serving Size 1 container (170.0 g)

Amount Per Serving	
Calories 110	Calories from Fat 0
	% Daily Value*
Total Fat 0.0g	**0%**
Saturated Fat 0.0g	**0%**
Trans Fat 0.0g	
Cholesterol 3mg	**1%**
Sodium 90mg	**4%**
Total Carbohydrates 20.1g	**7%**
Sugars 15.0g	
Protein 6.0g	
Vitamin A 15% •	Vitamin C 0%
Calcium 20% •	Iron 0%

*Based on a 2000 calorie diet

**54. The molarity of pure water is 55.5 M. Consider a 5.00-liter jug of purified water. How many moles of water does it contain? If you add 0.430 millimoles of sugar to this jug of water (and assume that the sugar does not change the volume), what is the molarity in millimolar units of the sugar solution you've just made? In other words, how many millimoles of sugar are in each liter of the solution?

55. INTERNET RESEARCH QUESTION

Choose an energy or sports drink that is not mentioned in this chapter. Find the ingredients in this drink by opening your search engine and typing in the name of this drink along with "ingredients" or "nutrition label." First, seek out the quantity of sodium, which will probably be given in units of milligrams sodium per liquid serving size. By using the following conversions, you can convert this amount to millimolar units, as this example shows:

$$\frac{115 \text{ mg sodium ions}}{240 \text{ mL per serving}} \times \frac{1 \text{ mmole sodium ions}}{22.99 \text{ mg sodium ions}} \times \frac{1000 \text{ mL}}{1 \text{ L}} = 21 \text{ mM sodium ions}$$

If it's available, locate the mass percent of sugar in your drink. Compare the values you obtained with the drink analysis shown in Table 9.2. How does the concentration of sodium ions in your drink compare? How does the amount of sugar compare?

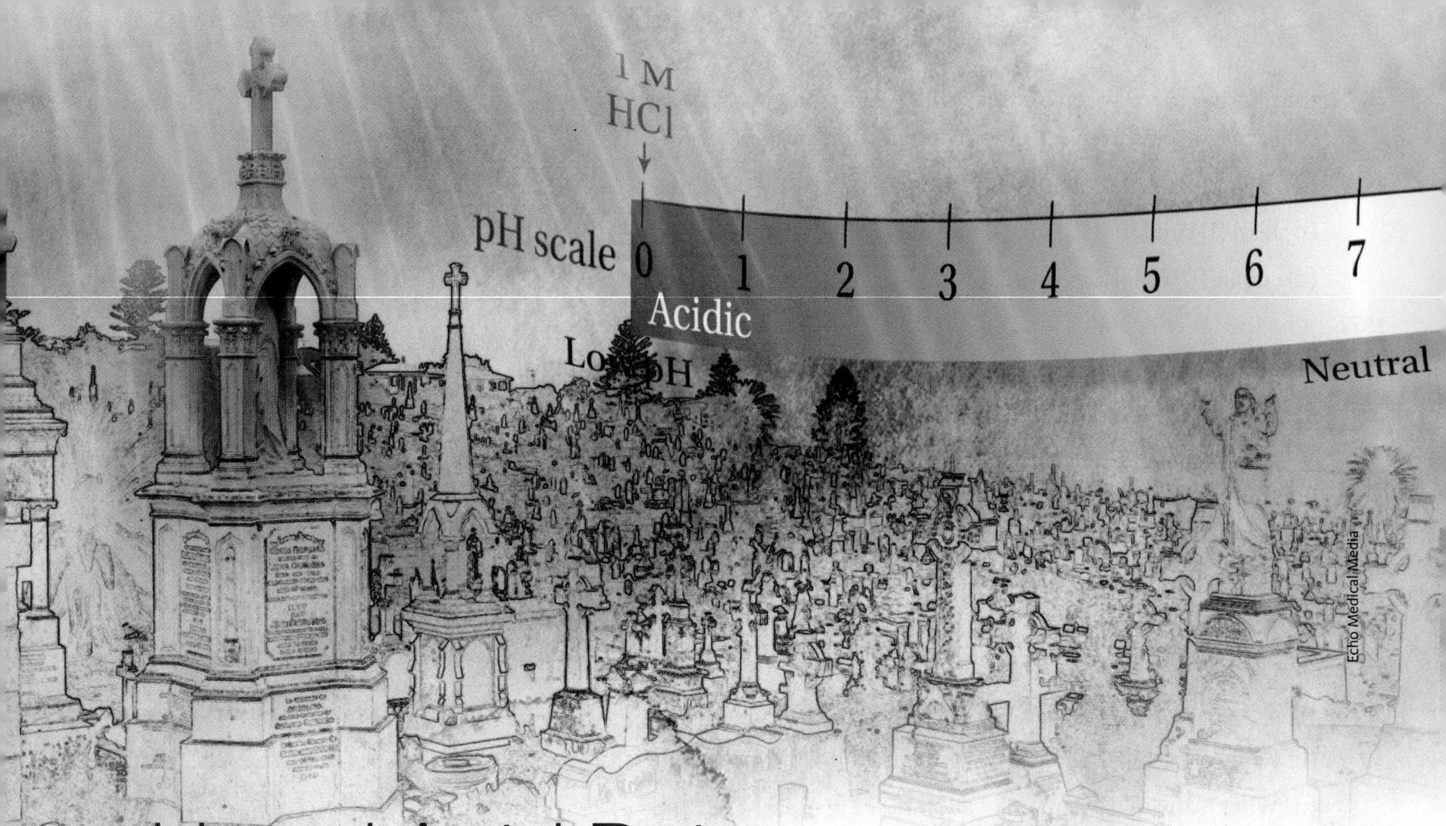

pH scale
0 1 2 3 4 5 6 7
Acidic
Low pH
Neutral
1 M HCl

Echo Medical Media

pH and Acid Rain

Acid Rain and Our Environment

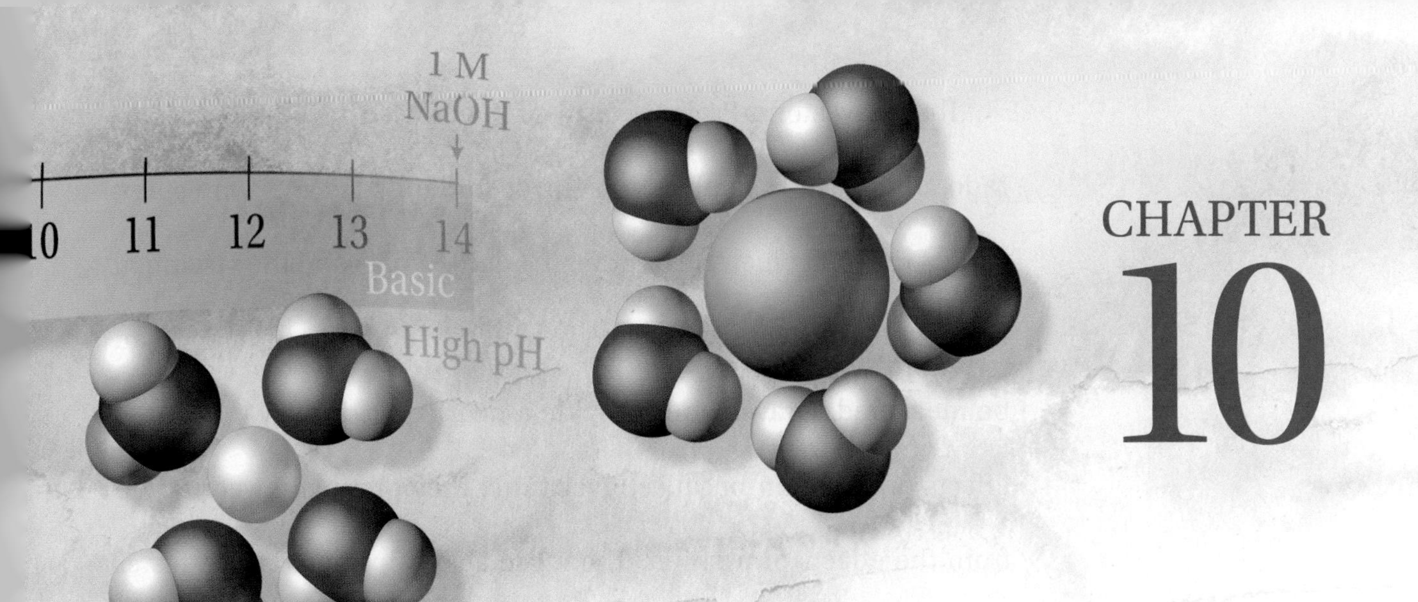

CHAPTER 10

It's true. Many chemists spend their days in a lab, wearing lab coats, doing chemistry. But some chemists spend their time out and about, with the world as their laboratory. Chemists refer to this outside-the-lab environment as "working in the field." Of course, a chemist could be working in an actual field—for instance, analyzing the soil on a baseball field—but working in the field means, generally, working outside of the laboratory.

A lot of laboratory equipment can be adapted for use in the field. So, we could find a chemist reaching an instrument out into the caldera of a volcano, analyzing air samples. Or, we could find a chemist leaning over the side of a boat, taking water samples. We could even find a chemist analyzing the air deep inside a mineshaft. But one place we might be surprised to find a chemist is in a cemetery.

The Gravestone Project was a project based in Boulder, Colorado, until 2011. It involved dozens of student-scientists who spent a lot of time in cemeteries, collecting data. What were they trying to figure out? They were interested in acid rain and how its effects can be measured. Gravestones provide a nearly perfect way to test the effects of acid rain because, over time, they gradually wear down from the effects of weather. The more acid rain there is, the faster they wear.

GSA EarthTrek 2010, Photo Gary Lewis.

Many gravestones are made of marble, one of the substances most easily broken down by acid rain. So gravestones are not only grave markers, they are also acid rain detectors. Each one includes the date the person died, which is usually when the gravestone was placed at the grave. The date is a reliable way to track the breakdown of the stone because it indicates about how long the stone has been standing out in the rain.

The mission of the Gravestone Project was to understand how acid rain damages the environment and where acid rain strikes hardest. This chapter has the same mission. But, before we can understand acid in rain, we have to understand what acids are. Then we'll move on to acid in rain and how it gets there. Finally, we can ask some practical questions, like, "How does acid rain damage the environment?" and, perhaps most important, "What can we do about it?"

10.1 The Autoionization of Water

There are two ways to depict protons in aqueous solutions.

Acids are complicated. They can be gases or liquids or solids. They can be single atoms, or ions, or molecules. Whole books have been written just about acids, but in this chapter we are primarily concerned with acid rain. Why are we focusing on acid rain? Because acid rain has been a persistent problem in the United States for decades. It's a significant cause of environmental damage, especially damage to forests and natural waters and the species of plants and animals that live in them. Acid rain also causes substances to degrade much faster than they ordinarily would. It degrades car paint, for example, as well as culturally significant materials such as gravestones and statues. Finally, the pollutants that cause acid rain also contribute to diseases in humans, especially heart and lung disease.

Because we are focusing on acid in rain in this chapter, we're interested primarily in acids that exist in water. Fortunately, this makes understanding acids much, much easier. However, before we can understand how acids behave in water, let's first review what you have already learned about water and about protons, which are both substances found in acid rain.

Recall from Chapter 2 that a proton, which has the symbol H^+, is a hydrogen atom that is missing its only electron. And, because most hydrogen atoms do not contain any neutrons either, the only particle in H^+ is a proton. In aqueous solutions, protons do not exist on their own; they're always surrounded by water molecules, just the way all ions in aqueous solutions are surrounded by water molecules.

There are two ways to depict a proton in aqueous solutions. First, we can use the symbol for a proton and follow it with the abbreviation for *aqueous,* like this: H^+ (*aq*). In this case, we still refer to it as "a proton." Alternatively, we can write a symbol that represents the proton combined with a water molecule, like this: H_3O^+. When we write a proton as H_3O^+, we refer to it as a **hydronium ion**. We can use the terms *proton* and *hydronium ion* interchangeably in most circumstances. As we will see, it's sometimes convenient to use one symbol in one situation and the other symbol in another situation. The two ways to depict a proton are summarized in **Figure 10.1**.

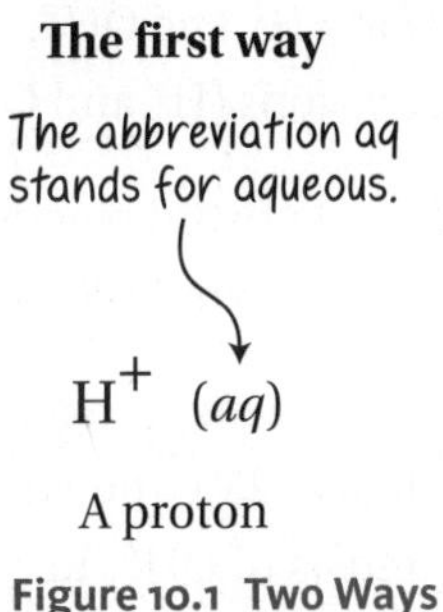

The first way
The abbreviation aq stands for aqueous.
H+ (aq)
A proton
The second way
This combines a proton, H+, with a water molecule, H2O.
H3O+
A hydronium ion

Figure 10.1 Two Ways to Depict a Proton in Water

Water molecules can break apart and form ions.

A beaker filled with high-purity water contains water molecules but, at any given time, a few of those water molecules are undergoing a chemical reaction known as **autoionization**. During autoionization, the water molecule splits and produces two ions. It happens like this:

$$H_2O \rightleftharpoons OH^- \,(aq) + H^+ \,(aq)$$

You may already recognize one of these ions. It's the proton surrounded by water: H^+ (*aq*). The other ion is a **hydroxide ion,** and it is an anion. And because the hydroxide ion is an ion in water, it is also surrounded by water molecules. We write it like this: OH^- (*aq*).

We can take a closer look at what happens when water autoionizes. Recall from Chapter 4 that the Lewis dot diagram for water is:

$$\mathrm{H-\ddot{\underset{\cdot\cdot}{O}}-H}$$

There are eight valence electrons in this molecule. Each line in the structure represents a single covalent bond, and a single covalent bond is composed of two electrons. So, in water, there are four valence electrons in its bonds and four more valence electrons in each of the two nonbonding pairs on the oxygen atom. This gives a total of eight electrons.

When water autoionizes, one of the two H–O bonds breaks, and both electrons in that bond go to the oxygen. The hydroxide ion that forms includes all eight electrons and a negative charge. This leaves the proton, which has no electrons and a positive charge, as shown in **Figure 10.2**.

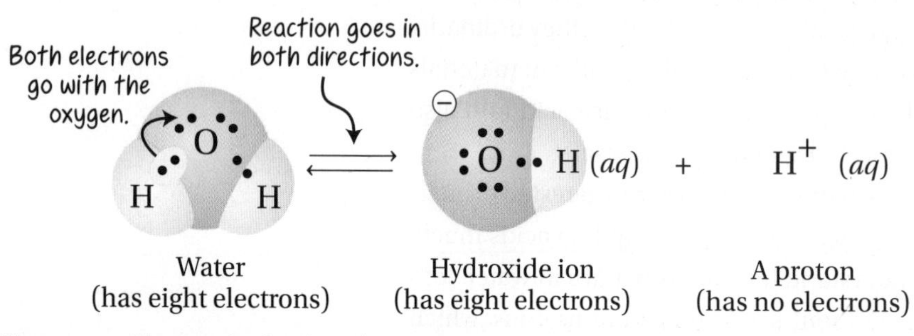

Figure 10.2 The Autoionization of Water

The eight electrons in water remain together in the hydroxide ion formed during autoionization. The proton that also forms has no electrons.

recurring theme in chemistry 4

Chemical reactions are nothing more than exchanges or rearrangements of electrons.

Notice that the autoionization reaction has a double arrow. Recall from the discussion of equilibrium in Section 9.3 that this double arrow means the reaction goes in both directions at once. We say that the system is at equilibrium when the rates of the two opposing reactions are equal. For the autoionization of water, the system is at equilibrium when the forward reaction, in which water is ionizing to form H^+ and OH^-, takes place at the same rate as the reverse reaction, in which the two ions (H^+ and OH^-) come together to form water:

Forward reaction $H_2O \rightarrow OH^- + H^+$

Reverse reaction $OH^- + H^+ \rightarrow H_2O$

Chemists have measured the extent of autoionization carefully, and we know that at 25°C, one liter of pure water contains 1.0×10^{-7} moles of H^+ and 1.0×10^{-7} moles of OH^-. Recall from Chapters 6 and 9 that the mole is a way to count very small things, like protons and hydroxide ions, and that one mole of something contains 6.02×10^{23} of that thing. We express the concentration of a substance dissolved in something else with *molarity,* which is the number of moles of a substance in one liter of solution (mol/L). Therefore, we know that the concentration of protons is 1.0×10^{-7} moles per liter of water. Another way to say this is that the concentration of protons is 1.0×10^{-7} molar. *Molar* is abbreviated with a capital letter M, so we express the concentration of protons this way: 1.0×10^{-7} M.

flashback

Read more about molarity in Section 9.4, "Concentration and Electrolytes."

In one liter of pure water, only about 1 in every 550,000,000 molecules autoionizes to form a hydroxide ion and a hydrogen ion. Thus, pure water is mostly water molecules, but it has a few ions as well. Why are we so concerned with a reaction that happens for only 1 in every 550,000,000 water molecules? Because it leads us to the definition of an acid: an **acid** is any substance that increases the concentration of protons in a watery environment. In one liter of pure water, there are

1.0×10^{-7} moles of H^+ ions. Water becomes an **acidic solution** when an acid is added to it. Because an acid increases the number of H^+ ions, one liter of an acidic solution contains more than 1.0×10^{-7} moles of H^+ ions.

wait a minute . . .

Is it just a coincidence that pure water has exactly the same number of H^+ and OH^- ions?

No, it is not a coincidence. In fact, this always is the case. Why? Because one water molecule always breaks into the same two parts. So each time auto-ionization takes place, it must result in the formation of the same number of H^+ and OH^- ions.

QUESTION 10.1 How many moles of H^+ are in two liters of pure water?

ANSWER 10.1 In one liter of pure water, there are 1×10^{-7} moles of H^+ ions. So, there are twice as many in two liters: 2×10^{-7} moles.

For more practice with problems like this one, check out end-of-chapter Question 7.

FOLLOW-UP QUESTION 10.1 True or false? Pure water is naturally acidic because it contains H^+ ions.

10.2 Acids, Bases, and the pH Scale

Acids ionize in water to produce protons.

One of the simplest acids is hydrochloric acid, HCl. It's simple because it contains only two atoms. HCl is a gas and, when it is bubbled into water, a solution of hydrochloric acid forms. The following chemical equation describes this reaction:

$$HCl\ (g) + H_2O \rightarrow H_3O^+\ (aq) + Cl^-\ (aq)$$

Notice a few things about this chemical equation. First, the equation tells us that when the HCl molecules are bubbled into water, they ionize and form ions, as we would expect from what we have learned in Chapter 9. The two ions formed when HCl ionizes, H^+ and Cl^-, are both surrounded by water molecules, as shown in **Figure 10.3**.

Second, it's significant that one of the products of this equation is a proton. Because HCl releases protons into water, we know that it must be an acid. Finally, notice that the reaction goes in one direction only. This means that all of the HCl that goes into the water makes ions. There is no HCl remaining, and no reverse chemical reaction takes place.

If we bubble HCl into one liter of pure water, the water becomes acidic. That means the concentration of protons in the solution changes from 1.0×10^{-7} M—the value in pure water—to a higher value. One liter of pure water can accommodate

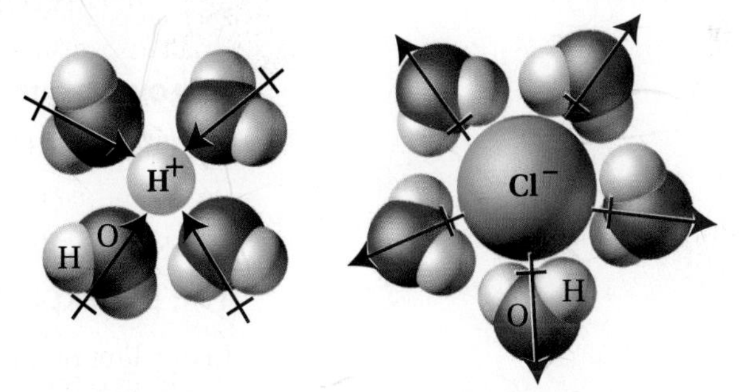

Figure 10.3 An Acid Ionizing in Water

In water, HCl makes protons and chloride ions. These ions are surrounded by water molecules. Notice that the negative end of the dipole within each water molecule points toward the cation, H^+, while the positive end of the dipole within each water molecule points toward the chloride anion, Cl^-.

about 12 moles of HCl, so the maximum molarity of the protons in the solution can be as much as 12 M. This solution contains 100,000,000 times more H^+ ions than pure water.

The term we use to describe this process, in which an acid makes ions in water, is **dissociation**. Hydrochloric acid dissociates in water completely. The arrow in the reaction points to the right only, not both ways, as we can see in the equation. When an acid dissociates in water completely, we call it a **strong acid**.

Other acids hardly dissociate at all. Still others dissociate somewhat, but not completely. When an acid does not dissociate completely, we refer to it as a **weak acid,** and we use a double arrow to show that dissociation is not complete. For example, the following chemical equation shows the dissociation of the weak acid, hydrogen fluoride (HF):

$$HF\ (aq) \rightleftharpoons H^+\ (aq) + F^-\ (aq)$$

wait a minute . . .

Are dissociation and ionization two words for the same thing?

No, not exactly. Ionization is a more generic term that's used when any substance becomes an ion. Dissociation is a special type of ionization. This term is usually used when substances such as salts and acids, which are neutral, break apart into their ions, which are charged.

QUESTION 10.2 A substance with the chemical formula $HC_2H_3O_2$ behaves in the following way in water:

$$HC_2H_3O_2 \rightleftharpoons H^+ + C_2H_3O_2^-$$

Is this substance an acid? How do you know? Is this substance a strong acid? How do you know?

ANSWER 10.2 Yes, it is an acid, because it releases protons into water. It is a weak acid because a double arrow shows that the equation is at equilibrium.

For more practice with problems like this one, check out end-of-chapter Question 2.

FOLLOW-UP QUESTION 10.2 What is the difference, if any, between the terms *proton* and *hydronium ion*?

The pH expresses proton concentration.

In one liter of water, it's common to have a very low concentration of H^+ ions, such as 10^{-14} M, or a very high concentration of H^+ ions, such as 12 M. This is a huge range of concentrations. Because these concentrations are so important and used so often, many years ago chemists devised a way to simplify how the numbers are reported—the pH scale. Rather than saying, "Pure water has a proton concentration of 1.0×10^{-7} M," we can say, "Pure water has a pH of 7.00."

To determine pH, we do some straightforward math. The **pH** is the negative logarithm of the molar concentration of protons:

$$pH = -\log[H^+]$$

The brackets around the H^+ indicate that we are talking about the molar concentration of protons. For example, suppose we want to know the pH of pure water. We start with the molar concentration of protons in it: 1.00×10^{-7} M. The pH is:

$$pH = -\log(1.0 \times 10^{-7}\ M) = 7.00$$

wait a minute . . .

How do I enter logarithms into my calculator?

Let's say we want to plug into a calculator the proton concentration in pure water, 1.00×10^{-7} M. Calculators usually do logarithms in one of two ways, but you likely won't know which method your calculator uses until you try it. A generic calculator is shown in **Figure 10.4**. You'll need to experiment to find out which way—method A or method B—to enter logarithms into your calculator.

To try method A, enter the number 1.00×10^{-7} into a calculator (step 1A). (If you don't know how to do this, review the "Wait a Minute" in Section 1.2.) Now press the LOG key (step 2A). Because there is a negative sign before the log in the equation shown above, you need to reverse the sign on the number (step 3A). So, if the calculator says −7.00, the pH is 7.00.

Method B is the reverse of the first method. Begin by pressing the LOG key (step 1B) and then enter the number 1.00×10^{-7} (step 2B). Change the sign to get the pH (step 3B).

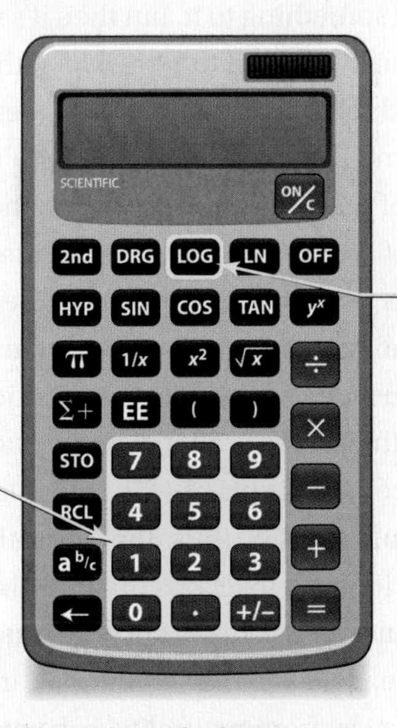

Figure 10.4 Finding a Logarithm with a Calculator
Calculators do logarithms in one of two ways; you have to experiment with your calculator to see which one it uses.

There is a shortcut to finding the pH of a solution when the concentration of protons is 1.0×10^{-x} M. Let's take a closer look at the following example—where x is 4—and the previous one, where x was 7, and see if we can figure it out:

$$\text{pH} = -\log(1.0 \times 10^{-4}\ \text{M}) = 4.00$$

We don't need a calculator to figure out the pH for a solution with a concentration of 1.0×10^{-x}. The pH of a solution is simply the value of x. So the pH of a 1.0×10^{-10} M solution of protons is 10.00. However, when the concentration is more complex, like 4.35×10^{-8} M, we will need a calculator. If we plug in that number, we can see that the pH of the solution is 7.361.

QUESTION 10.3 Without using your calculator, determine the pH of two solutions with the following proton concentrations: (a) $[H^+] = 1.0 \times 10^{-3}$ M (b) $[H^+] = 1.0 \times 10^{-5}$ M

ANSWER 10.3 (a) pH = 3.00 (b) pH = 5.00

For more practice with problems like this one, check out end-of-chapter Question 15.

FOLLOW-UP QUESTION 10.3 Using your calculator, determine the pH of two solutions with the following proton concentrations: (a) $[H^+] = 5.5 \times 10^{-4}$ M (b) $[H^+] = 3.0 \times 10^{-7}$ M

Acidic solutions have pH values less than 7.00.

What do we mean when we say that an aqueous solution is neutral? A **neutral** solution is one in which the hydrogen ion concentration and hydroxide ion concentrations are equal. Pure water is a neutral solution. When the pH of a substance is 7.00, the substance is a neutral solution. For pure water at 25°C, the only way to change the pH from 7.00 is to add something to it, but then it's no longer pure water. When an acid, a source of H^+ ions, is added to pure water, the concentration of H^+ ions increases because the acid releases them into the water.

Consider for a moment a glass of apple juice with a pH of 4.00. In this glass of juice, the concentration of protons is 1.0×10^{-4} M, and those protons come from the partial dissociation of weak acids in the juice. Recall from Section 10.1 that in the pure water that was used to make the juice, there is a tiny amount of H^+—1.0×10^{-7} M—but that amount is very small compared to the acid that's in the juice. For this reason, we usually neglect the contribution made by the acid in pure water. Its contribution to the acidity is almost always vanishingly small.

The glass of apple juice, with its pH of 4.00, contains much more acid than does pure water, which has a pH of 7.00. Thus, the lower the pH, the more acidic the solution. In fact, any solution with a pH less than 7.00 is **acidic** because it has more protons than pure water has. We use the pH of pure water at 25°C—7.00—as a reference point on the pH scale. Below pH 7.00, solutions are acidic.

When we deviate far away from neutrality, substances are potentially dangerous. For example, the 12-M solution of HCl that we discussed has a pH of −1.08, a value that's far away from 7.00. Concentrated acids, like this one, are corrosive and can cause severe skin burns. The same is true for solutions at the other end of the pH scale, as we are about to see.

Basic solutions have pH values greater than 7.00.

How is it possible to make the pH of a solution greater than 7.00? We know that because of autoionization, pure water contains hydroxide ions at a concentration of 1.00×10^{-7} M. If we add something that contains more protons—in other words, an acid—to the solution, its pH value is less than 7.00 and the solution is acidic. However, if we add more hydroxide ions, the pH value increases above 7.00 and our solution becomes a **basic** solution. We call anything that adds hydroxide ions to a solution a **base**. Many bases are not difficult to identify once we know how. For example, what do the following molecules, which are all bases, have in common?

$$NaOH, LiOH, Ca(OH)_2, Ba(OH)_2, KOH, RbOH$$

The answer is that they all contain the OH^- ion. These bases dissociate completely in water; for example:

$$KOH\,(aq) \rightarrow K^+\,(aq) + OH^-\,(aq)$$

Suppose we make a solution that contains 1.0×10^{-2} M hydroxide ions by adding a base to pure water. If we want to determine the pH of this solution, we start with the OH^- concentration rather than with the H^+ concentration. We calculate the pOH exactly the way we did pH. The pOH is the negative log of the molar concentration of hydroxide ions:

$$pOH = -\log[OH^-]$$

We can plug figures into the pOH equation exactly the way we did for pH. To find the pOH of our basic solution, we complete the following calculation:

$$pOH = -\log(1.0 \times 10^{-2}\,M) = -(-2.00) = 2.00$$

Because chemists are used to thinking in terms of pH rather than pOH, they use the following relationship to convert back and forth between pOH and pH:

$$pOH + pH = 14.00$$

Knowing this relationship, we can quickly determine the pH of our solution:

$$pH = 14.00 - pOH = 14.00 - 2.00 = 12.00$$

In our example, the pH changed from 7.00 (in pure water) to 12.00 when the base was added. Bases, by definition, increase pH by increasing the concentration of OH^-, and basic solutions have pH values greater than 7.00. A solution this basic should be treated with respect because, like a highly acidic solution, it can cause serious skin burns.

QUESTION 10.4 The pOH of several solutions are listed. For each, determine the pH and state whether the solution is neutral, acidic, or basic. (a) 3.4 (b) 6.9 (c) 1.0 (d) 12.5

ANSWER 10.4 (a) pH = 10.6; basic (b) pH = 7.1; basic (c) pH = 13.0; basic (d) pH = 1.5; acidic

For more practice with problems like this one, check out end-of-chapter Questions 5 and 17.

FOLLOW-UP QUESTION 10.4 Which of the following compounds are acids and which are bases? (a) HClO (b) NaOH (c) HBr (d) KOH

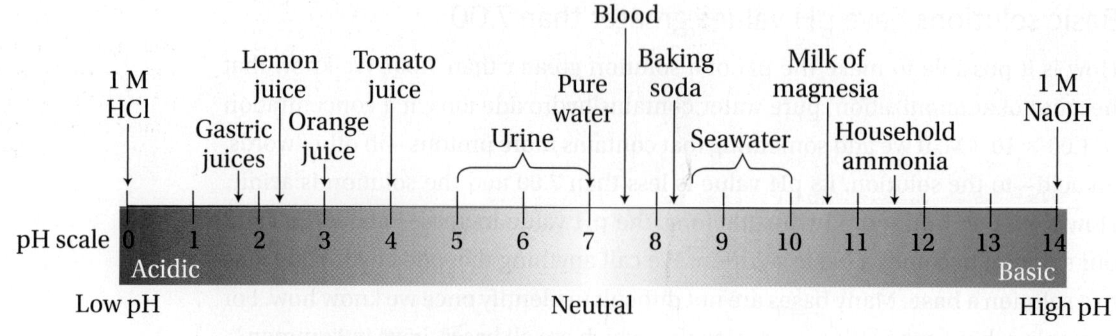

Figure 10.5 The pH Scale

Acidic substances all have pH values that are lower than the chemically neutral pH value of 7.00. Basic substances all have pH values higher than 7.00.

We use the pH scale to describe acid or base strength.

We now know that a solution with a pH less than 7.0 is an acidic solution, and a solution with a pH greater than 7.0 is a basic solution. The pH scale is shown in **Figure 10.5**. Notice that many substances that are slippery or caustic, such as a solution of baking soda or ammonia, are bases and have pH values greater than 7.0. Many substances that are tart, such as fruit juices, are acidic and have pH values less than 7.0. The pH scale shrinks a wide range of H^+ concentrations down to a convenient scale that typically ranges from 0 to 14.

We can see from Figure 10.5 that orange juice is more acidic than tomato juice, but how much more? We can estimate that difference by comparing their pH values:

Tomato juice $[H^+] = 1.0 \times 10^{-4}$ M
Orange juice $[H^+] = 1.0 \times 10^{-3}$ M

This result tells us that one liter of tomato juice contains 0.00010 mol of H^+ ions, and that one liter of orange juice contains 0.0010 mol of H^+ ions. Thus the acidity of orange juice is 10 times the acidity of tomato juice.

QUESTION 10.5 Determine which substance in each pair is more acidic: (a) orange juice/lemon juice (b) ammonia/blood (c) urine/baking soda.

ANSWER 10.5 (a) lemon juice; (b) blood; (c) urine.

For more practice with problems like this one, check out end-of-chapter Questions 18 and 20.

FOLLOW-UP QUESTION 10.5 A solution contains 1.0×10^{-6} mol of H^+ in one liter of water. (a) What is the pH of the solution? (b) What other substance has a pH value in this range (refer to Figure 10.5)?

What can pink polka-dotted airplanes tell us about the measurement of pH?

In the laboratory, pH can be measured with a device called a pH meter. But before pH meters were widely available, chemists measured pH using an indicator. An **indicator** is a brightly colored molecule that changes color as the pH of the solution containing it is changed. For example, the indicator phenolphthalein

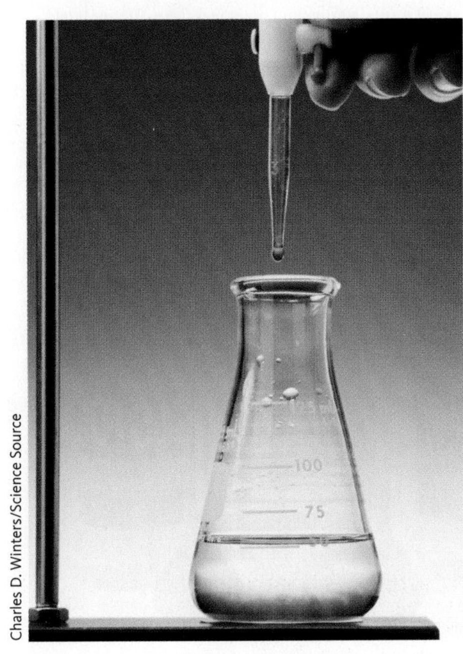

Charles D. Winters/Science Source

pH below 8.3

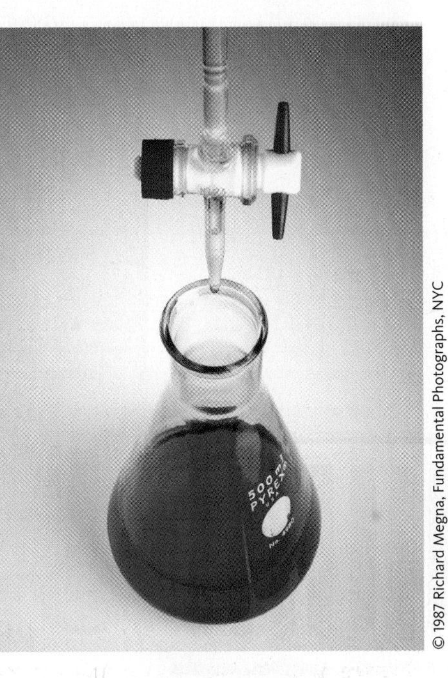

© 1987 Richard Megna, Fundamental Photographs, NYC

pH above 8.3

Figure 10.6 pH Indicators

Phenolphthalein changes color abruptly with changes in pH. At pH values below 8.3, a solution containing phenolphthalein is colorless. Once the pH surpasses 8.3, the solution turns bright pink.

(*feen-ol-THAY-leen*) is colorless at pH values below 8.3 but it is bright pink at pH values greater than 8.3, as shown in **Figure 10.6**. And, as we're about to see, sometimes old-school methods, like pH indicators, still work better for some purposes than sophisticated electronic gadgets, like pH meters.

Recently, the old-school indicator method was used by the airline industry to address a potentially dangerous problem: corrosion. **Corrosion** is a chemical reaction in which metals slowly deteriorate when they are exposed to the environment. Engineers who work on planes know that corrosion can cause serious damage. They also know that when a plane's fuselage (the main body of an airplane) corrodes, hydroxide ions are produced. Hydroxide ions, as we know, are bases, and bases increase pH. So, the engineers reasoned, why not use a pH test to look for corroding spots?

To do this, engineers painted planes with a colorless paint that included phenolphthalein indicator and waited. After a short time, pink spots began to appear on the plane. Wherever corrosion was just beginning, a bright pink spot showed up when the phenolphthalein changed from colorless to pink in the presence of hydroxide ions. The engineers who developed the new paint report that bright pink spots appear where a corroded spot is only 15 micrometers deep, a tiny crevice much too small to be seen with the naked eye.

This way of tracking corrosion has impressed a lot of people. It's much more sensitive than any of the high-priced instruments developed for the same purpose, and it is very inexpensive. In the future, planes coated with this new paint may become the norm.

Indicators are available for almost any pH range. In fact, a product called a *universal indicator* is a mixture of several different organic molecules, each one changing color at a different pH. As shown in **Figure 10.7a,** a universal indicator is

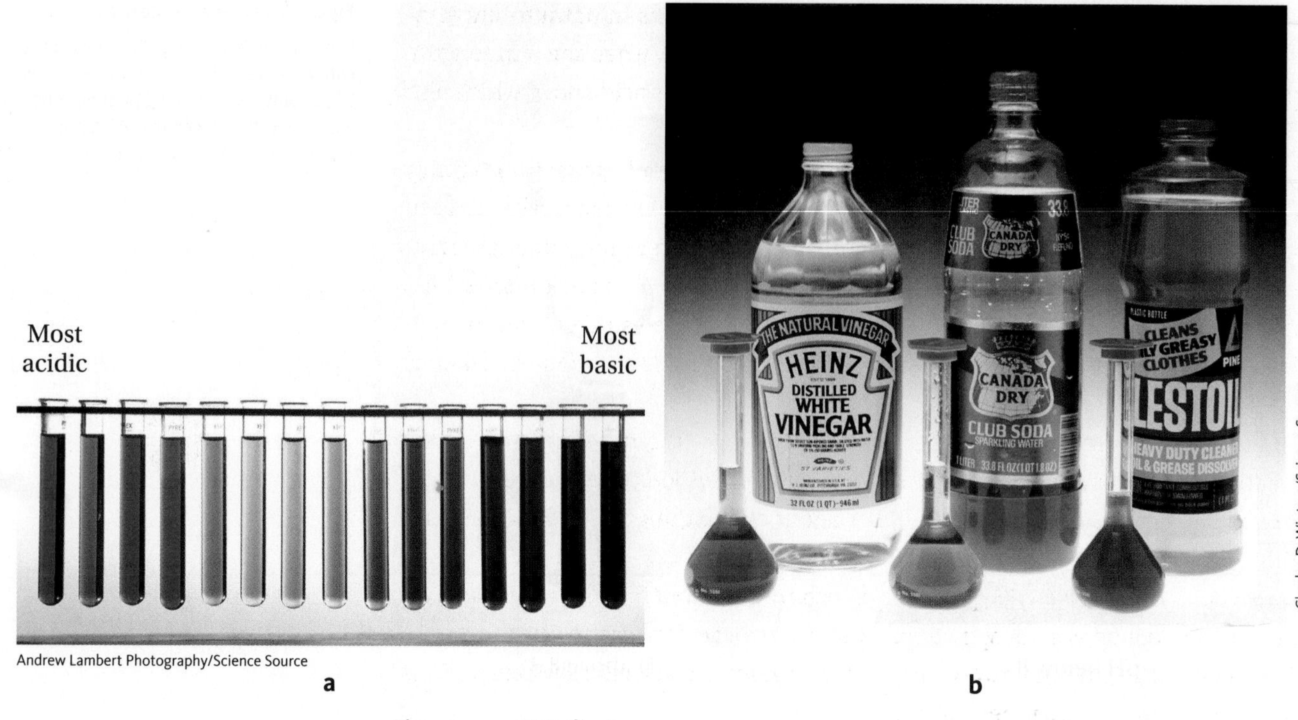

Andrew Lambert Photography/Science Source

a

b

Charles D. Winters/Science Source

Figure 10.7 pH Indicators

(a) A universal indicator is a mixture of indicators. Over a pH range from 1 to 12, the universal indicator changes color. (b) Different household products mixed with a universal indicator turn different colors. Vinegar has a pH of about 1, the pH value of club soda is about 4, and Lestoil stain remover has a pH of about 12.

sensitive enough to allow solutions that differ by one pH unit to be distinguished. **Figure 10.7b** illustrates how we can reveal the pH of different household products by adding a few drops of indicator to each one and comparing the colors to the pH values in Figure 10.7a.

10.3 Acid Rain Part I: Sulfur-Based Pollution

Compounds in air dissolve in water and change the water's pH.

Most natural waters, including lakes, rivers, and oceans, do not have a pH value of 7.0. That is, they are not chemically neutral. How do natural waters become acidic or basic? The answer to this question is exceedingly complex, because many things can affect pH. For example, minerals that ionize exist on the ocean floor, and aquatic life-forms remove nutrient molecules from water and leave other molecules behind. These and dozens of other factors, including temperature, salt content, depth, and so on, affect the pH of natural waters.

One of the most significant factors that determines the pH of a body of water is the air above it. Air is mostly nitrogen and oxygen molecules, in addition to small amounts of other naturally occurring gases. Human activities have also added several gases to the mix. Some of these gases form acids when they mix with water, and they are the gases responsible for acidifying rain. **Acid rain** is rainfall that has a lower pH than natural, unpolluted rain. The pH of acid rain is even lower than the pH of clean rain, which is weakly acidic and has a pH of about 5.5 to 6.0.

Acid rain is caused by atmospheric pollution, and it results in harm to the environment. We will discuss the reasons for this shortly. And, when the temperature drops below the freezing point of water, acid rain becomes acid snow, which also harms the environment.

The **Clean Air Act** of 1970 charged the **Environmental Protection Agency (EPA)** with setting limits on the amounts of gaseous pollutants that can be released into the air in the United States. These limits are reported in the National Ambient Air Quality Standards (NAAQS), which the EPA updates occasionally to address new challenges to the environment. In 2011, the EPA passed stricter limits on pollutants. **Table 10.1** lists the new allowable maximum levels of the most insidious air pollutants.

Let's look at these pollutants more closely. In Table 10.1, there are two concentration units: **parts per million (ppm)** and **parts per billion (ppb)**. These units are frequently used to express a very small amount of one thing mixed in with something else.

What is meant by the 75-ppb limit on ozone? This limit indicates that, in an air sample that contains a 75-ppb concentration of ozone, for every 1 billion (that is, 1,000,000,000) atoms and molecules of various gases in an air sample, at most 75 of them, on average, are ozone molecules.

Ozone, a molecule that naturally exists in Earth's upper atmosphere, is produced unnaturally when sunlight and smog react with oxygen molecules. Although the natural ozone in the upper atmosphere helps protect us from damaging ultraviolet rays, this same substance is toxic when formed near Earth's surface and can cause lung disorders in people who breathe it regularly.

Another gas listed in the table is carbon monoxide (CO), a potent greenhouse gas (recall from Chapter 5 that greenhouse gases trap heat in Earth's atmosphere and contribute to climate change and global warming). According to the NAAQS in Table 10.1, CO can have a concentration of 35 ppm. This means that, in an air sample containing this concentration of carbon monoxide, for every 1 million (that is, 1,000,000) atoms and molecules of various gases in an air sample, at most 35 of them, on average, can be carbon monoxide molecules. These two gases—ozone and carbon monoxide—are troublemakers. They are toxic pollutants that are present in air, especially in urban areas. But neither of these gases contributes to acid rain. So what chemicals *do* cause acid rain?

Table 10.1 Common Pollutants Found in Air

Pollutant Molecule	Concentrations of Common Pollutants That Exceed National Ambient Air Quality Standards (NAAQS)*
Ozone, O_3	75 ppb (8-hour average)
Carbon monoxide, CO	35 ppm (1-hour average)
Sulfur dioxide, SO_2	75 ppb (1-hour average)
Nitrogen dioxide, NO_2	100 ppb (1-hour average)

*These are the latest values from the EPA. Each limit is expressed in either ppm or ppb. The length of time given is the time elapsed for each measurement. For example, data measured with an 8-hour average is calculated by using a measurement performed over 8 hours.

QUESTION 10.6 According to Table 10.1, carbon monoxide, CO, is permitted to reach 35 ppm. At this concentration, how many CO molecules would you expect to find per 3 million molecules of air?

ANSWER 10.6 35 ppm is a concentration of 35 CO molecules per million molecules of various gases that make up air. Therefore, you would expect to find $3 \times 35 = 105$ CO molecules in 3 million molecules of air.

For more practice with problems like this one, check out end-of-chapter Question 33.

FOLLOW-UP QUESTION 10.6 Which of the gases listed in Table 10.1 has the highest allowed concentration in air?

Sulfur compounds are one of two major sources of acid rain.

The last two gases listed in Table 10.1 are the real culprits when it comes to the pollution of natural waters. In this section, we address sulfur dioxide, SO_2. When sulfur dioxide mixes with oxygen and water in the air, the product is sulfuric acid (H_2SO_4), a strong acid. In rainwater as well as the lakes, rivers, and oceans into which it falls, sulfuric acid is a problem because it releases protons. And, as we discussed in Section 10.2, those protons reduce the pH of the water. The following chemical equation shows this process:

$$H_2SO_4 \rightarrow H^+ (aq) + HSO_4^- (aq)$$

Natural, unpolluted rain has a pH of about 5.6. But in the United States, acid rain typically has a pH of about 4.5. This might not seem like a big difference, but the pH scale is logarithmic. So, a shift of one pH unit means a tenfold change in the concentration of protons. Therefore, acid rain with a pH of 4.5 has more than 10 times more protons than clean rain that has a pH of 5.6.

Where was the lowest recorded pH of acid rain? This dubious honor goes to Wheeling, West Virginia: at the height of its poorly regulated industrial activity in 1979, the city was home to rain with a recorded pH of 1.5. This pH falls somewhere between battery acid, with a pH of about 1, and concentrated lemon juice, with a pH of about 2. At these pH levels, it is not safe to go outdoors when it's raining.

The vast majority of sulfur dioxide released into the air comes from one source: coal-burning electrical power plants. **Figure 10.8** lists the sources for electrical energy in the United States. These data show that coal generates three times as much electricity as more renewable forms of energy do. There are several reasons for this. First, coal is cheap. It is the lowest-cost fuel listed in Figure 10.8. Second, coal is plentiful. There are coal deposits in two-thirds of the states, although the largest coal producers are Wyoming, West Virginia, Kentucky, Pennsylvania, and Texas (listed with the top producer first, in order of decreasing coal production). In fact, because coal is readily available, the United States exports coal to other countries.

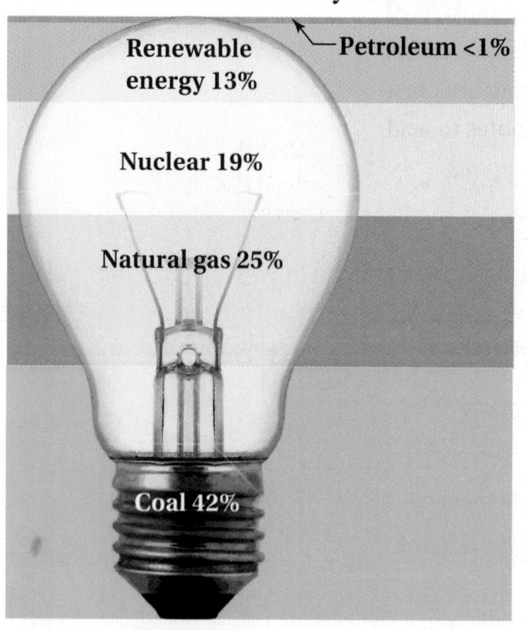

Figure 10.8 Sources of U.S. Electricity Generation in 2011

wait a minute . . .

Why does clean rain have a pH other than 7?

It's true. Natural, clean rain is acidic. It's acidic because the substances dissolved in it release protons into it. For example, sea spray and volcanic gases naturally contain sulfur dioxide, which contribute sulfuric acid to rain. Lightning is a source of nitrogen oxides, which also lower the pH of rainwater. Finally, the carbon dioxide that naturally exists in healthy, unpolluted air dissolves in natural waters to make a weak acid. All of these natural sources of acids—especially carbon dioxide—contribute to lowering the pH of rain to about 5.6.

Sulfur dioxide can be scrubbed from coal plant flue gases.

In 1990, the NAAQS reduced the limit for sulfur dioxide. But the new 2011 EPA guidelines for sulfur dioxide emissions impose stricter limits and require that most existing coal plants restrict what they release into the air as they burn coal. Coal plants that have installed equipment to remove some harmful pollutants are referred to as **clean coal plants**.

To remove SO_2 from air, coal-burning plants can scrub the gases—called flue gases—that leave their smokestacks. The term *scrub,* in this context, does not mean cleaning with a scrub brush. Rather, it means cleaning flue gases by removing an undesirable component—in this case, sulfur dioxide.

One of the most effective scrub methods is **flue gas desulfurization**. As previously noted, sulfur dioxide is present in flue gases. It's present because the sulfur contained in coal forms sulfur dioxide when it is burned in air. In the scrubber, this SO_2 gas comes into contact with jets of a watery limestone slurry, which contains calcium carbonate, $CaCO_3$, as shown in **Figure 10.9**. These two substances undergo the following chemical reaction to form solid calcium sulfate:

$$2CaCO_3 + 2SO_2 + O_2 \longrightarrow 2CaSO_4 + 2CO_2$$

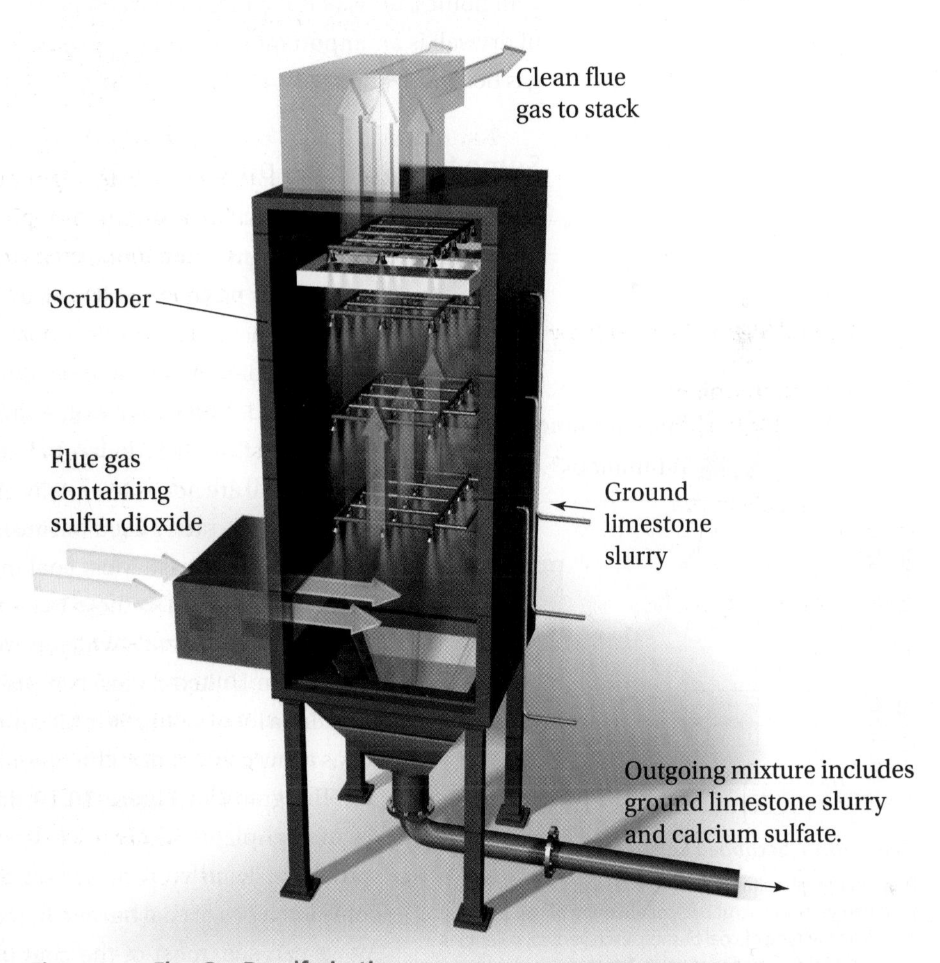

Figure 10.9 Flue Gas Desulfurization
Flue gases enter the scrubber from the lower left and move up through a spray of slurried limestone. In the process, sulfur dioxide in flue gases is converted to calcium sulfate (gypsum) and carbon dioxide.

Table 10.2 The Four Types of Coal Burned in the United States

Type of Coal	Percent by mass of coal used in the U.S. in 2010	Average energy produced from this coal type (kilojoules per kilogram)	Average percentage of sulfur by weight	Average cost per kilogram (cents)
Anthracite	<1%	35,300	0.7%	5.7
Bituminous	45%	31,000	2.3%	6.6
Sub-bituminous	47%	27,500	0.8%	1.3
Lignite	7%	15,000	0.4%	2.0

flashback
See *The Green Beat*, Chapter Five Edition, for more on the topic of carbon capture and sequestration (CCS).

In one day, a coal-burning power plant can potentially remove about 600,000 kilograms of sulfur dioxide from its flue gases, and that means a lot less sulfur dioxide is going into the air and forming acid rain. But it also means that these plants are producing a lot of calcium sulfate, as well as carbon dioxide, a greenhouse gas.

Another issue centers on what to do with the more than 1 million kilos of calcium sulfate also produced by coal-burning power plants. Luckily, one form of calcium sulfate, known as gypsum, is used to make plaster. Builders sandwich this plaster between two pieces of paper to form drywall (or Sheetrock®). In most American homes, drywall is the material of choice for interior walls. So the manufacture of drywall is an important use for the gypsum made from coal plants that use flue gas desulfurization.

Some types of coal produce more sulfur dioxide than others.

Not all coal is the same. Coal plants in the United States use four types of coal, and they differ in cost per kilogram, the amount of sulfur they contain, and how much energy they produce, as shown in **Table 10.2**.

Note the bargain price of all types of coal; 1 kilogram costs only pennies. Compare this to the cost of gasoline in the United States—about one U.S. dollar for 1 kilogram—and in Europe, where gasoline costs about three U.S. dollars per kilogram. Next, notice that the four types of coal are not used equally. Anthracite coal, which is the most energy rich, is rarely used because it's found only in small pockets in the U.S. Northeast. Most coal in U.S. mines is either bituminous or sub-bituminous, so these two types make up most of the coal we burn. Bituminous coal, which is mined mainly in the Midwest and the eastern United States, is more expensive than other types and contains a lot of sulfur. Sub-bituminous coal is less harmful because it has a low content of sulfur. It is also is very cheap.

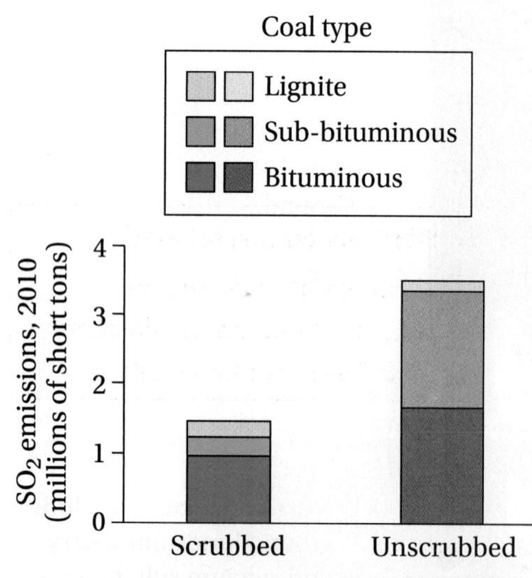

Figure 10.10 Scrubbed versus Unscrubbed Coal

These bar graphs show the 2010 emissions for SO_2 from scrubbed versus unscrubbed coal-burning power plants as well as the type of coal burned. Each vertical column is made up of three shades of color. The three shades represent the three types of coal, shown in the key.

The graph in **Figure 10.10** divides domestic coal plants into clean coal plants (those with scrubbers) and coal plants without scrubbers. Each vertical bar on the graph shows the three most common types of coal burned in plants with and without scrubbers. As we can see, most of the coal plants that scrub their flue gases burn bituminous coal. This makes sense: bituminous coal is the type of coal with the most sulfur. Therefore, plants burning this type of coal plant must remove the most sulfur dioxide in order to meet the new NAAQS requirements. As new, stricter requirements go into effect, some coal plants that do not comply with the new requirements are shutting their doors rather than adding costly clean coal equipment.

Even clean coal is a dirty, nonrenewable source of energy.

There is good news and bad news about coal and acid rain. The good news is that the limits on sulfur dioxide set by the EPA in 1990, and again in 2011, have dramatically affected the amount of sulfur dioxide released into the environment. From 1990 to 2008, sulfur dioxide emissions dropped by an impressive 70%.

Now the bad news: in 2010, the percentage of electricity generated by clean coal plants was only 58% of all energy generated by coal in the United States. The map in **Figure 10.11** features colored circles, each one representing a coal-fired power plant. The larger the circle, the more sulfur dioxide the plant emits. For this reason, in general, the green dots (which represent clean coal plants) are smaller than the red dots (which represent plants without scrubbers). The blue dots represent plants that are planning to switch to clean coal. Unfortunately, as the map shows, blue and green dots are dominated by red dots, especially in the eastern half of the map. More red dots mean more sulfur dioxide, and more sulfur dioxide means more acid rain.

Coal is dirty. Besides contributing to the pollution that causes acid rain, coal releases toxic substances, such as mercury, into the air and poisons the environment. And, although clean coal technology goes part of the way toward removing some of those toxins as well as SO_2, even clean coal plants release some toxins into our environment. As we noted, scrubbing also generates new by-products, such as carbon dioxide, that present their own environmental challenges.

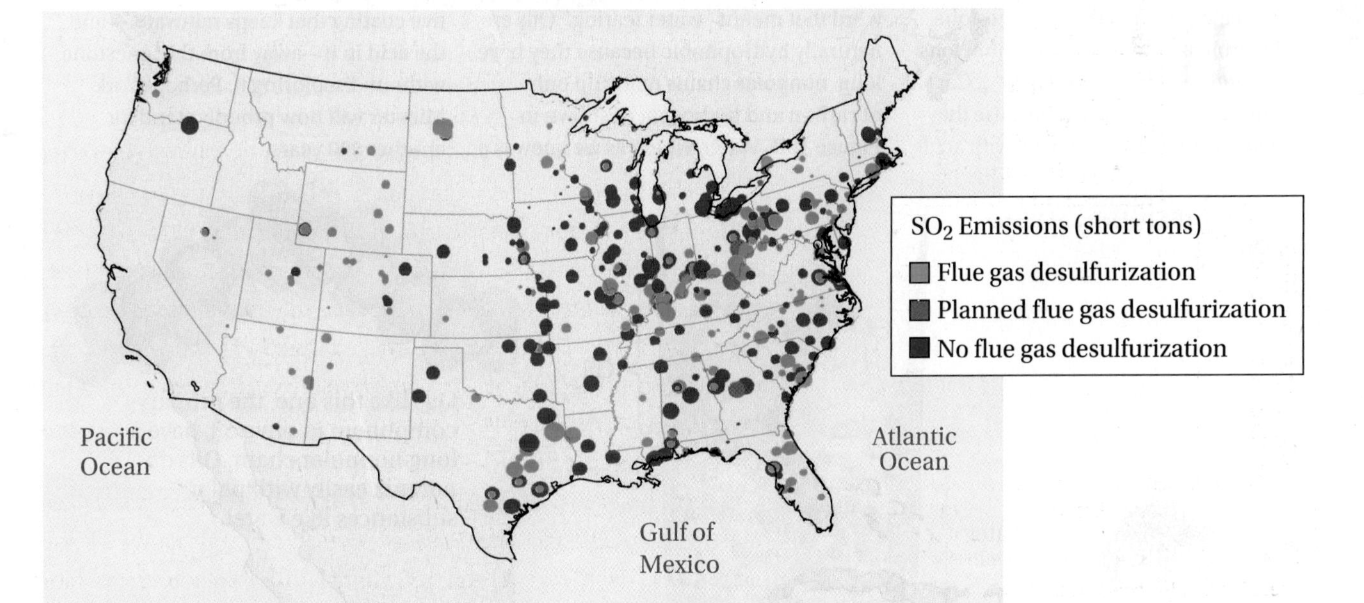

Figure 10.11 Sulfur Dioxide Emission by Coal Plant

Coal plants without scrubbers account for most of the U.S. SO_2 emissions. The circle size relates to the number of tons of sulfur dioxide released by a plant. The larger the circle, the greater the SO_2 emissions. The key for dot colors is shown to the right of the graph.

natureBOX • Cathedral Vinaigrette

York Minster cathedral in Great Britain (**Figure 10A**) is one of the best-known surviving examples of medieval European architecture, and one of the largest. Since it was built some 800 years ago, the cathedral has been pummeled by the elements and set on fire by a lightning strike, but it still stands. Part of York Minster's resilience is attributed to the material used to build it: limestone. Limestone is fine grained and sturdy so, throughout history, it has been used in monuments, statues, and buildings (and yes, gravestones). Limestone and its close cousin, marble, are both composed of calcium carbonate, $CaCO_3$ (also known as calcite), although marble is the denser and smoother of the two.

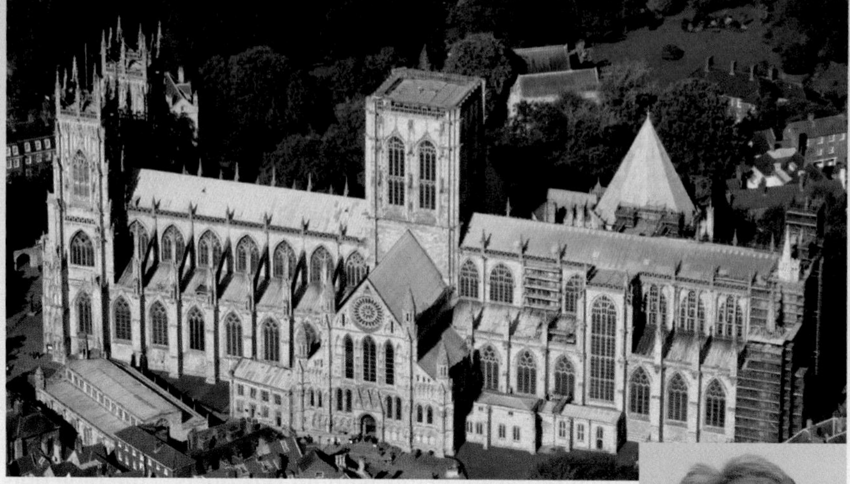

Figure 10A York Minster Cathedral in England (Inset: Dr. Karen Wilson)

Neil Mitchell/Shutterstock

Courtesy of Karen Wilson/Aston University

York Minster has withstood hundreds of years of punishment, but its greatest challenge has been pollution. Beginning in the industrial revolution, which spanned the years 1760 to 1850, and continuing to present times, impurities in the air have been embedding themselves into the limestone of York Minster, causing the stone to darken and weaken. Among all those formidable pollutants, though, the toughest adversary has been acid rain.

The calcite that makes up limestone and marble is composed of calcium ions (Ca^{2+}) and carbonate ions (CO_3^{2-}). Carbonate ions are basic, and because they are basic, carbonate ions react with acids and the calcite is eaten away. Thus, York Minster cathedral, constructed entirely of limestone, is gradually wearing away with each drop of acid rain that falls upon it.

But chemistry is coming to this cathedral's rescue. A group of chemists, led by Dr. Karen Wilson of the University of Cardiff in Wales (see inset, Figure 10A), is working to develop a protective coating that will defend the cathedral from the onslaught of acid rain. The coating needs to repel water, so the chemists sought out molecules that are **hydrophobic,** a word that means "water fearing." Oils are naturally hydrophobic because they have long, nonpolar chains made up only of carbon and hydrogen, as shown in **Figure 10B**. Water, which as we know is a very polar molecule, repels oil and does not mix with it.

The team of chemists tried a variety of oils and then settled on one: olive oil. It turns out that olive oil, when mixed with a Teflon-like substance, makes a protective coating that keeps rainwater—and the acid in it—away from the limestone without discoloring it. Perhaps York Minster will now proudly stand for another 800 years.

Figure 10B The Polarity of Water and Oil

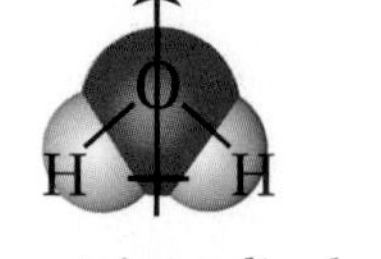

Water contains a dipole which makes it polar. It mixes easily with other polar substances.

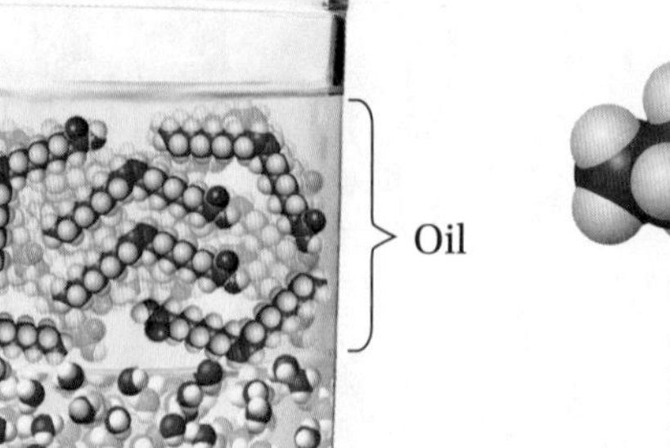

Magcom/Shutterstock

Oils like this one, the primary component in olive oil, have a long nonpolar chain. Oils do not mix easily with polar substances like water.

But the biggest problem with coal is that it's not a renewable resource. It is mined once and burned once. In Chapter 12, "Energy, Power, and Climate Change," we look at the future of renewable energy sources that will, hopefully, reduce our dependence on all coal, "clean" or otherwise.

recurring theme in chemistry 8

The Earth has a finite supply of raw materials that humans can use.

QUESTION 10.7 According to Figure 10.10, which type of coal is burned at most power plants that use clean coal technology?

ANSWER 10.7 Most plants that use clean coal technology burn bituminous coal.

For more practice with problems like this one, check out end-of-chapter Question 40.

FOLLOW-UP QUESTION 10.7 According to Table 10.2, which type of coal is least expensive per unit mass? Which type of coal contains the most sulfur?

10.4 Acid Rain Part II: Nitrogen-Based Pollution

Industrial and agricultural uses of nitrogen disrupt the nitrogen cycle.

Now we are ready to introduce the second culprit in our acid rain story—the final pollutant listed in Table 10.1—nitrogen dioxide, NO_2. But before we see how nitrogen compounds contribute to acid rain, let's first examine how nitrogen cycles through the environment and learn what can happen when that cycle is disrupted.

Although nitrogen moves through the air, the ground, and the water, as illustrated in **Figure 10.12,** most nitrogen exists in the air in the form of nitrogen molecules. When this inert nitrogen gas reacts to become part of other substances, we call this process **fixation**. Certain plants and bacteria are specially equipped to fix nitrogen and assimilate it into organic compounds. Lightning strikes can also fix nitrogen. Once it is fixed, the nitrogen that exists in soil can form different ions, such as NH_4^+,

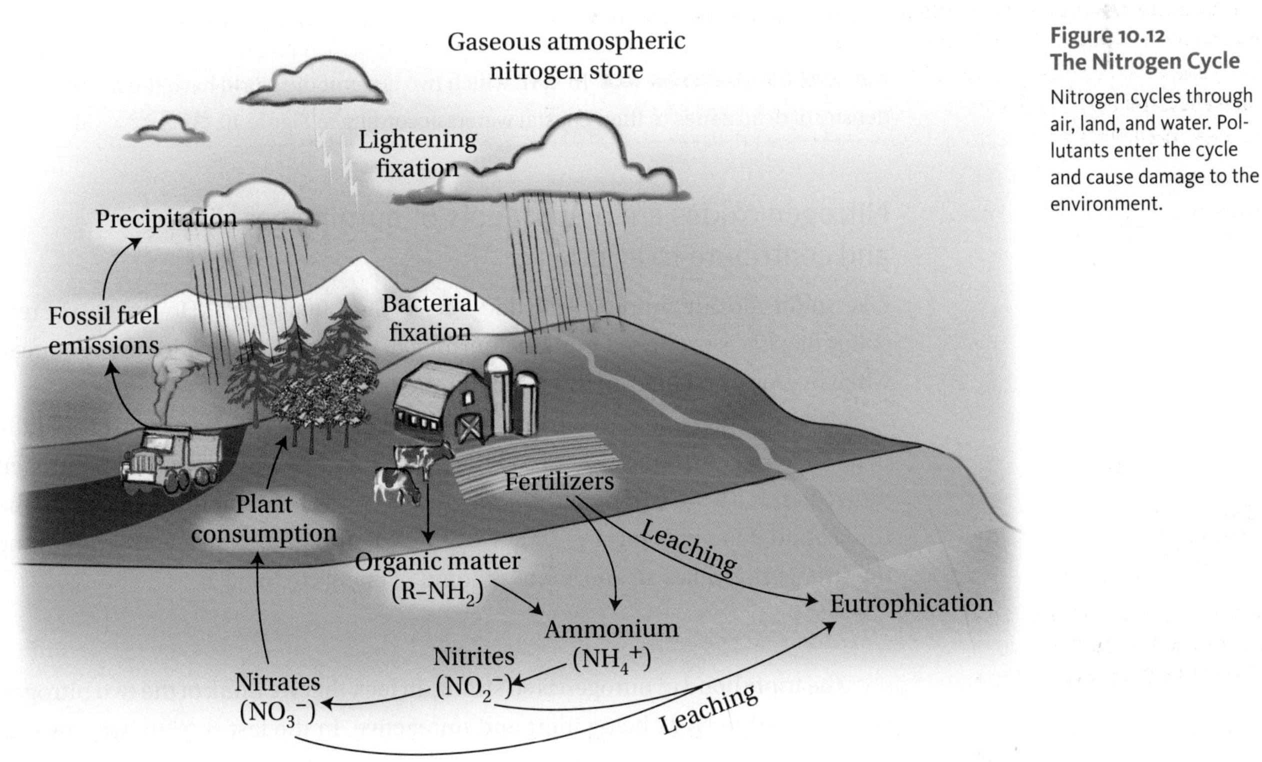

Figure 10.12 The Nitrogen Cycle
Nitrogen cycles through air, land, and water. Pollutants enter the cycle and cause damage to the environment.

NO_2^-, and NO_3^-. These forms of nitrogen—ammonium, nitrites, and nitrates, respectively—become nutrients for plants.

Nitrogen-containing compounds naturally occur in soil, but additional nitrogen is added to the soil in enormous quantities through agricultural activity. Most farming operations use fertilizers liberally, and most fertilizers are nitrogen-containing soil additives that increase crop growth by supplying nutrients to soil. In the United States, each year about 20 million tons of fertilizer are used in farms, lawns, and gardens. And, as Figure 10.12 shows, excess nitrogen from fertilizers leaches from the soil and drains into natural waters. In fact, the flux of nitrogen into coastal waters has doubled over the past five decades.

In natural waters, extra nitrogen promotes rapid growth of some organisms, such as plankton, and large blooms of these organisms require oxygen to live. When such organisms are present, they quickly consume the available oxygen in water, leaving little or no oxygen for the other organisms that need it. When a pollutant causes overgrowth of an organism and oxygen is consequently depleted from natural water, this is **eutrophication**.

When oxygen levels begin to fall in natural waters, the water first becomes hypoxic (oxygen deficient). As levels continue to drop, the eventual result is formation of a **dead zone,** an area where life—from fish to shellfish to corals—cannot thrive. Living things in these areas either die or move on to healthier environments. **Figure 10.13** shows a map of the world's dead zones. Areas in green and yellow are mild dead zones, and darker red areas are the most hypoxic. The number of dead zones is rising at an alarming rate in coastal waters worldwide.

QUESTION 10.8 According to Figure 10.13, do dead zones occur only in coastal areas adjacent to oceans, or do they also occur in inland lakes and rivers?

ANSWER 10.8 Some of the dead zones (red dots) in Figure 10.13 are not located in coastal areas, but most of the dead zones are near coasts.

For more practice with problems like this one, check out end-of-chapter Question 41.

FOLLOW-UP QUESTION 10.8 In 2011, which two areas of the world have the highest density of dead zones in their coastal waters according to Figure 10.13?

Nitrogen oxides are by-products of burning gasoline and contribute to acid rain.

Like sulfur dioxide, nitrogen dioxide forms when fuels burn. However, whereas sulfur dioxide is produced from the burning of coal, nitrogen oxides are produced when hydrocarbon fuels such as gasoline are burned.

How is nitrogen involved in the burning of gasoline, which contains very little of it? To answer this question, recall that the air we breathe is about 78% nitrogen molecules, N_2. If we compare the strengths of bonds between any two atoms, the triple bond that holds together the two atoms of nitrogen, shown below, is among the strongest. It takes an enormous amount of energy to break it.

$$:N \equiv N:$$

The triple bond of nitrogen is so strong, in fact, that we think of the two nitrogen atoms joined in N_2 as being inert and unreactive. In the last section, we saw that

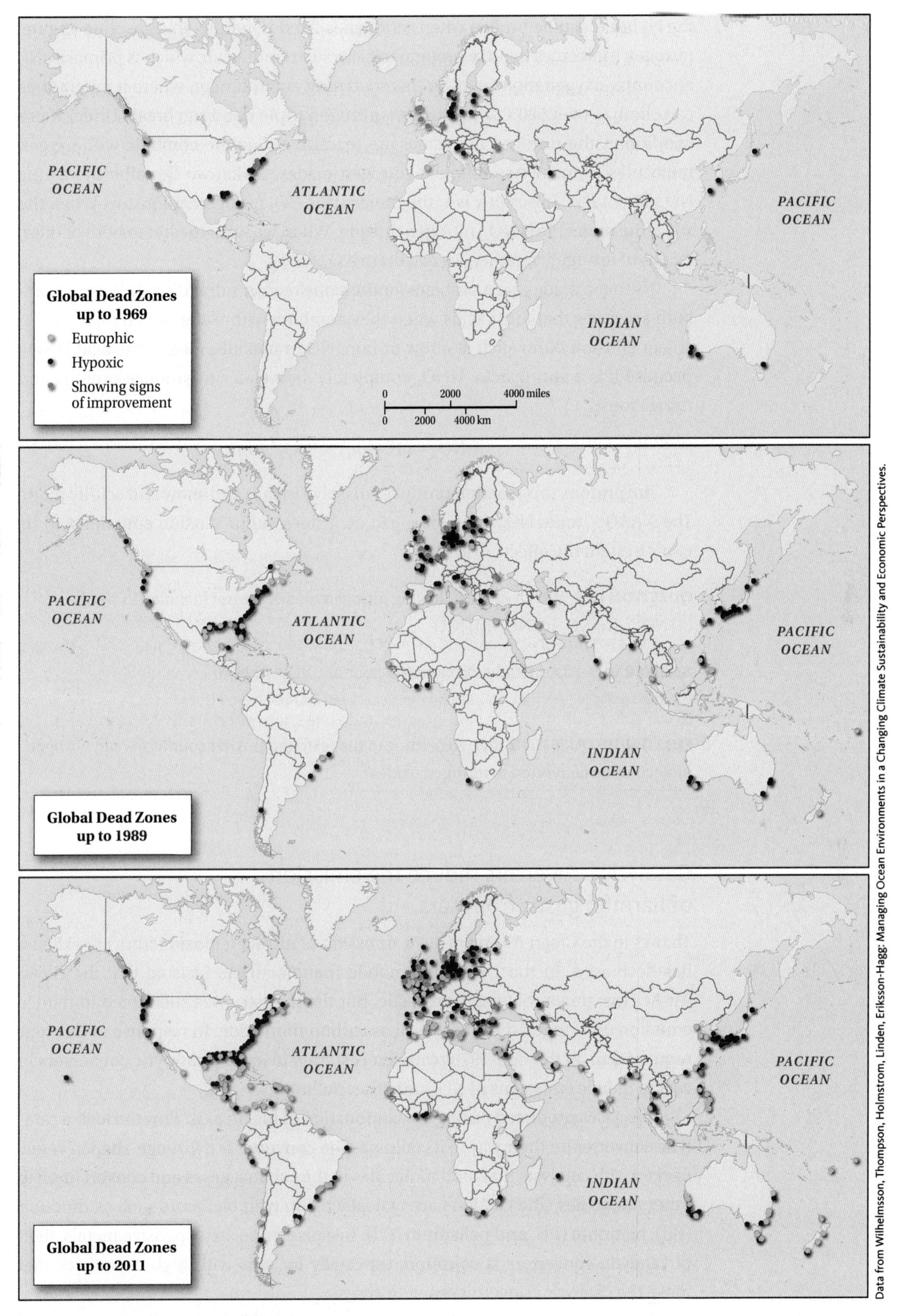

Data from Wilhelmsson, Thompson, Holmstrom, Linden, Eriksson-Hagg: Managing Ocean Environments in a Changing Climate Sustainability and Economic Perspectives.

Figure 10.13 Dead Zones around the World

The spots on the map are dead zones. Dead zones result from hypoxia due to excesses of nitrogen, from acid rain and fertilizer runoff, and from chemicals found in sewage and detergents.

the N_2 bond can be broken when nitrogen is fixed in nature. In a gasoline engine, nitrogen molecules that are brought in from surrounding air, which is primarily N_2, encounter oxygen molecules. In this scorching environment, where temperatures can climb to over 500°C, the nitrogen–nitrogen triple bond can break. Under these conditions, the nitrogen molecules are reactive enough to combine with oxygen molecules to make two different nitrogen oxides, which we describe by writing NO_x. When the subscript x is 1, the molecule is NO, nitrogen monoxide. When the x is 2, the molecule is NO_2, nitrogen dioxide. When we want to refer to both of these oxides of nitrogen together, we call them NO_x.

Nitrogen oxides are a problem for the same reason sulfur dioxide is a problem: both are gases that form acids when they combine with water. When dissolved in an aqueous medium such as snow or rain, NO_x molecules form nitric acid HNO_3. Because it is a strong acid, HNO_3 completely dissociates to produce protons and nitrate ions:

$$HNO_3 \rightarrow H^+ (aq) + NO_3^- (aq)$$

The protons this reaction produces dissolve in rain and snow and acidify them. The NAAQS limits NO_x emissions, just as it does sulfur dioxide emissions, in its effort to stem the effects of acid rain.

QUESTION 10.9 Which of the following nitrogen oxides are not included in NO_x?
(a) NO_3^- (b) NO (c) NO_2 (d) N_2O

ANSWER 10.9 NO_x includes NO and NO_2, but not NO_3^- or N_2O.
For more practice with problems like this one, check out end-of-chapter Question 37.

FOLLOW-UP QUESTION 10.9 According to the text, under what conditions can nitrogen molecules be converted to nitrogen oxides?

Catalytic converters reduce the emission of harmful gases from cars.

Thanks to the Clean Air Act of 1970, emission of nitrogen dioxide from car exhaust has decreased. In the 1970s, automobile manufacturers insisted that the Clean Air Act was untenable and unrealistic, but the industry was still forced to reduce emission of nitrogen dioxide as well as carbon monoxide. In response to this new requirement, in the mid-1970s carmakers started installing **catalytic converters** in cars to reduce emissions of NO_x and other pollutants.

The gases produced during the combustion of fuel in a car flow through a catalytic converter on their way to its tailpipe. The converter is a lozenge-shaped vessel packed with catalysts that bind molecules in the exhaust gases and convert them to other molecules. The catalysts are typically made with elements such as rhodium (Rh), platinum (Pt), and palladium (Pd). Because these are expensive metals, theft of catalytic converters is common, especially for cars with high clearances that make the catalytic converter easier to access.

flashback
Recall from Chapter 7 that a catalyst is a substance that facilitates a chemical reaction by making it faster.

The catalysts in a catalytic converter react with NO_x in the following chemical equation:

$$2NO_x \rightarrow xO_2 + N_2$$

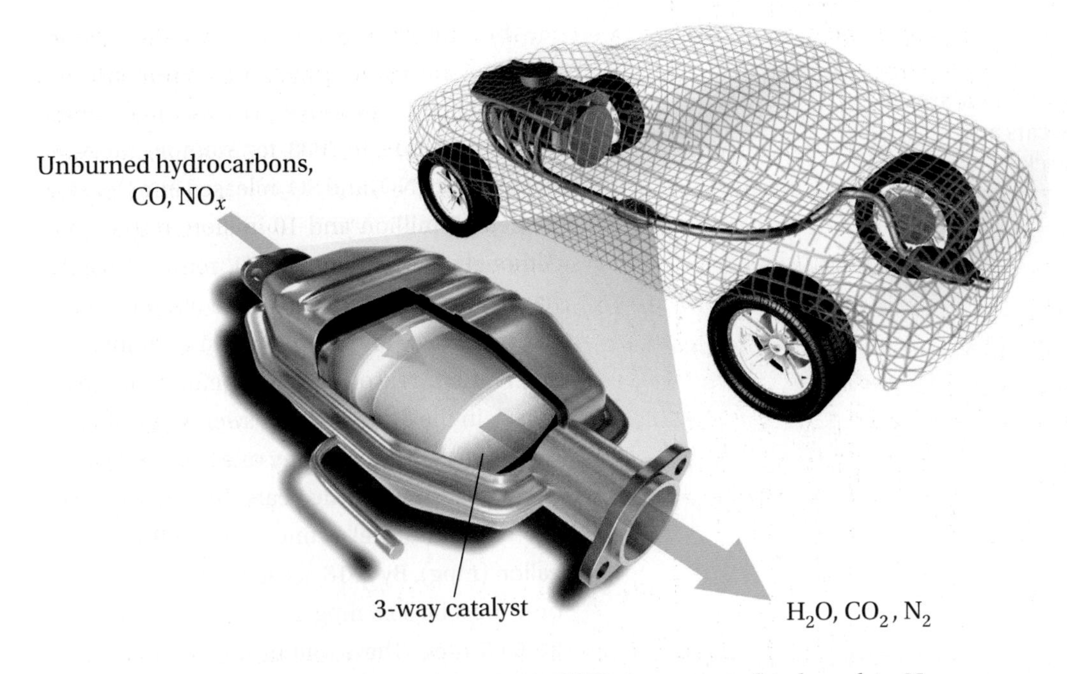

Figure 10.14 Catalytic Conversion

Catalytic converters are part of the exhaust system in cars. Unburned hydrocarbons, NO_x, and carbon monoxide are all converted to less harmful gases when they flow through a catalytic converter, which contains metal catalysts. Carbon dioxide, nitrogen molecules, and water are released at the other end of the converter.

Through catalytic conversion, about 90% of NO_x is converted to harmless N_2. Most modern catalytic converters are three-way, meaning that they also convert carbon monoxide (CO) and unburned fuel to carbon dioxide (CO_2), as indicated in **Figure 10.14**.

Despite their successes, catalytic converters are not the perfect solution to cleaning up car emissions. First, they produce carbon dioxide, a greenhouse gas. Second, they must be warmed up to work well. As a result, a typical car releases most of its pollution in the first five minutes of starting because the catalytic converter has not warmed up yet. Some car brands now have catalytic converter heaters that warm up the catalyst quickly so that more exhaust gases react with the catalyst, even when the car is just started.

NO_x emissions have decreased more slowly than SO_2 emissions.

Strict NAAQS requirements and innovations like the catalytic converter have led to decreased NO_x emissions over the past 30 years, as **Figure 10.15** shows. From 1980 to 2009, the levels of nitrogen oxides in air have been reduced by 48%. It's clear from these data that the Clean Air Act has significantly affected the levels of nitrogen dioxide levels in air.

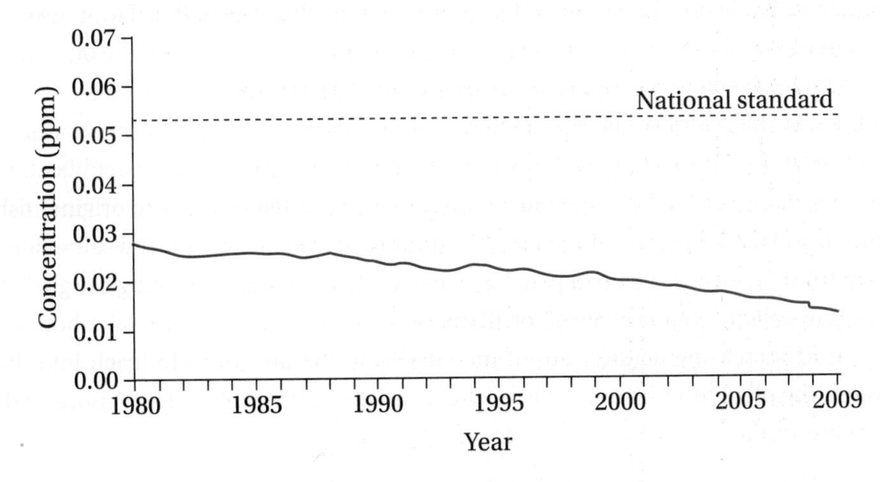

Figure 10.15 Nitrogen Dioxide Levels

This graph shows the concentration of nitrogen dioxide in air over a 29-year period. The concentration of nitrogen dioxide in ppm units is the vertical axis and the year is the horizontal axis. The horizontal dotted line shows the NAAQS standard for this gas, 0.053 ppm, when averaged over one year. (The value is not the same as the data in Table 10.2, because those data are averaged over one hour and these data are averaged over one year.)

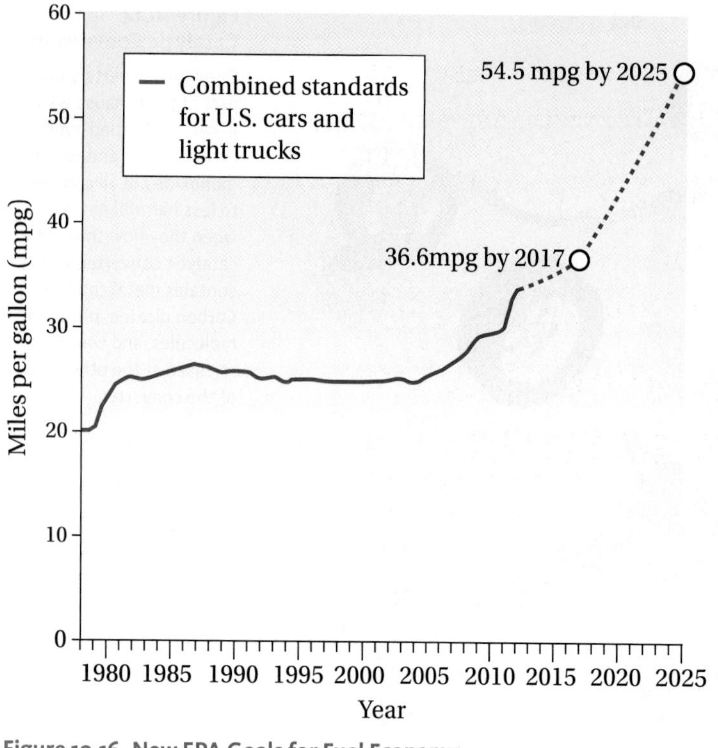

Figure 10.16 New EPA Goals for Fuel Economy

But the news is not all good. Even though levels of nitrogen oxides have dropped, they have dropped much more slowly than the levels of sulfur dioxide. In 2008, for example, the number of tons of NO_x and SO_2 released into U.S. skies totaled 17 million and 10 million, respectively. Although the amount of NO_x released per car has decreased significantly, the number of cars has increased, and it will continue to increase.

Is there reason to be hopeful about levels of NO_x in the future? The answer is yes. The EPA has released fuel economy rules that will go into effect over the coming years. In 2013, the average car had a fuel economy of 24.9 miles per gallon (mpg). By 2016, a car's average fuel economy must be 35.5 mpg. By 2025, that value must be 54.5 mpg. These values are summarized in **Figure 10.16**. Increased fuel efficiency translates into reduced car emissions, which will mean less NO_x and, presumably, less acid rain.

10.5 The Effects of Acid Rain

Natural waters have a limited tolerance for added acid or base.

A lake that's unaffected by the activities of humans has a pH between 6.0 and 7.0. When excessive amounts of acid rain produced by air pollutants lower the pH, the lake environment begins to change. When pH begins to drop, aquatic plants begin to die, and this adversely affects waterfowl that depend on those plants for food. As pH continues to drop, the bacteria that decompose dead organisms begin to die, leaving a buildup of organic matter. When pH drops below 4.5, no fish can live. **Figure 10.17** is a graph prepared by the EPA that shows how low pH levels can dip before certain species of aquatic life succumb to acid in their ecosystems.

Scientists worldwide use the scientific method as a basis for experimentation and deduction. The scientific method is based on the principles of peer review and reproducibility.

One of the most long-lasting, comprehensive acid rain studies ever done was carried out by limnologists (scientists who study lakes) at the University of Wisconsin at Madison. They obtained permission to acidify one half of Trout Lake in northern Wisconsin (**Figure 10.18**) and then compared that half with the other half of the lake, which was left untouched as the control experiment.

From 1984 to 1990, the pH in the half of the lake being acidified was gradually reduced from 6.1 to 4.7. Certain fish survived at the beginning of the acidification period, but as the pH dropped lower and lower, the offspring of the original fish did not survive. This half of the lake became very clear, and as a result, ultraviolet light from the sun was able to penetrate deeper than before, prompting the growth of algae called "elephant snot," or filamentous green algae, on the lake bottom. The ever-increasing acidity caused mercury from the lake floor to leach into the water. The plankton that once existed died, and they were replaced by a more acid-tolerant strain.

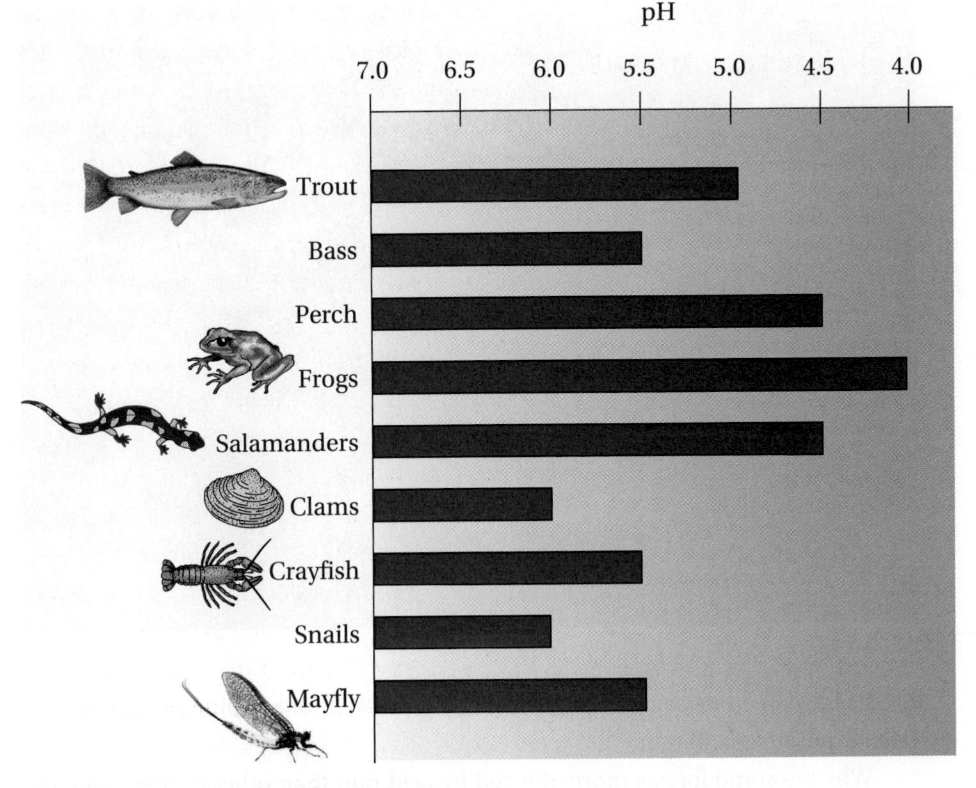

Figure 10.17 How Low Can They Go?
This bar graph shows the pH levels at which different species of aquatic life can survive.

After acidification was complete, the researchers studied the recovery of Trout Lake for the next decade. The pH and the concentrations of dissolved compounds returned to normal levels in only a few years. The biological recovery took longer; the restoration of aquatic plants and fish took several additional years. Nevertheless, this research gives us hope because it indicates that some types of damage done to natural waters by acid rain are not irreversible.

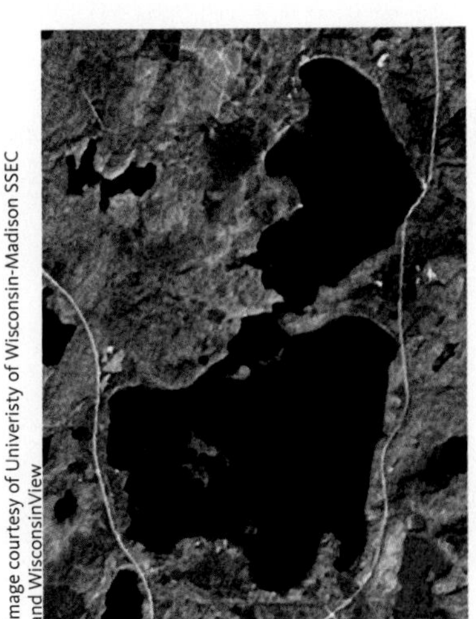

Image courtesy of Univeristy of Wisconsin-Madison SSEC and WisconsinView

Figure 10.18 Trout Lake in Northern Wisconsin
This satellite view of the lake shows the two halves separated by a narrow opening in the center of the lake.

QUESTION 10.10 True or false? The proton concentration in Trout Lake increased more than tenfold during the study performed by the University of Wisconsin.

ANSWER 10.10 True. Recall that a change in 1 pH unit means a tenfold change in proton concentration. The pH in the study was reduced from 6.1 to 4.7, a difference of 1.4 pH units. Thus, the change was more than tenfold.

For more practice with problems like this one, check out end-of-chapter Questions 19 and 23.

FOLLOW-UP QUESTION 10.10 In an imaginary lake where the pH decreases by 3 pH units, by what factor does the concentration of protons increase?

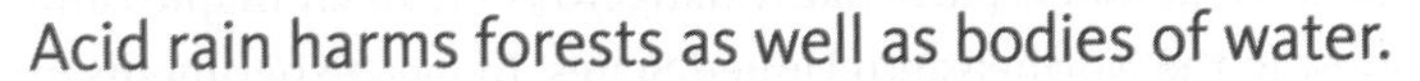

Acid rain harms forests as well as bodies of water.

Up to this point in the chapter, we have focused on the effects of acid rain on natural waters. But acid rain damages everything it falls upon, especially forests. For many years, scientists who study trees have noticed that certain trees are growing more slowly, or are losing their leaves or needles, or dying prematurely. This effect is not uniform: in some parts of the United States, trees are doing fine. In others,

Figure 10.19 Acid Rain Affects Trees

Tom McHugh/Science Source

forests have been devastated. **Figure 10.19** shows a photo of pine trees that have been exposed to acid rain.

Why are some forests more affected by acid rain than others? First of all, the pH of acid rain varies across the United States. In areas of heavy industrial activity, where SO_2 is pumped into the air by factories, and areas of high urban density, where cars add NO_x to the air, the pH of acid rain is lowest. But areas that do not have these stressors may still have rain with low pH values, because they are downwind of industrial or urban areas. This is especially true in the rural areas of the U.S. Northeast. In states such as Maine, New Hampshire, and Vermont, where there are few industrial or urban areas, the toll taken by acid rain is among the worst in the country. These areas feel the effects of acid rain created when factories and coal-fired plants in more urban and industrialized areas of the country pass on their dirty air to their neighbors.

Acid rain deprives soils of nutrients and releases toxins that damage trees.

How exactly does acid rain affect forests? In fact, the woody part of a tree is not harmed directly. Instead, acid rain takes away the tree's necessary nutrients from the soil and from its leaves. One element that's crucial for healthy tree growth is calcium. Acid rain leaches calcium from tree leaves and from the soil in which the tree grows, and the calcium is lost from the tree. A tree that doesn't get enough calcium gets sick. Sick trees are more susceptible to cold and disease in the same way that a sick person is. Acid rain is particularly dangerous for sugar maples and for red spruce that grow in the Northeastern United States. In fact, experiments in which calcium was added to an entire watershed affected by acid rain shows that reintroduction of this important element improved the health of the forest.

And there's more. Many soils have aluminum-containing minerals that are naturally present in the soil but, in a healthy ecosystem, do not dissolve and release aluminum. However, when the pH drops due to acid rain, the aluminum minerals

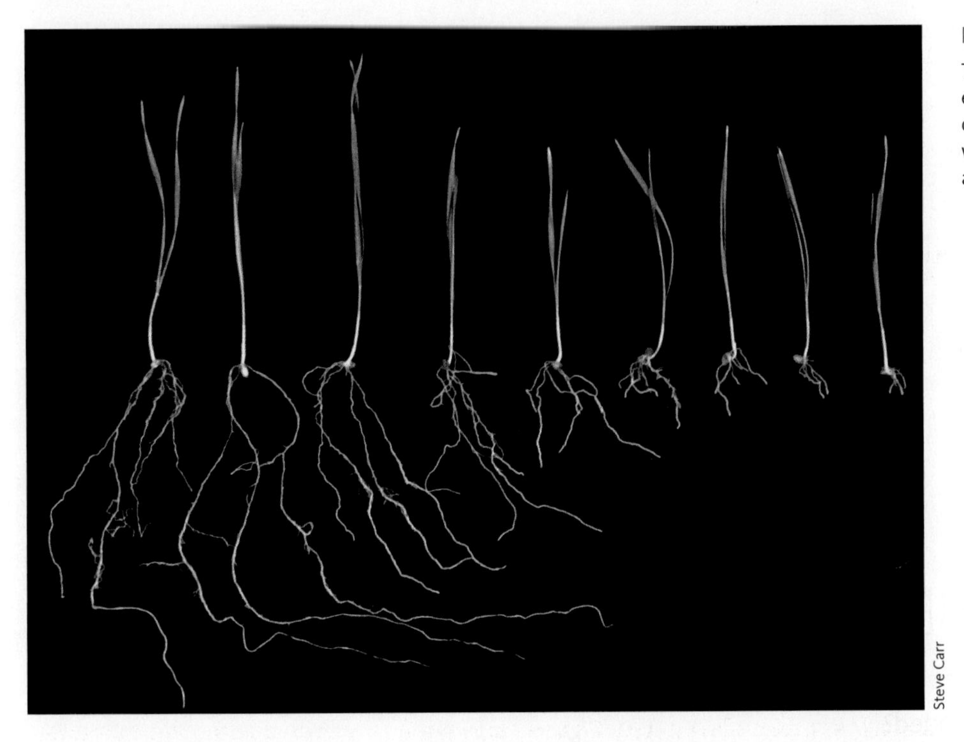

Figure 10.20 Aluminum Toxicity
These cotton plants demonstrate the effect of decreased pH on the growth of plant roots. The plants grown in soil with the lowest pH (at right) are damaged by high levels of aluminum.

begin to dissolve and toxic aluminum ions are released into the soil. Aluminum stunts the growth of healthy roots, as shown in **Figure 10.20** for cotton plants. It also reduces the amount of calcium the tree takes in through its roots, and it kills healthy bacteria in the tree.

In some places, the soils are naturally more resilient and can withstand the effects of acid rain longer. For example, soils in the Midwestern states of Nebraska and Illinois have soils that are buffered. These soils naturally contain compounds, called **buffers,** that react with excess acids in rain and reduce their effects on the forests grown in those soils. In the Northeastern United States, the soils are poorly buffered, and forests in this area feel the effects of acid rain more than forests in well-buffered areas do.

Acid rain continues to put stress on our environment. The drawing in **Figure 10.21** summarizes the effects of acid rain that we have discussed in this chapter. Fortunately, the hard work that has gone into restricting gaseous emissions from power plants, factories, and cars now seems to be paying off. Some parts of the country once had rain with a pH of 4.0. Now those areas have

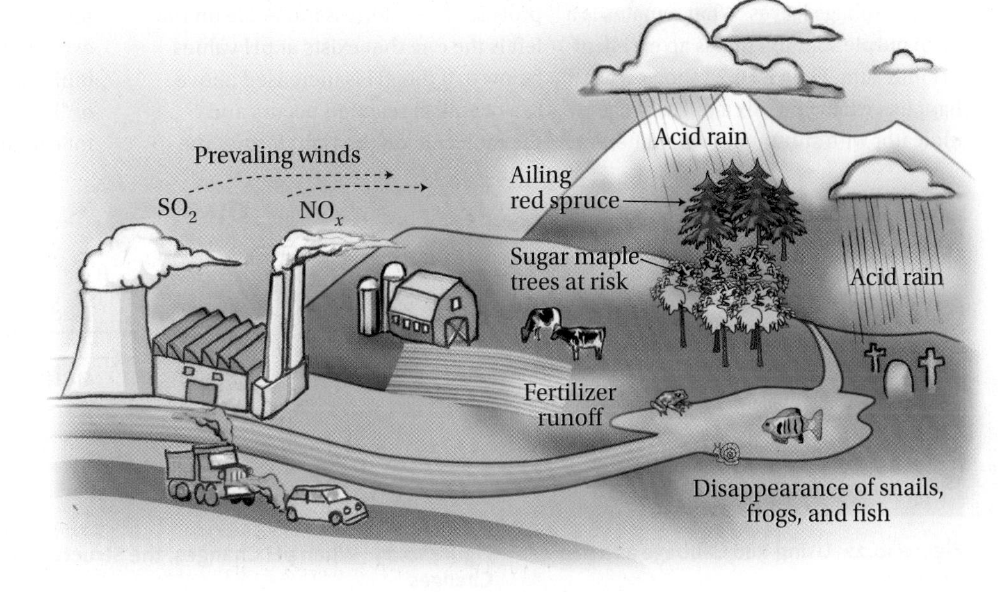

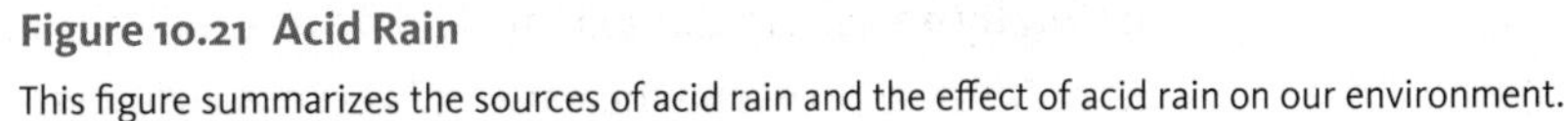
Figure 10.21 Acid Rain
This figure summarizes the sources of acid rain and the effect of acid rain on our environment.

levels nearly one pH unit higher. For every optimistic report, however, there are several reports of the damage acid rain continues to inflict. Our continued vigilance is needed.

QUESTION 10.11 Using Figure 10.21 as a reference, indicate which of the following is not a cause or an effect of acid rain: (a) emissions from car exhaust (b) decay of organic matter in soils (c) reduction of aquatic species in natural waters (d) sick trees

ANSWER 10.11 All of these are causes or effects of acid rain except (b). The decay of organic matter is a natural process that is neither a cause nor an effect of acid rain.
For more questions like this one, check out end-of-chapter Questions 44 and 47.

FOLLOW-UP QUESTION 10.11 True or false? Buffered soils are better able to resist the effects of acid rain than unbuffered soils.

THEgreenBEAT • news about the environment, chapter ten edition

TOP STORY **Caught Pink Handed?**

If you look through the cabinets in a typical kitchen, you are likely to find liquids like ammonia and milk, which are bases, as well as liquids like fruit juice and vinegar, which are acids. How do you know, though, if something is an acid or a base? One easy way is to use a pH indicator, which is a substance that changes color in solutions of different pH. You can make a pH indicator yourself by chopping up the leaves of a red cabbage, boiling them in water for about two minutes, and then straining off the cabbage leaves. What remains is a deep purple solution that is an excellent pH indicator. **Figure 10.22** shows what happens when you mix a red cabbage indicator with substances with different pH values. Acidic solutions will turn red cabbage indicator from purple to red or pink, and basic solutions will turn it blue, green, or yellowy-green.

How do pH indicators change color with changing pH? To answer this question we will look at an indicator that is simpler. The phenolphthalein molecule is colorless below about pH 8 and bright fuchsia above about pH 8. It changes color because the structure of the molecule changes when the pH changes. **Figure 10.23** illustrates this process. The colorless molecule on the left is the one that exists at pH values below 8. If the pH is increased above 8, a chemical reaction occurs and the molecule on the right forms. The molecule on the right is electric pink. This is always the case with pH indicators: as the pH changes, a change in the structure of the pH molecule takes place, which changes the color of the indicator.

Red cabbage is a complex mixture of indicators that give the spectrum of colors seen in Figure 10.22 as the pH is changed. Think about plants that you know that have intense colors. Blueberries, rhubarb, beets, and black tea, for example, have been used as pH indicators. Colored substances can also be extracted from a flower like rose and hibiscus to make pH indicators. Litmus, of "litmus test" fame, is a natural indicator extracted from lichens.

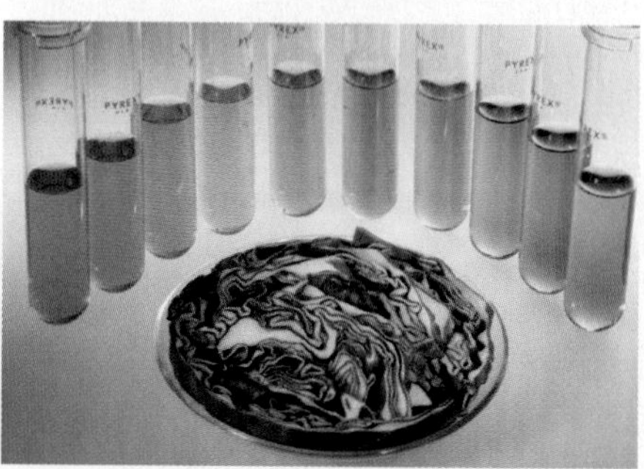

Figure 10.22 **Using Red Cabbage as a pH Indicator**

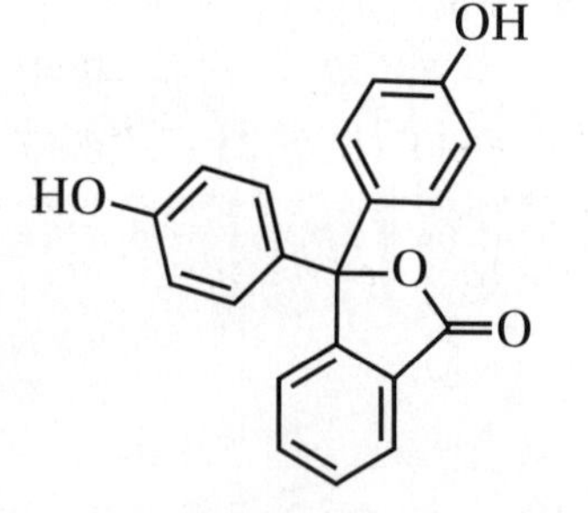

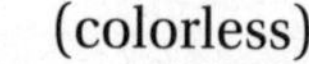

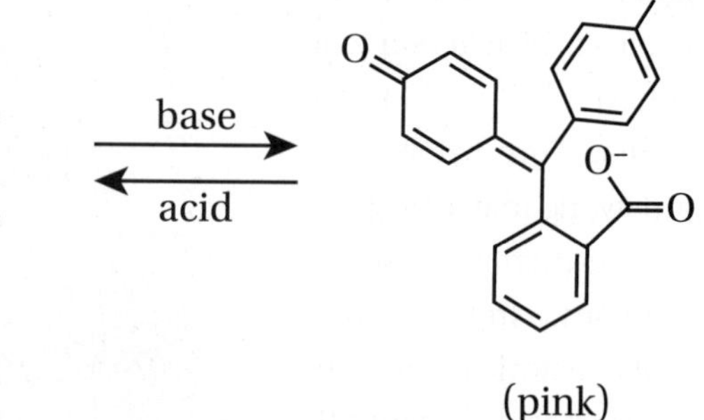

Figure 10.23 **When pH Changes, the Structure of Phenolphthalein Molecule Changes**

pH indicators can find uses outside of the chemistry lab. For decades, phenolphthalein, the molecule shown in Figure 10.23, has been used to prove guilt in cases of bribery. When it is a solid, phenolphthalein is a white powder. Law enforcement officers lightly coat paper money with this solid and then hand it off to someone who takes it as a bribe. To prove that this person handled the cash, the perpetrator's hands are rinsed with a weakly basic solution like sodium carbonate in water. This shifts the pH up above 8 and converts phenolphthalein to the fuchsia-colored molecule on the right side of Figure 10.23. This pink color may be admissible in court as evidence of bribery. Thus, if you live in a place where this test is still used and you are thinking of taking a bribe from a police officer, the best advice is this: remember to wash your hands!

Chapter 10 Themes and Questions

recurring themes in chemistry chapter 10

recurring theme 4 Chemical reactions are nothing more than exchanges or rearrangements of electrons. (p. 284)

recurring theme 8 The Earth has a finite supply of raw materials that humans can use. (p. 299)

recurring theme 1 Scientists worldwide use the scientific method as a basis for controlled experimentation and deduction. The scientific method is based on the principles of peer review and reproducibility. (p. 304)

A list of the eight recurring themes in this book can be found on page xv.

ANSWERS TO FOLLOW-UP QUESTIONS

Follow-Up Question 10.1: False. Pure water does contain protons, but those protons are balanced by an equal number of hydroxide ions, so water is not acidic. *Follow-Up Question 10.2:* The two words mean the same thing. *Follow-Up Question 10.3:* (a) pH = 3.26 (b) pH = 6.52. *Follow-Up Question 10.4:* Compounds (a) and (c) are acids, and compounds (b) and (d) are bases. *Follow-Up Question 10.5:* (a) The pH is 6.0. (b) This substance is in the same range as urine. *Follow-Up Question 10.6:* Carbon monoxide has the highest allowed concentration in air. *Follow-Up Question 10.7:* Sub-bituminous coal is least expensive; bituminous coal contains the most sulfur. *Follow-Up Question 10.8:* The northwestern coast of Europe and the eastern seaboard and Gulf of Mexico in North America have the highest density of dead zones in their coastal waters. *Follow-Up Question 10.9:* Under conditions of high temperature in a car engine, nitrogen molecules can be converted to nitrogen oxides. *Follow-Up Question 10.10:* The concentration of protons increases by a factor of 1000. *Follow-Up Question 10.11:* True.

End-of-Chapter Questions

Questions are grouped according to topic and are color-coded according to difficulty level: **The Basic Questions***, **More Challenging Questions****, and **The Toughest Questions*****.

Acids and Bases

*1. What is an acid and what is a base? How can you recognize each one?

*2. In your own words, explain the difference between a strong acid and a weak acid.

*3. Why does pure water at 25°C always have a pH of 7.00?

*4. Write a chemical equation for the dissociation of each of these acids in water:

(a) HCl (b) HI (c) H_2SO_4 (d) $HClO_4$

*5. Which of these compounds are acids and which are bases?

(a) $LiOH$ (b) $CsOH$ (c) HI (d) HNO_3

*6. Sketch the interaction between the following:

(a) a proton and several water molecules

(b) a hydroxide ion and several water molecules

**7. How many moles of H^+ are there in 8.0 liters of pure water?

**8. How many moles of hydroxide ion are present in 15 liters of pure water? Express your number in scientific notation.

**9. Determine if each substance is an acid, base, or salt and then write an equation that shows its ionization in water.

(a) $KClO_4$ (b) $RbOH$ (c) H_2CrO_4 (d) $Ca(CN)_2$ (e) H_3PO_4 (f) Li_2CO_3

**10. Determine if each substance is an acid, base, or salt and then write an equation that shows its ionization in water.

(a) KCl (b) HCl (c) $LiOH$ (d) $Sr(OH)_2$ (e) HI (f) $HClO_4$

**11. Nitric acid is a strong acid used in the manufacture of nitroglycerin, a powerful explosive. Write a chemical equation that shows the dissociation of nitric acid in water. What is the pH of a 4.5×10^{-4} M solution of nitric acid?

***12. The salt $Mg(OH)_2$ has limited solubility in pure water.

(a) If the maximum concentration of magnesium hydroxide in water is 1.4×10^{-4} M, how many moles of this salt can dissolve in 250 mL of water?

(b) Write a chemical equation that shows the dissociation of magnesium hydroxide in water.

(c) What is the pH of 1.0 L of a saturated solution of magnesium hydroxide?

(d) If you place 65.0 g of magnesium hydroxide in 2.0 L of pure water, will all of it dissolve, or will some of it remain as a solid in equilibrium with the solution?

The pH Scale

*13. (a) If you dissolve some NaOH in water, do you expect the pOH to be greater than or less than 7?

(b) Do you expect the pH to be greater than or less than 7?

*14. Indicate whether each of these pH values is acidic or basic.

(a) 4.3 (b) 9.0 (c) 11.2 (d) 5.3 (e) 7.2

***15.** Without using a calculator, determine the pH of solutions with the following proton concentrations:

(a) 1.00×10^{-6} M (b) 1.00×10^{-2} M

***16.** Identify the following pOH values as acidic, neutral, or basic.

(a) 3.00 (b) 0.01 (c) 12.2 (d) 7.00

***17.** The pH values of several solutions are given to you. For each, determine the pOH and state whether it is neutral, acidic, or basic:

(a) 4.0 (b) 8.1 (c) 11.0 (d) 1.3

***18.** Using Figure 10.5 as a guide, choose the one from each of the following pairs of substances that is more acidic.

(a) pure water and milk of magnesia

(b) seawater and baking soda

(c) gastric juice and tomato juice

****19.** If a sample of Fruit Juice A contains 100 times more H^+ than a sample of Fruit Juice B, which one is more acidic? What will the difference in pH be between the two juices?

****20.** Suppose a solution contains 1.0 mol of H^+ in one liter of water.

(a) What is the pH of the solution?

(b) What substance shown in Figure 10.5 has a pH value in this range?

****21.** Calculate the pH of a 1.25×10^{-2} M solution of HCl in water.

****22.** After looking at the table below, answer these questions.

(a) Which of the body fluids shown in this table has the highest pOH?

(b) Which fluid shows the greatest variation in pH?

(c) Which fluid contains the highest concentration of hydroxide ions?

(d) Which fluid has a neutral pH?

Body Fluid	pH
Blood plasma	7.4
Liquid outside living cells	7.4
Liquid inside living cells	7.0
Saliva	5.8–7.1
Gastric juice	1.6–1.8
Pancreatic juice	7.5–8.8
Intestinal juice	6.3–8.0
Urine	4.6–8.0
Sweat	4.0–6.8

****23.** What method mentioned in the chapter might be a viable option to adjust the pH of the acidic side of the lake in the Trout Lake study?

*****24.** Suppose you have two solutions, one made by adding 0.25 mol of NaOH to 1.0 L of water and one made by adding 0.25 mol of $Mg(OH)_2$ to 1.0 L of water.

(a) Which solution will have the higher pOH?

(b) Which solution will have the higher pH?

(c) Why are the pH values of these solutions different?

***25. A 2.0-L sample of water taken from a lake in an area where acid rain is prevalent contains 9.0×10^{19} H^+ ions. What is the pH of the sample?

Measuring pH

*26. A solution that has a volume of 250 mL contains 0.19 mol of sodium hydroxide. If a drop of phenolphthalein indicator is added, what color will the solution be?

*27. The indicator methyl orange is red below pH 4.0 and yellow above pH 4.5. Consider two separate solutions, one that contains KOH and one that contains $HClO_4$. When methyl orange indicator is added to each, one solution will be red and the other yellow. Which solution is which?

**28. Two methods for measuring pH are mentioned in this chapter. Describe each method.

**29. You may have heard the term *litmus test* in the news. In popular culture, this term refers to a single, important test or question used as a judgment. For example, you might hear something like, "Senator McKay's decision on this issue will be a litmus test for its future success." This term derives from chemistry, where a litmus test is a quick determination of pH using litmus paper. Litmus paper is blue or red paper that has been impregnated with an indicator dye. Blue litmus paper turns red in the presence of acid. Which of these solutions will turn blue litmus paper red?

(a) A solution that has a pH of 8.3

(b) A solution with $[H^+] = 2 \times 10^{-3}$ M

(c) A solution that has a pOH of 4.0

**30. The chemistry of corrosion had to be considered in the design of phenolphthalein paint for airplanes.

(a) How do water and oxygen contribute to the corrosion of metals?

(b) Why does phenolphthalein paint turn pink when corrosion is occurring?

Regulation and Measurement of Gaseous Pollutants

*31. In your own words, explain the link between the EPA and the Clean Air Act and the numbers in Figure 10.15. What event discussed in the text prompted the passage of the Clean Air Act?

*32. Over one year, the highest ozone concentration measured in a particular urban center is 0.22 ppm. Referring to Table 10.1, indicate whether this concentration meets the ozone standard set forth in the NAAQS.

*33. According to Table 10.1, sulfur dioxide (SO_2) is permitted to reach 75 ppb. At this concentration, how many SO_2 molecules would you expect to find per 2 billion molecules of the various gases that make up air?

**34. What are the top three sources of energy used to generate electrical power in the United States? Why is coal, which raises significant issues with toxic emissions, ranked high on this list?

**35. What change to cars took place in the 1970s after the Clean Air Act reduced the allowable levels of exhaust emissions? How did that change affect the levels of gaseous emissions from cars?

**36. What is flue gas desulfurization, and how does it help coal plants meet new NAAQS standards put forth by the EPA?

**37. Nitrogen dioxide is converted to nitric acid in air and then dissolves in water to form ions. If you begin with 4 mol of NO_2 gas, how many moles of protons result from these reactions?

***38. If a sample of gas from a coal plant's smokestack contains 800 molecules of sulfur dioxide per 5 billion molecules of gas, does this sample of gas exceed 75 ppb? What is its concentration in ppb units?

Acid Rain and Gaseous Pollutants

*39. According to Table 10.2, which type of coal is used most in U.S. coal-fired plants? Which type of coal contains the least sulfur?

*40. According to Table 10.2, which type of coal produces the most energy per unit mass in the United States? Why is this type of coal rarely used?

*41. In your own words, explain what factors contribute to the formation of a dead zone in an area of natural water.

*42. Which species of aquatic organism, listed in Figure 10.17, can survive at a pH of 4.5?

*43. In your own words, describe the Trout Lake study. How was the study designed? What did the chemists learn?

**44. The photograph on the left was taken in the 1920s; the photo on the right nearly 70 years later. Acid rain is the primary cause of the deterioration.

Monica Schroeder/Science Source

(a) What two atmospheric pollutants contribute most to acid rain?

(b) What strong acid is formed by each of these pollutants?

(c) Write chemical equations showing how each of these strong acids ionizes in water to produce corrosive protons.

**45. According to Figure 10.11, why is acid rain an enormous problem in New England (Pennsylvania and areas north), even though there are few coal plants north of Pennsylvania? What pollutant gas contributes to acid rain from these plants?

**46. In your own words, explain what types of pollutant are removed from car exhaust by a catalytic converter. What gases go into the converter, and what gases come out? What are the benefits and drawbacks of using catalytic converters?

**47. Review the naturebox about York Minster cathedral. In your own words, describe the problem that the scientists were trying to solve. What material was used to build the cathedral? Why is that material susceptible to pollutants?

**48. Look again at Figure 10.11. Comment on the size of the green dots versus red dots. Is one color larger than the other, in general? If so, why?

***49. Trout Lake contains 1.00×10^9 L of water. How many moles of protons (hydrogen ions) did it contain before it was acidified?

***50. A 1.0-mL sample of rainwater from an unpolluted area is analyzed and found to contain 3.00×10^{-9} moles of protons.

(a) What is the pH of this rain sample?

(b) Is this rain too acidic?

(c) Why does natural rain in the absence of pollution have a pH lower than 7?

*51. Swimming pools must be checked regularly to ensure that they have the proper pH and the proper concentration of added chemicals to prevent growth of microorganisms and keep the water safe for recreation. Pools are typically kept at a pH of 7.4 to 7.6. If the pH drops below this range, soda ash (sodium carbonate) can be added to increase the pH. If the pH rises above this range, muriatic acid can be added to decrease the pH. During a daily check, the pH in a certain public pool is found to be 8.3. What advice would you give to the pool manager to make the water safe for swimmers?

*52. HClO is a compound that can be used as a disinfectant for drinking water. It is made when chlorine gas is bubbled through liquid water:

$$H_2O\ (l) + Cl_2\ (g) \rightarrow HClO\ (aq) + H^+\ (aq) + Cl^-\ (aq)$$

(a) Is this chemical equation balanced? If not, balance it.

(b) Do you think that the pH of the water increases or decreases when this reaction takes place? How do you know?

**53. Evidence suggests that eating omega-3 fatty acids can help reduce the risk of heart disease and lower blood pressure. The molecule shown in the figure below is an example of an omega-3 fatty acid known as alpha-linolenic acid, or ALA. It is an acid because the COOH group on the right end of the molecule loses a proton, H^+. More specifically, it is also classified as an omega-3 fatty acid because it has a double bond at the position three carbon atoms from the left end of its chain (that is, between the third and fourth carbon atoms).

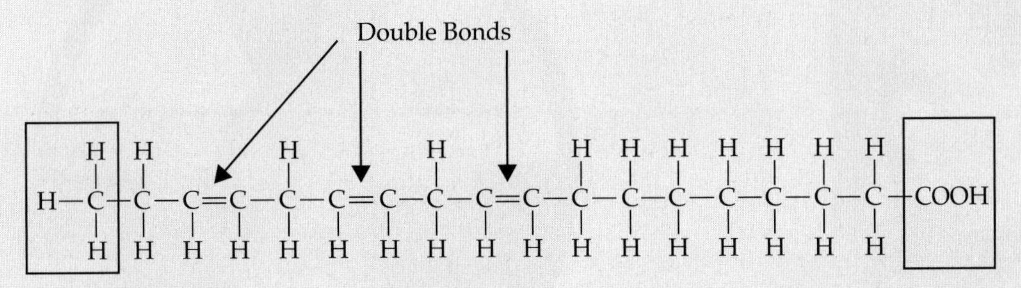

(a) Write a chemical formula for ALA.

(b) Calculate the molar mass of ALA.

(c) Consumption of omega-6 fatty acids has some positive and some negative health effects. Based on the definition of an omega-3 fatty acid, do you think that ALA is also an omega-6 fatty acid? How do you know?

***54. The figure shown below depicts several acids ranked in order of decreasing acid strength from left to right. The strength of an acid depends, in part, on the ability of the rest of the molecule to pull electron density away from the acidic hydrogen atom. Oxygen atoms connected to the central atom of the acid via a double bond are especially good at pulling electron density through the acid molecule and away from the hydrogen atom. This makes it easier for the molecule to release the hydrogen as a proton, H^+, and this, in turn, increases the acid's strength. Given these facts, examine the structures of the acids shown below. Does the order of acid strength make sense based on this information? Why or why not?

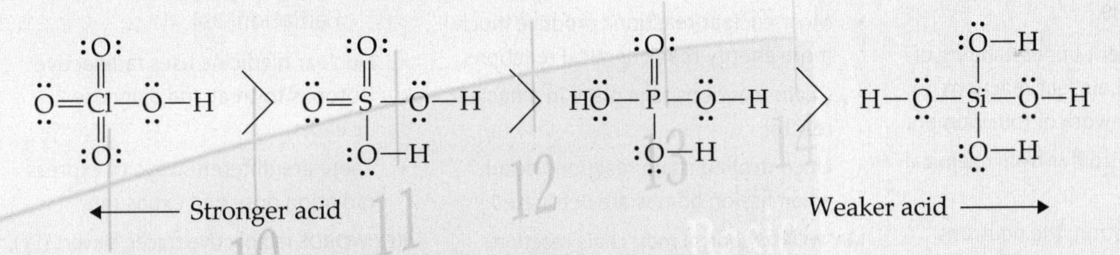

55. INTERNET RESEARCH QUESTION

It's not always easy to know if you live in an area affected by acid rain because many municipal water providers do not include pH in their water quality reports. However, it is possible to know if you live near coal-fired power plants, a source of acid-rain-producing sulfur emissions.

(a) Type "coal-fired power plants near me" into your Internet browser. Try to estimate how far you live from the nearest coal-fired power plant. Also, if you are able, determine whether you live in an area where there are many coal-fired plants, or relatively few.

(b) Try to learn more. For example, can you determine the type of coal that is burned in the coal-fired power plant nearest you? Types of coal are described in Table 10.2.

86
Rn
Radon
(222)

AP Images

Nukes

The Fundamentals of Nuclear Chemistry

Peteri/Shutterstock

CHAPTER 11

When a disaster strikes—whether it is natural or caused by humans—it can be difficult to know how to respond and how to help the people involved. In a crisis, communicating information is critical, so scientists have devised ways to notify the public about the severity of disasters. For example, the Richter scale is used to rank the severity of earthquakes on a scale from 1 to 10. (To date, the highest recorded quake was a 9.5 in Chile in 1960.) The International Nuclear Event Scale (INES) is a similar scale that ranks the severity of nuclear accidents. On that scale, a Level 1 is a minor event and a Level 7 is a catastrophe. The tsunami magnitude scale rates the potential for damage from a tsunami. It ranks severity on a scale of 1 to 6, where 6 is a tsunami that destroys hundreds of miles of coastline.

On March 11, 2011, all three of these scales suddenly became very relevant to people living on the coast 150 miles north of Tokyo.

On that day, an offshore earthquake struck at 2:46 P.M. The magnitude of that earthquake was 9.0 on the Richter scale, which puts it in a four-way tie for the fourth most powerful in history. Because it was offshore, the earthquake generated a powerful category 5 tsunami that made landfall about an hour later. Right in its path were several nuclear power plants. The Fukushima Daiichi plant took the full brunt of the tsunami, and its nuclear containments were breached. The result was a massive release of radiation into the air and the water. This nuclear disaster was given the highest possible INES ranking—7. Only one other nuclear accident has received an INES ranking of 7: the Chernobyl plant explosion, which took place in the Ukraine in 1986. The nearby photo is of burning reactors at the Fukushima Daiichi plant after the tsunami struck.

AP Images/KYDPL KYODO

This chapter is about the chemistry that happens when the nuclei of atoms undergo changes, as they do in nuclear power plants. The history of nuclear energy is a short one and we'll go back to its beginning with the discovery of radioactivity. We'll also learn how humans have used that discovery—for better and, mostly, for worse. Finally, we'll come full circle and see what lessons we can learn from the events in Japan on that fateful day in March 2011.

AP Images

11.1 The Nature of Nuclear Reactions

We owe our current understanding of radioactivity and nuclear reactions, in large part, to the work of four women.

Because human beings conduct scientific research, stories about important scientific discoveries sometimes seem like soap operas. The discovery of the unexpected goings-on in the atom's nucleus is such a story.

In the 1890s, Wilhelm Röntgen was just discovering X-rays and using them to take pictures of his wife's hand (see Chapter 2). And an equally important discovery was made during this period, quite by accident. A French scientist named Antoine Henri Becquerel, while studying how light affects specific rocks, found that certain rocks emit an unknown kind of powerful energy. What made this discovery so astonishing was that, whereas the creation of X-rays required an input of electricity, Becquerel's energy was being emitted with no input of anything. All on their own, without being heated or subjected to an electric current, Becquerel's rocks glowed in the dark. Becquerel published his peculiar findings about glow-in-the-dark rocks, but most people in scientific circles took little notice.

Someone who did notice was a young Polish woman studying for her doctoral degree in Paris. Marie Sklodowska Curie was one of only two women earning a PhD in Europe at the time. She was looking for something completely new and different to study for her dissertation research, and Becquerel's findings fit the bill perfectly. In a matter of months, Curie had characterized and measured these mysterious emissions, which she called *radioactivity*. Working with her husband, Pierre Curie, a respected French scientist, Marie began the first systematic study of these strange new materials. During her lifetime of research, she uncovered new and exciting information about the nuclei of atoms and discovered the new elements radium and polonium. **Figure 11.1** shows the couple, hard at work in the laboratory.

These bewildering discoveries set science on its ear. The notion that matter could give off energy on its own in the form of radiation was utterly new, and it seemed more like science fiction than fact. Everything known at the time about how substances behave was brought into question. And initially, as is often the case in scientific research, the discoveries made by Becquerel and the Curies raised more questions than they answered.

By the mid-1930s, scientists had begun to learn more about the atom and its contents. At this point, the two charged particles in the atom that we discussed in Chapter 2—protons and electrons—were known. The existence of the neutron had been documented. It was known that atoms of an element could have different numbers of neutrons,

AP Images

Figure 11.1 Marie and Pierre Curie

AP Images/MPG HO, undatiert

Figure 11.2 Lise Meitner

which were named isotopes. (Return to Section 2.3 for a more detailed discussion of isotopes.) Chemists and physicists had begun to use beams of neutrons to perform experiments on what we now know as **radioactive** materials, materials that release energy or particles when their nuclei undergo reactions.

One of these scientists, an Italian physicist named Enrico Fermi, decided to bombard every element, one at a time, with a beam of neutrons to see how each one responded. The most interesting result he obtained was with uranium, the element that has 92 protons in its nucleus. When Fermi bombarded uranium with neutrons, it produced a confusing mixture of unidentifiable products. Fermi's uranium experiment became one of the challenging puzzles for scientists at the time. Who would figure it out first?

One researcher working in the field at the time, German chemist Ida Noddack, suggested that—perhaps—the uranium atom was breaking apart and making two lighter elements of similar mass. This idea was not accepted by the physics community, and Noddack did not pursue experiments that could validate her proposal. In 1937, Marie Curie's daughter Irene Curie, a scientist like her mother, repeated Fermi's uranium experiments to try to figure out what the products might be. She found a product that had 57 protons. Curie was baffled by her findings, but she maintained that her results were valid.

In Germany, another group was working on this problem. The researchers included Otto Hahn, Fritz Strassmann, and Lise Meitner, the first woman professor of physics in Germany (**Figure 11.2**). Using the scientific method, these three scientists also began to work on unraveling the uranium data first collected by Fermi. They did not yet have an answer when Meitner, who was Jewish, was forced to sneak out of Germany to escape the Nazi regime. She quickly found work in Sweden and continued her contributions to the group by corresponding continuously with Hahn and Strassmann.

recurring theme in chemistry 1

Scientists worldwide use the scientific method as a basis for experimentation and deduction. The scientific method is based on the principles of peer review and reproducibility.

Upon hearing of the results obtained by Curie in France, Meitner turned to her nephew, physicist Otto Frisch, as a sounding board for her thoughts about the puzzling results. Working together in 1938, Meitner and Frisch imagined the uranium nucleus to be like a drop of water—an idea that other scientists had put forth earlier. When hit with a beam of neutrons, the water drop blasted apart into two pieces of similar size, as first suggested by Ida Noddack four years before.

Meitner believed that the two products were krypton (36 protons) and barium (56 protons) because together they contain 92 protons, the same number as uranium. She communicated her idea to Hahn and Strassmann in Germany. They performed experiments verifying that uranium (atomic number 92) does indeed split to produce krypton and barium. These results were published separately by Hahn and Strassmann and by Meitner and Frisch. **Nuclear fission,** the splitting of the atom's nucleus, had been discovered.

Upon announcement of the discovery, Meitner became an instant celebrity, especially in the United States. In 1944, however, Otto Hahn was awarded the Nobel Prize for "his discovery of the fission of heavy nuclei." No other names appeared on the award. In the years after World War II, the controversy surrounding the names on the 1944 Nobel Prize in Chemistry was researched, and Meitner's critical role in the discovery of nuclear fission has since been acknowledged. As a tribute to Meitner, element 109 was dubbed *meitnerium,* although she was not alive when the element was named in her honor in 1997.

wait a minute . . .

Is splitting the atom the same thing as nuclear fission?

During fission, the particles in the nucleus of the parent atom divide to make new atoms. All of the atoms involved in fission have protons, neutrons, and electrons, as all atoms do (except for some very small atoms and ions that may have no electrons or neutrons). Thus, all of the particles in the atoms are divided when fission takes place, not just the protons. So the phrase "splitting the atom" is a legitimate way to refer to fission, because all particles in the atom are divided during fission.

QUESTION 11.1 Suppose that an atom of plutonium, Pu, undergoes fission to make two atoms. One of those atoms is cerium, Ce. An atom of what other element also forms?

ANSWER 11.1 Plutonium has 94 protons. If it splits to make cerium, which has 58 protons, the atom of the other element must have $94 - 58 = 36$ protons. That element is krypton (Kr).

For more practice with problems like this one, check out end-of-chapter Question 3.

FOLLOW-UP QUESTION 11.1 True or false? When an atom undergoes nuclear fission, the number of protons in the two new atoms that are formed is equal to the number of protons in the nucleus of the parent atom.

Nuclear reactions differ from chemical reactions.

Chemical reactions typically involve electrons rearranging and chemical bonds being made and broken. As we know, chemical reactions do not involve changes in the number of protons and neutrons in the reacting atoms. In contrast, a **nuclear reaction** is one in which the number of protons and/or neutrons does change in some way. There are two major classes of nuclear reaction: fission, which we just introduced, and fusion.

recurring theme in chemistry 4

Chemical reactions are nothing more than exchanges or rearrangements of electrons.

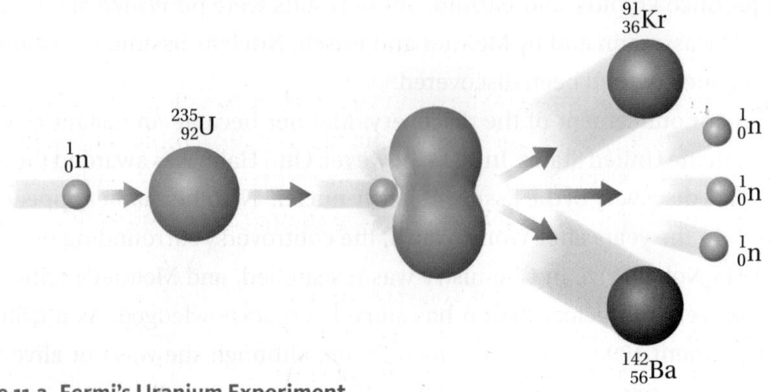

Figure 11.3 Fermi's Uranium Experiment

When a neutron smashes into uranium-235, the atom divides into krypton-91 and barium-142. Three neutrons are also produced.

The following nuclear fission equation represents Fermi's uranium experiment:

$$^{235}_{92}U + ^{1}_{0}n \rightarrow ^{91}_{36}Kr + ^{142}_{56}Ba + 3^{1}_{0}n$$

Fermi smashed uranium with a neutron, designated with $^{1}_{0}n$, as shown schematically in **Figure 11.3**. The products were ultimately identified by Hahn, Meitner, and Strassmann.

The equation shown here may look strange to you, because ordinarily we don't include superscripts or subscripts before chemical symbols in chemical equations. Let's discuss what these numbers mean. The superscript is the mass number, which is the sum of the number of protons and neutrons in the nucleus. In the case of uranium in Fermi's experiment, the mass number is 235. Because atoms of an element can exist as different isotopes, the mass number identifies the specific isotope of the element.

flashback

Recall from Chapter 2 that some elements have several isotopes and others have only one. There is no logical way to predict how many isotopes an element has. You just have to look it up.

The atomic number is shown as a subscript before the element symbol. The atomic number is the number of protons in the nucleus of the atom. In the case of uranium in Fermi's experiment, the atomic number is 92. For a given atom of an element, this number is always the same, because the number of protons defines the element. If we want to find the number of neutrons for a given isotope, we subtract its atomic number from its mass number, like this:

$$\text{mass number} - \text{atomic number} = \text{number of neutrons}$$

For example, the number of neutrons in uranium-235 is:

$$235 - 92 = 143 \text{ neutrons}$$

The neutron in this nuclear equation is depicted as an *n* with a superscript 1. Its mass number is 1 because it is just a neutron, so the sum of its protons and neutrons is 0 + 1. Its subscript is 0 because it contains no protons.

The other reason this nuclear equation looks strange is that it doesn't appear to be balanced. That is, the numbers of atoms of each element on each side of the equation are not equal: uranium is on the left, and krypton and barium are on the right. This is okay. Because elements can change into other elements in nuclear reactions, we should not expect atoms on each side to be balanced as they are in chemical equations.

In a nuclear equation, the neutrons and protons are usually balanced on each side.

Although they may not appear to be balanced, nuclear equations are balanced. In a balanced nuclear equation, even though the number of atoms is not balanced, the number of neutrons and protons on each side are equal in most cases. Let's count protons and neutrons in the equation for Fermi's uranium experiment and see if it is balanced.

For each symbol in this equation, the number of neutrons and protons is tallied in **Figure 11.4**. The total is the same on each side, so the nuclear equation is balanced because protons and neutrons are equal. Another way to check if a nuclear equation is balanced is to sum all of the superscripts on the left and sum all of the superscripts on the right and then sum all of the subscripts on the left and sum all of the subscripts on the right. If the sum of the superscripts on both sides and the sum of the subscripts on both sides are equal, the equation is balanced. These two methods are both a means of checking to see that the numbers of protons and neutrons are the same on each side.

$$^{235}_{92}U + ^{1}_{0}n \longrightarrow ^{91}_{36}Kr + ^{142}_{56}Ba + 3^{1}_{0}n$$

	$^{235}_{92}U$	$^{1}_{0}n$	$^{91}_{36}Kr$	$^{142}_{56}Ba$	$3^{1}_{0}n$
Number of neutrons	143	1	55	86	3
Number of protons	92	0	36	56	0
Total neutrons	**144**		**144**		
Total protons	**92**		**92**		

Figure 11.4 Is This Nuclear Equation Balanced?

This equation is balanced because both sides have the same number of protons and the same number of neutrons.

wait a minute . . .

Why is there a 3 in front of the neutron in the uranium equation in Figure 11.4?

Recall from Chapter 7 that, when we are balancing a chemical equation, we can place coefficients in front of any reactant or product as a multiplier. We can also use coefficients in nuclear equations. In this case, the number 3 tells us that there are three neutrons on the right side of the equation.

QUESTION 11.2 Sum the number of protons and neutrons on each side of this nuclear equation. Is this equation balanced?

$$^{209}_{83}Bi + ^{58}_{26}Fe \longrightarrow ^{266}_{109}Mt + ^{1}_{0}n$$

ANSWER 11.2 The equation is balanced because the sum of the mass numbers on the left (267) equals the sum of the mass numbers on the right (267) and the sum of the atomic numbers on the left (109) equals the sum of the atomic numbers on the right (109).

For more practice with problems like this one, check out end-of-chapter Questions 20 and 22.

FOLLOW-UP QUESTION 11.2 Indicate whether each nuclear equation is balanced.

(a) $^{1}_{1}H + ^{2}_{1}H \longrightarrow ^{3}_{1}H$

(b) $^{246}_{96}Cm + ^{12}_{6}C \longrightarrow ^{254}_{102}No + 4^{1}_{0}n$

(c) $^{14}_{7}N + ^{3}_{2}He \longrightarrow ^{17}_{8}O + ^{1}_{1}H$

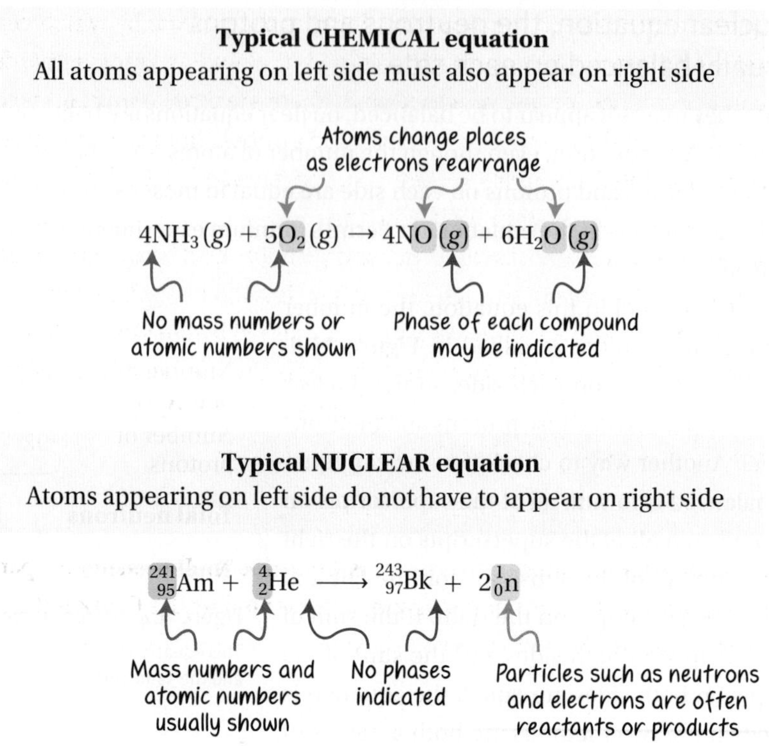

Figure 11.5 Chemical Equations Differ from Nuclear Equations

In a chemical equation, every atom shown on the left must also appear on the right. In a nuclear equation, the sum of the mass numbers (superscripts) on the left usually equals the sum of the mass numbers on the right. In this example, both sides have total mass numbers of 245. The same is true for the subscripts; both sides have 97 protons. Both of these equations are balanced.

Typical CHEMICAL equation

All atoms appearing on left side must also appear on right side

Atoms change places as electrons rearrange

$$4NH_3\,(g) + 5O_2\,(g) \longrightarrow 4NO\,(g) + 6H_2O\,(g)$$

No mass numbers or atomic numbers shown

Phase of each compound may be indicated

Typical NUCLEAR equation

Atoms appearing on left side do not have to appear on right side

$$^{241}_{95}Am + {}^{4}_{2}He \longrightarrow {}^{243}_{97}Bk + 2\,{}^{1}_{0}n$$

Mass numbers and atomic numbers usually shown

No phases indicated

Particles such as neutrons and electrons are often reactants or products

A typical chemical equation and a typical nuclear equation are shown in **Figure 11.5**. In the chemical equation, mass numbers are not normally shown, and the equation has the same elements on both sides. Remember, a typical chemical reaction involves only the rearrangement of the electrons of the atoms taking part in the reaction, so the nuclei stay the same and the atomic symbols stay the same, too. Chemical processes *never* involve the rearrangement of protons. In the nuclear reaction (shown at the bottom of the figure), however, atoms do change from one element to another because the number of protons in the nuclei change.

QUESTION 11.3 Which of the following equations represent nuclear reactions, and which represent chemical reactions?

(a) $^{246}_{96}Cm + {}^{12}_{6}C \longrightarrow {}^{254}_{102}No + 4\,{}^{1}_{0}n$ (b) $CH_4 + 2O_2 \longrightarrow CO_2 + 2H_2O$

ANSWER 11.3 Equation (a) represents a nuclear reaction because new elements are formed. Equation (b) represents a chemical reaction because there is no change in the identity of the atoms during the reaction. Rather, the same atoms are arranged in a new way. This example illustrates why atomic numbers and mass numbers are usually not written in chemical reactions. These numbers do not change, and they would add a great deal of clutter to the reaction.

For more practice with problems like this one, check out end-of-chapter Questions 14 and 15.

FOLLOW-UP QUESTION 11.3 Is the following equation a chemical equation or a nuclear equation?

$$^{2}_{1}H + {}^{2}_{1}H \longrightarrow {}^{3}_{2}He + {}^{1}_{0}n$$

There are three important types of radioactive decay.

Nuclear and chemical reactions differ in another way. Nuclear reactions produce more than just new elements; they also produce an energetic product that can be either a particle or radiation. **Radioactive decay** is the process by which certain isotopes give off those energetic products. Isotopes that undergo radioactive decay are radioactive isotopes (or just *radioisotopes*). There are three important types of radioactive decay, and they are given names and symbols from the first three letters of the Greek alphabet: alpha (α), beta (β), and gamma (γ). In this section, we consider each type of decay, very briefly, in turn.

In the following nuclear reaction, the energetic product is an alpha particle. The isotope on the left side of the equation, which is hassium-273, breaks apart to give seaborgium-269 and an alpha particle, represented by the Greek letter *alpha* in the equation. Another way to say this is that hassium, element 108, undergoes **alpha decay**. Or, we can say, "Hassium is an alpha emitter." The balanced nuclear equation for this process is shown here:

$$^{273}_{108}\text{Hs} \longrightarrow {}^{269}_{106}\text{Sg} + {}^{4}_{2}\alpha$$

The second type of radioactive decay is **beta decay**. The energetic particle emitted from this reaction is a beta particle, represented by the Greek letter *beta*. The following equation represents beta decay:

$$^{239}_{93}\text{Np} \longrightarrow {}^{239}_{94}\text{Pu} + {}^{0}_{-1}\beta$$

Neptunium-239 is radioactive, and it is a *beta emitter*. For beta emitters, the emitted particle is an electron, which is why there is a negative sign in the subscript of the symbol: ${}^{0}_{-1}\beta$.

The third and final kind of radioactive decay is the emission of **gamma radiation**. Alpha and beta radiation exist in the form of particles, but gamma radiation is a type of light. Recall from the discussion of the electromagnetic spectrum in Chapter 2 that gamma radiation is the most energetic type of light. Many nuclear reactions produce ultra-high-energy radiation in the form of gamma rays, which are usually given off along with alpha particles or beta particles. Gamma radiation is represented by the Greek letter *gamma* in the following balanced nuclear equation:

$$^{273}_{108}\text{Hs} \longrightarrow {}^{269}_{106}\text{Sg} + {}^{4}_{2}\alpha + \gamma$$

Nuclear chemists often do not include the gamma symbol, because it doesn't affect the balance of the nuclear equation. Therefore, this nuclear equation can be written with or without the gamma. Thus, it's understood that gamma radiation is often present even if it is not included in the equation.

Why is it important to know what type of radiation is given off by a particular radioactive substance? It's important because the three types of radiation affect the human body in different ways. In Section 11.4, we discuss how these different types of radioactive decay affect the human body.

QUESTION 11.4 How many protons and neutrons does an alpha particle contain?

ANSWER 11.4 An alpha particle contains two protons and two neutrons.

For more practice with problems like this one, check out end-of-chapter Questions 3 and 6.

FOLLOW-UP QUESTION 11.4 How many protons and neutrons does gamma radiation contain?

wait a minute . . .

Why aren't the number of protons and neutrons on each side of the beta decay equation equal?

The equation shown for beta decay is not balanced in the way the previous nuclear equations in this chapter have been balanced. A quick tally indicates that there are unequal numbers of protons and neutrons on each side of the balanced equation. This is why we qualify the statement in the text by stating that protons and neutrons are *usually* equal on both sides of a balanced nuclear equation.

The unique case of beta decay is an exception to this rule because, during beta decay, a neutron is converted into a proton and an electron. Thus beta decay is more complex than its balanced equation indicates. When we write the equation for beta decay, we include a -1 for the beta particle so that the superscripts and subscripts on the two sides of the equation are equal.

11.2 Energy from the Nucleus

Most nuclear reactions produce much more energy than chemical reactions.

A typical nuclear reaction produces much more energy than a typical chemical reaction. To get a sense of how they compare to one another, let's consider a process we have touched on in other chapters—the combustion of methane, also known as natural gas:

$$CH_4 + 2O_2 \rightarrow CO_2 + 2H_2O + \text{energy}$$

If one mole of methane reacts with oxygen as shown in this balanced chemical equation, this reaction gives off about 550 kilojoules (kJ) of energy. This is equivalent to about 130 Calories, which is roughly equal to the energy contained in a banana. Nuclear reactions, on the other hand, are typically about 1 million times more energetic than combustion reactions like this one. This is because, on one hand, nuclear reactions release enormous amounts of energy due to changes that occur in the atom's nucleus. A typical combustion reaction, on the other hand, releases energy through an exothermic chemical reaction in which chemical bonds are made and broken. Thus, in general, the amount of energy released in nuclear reactions is much greater than the amount released in chemical reactions.

QUESTION 11.5 The equations that follow are for two reactions: one releases 10,000,000 kJ of energy and the other releases 100 kJ of energy. Which is which?

$$C_3H_8 + 5O_2 \rightarrow 3CO_2 + 4H_2O$$

$$^{27}_{13}Al + ^{4}_{2}\alpha \rightarrow ^{1}_{0}n + ^{30}_{15}p$$

ANSWER 11.5 The first equation produces much less energy (100 kJ) because it is a chemical reaction. The second one produces more energy because it is a nuclear equation (10^7 kJ).

For more practice with problems like this one, check out end-of-chapter Question 18.

FOLLOW-UP QUESTION 11.5 Are the two equations in Question 11.5 balanced?

Chain reactions take place in a nuclear reactor.

As we have just seen, nuclear reactions can create a colossal amount of energy. And nuclear power plants based on those reactions are an obvious source of energy for human society, but they also produce radioactive waste, an enormous environmental challenge. Examples like the chapter-opening story about Fukushima Daiichi are reminders of what can happen when nuclear plants are compromised. Nevertheless, while some countries, such as Germany and Switzerland, are going nuclear-free, the United States is planning to ramp up its use of nuclear power.

For the United States, nuclear power is attractive because it presents a way to generate energy without relying on oil and coal, both resources that are present in limited quantities and increase greenhouse gas emissions. And, as we discussed in Chapter 10, "pH and Acid Rain," fossil fuels create environmental hazards of their own. It's important to note, however, that uranium is not an unlimited resource. A finite amount of uranium is left in the Earth for us to use.

recurring theme in chemistry 8
The Earth has a finite supply of raw materials that humans can use.

Figure 11.6 is an illustration of a typical nuclear reactor. The *reactor core* is located in a containment building. The hot, radioactive material is kept there. A fission reaction produces enormous amounts of heat energy in the core, and this energy converts the water passing over the fuel into steam. The pressure of the steam drives a turbine connected to an electric generator, which converts the energy of the rotating turbine to electrical energy. The steam leaves the plant at

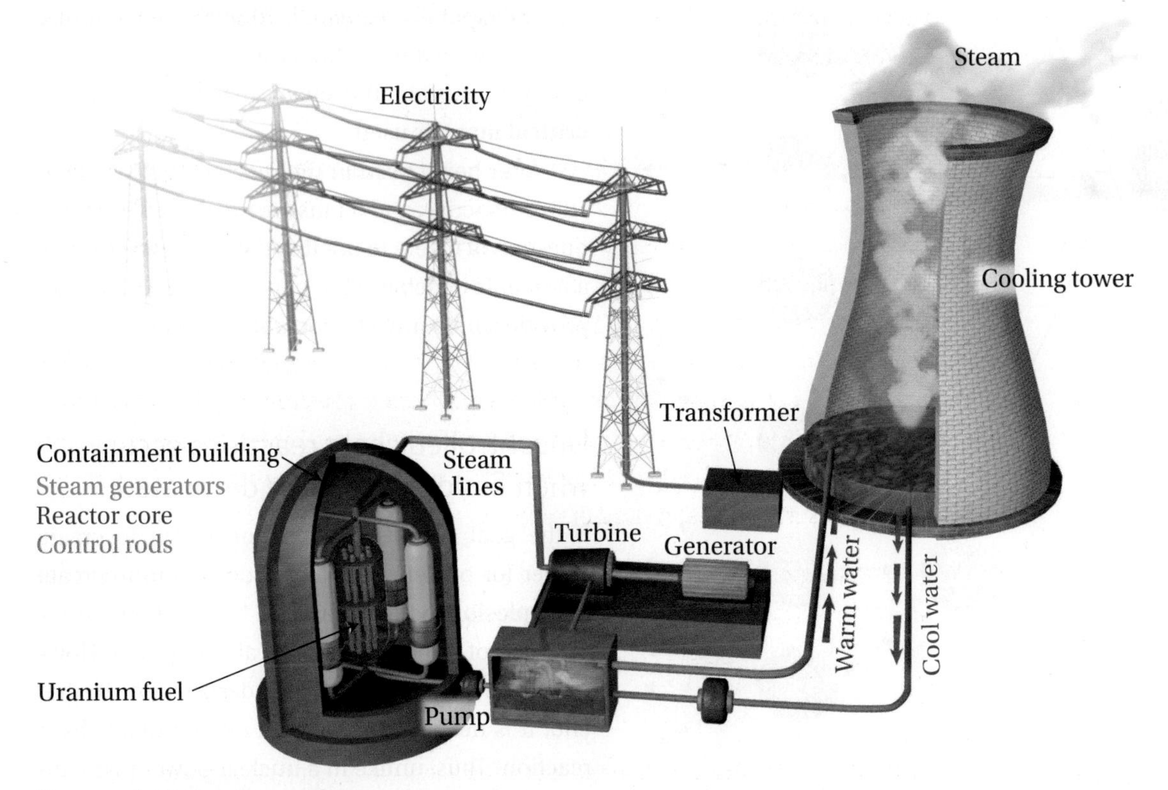

Figure 11.6 Nuclear Power Plant

The nuclear fuel (in this case uranium-235) is kept in a containment building. Water circulates through the containment building and is heated by the fission reaction. The steam produced turns a turbine, which generates electricity. Steam is released from the cooling tower.

the top of the cooling tower. Within the core, **control rods** can be lowered into the radioactive material to damp the nuclear reaction and prevent overheating of the reactor core.

We refer to the fuel in nuclear reactions as *fissionable material.* One fuel used commonly in fission reactors is uranium-235, the isotope studied by Meitner and her colleagues. We have already seen one way that uranium undergoes nuclear decay—the Fermi reaction, which is shown again here:

$$^{235}_{92}\text{U} + ^{1}_{0}\text{n} \longrightarrow ^{91}_{36}\text{Kr} + ^{142}_{56}\text{Ba} + 3^{1}_{0}\text{n} + \text{heat energy}$$

A collision between one neutron and one atom of uranium-235 initiates this reaction, which produces three neutrons. When these three neutrons are released, they smash into three more uranium-235 atoms, each one reacting to produce three more neutrons. Because this reaction amplifies itself by initiating more reactions, we refer to this series of fissions as a **chain reaction,** illustrated in **Figure 11.7**.

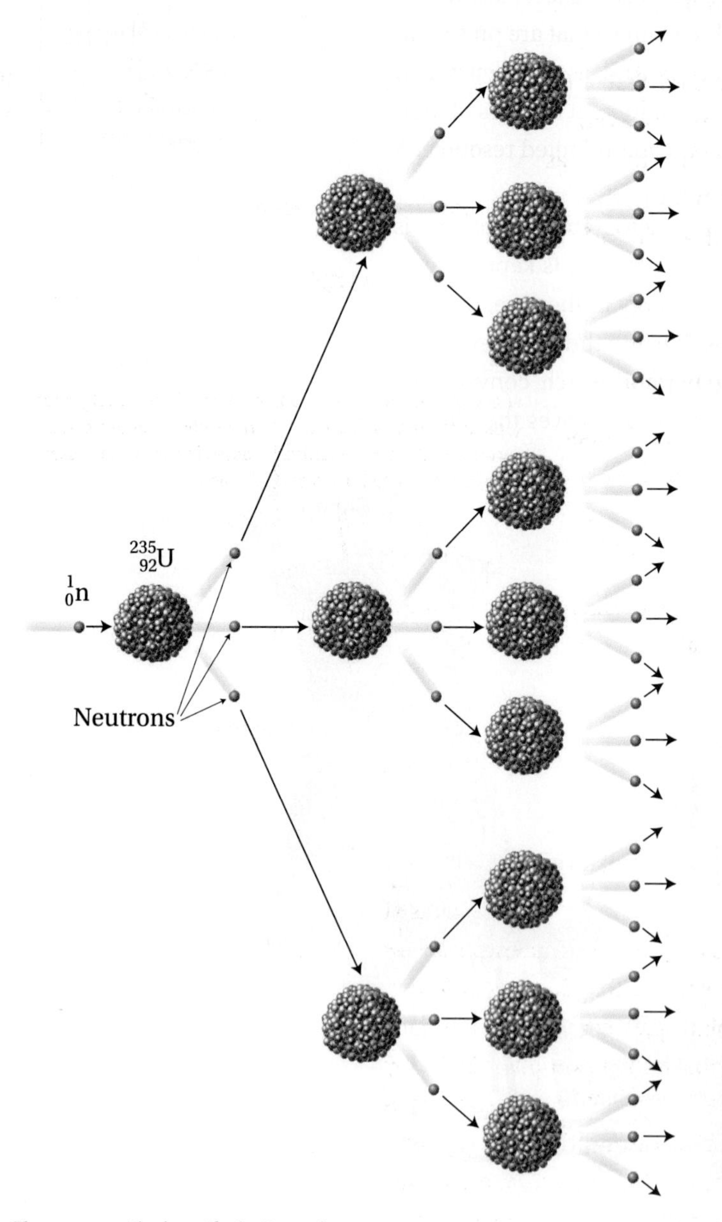

Figure 11.7 Fission Chain Reaction

When three neutrons are produced from one fission event, the subsequent reactions are multiplied and the reaction accelerates.

In looking at Figure 11.7, we can see how such a chain reaction could rapidly go out of control. The number of fission reactions quickly escalates and, because heat energy is produced in each step, the reactor can easily overheat if the reaction is not kept in check. If the acceleration is sufficiently fast, the result is an explosion. For this reason, the amount of uranium used in nuclear reactors is crucial and carefully regulated. To prevent explosions, the minimum amount of fissionable material necessary to sustain the chain reaction, called the **critical mass,** is used.

After being used in the reactor, nuclear fuel is still radioactive, and it must be put somewhere. The question of where to put it and how to get it there is a hot topic of debate. The naturebox in this chapter provides more information about that debate.

Uncontrolled chain reactions occur when fission bombs are detonated.

If the goal of a fission reaction is not to create power for our energy needs but instead to create an explosion, as in an atomic bomb, then a critical mass of radioactive fuel is also required. However, with a bomb, the goal is destruction. To that end, it is desirable to have an uncontrolled chain reaction. Thus, unlike in a nuclear power plant, no safeguards are in place to prevent the reaction in an atomic bomb from happening.

The key challenge that arises in creating a bomb with a critical mass is premature detonation. To circumvent this problem, some fission bombs hold two separate portions of fissionable material, each portion having a mass less than the critical mass, which is called a *subcritical mass*.

Figure 11.8 is a schematic drawing of a fission bomb. The two subcritical portions are placed one at each end of the bomb vessel, and a chemical explosive is placed behind one of the portions. Until the two subcritical masses are combined, no explosion occurs—at least in theory—and the bomb can be safely transported. When the bomb is discharged, the chemical explosive is detonated, forcing the two subcritical masses to smash violently into each other. The total mass of fissionable material becomes critical, and an explosive chain reaction is initiated.

The first fission bomb was dropped by the United States on Hiroshima, Japan, on August 6, 1945. The bomb's intensity was equal to that of 20,000 tons of TNT, and the devastation of Hiroshima was total. Every year on August 6, the citizens of Hiroshima gather to remember the victims of the bomb and to remind the world of the importance of peace.

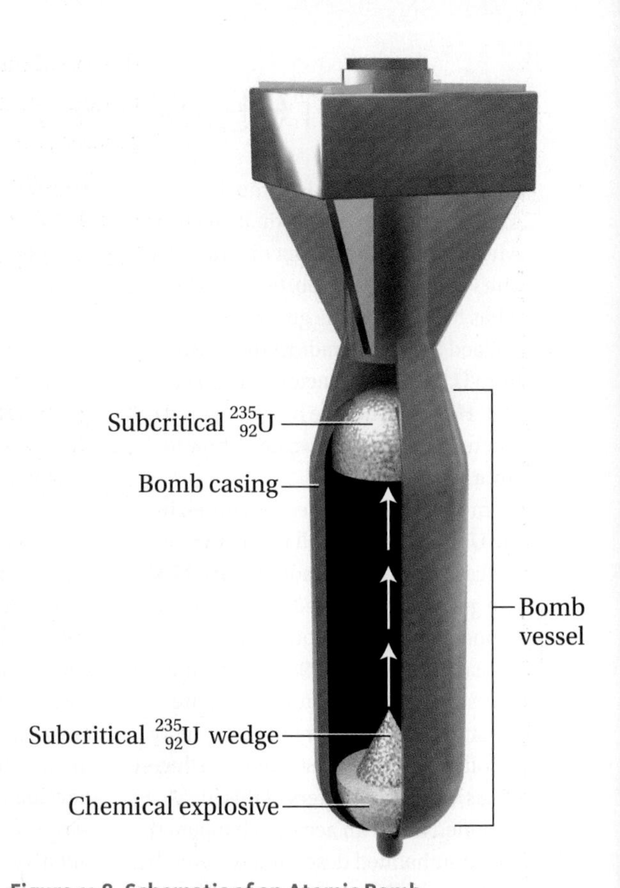

Figure 11.8 Schematic of an Atomic Bomb

Two subcritical masses of radioactive fuel are initially kept separated from each other. When the chemical explosive is detonated, the two subcritical masses in the bomb vessel slam together and form a critical mass.

QUESTION 11.6 True or false? Critical mass is the minimum mass of a fissionable material that must exist for a nuclear reaction to take place.

ANSWER 11.6 False. Nuclear reactions can occur in a sample of fissionable material for which the mass is less than critical. The critical mass refers to the mass required to sustain a chain reaction.

For more practice with problems like this one, check out end-of-chapter Questions 27 and 28.

FOLLOW-UP QUESTION 11.6 If the three neutrons produced in the fission of uranium-235 each initiate a subsequent fission reaction, how many neutrons are produced in the next step?

11.3 It's a Wonderful Half-Life

Radon gas is a natural form of radiation.

On a chilly December morning in 1985, the day began as usual for Stanley Watras of Colebrookdale Township, Pennsylvania. He drove to his job at the Limerick nuclear power plant at the usual hour and put in a full day's work. This day was different, though, because the radiation detectors at the plant exit screamed as Stanley passed through them on his way out. It's not so unusual for a power plant fueled by nuclear reactions to have an occasional radiation alarm. So, at first, it was assumed that Stanley had somehow been exposed to radiation during the workday, and the plant was scrutinized for leaks and contamination. But the radiation source wasn't located at his work site. After no leaks were found at the plant, the radiation levels in Stanley's house were checked and found to be higher than any reading ever recorded in a private residence. Indeed, the levels were so high that Stanley's

natureBOX • Who's Going to Take Out the Trash?

The U.S. government is in much the same position as the obstinate teenager whose job it is to take out the trash: No one wants to do the job, but somehow it has to be done. The government is obliged to remove radioactive waste from U.S. nuclear reactors and dispose of it. However, no one has yet figured out where to put nuclear waste or how to store it safely.

In 1982, a single site was chosen in the United States for all nuclear waste: Yucca Mountain, Nevada (**Figure 11A**). The government agreed to commence depositing all of the country's waste at Yucca Mountain in 1998, and the nuclear industry pitched in funds to prepare the site. Because the mountain is in a desolate place about 80 miles northwest of Las Vegas, supporters of this idea reasoned that if an accident happened, only uninhabited desert areas would be contaminated. This was the plan in 1982.

The Yucca Mountain plan was a hot-button issue for many Americans, especially Nevada residents, who did not want waste from all over the country traveling to their state. Environmental groups were vocal, too, about the threat to the natural beauty of Nevada as well as the hazards of transporting radioactive waste through 43 of the 48 contiguous states (**Figure 11B**). What's more, Yucca Mountain sits in an area prone to earthquakes, and the depth where the radioactive waste would be buried is close to the water table.

These potentially serious threats to the environment prompted President Bill Clinton to veto the bill for the Yucca Mountain project in April 2000. In July 2002, however, President George W. Bush reversed this decision. Then, in 2010, President Barack Obama cut most of the funding for the project. Now Yucca Mountain is little more than a five-mile-long exploratory tunnel and has yet to become home to any nuclear waste.

Where are the nuclear plants across the country currently storing their radioactive waste? About 50,000 tons of nuclear waste is stored temporarily at individual nuclear reactor sites, 90% of them east of the Mississippi River. And there the waste will remain until someone can figure out where it should go, how it will get there, and how it will be stored when it arrives.

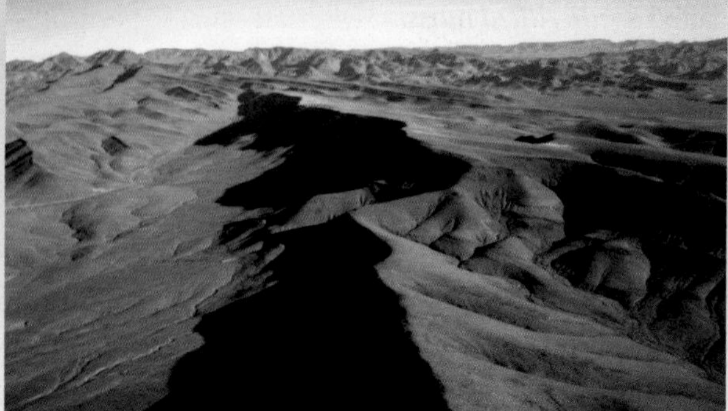

Figure 11A Yucca Mountain

Other countries have been more successful in dealing with nuclear waste. France, for example, gets more than three-quarters of its electricity from nuclear power, but its nuclear waste problem is minimal compared to ours. This is because the French have developed a way to recycle radioactive waste. The process produces some waste, but it's only a fraction of the 50,000 tons waiting for disposal in the United States. One French family using nuclear power over a 20-year period produces only about 25 milliliters of non-recyclable nuclear waste, a volume slightly smaller than an ice cube. What do the French do with nuclear waste that cannot be recycled? They put it into temporary underground storage repositories so that when scientists of the future figure out a way to recycle it, they can take it out and reuse it.

Meanwhile, back in the United States, lawsuits are piling up over nuclear waste buildup around the country. Utility companies that were promised a place to put their waste have had to pay huge amounts to store it themselves. Why doesn't the United States follow the French example? One reason is the government's concern that the radioactive isotopes produced in recycling could be stolen and used to make nuclear weapons that contain those isotopes. Additionally, some public figures believe that recycling nuclear material is too dangerous. At the time of this writing, Congress and the White House are locked in a debate over a plan to store nuclear waste temporarily. It seems that, at least for now, the Yucca Mountain storage plan will remain unrealized.

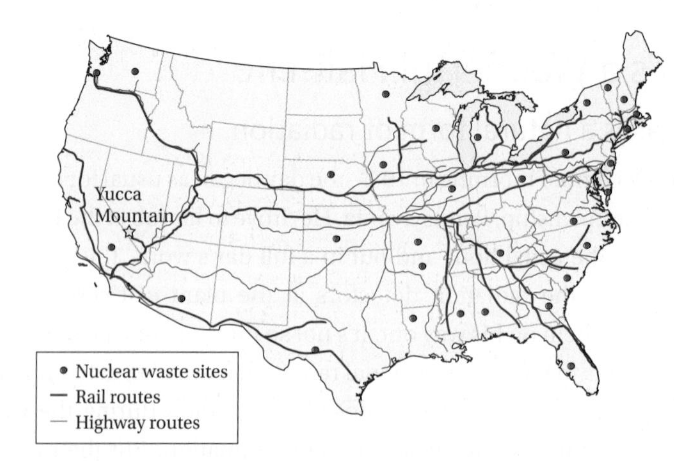

Figure 11B All Roads Lead to Yucca

By highway or by rail, the proposed transport routes for nuclear waste would pass through every state in the continental United States except Rhode Island, Delaware, North Dakota, South Dakota, and Montana. States that have at least one operating nuclear reactor are indicated with a red dot.

clothing tripped the radiation alarms at his job. Stanley was fortunate to work at a job where radiation levels are routinely checked. Otherwise, the extra-high levels in his house probably would have gone undetected.

The high levels of radiation in the Watras home were monitored by detectors that measure gamma radiation. Further testing in the house showed that the gamma radiation was due to the presence of the noble gas **radon,** Rn, atomic number 86. Radon gas does not interact chemically either with itself or with other substances. But radon is highly radioactive, and it decays to produce an alpha particle and gamma radiation:

$$^{222}_{86}\text{Rn} \longrightarrow {}^{218}_{84}\text{Po} + {}^{4}_{2}\alpha + \gamma$$

When we think of exposure to radiation, we normally worry about accidents at nuclear power plants or exposure from nuclear weaponry or X-ray machines, which are all sources of artificially produced radiation. However, radiation is also a natural phenomenon, and radioactive materials are found in Earth's crust. There is radioactive uranium in the ground in many regions across the United States. Areas with high uranium levels correlate with levels of radon found in homes: areas that have numerous deposits of uranium also have higher radon levels. **Figure 11.9** is a map of radon levels across the United States.

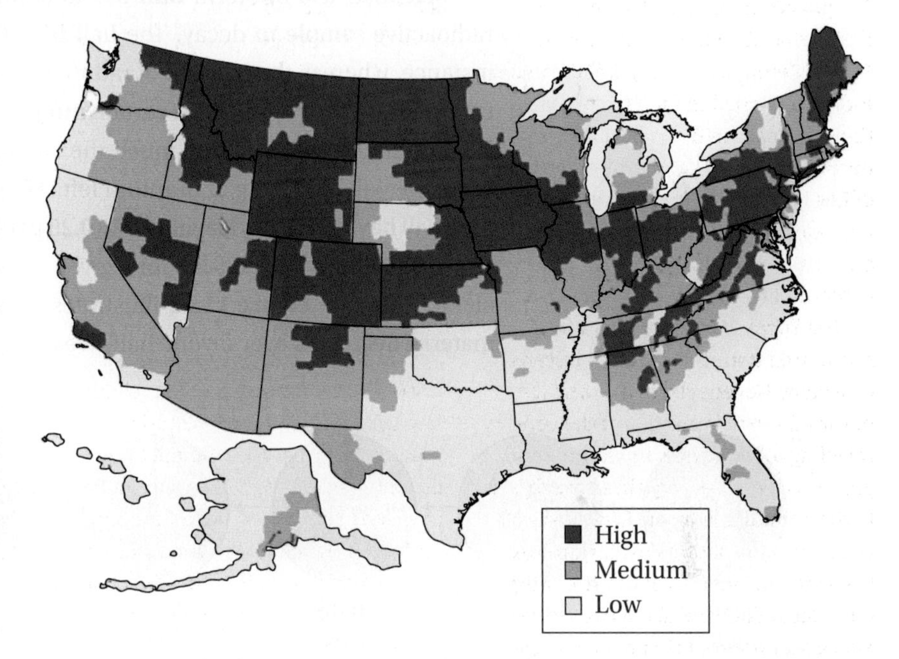

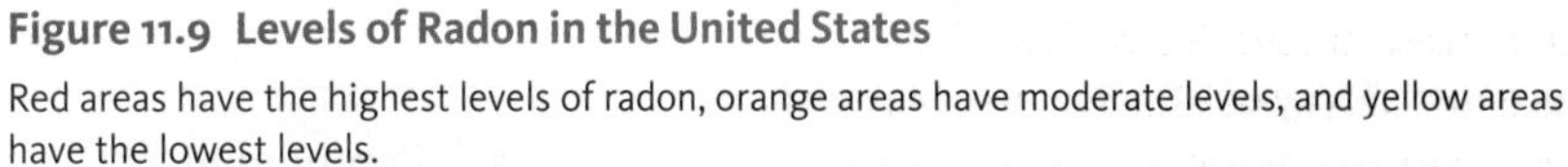

Figure 11.9 Levels of Radon in the United States

Red areas have the highest levels of radon, orange areas have moderate levels, and yellow areas have the lowest levels.

What does uranium have to do with radon? The answer is that radon is one of the elements that form as uranium-238 decays in a series of sequential nuclear reactions. Sequential strings of radioactive reactions are common because a radioactive isotope often forms another radioactive isotope, and the process continues until a stable isotope is reached. These sequential reactions are known collectively as a **radioactive decay series.** The radioactive decay series that leads to radon gas is shown in **Figure 11.10**. The production of alpha or beta particles in each step is shown above each arrow.

Once formed from uranium-238, radon-222 continues to emit radiation because it is itself radioactive. Seven more nuclear reactions occur until, at last, a stable isotope, lead-206, is formed.

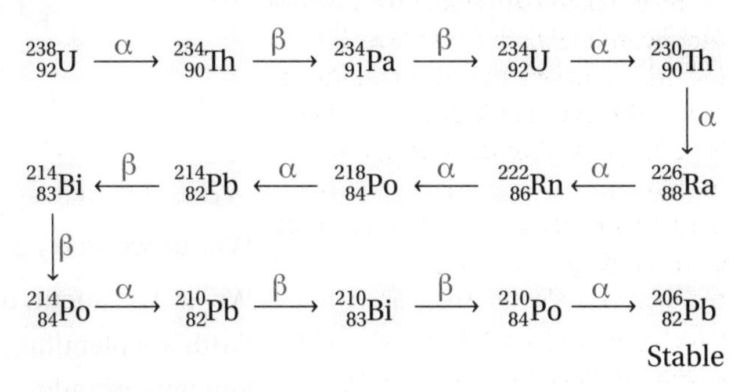

Figure 11.10 The Radioactive Decay Series That Produces Radon-222

Beginning with uranium-238, which exists in deposits in the Earth, a sequence of decay events leads eventually to radon-222, a gas. The radon-222 continues to decay until it becomes lead-206, which is not radioactive. Along the way, alpha and beta particles are emitted.

QUESTION 11.7 According to Figure 11.10, how many alpha particles and how many beta particles are produced in the radioactive decay series that produces lead-206 from uranium-238?

ANSWER 11.7 Eight alpha particles and six beta particles are produced in the entire radioactive decay series.

For more practice with problems like this one, check out end-of-chapter Question 11.

FOLLOW-UP QUESTION 11.7 What high-energy radioactive product that is not shown in the figure is made in some of the reactions shown in Figure 11.10?

Half-life is the time it takes for one-half of a radioactive sample to decay.

Uranium-235 and uranium-238 are highly radioactive, as is radon-222. And, according to Figure 11.9, some isotopes of lead are radioactive, but lead-206 is not. How do we keep track of how radioactive a substance is? Is there a practical way to determine how long a radioactive element will give off radiation? The answer is yes.

Scientists use the term **half-life** to define the time it takes for one-half of a radioactive sample to decay. The half-life of uranium-238 is 4.5×10^9 years, for instance, whereas the half-life of radon-222 is only 3.8 days. So, if we have 1 gram each of these two isotopes at a certain moment, then 3.8 days later we will have only 0.5 g of radon left but pretty much the whole 1 g of uranium-238. After 4.5×10^9 years, we will have 0.5 g of uranium left. After another 4.5×10^9 years, 50% of *that* 0.5 g will have decayed, leaving 25% (0.25 g) of the original amount remaining, and so on. The longer the half-life for a given radioactive element, the more slowly the element decays. **Figure 11.11** illustrates how the amount of a parent radioactive material decreases over several half-lives as it decays to make its daughter material.

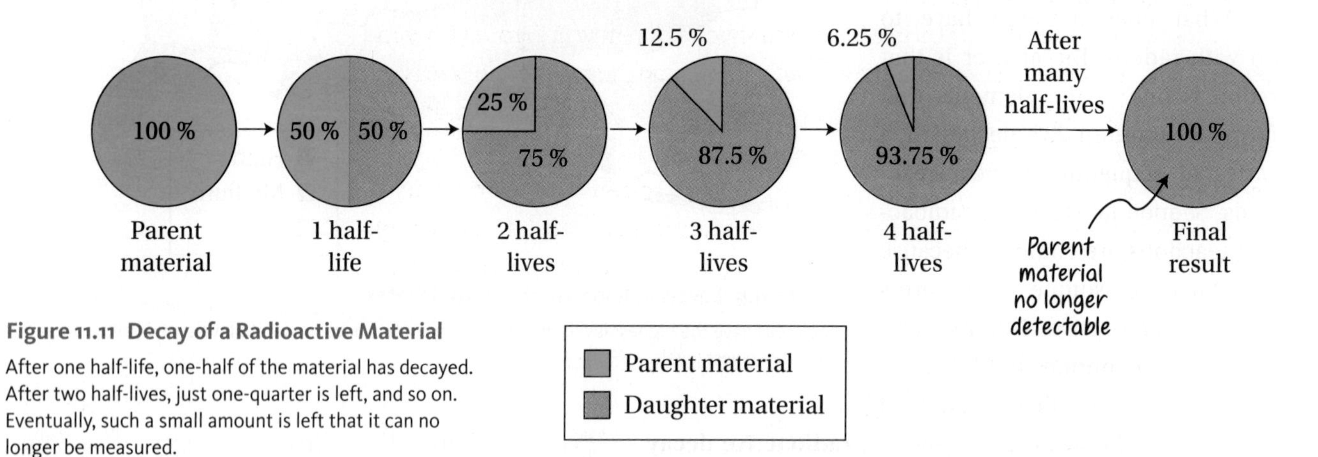

Figure 11.11 Decay of a Radioactive Material
After one half-life, one-half of the material has decayed. After two half-lives, just one-quarter is left, and so on. Eventually, such a small amount is left that it can no longer be measured.

wait a minute . . .

Why do we worry about uranium-238 decay if it decays so slowly?

We worry about uranium-238 decay because the uranium deposits on Earth are plentiful, and there is enough in the ground to release substantial amounts of radon even though the uranium-238 isotope has a half-life of 4,500,000,000 years. This uranium has been decaying for millions of years. So much uranium has decayed that a constant tide of radon-222 is emitted from the ground in areas where uranium deposits are found.

QUESTION 11.8 The half-life of iron-59 is 45 days. (a) How long does it take for one-half of a sample of iron-59 to decay? (b) How much of this iron isotope is left after 90 days?

ANSWER 11.8 (a) It takes 45 days for one-half of a sample of iron-59 to decay. (b) After 90 days, half of the half remaining after 45 days, or one-quarter, of the iron-59 remains. For more practice with problems like this one, check out end-of-chapter Questions 38 and 40.

FOLLOW-UP QUESTION 11.8 If a radioactive substance has a half-life of 100 years, how long does it take for three-quarters of a sample of the substance to decay?

Of all the isotopes in the uranium radioactive series, why do we worry most about radon-222? The answer is that only radon is a gas, and it seeps from uranium deposits in the ground and enters buildings through cracks and fissures in their foundations. Also, all the reactions from radon-222 to lead-210 in Figure 11.10 happen very fast. The radioactivity formed during these reactions, which includes four alpha particles and two beta particles, is released quickly as radon decays. If a person inhales radon and the radon decays before being exhaled, the decay products can damage lung tissue.

Should we be concerned about radon exposure? The answer is an unqualified yes. **Figure 11.12** shows a graph of radiation exposure for various sources of natural and artificial radiation. According to these data, more than half of the radiation an average person is exposed to in a typical year can come from radon. This number is greater than the exposure from all other types of radiation combined. Furthermore, it's estimated that nearly 10% of deaths due to lung cancer in the United States can be attributed to radon exposure.

Is it safe to assume that if you live in an area shaded in yellow in Figure 11.9, you are safe from radon exposure? Experts agree that although people living in areas of known uranium deposits are probably more likely to have high levels of radon in their houses, the values from house to house can vary. In areas where uranium is prevalent, it's common to find one house with a high level next door to a house with a very low level. This happens because the radon seepage rate into a building depends on the construction materials used, the age of the building, its ventilation,

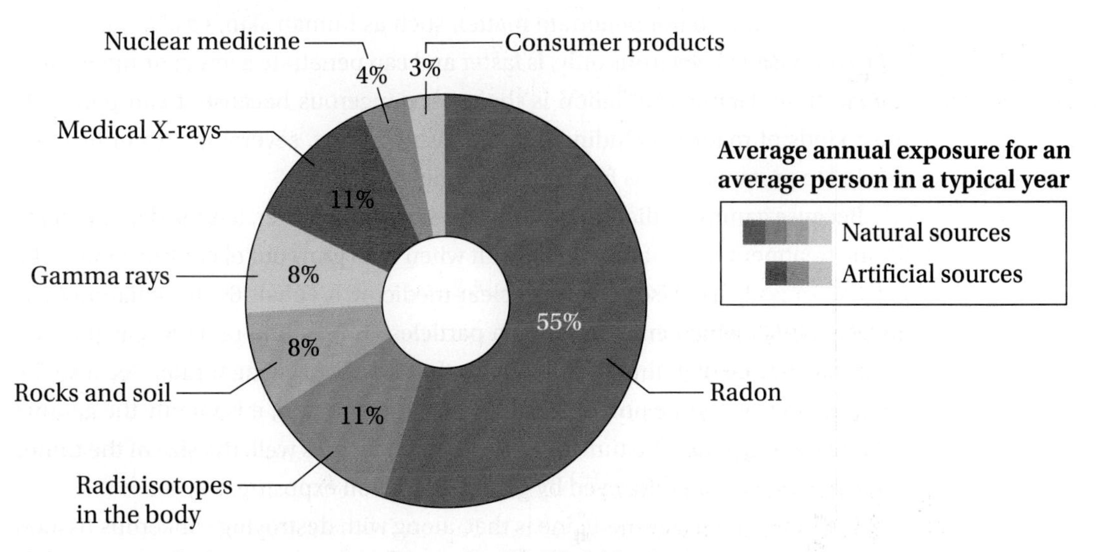

Figure 11.12 Radiation Exposure from Natural and Artificial Sources

and several other variables. And even for regions where uranium deposits are few and far between, it is possible to find buildings with very high radon levels.

The only way to be sure about the radon level in your own house is to have it tested. Stanley Watras and his family lived with radon levels 675 times higher than the allowable limit set by the U.S. government. The knowledge that radon levels were extraordinarily high probably saved their lives. The family moved out of the house immediately, and the Philadelphia Electric Company funded a remediation project in their house that included adding ventilation to the basement and sealing cracks in the basement floors and walls. These measures brought the levels of radon in the house down to acceptable levels, and they moved back in.

QUESTION 11.9 True or false? Artificial sources of radiation, such as medical X-rays, pose a greater threat to human health than natural sources of radiation.

ANSWER 11.9 False. According to Figure 11.12, natural sources of radiation, especially radon, present a significant health risk. A person's risk level depends on his or her exposure to each of the various sources.

For more practice with problems like this one, check out end-of-chapter Question 46.

FOLLOW-UP QUESTION 11.9 What artificial source of radiation causes the greatest exposure to humans annually?

11.4 Living Organisms and Radiation

Nuclear medicine uses radioactive isotopes to treat and diagnose diseases.

The three types of radiation we introduced in Section 11.1 differ. Gamma radiation is electromagnetic radiation of very high energy, whereas alpha and beta radiation consist of particles with positive and negative charge, respectively. Because they are dissimilar, these three types of radiation all interact differently with human tissue. **Figure 11.13** illustrates how each type of radiation has a different capacity for penetrating matter.

Alpha particles are comparatively large, slow-moving particles. Because they are so slow, they do not penetrate matter, such as human skin, easily. Beta radiation, composed of electrons only, is faster and can penetrate a few centimeters into human flesh. Gamma radiation is the most dangerous because it can penetrate most kinds of matter, including human tissue. In fact, several inches of lead are needed to protect humans from gamma radiation.

Because gamma radiation has strong penetrating power, it is used in medicine for the treatment of tumors, which result when cells grow out of control. One of the most commonly used isotopes in nuclear medicine is cobalt-60. In radiation therapy, cobalt-60, which emits both beta particles and gamma radiation, is put into a gun-like device that directs radiation to the site of a growing tumor. Because the beta particles penetrate only a short distance into the skin, it is mainly the gamma radiation that reaches the tumor. If the procedure works well, the size of the tumor decreases as cells are destroyed by gamma radiation exposure.

The irony of nuclear medicine is that, along with destroying cancerous tissues, the process kills healthy cells as well, making many patients undergoing radiation

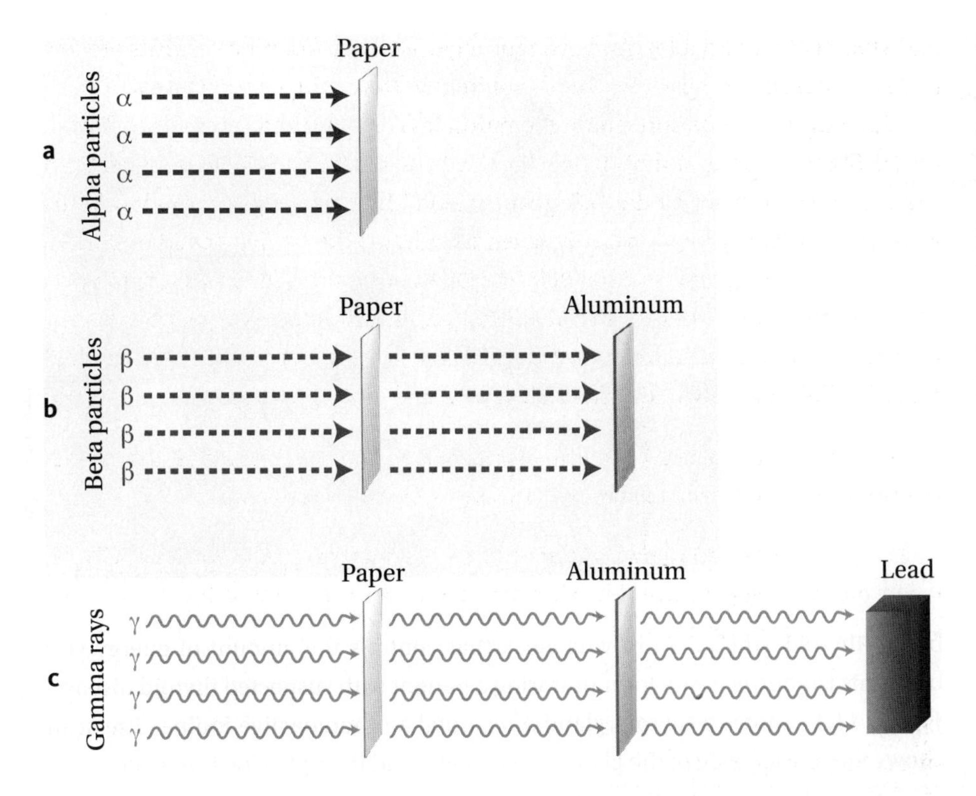

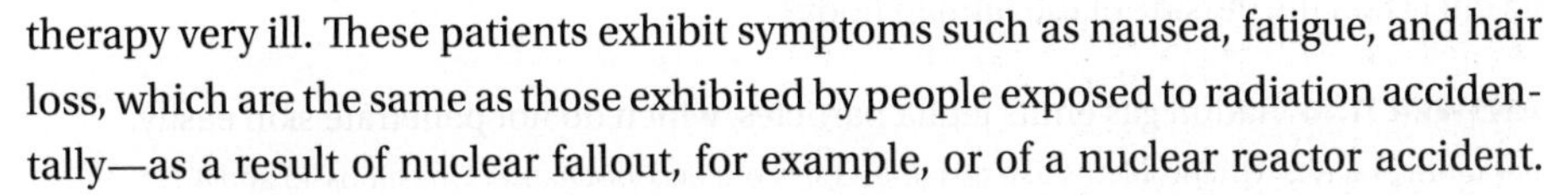

Figure 11.13 Relative Penetrating Powers of Radiation

(a) Alpha particles cannot penetrate paper, which means they also do not penetrate human skin. (b) Beta particles are more energetic and can penetrate paper. They are stopped by aluminum, however. Beta particles can penetrate skin to a depth of a few centimeters. (c) Gamma rays are able to penetrate paper and aluminum, but are stopped by a thick piece of lead. Gamma rays readily penetrate skin and are the most dangerous form of radiation.

therapy very ill. These patients exhibit symptoms such as nausea, fatigue, and hair loss, which are the same as those exhibited by people exposed to radiation accidentally—as a result of nuclear fallout, for example, or of a nuclear reactor accident.

Although they cannot easily penetrate skin, if the isotopes that emit these particles are ingested or inhaled into the body, the particles can damage healthy cells and tissues. Recall from Section 11.3 that radon undergoes a nuclear reaction that produces alpha radiation. Inside the body, alpha radiation is much more harmful than beta radiation, because the alpha particles interact with tissues in very localized areas and can irreparably damage the tissue in those small areas.

Although at times people absorb radiation inadvertently, some medical procedures require the patient to intentionally ingest a radioactive isotope called a **radioactive tracer,** which works inside the body in a controlled manner. Many tracers produce gamma radiation and beta particles, because potential damage from alpha particles is too difficult to control. For example, the tracer technetium-99 emits only gamma radiation. Its half-life is only 6 hours, which means a patient who ingests it will be nearly radiation-free in a few days.

After a patient ingests technetium-99, a specialized gamma camera scans the patient's body and detects gamma radiation. The camera moves around the body collecting images in three dimensions, images that a computer then converts into information that radiologists can use to diagnose abnormalities in the body. Technetium-99 moves quickly to the most active tissues in the body, and so it's especially useful for monitoring the brain and the heart as well as regions of arthritis or bone cancer.

Radioactive isotopes other than technetium-99 are also used in medicine. For example, the radioactive isotope iodine-131 is administered to patients in the form of radioactive sodium iodide, $Na^{131}I$. Because iodine is transported to and used

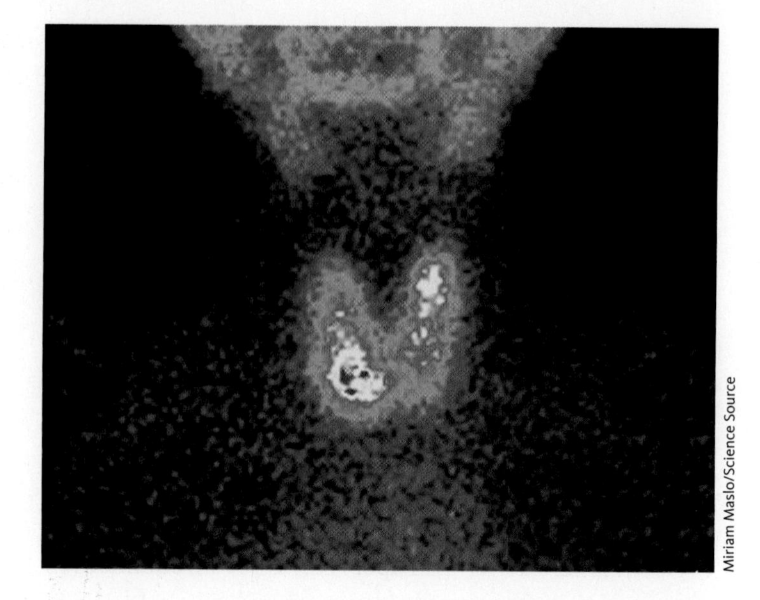

Miriam Maslo/Science Source

Figure 11.14 Radioactivity in Medicine

A healthy thyroid can be visualized using radioactive iodine.

by the thyroid—the gland in your neck that controls the amount of energy you use—this isotope is useful for monitoring a patient with suspected thyroid disease. **Figure 11.14** shows a thyroid gland after uptake of radioactive iodine. The scan shows that the left side of the gland is overactive and the right side is normal.

QUESTION 11.10 Radon gas is dangerous only when inhaled. Why don't we worry about its effects on the outside of the human body?

ANSWER 11.10 Radon gas emits alpha particles, which do not penetrate skin easily.

For more practice with problems like this one, check out end-of-chapter Questions 45 and 47.

FOLLOW-UP QUESTION 11.10 According to Figure 11.13, which types of radiation can penetrate a piece of paper?

There are different ways to express radiation dose and exposure.

The SI unit for the amount of radiation a human being is exposed to is the **sievert (Sv)**. In nuclear medicine, the commonly used unit for effective radiation dose is the **rem,** which stands for "*r*oentgen *e*quivalent for *m*an." By definition, 1 Sv = 100 rem. A typical person is exposed to about 0.3 rem every year, and symptoms of radiation sickness begin at doses of 25 rem. In other words, an individual would have to be exposed to about 80 times the normal annual dose of radiation to exhibit signs of mild radiation sickness.

Radiation affects the fastest-growing tissues in the body, because each new cell is exposed as it is created. Thus, the lymph nodes and the bone marrow—both places where cells multiply rapidly—are among the regions of the body hit hardest by high doses of radiation. Leukemia is the most common type of cancer associated with radiation exposure.

It's important to recognize that humans are constantly exposed to radiation from natural sources and artificial sources. Most of us will never accumulate enough radiation to cause radiation sickness, but it is advisable to limit exposure to low-level sources of radiation whenever possible.

QUESTION 11.11 Imagine that Homer Simpson, from the television series *The Simpsons,* is exposed to 0.50 Sv of radiation after an accident at the Springfield nuclear reactor. Will he exhibit symptoms of radiation sickness?

ANSWER 11.11 Homer has been exposed to 0.50 Sv $\times$ 100 rem/Sv = 50 rem of radiation. An analysis of Homer's blood might show that he has a temporary decrease in white blood cell count, but he is unlikely to exhibit other symptoms.

For more practice with problems like this one, check out end-of-chapter Question 50.

FOLLOW-UP QUESTION 11.11 Rank the following types of radiation in order of increasing ability to penetrate matter: gamma, alpha, beta.

THEgreenBEAT • news about the environment, chapter eleven edition

TOP STORY Are Our Nuclear Power Plants Safe Enough?

Every good real estate agent knows this mantra: location, location, location. Homebuyers often are drawn to waterfront property, or corner lots, or trendy neighborhoods. But there's one unlikely location that homebuyers are choosing more and more often: homes near nuclear power plants. In fact, the 2010 census showed that one in three Americans lives within a 50-mile radius of a nuclear power plant, and that number is up more than 6% from the previous census. Clearly, Americans seem to have confidence in their nuclear power plants (**Figure 11.15**).

Every nuclear plant is designed to withstand accidents. The design of the nuclear reactor includes control rods, which are inserted into the spaces in the core to dampen fission reactions. The plants have elaborate cooling systems to take heat away from the reactor core. The reactor core is kept inside a sturdy containment building. Used-up fuel is kept underwater in pools, and that water blocks radiation and keeps this spent fuel cool. And because many safety features of a nuclear power plant require electricity, the plant has backup generators that kick in if there is a power failure. When all of these systems are maintained and checked, a nuclear reactor should indeed be reasonably safe.

So what happened at Fukushima Daiichi in Japan? Was it a problem with the plants themselves, or was it just bad luck? It seems to have been a deadly combination of both. The problems started because Tokyo Electric (TEPCO), the utility company that runs the Fukushima plants, faced a public relations challenge. The Japanese public vocally opposed nuclear power, and new plants were not being approved. Therefore, TEPCO requested that its existing plants, many of them past their life spans, be extended and allowed to operate longer. TEPCO also wanted to save money, so the company skipped some safety inspections and stored excessive amounts of spent fuel at the Fukushima site. Japan's nuclear safety agency warned TEPCO about its aging diesel generators, which were cracked and vulnerable, and said that TEPCO did not carefully check and maintain the plant's cooling systems. TEPCO may have been guilty of shoddy maintenance on its reactors, but the company also was unlucky.

On March 11, 2011, when the tsunami struck the Fukushima Daiichi plants, the power went down. No power meant there was no circulating coolant. The backup diesel generators failed and water boiled away, leaving spent fuel rods

Flonline digitale Bildagentur GmbH/Alamy

Figure 11.15 In Our Back Yards

U.S. census data show that more and more Americans live near nuclear power plants than ever before.

(continued)

exposed and allowing the temperature at the reactor core to skyrocket. Explosions rocked the plants over the coming days, and the Japanese scrambled to get them under control. Radiation poured into the air and into the ocean, and contaminated air began to reach the United States and Canadian coastlines less than one week later. On April 4, 2011, a measurement of the milk supply in Hilo, Hawaii, showed levels of iodine-131, one of the fission products, at five times the EPA limit.

This example teaches us several things. First, nuclear power plants may be designed to be safe, but they're run by imperfect human beings whose first priority is not necessarily ensuring the safety of citizens near the plant. Second, nuclear power plants can fail under unlucky circumstances, such as a massive earthquake/tsunami. And third, one level of safety systems is not enough. When the stakes are as high as they are with a nuclear accident, these power plants must provide backups for backups for backups.

The Fukushima Daiichi plant was an American design, and nearly two dozen of the nuclear reactors in the United States are designed the same way. These stateside older-generation reactors may be a concern, especially if they are located in vulnerable areas such as earthquake zones.

AP Images/David Goldman

Figure 11.16 Safer Plants
A pair of modern reactors are being constructed near Augusta, Georgia.

New-generation reactors are currently being designed to include the concept of **passive nuclear safety**. Passive nuclear safety means that neither human nor electronic signals are required to initiate shutdown of the reactor. For example, if a problem arises and power is not available and no human being is nearby, the plant shuts down even though no computer tells it to do so and no human pushes the panic button. Passive safety also means the reactor is designed so that the fission reaction naturally stops when an emergency takes place. Now, as the United States is starting to build reactors again, these new-generation reactors may, one day, outnumber older-generation reactors like the ones in Fukushima. Of the five units currently under construction in the United States, four have modern designs that include passive safety features (**Figure 11.16**).

Chapter 11 Themes and Questions

recurring themes in chemistry chapter 11

recurring theme 1 Scientists worldwide use the scientific method as a basis for controlled experimentation and deduction. The scientific method is based on the principles of peer review and reproducibility. (p. 320)

recurring theme 4 Chemical reactions are nothing more than exchanges or rearrangements of electrons. (p. 321)

recurring theme 8 The Earth has a finite supply of raw materials that humans can use. (p. 327)

A list of the eight recurring themes in this book can be found on p. xv.

ANSWERS TO FOLLOW-UP QUESTIONS

Follow-Up Question 11.1: True. *Follow-Up Question 11.2:* Equations (a) and (b) are balanced; (c) is not balanced. *Follow-Up Question 11.3:* It is a nuclear equation because the element hydrogen is converted to the element helium. *Follow-Up Question 11.4:* Gamma radiation is a type of light, so it does not contain particles such as protons and neutrons. *Follow-Up Question 11.5:* Yes, they are both balanced. *Follow-Up Question 11.6:* Nine neutrons will result. *Follow-Up Question 11.7:* Gamma radiation is often produced along with alpha and beta particles. *Follow-Up Question 11.8:* 200 years. *Follow-Up Question 11.9:* Medical X-rays. *Follow-Up Question 11.10:* Beta particles and gamma rays can penetrate paper. *Follow-Up Question 11.11:* Alpha $<$ beta $<$ gamma.

End-of-Chapter Questions

Questions are grouped according to topic and are color-coded according to difficulty level: **The Basic Questions***, **More Challenging Questions****, and **The Toughest Questions*****.

The Nature of Nuclear Reactions and Radioactivity

***1.** List three ways that nuclear reactions differ from chemical reactions.

***2.** In general, which type of reaction is more energetic, a nuclear reaction or a chemical reaction? By roughly what factor?

***3.** Indicate the total number of neutrons, protons, and electrons in each isotope.

(a) $^{10}_{5}B$ (b) $^{90}_{38}Sr$ (c) $^{170}_{78}Pt$

***4.** According to Figure 11.10, in the radioactive series that begins with uranium-238 and runs all the way to the formation of polonium-218:

(a) How many alpha emissions are there?

(b) How many beta emissions?

***5.** Write the complete symbol (including subscript and superscript) for the isotope that contains the following:

(a) 56 protons and 82 neutrons (b) 35 protons and a mass number of 76

***6.** Write the complete symbol (including subscript and superscript) for the isotope that contains the following:

(a) 50 protons and 69 neutrons (b) a mass number of 84 and 47 neutrons

****7.** Write the complete symbol (including subscript and superscript) for the isotope that has the following characteristics:

(a) contains 52 neutrons and has an atomic number of 48

(b) has a mass number of 203 and an atomic number of 98

(c) is the isotope of ytterbium containing 84 neutrons

****8.** Lise Meitner (Figure 11.2) contributed much to our modern-day understanding of the atom. List her accomplishments in the field of nuclear chemistry.

AP Images

**9. We know that radon levels in buildings correlate somewhat with the presence of uranium in the ground under the buildings.

(a) How is uranium related to radon?

(b) How many nuclear reactions are required to convert uranium to radon? (Hint: Refer to Figure 11.10.)

(c) In these steps, how many alpha emissions occur?

(d) How many beta emissions occur?

***10. In general, are elements that have radioactive isotopes heavier elements or lighter elements? List three radioactive isotopes mentioned in this chapter to support your answer.

***11. Refer to Figure 11.10 to answer the following question: This radioactive decay series shows several elements, such as lead, more than once. How is this possible? Is the lead that appears in this decay series the same form of lead each time?

***12. New element 114, created in Russia in 1999, is just below lead on the periodic table. Suppose the owner of some manufacturing company has decided that this new element should be made by the truckload to use in place of lead in things such as batteries. As science consultant to the owner, what would you say to her about the validity of her proposal?

***13. Imagine that element 115, containing 145 neutrons, has just been created. Its name will be simpsonium, symbol Sp. (a) Write the complete symbol (subscript and superscript) for this isotope. (b) Under which element on the periodic table will this new element be placed?

Nuclear Equations

*14. Indicate whether each equation is a chemical reaction or a nuclear reaction.

(a) ${}^{174}_{77}Ir \rightarrow {}^{170}_{75}Re + {}^{4}_{2}\alpha$ (b) $H_2O \rightarrow H^+ + OH^-$ (c) $N_2 + 3H_2 \rightarrow 2NH_3$

*15. Indicate whether each equation is a chemical reaction or a nuclear reaction.

(a) $C_7H_{16} + 11O_2 \rightarrow 7CO_2 + 8H_2O$ (b) ${}^{12}_{6}C + {}^{4}_{2}He \rightarrow {}^{16}_{8}O$

(c) $CO + O \rightarrow CO_2$

*16. List three ways that a typical nuclear equation differs from a typical chemical equation.

*17. If you look at a typical nuclear equation, is it possible to tell whether the reaction produces gamma radiation? Why or why not?

*18. Which of these two reactions would you expect to produce more energy, as written?

(a) $C_3H_8 + 5O_2 \rightarrow 3CO_2 + 4H_2O$ + energy

(b) ${}^{1}_{0}n + {}^{235}_{92}U \rightarrow {}^{137}_{52}Te + {}^{97}_{40}Zr + 2{}^{1}_{0}n$ + energy

**19. In the mid-twentieth century, certain types of ceramic dinnerware were made with a surface glaze that contained uranium. If the isotope in the glaze is uranium-238, write a nuclear equation showing how this isotope decays via emission of one alpha particle. An alpha particle is designated as ${}^{4}_{2}\alpha$.

**20. Compare and contrast how nuclear equations and chemical equations are balanced. What must be equal on each side of a balanced nuclear equation?

**21. Polonium-218 is one of the isotopes formed from the decay of radon-222. When it decays, it produces one alpha particle and lead-214. Write a balanced nuclear equation to show this process. Recall that an alpha particle is represented as ${}^{4}_{2}\alpha$.

**22. Is each nuclear equation balanced or unbalanced?

(a) ${}^{40}_{19}K \rightarrow {}^{40}_{20}Ca$ (b) ${}^{1}_{1}H + {}^{11}_{5}B \rightarrow 3\,{}^{4}_{2}He$

**23. Is each nuclear equation balanced or unbalanced?

(a) ${}^{25}_{13}Al + {}^{4}_{2}He \rightarrow {}^{30}_{14}Si + {}^{1}_{1}H$ (b) ${}^{122}_{53}I \rightarrow {}^{122}_{54}Xe + {}^{1}_{0}n$

***24. Refer back to the discussion of the types of radioactive decay in Section 11.1 to answer this question. (a) Which type of radiation results in a change in the atomic number of the reactant but not a change in its mass number?

(b) Which type of radiation results in a change to both the mass number and the atomic number of the reactant?

***25. What is represented by the question mark in each equation?

(a) ${}^{222}_{86}Rn \rightarrow {}^{218}_{84}Po + ?$ (b) ${}^{14}_{7}N + ? \rightarrow {}^{17}_{8}O + {}^{1}_{1}H$

***26. Identify what the question mark stands for in each reaction.

(a) ${}^{64}_{29}Cu \rightarrow {}^{64}_{30}Zn + ?$ (b) ${}^{64}_{29}Cu + ? \rightarrow {}^{64}_{28}Ni$

Nuclear Power Plants

*27. About 4 kilograms (kg) of fissionable material constitutes a critical mass of uranium. If a bomb is designed to contain two portions of fissionable material, each with a mass of 6 kg, what engineering flaw might cause this bomb to function incorrectly? What undesirable event may occur?

*28. In your own words, explain what is meant by the term *critical mass.* Why must nuclear power plants carefully monitor the amount of nuclear fuel they employ?

*29. Refer to Figure 11.7 to answer this question. What is a chain reaction, and why, specifically, does the decay of uranium-235 create one?

**30. In the design of a nuclear power plant shown in Figure 11.6, how is electricity produced? Does the nuclear fission reaction produce electricity directly? Explain.

**31. What safety features are present in all nuclear reactors, even older-generation reactors?

**32. Define the term *passive safety.* Cite one example of a passive safety feature that will be used in new-generation reactor designs.

**33. Why is it necessary for nuclear plants to have complex cooling systems? In Fukushima, what happened when cooling systems were compromised?

***34. In your own words, describe the potential drawbacks and benefits to increasing the use of nuclear reactors as a major power source. Why might an environmentalist discourage the use of nuclear fuel as a source of energy in the United States? Why might an environmentalist support the use of nuclear power?

*****35.** The reaction in a typical nuclear reactor produces the isotopes krypton-91 and barium-142 plus some neutrons. Imagine that instead of breaking up into these two isotopes, 1 mol of uranium in the reactor breaks up into 2 mol of a single isotope plus at least one neutron and no other type of radiation.

(a) Using the minimum number of neutrons required, write a balanced nuclear equation showing this process.

(b) Would this reaction promote a chain reaction? Why or why not?

*****36.** In France, nuclear power has been used with success for several decades. Explain what the French have done to address some of the environmental challenges associated with using nuclear power.

Half-Life

***37.** In your own words, define the term *half-life.* What do we know about a radioactive isotope if we know its half-life?

***38.** The isotope tritium formerly was used to make watch dials glow. The half-life of tritium, $^{3}_{1}H$, is 12.3 years. What fraction of a sample of tritium remains after 24.6 years?

***39.** In Figure 11.11, what do the terms *parent material* and *daughter material* mean? List one example from the chapter of a pair of parent/daughter isotopes.

****40.** The half-life of polonium-214 is 1.6×10^{-4} s. How many microseconds will pass before 75% of a sample of this isotope has decayed?

****41.** The isotope potassium-40 has a half-life of 1,280,000,000 years.

(a) Express this half-life using scientific notation.

(b) How many minutes have passed when only 12.5% of a sample of this isotope remains?

*****42.** After 936 years, the amount of isotope X remaining after fission is 25% of the original mass. What is the half-life of this isotope?

*****43.** You have a 3.00-gram sample of $^{51}_{24}Cr$, which has a half-life of 27.8 days. How long will it take for this sample to decay until only 0.75 grams of $^{51}_{24}Cr$ remains?

Radiation and Living Organisms

***44.** According to Figure 11.12, what percentage of the radiation that a person receives comes from natural gamma radiation?

***45.** According to Figure 11.13, which types of radiation are blocked by a thick piece of lead?

***46.** Use the information from Figure 11.12 to answer these questions:

(a) What is the total percentage of radiation exposure from natural sources?

(b) Of that total, what percentage can be attributed to rocks and soil?

(c) What percentage of radiation exposure from artificial sources is attributable to medical X-rays?

****47.** Which type of radiation most easily penetrates human skin? In your own words, explain why this is important to human health.

****48.** In your own words, explain how radioactive isotopes are used in medicine. Cite one example from the chapter.

****49.** In your own words, explain why certain parts of the human body are affected by radiation before other parts are. What do the most quickly affected parts of the body have in common?

*****50.** Where do gamma rays fall on the electromagnetic spectrum? Compared to X-rays, do you think they are more or less harmful to humans?

****51.** Radioisotopes have been used as murder weapons. Alpha particles, once ingested, can do considerable damage to the body, but the radioactivity is not detectable outside of the body. In one case, a man who was poisoned with a radioisotope did not give off any radioactivity, so the identity of the poison was not immediately determined. Eventually, it was determined that the culprit was polonium-210, which is an alpha emitter with a half-life of 138 days. If a murder victim is given a 20.00-milligram dose of polonium on January 1 and if the victim dies the same day, roughly how much of the radioactive poison remains in the corpse on May 18?

****52.** Radioactive fuel is ranked according to its potential for use in a nuclear weapon. "Super weapons grade fuel" is mostly plutonium-239, but it also contains less than 3% plutonium-240, an unwanted contaminant. "Weapons grade fuel" contains up to 7% plutonium-240. "Fuel grade fuel" contains 7–19% plutonium-240 and "reactor grade fuel" contains more than 19% plutonium-240. A sample of radioactive fuel has a mass of 64.53 grams of plutonium-239 and 4.30 grams of plutonium-240. What is the grade of this mixture?

*****53.** This question is a continuation of Question 5.52. Plutonium-239 and plutonium-240 have half-lives equal to 87.74 years and 6500 years, respectively. Consider the sample discussed in Question 5.52. If the decay of the plutonium-240 in the sample is negligible, what is the mass of the whole sample after 87.74 years?

*****54.** Many modern spacecraft use Radioisotope Thermoelectric Generators or RTGs, which are a type of nuclear-powered battery. These have worked well because they are simple and don't often require repair. They also work better than solar panels in deep space where sunlight is limited. RTGs take advantage of the decay of plutonium-238, an alpha and gamma emitter.

(a) Which of the two types of plutonium-238 emission is light, and which is particulate in nature?

(b) Write a nuclear equation for the decay of Pu-238 via alpha emission. What radioisotope is produced?

(c) If Pu-238 has a half-life of 87.74 years, how long will it take for a 1.00-gram sample of it to decay to 12.5-grams?

55. INTERNET RESEARCH QUESTION

Go to your search engine and type the words, "nuclear power plant near me." How many miles is the nearest nuclear power plant to your current location? Most nuclear plants are one of two types: BWR or PWR. "BWR" stands for "boiling water reactor." In this type of power plant, the reactor core heats water to steam, and the steam turns a turbine, which generates an electric current. "PWR" stands for "pressurized water reactor." These plants operate with water that is heated, but not to boiling. Investigate the power plant closest to you. Can you determine which one it is? Can you determine if your nearest plant has passive controls installed?

GH1/Pavegen/WENN.com/Newscom

Energy, Power, and Climate Change

New Ways to Generate Power and Store Energy

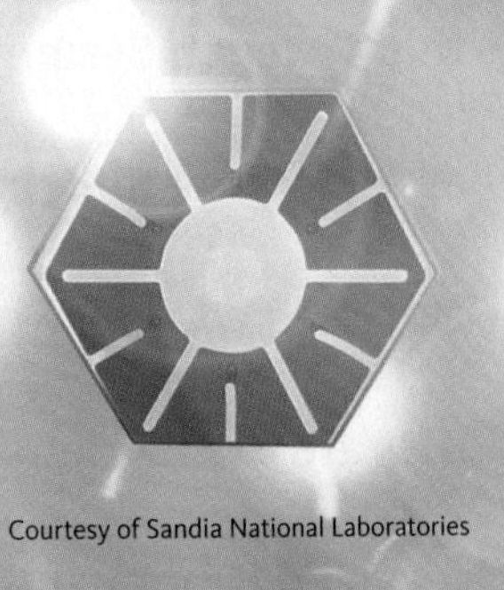

Courtesy of Sandia National Laboratories

CHAPTER 12

The Simon Langton Grammar School for Boys sits in the countryside near the medieval town of Canterbury, England. In many ways, it's just like other private schools in Great Britain. What sets it apart is one particular hallway—the most heavily trod on the whole campus. In that hallway, the lighting is not connected to the electrical grid. Instead, it's connected to unique floor tiles that pepper the hallway from end to end. These revolutionary tiles use the energy from student footfalls to power the hallway lights as well as small lights in the tiles that are illuminated when someone steps on them.

These tiles are the brainchild of Laurence Kimball-Cook, a twenty-something British inventor and designer who founded the company, Pavegen Systems, that sells the tiles. When a foot strikes a Pavegen tile, the tile depresses about 5 millimeters, a distance barely discernible to the walker. The friction generated by this tiny movement is converted to electricity, a flow of electrons, by a still-secret method that Pavegen Systems has not revealed. About 5% of the energy generated from footfalls goes to lighting the tile itself, and the rest can be stored in a battery or used for anything that requires electricity. The tiles are earth-friendly and practical in other ways: they're made from recycled rubber and can be used outdoors in any weather.

Pavegen Systems may have a patent on its own floor tiles, but it does not have a monopoly on this groundbreaking idea. A company in the Netherlands, Enviu, has developed a spring-loaded dance floor that converts dance into power. Nightclubs with bone-jarring speakers and dazzling light shows are notorious power hogs. But this new technology will eventually allow dance power to offset some, or perhaps all, of the power usage at night clubs.

GH1/Pavegen/WENN.com/Newscom

One goal of this chapter is to discuss exciting, original sources of energy like this one. We look at several examples of new ways to store energy and generate power. But first we'll talk about the terms *energy* and *power,* to get a deeper understanding of what they are and how they differ. We'll also consider the predominant source of energy that we now use, fossil fuels, and see why it's imperative that we start to focus on other sources. Finally, we'll turn our attention to new energy ideas, especially those that tap our most underutilized energy resource, the sun.

12.1 Energy and Power

Energy can be converted from one form to another.

In the simplest terms, **energy** is the ability to apply a force over a distance, and a **force** is a push or pull on an object.

Here's an example that will help clarify the nature of energy. Imagine an enormous lake that has a dam, which is closed to hold the water in the lake. When a sluice gate is opened in the dam, water pours out of the lake onto a water wheel and strikes its paddles (**Figure 12.1**). Before the water begins to flow, the water wheel is still; but the wheel begins to turn when the force of the moving water hits its paddles. The force of the water on the paddles moves them around and around, over a distance. According to our definition, the moving water is a source of energy. And we can say that the energy of this moving water as it strikes and turns the wheel is "converted to" or "transferred to" the energy of the moving wheel.

From this example, we can see how it's possible to convert one type of energy to another type of energy. In fact, this happens constantly in the world around us: a blender takes electrical energy and converts it to the energy of chopping blades; a mechanic takes energy from the food she has eaten and converts it to the energy of a turning screwdriver; or a car takes the energy stored in its fuel and converts it to the energy of movement.

> **flashback**
> Recall from Chapter 5 that the metric unit for energy is the joule, represented with the capital letter J.

We can put the various types of energy that exist in the world into categories such as nuclear, light, electrical, mechanical, thermal, and chemical. **Figure 12.2** shows examples of these various types of energy as well as how energy can shift from one form to another.

Snehitdesign/iStock/Getty Images

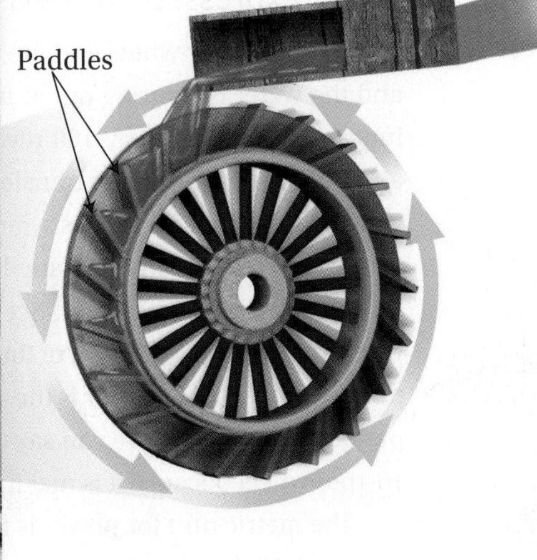

Figure 12.1 A Water Wheel
A water wheel converts the energy of falling water into the energy of the moving wheel.

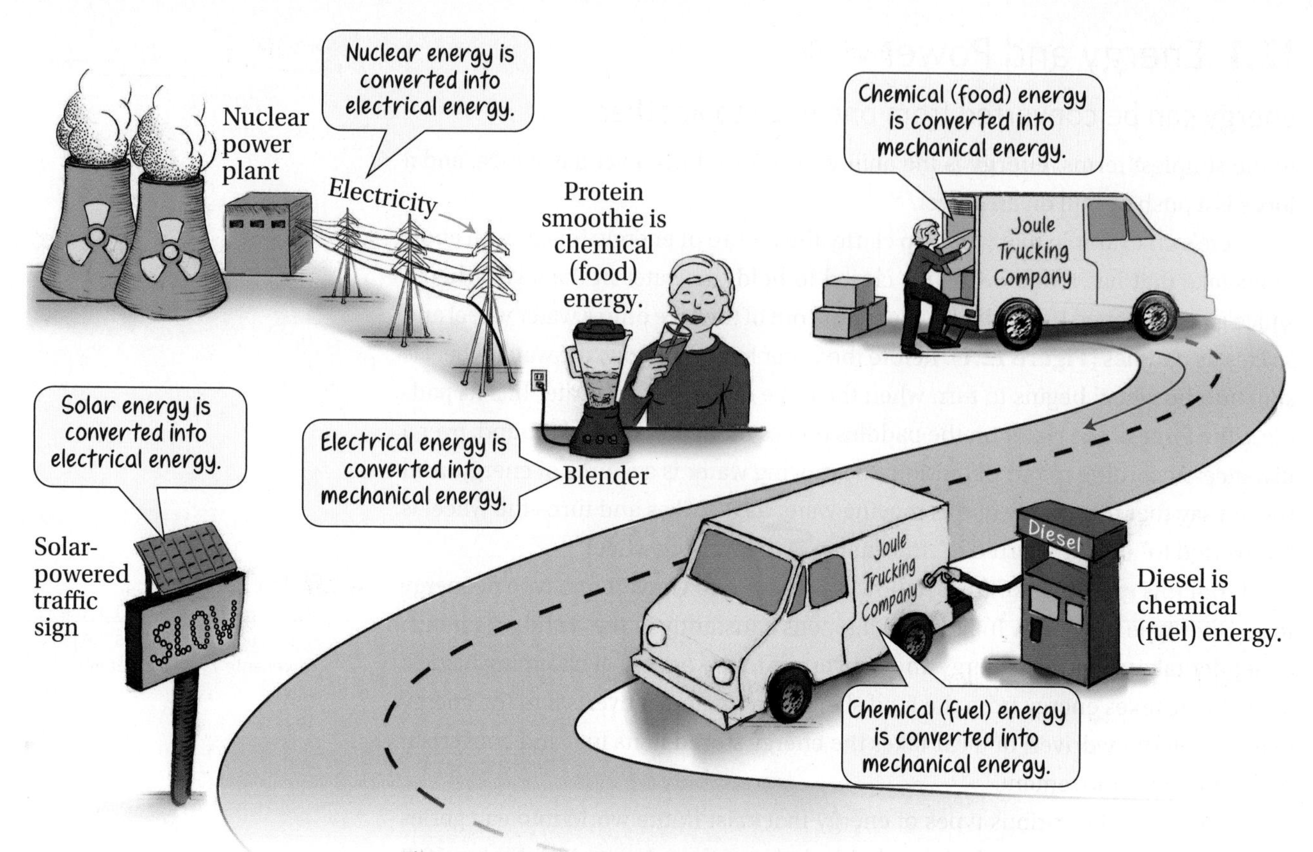

Figure 12.2 Energy Can Be Transferred from One Form to Another

Energy generated in a nuclear power plant is converted to electrical energy, which is used to power a blender. The blender converts the electrical energy to mechanical energy, the energy of work being done, such as blender blades turning. The protein smoothie made in the blender provides chemical energy in the form of food to the person who drinks it. The person converts that chemical energy to mechanical energy and moves boxes onto a truck. The truck moving the boxes stops for diesel fuel, which contains chemical energy. The truck burns the fuel and converts its chemical energy into the mechanical energy of moving wheels and turning axles. Finally, energy from the sun shining on a solar panel is converted to electrical energy, which is used to power emergency warning lights on the highway.

Power expresses energy use over time.

Let's return to our wheel and the water that turns it. Suppose the dam is fully open and the water is pouring out at full force, so that the water wheel is turning at an impressive rate, let's say 100 revolutions per minute (rpm). Clearly, a significant amount of energy is being transferred from the moving water to the turning wheel. Now imagine what happens when the water flowing through the sluice is significantly reduced. Much less water hits the paddles of the wheel, and the wheel turns more slowly, let's say 10 rpm. Because we are now thinking about the energy of the wheel over a specific period of time—in this case, one minute—we are defining the power of the wheel. **Power** is the amount of energy generated or used in a specified period of time. Thus, we can say that the 100-rpm wheel has more power than the 10-rpm wheel because it is making more turns per minute (**Figure 12.3**).

The metric unit for power is the **watt,** designated with the symbol W. A watt is the number of joules of energy transferred per second. For example, whenever it is switched on, a 100-watt light bulb uses 100 joules of energy every second. Unlike a 100-watt light bulb, your laptop might use 30 watts of power when it's sitting idle, but 60 watts when you're playing Grand Theft Auto with your monitor at full brightness.

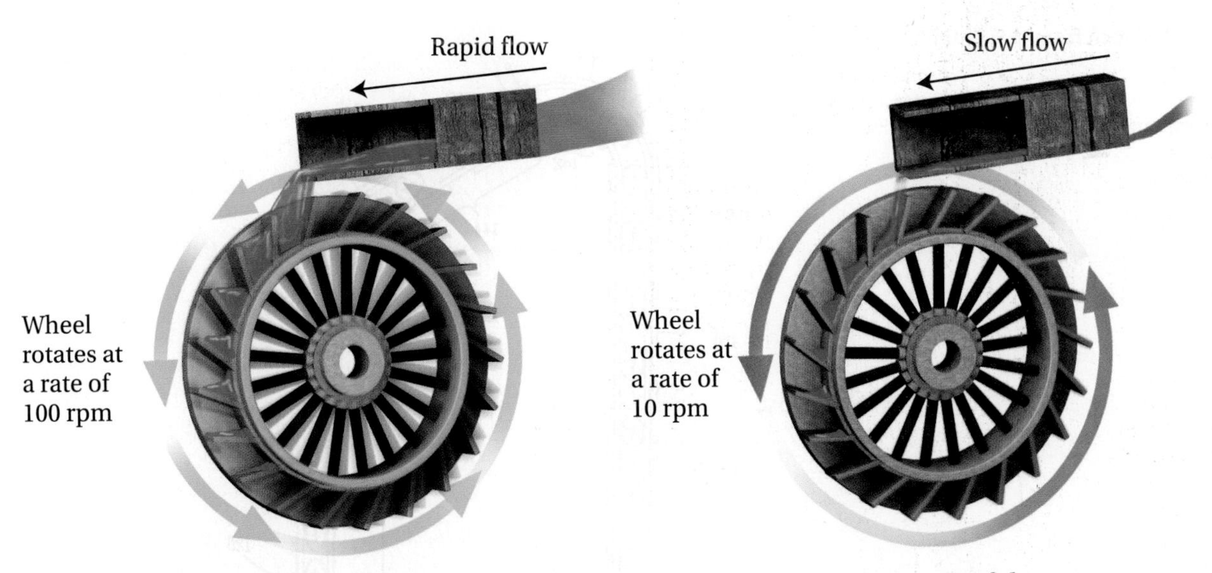

Figure 12.3 Two Water Wheels
The wheel that turns faster is more powerful than the wheel that turns more slowly.

QUESTION 12.1 A light bulb uses 300 joules (J) every two seconds. What is the wattage of the light bulb?

ANSWER 12.1 The number of watts is equal to joules of energy per second. In this example, 300 joules are used every two seconds, so the light bulb uses 150 joules each second. It is a 150-watt light bulb.

For more practice with problems like this one, check out end-of-chapter Question 6.

FOLLOW-UP QUESTION 12.1 Which light bulb uses more power, one that uses 75 watts of power or one that uses 500 joules every 5 seconds?

Energy cannot be created or destroyed.

Anyone who has ever had to change an old-fashioned incandescent light bulb knows that these bulbs produce more than light. Higher wattage incandescent bulbs produce so much heat, they cannot be touched when they are lit. These very hot bulbs are not very efficient at converting electrical energy into light energy. In fact, these bulbs convert only about 2% of their electrical energy to light; the remaining 98% becomes heat, or *thermal energy*. Light-emitting diode bulbs (called LEDs) do a bit better. They convert up to about 18% of the electrical energy they receive into light, and the rest becomes heat.

All light bulbs everywhere have one thing in common: 100% of the energy that goes into the bulb is converted to other forms of energy (**Figure 12.4**). That's because one form of energy can be converted to other forms of energy, but energy always persists—it doesn't just disappear. In the case of the light bulb, part of the electrical energy it receives may be converted to heat or to light, but all of it turns into some form of energy and goes somewhere.

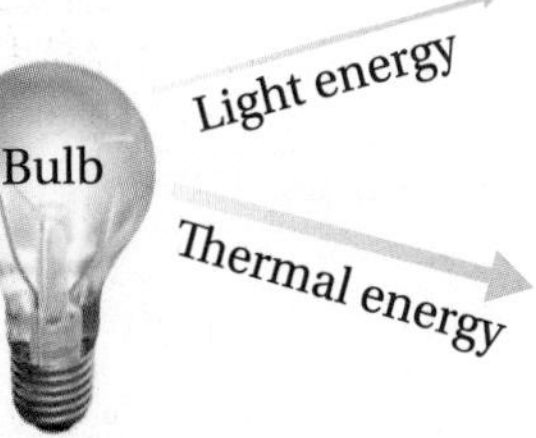

Figure 12.4 All of the Energy Going into a Light Bulb Leaves as Light or Thermal Energy

Figure 12.5 **A Sketch from U.S. Patent 6,734,574 B2**

The inventor of this device, which was patented in 2004, claims that it produces more energy than it consumes.

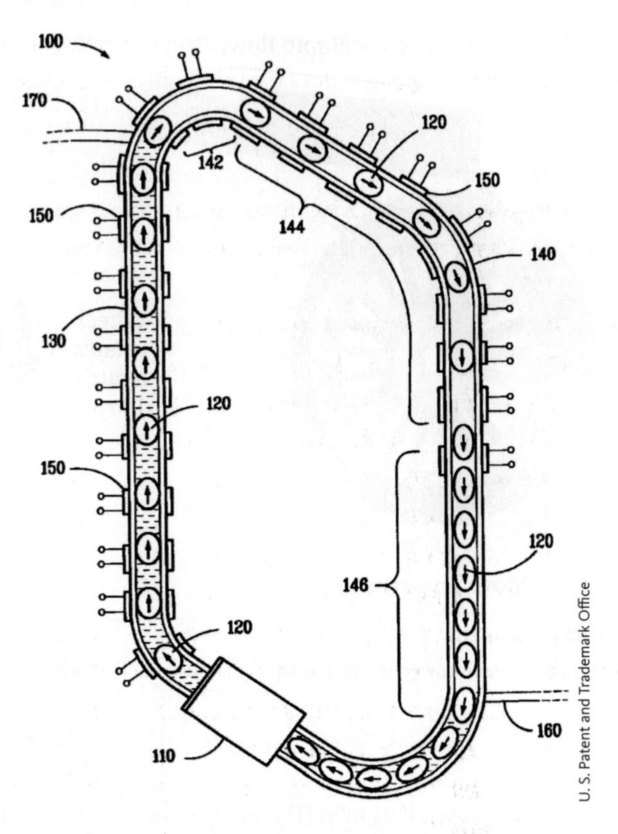

The behavior of energy and power are governed by one fundamental, absolute law of nature, the **first law of thermodynamics,** which states that energy cannot be created or destroyed. Despite this well-known fact of nature, a search of U.S. patents will generate a list of hundreds of gadgets that claim to produce energy from nothing. For example, the device shown in **Figure 12.5** claims to use a system of floating magnets to generate more energy than is put into the system. Again and again, machines purported to create new energy have been debunked, and no gizmo or process has ever successfully challenged the first law of thermodynamics.

wait a minute . . .

If energy cannot be created, then what do electrical generators do?

An electrical generator is a device that creates a useful form of energy, electricity, from some other form of energy. (We discuss electrical energy in Section 12.5.) For example, a steam turbine takes the energy of steam and uses it to turn a turbine, a wheel-like device similar to the water wheel described earlier in Section 12.1. The energy of the turning turbine can be converted into electrical energy. This machine is considered an electrical generator because it makes electricity, but it doesn't create energy from nothing. Rather, it converts a less usable form of energy—the energy of steam—into more usable electrical energy. The steam comes from water that was heated using the energy from a burning fuel. Thus, despite what their names seem to suggest, generators do not violate the first law of thermodynamics.

QUESTION 12.2 If light bulbs convert only a smaller percentage of the electrical power they receive into light energy, how does the light bulb obey the first law of thermodynamics?

ANSWER 12.2 According to the first law of thermodynamics, energy cannot be destroyed, so the energy that goes into the light bulb must go somewhere. As noted in the text, the remaining energy is emitted in the form of thermal energy (heat).
For more practice with problems like this one, check out end-of-chapter Question 7.

FOLLOW-UP QUESTION 12.2 A new light bulb converts 10% of the electrical energy it receives into light. If the light bulb receives 1000 J of energy, how many joules of light does it emit?

12.2 Fossil Fuels: What They Are and Where We Get Them

Fossil fuels are hydrocarbon mixtures.

Over the last 200 years, there has been a radical shift in energy use in the United States. **Figure 12.6** is a graph that plots the year on the horizontal axis and energy, in units of exajoules (EJ), on the vertical axis. The prefix *exa-* is a metric prefix that indicates its base unit is multiplied by 10^{18}. In other words, there are 1,000,000,000,000,000,000 joules in one exajoule. This graph breaks down energy consumption into the major energy sources that people in the United States have used over the years, including everything from wood to petroleum.

In the 1700s and 1800s, the primary source of energy in the United States was wood. During the industrial revolution of the late 1800s, machines began to replace human-powered production. Early machines were powered first by horses or other

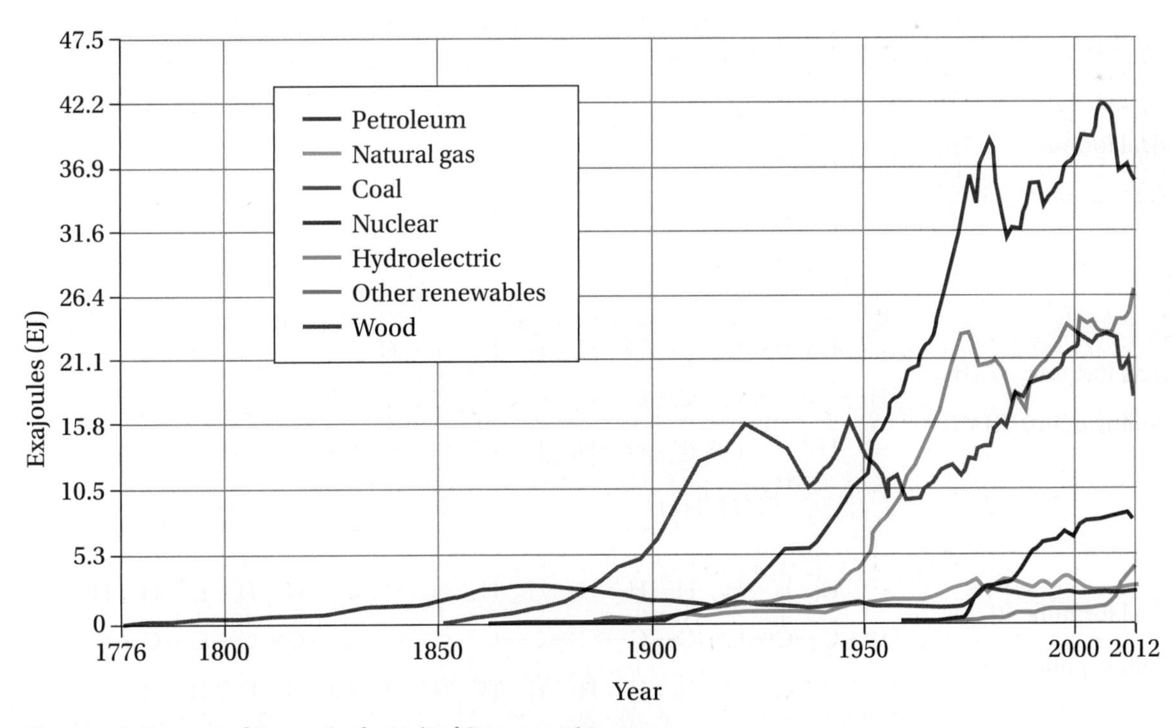

Figure 12.6 Sources of Energy in the United States, 1776 to 2012
This diagram shows how the United States has shifted its use of energy resources over the past 200-plus years. The vertical axis is given in exajoules (EJ). One exajoule (EJ) is equal to 10^{18} joules.

draft animals, then by flowing water, then by steam, and finally by coal. Coal was a dominant energy source in the early 1900s until it was surpassed by petroleum and natural gas, which is composed primarily of methane, CH_4. Coal had a resurgence when the technology was developed to create electrical energy in coal-fired power stations. Today petroleum, natural gas, and coal are still dominant, while nuclear energy and other sources such as hydroelectric (electricity generated by falling water), wind, and solar lag far behind.

Figure 12.6 presents some indisputable facts. First, the United States has dramatically increased its energy consumption over the past 200 years. Second, since the 1950s, petroleum, natural gas, and coal have dominated the U.S. energy profile, and there does not seem to be a strong trend toward any of the other energy sources listed. The big three—petroleum, natural gas, and coal—are all examples of **fossil fuels**. Fossil fuels are derived from organic matter, including plant and animal fossils, that has decayed over eons to make a mixture of carbon-based molecules. Within Earth, those carbon-based molecules became compressed over millennia and formed deposits of coal and oil.

Examples of molecules that you would find in various types of fossil fuels are shown in **Figure 12.7**. What do all of the molecules in this figure have in common?

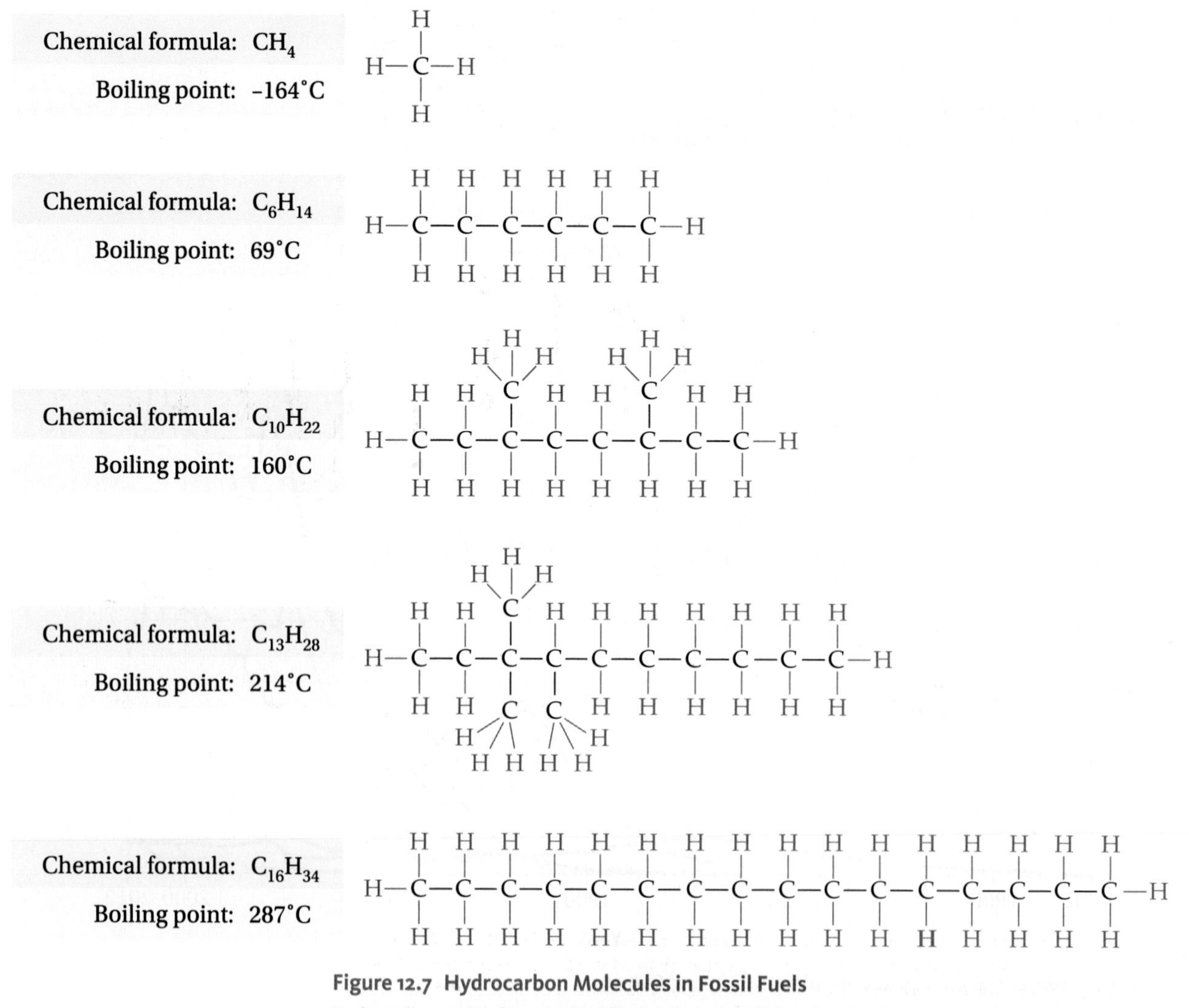

Figure 12.7 Hydrocarbon Molecules in Fossil Fuels

Hydrocarbon molecules come in different sizes. For each molecule, the structure, chemical formula, and boiling/condensation point, in degrees Celsius, are shown.

First, the atoms in the molecules are connected by lines, so we know that they are molecules and that the bonds are covalent. Second, all the molecules are hydrocarbons, because they are comprised of just two elements, carbon and hydrogen. Hydrocarbons come in different shapes and sizes. They can have chains that are straight or branched. They can even form rings.

recurring theme in chemistry 6

The structure of carbon-containing substances is predictable because each carbon atom almost always makes four bonds to other atoms.

QUESTION 12.3 According to Figure 12.7, is there a relationship between the number of carbon atoms and the boiling points of the molecules shown in the figure?

ANSWER 12.3 For the molecules shown in the figure, the lower the number of carbon atoms, the lower the boiling point of the hydrocarbon.

For more practice with problems like this one, check out end-of-chapter Questions 12 and 13.

FOLLOW-UP QUESTION 12.3 Which type of non-fossil-fuel energy is the most widely used in the United States today, according to Figure 12.6?

Refineries separate crude oil into usable fractions.

Fossil fuels, including natural gas, crude oil, and coal, are mixtures consisting mainly of hydrocarbon molecules. When crude oil is pumped out of the ground, it is a dark, viscous liquid, and it must be refined. Refining involves separating oil's various components on the basis of how many carbon atoms each component contains, as shown in **Figure 12.8**. The process that accomplishes this is called **fractionation**. In the fractionation process, the mixture of molecules in the crude oil is

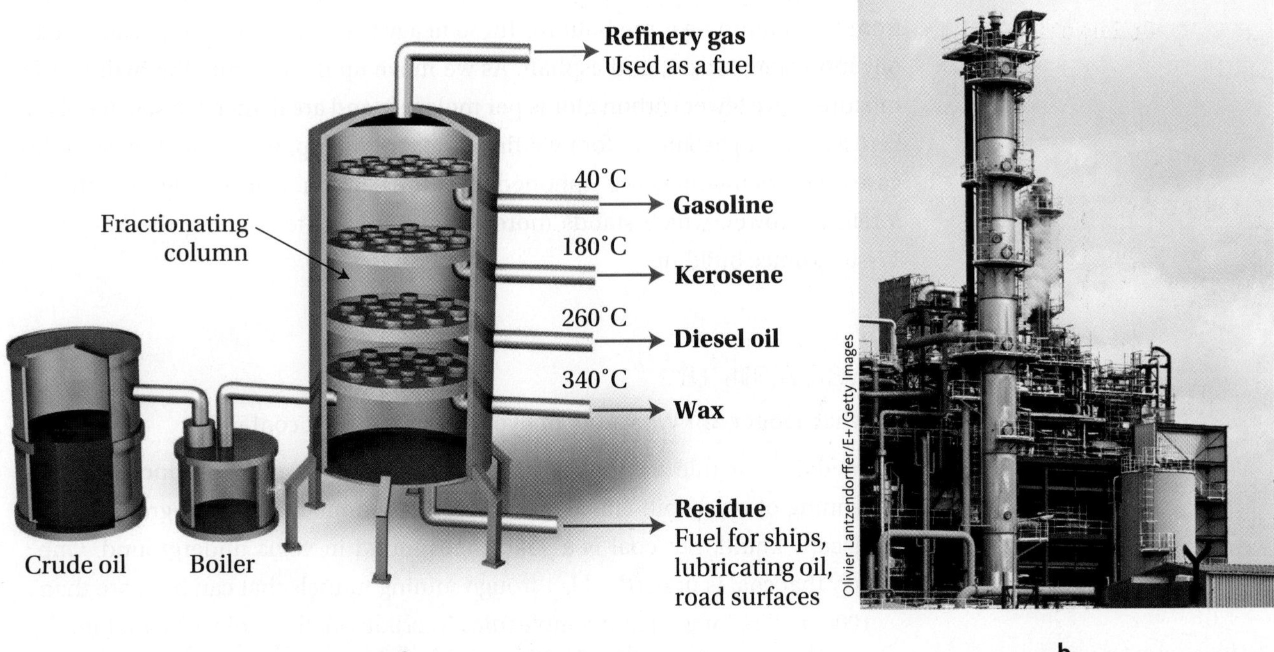

Figure 12.8 How a Fractionating Column Works

(a) The mixture of crude oil is heated to a temperature that converts most of the hydrocarbon molecules in the mixture into the gas phase. As the mixture travels up, it moves into progressively cooler regions of the column. When a given hydrocarbon molecule reaches the temperature at which it condenses, it returns to the liquid phase and is removed from the column along with other hydrocarbons with a similar carbon count. (b) Fractionating columns are contained within towers at oil refineries. Towers like these can be 30 stories high.

heated to a high temperature and then allowed to travel in the gas phase up a tall column. The *fractionating column* separates the hydrocarbons into fractions containing hydrocarbons with similar masses and physical properties.

Every hydrocarbon molecule has a boiling point, the point at which it turns from a liquid to a gas. This is also the condensation temperature of that hydrocarbon, the temperature at which it changes from a gas into a liquid. In Figure 12.7, the boiling points for the various hydrocarbons are listed. Is there a relationship between the number of carbon atoms in the hydrocarbon and its boiling point? Yes, there is. Methane has the lowest boiling point, and it has one carbon atom. The hydrocarbon with six carbon atoms has the next highest boiling point, followed by the hydrocarbons with 10, 13, and then 16 carbon atoms.

Why do hydrocarbon molecules with more carbon atoms generally have higher boiling points? Because the more carbon atoms a molecule has, the more surface area the molecule has. The more surface area the molecule has, the more it interacts with neighboring molecules like itself. As a result, a higher temperature is required to disrupt interactions between molecules if a hydrocarbon contains 16 carbon atoms than if it contains four carbon atoms.

The fractionating column gets progressively cooler toward the top, and this drop in temperature causes the hydrocarbons that distill off of the crude oil to condense back into the liquid phase. The heavier the hydrocarbon, the shorter its trip up the column, and the sooner it is converted from gas to liquid. Smaller hydrocarbons travel farther up the column before condensing, because they change from gases to liquids at colder temperatures. Methane travels all the way to the top of the column, where it is collected as a gas and sold as natural gas.

As Figure 12.8a shows, hydrocarbons with many carbon atoms are removed from the column near the bottom. These heavy hydrocarbon mixtures are used as oily lubricants, waxes, and asphalt. As we move up the column, the hydrocarbon mixtures have fewer carbon atoms per molecule and are lighter. We see diesel fuel, kerosene, and gasoline before we finally reach the top, where light hydrocarbon gases such as methane are siphoned from the column. Figure 12.8b is a photo of a refinery tower, which stands more than 50 meters tall—about the height of a 17-story office building.

wait a minute . . .

What about coal? What kind of hydrocarbons does it contain?

We didn't include coal on the list of hydrocarbons that are produced by the refining of crude oil. This is because oil is pumped out of the ground as a viscous liquid, but coal is a solid that's found in veins underground. One way that coal is unearthed is through mining tunnels that can be more than 1000 meters long. Like the molecules in crude oil, the molecules that make up coal are composed primarily of carbon and hydrogen. However, the molecules in coal are enormous and can contain more than 100 carbon atoms. There are several varieties of coal, and each has a different mixture of complex hydrocarbon molecules. In Section 10.3, we discussed the varieties of coal and coal-fired plants in more detail.

QUESTION 12.4 Which type of crude oil has a higher boiling point: a fraction that is taken from the top of the fractionating column or a fraction that is taken from the bottom?

ANSWER 12.4 The fraction from the bottom has a higher boiling point. The liquid mixture of hydrocarbons in crude oil is heated enough to make most components into gases. A fraction that travels only a short way up the column reaches its boiling/condensation point lower on the column, where temperatures are higher.

For more practice with problems like this one, check out end-of-chapter Question 17.

FOLLOW-UP QUESTION 12.4 Which of the following is not produced from crude oil? (a) kerosene (b) coal (c) wax (d) jet fuel

12.3 Fossil Fuels and Climate Change

Fuels combust in the presence of oxygen molecules.

Now that we know where fossil fuels come from, let's consider how they provide energy. Fossil fuels are fuels because energy is released when the fuels undergo a specific chemical reaction. For that reaction to occur, oxygen must be present. In the presence of oxygen, the molecules in fossil fuels combust to make carbon dioxide and water. The combustion of the hydrocarbon methane occurs according to the following balanced chemical equation:

$$CH_4\,(g) + 2O_2\,(g) \rightarrow CO_2\,(g) + 2H_2O\,(g) + \text{heat}$$

For example, when a person lights the gas stove in a kitchen, methane gas pours out of the burner and a spark initiates this combustion reaction. The reaction in the burner produces heat as long as gas continues to flow and air is available. We don't normally write the word *heat* in a chemical equation. For this combustion reaction, though, we include it because we want to emphasize that this reaction produces heat.

In a car engine, another fossil fuel, gasoline, undergoes the same type of reaction. **Gasoline** is a mixture of hydrocarbons that have, on average, about six to eight carbon atoms per molecule. When gasoline combusts, the following products are obtained: carbon dioxide, water, and lots of heat. (Side products such as carbon monoxide are also produced.) In a car engine, the heat produced by combustion causes the gases it produces to expand quickly, and the pressure in the engine sharply increases. This pressure pushes pistons, the pistons turn a crank, the crank turns the axles, and the tires start to roll.

flashback
Combustion was discussed in Chapter 7.

flashback
In Chapter 7, we learned how to balance chemical equations. A balanced chemical equation has equal numbers of each type of atom on each side. Note that in the equation for the combustion of methane, there are four oxygen atoms, one carbon atom, and four hydrogen atoms on both sides of the equation.

A car's fuel economy is related to its power.

Recall from Section 12.1 that energy can be transferred from one form to another. A car moves when the energy provided by the combustion of gasoline is converted to the energy of movement. How fast this happens depends on how powerful the car is. For cars, a word sometimes used to describe power is *pickup*. A car with good pickup has a great deal of power because it can transmit its fuel's chemical energy into the energy of turning wheels in a short amount of time. A car's power is usually expressed in units of horsepower, where one horsepower is equal to about 746 watts.

Figure 12.9 **Two Cars, Two Fuel Economies**

The information in this figure compares particular specifications for the 2013 Smart car ForTwo and the 2013 Dodge Viper SRT-10.

Specification	2013 SmartForTwo	2013 Dodge Viper SRT-10
Trunk volume (cubic feet)	12	14.7
Price (base model)	$12,490	$97,395
Top speed (miles per hour)	84	206
Mass (pounds)	1610	3300
Horsepower (at 5800 rpm)	70	570
Gas consumption (miles per gallon, city driving)	34	12
Second from 0 to 60 mph	15.5	4.0

For cars powered by fossil fuels, there's an important link between the power a car has and the amount of fuel it uses. Most powerful cars have low **fuel economy**; the distance they can travel on a given volume of fuel is not very far. In the United States, we typically express fuel economy in units of miles per gallon (mpg). (In this book, we use metric system units almost exclusively. But because the domestic auto industry and the U.S. government use English units, we use miles per gallon in our discussion of gas mileage in this chapter.)

Let's compare the power and fuel economy of two different cars that are available on the market in the United States.

Figure 12.9 shows selected specifications for the 2013 Smart ForTwo and the 2013 Dodge Viper SRT-10. What do these cars have in common? They both hold two people and have small trunks. Beyond that, these cars couldn't be more different. The humble 70-horsepower Smart car can go from 0 to 60 miles per hour (mph) in 15.5 seconds, and its maximum speed is 84 mph. The powerful 570-horsepower Dodge Viper SRT-10 is a muscle car that can go from 0 to 60 mph in 4.0 seconds and has a maximum speed of 206 mph. (But the Dodge Viper can't accelerate faster than a cheetah, which can go from 0 to 60 mph in about 3.5 seconds!)

Clearly, these two cars attract very different buyers. If you can afford the $97,395 price tag and want to choose a very powerful car, then maybe the Viper is for you. But, more and more, people are basing their car-buying decisions not just on price and style, but on how driving that car will affect the environment. The Smart ForTwo travels nearly three times as far as the Viper on one gallon of gas. And, for each mile driven, the Smart car releases about 260 grams of carbon dioxide, whereas the Viper releases about 740 grams. When you consider that fossil fuels exist in finite quantities on Earth and contribute to climate change and pollution problems in a number of ways, it's easy to see where the Smart car gets its name.

recurring theme in chemistry 8

The Earth has a finite supply of raw materials that humans can use.

QUESTION 12.5 How many watts of power does a Dodge Viper use?

ANSWER 12.5 We know that the horsepower of the Viper is 570. We also know the conversion factor that relates horsepower to watts, so we can convert one to the other:

$$570 \text{ horsepower} \times \frac{746 \text{ watts}}{1 \text{ horsepower}} = 430{,}000 \text{ watts}$$

Because there are 1000 watts in one kilowatt, we could also express this answer as 430 kilowatts.

For more practice with problems like this one, check out end-of-chapter Question 18.

FOLLOW-UP QUESTION 12.5 Which of the following molecules is most likely to be a component of the mixture we know as gasoline? (a) C_3H_8 (b) C_7H_{16} (c) $C_6H_{12}O_6$ (d) $C_{14}H_{30}$

Climate change is the result of global warming.

The climate of our planet is changing. Measurements show changes in the prevalence of droughts and extreme weather, the amount of precipitation, and the number of heat waves, among other things. These changes are known, collectively, as *climate change*. Climate change is the result of one important phenomenon: the planet is getting warmer at a faster rate than it has warmed in the past. A vast majority of climate and environmental scientists agree that this change is due to human activities, especially the burning of fossil fuels, including coal, natural gas, and petroleum.

The burning of fossil fuels makes the planet warmer because carbon dioxide is released into the atmosphere and traps heat from the sun. This process is illustrated in **Figure 12.10**. The light from the sun that passes into Earth's atmosphere

flashback
In Chapter 1, we introduced *climate change* as a term that describes the effects of *global warming*, such as rising sea levels and disappearing islands.

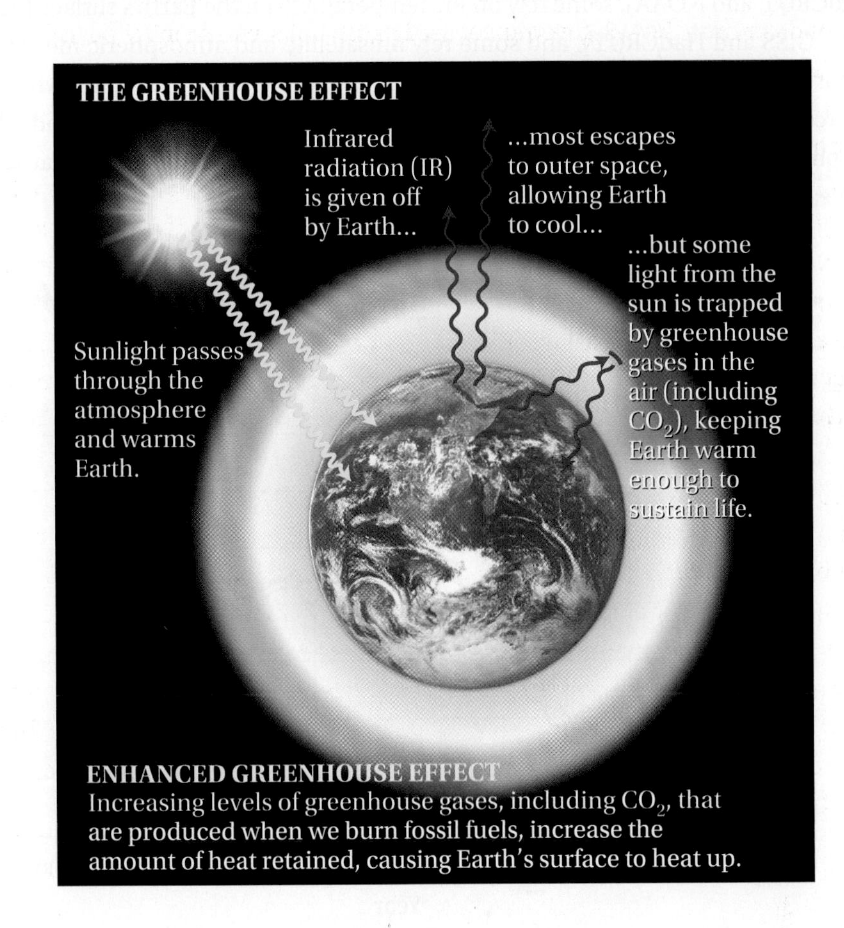

THE GREENHOUSE EFFECT
Infrared radiation (IR) is given off by Earth...
...most escapes to outer space, allowing Earth to cool...
...but some light from the sun is trapped by greenhouse gases in the air (including CO_2), keeping Earth warm enough to sustain life.
Sunlight passes through the atmosphere and warms Earth.
ENHANCED GREENHOUSE EFFECT
Increasing levels of greenhouse gases, including CO_2, that are produced when we burn fossil fuels, increase the amount of heat retained, causing Earth's surface to heat up.

Figure 12.10 The Greenhouse Effect
The greenhouse effect occurs naturally and makes Earth's atmosphere warm enough to support life. However, greenhouse gases produced by the burning of fossil fuels cause an enhanced greenhouse effect, where excess heat is trapped within Earth's atmosphere and the planet becomes unnaturally warmer.

can be absorbed by the surface of Earth, or it can escape back into space. Some of that energy is trapped by certain gases, called greenhouse gases, that help to heat Earth's atmosphere. This natural process is known as the **greenhouse effect**; it maintains Earth's temperature in a range that can sustain life.

In the 1800s, humans began to use fossil fuels as a source of energy. By the early to middle 1900s, fossil fuels, including coal, natural gas, and petroleum, were our dominant energy source (see Figure 12.6). All fossil fuels produce carbon dioxide and water when they combust, and carbon dioxide is a greenhouse gas. Recall from Chapter 5 that greenhouse gases are gases that trap heat in Earth's atmosphere and contribute to *global warming*, which results in climate change. Thus, it's reasonable to deduce that the enormous amounts of carbon dioxide that have entered the atmosphere due to fossil fuel use have caused the planet to get warmer. This phenomenon is known as the **enhanced greenhouse effect**.

flashback

In Chapter 1, we introduced *global warming* as a term describing the increasing temperatures on Earth that result from human activities such as the burning of coal to heat homes or the burning of gasoline in cars.

Figure 12.11 shows data about global warming, the increase in temperature of Earth as a result of human activity. The data supporting the evidence in the figure have been collected by scientific agencies, including the NASA Goddard Institute for Space Studies (GISS), a collaborative effort between NASA and Columbia University; HadCRUT, a set of temperature data compiled by a British agency based at the Climatic Research Unit (CRU) of the University of East Anglia; the U.S. National Oceanic and Atmospheric Administration (NOAA) based in Boulder, Colorado; RSS, a private company that performs calculations using NASA satellite data; and the University of Alabama at Huntsville (UAH).

recurring theme in chemistry 1

Scientists worldwide use the scientific method as a basis for experimentation and deduction. The scientific method is based on the principles of peer review and reproducibility.

Some data rely on ocean temperature measurements (data from GISS, HadCRUT, and NOAA), some rely on the temperatures on the Earth's surface (data from GISS and HadCRUT), and some rely on satellite and atmospheric measurement (data from GISS, NOAA, RSS, and UAH). All of the data, no matter what its source, are collected, analyzed, and scrutinized using the scientific method. The graph indicates that since 1890, Earth has been getting warmer. In fact, the warmest 10 years on record all occurred between 2000 and 2012.

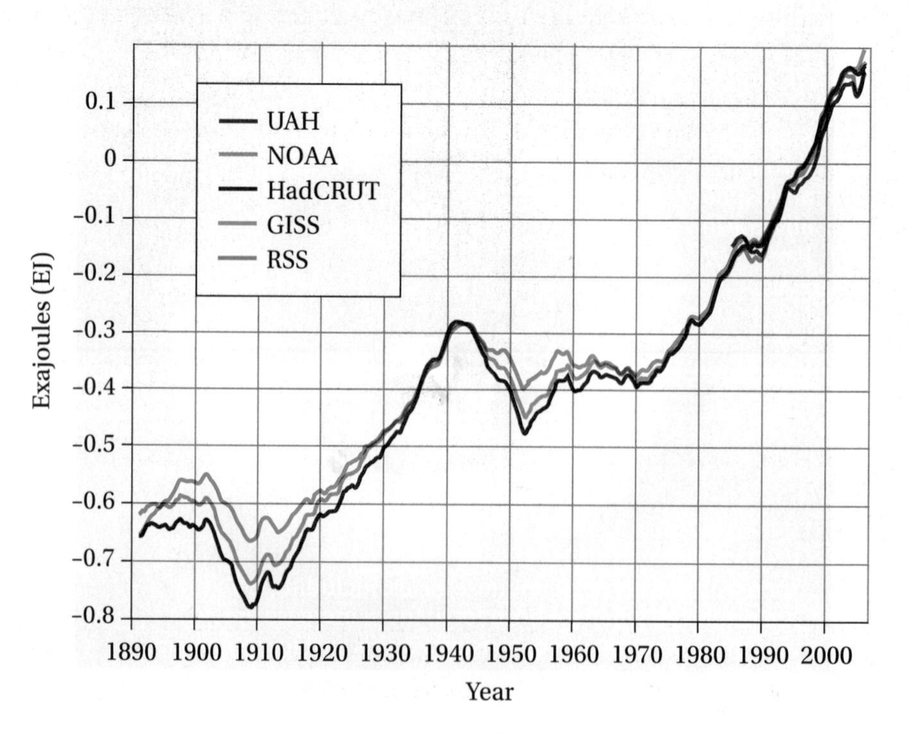

Figure 12.11 Measurements of Global Temperature by Five Scientific Agencies

Global temperatures have been tracked by various scientific agencies. Please see the text discussion for more information on these agencies.

Evidence that the climate is changing because of these warmer temperatures is remarkable and well documented. There is evidence that sea levels have risen (see *The Green Beat,* Chapter One Edition). Ice sheets have been shrinking, and the ice in the Arctic has become thinner. Glaciers are disappearing worldwide (see *The Green Beat,* Chapter Five Edition). Despite overwhelming evidence, though, some people still claim that climate change is not taking place. But no one has yet presented credible, peer-reviewed scientific data that contradicts or raises doubts significant enough to rival the immense volume of evidence in support of climate change due to global warming.

wait a minute . . .

But aren't global warming and climate change the same thing?

No, global warming and climate change are not the same thing. Global warming is the warming of the planet as a result of human activity. Global warming is the cause, and climate change is the effect. Climate change is a broad term that encompasses all the effects of a warming planet, such as changing climate patterns, melting ice caps, and rising sea levels.

Atmospheric carbon dioxide levels reached 400 ppm in 2013.

May 20, 2013, was an important date for people who are interested in global warming and climate change. On that date, the measured levels of carbon dioxide in Earth's atmosphere exceeded 400 parts per million (ppm). Why was the number 400 such a big deal? **Figure 12.12** provides the answer. Throughout the history of our planet, carbon dioxide levels have gone up and down, but they have stayed mostly in the 180- to 300-ppm range. The 300-ppm mark was passed before the 1950s, and the next milestone was the 400-ppm mark, the next round number. It's like turning 30 years of age; you're the same person you were when you were 29, but you've reached an important milestone and then you set your sights on the next one. The 400-ppm mark was a big deal because we reached it very quickly, much faster than scientists expected.

flashback

Recall from Chapter 10 that parts per million is a way to express the concentration of one thing in another. For example, "400 ppm carbon dioxide" means there are 400 grams of carbon dioxide per million grams of air.

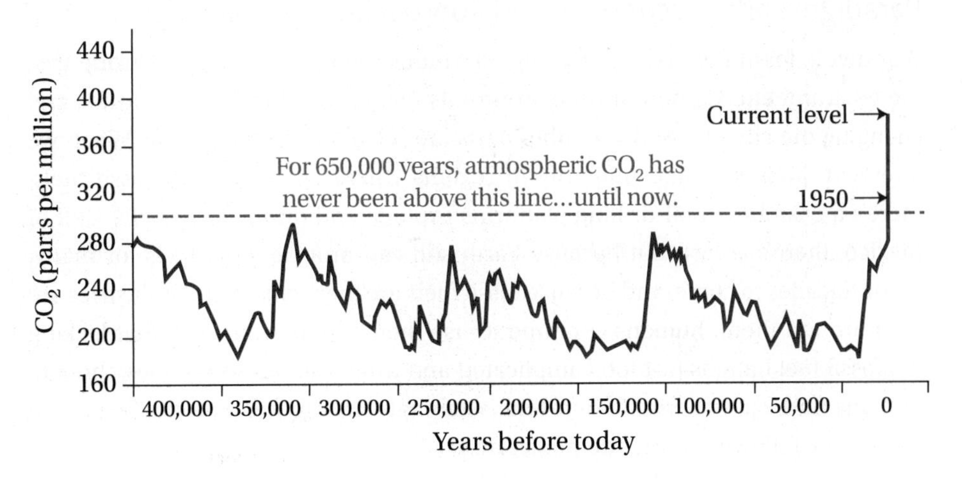

Figure 12.12 Carbon Dioxide Levels over the Last 400,000 Years

Now that we've reached the 400-ppm mark, is it too late to turn back? Can humans easily reverse the effects of global warming? Recent studies have suggested that yes, it is too late to turn back; and, no, we cannot easily reverse the effects of global warming. These studies explain that even if we reduce carbon dioxide emissions in a radical way, right now, climate change would continue to play out—not just for hundreds of years, but perhaps for a thousand years or more.

wait a minute . . .

Aren't there other gases besides carbon dioxide that are greenhouse gases?

Yes, many gases in our atmosphere are greenhouse gases. For example, water is a natural greenhouse gas that contributes significantly to the greenhouse effect. But many greenhouse gases are generated by humans. For example, cars release carbon monoxide, nitrous oxide, and methane. The last two are more potent greenhouse gases than carbon dioxide—per molecule, they trap heat in the atmosphere more than carbon dioxide does. However, because carbon dioxide is present in much greater quantities, CO_2 is the human-generated greenhouse gas that contributes most to global warming. Methane is a close second, and it's discussed in *The Green Beat,* Chapter Five Edition.

QUESTION 12.6 Which of the following molecules is *not* a greenhouse gas according to the text discussion? (a) radon gas (b) carbon dioxide (c) methane (d) nitrous oxide

ANSWER 12.6 (a) Radon is not a greenhouse gas. The others were identified as greenhouse gases in the text.

For more practice with problems like this one, check out end-of-chapter Question 22.

FOLLOW-UP QUESTION 12.6 According to Figure 12.12, which one of the following carbon dioxide levels has recently been surpassed for the first time in 400,000 years? (a) 200 ppm (b) 380 ppm (c) 410 ppm (d) 220 ppm

12.4 Meeting New Environmental Standards

Paradigm shifts occur when old approaches are challenged.

Worldwide fossil fuel use is releasing enormous amounts of carbon dioxide into the environment. Carbon dioxide, in turn, is increasing global temperatures and changing the climate. And still other costs are related to fossil fuel use. When we intervene politically and militarily in regions where we obtain our fossil fuels, there's a cost. When oil-drilling platforms explode and dump oil into the Gulf of Mexico, there's a cost. But because Earth still can provide fossil fuels for many, many decades to come, and because fossil fuels are inexpensive and industries are built around them, humans continue to use them. Some would say that kicking the fossil fuel habit is just too complicated and troublesome. Sometimes, though, humans find ways to work around the seemingly insurmountable challenges we face, as the following stories demonstrate.

natureBOX • Ice-Core Measurements

Much of the evidence for climate change has come from measurements of carbon dioxide levels from the past 400,000 years (see Figure 12.12). How is it possible to accurately measure levels of gas in the atmosphere that existed on Earth before the evolution of modern humans? Much of what we know about Earth's natural history this long ago has come from ice-core samples.

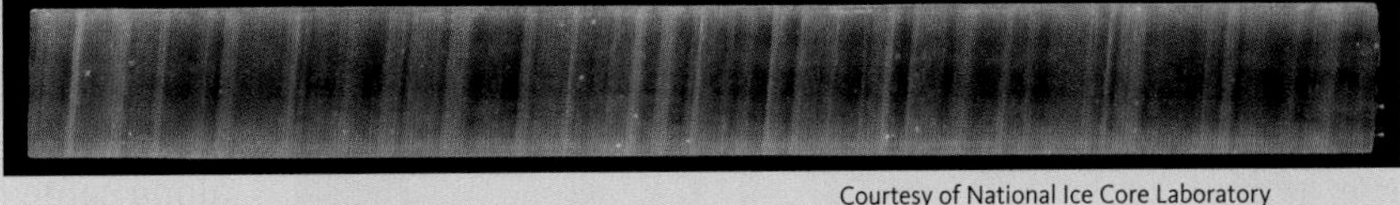

Courtesy of National Ice Core Laboratory

Figure 12A An Ice Core Collected in Greenland

This 1-meter section of ice core was collected from a depth of 1837 meters within the Greenland ice sheet. The oldest ice in this core was formed 16,250 years ago

Wherever we can find very, very deep ice, there's an opportunity to learn more about what was happening in that spot thousands of years ago. Ice cores are collected by using a long pipe with a sharp end that extends down vertically into the ice. When the pipe is raised, it contains a pipe-shaped core of ice that's a lot like a time capsule because it can tell scientists what was happening when those layers of ice formed.

Figure 12A shows an ice core that was collected from a depth of 1837 meters within the Greenland ice sheet. Each striation along the ice core represents one year. This is because the snow that falls year-round changes from summer, when the sun shines 24 hours per day near the poles, to winter, when it is always dark. As the seasons dramatically change, so does the density and texture of the snow that falls. That seasonal variation is visible as lines within the ice core.

The thickness of each striation within the ice core tells scientists how much snow fell during a given year. A chemical analysis of the ice reveals the presence of trapped dust, pollen, and volcanic ash. These particulates can reveal how windy it was that year, or whether there was a major volcanic eruption. Within the ice, trapped bubbles contain a tiny sample of the atmosphere that existed that year. For example, measurements of methane in ice-core bubbles reveal how much of Earth was covered with wetlands each year, because methane-producing bacteria live in wetlands and release the methane into the atmosphere.

The tiny bubbles within ice cores also contain carbon dioxide, which is a natural component of air. Through painstaking analysis, the carbon dioxide data in Figure 12.12 were collected from ice cores that spanned 400,000 years. These data give incontrovertible evidence supporting the role of carbon dioxide emissions in global warming in the twenty-first century.

What happens to ice cores once they are pulled from Earth? They must be carefully transported, in well-insulated containers, to the National Ice Core Laboratory in Denver, Colorado. There, scientists can analyze ice-core samples in specialized laboratories that are kept at a chilly −35°C. The facility stores more than 14,000 meters (that's more than eight miles) of ice-core samples collected from all over the world.

The scientists who study ice-core samples are called paleoclimatologists because they study the history of climate change. In most photos of these people, they all have one thing in common: they always wear hats. It takes an especially dedicated person to work in bitterly cold climates where ice exists year-round. **Figure 12B** shows such a person, Gifford Wong, a graduate student in paleoclimatology. In this photo, he points to striations on the wall of an ice cave.

Courtesy of Gifford Wong

Figure 12B Gifford Wong, Graduate Student in Paleoclimatology

Gifford Wong stands in a backlit snow pit, where striations within the snowpack are evident.

Environmental problems are not a new phenomenon, and they don't always involve machines, such as cars, threatening nature. Sometimes, it's just the opposite. In the 1800s, cities around the world were facing a devastating problem: horse dung. Because horses were the primary mode of transport in cities, cities had to find a way to cope with horse dung. Lots of it. International meetings were convened to discuss this seemingly insurmountable problem, and many predicted

AP Photo

Figure 12.13 No Gas to Sell
During the oil embargo of 1973, gasoline stations often had no gas to sell.

doom as horse droppings piled up in empty city lots to a height of more than 12 meters. It was predicted that, by the year 1950, the dung would reach third-story windows in New York City!

For people of the time, the problem seemed intractable. When the next century arrived, though, along came the electric streetcar and the automobile, which offered a paradigm shift in urban transportation. The horse dung controversy was resolved when horses were replaced with other modes of transportation, and that particular environmental crisis was over.

As cars replaced horses on the road, new conflicts over transportation arose. From its very beginnings, the auto industry resisted attempts to regulate the fuel economy and safety of its cars. When seat belt laws were enacted, Henry Ford predicted that the new safety laws would cause his company to shut down.

In the year 1900, there were 8000 cars in the United States. By 1970, more than 100 million cars were registered in the United States, and another environmental crisis was about to play out. Gas-guzzling cars ruled the road, and the average fuel economy was below 15 miles per gallon (mpg). In 1973, political tension in the Middle East resulted in an oil embargo, and the United States was cut off from most Middle Eastern oil. Suddenly, gas prices were 30% higher, and supply was limited. Rationing was imposed, and many gas stations had no gas to sell (**Figure 12.13**).

U.S. CAFE laws set minimum levels on fuel economy for cars and trucks.

In the wake of the oil embargo, in 1975, the U.S. Congress passed the **Corporate Average Fuel Economy (CAFE) laws** that required a doubling of fuel efficiency within 10 years to a value of 27.5 mpg. Automakers predicted certain doom for the auto industry. But within 10 years, the industry persevered and the new standard was met.

Figure 12.14 shows fuel economy for U.S. cars and trucks beginning in 1975. After CAFE laws brought improvements to fuel economy from 1975 to 1985, average mpg levels stagnated for the next two decades. Modest increases were seen from 2005 to 2010 after new fuel economy rules were put in place in 2005. Then, in 2010, the Obama administration passed two new CAFE laws. The laws mandate that average mpg values (for cars and trucks combined) must increase from 25.3 mpg in 2010 to 34.1 mpg in 2016 and to 54.5 mpg by 2025.

In a *Forbes* article titled "New CAFE Rules Anger Just About Everyone" (September 4, 2012), the writer said, "Not surprisingly, automakers are not happy with the new rules." Will automakers rise to this new challenge? Time will tell, but history predicts that, yes, somehow, they will.

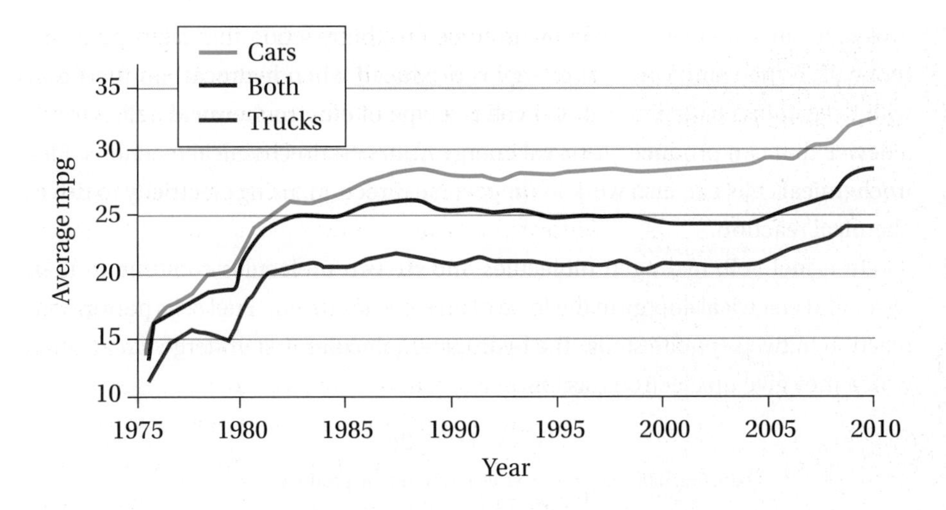

Figure 12.14 The Fuel Economy of Cars in Miles per Gallon (mpg) as a Function of the Year They Were Produced

The CAFE laws passed in the 1970s resulted in improvements in fuel economy into the 1980s. Fuel economy remained flat until about 2005, when new fuel economy standards were enacted. A series of new CAFE laws should produce sharp increases in mpg over the coming decade. In the figure, *Both* refers to both cars and trucks.

QUESTION 12.7 Which of these statements about the data in Figure 12.14 is not true? (a) Cars generally have better fuel economy than trucks. (b) Fuel economy rose sharply at the start of the twenty-first century. (c) During some periods in recent history, the average fuel economy of trucks has decreased. (d) Fuel economy in this figure is expressed in units of miles per gallon.

ANSWER 12.7 (b) The graphed data for the year 2000 in Figure 12.14 show no sharp rise in fuel economy. This statement is false.

For more practice with problems like this one, check out end-of-chapter Question 29.

FOLLOW-UP QUESTION 12.7 True or false? U.S. laws that mandate improved fuel economy have set unrealistic goals for car makers.

12.5 Storing Energy in Hydrogen Molecules

Fuel cells use redox reactions to produce electricity and water.

Wind is an energy resource that can provide power to people who happen to live in windy places. One such place is the tiny island of Utsira, which sits just off the coast of southwestern Norway. Utsira is home to about 240 people and hundreds of sheep. It is also home to one of the world's first hybrid wind-hydrogen power stations.

Utsira has been using wind power for a long time. In fact, the island's wind turbines have been producing more electricity from wind than inhabitants can use, but the island has had no way to store the energy, so the power was wasted. To solve this problem, the island recently added an electricity generator that splits water according to the following chemical equation:

$$2H_2O + \text{electrical energy} \longrightarrow 2H_2 + O_2$$

This process is known as **electrolysis** because it uses electricity (*electro-*) to split (*-lysis*) water molecules. Electrolysis is currently the most commercially viable way to produce hydrogen gas from water, but it requires a tremendous amount of electricity. In Utsira, the excess wind power is converted into electrical power and harnessed to electrolyze water. The hydrogen gas that is produced via electrolysis is stored for later use.

Why is hydrogen gas—represented by H_2 in the electrolysis equation—a fuel? It's a fuel because energy is produced when hydrogen molecules react with oxygen

recurring theme in chemistry 4
Chemical reactions are nothing more than exchanges or rearrangements of electrons.

molecules in a fuel cell. Energy is produced in the reaction that takes place in a **fuel cell** in the same way that energy is produced when hydrocarbon fuels react with oxygen in a car engine. A fuel cell is a type of **electrochemical cell,** which is a device that can produce electrical energy from certain chemical reactions. Electrochemical cells can also work in the reverse direction, using electricity to drive a chemical reaction.

In a fuel cell, hydrogen molecules and oxygen molecules combine to make water and electrical energy in the form of an electric current. Fuel cells perform this reaction in two separate steps. The hydrogen molecules first undergo a reaction in which they give up electrons, as follows:

$$H_2 \longrightarrow 2H^+ + 2e^-$$

Oxidation half-reactions have electrons on the products (right) side.

flashback
Recall from Chapter 2 that we use the word *proton* to describe the positively charged particle in the nucleus of an atom. It's also one of the terms we use for H^+, because this ion contains one proton and no electrons. In this discussion, we use *proton* in the second sense.

This is a special type of chemical equation called a half-reaction, because it represents one-half of a larger process. You can recognize half-reactions because they always show electrons as either a reactant (if they are on the left side) or a product (if they are on the right side). When electrons are the product in a half-reaction, we call that reaction an oxidation reaction. **Oxidation** is a *loss* of electrons. In this half-reaction, hydrogen molecules are oxidized to make protons, H^+, and electrons, represented with the symbol e^-.

When something is oxidized, the electrons produced take part in the other half of the process, reduction. **Reduction** is a *gain* of electrons. In a fuel cell, the oxygen molecule is reduced when it combines with protons (H^+) to form water as follows:

$$O_2 + 4H^+ + 4e^- \longrightarrow 2H_2O$$

Reduction half-reactions have electrons on the reactants (left) side.

QUESTION 12.8 Indicate whether each of the following half-reactions is an oxidation reaction or a reduction reaction: (a) $Cu \longrightarrow Cu^{2+} + 2e^-$ (b) $Ni^{3+} + e^- \longrightarrow Ni^{2+}$

ANSWER 12.8 The reaction in (a) is an oxidation half-reaction because electrons are on the products side. The reaction in (b) is a reduction half-reaction because electrons are on the reactants side.

For more practice with problems like this one, check out end-of-chapter Question 42.

FOLLOW-UP QUESTION 12.8 True or false? Oxidation is a gain of electrons.

Redox reactions combine oxidation and reduction half-reactions.

Oxidation and reduction half-reactions must be paired together to make a whole reaction, because one must occur for the other to happen. Oxidation half-reactions produce electrons, and those electrons are then used up in a reduction reaction. An oxidation-reduction or **redox reaction** is what we get when we combine them in a way that cancels electrons on both sides. To see how this is done, consider that the two half-reactions we looked at in the previous section include different numbers of electrons. To make them equal, we multiply everything in the oxidation reaction by a factor of two, with the following result:

$$2H_2 \longrightarrow 4H^+ + 4e^-$$

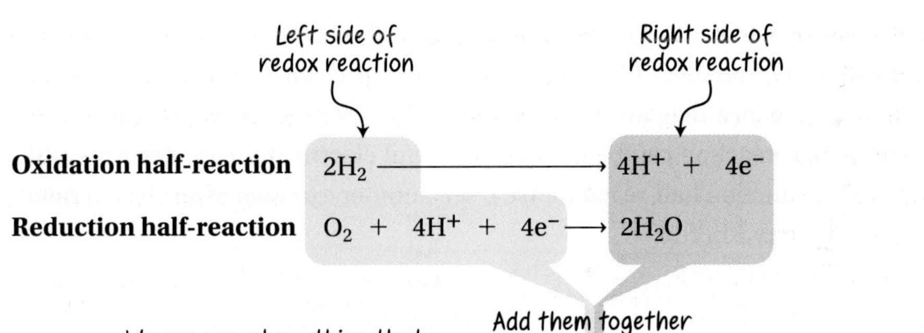

Figure 12.15 Half-Reactions in a Fuel Cell

Now that each of our half-reactions has four electrons in it, we can add them together to make a redox reaction. We do this by combining everything on the left side of each half-reaction and everything on the right side of each half-reaction and putting them into a single new chemical equation, as shown in **Figure 12.15**. Then we cancel anything that appears on both the left and right sides of the chemical equation. After canceling out the protons, H^+, and the electrons, e^-, we have a redox reaction that summarizes the chemistry that takes place in a fuel cell.

Figure 12.16 is a schematic drawing of a fuel cell. The hydrogen gas flows into the cell from the left, and it is oxidized according to the oxidation half-reaction shown earlier. Oxidation occurs at the **anode**. The electrons produced from the oxidation half-reaction at the anode leave the cell as **electric current,** a flow of electrons, also known as electricity. Current flowing at a specific rate through the wire can provide electrical power to a device that is powered by electricity, such as an electric car or a kitchen appliance on the island of Utsira.

The flow of electrons returns to the fuel cell at the **cathode,** which is where the reduction half-reaction takes place. This is where protons (H^+), electrons, and oxygen molecules come together to form water. When we look at this process

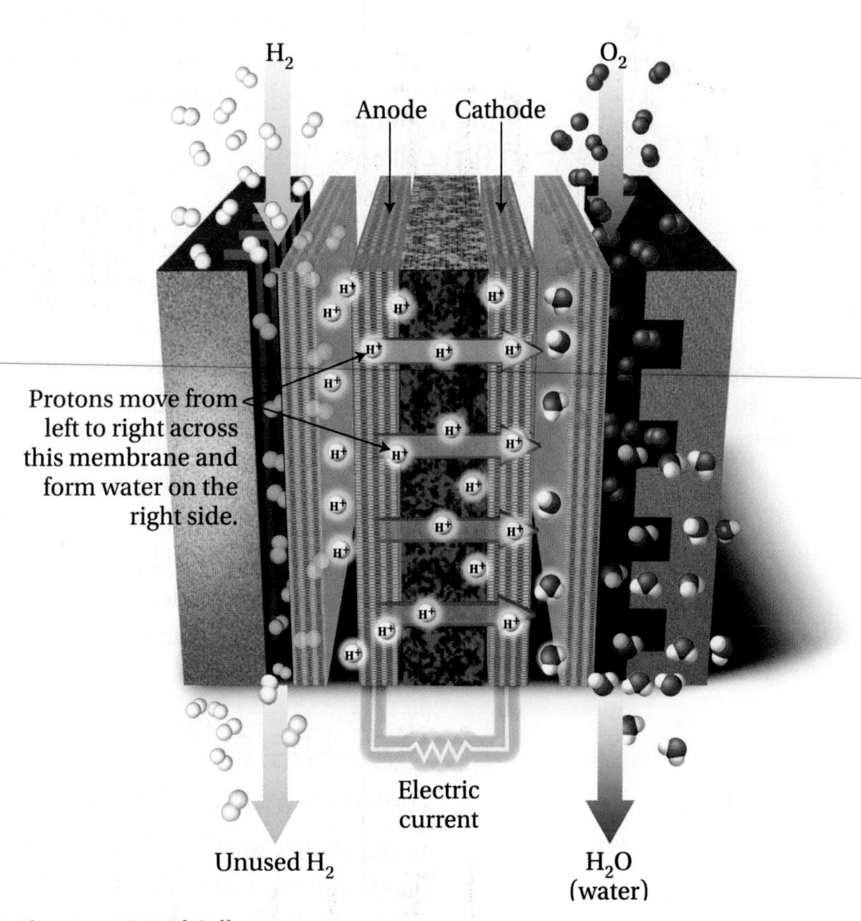

Figure 12.16 Fuel Cell

In a fuel cell, hydrogen molecules (H_2) flow into the cell, where they are converted to H^+ ions (protons) at the anode through an oxidation half-reaction. Oxygen molecules (O_2) flow in at a different inlet into the cell. When combined with H^+ produced at the anode, the oxygen molecules are converted to water (H_2O) through a reduction half-reaction at the cathode.

as a whole, we can see that an electrochemical cell can generate a current because it separates a redox reaction into two half-reactions, which are physically separated in the cell. Because they are physically separated, the electrons produced in the oxidation half-reaction can be diverted as useful electric current before they take part in the reduction half-reaction. We'll see another example of an electrochemical cell, a battery, shortly.

QUESTION 12.9 Add together the following half-reaction equations, performing cancellation if needed:

$$Cu \longrightarrow Cu^{2+} + 2e^-$$
$$Ni^{2+} + 2e^- \longrightarrow Ni$$

ANSWER 12.9 The resulting equation—with electrons canceled—looks like this:

$$Ni^{2+} + Cu \longrightarrow Cu^{2+} + Ni$$

For more practice with problems like this one, check out end-of-chapter Question 41.

FOLLOW-UP QUESTION 12.9 In the reduction half-reaction that occurs within a fuel cell, which three things combine to make water?

Fuel cells are limited by the availability of hydrogen gas.

Fuel cells seem like a great way to generate energy. They take hydrogen gas and oxygen gas and make water and electric current. If they are such a great idea, though, why don't we all have fuel cells in our homes and in our cars? The answer is that hydrogen gas is hard to come by. It is rare in nature, and it's not easy to make or store. In Utsira, they are fortunate to have excess energy from wind power that they use to electrolyze water to produce hydrogen. But the electrolysis of water requires enormous amounts of electricity, and right now there are no other commercially viable ways to make hydrogen.

The cartoon in **Figure 12.17** is a tongue-in-cheek look at fuel cell cars that run on hydrogen gas. It might seem like a great idea to drive a "clean energy" car that emits zero carbon dioxide, but if the hydrogen you put into the car is made using nuclear energy or fossil fuels, then, at the end of the day, your car may not be very clean after all.

Figure 12.17 Is Hydrogen a Clean Fuel?

12.6 Energy from the Sun

The sun is a virtually limitless source of energy.

As the world's population grows, so does the world's demand for energy. That energy must come from somewhere. And, as we have discussed, right now most of it comes from burning fossil fuels, such as coal, natural gas, and petroleum. But fossil fuels are taking a disastrous toll on the environment, and supplies are not unlimited. Depending on who you ask, it is estimated that our fossil fuel reserves will last for another 75 to 500 years.

There is, however, one virtually unlimited source of energy: the sun. Ultimately, it's the sun that gives us the energy we need to sustain our bodies. Through the extraordinary process of **photosynthesis,** plants convert the sun's energy to chemical bond energy. The following chemical equation is a simplified version of the photosynthesis process:

$$6CO_2 + 6H_2O \text{ (+ sunlight energy)} \rightarrow C_6H_{12}O_6 + 6O_2$$

Carbon dioxide Water Glucose Oxygen

Through photosynthesis, plants are able to combine sunlight, carbon dioxide, and water to produce glucose and oxygen. The glucose becomes food for the plant, and the oxygen is released into the air. That oxygen becomes part of the supply of oxygen that humans breathe. Therefore, plants are important sinks for carbon dioxide, a greenhouse gas, as well as producers of oxygen molecules. This is the reason for serious concern about **deforestation,** the removal of trees to allow the land to be used for another purpose.

Sunlight is the ultimate source of energy for most of the organisms on Earth. Vegetarians consume plants as a source of energy, and those plants consumed sunlight as their source of energy. Carnivores consume animals, those animals consumed plants, and those plants consumed sunlight. Thus, no matter what animals—such as humans—eat, the chemical energy in our food ultimately comes from sunlight. We don't need to worry about the limitations of this energy source, because the sun will continue to shine for the next few billion years.

Photovoltaic cells convert sunlight into electrical energy.

The sun that shines on the surface of Earth is a free and virtually unlimited resource, if we can capture it and use it. This is tricky, because the sun doesn't shine 24 hours a day. But, if we collect and store energy from the sun—**solar energy**—when the sun is shining, then we can use it to power our homes, buildings, and cars after the sun sets.

Solar energy is collected using solar panels that are made up of **photovoltaic cells,** a type of device that converts light energy to electrical energy. Most photovoltaic cells are comprised of sheets of the element silicon, a metalloid element. When silicon atoms form Si–Si bonds, those bonds extend in a crystalline solid, like carbon does in diamond. Some of the valence electrons in the silicon atoms in the network are occasionally able to move across a gap from a lower energy level to a higher one (**Figure 12.18**). This jump to a higher energy level requires an outside source of energy, such as light energy. When silicon in a photovoltaic cell is exposed to light from the sun, some of the electrons absorb that solar energy and are promoted to a higher energy level.

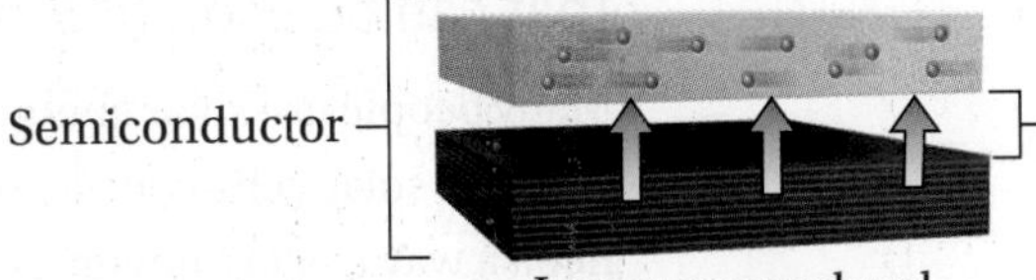

Figure 12.18 How a Semiconductor Works

A semiconductor possesses two levels of energy separated by a gap. When excited by light, for example, an electron can move from the lower to the higher energy level. The electrons in the higher energy level are mobile and form an electric current through the semiconductor. If the source of excitation is light, then the current flows as long as light is shining on the semiconductor.

An electron moving to the higher energy level leaves behind a "hole" in the crystalline solid. And because this hole represents a "missing" negative charge, we consider the hole to have a positive charge. The positively charged hole attracts an electron from someplace else in the network, and that electron, as it migrates to the hole, leaves a fresh hole behind at its original location. As a result, there is a great deal of "traffic" within the silicon network. As we learned earlier in this chapter, moving electrons constitute an electric current. Thus silicon is used in photovoltaic cells because it can conduct an electric current when light shines on it. We refer to silicon as a **semiconductor** because it conducts an electric current only under these specific conditions.

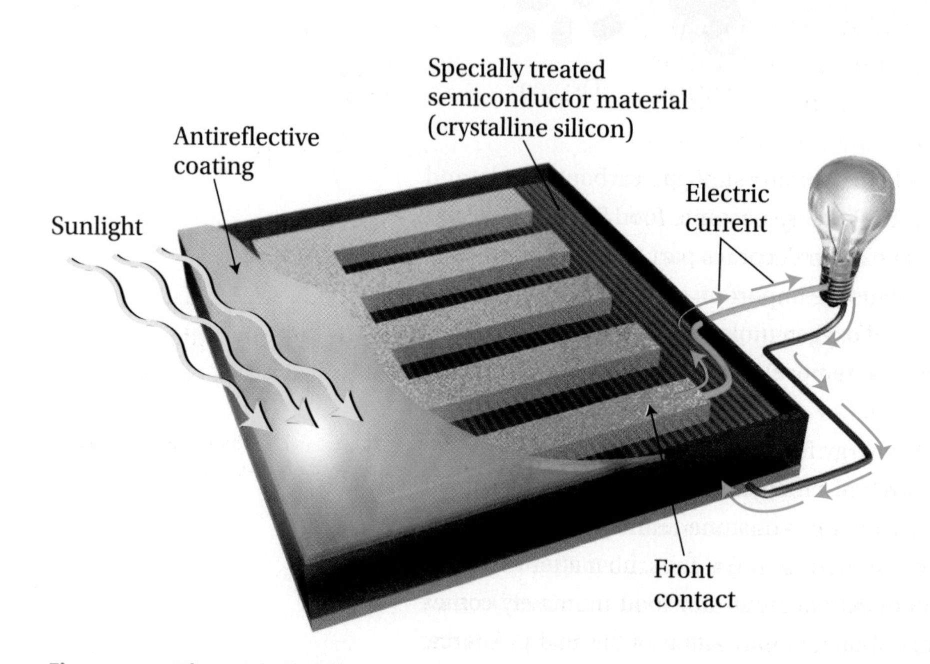

Figure 12.19 Photovoltaic Cell

Sunlight hits the antireflective coating on the surface of the photovoltaic cell. In the specially treated silicon inside the cell, an electrical current is generated. Electricity flows into the contacts and wires and can be used to power a light bulb, an appliance, or a car. This electricity also can be stored in a battery.

On the top of the photovoltaic cell in **Figure 12.19** is an antireflective coating. This minimizes the reflection of light off the cell's surface, so that as much light as possible goes into the cell. Under the coating is a grid of electrical wires and contacts that carry electric current generated within the cell. Under these wires is a specially treated form of crystalline silicon. When light strikes this crystalline silicon, electrons flow to the front contact and then out of the cell. That electric current can be used, for example, to run a refrigerator or power up an electric car, or it can be stored in a battery.

QUESTION 12.10 In a solar cell, what is the role of the antireflective coating on the cell's surface?

ANSWER 12.10 The antireflective coating reduces reflection of sunlight, so that the amount of light hitting the silicon semiconductor is maximized.

For more practice with problems like this one, check out end-of-chapter Question 43.

FOLLOW-UP QUESTION 12.10 How many valence electrons does a silicon atom possess?

Rechargeable batteries are electrochemical cells that can be used to store solar energy.

Many people use solar energy to collect 100% of the energy they need in their homes. Because solar cells collect solar energy only when the sun is shining, these people need a way to store energy collected during the daytime for use at nighttime. Batteries are a common way to store solar energy, and other types of energy, too.

Like fuel cells, a **battery** is an electrochemical cell that produces electricity. Batteries are based on two half-reactions. As in the fuel cell, one half-reaction in a battery is an oxidation half-reaction in which electrons are the product (on the right side). The other half-reaction is a reduction half-reaction in which electrons are the reactant (on the left side).

Figure 12.20 illustrates the two half-reactions that occur in a modern nickel-metal hydride (NiMH) battery, the battery of choice for electric cars. The reduction half-reaction uses a mixture of metals represented by the letter *M*. In the oxidation half-reaction, a nickel ion is oxidized and one electron is produced. These two half-reactions are physically separated within the battery so that electrons flow between the two halves of the battery. If the job of the battery is to light a flashlight, for example, that current is diverted to a light bulb in the flashlight and the light bulb glows.

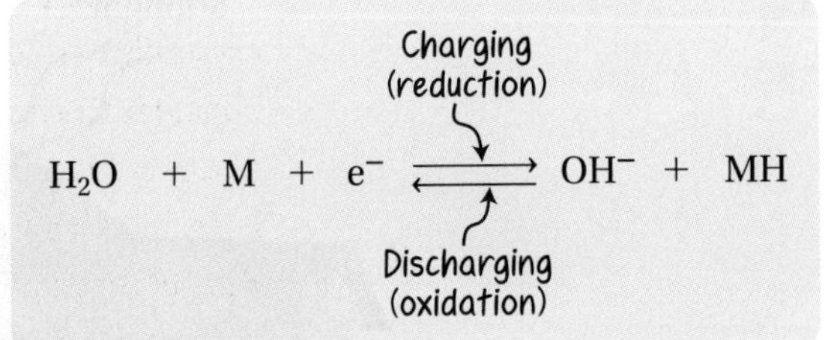

One half-reaction occurs at one electrode...

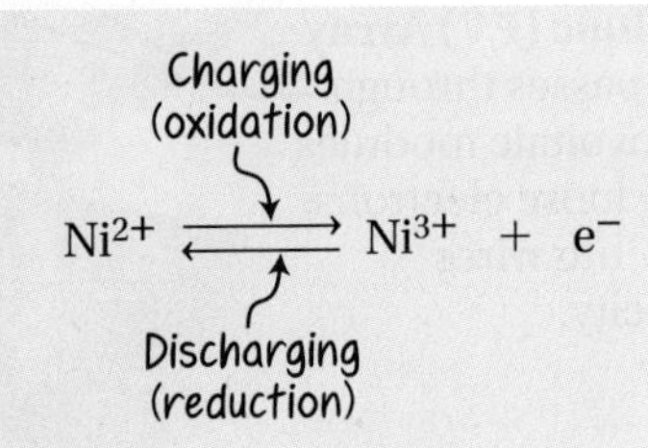

...and the other half-reaction occurs at the other electrode.

Figure 12.20 Half-Reactions in a Rechargeable NiMH Battery

In a rechargeable battery, the two half-reactions can go in two directions, one for charging and one for discharging.

NiMH batteries are special because the half-reactions in them are reversible chemical reactions. They can proceed either in the forward direction or in the reverse direction. We indicated this in Figure 12.20 with double arrows. Compare these half-reactions to those in the fuel cell in Figure 12.15. For a fuel cell, the arrows point only in the forward direction and are not reversible.

Why do some electrochemical cells employ reversible half-reactions and others employ irreversible half-reactions? The answer lies in the cost required to make the battery and in the job that the electrochemical cell is performing. For a fuel cell, the half-reactions need to go in the forward direction only, because the purpose of the fuel cell is to combine hydrogen and oxygen to make water and electrical energy. That is the only thing it's designed to do. Likewise, traditional flashlight batteries use irreversible half-reactions because their job is always to produce an electric current that lights the flashlight. That one job requires the half-reactions to go in one direction only.

Some batteries, though, have two jobs. One such battery is the NiMH battery that's often used in electric cars. When the NiMH battery is being charged, electrical energy is flowing into it, and the half-reactions shown in Figure 12.20 both proceed in the forward charging direction. When the battery is fully charged, you can take the car for a spin. Driving the car is a job that requires electric current from the battery to power the car. When the car is being driven, the half-reactions proceed in the reverse direction. In Figure 12.20, this is the discharging direction. Because the battery can be charged and discharged again and again, the NiMH battery is called a *rechargeable battery*.

Rechargeable batteries are used in solar homes and buildings to store electrical energy that is collected when the sun is shining. **Figure 12.21** shows one way to design a solar home. During the day, some of the electrical current generated by the solar cells powers the home. Extra electrical energy is stored in rechargeable batteries. During charging, the half-reactions in the battery move in the charging direction. When the sun goes down, no more electrical energy is being generated from sunlight, and power for the home comes back out of the batteries. The half-reactions in the batteries switch to the discharging direction, and electrical energy flows out of them. This process repeats when the sun rises again.

Solar-powered buildings and homes use solar cells that are connected together and mounted into a solar panel. Solar panels can be combined into arrays, and arrays can be mounted onto motorized platforms that follow the sun throughout the day.

The home shown in Figure 12.21 is attached to the electrical grid, the network that carries electricity from power plants to consumers. If a home can get 100% of

Sunlight

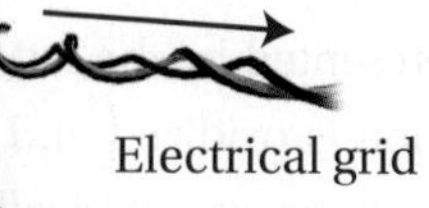

Electrical grid

Photovoltaic (PV) Array
Sunlight passes through the photovoltaic modules, knocking loose electrons, that flow into wires as electricity.

The photovoltaic modules are wired together.

Charge Controller
The controller regulates the electrical energy going into the rechargeable batteries.

Batteries
Batteries store the energy generated by the system.

Electrical Panel and Meter
Grid power consumption and the power production from the photovoltaic array are metered here. Net metering means the consumers in the solar home only pay the difference between what they use and what they produce.

Figure 12.21 One Way to Design a Solar Home

In a solar home, the photovoltaic (pv) array on the roof collects solar energy, converts it to electrical energy, and sends that energy into the home. Some of the energy powers the home during the day, and the rest is stored in rechargeable batteries. Homes can also be connected to the electrical grid and sell electricity to the utility company through net metering.

its energy from sunlight, why is it attached to the electrical grid? It's attached to the grid because if the home collects more energy from the sun than the total amount the home consumes, that excess energy can be put back into the grid. In many locations, the utility company will credit the homeowner for that energy through a payback system known as **net metering**. It's estimated that in the sunniest spots in the United States, solar homes can give about 20 to 40% of the energy they produce back to the community through net metering.

QUESTION 12.11 In the redox reaction for a NiMH battery shown in Figure 12.20, what does the letter *M* represent?

ANSWER 12.11 It represents a mixture of metals.

For more practice with problems like this one, check out end-of-chapter Question 48.

FOLLOW-UP QUESTION 12.11 In Figure 12.21, the solar home is connected to the electrical grid. In your own words, explain why a house that gets 100% of its power from solar energy is connected to the electrical grid.

Solar energy can be used anywhere the sun shines.

In some parts of the world, it is more difficult to use solar power than in others. But, at least in theory, solar energy can be used anywhere the sun shines. In general, in places that get four or more peak sun hours per day, solar power is an obvious alternative. In the solar industry, a peak sun hour is one hour of sunlight at a power of 1000 watts shining on one square meter of Earth. Sunlight energy for any location on

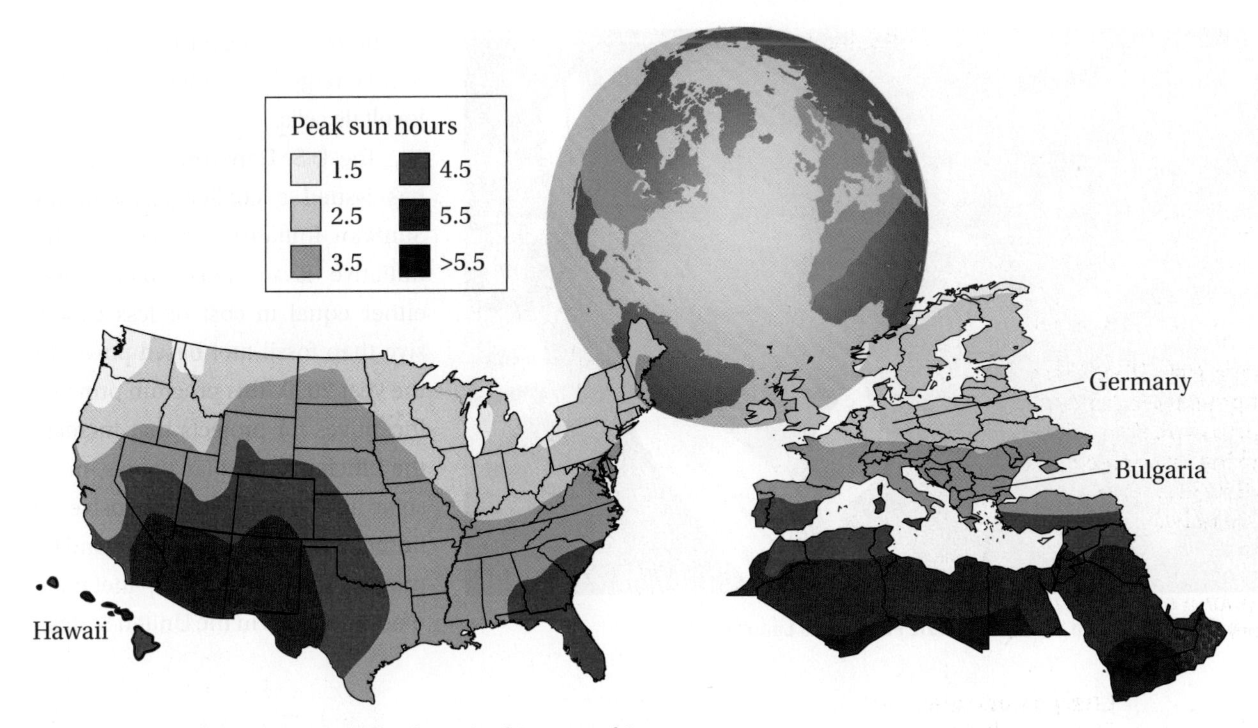

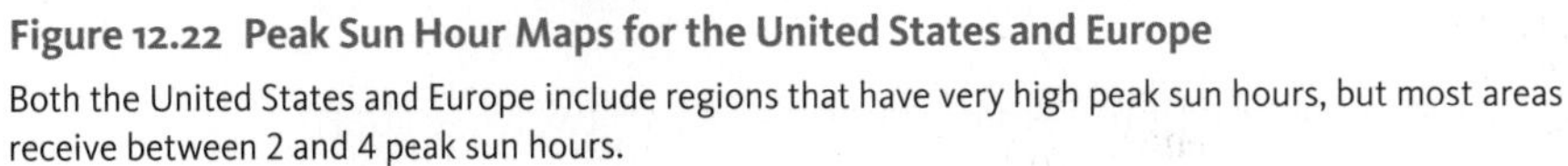

Figure 12.22 Peak Sun Hour Maps for the United States and Europe
Both the United States and Europe include regions that have very high peak sun hours, but most areas receive between 2 and 4 peak sun hours.

Earth can be referenced to that value, and the number of peak sun hours per day is the measure used to determine how easy it is to use solar power in a particular location.

The maps in **Figure 12.22** show the peak sun hours of the United States and Europe. In the United States, the places that receive the most solar energy per day are in the desert Southwest and Hawaii. Most of Europe receives the same amount of solar energy as the northern and midwestern United States. Worldwide, the countries that use the most solar energy per capita are Germany and Bulgaria, which get between 2 and 4 peak sun hours. This means that it's possible for nearly all of the states in the United States to use solar energy as a significant source of power, just like power companies do in Germany.

Hawaii is a state that pays very high prices for the fossil fuels its people use, because it's expensive to transport these fuels. Because Hawaii has this incentive to find a more affordable source of energy, and because Hawaii has high peak sun hours, it is one of the biggest solar power users in the United States.

One common criticism of solar power is the inefficiency of solar panels. The best commercially available solar panels have an efficiency of about 20%. This means they convert 20% of the light that shines on them into electrical energy. But solar panel efficiency has reached this point after steadily increasing over the years, so it's expected to keep increasing gradually. A massive research effort is aimed at increasing the efficiency of solar panels, and the best experimental panels now have efficiencies of more than 50% (**Figure 12.23**). As efficiencies increase, homes and buildings in more and more places on Earth will be able to use solar energy as a sole source of electricity.

Some of the newest solar panels still in the research stage work by concentrating sunlight onto photovoltaic cells. These panels can dramatically increase peak

AP Images/Peter Endig/picture-alliance/dpa

Figure 12.23 Research on Photovoltaic Panels in a Laboratory

sun hours by using mirrors that focus more energy from sunlight onto photovoltaic cells.

The U.S. Department of Energy has issued a challenge called the SunShot Initiative. The goal of this initiative is to make solar power either equal in cost or less expensive than fossil-fuel-based power by the year 2020. This program provides incentives for projects that increase the efficiency of solar panels, make solar installations less expensive for homeowners and businesses, and for projects that create solar panel manufacturing jobs in the United States.

THEgreenBEAT • news about the environment, chapter twelve edition

TOP STORY: Energy Use by Internet Server Farms

In October 2010, Mount Everest finally entered the twenty-first century when a Nepalese telecom company made high-speed Internet access possible on the mountain's summit. Since that day, it has been possible to post a photo to Facebook or make a live Skype call from the roof of the world. Indeed, there are few places on Earth where Internet access is not available.

Figure 12.24 is a graph of Internet users compared to world population. At some point during this decade, one-half of the world's population will be Internet users. We've seen the adoption of new technologies on a grand scale before, such as when automobiles were first available or telephones began to appear in individual homes. However, the Internet revolution is a bit different from previous technological revolutions. For one thing, there are more people on Earth today than there were in 1900, when everyday citizens became telephone owners for the first time.

The Internet is also different because it is less tangible. Unlike a telephone, with its wires, telephone poles, and operators, most of the apparatus of the Internet is invisible to its users. We can see our smartphones and laptops and tablets, but the infrastructure that exists beyond each person's device is mostly kept out of view. Data that do not reside on our own hard drives are kept in a cloud. The very word *cloud* conjures up a floating, ephemeral puff of megabytes that exist in some virtual place. But our data must be somewhere, and it's not kept in real clouds or in anything resembling a cloud. It is kept in *server farms*, which are also known as data centers.

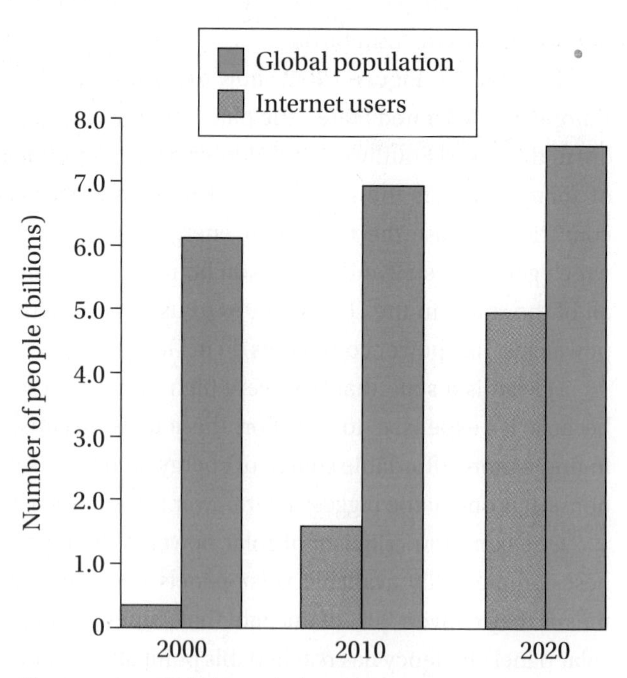

Figure 12.24 Global Population versus Internet Users for the Years 2000 to 2020

At some point between 2010 and 2020, one-half of the world's population will be Internet users.

Server farms are storehouses for servers, which are powerful computers that lack peripherals such as printers, monitors, and keyboards. Each server farm can

THEgreenBEAT *continued*

contain 25,000 servers or more. About 3 million server farms exist worldwide, and the number is growing steadily as demand for data storage continuously increases.

Every time we use the Internet to send an e-mail message or to search for a guacamole recipe, we put a server somewhere to work. An e-mail message sent to a next-door neighbor must travel through miles of Internet infrastructure and at least one data center to reach a neighbor 15 meters away. Servers also are the repository of all our digital stuff. The photos taken five years ago at Aunt Sue's wedding, lists of favorite websites, old e-mails and term papers, fantasy football statistics; it's all stored somewhere on a server in a server farm.

Server farms are massive warehouse buildings that usually do not advertise whether they belong to Microsoft, Google, Yahoo, or some other Internet-based business. **Figure 12.25** is a map of server farm locations in the United States. The ideal spot for such a warehouse is a place where energy is cheap and the air is cold, because servers produce a lot of heat and use tremendous amounts of energy. Servers must be kept cool, and the chillers used to cool servers in a server farm make up a significant chunk of their massive energy demand.

There's a misperception that the Internet, and the businesses that use it, are clean, green, and efficient. But in an investigative report published in 2012 by the *New York Times* and entitled "Power, Pollution and the Internet," the writer pointed out that Internet-based industry takes a significant toll on the environment. Worldwide, server farms use about 76 billion kilowatts of power, and a single server farm can use the same amount of electrical power as a medium-sized town.

Consider that in the United States in 2011, about 67% of electrical power was generated in coal-burning power plants or with natural gas, both fossil fuels. About 19% came from nuclear power plants, and only about 13% came from other sources such as hydroelectric, solar, and wind. That means that, depending on where it's located, a server farm likely is contributing either to carbon dioxide release through the burning of fossil fuels or to the generation of nuclear waste.

But there's more. Because Internet consumers do not tolerate server downtime, server farms must ensure that their servers are always available. To guarantee this, on average, server farms use only a small fraction of their available servers at any given time. According to the *New York Times,* only about 6 to 12% of server time is spent computing. The rest is idle time. This system guarantees that in the event of a surge in Internet demand, no server farm will ever crash. In addition, many server farms have numerous backup generators that use diesel fuel. When the backup generators are running or being tested, they too contribute to carbon dioxide release into the atmosphere.

How does an environment-conscious consumer reduce the negative effects that server farms have on the environment? First, we can reduce the amount of digital content we put on the Web and send electronically. Second, we can seek out Internet companies that use alternative energies to run their server farms. As the public becomes more aware of the impact of Internet use on the environment, hopefully more and more Internet-based companies will move to greener energy sources.

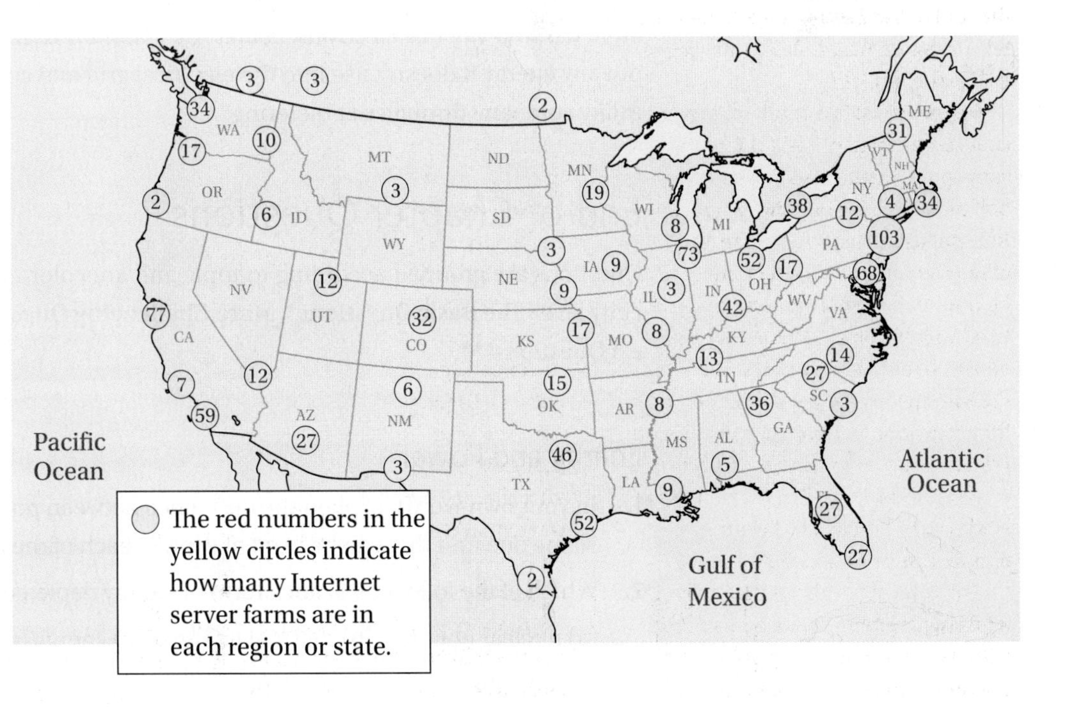
Pacific Ocean
Atlantic Ocean
Gulf of Mexico
The red numbers in the yellow circles indicate how many Internet server farms are in each region or state.

Figure 12.25 Server Farms in the United States

Chapter 12 Themes and Questions

recurring themes in chemistry chapter 12

recurring theme 6 The structure of carbon-containing substances is predictable because each carbon atom almost always makes four bonds to other atoms. (p. 353)

recurring theme 8 The Earth has a finite supply of raw materials that humans can use. (p. 356)

recurring theme 1 Scientists worldwide use the scientific method as a basis for controlled experimentation and deduction. The scientific method is based on the principles of peer review and reproducibility. (p. 358)

recurring theme 4 Chemical reactions are nothing more than exchanges or rearrangements of electrons. (p. 364)

A list of the eight recurring themes in this book can be found on p. xv.

ANSWERS TO FOLLOW-UP QUESTIONS

Follow-Up Question 12.1: The light bulb that uses 500 joules every five seconds is a 100-watt bulb, which uses more power than a 75-watt bulb. *Follow-Up Question 12.2:* It can emit 100 joules of light. *Follow-up Question 12.3:* Nuclear energy. *Follow-Up Question 12.4:* Coal is not produced from crude oil. *Follow-Up Question 12.5:* The answer is (b). Options (a) and (d) are hydrocarbons with 3 and 14 carbon atoms, respectively, and (c) is not a hydrocarbon. *Follow-Up Question 12.6:* (b) 380 ppm is the only value that has existed recently but did not exist in the previous 400,000 years. *Follow-Up Question 12.7:* False. Car makers are already producing cars that meet new fuel economy standards. *Follow-Up Question 12.8:* False. Oxidation is a loss of electrons. *Follow-Up Question 12.9:* Oxygen molecules, protons, and electrons combine to form water. *Follow-Up Question 12.10:* A silicon atom has four valence electrons. *Follow-Up Question 12.11:* A home like this can put any energy it doesn't use into the electrical grid and collect payment from a utility company through net metering.

End-of-Chapter Questions

Questions are grouped according to topic and are color-coded according to difficulty level: **The Basic Questions***, **More Challenging Questions****, and **The Toughest Questions*****.

Energy and Power

***1.** In your own words, explain the difference between power and energy. Name one unit that can be used to express each of these.

***2.** Which of the following is not a form of energy depicted in Figure 12.2?

(a) nuclear energy (b) thermal energy (c) energy from food

(d) sunlight (e) electrical energy

*3. In the water wheel example discussed in the text, how is energy transferred from one form to another?

*4. True or false? Energy is a force applied over a distance.

*5. Name one unit used for power besides the watt.

**6. A light bulb uses 400 joules of power every 10 seconds. How many watts of power does the light bulb use?

**7. In your own words, state the first law of thermodynamics. Give a real-life example of this law.

**8. If energy cannot be created or destroyed, how is it possible for electrical generators to exist?

***9. The unit for energy, the joule, is the same as one newton meter (Nm), where a newton is a unit of force. For example, if four newtons of force are required to push a box two meters across the floor, then eight joules of energy are used to push the box. Given this, how many joules are required to push the same box five meters? If the pushing takes three seconds, what power, in watts, is used for this process?

Fossil Fuels

*10. Using chemical formulas, rank the following molecules according to the distance they will travel up a fractionating column. Start the ranking with the one that travels farthest:

(a) $C_{20}H_{42}$ (b) C_2H_6 (c) C_7H_{16} (d) C_5H_{12} (e) $C_{16}H_{34}$

*11. Which of the following is not a type of fossil fuel?

(a) gasoline (b) radioactive material

(c) methane (d) diesel fuel

*12. Which of the following is/are not hydrocarbon molecule(s)?

(a) C_4H_8 (b) CH_4 (c) H_2O (d) $C_6H_{12}O_6$

*13. According to Figure 12.8a, which product of crude oil refining has the lowest average number of carbon atoms?

(a) diesel fuel (b) gasoline (c) wax (d) lubricating oil

*14. According to Figure 12.6, which type of fuel was most often used to provide energy in the United States in 1800? In 1920?

**15. The fossil fuels that we use, such as diesel fuel and gasoline, are mixtures of hydrocarbons. Based on what you know about the refining process and Figure 12.8a, explain why they are mixtures.

**16. Figure 12.8a is a diagram of a fractionating column. Why is coal not mentioned in this diagram?

**17. Which type of crude oil has a lower boiling point, a fraction that is taken from the top of the fractionating column or one taken from the bottom?

**18. Consider Figure 12.9, and use a conversion factor to figure out how many watts of power a SmartForTwo uses.

***19. Using Figure 12.7 as a guide, roughly approximate the boiling point of the following hydrocarbons:

(a) C_8H_{18} (b) $C_{18}H_{38}$

***20. You are put in charge of designing a new oil refinery. Your plans call for a fractionating column that is eight feet in height. Once built, your column does not provide fractions from crude oil like those shown in Figure 12.8a. What must you change to improve the performance of your column?

***21. According to Figure 12.6, by *roughly* what factor did the total energy consumption in the United States increase from 1915 to 2012?

Global Warming and Climate Change

*22. Which one of the following greenhouse gases contributes most to global warming?

(a) ozone (b) carbon dioxide (c) butane (d) helium gas

*23. According to Figure 12.12, what is the lowest carbon dioxide level reached in the past 400,000 years on Earth?

(a) 280 ppm (b) 595 ppm (c) 260 ppm (d) 180 ppm

*24. True or false? Global warming and climate change are two words that describe the same thing.

**25. What type of experimental evidence is used to study atmospheric conditions on Earth over hundreds of thousands of years? What are scientists who work in this field called?

**26. Some greenhouse gases, such as methane, cause more atmospheric heating per molecule than carbon dioxide does. Given this, why is carbon dioxide the greenhouse gas that is blamed most for global warming?

**27. Recount one story from the text that tells how society has risen to meet a challenging environmental problem.

**28. In your own words, explain what CAFE laws are. Does Figure 12.14 support or refute the use of CAFE laws in raising fuel efficiency for cars and trucks?

***29. Consider Figure 12.14. From 1975 until the 1990s, the line labeled "Both" is much closer to the car line than to the truck line. What is one possible explanation for this?

***30. Your friend Joe is skeptical of global warming. He cites a recent cold snap as evidence that global warming is a hoax. What explanation could you offer to your friend to explain this contradiction?

Courtesy of Sandia National Laboratories

Hydrogen and Fuel Cells

*31. In an electrochemical cell, which half-reaction occurs at the anode and which occurs at the cathode?

*32. True or false? When water is electrolyzed, a great deal of energy is needed to produce hydrogen and oxygen molecules.

*33. How many bonds connect the two atoms of hydrogen in a hydrogen molecule? How do you know?

*34. Why did the people of Utsira, Norway, electrolyze water?

**35. A fuel cell performs the following redox reaction:

$$2H_2 + O_2 \rightarrow 2H_2O$$

In this redox reaction, what is reduced and what is oxidized?

**36. Why are electrochemical cells based on half-reactions? What function does this allow the cell to perform? Why is hydrogen gas—represented by H_2 in the electrolysis equation—a fuel?

**37. In one part of the United States, a car fueled by hydrogen gas may contribute to climate change more than a gasoline-powered car in another part of the United States. Explain how this could be true.

**38. Why aren't fuel cells widely used as a source of electrical power?

**39. True or false? Purchasing a hydrogen-fueled car is always an environmentally friendly choice.

**40. Why is the term *proton* used to refer to H^+?

***41. Add the following half-reaction equations, performing cancellation if needed:

$$Cu^{2+} + 2e^- \rightarrow Cu$$
$$Al \rightarrow Al^{3+} + 3e^-$$

GH1/Pavegen/WENN.com/Newscom

Solar Power and Batteries

*42. Indicate whether each half-reaction is an oxidation reaction or a reduction reaction.

(a) $C + 2O^{2-} \rightarrow CO_2 + 4e^-$ (b) $Na^+ + e^- \rightarrow Na$

*43. What is the purpose of the wires and contacts that lie on top of the silicon in a solar cell?

*44. True or false? Solar panels are much too inefficient for use in most climates around the world. Explain your answer.

*45. According to the map in Figure 12.22, only three countries in Europe experience just 1–2 peak sun hours. Name one of them.

**46. According to the text, what new modification to solar panels might dramatically increase their efficiency in converting sunlight to electrical power?

**47. Add the following half-reaction equations, performing cancellation if needed:

$$Fe^{2+} + 2e^- \rightarrow Fe$$
$$Zn \rightarrow Zn^{2+} + 2e^-$$

**48. In a NiMH battery, what ion is oxidized at the anode? Is this oxidation half-reaction a charging process or a discharging process?

***49. If the solar array on the roof of a home generates 15,000,000 joules of energy in one day, and the home requires 10,000,000 joules per day for lighting, appliances, and so on, what number of joules is contributed to the electrical grid for that day? If the cost per 1 million joules is 33 cents, how much does the homeowner earn through net metering on that day?

***50. In your own words, explain the difference between an electrochemical cell that uses reversible half-reactions and one that uses irreversible half-reactions. Cite one example of each and explain why, in each case, the choice of reversible or irreversible makes sense.

Integrative Questions

*51. You own a truck that was manufactured in 1977. According to Figure 12.14, how many miles should that truck take you on one gallon of gasoline? If, instead, that truck had been produced in the year 2010, how many more miles would the newer truck travel on one gallon of gasoline? Convert this answer to kilometers using dimensional analysis.

**52. Figure 12.14 in this chapter shows U.S. fuel economy data beginning in 1975. The figure shown below presents Australian fuel economy data from a different perspective. In this figure the lines slope downward rather than upward. Compare the two figures. Why does this one trend downward? Has the fuel economy of Australian vehicles been decreasing, while the fuel economy of vehicles in the United States has been increasing? What is being represented here?

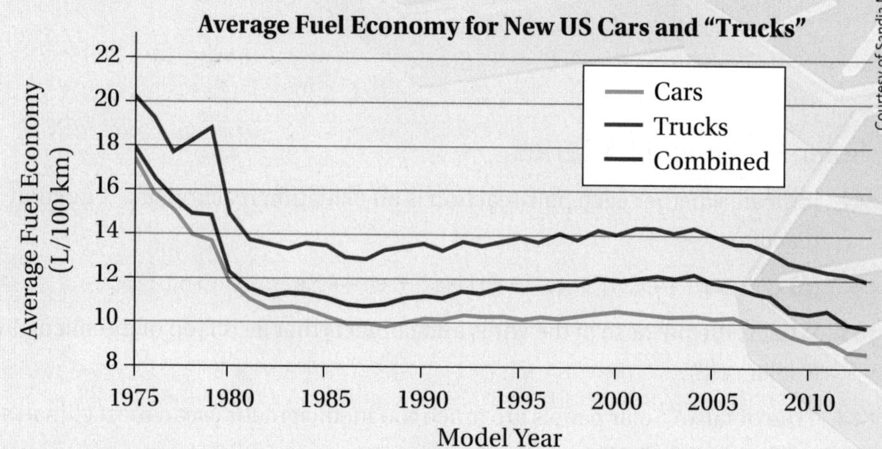

Courtesy of Sandia National Laboratories

**53. Peak sun hours across the United States are shown in Figure 12.22. Return to that figure, and determine the number of peak sun hours in your area. The number of peak sun hours is roughly equivalent to the number of kilowatt hours (kWh, a common unit for residential electrical power) per square meter of solar panel surface area per day. Therefore, if you have 5.0 peak sun hours and a solar panel with an area of one square meter, then you should collect roughly 5.0 kilowatts of power per day. Now imagine that you have a refrigerator that uses 1095 kWh per year. Calculate the number of kWh it uses in one day (365 days = 1 year). Ask yourself "Will my one, 1-square-meter-area solar panel power this refrigerator, or will I need more than one?" (Note: We do not include reductions in solar power due to panel efficiency in these calculations.)

***54. Batteries combine a reduction and an oxidation reaction to create a current. Old alkaline batteries were based on the two half-reactions shown as (1) and (2) below.

$$4e^- + 2H_2O + 3MnO_2 \rightarrow Mn_3O_4 + 4OH^- \qquad (1)$$

$$4OH^- + 2Zn \rightarrow 2ZnO + 2H_2O + 4e^- \qquad (2)$$

With reversible batteries, both of these reactions must run efficiently in the reverse direction during recharging. Alkaline batteries were not easily recharged because the reverse of these reactions did not happen efficiently. Upon recharging, an unwanted side reaction (see (3) below) could also take place, making the recharging process potentially dangerous:

$$2e^- + 2H_2O\,(l) \rightarrow H_2\,(g) + 2OH^-\,(aq) \qquad (3)$$

This old alkaline technology is still used in batteries today, but these types of batteries have been modified so that they recharge more efficiently, although they are not as efficient as modern rechargeable batteries like NiCad batteries.

(a) Add together reactions (1) and (2) to find the net redox reaction for an alkaline battery.

(b) What species in reactions (1) and (2) shows you that the battery is "alkaline"?

(c) If you buy a rechargeable alkaline battery, you may notice that it has little vents. Looking at reaction (3), what do you think the purpose of these vents might be? Why do you think the recharging process may be potentially dangerous?

55. INTERNET RESEARCH QUESTION

Perform research on any car or truck of your choosing, except the 2013 Smart for Two and the 2013 Dodge Viper. Refer to Figure 12.9 and imagine that you will create a column in this table for your car or truck of choice. Look up the seven specifications listed in the figure for your car or truck and compare your new data to the two cars shown. In terms of fuel economy, how does your car compare? How does your car compare for top speed and horsepower? Price? After inspecting the specifications for all three cars, which one would you purchase?

Courtesy of Sandia National Laboratories

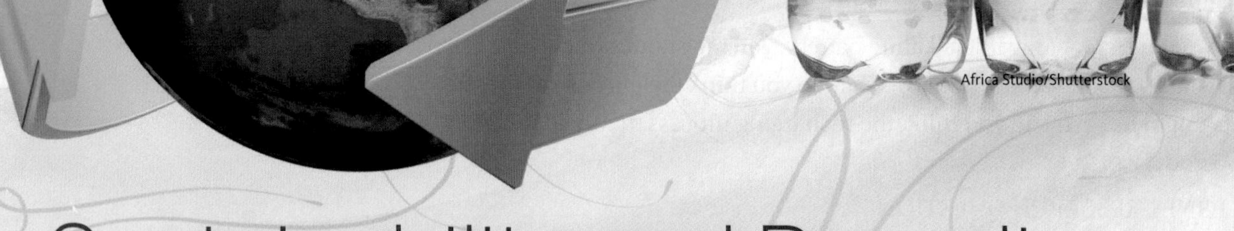

Africa Studio/Shutterstock

Sustainability and Recycling

Finding Better Ways to Use (and Reuse) Our Resources

13.1 What Is Sustainability? 383

- Sustainability is the capacity to endure.
- A life-cycle assessment accounts for the energy and materials that go into making a product.
- It is possible to estimate the environmental impact of a product or process.
- Cradle-to-cradle design includes a plan for product reuse.

KEY WORDS biodegrade, sustainability, life-cycle assessment (LCA), cradle-to-cradle, bisphenol A (BPA)

naturebox Landfills, Paper, and the Menace of Single-Use Water Bottles 390

13.2 What Is Plastic? 392

- Plastics are made from polymers, which are large organic molecules.
- Natural rubber is a polymer called polyisoprene.

KEY WORDS monomer, polymer, plastic

13.3 The Physical Properties of Polymers 394

- The structure of a polymer often dictates its physical properties.
- Polymers can be designed for rigidity and toughness.
- Crystallites can make polymers more rigid.

KEY WORDS hydrophobic, hydrophilic, amide, cross-links, crystallite

13.4 Recyclable and Sustainable Plastics 398

- We can categorize plastics as thermoplastic polymers or thermosetting polymers.
- Plastics are recycled according to their resin ID codes.
- Recycled plastics can be made into new products and structures.
- Waste plastics that are not recycled end up in the landfill or the incinerator.
- The structure of a polymer dictates its biodegradability.

KEY WORDS thermoplastic polymers, thermosetting polymers, resin ID code, VOC, recyclable, bioplastic, lightweighting

THEgreenBEAT News about the Environment: The Arrival of Sunshine-Fueled Batteries 406

Africa Studio/Shutterstock

stable/Shutterstock

CHAPTER 13

In the distant past, telephones, like the one in the photo on the next page, had cords. These old phones weren't designed to be mobile; they weighed about 100 times more than a modern cell phone. They were clunky, yes, but they had one clear advantage over a cell phone: they were virtually indestructible.

The telephone in the photo is made of Bakelite, the first purely artificial material made without any natural substances. Bakelite is a type of plastic that once was used to make everything from bracelets to telephones to chess pieces. Early plastics, such as Bakelite, were touted for their durability. Once a manufacturer made Bakelite into a phone shape, it probably always would be phone-shaped. It was incredibly tough stuff. But an indestructible product presents a tricky problem: What do you do with it when you don't want it anymore? Luckily, this was never a big concern, because Bakelite was

Africa Studio/Shutterstock

not made for long—only from about the 1920s to the 1940s. Now it's a rare and valuable commodity. A phone similar to the one shown here recently sold for $60 on eBay.

Luckily, twenty-first-century cell phones are not made of Bakelite. But even though our phones are not made out of indestructible plastic, discarding them is still a challenge. New phones contain toxic elements such as cadmium (Cd), lead (Pb), beryllium (Be), and mercury (Hg). *The Green Beat,* Chapter Three Edition, discusses the difficulties posed by the accumulation of discarded cell phones—known as e-waste—and the effects their toxic components can have on the environment.

stable/Shutterstock

This chapter is all about our stuff, where it comes from, how long it lasts, and what we do with it when we no longer want it. It's about how we design the materials we use to make our stuff. Are our materials extra durable, like Bakelite, or less durable and potentially reusable? What kind of trash will this material become? We'll begin in the place where it's easy to see what becomes of what we discard: the ocean.

13.1 What Is Sustainability?

Sustainability is the capacity to endure.

If we were to sail directly from San Francisco to Hawaii, we would pass through an area called the North Pacific Gyre. At the gyre, which covers an area greater than the state of Texas, several ocean currents pass each other while moving in different directions. The result is a huge, spiraling circle of water—a *vortex*—that continuously moves clockwise (**Figure 13.1**). Because water from all over the Pacific Ocean flows into and out of the gyre, and many ocean currents flow toward its location, this is where debris in the ocean collects. In other words, it's the final destination for much of the trash that goes into the ocean.

About 20% of the millions of tons of plastic made around the world each year winds up in the ocean. Much of this trash makes its way into the ocean from the shore—it's either swept off beaches or carried with river currents or sewage—but a surprisingly large amount of plastic comes from toppled shipping containers. The most infamous lost cargo was a container holding 28,000 toy rubber ducks that fell overboard in the Pacific Ocean in 1992. The ducks ended up all over the world, and their locations have been used as a way to follow ocean currents. A large number of the ducks became trapped in the North Pacific Gyre.

The vast majority of the trash at all depths of the ocean is plastic. And, of all the plastic that finds its way into the ocean, about 70% sinks; the rest floats. Since the 1950s, when plastic was first widely used, it has been accumulating in landfills

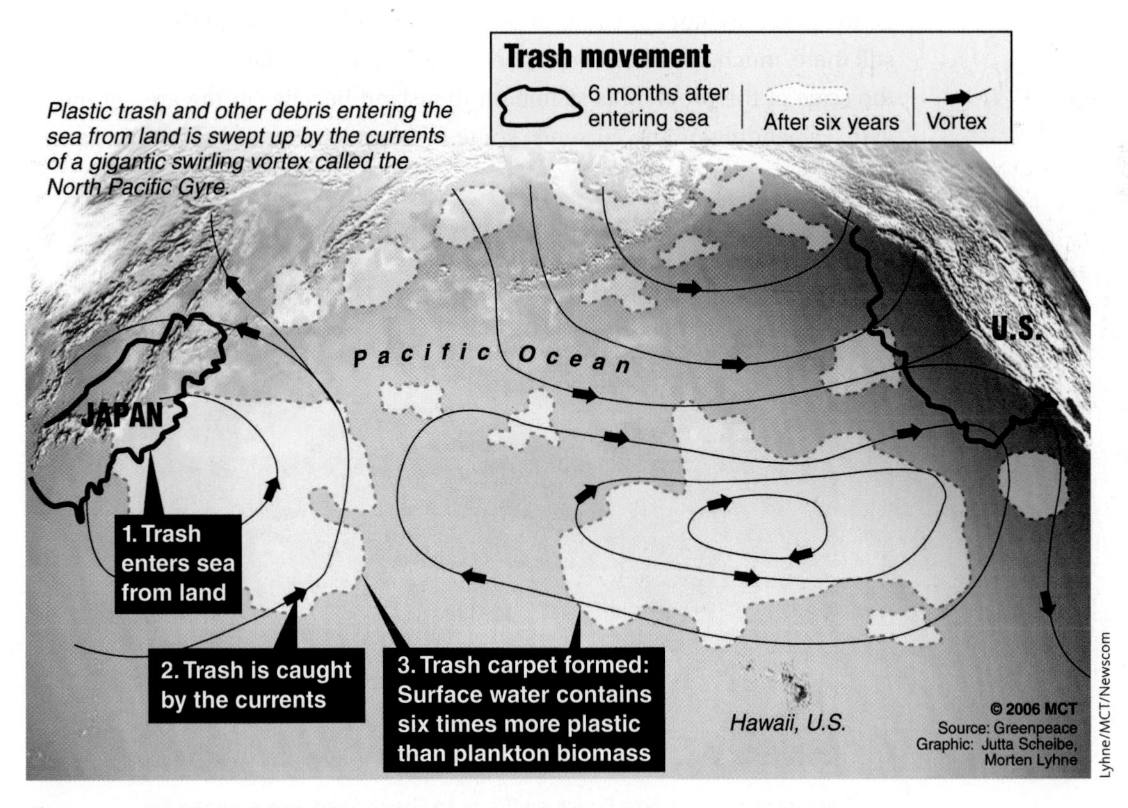

Trash movement
6 months after entering sea
After six years
Vortex
Plastic trash and other debris entering the sea from land is swept up by the currents of a gigantic swirling vortex called the North Pacific Gyre.
Pacific Ocean
JAPAN
U.S.
1. Trash enters sea from land
2. Trash is caught by the currents
3. Trash carpet formed: Surface water contains six times more plastic than plankton biomass
Hawaii, U.S.
© 2006 MCT
Source: Greenpeace
Graphic: Jutta Scheibe, Morten Lyhne
Lyhne/MCT/Newscom

Figure 13.1 The North Pacific Gyre

The gyre we discuss in this section is the larger gyre shown directly above Hawaii. Ocean trash also accumulates in a smaller gyre off the east coast of Japan.

and in oceans. Over time, plastic in the ocean **biodegrades**—breaks down naturally into smaller and smaller pieces—but the process is much slower than it is on land, so plastic in the ocean tends to remain there for decades. And it causes other problems. For example, in many cases, seabirds mistake small pieces of plastic for food. It is estimated that nearly half a million albatross die each year because their stomachs are full of plastic that they cannot digest. When scientists perform autopsies on these dead albatross, they find anything from Legos to bottle caps to disposable lighters in the birds' stomachs. It is an overwhelming and intractable problem.

This ocean trash problem stems from unsustainable methods being used on a grand scale. If humans continue to produce millions of tons of plastic with no plan to dispose of them or recycle them, then plastic waste will build up—and up—and up. This is a one-way process: plastics are made, and they are used, and then most

wait a minute . . .

Does the North Pacific Gyre really look like a garbage dump?

When we imagine a garbage dump, we see mountains of trash, much of it plastic. The North Pacific Gyre does indeed contain millions of tons of plastic, but it doesn't look like a floating garbage dump. Much of the plastic in the ocean exists in very small pieces, like confetti. These small pieces are produced when larger pieces are broken down slowly by the action of wind and waves and by chemical reactions that are sometimes hastened by sunlight. So the gyre does not appear to be a floating island of trash, but the plastic is still there, much of it bobbing just below the surface. In Hawaii, for example, the scale of the problem is visible on the island beaches in the area of the gyre. Depending on the currents, some beaches may indeed resemble garbage dumps (**Figure 13.2**).

Courtesy of Barbara Zazzi

Figure 13.2 Trash on a Beach en Route to Green Sand Beach on the Big Island of Hawaii

of them become trash. Currently, we do not reuse a significant portion of what we make. We use plastics in a linear way—we use them and discard them—rather than in a cycle where we use and discard, then reuse and discard, and so on. Worldwide, people recycle only about 10% of the plastics that are produced. This process is *unsustainable,* because it cannot be maintained. We will reach a limit—some would say we already have—where Earth's oceans cannot accommodate any more plastic trash.

Sustainability is a broad term that refers to the capacity to endure. It is best explained with the examples that appear throughout this chapter. For example, for the ocean gyre problem, a sustainable solution would be to recycle 100% of our plastics so that we can dramatically reduce the amount of new plastic made each year. This solution increases the ocean's capacity to endure because it puts plastics into a cycle and reduces the amount of plastic that becomes ocean trash. Another sustainable approach would be to create plastics that are less durable. Plastics that would break down more readily would not persist for decades. This solution increases our capacity to endure because it reduces the millions of tons of plastic trash that we generate each year.

A life-cycle assessment accounts for the energy and materials that go into making a product.

We can begin to understand sustainability with a familiar example. Consider a company that wants to begin manufacturing disposable drinking cups. The company is trying to decide which to make: plastic or paper cups? For each alternative, company executives want to consider every step of the production process and the price they will charge to consumers. **Figure 13.3** shows the steps that most products go through—from raw materials to the production of the product, to the distribution

Figure 13.3 The Life Cycle of Manufactured Products
Most products that we buy go through this type of process to reach a store: raw materials are harvested and make their way to a factory, where they are made into something. That thing travels to a store, where you buy it, take it home and use it, and ultimately discard it or its packaging. Finally, many products go to the incinerator.

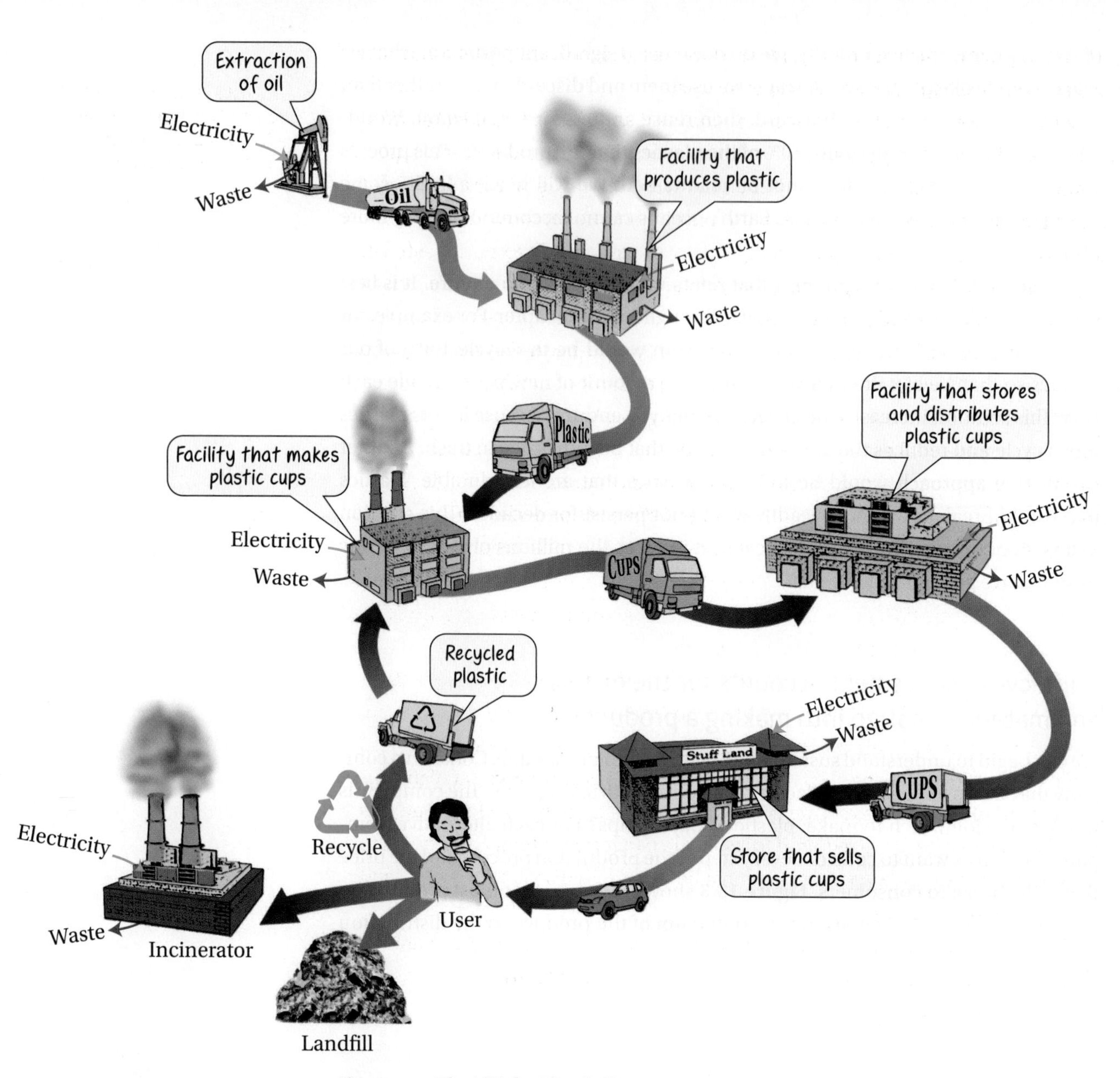

Figure 13.4 The LCA for Plastic Cups

A plastic cup is made from oil. Oil that is extracted from Earth is made into plastic, and the plastic is made into cups. The cups are distributed to warehouses and then to stores where consumers buy them. In each of these steps, energy is consumed and waste is produced. The used cup usually has one of three fates: incineration, a landfill, or recycling.

and sale of the product, to its use and disposal. For each step along the way, the company can try to figure out what it will cost to get the product to the next step.

Our cup maker performs this sort of analysis for plastic cups versus for paper cups, so the company can reach an informed decision about which type of cup to make. This type of investigation is sometimes called a cradle-to-grave analysis, because it tracks a product from its raw materials (the "cradle") all the way to its disposal (the "grave"). A more modern term for this analysis is **life-cycle assessment,** or **LCA**. An LCA accounts for all the energy and materials that go into making a product and disposing of it. This same type of analysis is also used to determine the carbon footprint or the water footprint of processes. (Carbon and water footprints are discussed in Chapters 5 and 8, respectively.)

flashback

In Chapter 12, we discussed fossil fuels and their relationship to global warming and climate change in detail.

Let's consider the LCA for plastic cups. As we'll see later in this chapter, most plastic is made from compounds in petroleum, a fossil fuel. To make plastic cups, a manufacturer must obtain crude oil from an oil well. The crude oil must be refined so that the materials needed to make the plastic can be obtained. The oil drilling and refining processes require energy, a cost that also has to be considered. The petroleum by-products used to make plastic must be transported to a factory where the plastic cups will be made. The factory's energy use must also be counted.

Once the cups are made, they must be put into a package. This step requires either more plastic (if the package is a plastic bag) or cardboard (if the package is a box). The raw materials used to make the packaging, and all of the energy required to obtain them and make the package, must be considered. Finally, our packaged cups must be transported to a distribution center and then transported to stores. People buy the cups, use them once or twice, and discard them. If the cups are not recycled, they make their way to a landfill where, depending on the type of plastic they are made from, they are put directly into the landfill or first incinerated and then put into the landfill. The LCA for plastic cups is summarized in **Figure 13.4**. We discuss the fate of recycled plastics in Section 13.4.

We can also conduct an LCA for paper cups. The paper comes from trees, which get their energy from sunlight. Trees need land and water, and machines are needed to cut down the trees and remove their bark. The trees must be transported on roads to a paper mill to be chipped into 1-inch squares and cooked at high temperatures. All of these steps require energy. Enormous amounts of clean water, as well as harsh chemicals, are required to make the paper that is then fashioned into cups. Like the plastic cups, paper cups must be taken to a distribution center and transported to retail stores.

When all the costs of each of these multiple steps for each type of cup are tallied, the company can use that data as a basis for deciding to invest in plastic or paper cup production.

It is possible to estimate the environmental impact of a product or process.

More and more companies are using LCAs, not just as a way to determine the cost of a process, but also to determine the environmental impact of a process. For our cup example, the company may want to consider the final cost of making both cups, but an equally important consideration is how much the production of each type of cup affects the environment. The plastic cups, for example, may require fewer transportation steps than the paper cups. Assuming that the mode of transportation uses fossil fuels, as most do, these steps result in the release of carbon dioxide and nitrogen oxides into the environment. The first substance contributes to global warming, and the second one to the formation of acid rain. Likewise, if the plastic-cup-making factory gets its energy from a coal-burning power plant, then the production of carbon dioxide and sulfur oxides must be considered in the environmental impact of the cups. **Table 13.1** (on p. 388) is a summary of key pollutants.

Table 13.1 Key Pollutants

This/these pollutant(s) . . .	is/are released into the environment mostly through . . .	and harm(s) the environment by causing . . .	Section of this book where you can find more information
Sulfur oxides	the burning of coal	acid rain.	10.3
Nitrogen oxides	the burning of fossil fuels	acid rain.	10.4
Carbon dioxide	the burning of fossil fuels	global warming and climate change.	12.3
Methane	agricultural activity	global warming and climate change.	12.2

Table 13.2 summarizes some of the environmentally relevant findings of a group of scientists who performed an LCA for plastic and paper cups. First, note that the manufacturing process for both types of cup requires significant electrical power. Both processes use natural resources, such as crude oil and wood. They both contribute to pollution. The plastic cup has a greater carbon footprint, meaning that the production of one plastic cup releases high levels of carbon dioxide into the air. The production of the paper cups also causes pollution, but paper production pollutes water more than air.

flashback

The naturebox in Chapter 5 introduced carbon footprints—the amount of carbon dioxide, or its equivalent, produced by a process or the manufacture of a product over a specified period of time.

Which cup will the company decide to manufacture, disposable paper cups or plastic cups? The firm will base that decision on the LCA results, which will reveal the most cost-effective process. Many companies also now consider the carbon and water footprints of their products as well as the energy requirements for producing the product. So, is it better for the environment to make plastic cups or paper cups? The answer depends on the priorities of the manufacturer.

If you want to make the best choice for the environment, which should you choose, paper or plastic? If you happen to live in an area where paper companies clear-cut forests and pollute the water, then you might choose plastic. If you want to choose the cup with the smallest carbon footprint, then you might choose to buy the paper cup. There is no easy answer. Each type of cup has its pros and cons, and consumers must make their choices in the context of their own lives and their own priorities.

QUESTION 13.1 For each of the following pollutants, name the type of cup manufacturing (plastic or paper) that produces more of that pollutant. (a) carbon dioxide (b) water pollution

Table 13.2 Comparison of LCAs for Plastic versus Paper Cups

	Plastic	Paper
Resources used in production	natural gas, crude oil, electrical power	sunlight, timberland, wood, electrical power
Price per cup for manufacture	one cent	12 cents
Carbon footprint (mass of carbon dioxide emitted per cup)	high	low
Biodegradability	low	medium
Contribution to water pollution	low	high

ANSWER 13.1 (a) The plastic cup has a larger carbon footprint; its production results in more carbon dioxide emissions. (b) Paper cup production causes more water pollution.

For more practice with problems like this one, check out end-of-chapter Question 11.

FOLLOW-UP QUESTION 13.1 True or false? An LCA is a method used solely to determine the cost of making a particular product.

Cradle-to-cradle design includes a plan for product reuse.

An LCA can also take into account the sustainability—the capacity to endure—of plastic versus paper cups. Plastic cups are made from an unsustainable resource: fossil fuels. A finite amount of fossil fuels exist on Earth and, as time goes on, they are increasingly difficult to find and extract. Any product that relies on fossil fuels and is not conscientiously recycled is unsustainable, because the raw material is in finite supply. A sustainable process does not deplete a resource. Rather, a sustainable process uses resources that are part of a cycle.

recurring theme in chemistry 8
The Earth has a finite supply of raw materials that humans can use.

Sustainability relates to more than just raw materials: it also relates to the end of a product's life. Product manufacturers lately are thinking more about what will happen to their products when they're ready to be discarded. For example, Nike and Puma both manufacture sneakers that can be recycled. When a consumer is done using the sneakers, the company can make the used shoes into something else.

Nike takes a sneaker and dismantles it into its rubber, foam, and fabric components to make Nike Grind (**Figure 13.5a**). To make Nike Grind, the sneaker company grinds up the shoe parts to create a premium-grade raw material with many innovative applications. It is an ideal component for high-performance Nike footwear and apparel as well as high-quality sports surfaces including courts, turf fields, tracks,

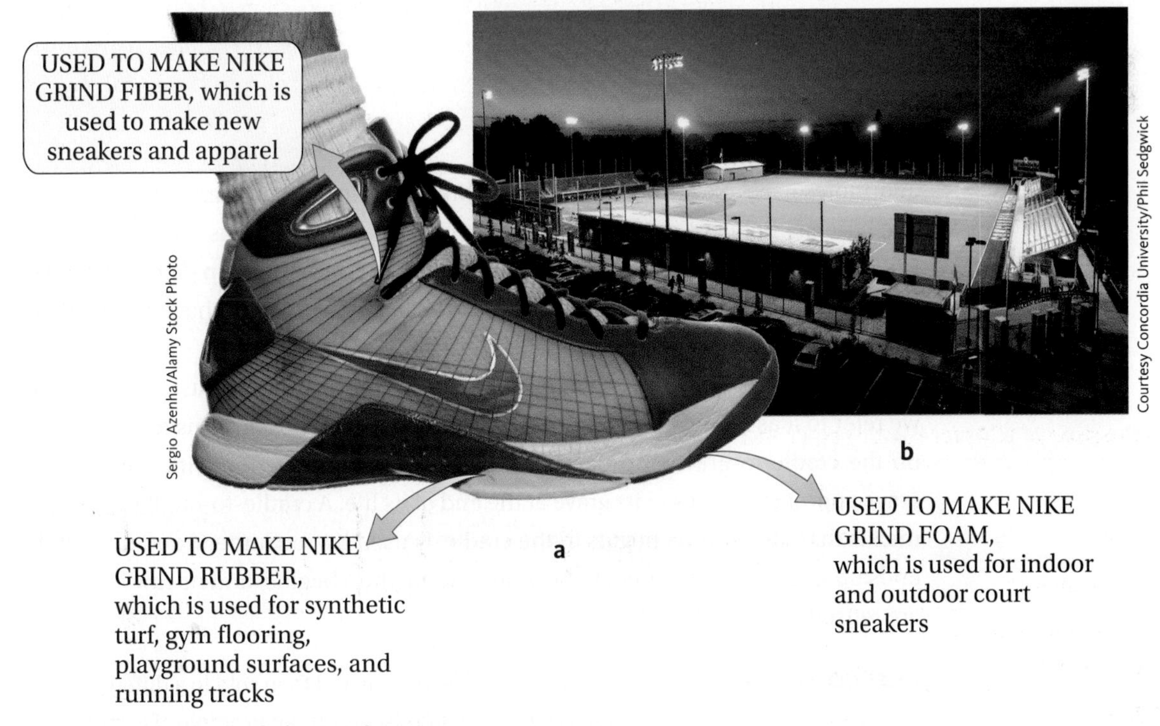

USED TO MAKE NIKE GRIND FIBER, which is used to make new sneakers and apparel
USED TO MAKE NIKE GRIND RUBBER, which is used for synthetic turf, gym flooring, playground surfaces, and running tracks
USED TO MAKE NIKE GRIND FOAM, which is used for indoor and outdoor court sneakers
a
b
Sergio Azenha/Alamy Stock Photo
Courtesy Concordia University/Phil Sedgwick

Figure 13.5 A Second Life for Sneakers

(a) Nike dismantles used sneakers and repurposes each part to make a different material. (b) The Concordia University athletic field is made of Nike Grind.

natureBOX • Landfills, Paper, and the Menace of Single-Use Water Bottles

For many of us, garbage is something we reluctantly drag to the curb or to the dumpster. Most of us don't think about what happens to our garbage once it leaves the house, but that is not true for some scientists at the University of Arizona (UA). The Garbage Project, begun at UA in 1973, was a data collection effort that spanned more than two decades, and some of its findings may surprise you.

When we think of landfills, many of us conjure up an image of discarded baby diapers and plastic wrappers, imagining that anything made from organic materials decomposes relatively quickly. The data show, however, that plastic items make up only about one-fifth of an average landfill. What the UA researchers did find in great quantities was paper, which accounted for about 40% of the volume. They found buried newspapers, dozens of years old, whose type was still clearly legible. Apparently, paper doesn't biodegrade as quickly as we thought. In fact, the researchers were surprised to find that the landfills that had been degrading undisturbed for 50 years, still contained undegraded organic materials, including paper. Food scraps and grass clippings were the only things they found that did degrade reliably.

The chart in **Figure 13A** gives the approximate percentages of various materials, by weight, in a typical landfill.

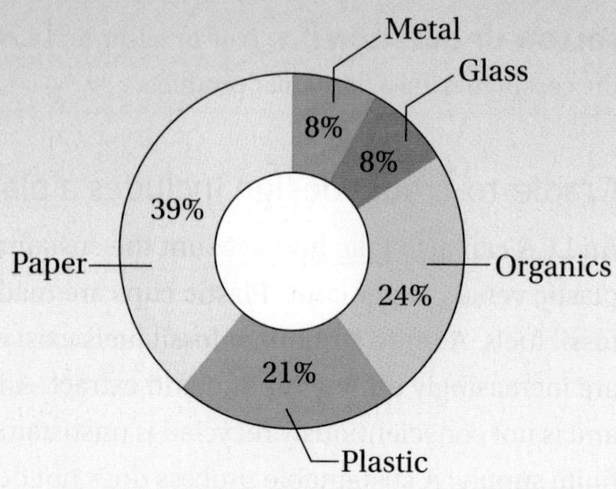

Figure 13A Percentages of Various Materials in a Typical Landfill
In the chart, the term *organics* refers to yard clippings and food waste.

The Garbage Project also revealed that the mass of plastics at landfills stayed pretty much the same over the 32 years of the study. That's surprising, given the radical increase in the use of plastics during the same period. This phenomenon is the result of **lightweighting,** the process of reducing the mass of plastic used to make a container to hold a given volume of something. For instance, a typical half-liter (16.9-ounce) water bottle had a mass of 18.9 g in 2000 but only 12.7 g in 2008. And your paper grocery bags fall apart more often because they're made from thinner paper—it has gone from 33 μm in 1976 to less than 13μm today.

Despite the surprisingly small piece of the pie allotted to plastics in Figure 13A, plastics, especially single-use water bottles, are piling up in landfills.

---

and more. Each playground surface requires outsole rubber from 2500 sneakers. The baseball field at Concordia University in Portland, Oregon, is made entirely of Nike Grind. The artificial field, which is made of more than 6 million pairs of recycled shoes, saves the university the thousands of dollars that it would have to spend on watering and maintaining a grass field (**Figure 13.5b**).

When a company's original design incorporates plans for reusing a product, we refer to it as a **cradle-to-cradle** approach to product design. This term is a play on the cradle-to-grave approach. When a product is designed with a cradle-to-grave approach, it goes to its grave at the end of its life. A cradle-to-cradle approach means that the product begins in the cradle, is used by consumers, and ultimately ends up in a different cradle. Its second lifetime then begins, and its raw materials are reused in a new product.

QUESTION 13.2 A Bakelite phone is used by one person and then sold to a different person. That person uses it and sells it to someone else. Is this an example of a cradle-to-cradle process?

In the United States each year, we use and discard more than 50 billion single-use water bottles. We recycle about 23% of those plastic bottles, meaning that the other 77%—that's 3.85×10^{10} bottles—go to the landfill every year. And because those bottles are made from petroleum, plastic bottles translate into increases in our demand for and reliance on crude oil. On average, in one year, every American uses 167 disposable plastic bottles but recycles only 38.

Why do we drink so much bottled water? For one thing, it's convenient. For another, many people believe that filtered tap water is less pure than bottled water. But most bottled water does not come from clean mountain streams; it's simply filtered municipal water. Furthermore, according to Dr. Gina Solomon, a senior scientist at the Natural Resources Defense Council, recent studies on filtered tap water versus bottled water assert that bottled water is not healthier than most filtered tap water. Certainly, when you factor in the environmental impact of 50 billion water bottles, the choice is a no-brainer.

So what should we do about the massive number of plastic bottles that continue to accumulate in landfills? Octogenarian Jean Hill (**Figure 13B**) of Concord, Massachusetts, had an idea: ban single-use water bottles. She put it this way: "We're a smart community of people who cannot be tricked by clever marketing. And we are not willing to put convenience ahead of our concern for the near and long-term consequences of bottled water. . . . I'm going to work until I drop on this." And work she has. Thanks to Jean's hard work and the efforts of other advocates, at the beginning of 2013, Concord became the first U.S. city to ban single-use water bottles. Since then, dozens of places around the United States have enacted full or partial bans on single-use water bottles. The website www.banthebottle.net provides facts about single-use water bottles and tracks campaigns to ban them.

Matthew J. Lee/The Boston Globe via Getty Images

Figure 13B Jean Hill of Concord, Massachusetts, and the Bottle Ban Logo

ANSWER 13.2 No. Once a Bakelite product is made, it serves only one function because Bakelite cannot be recycled or reused. In a cradle-to-cradle process, a product spends one lifetime performing a function and then is remade and reused for a different purpose.

For more practice with problems like this one, check out end-of-chapter Questions 1 and 3.

FOLLOW-UP QUESTION 13.2 Which of the following resources are sustainable and which are unsustainable? (a) petroleum (b) Bakelite (c) sunlight

13.2 What Is Plastic?

Plastics are made from polymers, which are large organic molecules.

Think about words beginning with *poly-*. The prefix *poly-*, as in *polyglot, polygamist,* and *polygraph,* multiplies the root word. A polyglot is someone who speaks many languages. A polygamist has many spouses. A polygraph, commonly known as a lie detector, measures many things in its efforts to detect liars—blood pressure, respiration, heart rate—all at once. In chemical terms, the prefix *poly-* has the same meaning: "many." Thus, a polysaccharide contains many saccharide molecules, polystyrene contains many styrene molecules, and polyester contains many ester molecules.

An enormous list of chemicals have names that begin with the prefix *poly-*. These molecules are typically made up of small organic molecules, called **monomers,** that are connected to form a long chain, or **polymer**. There are no hard-and-fast rules about how many monomers must be linked together to constitute a polymer, although most polymer chains contain at least 100 monomers and often many thousands.

The variety of polymers is nearly as vast as the number of monomers they contain. Polymer chains can be long or short, branched or straight. Their physical properties also vary widely. Some polymers are tougher than steel. Other kinds of polymers are stretchy, water resistant, biodegradable, or even antibacterial.

Figure 13.6 shows some everyday items made from polymers. Perhaps the most common way to describe all these items is to say they're **plastic**. The word *plastic* can be used as an adjective meaning "mutable, or capable of change." Most items made from plastic are changeable and can be shaped in many ways. (One notable exception is the first true plastic, Bakelite, which, as we noted, was not easily changed.)

From a plastic material, a manufacturer can fashion a food container or a cell phone or a weather-resistant coat. Thus, plastics are indeed *plastic*—capable of being molded, stretched, or flattened into any conceivable shape. But to understand how plastics are able to be made into different shapes, we must first understand their molecular structures.

Natural rubber is a polymer called polyisoprene.

Figure 13.7a is an example of a typical polymer molecule, drawn as a line structure. This molecular structure is enormous, but not as complex as it may seem at

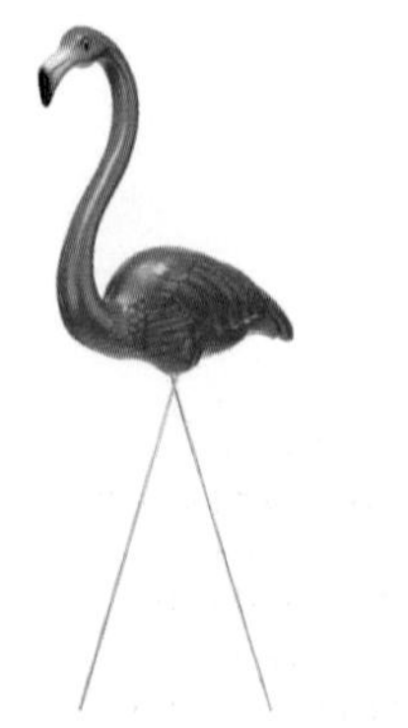

Michael Kraus/Shutterstock

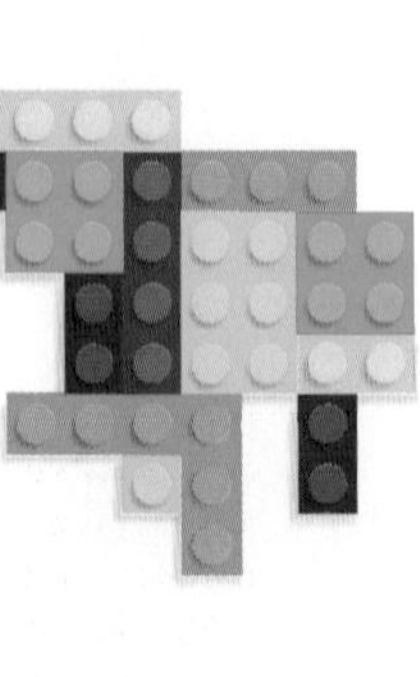

fjmjcreativo/Shutterstock

kostakirov/Shutterstock

Nadezda/Shutterstock

lucadp/Shutterstock

Figure 13.6 Items Made from Polymers

The items shown in this figure are made mostly (or entirely) from polymers.

first. If you look closely at the structure, you can see it contains a motif that repeats over and over again. This motif is highlighted here and there in yellow. If you trace the structure, you can count the number of times this motif is repeated: 36 times. Now, imagine that you are a polymer chemist who works with these molecular structures every day. It might be much easier just to write the repeating motif once, and then indicate how many times it is repeated. This is exactly what chemists do.

The brackets around the motif in **Figure 13.7a** indicate that the repeating motif connects to itself, end to end, at the points where the brackets are located. The subscript 36 indicates that the motif is repeated 36 times. Quite often we don't know how many times the motif is repeated. In that case, the number is not indicated. Instead, we write the letter *n* to show that the motif is repeated many times.

The polymer in Figure 13.7a is polyisoprene, and its monomer—its repeating motif—is isoprene. Polyisoprene is the primary component of natural rubber, and it is also used to make chewing gum. Natural rubber is made from a white liquid called *latex*, which is tapped from a tropical tree called *Hevea braziliensis* (**Figure 13.7b**).

flashback

Recall from Chapter 5 that line structures are an easy way to draw complex organic molecules. In a line structure, we don't use letters to depict atoms. Instead, the end of each line and the angle in each line represent carbon atoms. Hydrogen atoms are also not shown but implied.

recurring theme in chemistry 6

The structure of carbon-containing substances is predictable because each carbon atom almost always makes four bonds to other atoms.

Figure 13.7 Natural Rubber

(a) A drawing of the complete line structure of the natural rubber polymer. It is drawn in a circular way here, so it fits onto the page, but the structure is flexible and so can be drawn in different ways. One monomer unit is highlighted in yellow. Its structure is shown again in abbreviated form. It is repeated 36 times. (b) Latex drips from a *Hevea braziliensis* tree.

The Mayans knew about this tree and knew how to extract rubber from it. They used the rubber as weatherproofing for shoes and clothing. They also knew that rubber could make things bouncy. They used rubber to coat round things—even human heads—to make balls that they used to play a game similar to soccer.

When the Mayans began harvesting latex, the Europeans had no such bouncy material. The British used leather-covered pig bladders or animal skulls as balls, but neither of them was very bouncy. When Spanish conquistadors arrived in America in the 1500s, they took some bouncy balls with them back to Europe. The Europeans also took latex-producing *H. braziliensis* seedlings and planted them in tropical climates such as present-day Sri Lanka and Malaysia.

QUESTION 13.3 Polyvinylchloride (PVC) is a polymer used to make sewage pipes. Its structure is shown here. What is the repeating motif in this polymer?

$$-\!\!\left[\begin{array}{cccccccccccccccccccc} H & H & H & H & H & H & H & H & H & H & H & H & H & H & H & H & H & H & H & H \\ | & | & | & | & | & | & | & | & | & | & | & | & | & | & | & | & | & | & | & | \\ C- & C- & C- & C- & C- & C- & C- & C- & C- & C- & C- & C- & C- & C- & C- & C- & C- & C- & C- & C \\ | & | & | & | & | & | & | & | & | & | & | & | & | & | & | & | & | & | & | & | \\ H & Cl & H & Cl & H & Cl & H & Cl & H & Cl & H & Cl & H & Cl & H & Cl & H & Cl & H & Cl \end{array}\right]\!\!-$$

ANSWER 13.3

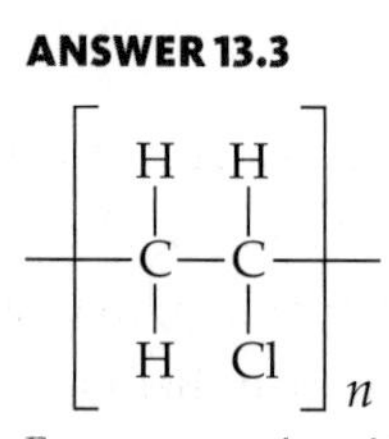

For more practice with problems like this one, check out end-of-chapter Questions 12 and 13.

FOLLOW-UP QUESTION 13.3 The string of letters shown here has a repeating motif like polymers do. What is the repeating motif? How many times does it repeat?
AABCBBAABCBBAABCBBAABCBBAABCBBAABCBBAABCBBAABCBB

13.3 The Physical Properties of Polymers

The structure of a polymer often dictates its physical properties.

recurring theme in chemistry 7
Individual bonds between atoms dictate the properties and characteristics of the substances that contain them.

flashback
In Section 4.6, we learned that when a bond forms between atoms of two elements, the polarity of the bond is dictated by how close the elements are to one another on the periodic table. Elements that are far apart make the most polar bonds. Elements that are close together make bonds that are the least polar. In general, covalent bonds are nonpolar or polar while ionic bonds are very, very polar.

We can often trace the properties of a substance back to its molecular structure. For example, we know that natural rubber is a huge molecule made up of only carbon and hydrogen. Recall from Section 4.6 that covalent bonds can range from very polar to nonpolar. Hydrocarbons are organic molecules that contain only carbon–carbon bonds and carbon–hydrogen bonds, and both of these types of bonds are quite nonpolar. Thus, like most hydrocarbon polymers, polyisoprene is a highly nonpolar polymer.

Compared to polyisoprene molecules, water molecules are extremely polar. And, because their polarities differ, polyisoprene and water repel one another. In general, polar substances mix well together and nonpolar substances mix well together, but polar and nonpolar substances, such as polyisoprene and water or oil and vinegar, do not. We can summarize this fundamental idea with the adage, "Like mixes with like." Because of its properties, polyisoprene makes a great waterproofing substance. This type of polymer is described as **hydrophobic,** meaning "water-fearing."

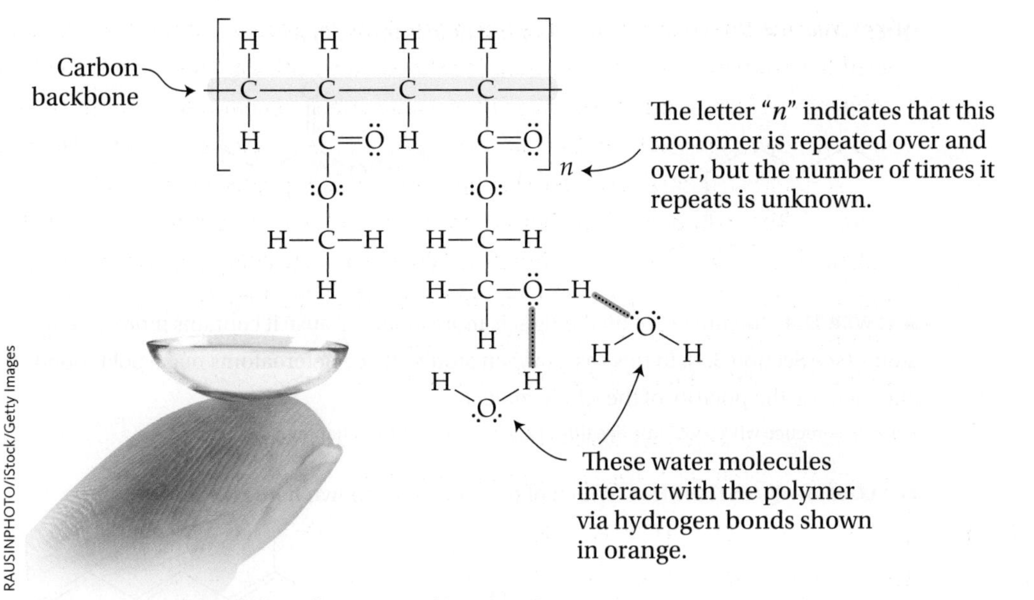

Carbon backbone
The letter "n" indicates that this monomer is repeated over and over, but the number of times it repeats is unknown.
These water molecules interact with the polymer via hydrogen bonds shown in orange.
RAUSINPHOTO/iStock/Getty Images

Figure 13.8 A Hydrophilic Polymer Found in Some Contact Lenses

We have just seen that a rubbery hydrocarbon polymer makes good waterproofing material. But not all polymers are made only of carbon atoms and hydrogen atoms. What if a manufacturer needs a polymer with different properties, like, say, a soft contact lens? What properties would be desirable in that type of polymer? A soft contact lens should be **hydrophilic**—a word that means "water loving"—because it must be accepted comfortably onto the surface of the eyeball, which is coated with an aqueous solution. Thus, rather than using a hydrocarbon, a contact lens manufacturer needs a polymer that is water-friendly yet does not dissolve in water.

A contact lens should also be flexible, because it needs to conform to a person's eyeball. In this sense, the contact lens should be like natural rubber, which is a flexible molecule with a bendable backbone. It is bendable because many of the bonds in its backbone are carbon–carbon single bonds. These bonds give natural rubber flexibility.

Consider the polymer shown in **Figure 13.8**. This polymer is composed of an unknown number of linkages of the monomer, indicated with the letter *n*. Notice that this polymer is not a hydrocarbon. It contains many oxygen atoms, which give the polymer more polarity. In the figure, water molecules interact with the polymer via hydrogen bonds. Recall from Chapter 8 that hydrogen bonds can be made between water molecules and certain atoms—in this case, an oxygen atom attached to a hydrogen atom. This type of interaction between the polymer and water makes the polymer hydrophilic and guarantees that it will interact easily with water. It will not repel water like rubber does.

The backbone of this polymer is made up entirely of carbon–carbon single bonds. This means that the polymer is flexible and bends easily, whereas a polymer with double bonds would be less flexible. These properties together ensure that this polymer is compatible with the eyeball. Thus, these features can be included in the design of contact lenses, and the polymers can be created in a factory by mixing together the monomers in specific ways.

QUESTION 13.4 Which of molecules shown here is most polar?

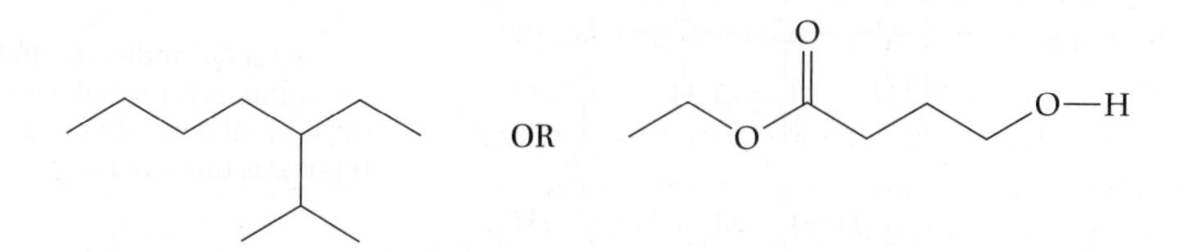

ANSWER 13.4 The molecule on the right is more polar because it contains more heteroatoms (see Section 5.3), in this case, oxygen atoms. These heteroatoms make polar bonds and increase the polarity of the whole molecule.

For more practice with problems like this one, check out end-of-chapter Question 23.

FOLLOW-UP QUESTION 13.4 Which of the molecules shown here is more flexible?

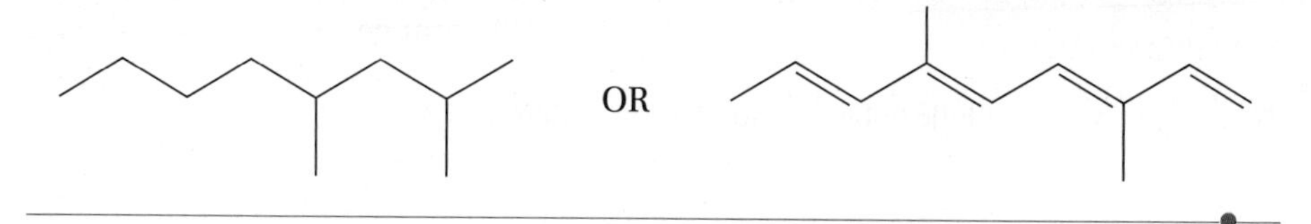

Polymers can be designed for rigidity and toughness.

Polymers can also be designed for strength, resilience, and *in*flexibility. This was the challenge put to chemist Stephanie Kwolek, who worked as a polymer expert at I.E. DuPont de Nemours for 45 years (**Figure 13.9a**). During this time, she experimented with new polymers that had exceptional strength. She was particularly interested in polymers that were rigid enough to pack together as crystals, in much the same way the ions in solid salts pack together into a crystalline lattice (see Chapters 4 and 9). Dr. Kwolek came up with the polymer shown in **Figure 13.9b**. It contains an alternating pattern of benzene rings, which make the polymer less flexible, and a special grouping of atoms that contain one carbon atom, one nitrogen atom, and one oxygen atom. This functional group, known as an **amide,** is highlighted in yellow in the

flashback

Recall from Chapter 5 that a functional group is a grouping of atoms within an organic molecule that gives the molecule characteristic properties.

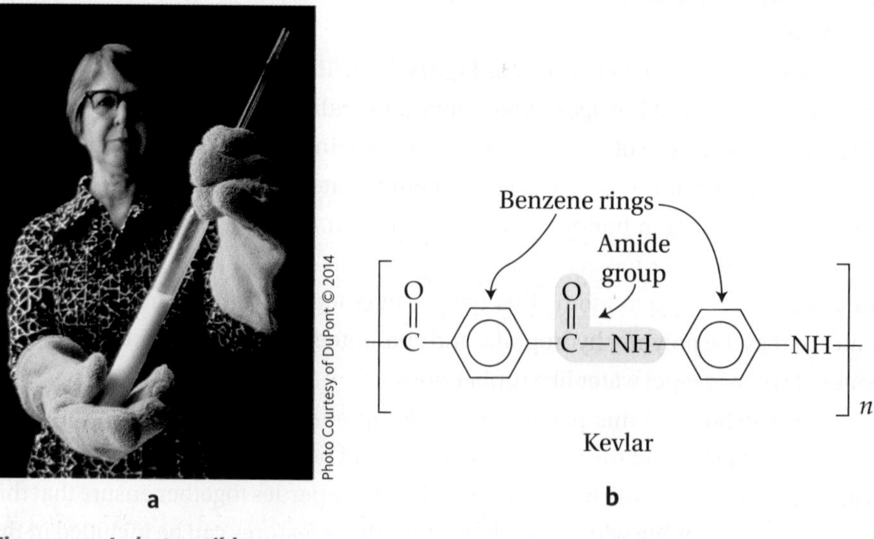

Figure 13.9 Indestructible

(a) Stephanie Kwolek, a polymer chemist, led the development of the Kevlar polymer. (b) Kevlar is a polymer of amides, which are highlighted in yellow.

figure. Amides are important in plastics and in other polymers such as proteins. (We'll learn about proteins and the importance of their amide groups in Chapter 14, "Food.")

For reasons we will not explain here in detail, the bond between the nitrogen atom and the carbon atom in the amide group is relatively inflexible. This bond and the benzene ring together made Kwolek's polymer very rigid. The polymer flexes very little, so each strand is relatively straight. And because the strands are straight and rigid, they can pack together the way uncooked spaghetti can be stacked together and fit into a box. These tightly packed strands make a regular three-dimensional crystalline matrix.

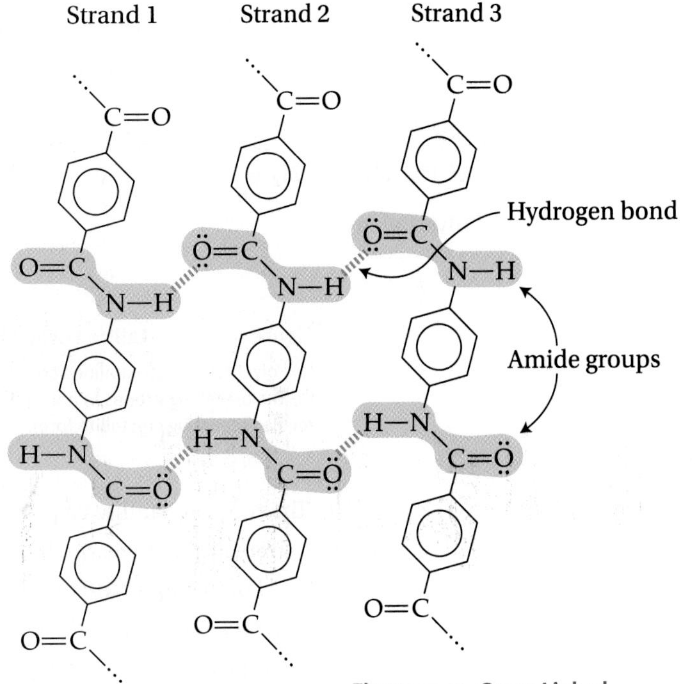

Figure 13.10 Cross-Linked and Rigid
The molecular structure within the Kevlar polymer make it an extraordinarily heavy-duty, durable polymer. Amide groups are highlighted in green and hydrogen bonds are shown in orange.

When they pack together closely, the individual strands of this polymer interact with one another via hydrogen bonds. These hydrogen bonds, which are shown with orange dashes in **Figure 13.10,** act as **cross-links** between strands and make the material even stronger. In this polymer, they occur between amide groups, highlighted in green in the figure, on neighboring strands. This three-dimensional crystalline structure gives the polymer extraordinary strength. In fact, this polymer, which is known as *Kevlar*, is five times stronger than steel. It is used to make cables for suspension bridges, bulletproof vests, protective helmets, and sails, among other things.

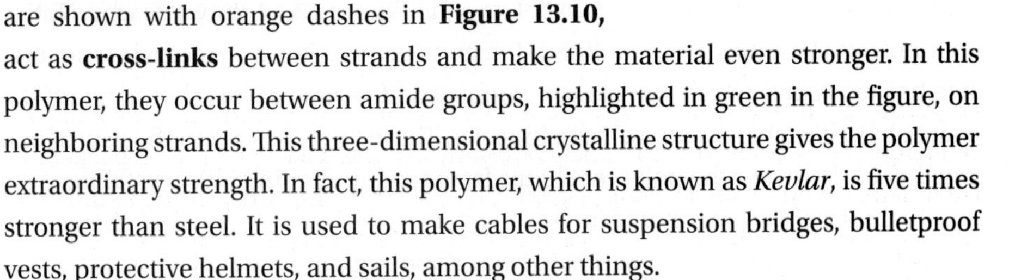

QUESTION 13.5 Which of the following molecules contains an amide functional group?

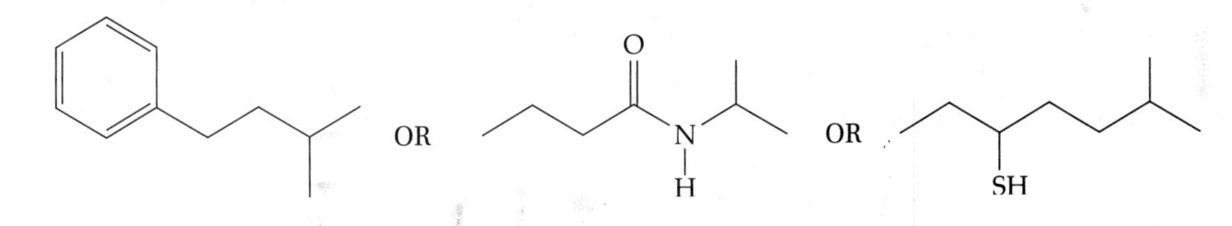

ANSWER 13.5 The structure in the middle contains an amide functional group—a specific grouping of one carbon atom, one oxygen atom, and one nitrogen atom, along with hydrogen atoms as needed.

For more practice with problems like this one, check out end-of-chapter Question 28.

FOLLOW-UP QUESTION 13.5 Which of the following does not contribute to the strength and rigidity of Kevlar? (a) a carbon–carbon chain composed mostly of single bonds (b) rigid benzene rings in the backbone (c) cross-linking between strands (d) inflexible amide groups

Crystallites can make polymers more rigid.

Kevlar owes its extraordinary strength, in part, to the ability of the polymer strands to stack together in a crystalline pattern. This phenomenon occurs in many other polymers. Any highly ordered, crystalline region in a polymer is called a **crystallite.**

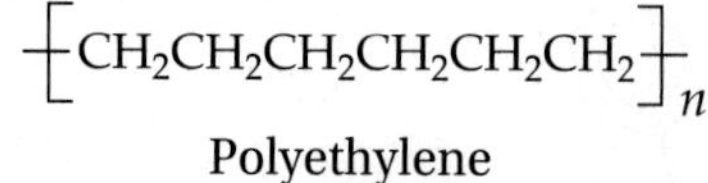

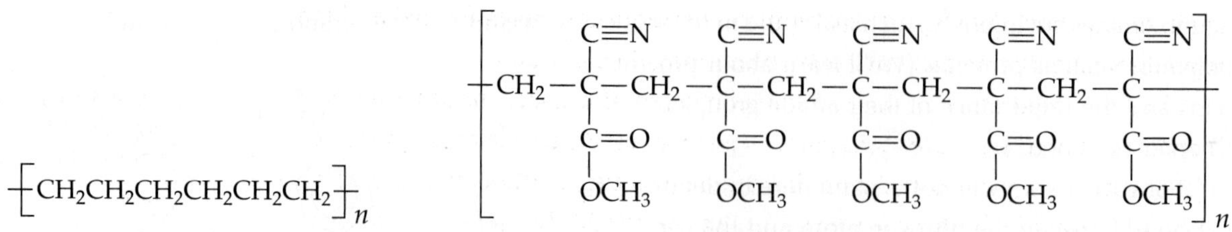

Polyethylene

a

Polymethylcyanoacrylate

b

Figure 13.11 Crystallite Formation

(a) Polyethylene's streamlined structure allows for the formation of crystallite structures in the polymer. (b) The branching groups in polymethylcyanoacrylate make it difficult for its polymers to pack into a regular, repeating crystallite form.

The more crystallites a polymer contains, the harder it is and the more resistant to external influences, such as increases in temperature.

For example, the number of crystallites in a given volume of the relatively rigid polymer called *polyethylene* is much larger than the number in the same volume of the liquid polymer known as *polymethylcyanoacrylate* (the polymer in superglue). A look at the structures of these two polymers in **Figure 13.11** shows us why. Notice that polyethylene has a streamlined, linear structure. But polymethylcyanoacrylate has branches, which are groups composed of several atoms that hang off the main chain like charms on a charm bracelet. These branches get in the way of one another and prevent the polymer chains from readily aligning into crystallites.

In this brief survey of some interesting polymers, we have seen that it's possible to incorporate specific characteristics into the design of new polymers. Scientists can, for example, design a water-resistant polymer by using a purely hydrocarbon structure or a hydrophilic polymer to which they then add groups that can take part in hydrogen bonding. Polymers can also be designed to be flexible or rigid, and cross-links can be included to add strength. And, as we'll see in the next section of the chapter, the structural features of a polymer often dictate whether that polymer can readily be recycled.

13.4 Recyclable and Sustainable Plastics

We can categorize plastics as thermoplastic polymers or thermosetting polymers.

When heated, some polymers gradually change composition as the energy supplied by the added heat breaks crystallites apart and changes them to more fluid regions that lack specific shape and order. The lower the percentage of crystallites in a polymer, the more pliable and more malleable the polymer is. These useful characteristics allow the hot polymer to be molded into a comb, say, or a toothbrush. When cooled, **thermoplastic polymers** re-form crystallites and therefore retain their new shape.

Most of the plastic things we encounter every day—including plastic forks, cell phones, pens, and cases for contact lenses—are made of thermoplastic polymers. If we heat one of these items, it melts and its crystallites will become fluid. That fluid mass can take on a new shape and retain the new shape when cooled. As we'll see later in this section, the ability to manipulate thermoplastic polymers makes them **recyclable,** meaning we can use these polymers again and again.

What if instead of being shaped into combs, toothbrushes, or other molded products, plastic is needed to make parts for jet engines? An important requirement for jet engine parts is that they have to be stable no matter how hot the engines run. Not even the toughest thermoplastic polymers are suitable because, by definition, a thermoplastic polymer becomes fluid at high temperatures. For jet engines, the manufacturer needs a polymer that will maintain its shape when heated. **Thermosetting polymers** are extremely tough polymers that do not change shape when heated. Bakelite is an excellent example of a thermosetting polymer.

flashback
Recall from Section 8.3 that melting is a transition from the solid phase to the liquid phase.

QUESTION 13.6 Which of these products should not be made from a thermoplastic polymer? (a) a cup for cold beverages (b) a highlighter pen (c) a part in a car engine (d) a cup holder inside a car's interior

ANSWER 13.6 Choices (c) and (d) should not be made out of thermoplastic polymers. When a plastic will encounter very high temperatures, like those in a car engine, or even just somewhat high temperatures like those inside a closed car parked in the sun, a thermoplastic polymer is unsuitable.

For more practice with problems like this one, check out end-of-chapter Question 29.

FOLLOW-UP QUESTION 13.6 True or false? Polymers with higher percentages of crystallites are less malleable than polymers with lower percentages of crystallites.

Plastics are recycled according to their resin ID codes.

In contrast to thermosetting polymers, thermoplastic polymers are easily reshaped and reused. The word *thermoplastic* means "heat-moldable." Therefore, all thermoplastic polymers are, by definition, capable of being melted; but this doesn't mean they all have the same melting point. Small differences in structure give each polymer a different melting point. These differences in melting point can also be exploited to create plastics that serve different purposes.

The six thermoplastic polymers that make up most modern plastics are listed in **Table 13.3** (on p. 400). In the first column of the table for each entry is a number known as the **resin ID code**. These codes indicate how easily a thermoplastic polymer melts and therefore how easily recycled each polymer is. The numbers range from 1 through 6, where 1 indicates the most readily recycled. Note that LDPE and HDPE are two polymers that have the same monomer but different resin ID codes. The difference between the two is in how they are manufactured. Two different manufacturing methods produce polymers with two different densities. A miscellaneous category, resin ID code 7, includes other thermoplastic polymers and plastics that are a blend of more than one polymer.

Take a few moments to look at some of the plastic items around you. If you're reading in the park, look at a plastic Frisbee or a plastic soda bottle if you can find one. If you're in the library, consider a plastic waste basket, stapler, or pencil holder. If you're at home, look for plastic containers of shampoo, cleanser, saline, milk, or detergent. Many of these items have a resin ID code stamped on the bottom, surrounded by a triangular arrow indicating that the item is recyclable. Can you find examples of all seven resin ID codes?

Table 13.3 Resin ID Codes for Recyclable Polymers

Resin ID Code	Polymer Name	Monomer Structure	Used for
1	PET (polyethylene terephthalate)	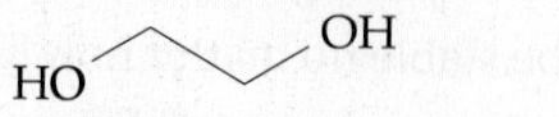 HO–CH₂CH₂–OH Ethylene glycol *and* 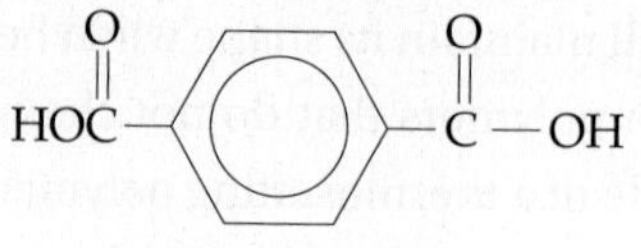 HOC(=O)–C₆H₄–C(=O)–OH Terephthalic acid	Soft drink, salad dressing, and peanut butter containers, fleece clothing, luggage
2	HDPE (high density polyethylene)	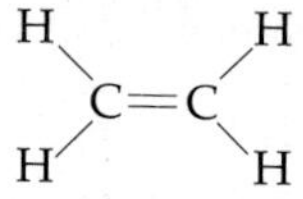 $H_2C{=}CH_2$ Ethylene	Milk, juice, liquid detergent, and water containers, toys
3	PVC (polyvinyl chloride)	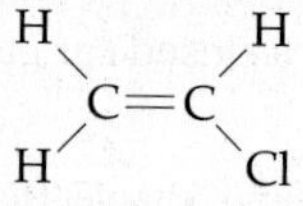 $H_2C{=}CHCl$ Vinyl chloride	Raincoats, packaging film, shower curtains, credit cards, loose-leaf binders, mud flaps, traffic cones, pipes
4	LDPE (low density polyethylene)	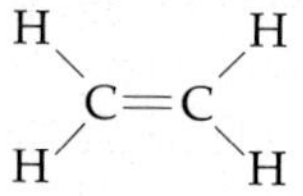 $H_2C{=}CH_2$ Ethylene	Bags for bread, grocery bags, zipper seal bags, shipping envelopes, compost bags
5	Polypropylene	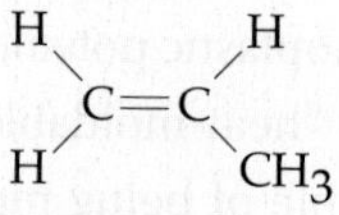 $H_2C{=}CH(CH_3)$ Propylene	Food packaging, injection molding, automobile interior parts, ice scrapers
6	Polystyrene	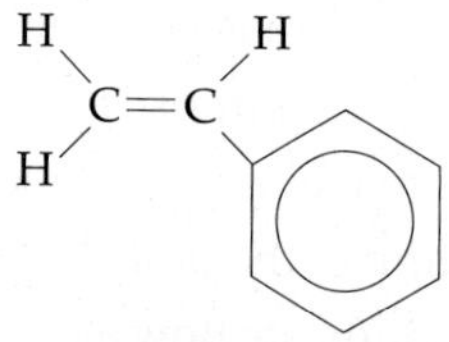 $H_2C{=}CH(C_6H_5)$ Styrene	Styrofoam, fast-food containers, CD cases, cafeteria trays, insulation
7	Mixture	Mixture	Plastic lumber, bottles, water carboys, citrus juice bottles, ketchup bottles

Because most municipal recycling programs allow residents to deposit all their used plastic items in one container, the first step at a recycling facility is usually to separate everything into the seven resin ID code categories (**Figure 13.12**). Once the items have been separated, they are sent to a chipper, where they're cut into small flakes that are then thoroughly washed to remove contaminants and labels. After the flakes have been sterilized, the sterile, label-free plastic in each category is sold to a plastics manufacturer who then melts it down and re-forms it into some new shape.

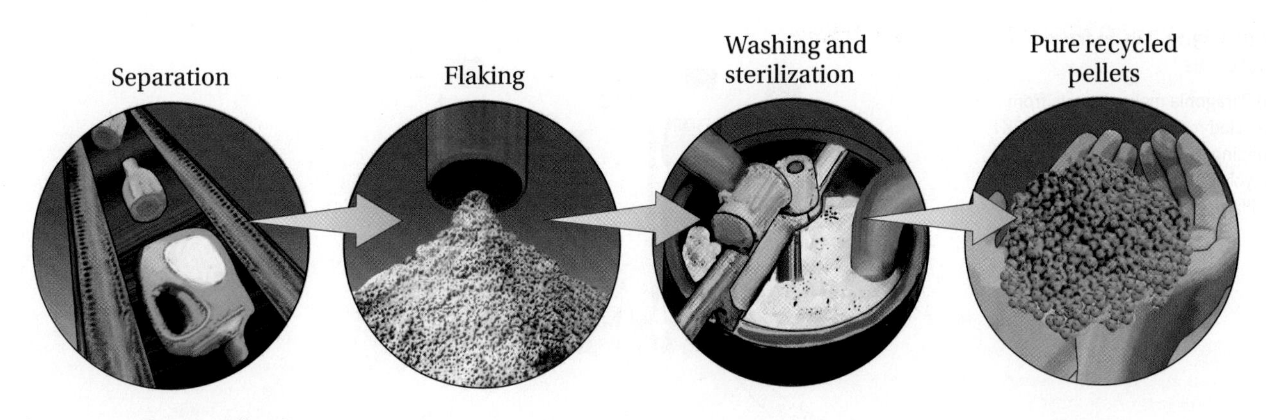

Figure 13.12 The Recycling Process

After separation, plastic items with the same resin ID code are chipped into small flakes, washed, and sterilized to produce recycled plastic pellets.

QUESTION 13.7 Rank the following polymers from Table 13.3 for ease of recycling. List the easiest to recycle first. (a) PET (b) polystyrene (c) LDPE (d) PVC

ANSWER 13.7 PET is the easiest to recycle, followed by PVC, LDPE, and polystyrene, in that order.

For more practice with problems like this one, check out end-of-chapter Question 43.

FOLLOW-UP QUESTION 13.7 A child's juice box contains three kinds of plastic. Under which resin ID code will it be recycled?

Recycled plastics can be made into new products and structures.

During the 1980s, only a small percentage of people in the United States recycled plastics. The process was time-consuming because no central infrastructure was in place to collect and handle plastics for recycling. Today, most cities in the United States have recycling programs, and non-recyclers are in the minority. The success of the recycling effort can be attributed in part to legislation at the state level requiring that rigid plastic containers be manufactured with a certain minimum percentage of recycled plastic. Several U.S. states do not accept recyclable materials in their landfills (Wisconsin, Minnesota, Michigan, and North Carolina). Which city recycles the most per capita? San Francisco.

As recycling rates increase, it is not uncommon for the plastic in a container to be used five times. The plastic in a shampoo bottle may well have been part of a motor oil container or a soda bottle in a previous life. And, because the raw material for creating thermoplastic polymers is petroleum, when our demand for plastic decreases as a result of recycling, so does our dependence on petroleum.

Recycled plastics are being used in innovative ways to make new products from old plastic. We saw that Nike Grind is used to make athletic fields and clothing from recycled sneakers. Patagonia, one of the first U.S. companies to use recycled plastic, makes a fleece jacket. Each jacket is made from about 25 recycled 2-liter plastic soda bottles (**Figure 13.13a**). Through the company's Common Threads Partnership, Patagonia will take back that jacket from a customer who is done with

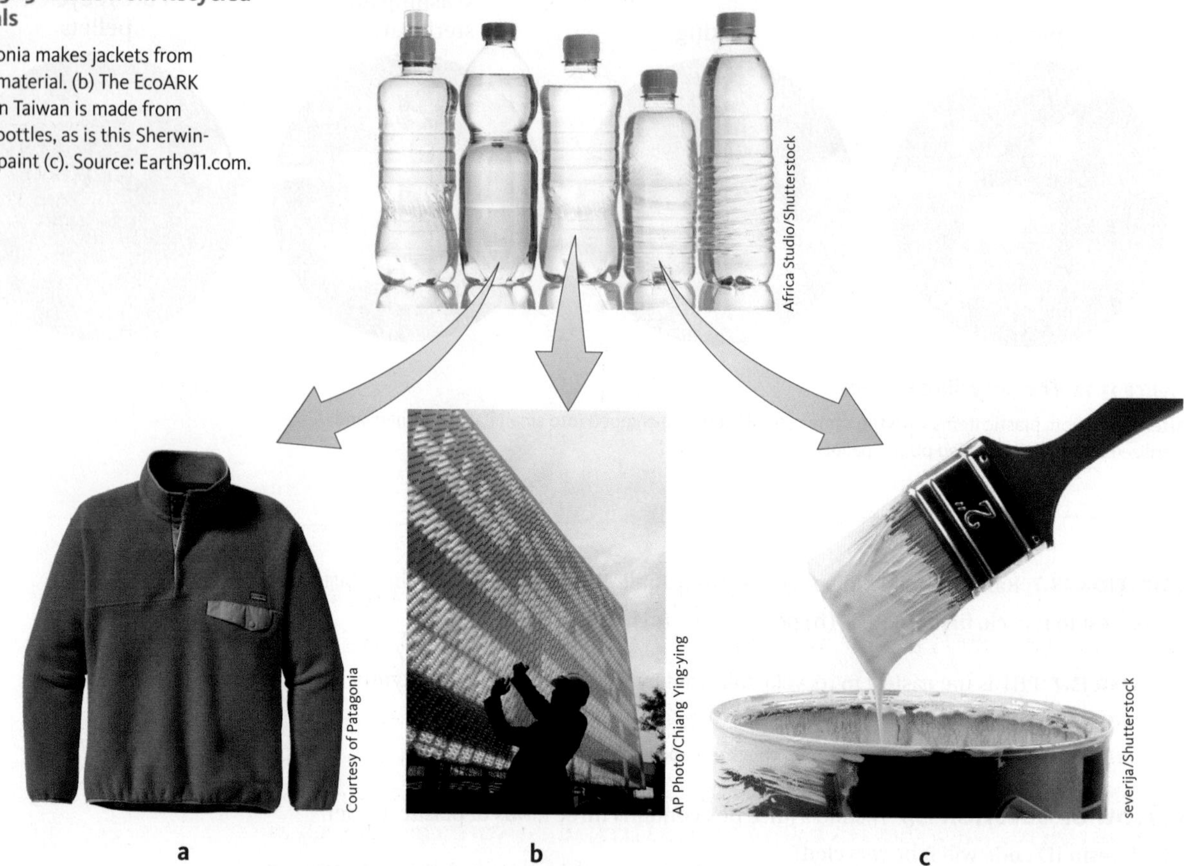

Figure 13.13 Made from Recycled Materials

(a) Patagonia makes jackets from recycled material. (b) The EcoARK building in Taiwan is made from recycled bottles, as is this Sherwin-Williams paint (c). Source: Earth911.com.

it and use it to make new plastic items such as eBook reader covers. The EcoARK building in Taiwan is built from 1.5 million recycled plastic bottles (**Figure 13.13b**). The building's purpose is to encourage recycling in a country where the recycling rate is only 4%. Finally, Sherwin-Williams now has a paint made, in part, from recycled bottles (**Figure 13.13c**). As a bonus, this paint has low levels of volatile organic compounds, or **VOCs**. Because they are volatile, VOCs enter the gas phase easily. They are found in many paints, cleaning supplies, and glues, can cause respiratory issues, and can harm kidney and lung tissues.

Waste plastics that are not recycled end up in the landfill or the incinerator.

Different pieces of plastic have different fates, and their fates depend in part on their resin ID codes. If, like the vast majority of the plastics we use, the plastic has a resin ID code of 1 through 6, it is **recyclable**. But unfortunately, even though most plastics could be recycled, only about 10% of them are recycled in the United States. The rest go to the landfill, are incinerated, or make their way into the ocean. At the landfill, those non-recycled pieces of plastic persist, because plastics with resin ID codes 1 through 6 do not degrade quickly. Rather, these thermoplastic polymers can survive in the environment for decades.

Plastics with resin ID codes 1 through 6 are not sustainable, because they usually come from petroleum. If they aren't recycled, they make their trip from the cradle (petroleum) to the grave—the landfill, or the incinerator. During incineration, plastics undergo a chemical reaction that degrades the polymer to small

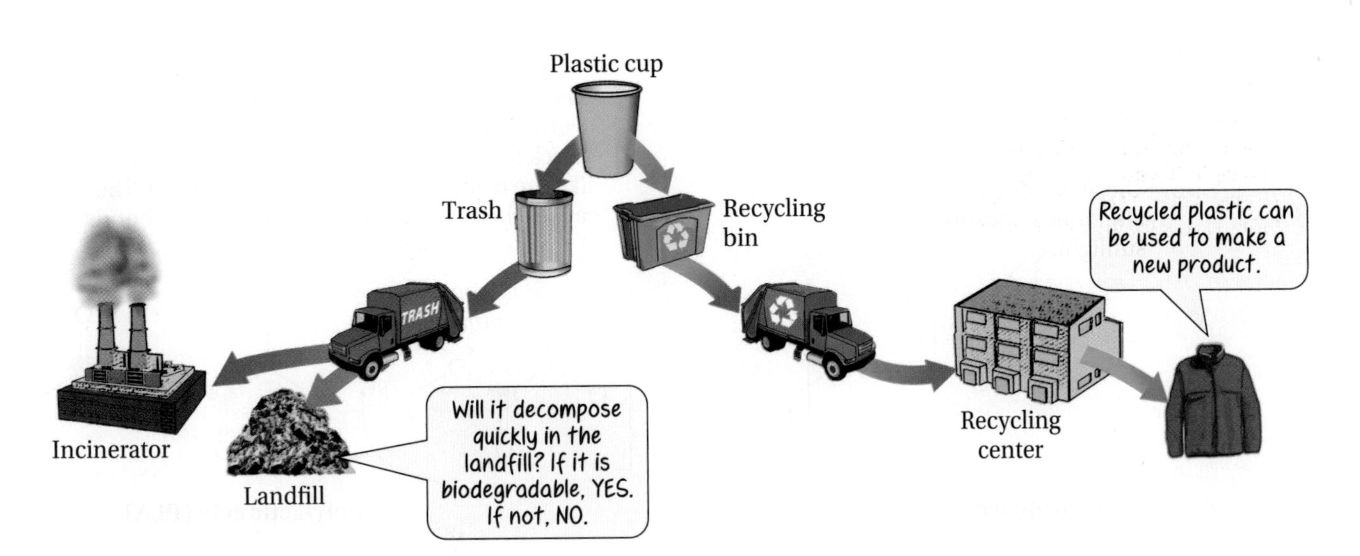

Figure 13.14 The Possible Fates of Plastic Waste

molecules including carbon dioxide and water. Carbon dioxide produced in this process contributes to global warming and climate change. Incineration also can produce extremely toxic compounds such as dioxin, a molecule that damages the human immune and endocrine systems and is linked to various cancers.

For this reason, today the push is to design new polymers that are biodegradable. As you learned earlier, biodegradable materials break down (degrade) into smaller pieces when they're exposed to something in the environment, such as microorganisms and sunlight. A piece of biodegradable plastic that makes its way to the landfill will decompose over a relatively short period of time—perhaps less than 100 days.

What happens to a polymer as it biodegrades? Biodegradation is very different from the reshaping that takes place when recycled plastic pellets are melted and made into something new. When a plastic is melted, it is still composed of long polymer chains, and these chains move fluidly in liquid form at a higher temperature. When cooled, the polymer reverts to the solid phase and takes on any new shape it has been given. Thus, the recycling process does not change the polymer's identity.

But when a polymer is biodegraded over time, its chains are broken, and smaller pieces result. If the process continues long enough, it's possible to convert the polymer back to its constituent monomers or still smaller molecules, as we'll discuss in the next section. **Figure 13.14** summarizes the possible fates of plastics.

QUESTION 13.8 True or false? When plastics are recycled, they are heated, but not broken down into monomers.

ANSWER 13.8 True. As we will see later in this section, in some special cases plastics are converted to monomers when recycled. But the vast majority of polymers, including those with resin ID codes 1 through 6, are simply heated to the point where they can be reshaped into a new product.

For more practice with problems like this one, check out end-of-chapter Question 39.

FOLLOW-UP QUESTION 13.8 According to the text, which of the following is not a place where a significant amount of waste plastic ends up? (a) in the ocean (b) on farmland (c) in the incinerator (d) in the landfill

These bonds can break after long exposure to sunlight.

Polypropylene

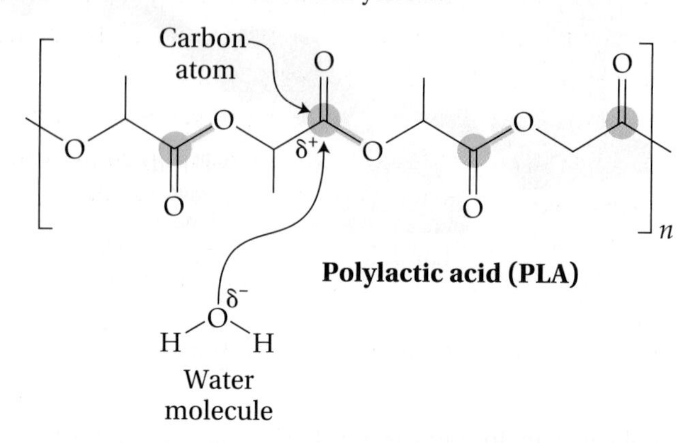

Because they have a partial positive charge when attached to an atom like oxygen, carbon atoms (shown in green) are susceptible to attack by atoms with a partial negative charge, like the oxygen atom in a water molecule. This causes the polymer chain to break at the bonds shown in yellow.
Carbon atom
O
δ+
δ−
H
Water molecule
n
Polylactic acid (PLA)

Figure 13.15 Comparing How Two Polymers Degrade

recurring theme in chemistry 7

Individual bonds between atoms dictate the properties and characteristics of the substances that contain them.

flashback

Recall from Section 5.3 that a heteroatom is an atom in an organic molecule that is neither carbon nor hydrogen.

The structure of a polymer dictates its biodegradability.

Why do some polymers degrade easily while others do not? We'll answer this question by looking at two examples of very different polymers, polypropylene and polylactic acid (PLA), shown in **Figure 13.15**.

Polymers that degrade most easily have vulnerable places in their structures. Recall from Chapter 4 that bonds within organic molecules can be very polar bonds or very nonpolar bonds, or anything in between. The bonds between two carbon atoms or a carbon atom and a hydrogen atom are quite nonpolar. Because the bonds are nonpolar, nowhere in the molecule is there an excess of charge or a deficit of charge.

When an organic molecule contains heteroatoms, which are typically oxygen, nitrogen, or sulfur, these atoms pull electron density toward themselves. Thus, when one of these heteroatoms makes a bond to a carbon atom, that bond is polar. For this reason, the presence of such an atom in an organic molecule can make that molecule vulnerable to attack by another polar molecule, such as water.

Let's first consider a polymer with no heteroatoms. Polypropylene is a type of plastic found in everything from food containers to car parts to packages and labels. Its resin ID code is 5, indicating that it's not as easily recycled as polymers with lower resin ID codes. If we examine Figure 13.15a, we can see that polypropylene is a hydrocarbon and possesses no heteroatoms where degradation might easily be initiated. In general, hydrocarbon polymers are inert to chemical degradation, in contrast to polymer chains that contain heteroatoms. This is true of polypropylene. But over a very long period of time—perhaps millennia—the carbon-carbon bonds indicated in yellow in polypropylene break when exposed to sunlight.

Our second example is polylactic acid (PLA), a plant-based polymer or **bioplastic** that can be made from corn, tapioca, or sugarcane. These plants provide the monomers that join to make the PLA polymer shown in Figure 13.15b. The structure of PLA includes several oxygen atoms. Each oxygen atom pulls electron density from surrounding atoms toward itself. This means that the carbon atom shown in green is electron deficient and possesses a partial positive charge. Because opposite charges are attracted to one another, the partially positive carbon atom is vulnerable to attack by molecules with a negative charge.

In nature, the oxygen atom of the water molecule is often the attacker. Recall from the discussion in Chapter 8 that the oxygen atom in the polar water molecule carries a significant partial negative charge, so it can attack the vulnerable carbon and cause the polymer chain to break. The polymer then degrades. This is often how chemical reactions begin: a positive region of one molecule interacts with the negative region of another molecule.

Because it can be broken down easily, PLA is a highly biodegradable polymer. It's also a sustainable polymer because it comes from fast-growing plants (as opposed to very slow-growing trees, which would be unsustainable). Moreover, PLA is recyclable: it has a resin ID code of 7.

recurring theme in chemistry 8
The Earth has a finite supply of raw materials that humans can use.

Table 13.4 summarizes the plastics terminology introduced in this chapter. A specific plastic can be one or more of these things, or none of them.

wait a minute . . .

Why does a biodegradable plastic like PLA have a resin ID code of 7?

Resin ID codes indicate how easy it is to recycle different polymers. The code assigned to a polymer depends largely on how well it fits into the existing recycling process. Recall that the polymers shown in Table 13.3 (on p. 400) are not degraded into monomers during recycling. Rather, they follow the process shown in Figure 13.14 (on p. 403), which leaves the polymer chains intact. PLA, in contrast, must be degraded to its monomers when it's recycled. The monomers are then remade into virgin polymer chains of PLA. For this reason, PLA falls into the miscellaneous category of plastics that also includes materials composed of a mixture of plastics. These plastics usually are destined for the landfill. So, even though PLA is biodegradable, it's not easily recycled using the recycling infrastructure in place today.

Bioplastics such as PLA biodegrade much more rapidly in landfills than petroleum-based plastics do, but they are difficult to recycle. This is because our recycling facilities have been designed to recycle petroleum-based plastics. More bioplastics recycling centers are being built specifically to handle these new, sustainable polymers. And because bioplastics are made from sustainable resources, in the long run, they are a better bet than plastics that come from petroleum. As bioplastic technology continues to improve, these new plastics will become cheaper and, someday, may eventually replace unsustainable plastics.

Table 13.4 Plastics Terminology

A polymer can be . . .	which means that it . . .	Example of this type of polymer from the text
Degradable (or biodegradable)	can be broken down into monomers either by a chemical reaction or by bacterial action, respectively.	PLA
Recyclable	can be recycled and used to make a different plastic product.	polypropylene
Sustainable	is made from a sustainable resource such as plants.	PLA

QUESTION 13.9 Which of the following molecules will be more susceptible to attack by a water molecule?

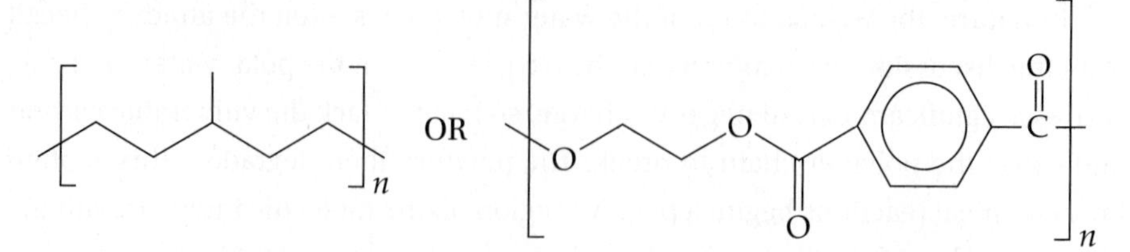

ANSWER 13.9 The molecule on the right has heteroatoms that create vulnerable sites where water can attack.

For more practice with problems like this one, check out end-of-chapter Question 44.

FOLLOW-UP QUESTION 13.9 Why is water often an attacking molecule that begins the degradation process? Choose the best answer: (a) because the oxygen atom in a water molecule has a partial positive charge (b) because water can form hydrogen bonds (c) because water is nonpolar (d) because the oxygen atom in water has a partial negative charge

THEgreenBEAT • news about the environment, chapter 13 edition

TOP STORY The Arrival of Sunshine-Fueled Batteries

The world's most abundant fossil fuels include gasoline, diesel fuel, methane, and other hydrocarbons derived from underground oil reserves. These reserves were created over millions and millions of years from the remains of decayed organisms. The problem with fossil fuels, however, is that they release heat, water and carbon dioxide when burned, and the carbon dioxide cannot be easily captured and recycled. It's a one-way system: Fuel is spent; heat is released; nothing is recovered. This chapter covers alternatives to this system, and modern energy research aims to create energy systems that are "closed."

Closed systems are loops that cycle through the steps of a process. A simple battery is a closed system. Within a battery that produces an electrical current, two reactions take place, one in which electrons are produced and one in which those electrons are used. In between, a current flows. The general problem with batteries is that the chemicals they use are often toxic and/or unsustainable metals, and they may also contain membranes and other components that wear out.

What if there were a battery that's simpler, which doesn't require charging, doesn't wear out, and doesn't even produce an electric current? What if we simply wanted a thermal battery that stores heat in a closed system in a sustainable way? Chemist Dhandapani Venkataraman and his research group at the University of Massachusetts at Amherst are working on such a battery.

The Venkataraman group designed a molecule that changes shape when exposed to sunlight through a process called *photo-excitation* (see **Figure 13.16**). Before it's excited, the azobenzene molecule is stretched out (the "trans" version). After excitation, the azobenzene molecule folds over (the "cis" version). The newly-bent molecule has stored solar energy by undergoing a shape change that doesn't happen without sunlight. Thus, this system acts like a heat sink, holding onto solar energy until the time comes to release it.

In order to make this process useful for trapping and storing solar energy, the Venkataraman group attached these molecules along a long polymer chain. The azobenzene groups dangle off of the polymerized chain like bulbs on a strand of holiday lights. When they are attached to a chain like this, the groups align and stack with each other. When the parts of the chain pack tightly together, the whole chain with its appendages clicks together into the new shape in the presence of light. Without light, the whole system can click off again, releasing its stored energy in the form of heat.

This process is shown in **Figure 13.17**. On the left, the azobenzene groups

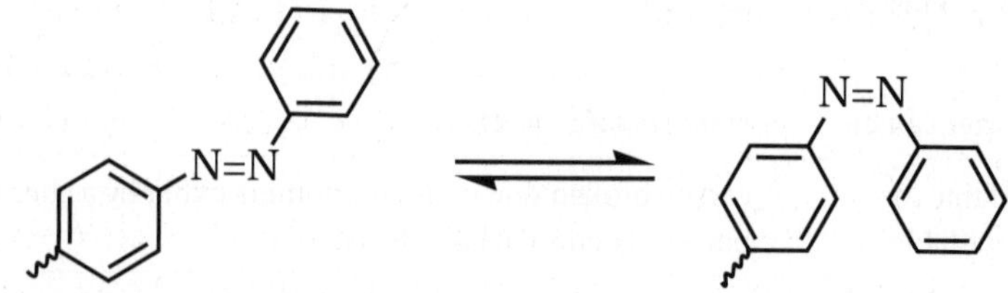
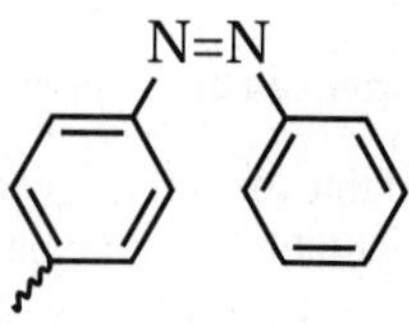

Figure 13.16 The Process of Photo-Excitation

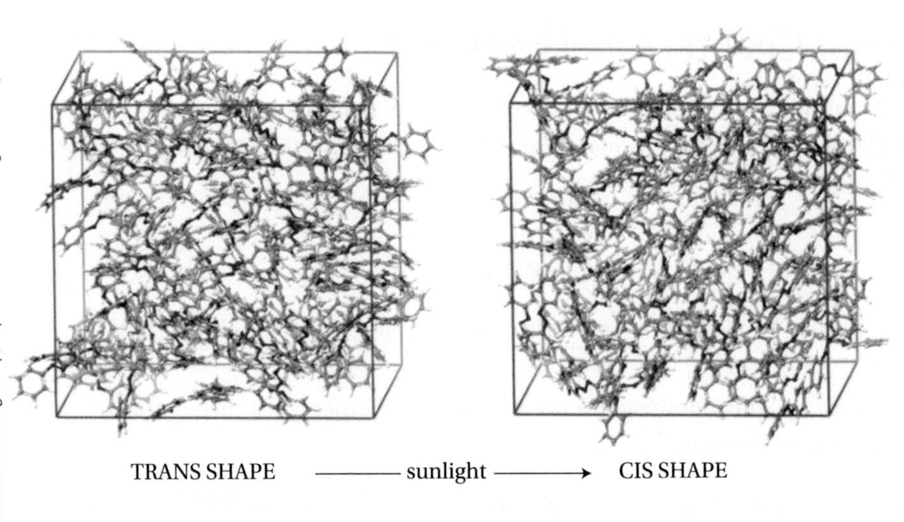

Figure 13.17 Trans and Cis Shape

attached to the polymeric chain are all in the trans state. In the presence of sunlight, the entire polymer has shifted to the shape on the right, in which all of the azobenzene groups are in the cis state.

Chapter 8 discusses passive solar homes in which water is used as a heat sink, a place to store heat energy. The water absorbs heat from the sun during the day and releases it into the house when the sun sets. Venkataraman's photothermal "battery" does the same thing, but better. His polymer can hold much more heat energy than the comparable amount of water. In fact, it can actually hold more energy than a typical lithium ion battery. And because the shift in shape of the long polymer chain is subtle—only the azobenzene appendages change shape—this back-and-forth can happen over and over again. This is the ultimate closed system: Light goes in; heat goes out. There are no substances that enter or leave, no waste products, and no need to recharge anything. These new materials are so good at trapping and holding heat from the sun that they can even be used as solar cooktops. Put your polymer outside in the sun, then bring it in and use it to warm up your dinner.

Chapter 13 Themes and Questions

recurring themes in chemistry chapter 13

recurring theme 8 The Earth has a finite supply of raw materials that humans can use. (pp. 389, 405)

recurring theme 6 The structure of carbon-containing substances is predictable because each carbon atom almost always makes four bonds to other atoms. (p. 393)

recurring theme 7 Individual bonds between atoms dictate the properties and characteristics of the substances that contain them. (pp. 394, 404)

A list of the eight recurring themes in this book can be found on p. xv.

ANSWERS TO FOLLOW-UP QUESTIONS

Follow-Up Question 13.1: False. An LCA can emphasize factors such as environmental impact. *Follow-Up Question 13.2:* Petroleum is unsustainable. It is used to make Bakelite, which therefore is also unsustainable. Sunlight is a sustainable resource because it is not depleted by being used. *Follow-Up Question 13.3:* The pattern AABCBB repeats eight times. *Follow-Up Question 13.4:* The molecule on the left is more flexible because it has more carbon–carbon single bonds. *Follow-Up Question 13.5:* (a) Carbon–carbon single bonds are flexible and not rigid. *Follow-Up Question 13.6:* True. The lower the percentage of crystallites in a polymer, the more pliable and more malleable the polymer is. *Follow-Up Question 13.7:* Any item that contains a mixture of plastics is classified as resin ID code 7. *Follow-Up Question 13.8:* (b) There is no issue with waste plastics polluting farmland. *Follow-Up Question 13.9:* (d) because the oxygen atom water has a partial negative charge.

End-of-Chapter Questions

Questions are grouped according to topic and are color-coded according to difficulty level: **The Basic Questions***, **More Challenging Questions****, and **The Toughest Questions*****.

Sustainability and Life-Cycle Analysis

***1.** Cite one example each of a cradle-to-grave process and a cradle-to-cradle process.

***2.** For each of the following resources, indicate whether it is a sustainable or unsustainable resource.

(a) trees (b) corn (c) sunlight

***3.** In your own words, explain the difference between a cradle-to-grave process and a cradle-to-cradle process.

***4.** Using the figures cited in the text, estimate how many 2-liter plastic bottles Patagonia needs to make 10 fleece jackets?

****5.** What type of polymer is Bakelite? Why don't manufacturers still make items out of Bakelite?

****6.** In your own words, describe the North Pacific Gyre. Why is trash concentrated in this area? Where is the gyre located?

****7.** Describe the issues at stake when a consumer is deciding whether to use plastic or paper cups. Which type of cup has a larger carbon footprint? Which type of cup contributes more to water pollution?

****8.** How is it possible to use all the parts of an athletic shoe to make new products? Give an example of a product mentioned in the text that is made from recycled shoes.

****9.** What is BPA? Why does it contaminate recycled paper? How and why does it affect the human body?

*****10.** According to the text, what percentage of plastic made around the world ends up in the ocean each year?

*****11.** Producing a particular kind of paper cup requires several transportation steps that use fossil fuels, but the cup is made in a factory that is fueled by a nuclear power plant. A different kind of plastic cup is made in a factory that gets its energy from a coal-fired plant, but the product is transported with electric vehicles that also are powered by that plant.

(a) Which cup is likely to produce more pollution in the form of nitrogen oxides?

(b) Which cup is likely to produce more pollution in the form of sulfur oxides?

Polymeric Structures

***12.** The string of letters shown here has a repeating motif just like polymers do. What is the repeating motif? How many times does it repeat?

XYZZYXXYXYZZYXXYXYZZYXXYXYZZYXXYXYZZYXXYXYZZYXXY

***13.** The polymer shown here is known as PET. Circle the repeating motif in this polymer.

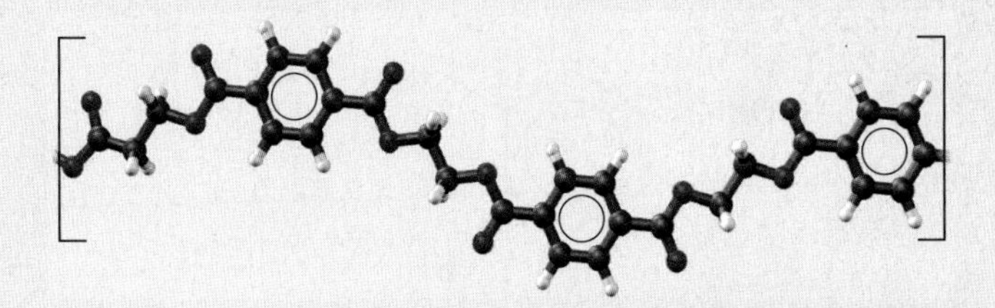

***14.** The subscript *n* in the structural formula of a polymer indicates the number of monomers in the polymer chain for artificial polymers. True or false? In nature, the value of *n* is always a fixed number for a given polymer.

***15.** What is the resin ID code and three-letter abbreviation for the polymer polyethylene terephthalate?

***16.** Write down five words not mentioned in the text that begin with the prefix *poly–*. For each, explain what *poly–* means in that word.

***17.** Can the same polymer exist in various lengths? Is there any requirement for the length of a polymer?

****18.** In your own words, explain what the word *plastic* means, both as a noun and as an adjective. Why is this word suitable for many polymers? Are there polymers that cannot be described with the adjective *plastic*? If so, name one.

****19.** Draw a polymer chain that has five repetitions of the structure shown here. Brackets should be kept on the ends of the structure.

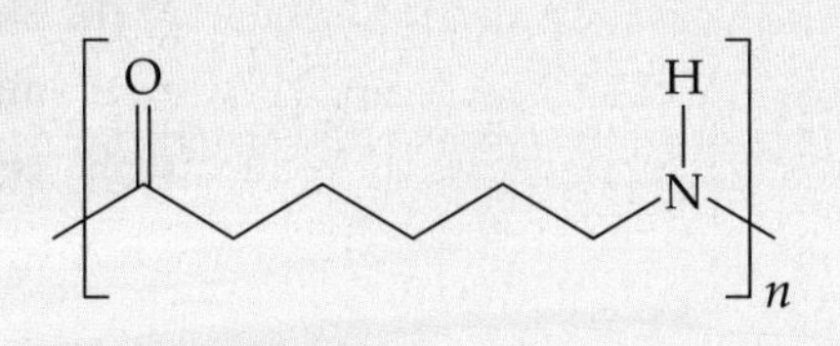

****20.** Draw a polymer chain that has four repetitions of the structure shown here. Brackets should be placed on the ends of the structure.

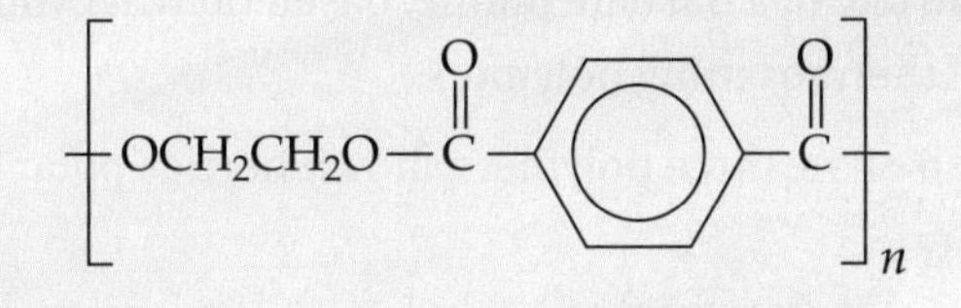

*****21.** Look at the structure of these two molecules and then answer the following questions:

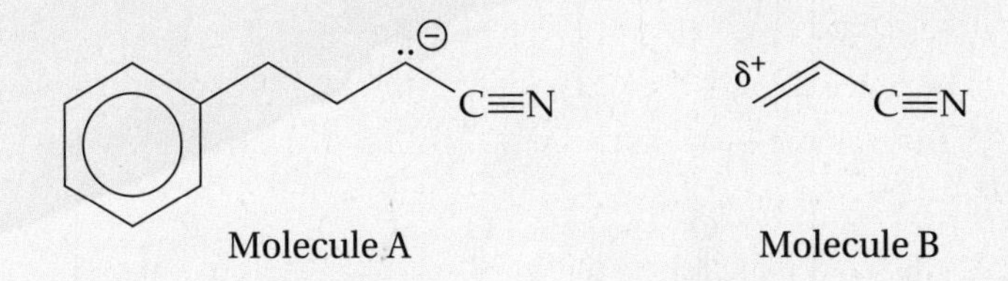

(a) Which area of molecule A is likely to interact with which area of molecule B?

(b) Are two molecules of A likely to react with each other?

(c) Are two molecules of B likely to react with each other?

*****22.** If an 897-g Bakelite telephone were composed of one polymeric molecule, what would be the molar mass of that molecule in grams per mole? (Hint: Recall that Avogadro's number is 6.02×10^{23} molecules in 1 mole.)

The Properties of Polymers

***23.** Which of the following molecules is most polar?

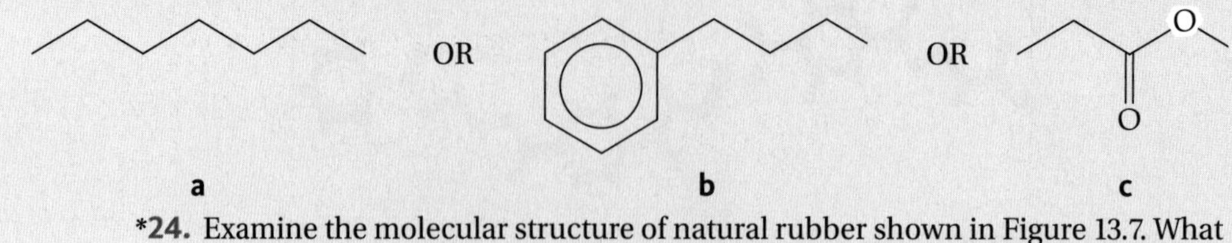

***24.** Examine the molecular structure of natural rubber shown in Figure 13.7. What features of its structure indicate that it is a good waterproofing material?

***25.** In your own words, explain what a sustainable polymer is. Give one example from the text of a polymer that comes from a sustainable resource.

***26.** Name one way that states in the United States are increasing the amount of recycling in their communities.

****27.** Thermoplastic polymers and thermosetting polymers are structurally very different from each other.

(a) Which type of polymer is more easily recycled, and why?

(b) Name one example of a thermosetting polymer.

(c) Does the polymer you listed to answer part (b) have a resin ID code? If so, what is it?

****28.** Sketch an amide functional group. What property does it impart to Kevlar, an amide-containing polymer?

****29.** Which of these items is probably not made from a thermosetting polymer?

(a) a plastic rotor in a jet engine
(b) a plastic hairbrush
(c) a Frisbee
(d) a cell phone

****30.** Describe the fate of a Bakelite phone, based on what you know about the structure of thermosetting polymers.

****31.** Classify each of the three polymers shown here as polar or nonpolar.

a

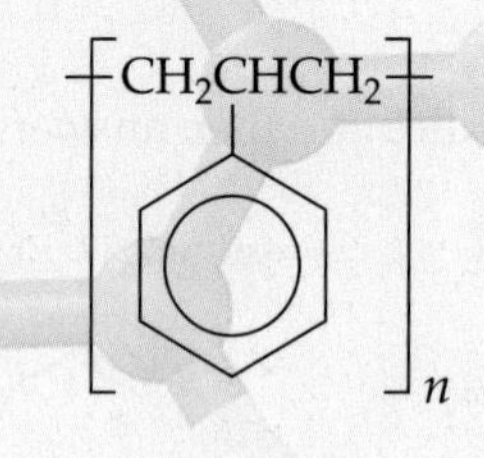

b

$$\left[-O-\overset{O}{\overset{\|}{C}}-(CH_2)_2-\overset{O}{\overset{\|}{C}}-NH-(CH_2)_2-NH- \right]_n$$

c

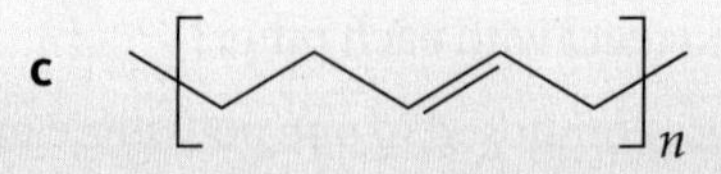

*****32.** Liquid latex straight from the tree is allowed to sit undisturbed for a period of time, until it reaches a creamy consistency. What do you think is happening during that time to change the consistency of the latex?

***33. (a) Draw the line structure of Kevlar, showing the smallest unit of the polymer repeated three times.

(b) What is the molar mass of the smallest unit that makes up Kevlar?

(c) Cite two reasons that Kevlar is so much stronger than most other polymers.

Materials Recycling

*34. According to the *Green Beat* in this chapter, which of the following is found in the greatest quantities (by mass) in a typical landfill?

(a) paper (b) plastic (c) metal (d) glass

*35. What myths about the contents of landfills were dispelled by the Garbage Project? Did this project paint an optimistic or pessimistic view of our efforts to manage our landfills?

*36. Why are single-use water bottles referred to as a menace in the *Green Beat* in this chapter? Why do people use so many single-use bottles rather than filtered tap water?

*37. Which one of these attributes might make a polymer biodegradable?

(a) a hydrocarbon chain

(b) a benzene ring

(c) one or more heteroatoms

(d) a chain with more than 500 monomers in it

*38. Of all the polymers shown in Table 13.3, which one contains atoms that might be susceptible to attack by water molecules?

**39. What is the difference between the process of reshaping a recycled polymer and the process of biodegradation that a polymer experiences over time in a landfill? What are the end products of each process?

**40. What is lightweighting? How much did the average mass of a water bottle that contains 16.9 ounces of water decrease from 2000 to 2008? Give one other example of a plastic item that has been lightweighted.

**41. Outline the steps that a typical bin of recycling goes through after it leaves a house and arrives at the recycling plant.

**42. Describe the differences between a polymer that can be a recycled and a polymer that is biodegradable.

**43. Rank the following polymers from Table 13.3 in order of ease of recycling, with the easiest to recycle first. (a) HDPE (b) PVC (c) polypropylene

**44. Which of these two molecules is more susceptible to attack by a water molecule?

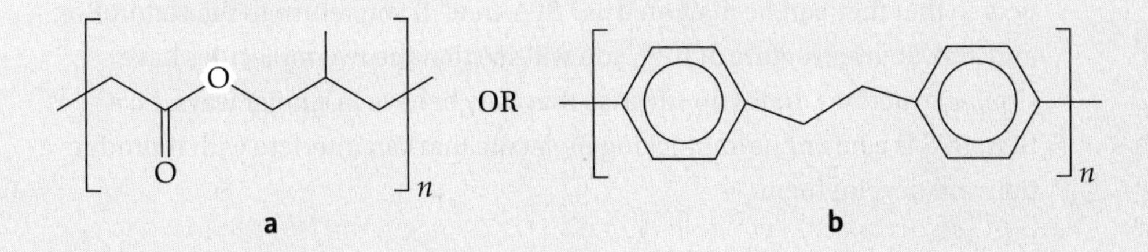

**45. It is possible to buy very inexpensive recycled plastic that contains a mixture of thermoplastic polymers. Why do you suppose this mixed type costs less than the same amount of a recycled plastic containing only one type of polymer?

**46. Recycling has the potential to dramatically reduce the amount of crude oil our country uses each year.

(a) How is recycling linked to the use of crude oil?

(b) What happens to recyclable plastics that are thrown away and not recycled?

**47. A milk jug at a recycling plant sits in a pile with other milk jugs. All have their caps on. Which step in the recycling process have these milk jugs already been through? Which steps have they not yet undergone?

**48. A certain type of milk jug has a resin ID code of 2.

(a) What type of polymer was used to make the jug?

(b) Based on the structure of this polymer, would you expect milk jugs to be hydrophilic or hydrophobic? Why?

**49. If recycling is becoming increasingly successful, why are efforts being made to make biodegradable plastics?

***50. (a) Why is PLA recycled in a different way from other plastics?

(b) What process does it undergo during recycling?

(c) Where are the monomers for PLA obtained when virgin PLA is made?

(d) What feature of PLA's structure makes it biodegradable?

Integrative Questions

*51. Collect three items made of plastic that have resin ID codes stamped on them.

(a) Turn to Table 13.3, and identify the polymer that each item contains. Are any of your items hydrophilic like the one shown in Figure 13.8?

(b) Would the plastic in your items be suitable for making contact lenses? Why or why not?

(c) What structural feature of the contact lens polymer makes it ideal for use in the eye?

*52. Think of three nonscientific words that contain the prefix "poly-." Consider what this prefix means in the context of each of these words. Why do so many plastics have names that begin with the prefix "poly-"?

***53. The bisphenol S (BPS) molecule, shown below, is sometimes used in products so that they can be marketed as "BPA-free." If you return to the NatureBox and look at the structure of BPA, you will see that the two molecules have similar structures, so it's no surprise that they behave in similar ways. Like BPA, BPS is a hormone-mimicking molecule that can interfere with reproduction and development.

(a) Sketch a full structure of BPS that shows every bond and atom. Write a chemical formula for it and determine its molar mass.

(b) Comparing the two structures, is it reasonable to surmise that BPS may bind to hormone molecules in a similar way to BPA? Provide support for your answer.

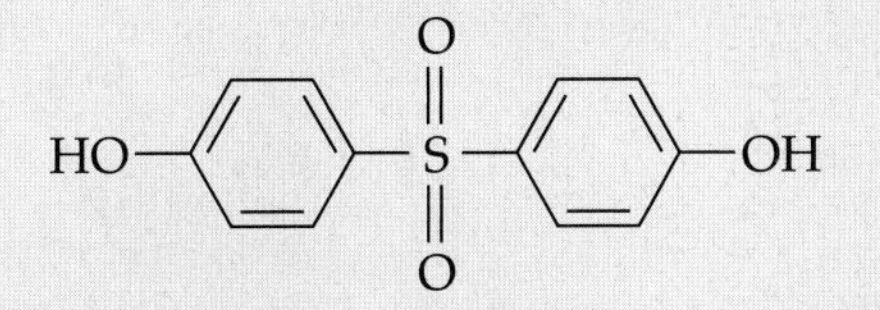

***54. Modern, air-filled automobile tires are made from different types of polymers, including styrene-butadiene, shown below.

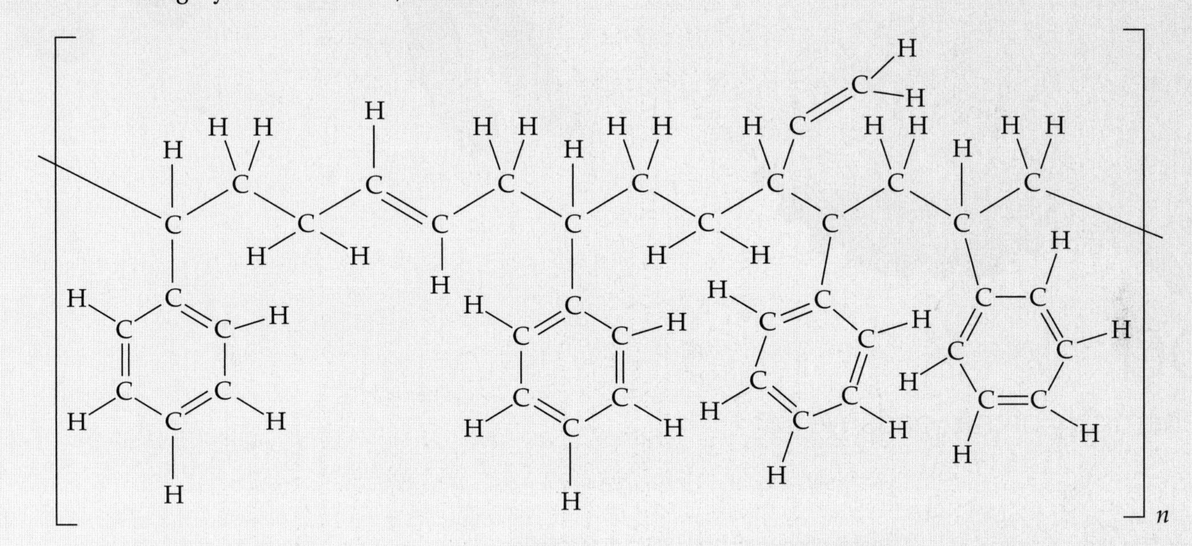

(a) Looking at this structure, would you expect this polymer to be waterproof? What makes you think so?

(b) Draw the line structure for the monomer shown. What is the chemical formula for this monomer?

(c) Styrene-butadiene is a *copolymer*, meaning that it is made from the polymerization of two different starting materials, in this case styrene and butadiene, shown right. How many of each of these two molecules would make up one monomer unit of styrene-butadiene, shown above?

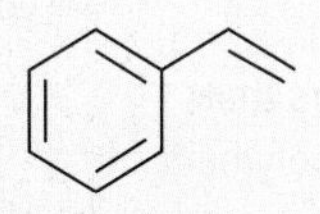

styrene

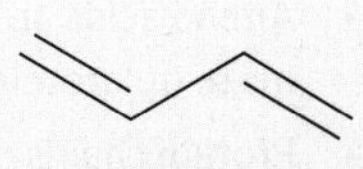

butadiene

55. INTERNET RESEARCH QUESTION

Choose from among these everyday items that you might find on a typical desk: Post-It notes, a paper clip, a pencil, a stapler. In your Internet search engine, type in "LCA for ___________" and insert the name of the item you just chose. Investigate the life cycle of this item and try to answer as many of the following questions as you can: What raw materials are used to make it, and where do they come from? What steps are used in its manufacture, and what types of energy are required? Can this item be recycled or repurposed? Where is this item discarded? Is there a way that this item could be produced or used more sustainably?

Food

The Biochemistry of the Foods We Eat

14.1 Protein: The Most Critical Nutrient 417

- The human body needs a mixture of fuels for optimal health.
- Both micronutrients and macronutrients are part of a balanced diet.
- Amino acids are the monomers that make up proteins, which are polymers.
- Protein chains can fold into globular proteins.
- Structural proteins play mechanical and structural roles.

KEY WORDS nutrient, macronutrient, biomolecule, micronutrient, malnutrition, protein, amino acid, peptide bond, disulfide bond, globular protein, enzyme

14.2 How Proteins Are Made 425

- DNA is a polymer of nucleotides.
- Genes within DNA are transcribed into RNA, which is translated into protein.

KEY WORDS deoxyribonucleic acid (DNA), nucleic acid, nitrogenous base, nucleotide, complementary base pair, double helix, gene, ribonucleic acid (RNA), transcription, genetic code, triplet, translation

14.3 Genetic Engineering and GMOs 430

- Genetic engineering alters the DNA of food crops.
- Genetic engineering is used to grow herbicide- or insect-resistant food crops.
- The use of certain GMOs is controversial.

KEY WORDS genetic engineering, genetically modified organism (GMO), transgenic organism

naturebox The Demise of Natural Orange Juice 432

14.4 Carbohydrates 434

- There is an obesity epidemic among children and adults in the United States.
- Carbohydrates can be simple or complex, and grains can be refined or whole.
- Digestion breaks down complex carbohydrates into monosaccharides and disaccharides.

KEY WORDS body mass index (BMI), carbohydrate, saccharide, sugar, complex carbohydrate, whole grain, dietary fiber

14.5 Fats 439

- Human beings store energy in the form of fat.
- Lipids are hydrophobic molecules that are not polymeric.
- Many properties of fatty acids depend on the number of double bonds the fatty acids contain.
- Fatty acids are stored in the form of triglycerides.

KEY WORDS lipid, hydrophobic, fatty acid, triglyceride, trans fat, hydrogenation, biodiesel

THEgreenBEAT News about the Environment: Indoor Plantations: The Answer to the World's Food Shortage? 444

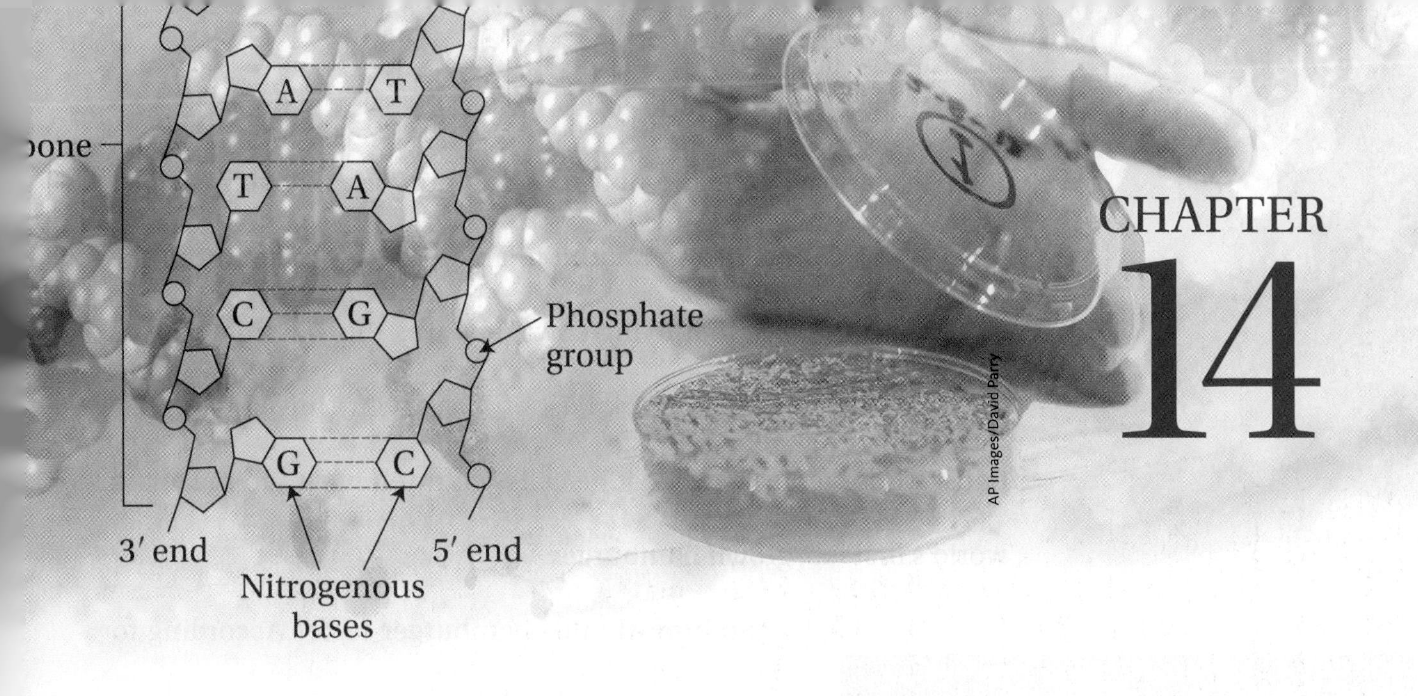

CHAPTER 14

*On August 5, 2013, a television stage was prepared in London in antici-*pation of the biggest culinary event of the year. The usual cooking-show set was assembled: the kitchen, the chef in a white hat and jacket, the pan on the stove, the ingredients measured out in advance. Accompanied by the sound of dozens of clicking cameras in the hands of news photographers, three onlookers took their seats at the kitchen stools and watched as the chef worked his magic at the stove.

What was on the menu for this event? It was a hamburger, which the chef sautéed with the usual seasonings and butter. But this was no ordinary hamburger. This hamburger cost $330,000. And as this remarkable hamburger cooked in the pan, the world waited to find out the answer to this question: how does it taste?

The press came out to see this extraordinarily expensive hamburger because it was made from a cow that was still alive and well on the day of the big taste test. And no, this isn't science fiction: the hamburger was made not from ground beef, but from special cells—the smallest structural unit in an organism—taken from the shoulder muscle of that living cow. How was this possible? It was due to the efforts of Dr. Mark Post of Maastricht University, a scientist whose team grew what he calls "cultured meat," to make the world's first lab-grown hamburger.

AP Images/David Parry

So how did this hamburger taste? According to the tasters, it had the correct consistency, aroma, and flavor. But unlike all great hamburgers, it lacked juiciness—an inevitable shortcoming in a hamburger that contained no fat and blood. Even so, the first lab-grown hamburger was edible, and that was a huge step forward. Encouraged, Post returned to his laboratory-kitchen to work on creating a better cultured hamburger.

The efforts made by Post and others to develop artificial meat may indeed seem like science fiction, but they're doing important work. Raising animals for food requires 30% of our available land, uses 8% of our fresh water, and accounts for 40% of emitted methane, a potent greenhouse gas. And for every 100 grams of vegetable-based protein we feed our cows, they produce only 15 grams of meat and use an enormous amount of energy and water, a very inefficient process. So it's clear that, as our human population grows, the practice of raising animals for food is not sustainable. When we consider the impact of the methane on our environment, and the amount of energy and water we use to raise livestock, many would say that we cannot afford it even now.

This chapter is all about our food and what's in it. We'll explore the three main constituents of the foods humans eat: protein, carbohydrates, and fat. We'll also consider where our food comes from, how modern crops are grown, and how new agricultural practices—and new technologies—might change the way we eat. Finally, we'll consider the artificial foods that we already eat and how they affect our health and our environment. Let's begin with the most critical part of the human diet: protein.

14.1 Protein: The Most Critical Nutrient

The human body needs a mixture of fuels for optimal health.

Like a car, the human body needs fuel to function. But, whereas most cars use only one type of fuel, the human body needs a mixture of fuels for optimal functioning. That mixture can vary, though, within limits and it depends on personal choice. For human bodies, the word **nutrients** is commonly used in place of *fuels* to describe the substances that provide nourishment which the body needs for growth, for the maintenance of life, and for optimal health.

The human body is a collection of fluids and *cells.* The body of an adult human contains about 37 trillion cells. That number includes dozens of cell types, each having a specific function and needing a specific balance of nutrients. Cells make up *tissues* that in turn compose *organs,* the structural units in the body that serve a particular function. Different organs, such as the heart, bone, and brain, have different nutritional requirements, and the nutritional requirements for the whole body reflect the needs of its individual parts.

There is no shortage of advice on what human beings should eat and what makes a healthy diet. Popular diets come and go—the Mediterranean diet, the paleo diet, the low-carb diet—but all of them have one thing in common: they all include some combination of the three major **macronutrients,** substances required by the human body in large amounts. These macronutrients, which are sometimes referred to as *major food groups,* are protein, carbohydrate, and lipid (also known as fat). Although various diets may recommend that we eat different amounts of each of these macronutrients, all three are essential to a healthy diet.

Protein, carbohydrate, and fat are examples of **biomolecules,** the molecules that support life. Biomolecules are just like the other molecules you've seen throughout this book, only larger.

QUESTION 14.1 Which of the following are macronutrient(s) for humans?
(a) fat (b) carbohydrate (c) water (d) protein (e) iron

ANSWER 14.1 Choices (a), (b), and (d) are macronutrients. Choices (c) and (e) are not.
For more practice with problems like this one, check out end-of-chapter Question 1.

FOLLOW-UP QUESTION 14.1 True or false? Diets that exclude one of the three macronutrients are not healthy.

Both micronutrients and macronutrients are part of a balanced diet.

About one in eight people on Earth do not get enough to eat. And while about 80% of those people live in only 20 countries, primarily in Asia and Africa, hunger does exist within the United States. In fact, in nearly 15% of U.S. households, food insufficiency is a problem.

Sometimes hunger is not a matter of the *quantity* of food, but the *quality* of food that people can obtain in their diets. As we noted at the start of this chapter, a healthy diet includes the correct balance of the macronutrients protein, carbohydrate, and fat. (Humans also need clean water, and water is the topic of Chapter 8.) A healthy diet also includes micronutrients. **Micronutrients,** such as vitamins (like niacin and vitamin B_{12}) and minerals (like iron and calcium), are substances required in small amounts in the human diet. (Vitamins are organic substances, whereas minerals are inorganic substances.)

recurring theme in chemistry 8
The Earth has a finite supply of raw materials that humans can use.

A severe deficiency in any micro- or macronutrient can result in **malnutrition,** the lack of proper nutrition. According to the medical journal *The Lancet,* malnutrition is the cause of death for 45% of the children who die worldwide. Among the macronutrients, worldwide, protein inevitably is the hardest to find, and protein deficiency is the most common cause of malnutrition.

Protein is present in meat and dairy products, and it's also present in vegetarian foods such as beans, legumes, soy, and nuts. But in areas where malnutrition is omnipresent, these common meat and vegetarian sources of protein are often extremely limited. For this reason, the arm of the United Nations devoted to issues of hunger and malnutrition recently released a new directive that its advocates feel could help reduce malnutrition. What source of protein did the UN recommend? Insects.

According to the UN report, insects provide as much protein as meat and dairy but a lot less fat. Insects can be raised much more efficiently than livestock and without significant environmental impact. In fact, more than 2 billion people on Earth already eat insects as a regular source of protein. **Figure 14.1** shows an energy bar made from crickets and a plate of insects prepared as food.

QUESTION 14.2 Identify each of the following as a macronutrient or a micronutrient:
(a) iron (b) fat (c) protein (d) vitamin C (e) carbohydrate

p.studio66/Shutterstock

a

Courtesy of EXO

b

Figure 14.1 Insect Cuisine

(a) A plate of snacks in Thailand may include seasoned, grasshoppers, crickets, beetles, or larvae.
(b) The Exo Bar is a protein bar made from cricket flour.

ANSWER 14.2 Macronutrients include fat (b), protein (c), and carbohydrate (e). Other nutrients that are required in smaller amounts, such as iron (a) and vitamin C (d), are micronutrients.

For more practice with problems like this one, check out end-of-chapter Questions 4 and 5.

FOLLOW-UP QUESTION 14.2 According to the text, why are insects a healthy choice for people and for the planet?

Amino acids are the monomers that make up proteins, which are polymers.

In the future, as Earth's resources become more depleted, perhaps more of us will be eating insects or cultured beef as our source of complete protein. What do we mean when we say that a food is a source of complete protein? **Proteins** are polymers of **amino acids**. Amino acids are small organic molecules that connect together to form long, chainlike protein polymers. When we talk about complete protein, we mean that a given source of protein includes a wide variety of certain amino acids, the building blocks of protein.

Figure 14.2 shows the molecular structure of a protein chain and the amino acids within it. Several things about this protein's structure are important. First, note that the amino acids have similar structures. They differ from each other only in the identity of the side chain they contain. Second, each amino acid is bonded to the next, one after the other, at the point where a carbon atom attaches to its neighboring nitrogen atom (indicated with a green slash in the figure). These links, called **peptide bonds,** can be made and broken. In this way, amino acids—monomers—can be assembled into proteins, which are polymers, and proteins can be disassembled back into amino acids.

flashback

Recall from Section 13.2 that polymers are long, chainlike molecules that are composed of monomers. Polymers can be made by humans (like Kevlar and nylon) or they can form naturally (like proteins).

Figure 14.2 The Structure of Protein

Part of a protein chain. The backbone of the protein chain is highlighted in yellow; the locations where amino acids connect—the peptide bonds—are indicated with green slashes. Pink circles highlight the side chains that branch off from the main protein chain. Twenty different side chains are possible, which means there are 20 different amino acids. Six side chains are shown in this figure.

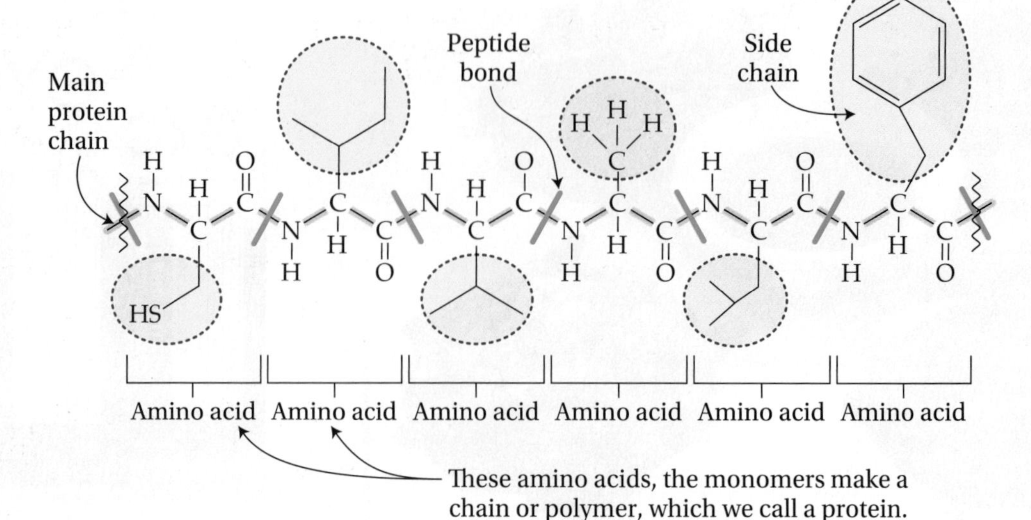

recurring theme in chemistry 6

The structure of carbon-containing substances is predictable because each carbon atom almost always makes four bonds to other atoms.

Twenty different amino acids are commonly found in proteins. **Figure 14.3** depicts a sequence of amino acids in a protein. In the figure, each amino acid is identified with its three-letter designation. This specific sequence is the sequence for *insulin*, a protein that helps regulate the metabolism of fats and carbohydrates in the body. Notice that this protein actually contains two individual protein chains. They are held together by **disulfide bonds,** shown in the figure as S–S. Disulfide bonds are composed of two sulfur atoms. They are a common feature of proteins, and they're important because they connect protein chains together. All insulin molecules have two chains and this specific sequence of amino acids.

Amino acids can be divided into two groups, which are identified in **Figure 14.4**. First, there are the 11 *nonessential* amino acids, highlighted in yellow in Figure 14.4. The adult human body can produce these amino acids on its own. Second, there are the nine *essential* amino acids. They are deemed essential not because they play a more important role in proteins than do the nonessential amino acids, but because they must be included in the adult human diet for humans to survive. The human body doesn't know how to make these amino acids, so we must consume them instead. (The list of essential versus nonessential amino acids varies slightly for children.) In Figure 14.4, each amino acid's side chain is highlighted in green.

Figure 14.3 The Amino Acid Sequence of Insulin

The two chains of insulin are shown as sequences of three-letter amino acids. The two chains are connected with disulfide bonds in two places. Disulfide bonds always connect at the amino acid abbreviated with the letters *Cys*, known as cysteine.

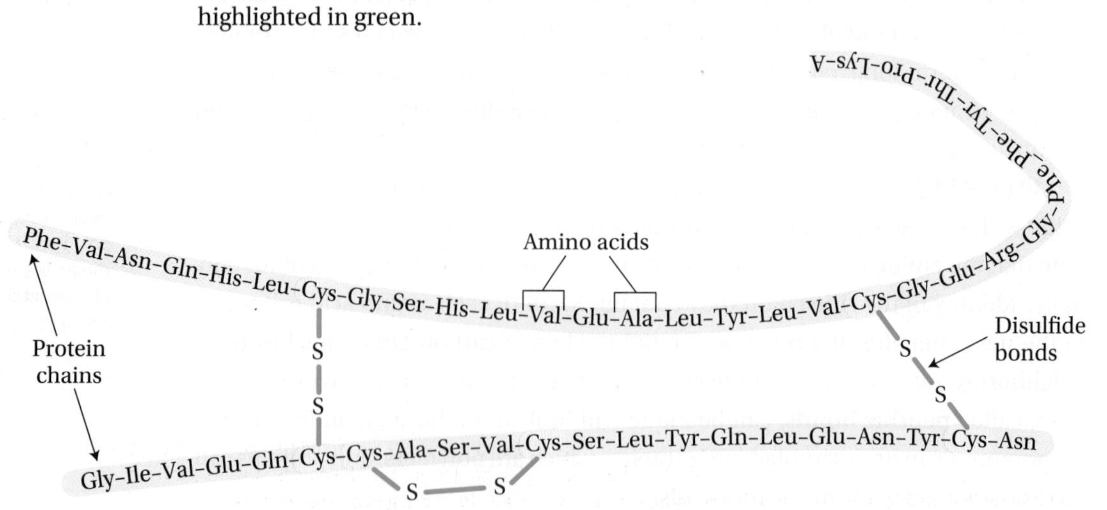

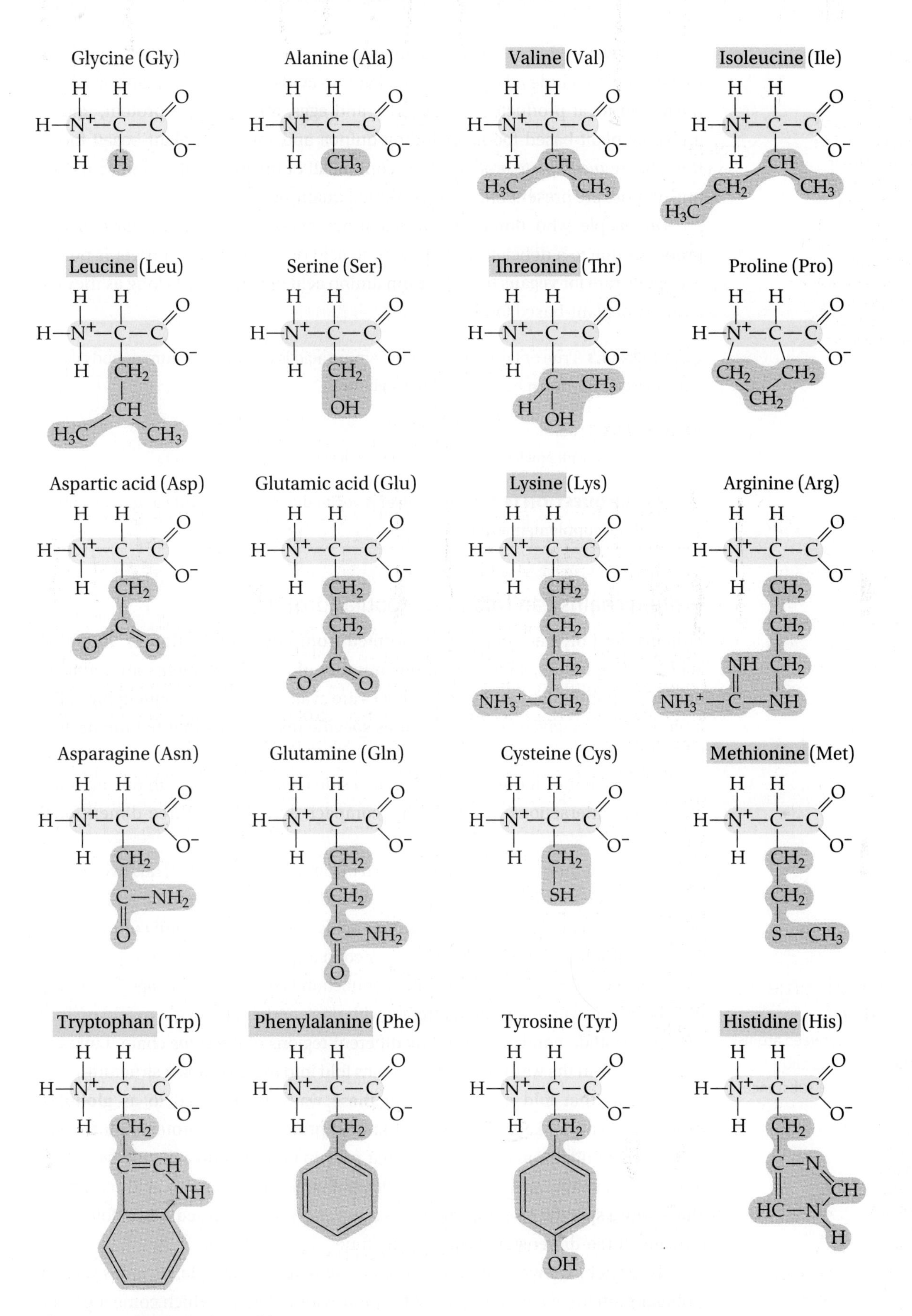

Figure 14.4 The 20 Amino Acids

In the 20 amino acids, shown here, the part shown in pink is the side chain that extends from the protein backbone. The atoms of each amino acid that become part of the backbone are highlighted in yellow. The names of the essential amino acids are highlighted in green. The three-letter abbreviation for each amino acid is given in parentheses.

When we say that a diet should include a source of complete protein, we mean that it should include all of the essential amino acids in adequate quantities. Most meat, and meat products such as dairy and eggs, are complete protein, and so are some plant-based foods, including quinoa and soy. Most plant-based foods, though, are *incomplete proteins;* they contain all of the essential amino acids, but one or more are present only in very limited quantities.

Do people who don't eat animal products—that is, vegans—have health problems due to a deficiency of one amino acid or another? The answer is no. It's extremely rare for vegans to suffer from amino acid deficiencies as long as they eat a variety of plant-based foods.

QUESTION 14.3 True or false? An essential amino acid is one that the human body cannot make and that must be obtained through diet.

ANSWER 14.3 True.

For more practice with problems like this one, check out end-of-chapter Question 13.

FOLLOW-UP QUESTION 14.3 True or false? A vegan diet cannot provide adequate amino acids without supplementation.

Protein chains can fold into globular proteins.

Proteins are polymers that can be formed from any combination of 20 amino acid building blocks, so an enormous number of different proteins are found in nature. An impressive variety of sequences are available, and the sequencing of the amino acids in a given protein provides specific instructions that tell the protein how to fold into a unique three-dimensional shape. Mature proteins are not just long, floppy chains. Rather, the amino acid side chains interact with one another, and those interactions stabilize the compact and folded three-dimensional structure.

Notice that the insulin molecule shown in Figure 14.3 (on p. 420) has two chains. The one on the bottom is made up of 21 amino acids linked together, and the one on the top is made up of 30 linked amino acids. The two chains are connected by two pairs of sulfur atoms. Each pair is located in a specific place. In both places, the amino acid Cys (cysteine) is where the connection is made. This linkage is one way that the chains within a protein interact with one another and begin to fold up into shapes. Disulfide bonds can also link different regions of the *same* chain. Linkages play a big role in the way long protein chains fold into more compact structures.

recurring theme in chemistry 7

Individual bonds between atoms dictate the properties and characteristics of the substances that contain them.

Proteins that fold and curl up in these ways are referred to as **globular proteins**. If you looked at a list of all known proteins, most proteins on the list would be globular proteins. Each globular protein folds up into a distinctive three-dimensional shape, and that shape is dictated largely by its amino acid sequence. That's why a specific protein, which has a specific amino acid sequence, folds into the same three-dimensional shape every time.

The protein known as *human exosome complex*, for example, includes several globular proteins, each more than 200 amino acids in length, which come together to make the larger megaprotein molecule. Its job is to dismantle unneeded molecules in human cells. **Figure 14.5** is a space-filling model of this enormous molecule. This type of drawing is designed to show you the surface of the protein—

its outside contour—rather than the protein chain itself and the details of its amino acid sequence. In this view, you can see that this protein is roughly spherical—a round lump. The term *globular protein* suggests that the three-dimensional shape is somewhat spherical, like the shape of bovine chymotrypsinogen, but other globular proteins can have an oblong, ellipsoidal, or flattened shape.

Globular proteins exist in a variety of shapes. Because it's often the protein surface that determines what the protein does, this means that proteins can serve a variety of functions. For example, many globular proteins have surface indentations where they can bind other molecules. Consider *ribonuclease III*, the protein molecule illustrated in **Figure 14.6**. It has a deep crevice where another molecule, shown in red in the figure, fits exactly. The *ribonuclease III* molecule cuts this red molecule, releases it, and then binds the next molecule that it will cut.

In living systems, globular proteins such as *ribonuclease III* do most of the grunt work. When they facilitate or *catalyze* chemical reactions, as *ribonuclease III* does, we call them **enzymes**. When they are chemical messenger molecules that travel through the blood, we call them *hormones*. When they are produced and sent out in response to an infection, we call them *antibodies*. The biomolecules that carry and store oxygen in our blood—*hemoglobin* and *myoglobin*, respectively—are globular proteins, too.

Most of the protein we eat is globular protein. Most dietary proteins are broken down into amino acids in our bodies through the process of digestion, and our bodies use those amino acids to build new proteins. Thus, proteins are our most critical nutrient because they play so many vital roles in the human body.

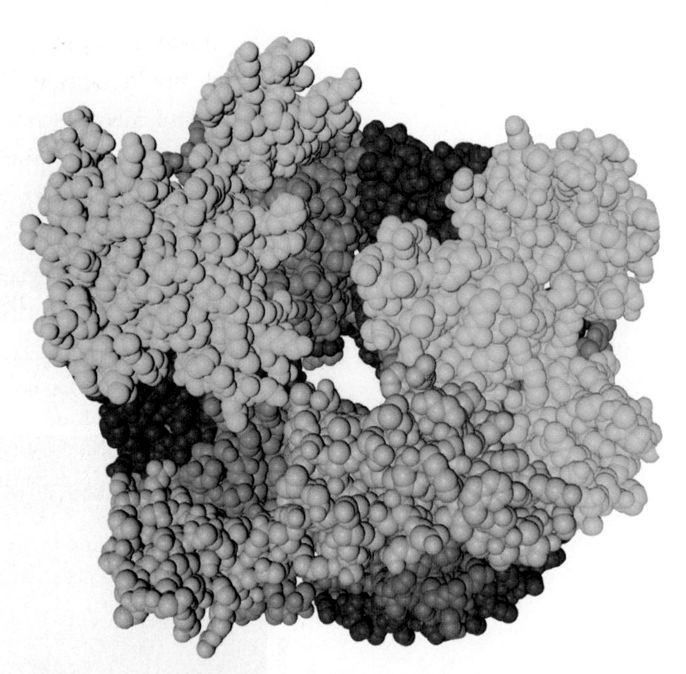

Figure 14.5 A Space-Filling Model of Bovine Chymotrypsinogen

This view of a protein shows its surface and may provide clues to its function. The contours of proteins, for example, often serve as binding sites for other molecules.

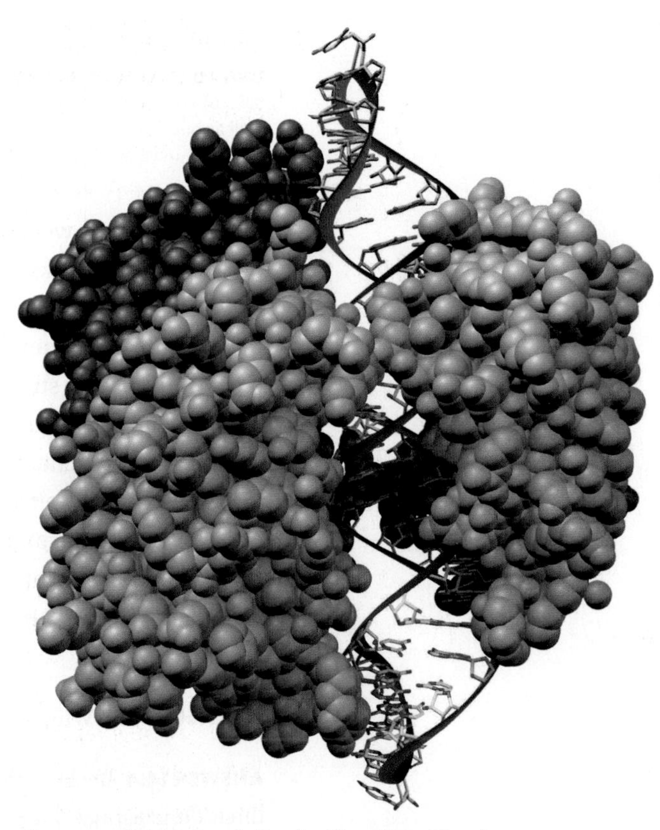

Figure 14.6 A Molecule Docked in a Deep Crevice on the Surface of a Protein

The protein *ribonuclease III* is shown as a blue space-filling model. In the model, the red molecule docks in the blue *ribonuclease III* protein molecule. After it docks, the protein cuts it into two pieces and then releases it. Proteins like *ribonuclease III*, which catalyze chemical reactions, are enzymes.

Structural proteins play mechanical and structural roles.

Proteins can be divided into two broad categories. Globular proteins, which we just discussed, make up one of those categories. The other category is *structural proteins,* which have mechanical roles. In living organisms, structural proteins have protective jobs. Skin and fingernails, for example, are made up of structural proteins. Structural proteins also are

Figure 14.7 Nature's Scaffolding

(a) The hides of animals, such as the bison shown here, as well as human skin, are protective barriers made of pliable structural proteins. (b) Structural proteins give bones the strength needed to support the body. (c) A spider's web is made possible by the extraordinary strength of the structural proteins in spider silk. (d) The toughness of structural proteins is manifested in the antlers and horns of some animals as well as our own fingernails.

the basis of hard-wearing materials such as the horn of a rhinoceros and the ivory of an elephant's tusks (**Figure 14.7**).

Structural proteins also can be used to do mechanical work, such as in contractile systems, which are the basis for muscle movement in humans and other animals. They can be stretchy and flexible as in *resilin,* the biomolecule found in the hinges of insect wings. Nature has found a way to make a structural protein suited to almost any mechanical or structural application.

Structural proteins do not have a globular shape. Instead, they're often fibrous or ropelike. For example, silk, which is made by silkworms, is made up of chains of amino acids that fold upon themselves and interdigitate with other chains, forming the interlocking pattern shown in **Figure 14.8**. As a result, silk makes a very strong fiber.

QUESTION 14.4 Why is the sequence of amino acids in a globular protein so important?

ANSWER 14.4 Because the amino acid sequence dictates how each protein folds into a three-dimensional shape.

For more practice with problems like this one, check out end-of-chapter Question 17.

FOLLOW-UP QUESTION 14.4 How does the zigzag pattern of protein chains strengthen silk?

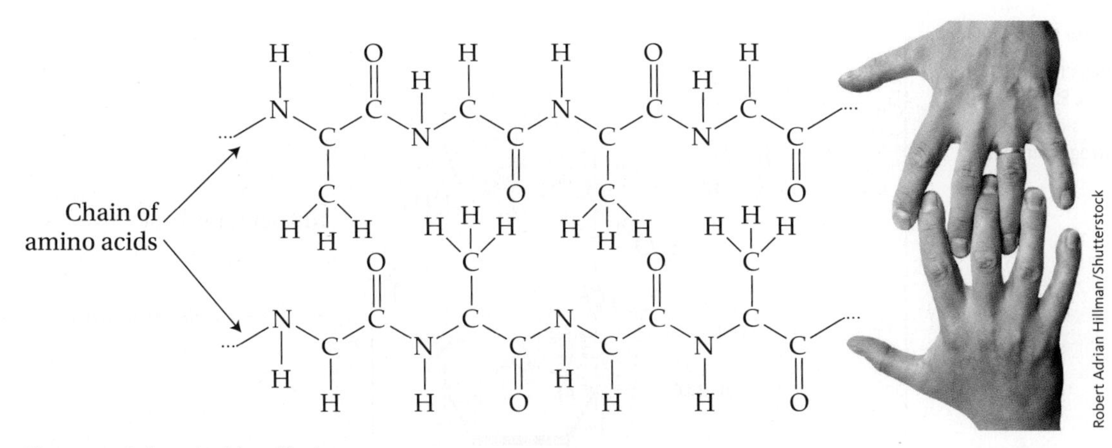

Robert Adrian Hillman/Shutterstock

Figure 14.8 Interlocking Chains

The chains of the silk protein form a zigzag shape that allows neighboring strands to interlock with each other, similar to the way that people's fingers can interlock.

14.2 How Proteins Are Made

DNA is a polymer of nucleotides.

Now that we know a bit about what proteins look like, let's consider how a protein is made. This process is exceedingly complex, and scientists' current understanding of it is incomplete. We do know that the instructions for making every protein are stored in molecules of **deoxyribonucleic acid,** which is called **DNA**. DNA and protein are both polymers. They are alike in some ways and different in others.

DNA belongs to the class of biological molecules called **nucleic acids**. DNA is similar to protein in that it has a backbone with appendages branching off it. In proteins, the appendages are the side chains of amino acids. And as we have discussed, there are 20 different amino acids (Figure 14.4). In DNA, each appendage is a **nitrogenous base,** sometimes referred to simply as a "base." There are four different bases in DNA, each one represented with a different one-letter symbol. The four bases are guanine (its symbol is G), cytosine (C), thymine (T), and adenine (A). Thus, DNA contains only four different nucleotide building blocks, as compared to proteins, which have 20 amino acid building blocks.

Both DNA and protein have backbones, but the backbones are composed of different kinds of molecules. The backbone of proteins is made up of amino acids linked together. The backbone in DNA is made up of units called **nucleotides**. Each nucleotide has three parts. A phosphorus-containing group and a sugar part make up the backbone and, as we already mentioned, nitrogenous bases dangle from the backbone (**Figure 14.9**). The sugar in DNA is deoxyribose, and this word is part of the full name of DNA: *deoxyribo*nucleic acid.

DNA is made up of not one strand, but two. How do two strands come together to form the double-stranded DNA molecule? The four bases on one strand—G, C, T, and A—interact with the four bases on the other strand. Here's how it works: Each time a G appears on one strand, it pairs with a C on the other strand. Each time a T appears on one strand, it pairs with an A on the other strand. We refer to G and C as well as A and T as **complementary base pairs**. The bases in base pairs are

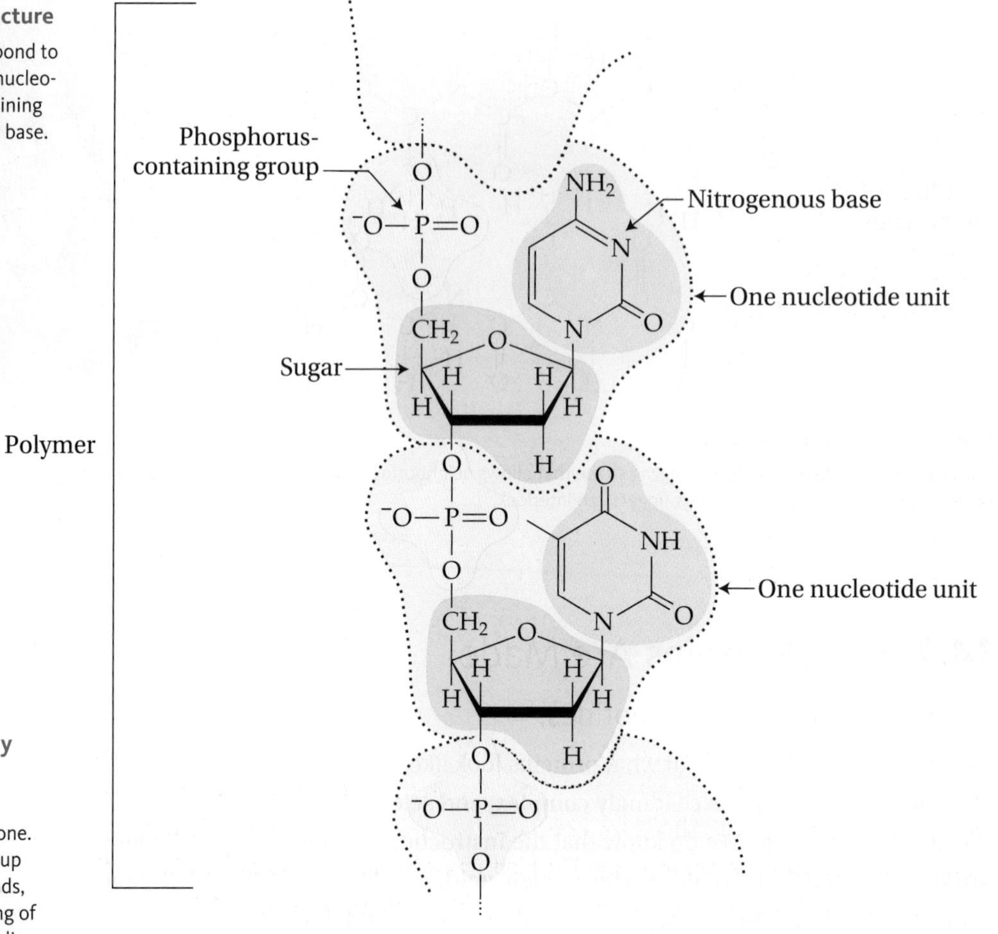

Figure 14.9 Nucleotide Structure

Nucleotides are monomers that bond to form DNA polymer chains. Each nucleotide includes a phosphorus-containing group, a sugar, and a nitrogenous base.

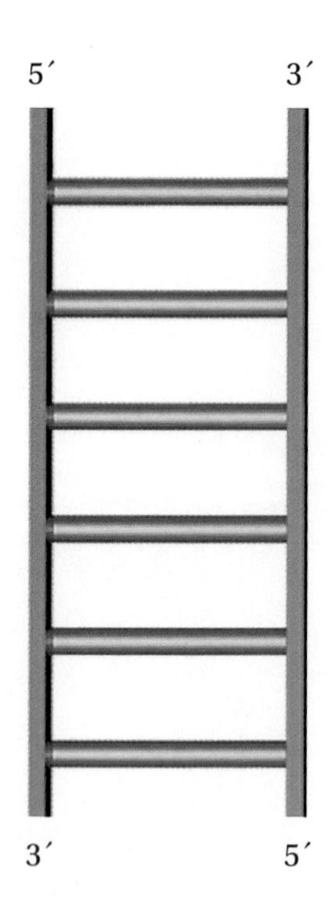

Figure 14.10 Complementary Base Pairing

The bases in a DNA molecule are attached to the molecule's backbone. When complementary bases pair up with each other via hydrogen bonds, the result is a long chain consisting of two connected backbones resembling a ladder. The bases act as the rungs of the ladder, and the backbones act as the two vertical pieces.

connected to each other by hydrogen bonds and, in this way, two strands are connected to each other.

Each of the two strands of DNA resembles one vertical piece of a ladder. When the two strands come together, the bases on the two chains form hydrogen bonds that create "rungs" of the ladder, as illustrated in **Figure 14.10**.

The ladder analogy is a useful way to visualize a double helix. However, a ladder is an oversimplification because, unlike a ladder, the two strands that make up the double helix each have different ends. We call one end the *five-prime end* (often written 5′ end) and the other end the *three-prime end* (or 3′ end). These ends are labeled on both DNA strands in Figure 14.10.

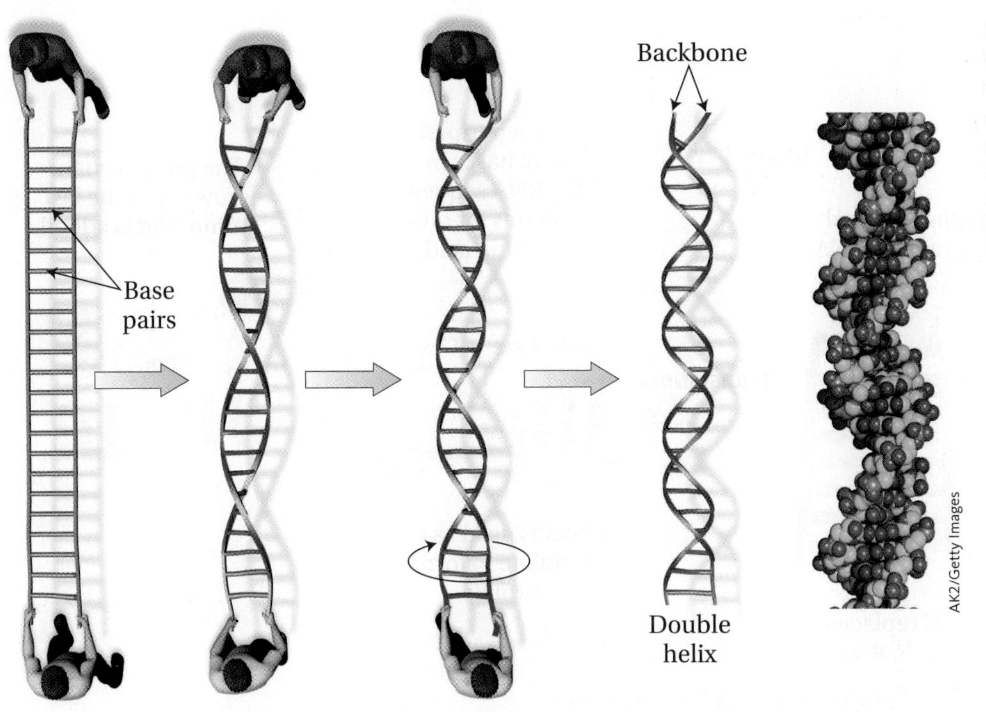

Figure 14.11 Ladder with a Twist

Two people holding the ends of a flexible ladder can create a structure like a DNA double helix by twisting the ladder to create a right-handed spiral. The structure on the far right is a space-filling model of a DNA molecule.

Notice that the two strands in Figure 14.10 run in opposite directions. The one on the left goes from 5′ to 3′ (top to bottom) while the one on the right goes from 3′ to 5′ (again, top to bottom). We indicate the direction that the strand is going by adding these numbers to the sequence. For example, if the left strand in Figure 14.10 is 5′-CTATCG-3′ then its complementary strand, the one on the right, would be 3′-GATAGC-5′.

Let's consider another example of complementary base-pair sequences. If one strand of DNA has a sequence that reads 5′-GGCCTATC-3′, then its complementary strand will read 3′-CCGGATAG-5′. G pairs with C and T pairs with A. So, not only are the individual bases complementary to each other, but one entire strand is the complement of the other.

Imagine for a moment that the ladder in Figure 14.10 is very flexible. You and a friend stand at opposite ends of the ladder, each taking one end, and start twisting. The resulting shape is the shape of DNA, called a **double helix (Figure 14.11)**.

QUESTION 14.5 A single strand of DNA has the sequence 3′-AGGTACCTGGTA-5′. What is its complementary sequence?

ANSWER 14.5 Because G is the complement to C and T is the complement to A, the complementary sequence is 5′-TCCATGGACCAT-3′.

For more practice with problems like this one, check out end-of-chapter Question 19.

FOLLOW-UP QUESTION 14.5 A single strand of DNA has the sequence 3′-CGCGTTCGCGAATAG-5′. What is its complementary sequence?

Genes within DNA are transcribed into RNA, which is translated into protein.

The DNA molecule is a dynamic structure. Each bond *within* each DNA strand is covalent and strong, but the hydrogen bonds holding the two strands together are much weaker and more easily broken. Because weaker bonds connect the

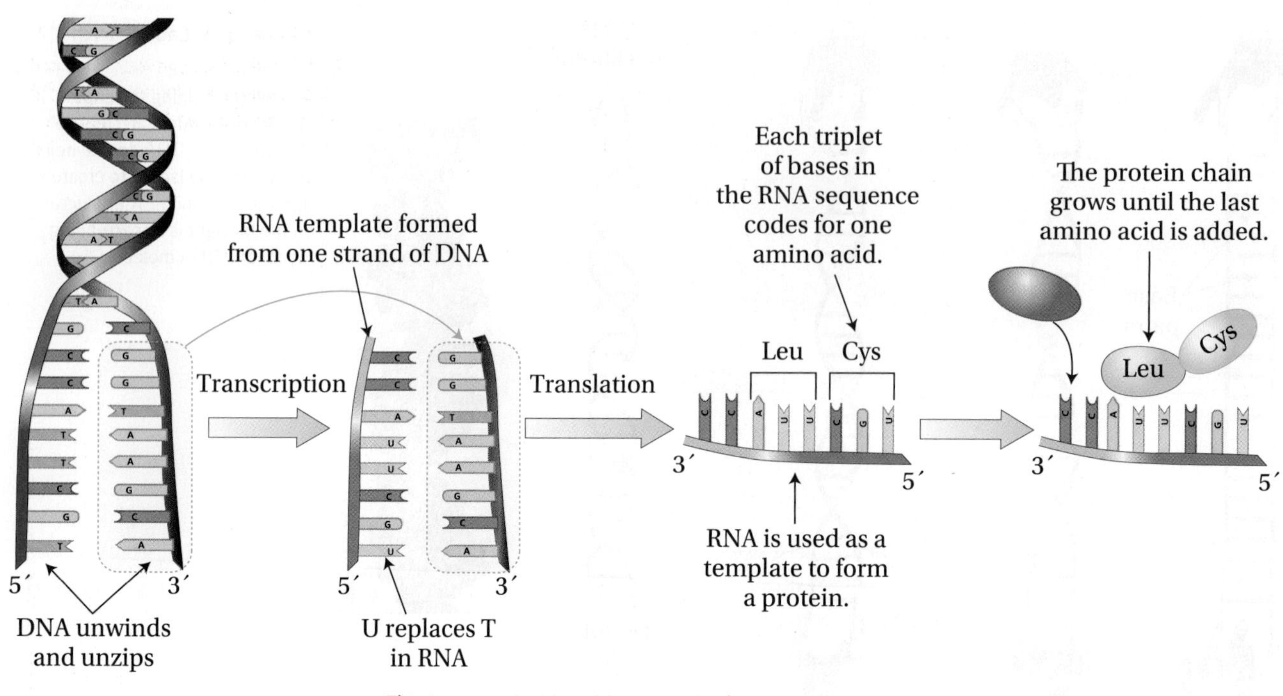
RNA template formed
from one strand of DNA
Transcription
Translation
Each triplet
of bases in
the RNA sequence
codes for one
amino acid.
Leu
Cys
The protein chain
grows until the last
amino acid is added.
3′
5′
RNA is used as a
template to form
a protein.
DNA unwinds
and unzips
U replaces T
in RNA

Figure 14.12 DNA to RNA to Protein

(a) An unzipped section of DNA is used as a template to create a strand of RNA in a process called transcription. (b) RNA is used as a template to form a protein. (c) Each triplet of bases in the RNA sequence codes for one amino acid. The protein chain grows until the last amino acid is added.

two strands together, the two strands can zip and unzip. As we'll see, this exposed strand contains the information needed to make new proteins. When the strands unzip, it is possible for the exposed bases of one strand to act as a template for the formation of a new strand.

How does a strand of DNA become a protein chain? The nucleotide sequence of DNA is the key. The long nucleotide sequence on any strand of DNA is divided into small segments called **genes,** each of which provides instructions for making a specific protein needed by the body. Every cell of every living organism contains DNA, and that DNA includes enough genes to create nearly 100,000 different proteins.

Unzipped DNA has an exposed code of bases, and that code eventually becomes a protein sequence. Before that can happen, though, the exposed strand of DNA is used as a template to make a strand of a different nucleic acid, called **ribonucleic acid** or **RNA**. RNA has almost the same structure as DNA, but with two differences. First, the sugar in RNA is *ribose,* and this term is part of the full name of RNA: *ribo*nucleic acid. Second, RNA includes the base uracil (U) in place of T. When RNA is made from a DNA template in a process called **transcription,** the A on DNA makes a base pair with the U on the new RNA strand, as shown in **Figure 14.12a**.

The new RNA strand is now ready to make a protein. For example, let's say your body needs a specific protein molecule to fight off an infection. This process takes place within cells of your body that are responsible for fighting infection. When a protein that fights infection, or any protein, is made, the strands of DNA located in the infection-fighting cell's nucleus separate to expose the DNA sequence for the specific gene that makes the protein you need. As the DNA strands separate, one of those strands serves as a template to create a new strand of RNA, as shown in **Figure 14.12b**.

Codon	Amino acid	Codon	Amino acid	Codon	Amino acid	Codon	Amino acid
UUU	Phe	UCU	Ser	UAU	Tyr	UGU	Cys
UUC		UCC		UAC		UGC	
UUA	Leu	UCA		UAA - STOP		UGA - STOP	
UUG		UCG		UAG - STOP		UGG	Trp
CUU		CCU	Pro	CAU	His	CGU	Arg
CUC		CCC		CAC		CGC	
CUA		CCA		CAA	Gln	CGA	
CUG		CCG		CAG		CGG	
AUU	Ile	ACU	Thr	AAU	Asn	AGU	Ser
AUC		ACC		AAC		AGC	
AUA		ACA		AAA	Lys	AGA	Arg
AUG	Met	ACG		AAG		AGG	
GUU	Val	GCU	Ala	GAU	Asp	GGU	Gly
GUC		GCC		GAC		GGC	
GUA		GCA		GAA	Glu	GGA	
GUG		GCG		GAG		GGG	

Figure 14.13 The Genetic Code

Each triplet of nitrogenous bases codes for an amino acid. The triplets labeled "stop" signal the end of the protein chain.

The question of how you get from RNA, a molecule built from various combinations of only four bases, to protein, a molecule built from various combinations of 20 amino acids, stumped scientists for decades. In the 1960s, the answer was found. To translate a message that uses four "letters" (bases) to one that uses 20 (amino acids), the cell uses what is called the **genetic code**. The code works by equating each group of three RNA bases, called a **triplet** (also known as a *codon*), with a specific amino acid. For every RNA triplet there's a corresponding amino acid, as shown in **Figure 14.13**. For example, either of the two triplets UUU or UUC "code for" the amino acid phenylalanine, which has the three-letter abbreviation Phe.

The base sequence of the RNA strand is made into a sequence of amino acids that reflects the genetic code, as shown in **Figure 14.12c**. For every three nucleotides of the RNA strand, one specific amino acid is created and added to a growing protein chain in which the amino acids are connected with peptide bonds. When one of the "stop" triplets in Figure 14.13 is reached, the protein is complete and no more amino acids are added. This process of **translation** takes one sequence—that of RNA nucleotides—and translates it into another sequence—that of the amino acids in a protein. As the new protein begins to form, it changes from an unfurled, floppy chain into a tightly folded protein. In this way, hundreds of times every second, your cells produce new protein from nonessential amino acids made in the body and from essential amino acids derived from the food you eat.

QUESTION 14.6 The following steps are involved in the production of a protein beginning with a sequence of DNA. Put them in the order they occur. (a) The triplets in RNA are translated. (b) DNA unzips. (c) A new protein chain is made. (d) A small piece of RNA is created from the DNA template.

ANSWER 14.6 The sequence is (b)-(d)-(a)-(c).

For more practice with problems like this one, check out end-of-chapter Question 23.

FOLLOW-UP QUESTION 14.6 An RNA sequence is UUU AAA. What two amino acids are coded for by this sequence (see Figure 14.13)? List the three-letter abbreviation for each.

14.3 Genetic Engineering and GMOs

Genetic engineering alters the DNA of food crops.

In the United States, consumers are used to finding high-quality produce at the grocery store. Although most fruits and vegetables grow only during one part of the year in most regions of the country, we can buy many crops year-round because produce is shipped to us from around the world. And our produce looks good. We demand tomatoes that are plump, red, and shiny, and we like apples without bruises (**Figure 14.14**). Thus, our modern produce has to withstand shipping and changing temperatures. It also has to have a long shelf life and harbor no insects.

Maks Narodenko/Shutterstock

Figure 14.14 Is It Real?
Modern produce is so perfect that it sometimes looks artificial.

It wasn't always this way. The quality of our produce has changed greatly over the past 50 years, and some of that progress is owed to **genetic engineering,** technology that allows scientists to manipulate the DNA of organisms—plants or animals. A **genetically modified organism** (**GMO**) has altered DNA. This alteration affects the proteins the DNA makes. For example, a genetically modified piece of fruit may contain altered DNA that allows it to produce a new, beneficial protein. Such a protein might extend the shelf life of the fruit or reduce its bruising.

If scientists want a plant to contain a specific protein that it does not naturally contain, how do they go about making that happen? The secret behind any kind of genetic engineering is cutting and pasting pieces of DNA from one organism into another organism. Technology gives scientists the capacity to cut a strand of DNA into fragments and then reattach the fragments in a different way. Any organism containing one or more genes that came from a different organism is referred to as a **transgenic organism**. The terms *GMO* and *transgenic* are often used interchangeably.

Genetic engineering is used to grow herbicide- or insect-resistant food crops.

Two genetic modifications are frequently employed in modern agriculture. The first of these involves splicing a gene that confers a resistance to insects into a plant's DNA. The gene comes from a bacterium called Bt (which stands for *Bacillus thuringiensis*) that naturally lives in soil. This bacterium produces a protein that's toxic to certain insects. GMO plants that are Bt-modified have the gene for that toxic protein spliced into their DNA, and the transgenic plant is protected over its life span from the insects that are killed by the toxic protein.

The second common genetic modification confers tolerance to herbicides. Herbicide-resistent GMOs are often labeled with the letters *HT*. In the United States, for example, most soybeans are HT soybeans. These transgenic plants contain a gene for herbicide resistance built into their genetic material. When HT plants grow, they—like many plants—are often overcome by weeds, but the HT crops can be sprayed with a herbicide that kills all plants *except* those with the resistance gene. **Figure14.15a** shows a soybean crop that has not been genetically modified for herbicide resistance.

One herbicide that many plants have been designed to tolerate is Roundup, a broad-spectrum herbicide made by the Monsanto Corporation. The active ingredi-

a

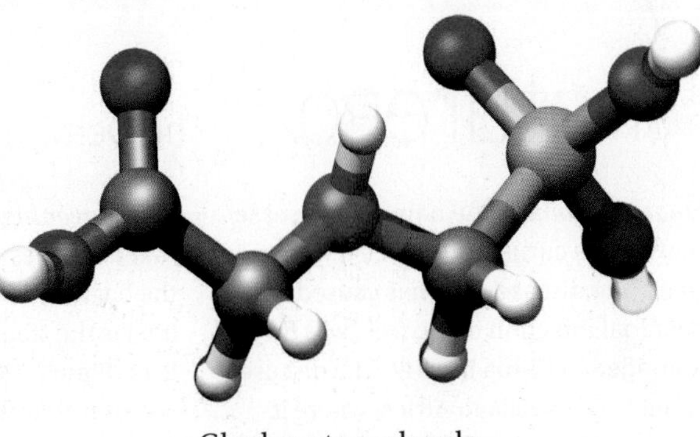

b

Figure 14.15 Roundup Herbicide from Monsanto

(a) This photo shows weed growth in a weed-filled soybean crop that is not genetically modified. (b) The glyphosate molecule is the active ingredient in the herbicide marketed by Monsanto under the name Roundup.

ent in Roundup is a chemical called glyphosate. The glyphosate molecule, shown in **Figure 14.15b,** inhibits one of the steps in the pathway that plants use to make certain amino acids. These amino acids are essential to plant growth, so once the process is blocked, the plants cannot grow. Because this pathway occurs only in plants and bacteria, glyphosate is harmless to animals, at least in theory.

When Roundup is applied to a soybean crop, it kills the weeds and leaves the HT soybeans untouched. The difference in treated and untreated plants is clear and undeniable. The development of herbicide-tolerant and pest-resistant crops has markedly improved crop yields. Furthermore, farmers growing pest-resistant transgenic crops need to use fewer conventional pesticides, which are usually toxic to animals, contaminate the environment, and often are linked to long-term health problems in humans.

GMO crops have become more and more prevalent in the United States since being introduced in the 1990s. **Figure 14.16** is a graph of the percentage of various crops that are genetically modified today. Bt cotton, for example, made up about 35% of the U.S. cotton crop in 2000 and 76% in 2013.

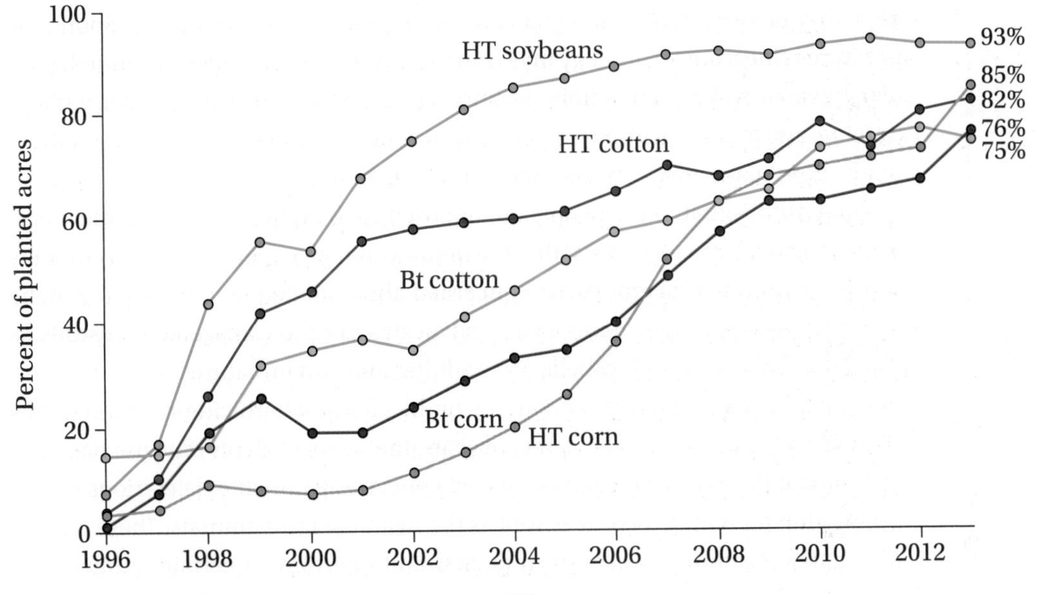

Figure 14.16 GMO Crops in the United States

natureBOX • The Demise of Natural Orange Juice

Huanglongbing is the name of a disease that infects citrus fruit, especially orange trees. The disease, which is caused by a bacterial infection of the tree, was first identified in China in 1942. The disease spread across Asia to Africa, where it decimated orange crops.

Bacterial infections can be transferred among humans in various ways. For example, a person on a bus sneezes and infected mucus spreads to people standing nearby, who then carry it with them when they get off the bus. But how does an organism like an orange tree, which doesn't ride a bus or move at all, become infected with a bacterium? The answer is that insects carry the infection from one tree to another. Huanglongbing disease spreads among orange trees when the Asian citrus psyllid (**Figure 14A**) drinks the sap from an infected tree and carries the infection to the next tree, where it then infects that tree by drinking its sap.

Shortly after the turn of the millennium, much to the dismay of U.S. orange growers, huanglongbing showed up in Florida citrus groves. In the United States, it is called by a new name—*citrus greening*—that describes how the bacterium affects the trees' fruit. **Figure 14B** shows fruit from an infected tree. The fruit is green on the bottom and tastes bitter.

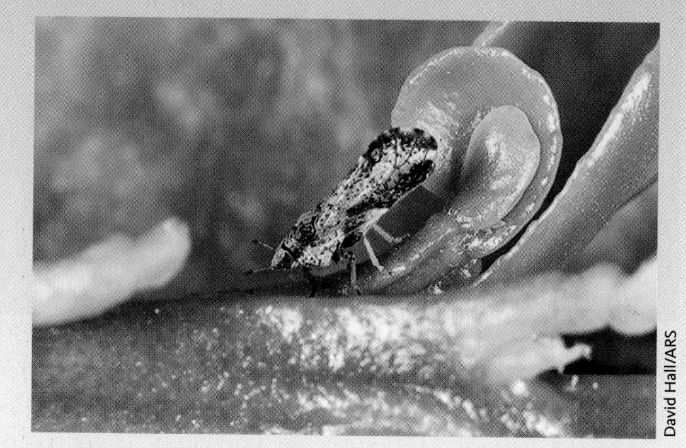

Figure 14A The Asian Citrus Psyllid
This insect carries the bacterium that infects trees with huanglongbing disease, also known as citrus greening.

Citrus greening has now spread across the United States to California. If left unchecked, the disease may make orange juice—long a staple on most American breakfast tables—a thing of the past. As this stark reality looms, the U.S. orange juice industry faces a tough decision. Should orange growers continue to increase the amount of pesticide they spray on orange trees, an approach that has had limited success in battling huanglongbing? Or, should they genetically modify their trees to resist the bacterial infection? The second option is a serious step to take for an industry that prides itself on producing natural, unadulterated juice from a wholesome piece of fruit.

One of the biggest orange growers in the United States is Southern Gardens Citrus, a Florida-based company that supplies orange juice for the brands Florida's Natural and Tropicana. In 2008, the president of Southern Gardens Citrus, Ricke Kress, began to address the problem by first scouring the globe

The use of certain GMOs is controversial.

It's a rosy picture. Better crop yields mean more food for the hungry. Soon new genetic engineering technology may bring us canola oil with higher vitamin E content, bananas spiked with edible vaccines against tropical diseases, coffee without caffeine, and tobacco without nicotine. Who could be opposed to that? Won't these innovations make all of our lives happier and healthier?

The truth is that many people are opposed to the growth of transgenic crops. Some religious protesters insist that the modification of genetic material is morally wrong. Environmental groups are concerned about inadequate monitoring of the use of transgenic plants. Some health advocates refer to transgenic foodstuffs as "frankenfood," because they believe it's adulterated and unnatural.

Still others are concerned about the herbicides that herbicide-resistant plants allow farmers to use so freely. In recent years, the safety of glyphosate use has been questioned. Reports have been published concerning possible deleterious effects of Roundup on human health as well as the health of farm animals. These reports suggest that Roundup causes birth defects and infertility in certain animals that are exposed to it, such as cattle, chickens, frogs, and pigs. Another study showed

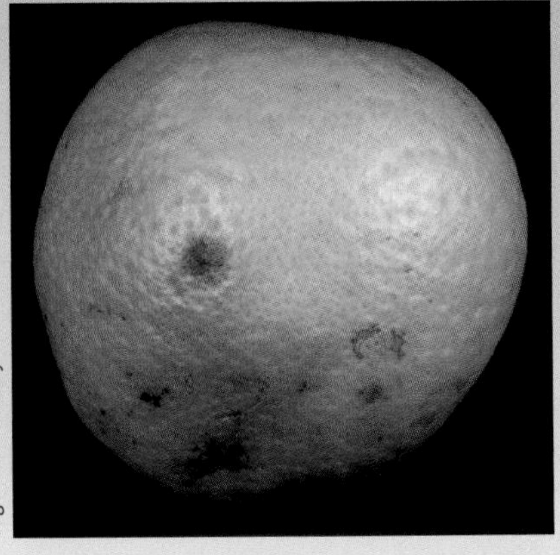
Nigel Cattlin/Alamy

Figure 14B The Effect of Citrus Greening

Fruit from infected trees is green on the bottom and has a bitter taste.

to find an orange tree with natural resistance to the infection. No such tree existed. In a last-ditch effort to help his company survive, he decided to invest millions of dollars in the development of genetically modified orange trees.

Kress chose five scientists, each of whom headed up a research group that worked on the genetic modification of plants. Four of these research groups chose a different gene from a different organism to insert into the DNA of the orange tree to confer resistance to the citrus-greening bacterium. The genes came from spinach, onion, a pig, and a virus. The fifth group decided to design a gene from scratch. To date, the most promising results have been obtained from the spinach and pig genes. But juice from these modified trees is still far from being ready to test, because it takes years for new trees to mature and grow fruit.

Now the question on the minds of citrus growers is, "If this approach works, will the American public drink genetically modified orange juice?" This is a serious concern, because the American public generally is not well educated about GMOs. When asked about genetically modified juice, many people express concern that orange juice modified with a gene from a pig might taste like pork. They do not understand that only one gene from the pig—one that has nothing to do with how pork tastes—is being used to modify the orange. Others believe that genetic modification is unsafe and could make us sick.

The GMO question is complex, and behind each GMO is a unique story that must be considered. For example, Monsanto Corporation developed HT-modified crops. These crops have been modified to resist an herbicide that Monsanto also sells and that farmers apply to their HT-modified crops in copious amounts to reduce the growth of weeds. The use of this GMO is thus linked to the application of a specific chemical to crops. Application of the chemical benefits Monsanto, the company that developed the GMO, and that chemical may have serious health consequences for the American public.

The story is different for the case of citrus greening. To fight this disease, citrus growers have been forced to dramatically increase the amount of pesticide they use to kill the Asian citrus psyllid, which carries the citrus-greening bacterium. Genetic modification would, ideally, give the tree resistance to the bacterium and reduce the amount of pesticide needed. And the orange growers may be out of business without the help of GMOs.

The battle will soon play out on the public stage as more and more orange trees succumb to citrus greening and Kress's GMO trees mature and begin to produce fruit. According to one Florida scientist, "People are either going to drink transgenic orange juice or they're going to drink apple juice."

that farmers and their families carry higher levels of glyphosate in their blood, and elevated glyphosate has been linked to increased risk for certain serious diseases, such as multiple myeloma, in humans.

What's worse, reports have shown that Roundup-resistant weeds are now growing on more than 10 million acres of farmland in the United States. So, even though farmers are using more than 35 million pounds of Roundup on crops each year, some weeds are growing anyway. In short, GMOs may not be harmless to humans or to the environment; they may not even be safe.

Many people, however, do not balk at the idea of corn that has been modified for insect or herbicide resistance. Transgenic corn is now an ingrained feature of food production in the United States. According to Figure 14.16, at least 80% of our corn crops are transgenic corn. And products made from corn, such as high fructose corn syrup, are used to produce a large percentage of processed foods such as baked goods, breakfast cereals, and sodas. According to food writer Michael Pollan, sweeteners that come from corn now make up more than 10% of the average diet in the United States. Whether we're aware of it or not, most of us regularly eat products derived from GMO plants.

Figure 14.17 Size comparison of an AquAdvantage® Salmon (background) versus a non-transgenic Atlantic salmon sibling (foreground) of the same age.

Courtesy of AquaBounty Technologies, Inc.

What about GMO animals? Would we want to eat a fish that's modified to grow faster? At Aqua Bounty Farms in Prince Edward Island, Canada, biotechnologists have created such fish. These fish have one gene from another fish spliced into their DNA. This splice allows these fish to produce growth hormones year-round, in contrast to non-transgenic fish that produce growth hormones only during warm-weather months. As a result, GMO fish grow twice as fast and are ready for market in a shorter amount of time.

Figure 14.17 shows a transgenic fish from AquaBounty (back) and a non-transgenic fish (front). These two fish are siblings that are exactly the same age, and both will reach the same mass at maturity. The transgenic fish, though, will get there in half the time. This transgenic fish is awaiting approval for sale in the United States.

A lively and heated debate over GMOs continues in the United States and in many other countries. Should we use GMOs, or shouldn't we? Can we feed the world without them? In this section of Chapter 14, we have presented a basic explanation of transgenic organisms and how they are developed. After reading this chapter, you can seek out opinions from known experts in the field and alternative viewpoints and evaluate them. Then, when you walk into the voting booth—or the supermarket—you will be able to make an informed decision based on your own opinions about these important questions.

QUESTION 14.7 According to Figure 14.16, for which food crop did farmers harvest about four times more HT-resistant variety in 2013 than they did in 2004?

ANSWER 14.7 According to Figure 14.16, HT corn was 20% of the total corn crop in 2004. In 2013, it was 85% of the total corn crop in the United States.

For more practice with problems like this one, check out end-of-chapter Question 33.

FOLLOW-UP QUESTION 14.7 Which of these statements about the two fish shown in Figure 14.17 is true? (a) They have the same mass. (b) They are the same age. (c) They both produce growth hormones year-round. (d) They both grow at the same rate.

14.4 Carbohydrates

There is an obesity epidemic among children and adults in the United States.

The United States has one of the highest obesity rates in the world. Roughly 32% of Americans are obese—a percentage exceeded only by Mexico, where 32.8% of the

BMI Ranges

<18.5	underweight
18.5–24.9	normal
25.0–29.9	overweight
≥ 30.0	obese

Weight: 190 lb (86 Kg)

Height: 65 inches (1.6 meters)

Sample BMI calculation

$$\text{BMI} = \frac{(190)(703)}{(65)^2} = 32$$

Figure 14.18 Sample BMI Calculation and BMI Ranges

adult population is obese. Obesity is determined from **body mass index (BMI)**, a calculation based on a person's height and weight:

$$\text{BMI} = \frac{(\textit{body mass in pounds})(703)}{(\textit{height in inches})^2}$$

Figure 14.18 provides an example using this BMI equation and shows how the result is interpreted. (In this book, we use metric system units almost exclusively; but in this chapter, because the people in the United States government use English units, we use inches and pounds in our discussion of BMI.) A person who is 65 inches tall (that is, 5 feet 5 inches tall) and weighs 190 pounds has a BMI of 32. According to Figure 14.17, that person is obese. When applied to children in the United States, BMI calculation reveals that about 17% are obese today compared to 5% in the 1960s. It's clear that dietary habits are changing in the wrong direction.

The U.S. Department of Agriculture (USDA) and the Department of Health and Human Services (HHS) puts out nutritional guidelines for U.S. citizens every five years in an effort to help citizens eat in a healthy way. For many years, these agencies summarized healthy eating guidelines with a food pyramid, such as the one shown in **Figure 14.19a**. To eat according to the pyramid, a person would eat the items at the top sparingly and those at the bottom liberally. At the top are the unhealthiest foods, such as butter and salt and candy, and at the bottom are bread products and rice and pasta.

In recent years, the food pyramid has been phased out and replaced with MyPlate. This change reflects evolving trends about what government nutritional experts think we should eat. The MyPlate diagram is shown in **Figure 14.19b**. If we compare MyPlate to the food pyramid, it's clear that the recommended portions of some food have shifted. Dairy is now represented by a glass of milk, but other dairy products, such cheese and eggs, can make up the protein section on the plate. Fruits and vegetables now take up half of the plate. All of the bread, rice, and pasta products, which made up the largest segment of the food pyramid, are now represented by the quadrant of the plate labeled "grains."

As MyPlate demonstrates, a gradual shift is taking place away from recommendations that we eat more bread

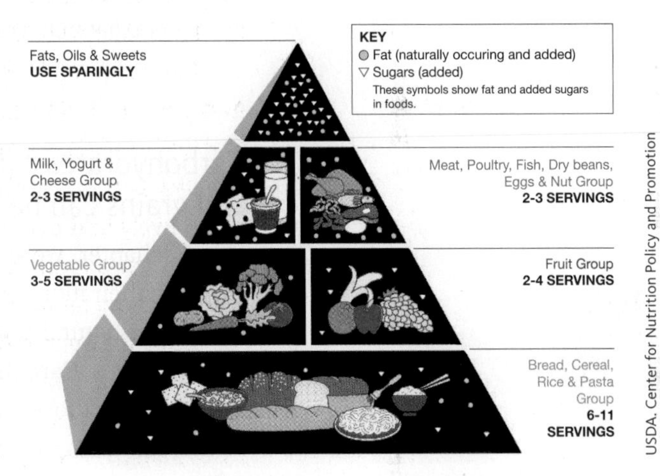

a

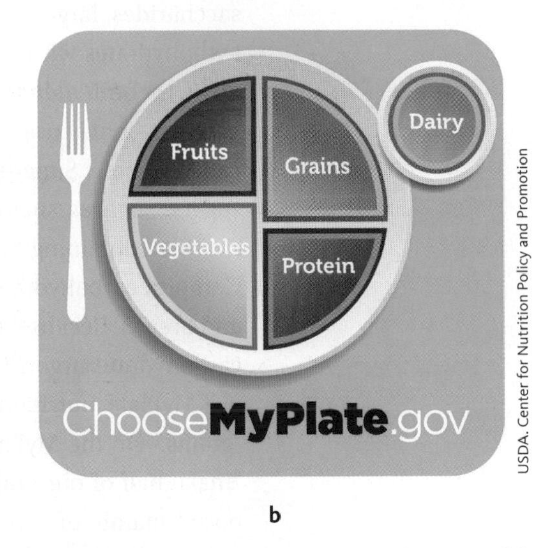

b

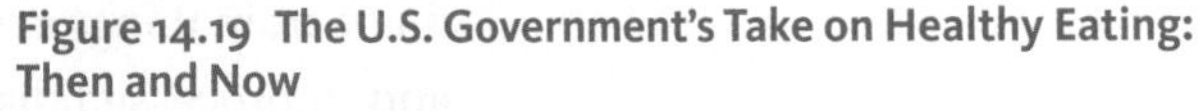

Figure 14.19 The U.S. Government's Take on Healthy Eating: Then and Now

(a) The food pyramid was promoted by the U.S. government in the 1990s. (b) The food pyramid has since been replaced by the MyPlate diagram, which illustrates what a healthy meal should look like.

and grain products, as the food pyramid suggested, and toward recommendations that we eat less. But even more important is the type of bread and grain products that we eat. In common usage, we refer to bread and rice and other products made from grains as being composed mainly of carbohydrates. The term *carbohydrate* is complicated, because it is the primary component of everything from cookies and chips to whole-grain bread and brown rice.

Which carbohydrates are complex? Which are healthy? In this section, we explore the different kinds of carbohydrate. To understand the diversity of carbohydrates and why some are healthy while others are not, we must understand what type of molecules they are.

QUESTION 14.8 Rank the following foods (from least to most) to show the amounts a person should eat according to MyPlate: sweets; grains; vegetables; fruit

ANSWER 14.8 sweets < fruit < grains< vegetables

For more practice with problems like this one, check out end-of-chapter Questions 8 and 9.

FOLLOW-UP QUESTION 14.8 Which of these foods is not composed primarily of carbohydrate? (a) hamburger meat (b) cookies (c) pasta (d) brown rice

Carbohydrates can be simple or complex, and grains can be refined or whole.

In this chapter, we've seen that proteins are polymers, as are DNA and RNA, although their monomers are different. Most **carbohydrates,** which are macronutrients, are natural polymers, too. **Saccharides** are the monomers that make up carbohydrates. Carbohydrates made up of only one monomer unit are *monosaccharides,* and those containing two monomer units are *disaccharides.* For molecules containing more monomers, we use the term *polysaccharide.* A polysaccharide can contain thousands of monomers. The name *carbohydrate* refers to all saccharides, large and small. We give a special name, **sugar** (or *simple sugar*), to carbohydrates with only one or two monomer units. So the term *sugar* is a synonym for both *monosaccharide* and *disaccharide.*

In everyday usage, the term *sugar* refers to table sugar, a specific disaccharide called *sucrose.* Super-sweet foods, such as candy, are composed mainly of simple sugars, such as sucrose. The term **complex carbohydrate** refers to a polysaccharide containing hundreds or thousands of individual saccharide monomers. Complex carbohydrates are the primary ingredient in foods such as potatoes, rice, and breads. Because this terminology can be confusing, **Figure 14.20** summarizes carbohydrate jargon.

MyPlate nutritional guidelines encourage us to eat carbohydrates that are "grains." On the MyPlate website, there's also a recommendation that we should "make half of our grains whole." The term "grains" refers to foods that are composed mainly of complex carbohydrates (polysaccharides). This includes foods made from things such as wheat, rice, barley, and cornmeal. **Whole grain** is intact and unprocessed, like the grain of wheat shown in **Figure 14.21**. A single whole grain includes parts called the bran, the germ, and the endosperm. When you eat

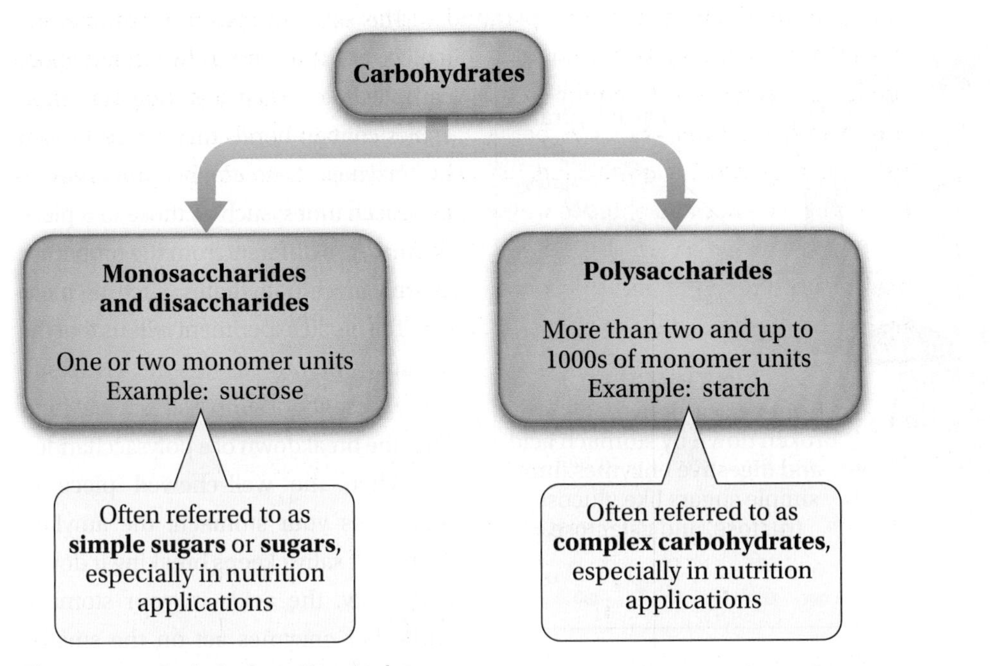

Figure 14.20 Carbohydrate Terminology

food that contains whole grain, such as whole wheat bread or oatmeal or brown rice, it includes all of these parts. The bran and the germ are nutritious and help maintain a healthy digestive tract.

When a grain is processed, only the endosperm remains. Foods made from the endosperm alone, called refined grains, have a softer texture and are more easily processed. They include white pasta, white bread, and many kinds of crackers, noodles, and tortillas. These foods lack the nutritional content of whole grains because they have no bran and no germ. Thus, the saying, "make half of our grains whole" is reminding us about the higher nutritional content of whole grains.

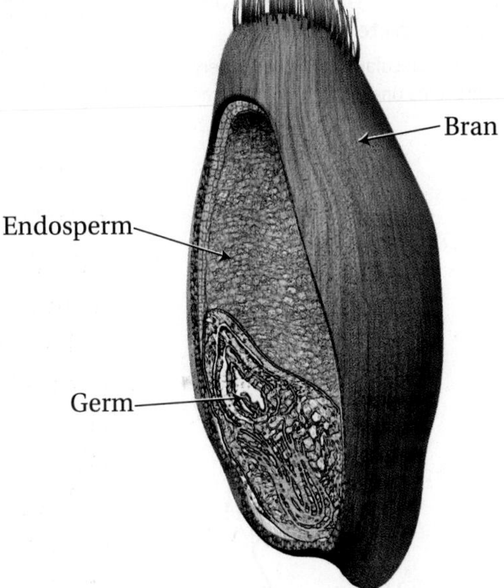

Figure 14.21 A Grain of Wheat

Digestion breaks down complex carbohydrates into monosaccharides and disaccharides.

When you take a bite of carbohydrate and begin to chew, the molecules in the food begin to break down as soon as the food hits the saliva in your mouth. Saliva contains an enzyme, called *amylase*, that takes polysaccharides and cuts them up into small pieces. Try this experiment the next time you're eating a piece of bread: Chew the bread much longer than you normally would, giving it ample opportunity to mix thoroughly with the saliva in your mouth. Do you notice any change in the flavor of the bread? Specifically, does it seem to get sweeter? If you do notice such a change, it's due to the action of amylase, which chops the long-chain polysaccharides in the bread into small pieces. Some of the small pieces produced are disaccharides. The disaccharide you taste after chewing bread is called maltose. This common simple sugar tastes sweet, as do many other disaccharides and most monosaccharides.

flashback

Recall from Section 14.1 that enzymes are proteins that perform chemical reactions. Salivary and digestive enzymes are special enzymes that break down various large molecules in foods into smaller molecules.

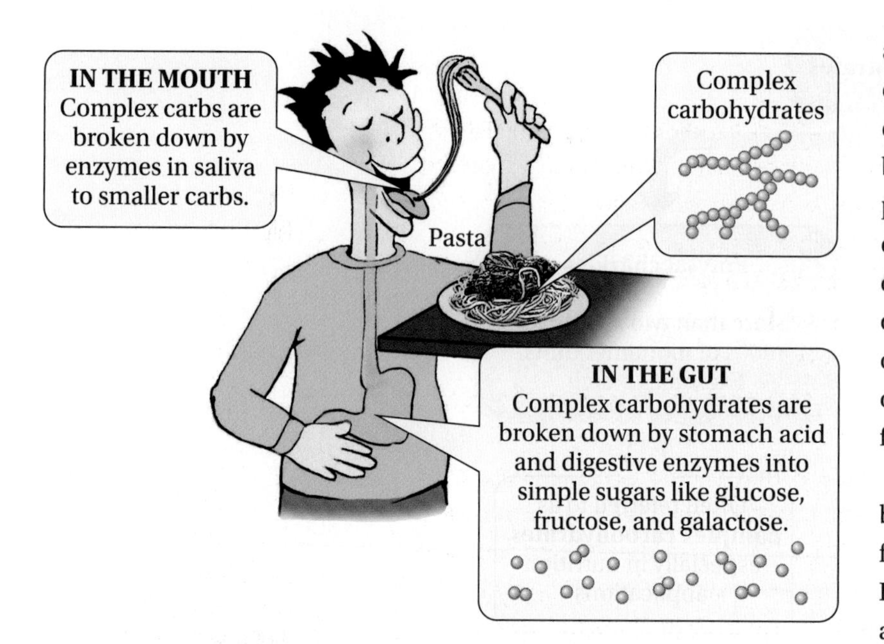
IN THE MOUTH
Complex carbs are broken down by enzymes in saliva to smaller carbs.
Complex carbohydrates
Pasta
IN THE GUT
Complex carbohydrates are broken down by stomach acid and digestive enzymes into simple sugars like glucose, fructose, and galactose.

Figure 14.22 Digesting a Complex Carbohydrate
In the molecular views, each sphere represents one saccharide monomer.

The simple experiment of chewing a piece of bread tells us two things about carbohydrates. First, carbohydrate molecules contain bonds that can be broken by enzymes. Second, the properties of polysaccharides, such as those in a piece of bread, are different from the properties of smaller carbohydrates, such as maltose. The bread experiment tells us that the quality we perceive as sweetness is a result of the presence of simple sugars that form from the breakdown of a polysaccharide.

When the well-chewed piece of bread hits your stomach, the amylase from your saliva keeps breaking it down. Eventually, the acids in your stomach and other enzymes act on the smaller and smaller remnants of the original polysaccharide. These enzymes are capable of breaking down disaccharides to monosaccharides. This process is illustrated in **Figure 14.22**.

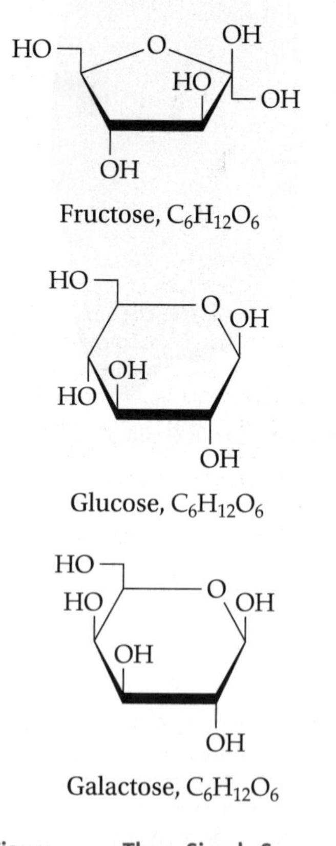
HO
O
OH
HO
OH
OH
Fructose, $C_6H_{12}O_6$
HO
O
OH
OH
HO
OH
Glucose, $C_6H_{12}O_6$
HO
HO
O
OH
OH
OH
Galactose, $C_6H_{12}O_6$

Figure 14.23 Three Simple Sugars
Shown here are the structures of the three most common monosaccharides derived from the human diet: fructose, glucose, and galactose.

When complex carbohydrates such as pasta break down into smaller saccharides during digestion, the most common monomers that result are the monosaccharides fructose (also known as *fruit sugar*), glucose, and galactose. These monosaccharides are typical carbohydrates; their structures are shown in **Figure 14.23**. Notice that the basic carbohydrate structure contains carbon, hydrogen, and oxygen. In fact, all three of the structures in this figure have the same molecular formula: $C_6H_{12}O_6$. If we rearrange this formula, we can write it as $C_6(H_2O)_6$. Thought of this way, it makes sense that the name for these molecules is *carbo- + -hydrate,* which loosely translates as "carbon and water." Today we know that the structure does not actually include water molecules, but the name has stuck nevertheless. During digestion, these tiny carbohydrates are small enough to be absorbed into the cells that line the intestinal walls.

Some carbohydrates are not broken down easily by the human digestive tract. For example, cellulose, which is found in the cell walls of fruits and vegetables, is not digested. Nevertheless, these carbohydrates play an important role in the human body. Carbohydrates that are not digested make up **dietary fiber**. Dietary fiber, such as the bran obtained from whole grain, is solid, undigested food that helps maintain the health of the digestive tract. It does this by keeping moisture in the gut and helping to move waste through to its final destination.

QUESTION 14.9 Put the following digestive processes in the order they occur: (a) Complex carbohydrates are broken down by acids and enzymes. (b) Monosaccharides and disaccharides are absorbed by intestinal walls. (c) Carbohydrates are chewed. (d) Salivary enzymes break down complex carbohydrates.

ANSWER 14.9 First (c), then (d), then (a), then (b).

For more practice with problems like this one, check out end-of-chapter Question 42.

FOLLOW-UP QUESTION 14.9 How did carbohydrates get their name?

14.5 Fats

Human beings store energy in the form of fat.

When you consider the way lifestyles in the United States have changed over the past few decades, it's not surprising that bathroom-scale readings are creeping higher and higher each year. Twenty-first-century Americans enjoy diversions that require little or no physical activity, and the Internet, videogames, and television make it easy to find sedentary entertainment that burns few calories. Couple this low level of activity with the convenience of fast-food, fat-laden meals, and it's no wonder about one-third of U.S. adults are obese.

The government invests substantial resources into collecting data on the weight of its citizens, but why should the government care about the weight of the U.S. public? Why are our bodies, and the mass of matter they contain, anybody's business but our own? The answer is that obesity takes a significant toll on our health care system. Estimates suggest that as much as 10% of our annual health care budget is spent on treating problems related directly to obesity and that the risk of death for obese people is 50 to 100% higher than for the normal-weight public.

For those who are gaining weight, the energy they consume each day in the form of food is greater than the energy they expend through activity in one day. Our bodies require a certain amount of energy to fuel a certain level of activity. To maintain a steady weight, a person should consume exactly that amount of energy in the form of food. The problem arises when a person consumes excess calories, and those extra calories most often take the form of fats and carbohydrates. When excess fat is consumed, the portion not used as fuel is stored in adipose (fatty) tissue. When excess carbohydrates are consumed, the leftovers are converted to fat. This fat also goes into long-term storage in adipose tissue.

Lipids are hydrophobic molecules that are not polymeric.

Most of the food we eat is some combination of three biomolecules, the macronutrients we introduced in Section 14.1: protein, carbohydrate, and fat. In addition, food contains small amounts of nucleic acids. Of those four major classes of biomolecules—nucleic acids, proteins, carbohydrates, and fats—fats are the only ones that are not polymers. That is, they are not made up of repeating units of a smaller molecule. The word *fat* is colloquial, and it is used to describe the subset of biological molecules called **lipids,** which are defined as biomolecules that dissolve readily in solvents that are not polar. Lipids are generally **hydrophobic,** or water-fearing, because they are not polar like water. Rather, they contain long hydrocarbon chains that are very nonpolar. **Figure 14.24** shows the differences between the structures of these two types of molecule.

flashback
Recall from Section 8.2 that water is a very polar substance because it contains a dipole. The presence of a dipole means that water has a region of negative charge and a region of positive charge. Unlike water, nonpolar substances do not have regions where charge is concentrated.

NONPOLAR

A hydrocarbon chain is nonpolar because it has no regions where positive or negative charge is concentrated.

POLAR

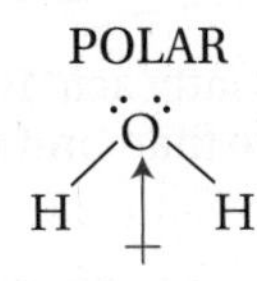

Water is a very polar molecule. The dipole drawn in red indicates that there is excess negative charge centered on the oxygen atom.

Figure 14.24 A Nonpolar Molecule versus a Polar Molecule

flashback

Recall from Section 5.4 that a functional group is a group of atoms within an organic molecule that has specific characteristics.

The simplest lipids are **fatty acids**. Every fatty acid molecule is composed of a long hydrocarbon chain with a carboxylic acid functional group (-COOH) attached at one end (**Figure 14.25**). The presence of a carboxylic acid group is usually a signal that a molecule is polar, because this functional group contains electronegative atoms that cause the electron density in the molecule to be lopsided. Conversely, molecules that contain only hydrocarbon groups are decidedly nonpolar. Why are fatty acids classified as nonpolar lipids if they have both polar and nonpolar character? It's because the hydrocarbon chains in fatty acids are so long that the nonpolar character of the chain dominates the polar character of the carboxylic acid functional group. Because the nonpolar character is dominant, fatty acids dissolve in nonpolar solvents and are classified as lipids.

Figure 14.25 The Structure of a Typical Fatty Acid

Palmitic acid is a fatty acid. Its carboxylic acid functional group is indicated in blue.

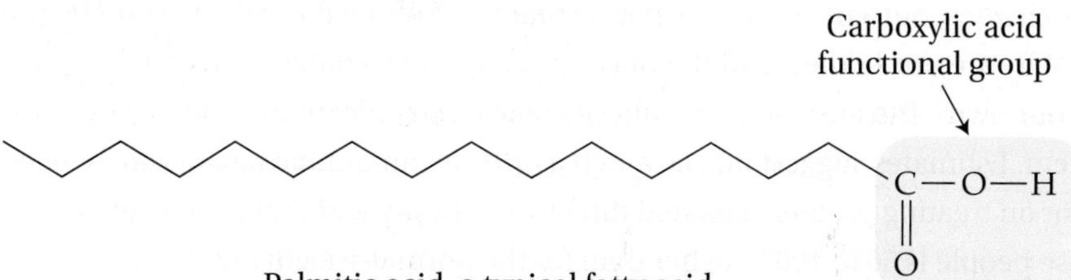

Palmitic acid, a typical fatty acid

Fatty acids can be classified as saturated or unsaturated. These identifiers tell us whether a fatty acid molecule contains any carbon–carbon double bonds, as shown in **Figure 14.26**. Fatty acids with only single bonds between carbon atoms are referred to as saturated; those that contain one or more carbon–carbon double bonds are unsaturated. Monounsaturated fatty acids contain one double bond, whereas polyunsaturated fatty acids contain more than one double bond.

Figure 14.26 Saturated and Unsaturated Fatty Acids

O

C—O—H

This fatty acid is *saturated* because it contains no double bonds.

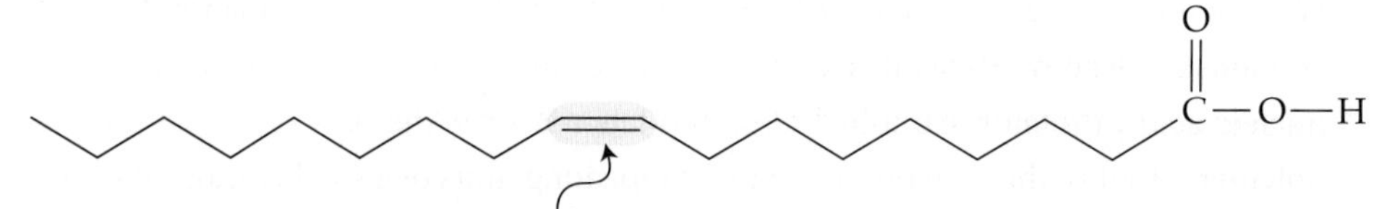

This fatty acid is *monounsaturated* because it contains one double bond.

O

C—O—H

This fatty acid is *polyunsaturated* because it contains more than one double bond.

Saturated and unsaturated fats are not equally healthy. According to the American Heart Association, eating foods that contain high levels of saturated fats significantly increases the risk of heart disease and stroke. This is why dietary

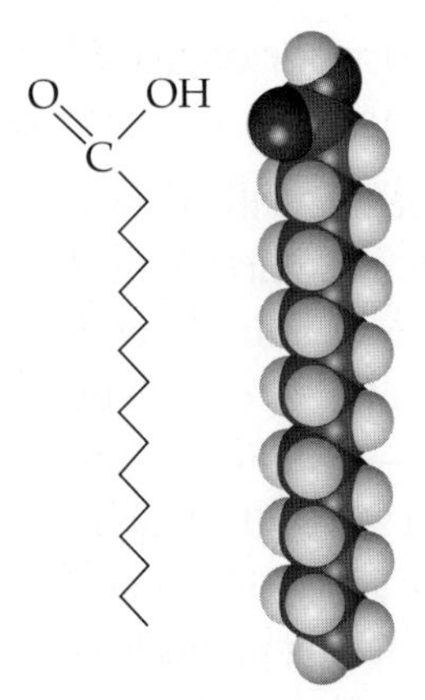

Palmitic acid
Carbons: 16
C=C bonds: 0
Melting point: 63°C

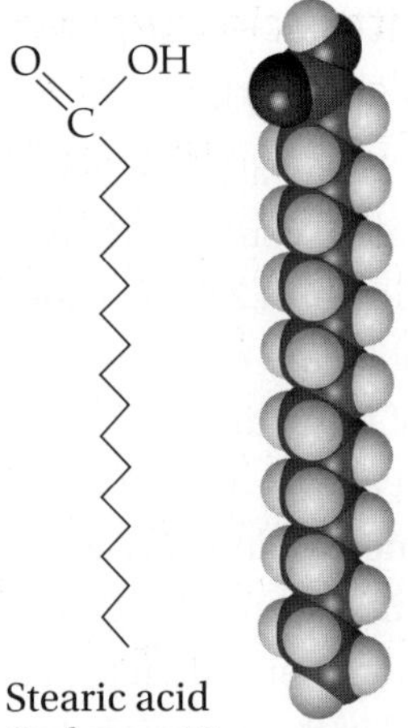

Stearic acid
Carbons: 18
C=C bonds: 0
Melting point: 70°C

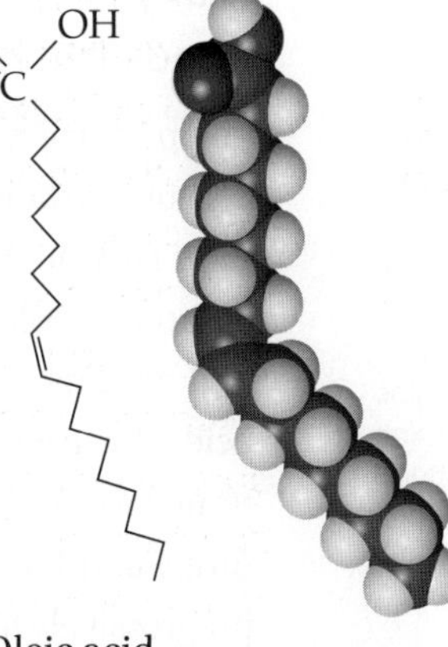

Oleic acid
Carbons: 18
C=C bonds: 1
Melting point: 13°C

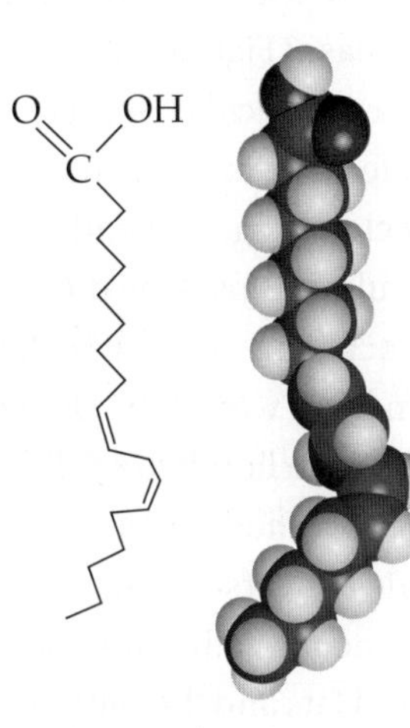

Linoleic acid
Carbons: 18
C=C bonds: 2
Melting point: -5°C

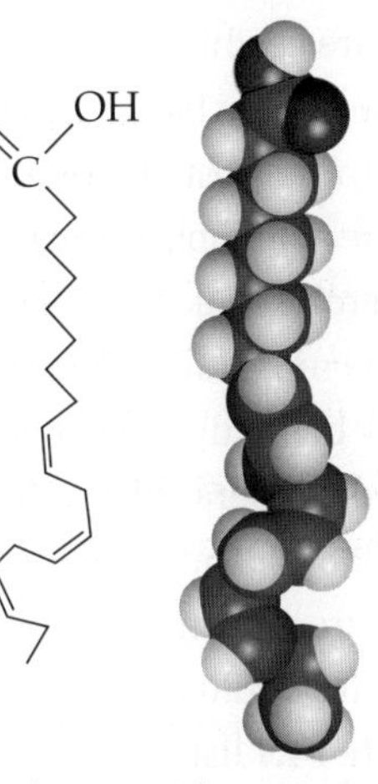

Linolenic acid
Carbons: 18
C=C bonds: 3
Melting point: -11°C

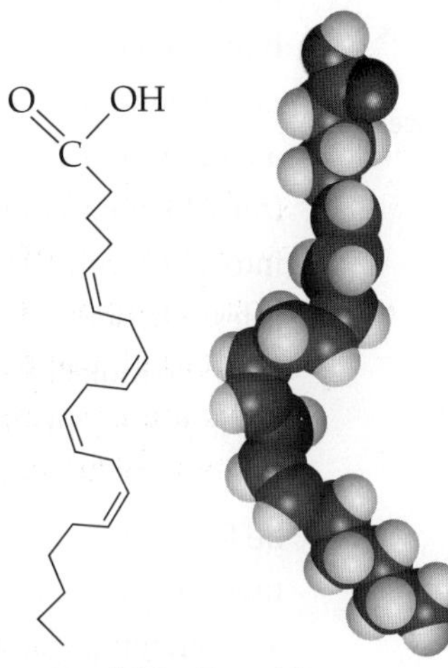

Arachidonic acid
Carbons: 20
C=C bonds: 4
Melting point: -50°C

Figure 14.27 Selected Fatty Acids

Line structures and space-filling models are shown for six fatty acid molecules.

guidelines commonly recommend choosing fats that contain fewer saturated fats and more unsaturated fats. **Figure 14.27** shows some of the fatty acids found in the human diet.

QUESTION 14.10 For each fatty acid in Figure 14.27, indicate whether it is saturated, monounsaturated, or polyunsaturated.

ANSWER 14.10 Palmitic and stearic acid are saturated. Oleic acid is monounsaturated, and the rest are polyunsaturated.

For more practice with problems like this one, check out end-of-chapter Question 43.

FOLLOW-UP QUESTION 14.10 True or false? Polyunsaturated fatty acids are fatty acids that do not contain a carboxylic acid functional group.

Many properties of fatty acids depend on the number of double bonds the fatty acids contain.

flashback

Recall from Section 3.5 that for any substance, the melting point and the freezing point are the same thing. We use two terms to describe liquids turning to solids (freezing) and solids turning to liquids (melting).

The physical properties of a fatty acid are dramatically affected by the number of carbon–carbon double bonds it contains. Consider the melting points for the common fatty acids listed in Figure 14.27. The figure also lists the number of carbon–carbon double bonds. Notice the correlation between number of double bonds and melting point. Both of the saturated fatty acids shown—palmitic and stearic acid—have a melting point above room temperature (about 25°C). The four unsaturated examples—oleic, linoleic, linolenic, and arachidonic acids—have one, two, three, and four double bonds, respectively. Note that the melting points become progressively lower as the number of double bonds increases. The melting point of every unsaturated fatty acid in the figure is below room temperature, which means they all are liquids at that temperature.

recurring theme in chemistry 7

Individual bonds between atoms dictate the properties and characteristics of the substances that contain them.

It's clear that as the number of double bonds in a fatty acid increases, the melting/freezing point decreases, but why does this correlation exist? How do double bonds affect the temperature at which a fatty acid changes phase? Figure 14.27 gives us the answer. As we can see in the figure, a double bond in a fatty acid chain causes a kink. When a fatty acid (or almost any other substance, for that matter) changes from the liquid phase to the solid phase, the molecules get closer together and pack into a firm solid. When there's a kink in the fatty acid molecule, this packing process is less efficient. Usually, the more kinks there are in a fatty acid, the more difficult it is for the molecules to pack in tightly against one another. In other words, the kinkier the fatty acid, the lower the temperature at which it converts from liquid to solid.

Butter, margarine, salad oils, and the other lipids we commonly consider to be food all contain a mixture of fatty acids. In nonscientific conversations, we call these mixtures oils if they are liquids at room temperature and fats if they are solids at room temperature. **Table 14.1** lists some common oils and fats and the saturated fatty acids they contain. As the numbers in the table indicate, the firmer the product is, the higher its percentage of saturated fatty acids. For example, the percentage of saturated fatty acids is lowest for the vegetable oils, which are completely liquefied at room temperature. This trend is the direct result of molecular structure. As we have seen throughout this book, the molecular structure of a substance often dictates its physical properties.

QUESTION 14.11 Have you ever noticed that some oily substances become solid when cooled in your refrigerator? This happens because, when an oil is cooled to below its melting point, it becomes a solid. Say your refrigerator is set at 3°C. Which of the unsaturated fatty acids in Figure 14.27 will be solids in this refrigerator?

Table 14.1 Fatty Acid Content of Selected Fats and Oils

Product	Saturated (%)	Monounsaturated (%)	Polyunsaturated (%)
Butter	68.2	27.8	4.0
Hard (stick) margarine	18.2	50.5	31.3
Soft (tub) margarine	17.1	36.5	46.4
Vegetable oil	10.8	20.4	68.8

ANSWER 14.11 Any fatty acid that has a melting point above the temperature of the refrigerator is a solid at this temperature. Therefore, oleic, stearic, and palmitic acid are solids in a refrigerator set to 3°C.

For more practice with problems like this one, check out end-of-chapter Question 46.

FOLLOW-UP QUESTION 14.11 Which of the unsaturated fatty acids in Figure 14.27 are solids in a freezer set at −22°C?

Fatty acids are stored in the form of triglycerides.

In the human body, lipids are stored in the form of *triacylglycerols,* also known as **triglycerides** or *TAGs.* Each TAG is made up of one molecule of glycerol, shown in **Figure 14.28a,** combined with three fatty acid molecules (**Figure 14.28b**). TAGs are the primary component of *adipose* cells, the cells in the body where fat is stored.

Adipose tissue serves several functions in an animal's body. For example, it acts as an insulator for animals that live in very cold environments. When energy is needed in the body, the links between glycerol and fatty acids in a TAG molecule can be broken, releasing the fatty acids, which can then be used as fuel for the body. Thus, TAGs represent an efficient way to package and store fatty acids.

In this chapter, we discussed the three major macronutrients: protein, carbohydrate, and fat. We also discussed the nutritional guidelines set forth by the U.S. government. These guidelines are based on well-established facts that link health

$CH_2-O-\overset{O}{\overset{\|}{C}}-R$

$CH-O-\overset{O}{\overset{\|}{C}}-R'$

$CH_2-O-\overset{O}{\overset{\|}{C}}-R''$

a

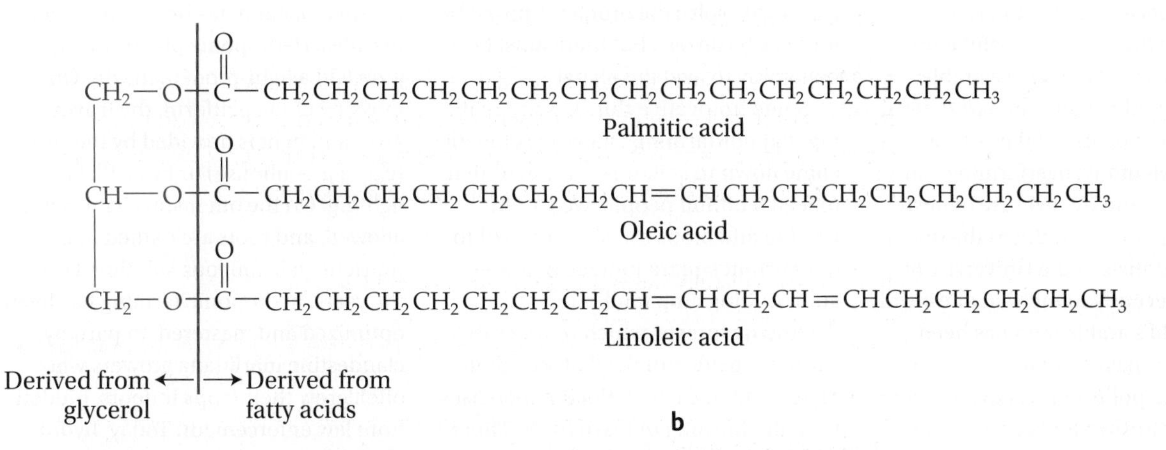

Figure 14.28 The Structure of Triglycerides (TAGs)

(a) This generic molecule shows how a triglyceride is a combination of one glycerol molecule (red) and three fatty acid molecules (black). (b) The triacylglyceride made up of palmitic acid, oleic acid, and linoleic acid.

to good nutrition. For example, it's a fact that whole-grain breads are healthier than breads that contain processed flour. And it's a fact that eating excess saturated fat can lead to heart disease. It's also a fact that obesity increases the risk for certain types of illness such as diabetes. But despite these available facts, in the United States, the percentage of overweight and obese adults and children continues to trend upward.

With so much information about nutrition available, how is it possible to have so many overweight and obese Americans? The answer lies, in part, in the confusion of good information with bad. The American public is bombarded with conflicting information about nutrition. For example, although MyPlate recommends eating reasonable portions of fats and carbohydrates, some people still believe that all fat is bad, or that all carbohydrate is bad.

We are also at the mercy of food companies that want us to eat their products. A shopping trip through a typical grocery store in the United States can be a confusing and frustrating experience. Boxes of food are covered with enticements. Even savvy consumers may question what they think they know in the face of an overwhelming number of choices and advertisements. Hopefully, the information in this chapter will help you, the reader, be a more informed consumer of nutrition information.

THEgreenBEAT • news about the environment, chapter fourteen edition

TOP STORY **Indoor Plantations: The Answer to the World's Food Shortage**

Most experts agree that our agricultural systems are stretched to their limits; we are turning to more extreme farming methods to feed the Earth's 7.5 billion people. For example, traditional farming practices called for arable land to have recovery time between crops; that is, time provided to replenish nutrients and high-quality soil to produce the next successful crop. Today, however, we can use arable land constantly, thanks in part to fertilizers, sources of artificial nutrition. It's a double whammy: overfarming causes soil erosion, and the overuse of pesticides and pollution reduces the quality of soil. Scientists at the University of Sheffield recently found that one-third of the world's arable land has been lost over the past 40 years due to soil erosion and pollution. Today, roughly 70% of our freshwater is used for agricultural purposes, and the effects of climate change will likely reduce the volume of freshwater that we can use to grow food.

Clearly, a different approach to farming is needed in order to feed 9 billion people by the year 2050. Some innovative proposals involve further tweaking our strained agricultural system. For example, some support the use of human waste, referred to as "night soil," for use as a natural fertilizer. This would potentially solve the problems posed by artificial fertilizers, but there must be a better way to feed the planet.

Some innovative thinkers are realizing that our farming challenges mostly come down to soil. It is estimated that to feed 9 billion people, we would need to add crops the size of Brazil to our current square footage of arable land. Adding crop land often means destroying forests, which reduces the carbon dioxide uptake that we count on to combat carbon dioxide increases from the burning of fossil fuels. Therefore, it is thought that if soil is the crux of the problem, that is, how much we have and its quality, then perhaps soil isn't needed to grow food. This radical idea may provide a sustainable, long-term solution to our food problem.

Soil warms and stabilizes the plant so that it can grow upright. It also circulates nutrients, including water and oxygen. What if we could provide these to a plant without soil? In fact, gardening without soil is known as *hydroponics* and has been around for decades. Hydroponic plants are often grown in a light-proof platform. On one side of the platform, the leaves grow and light is provided by the sun (e.g., a greenhouse) or by artificial lighting. On the other side, light is not allowed, and roots are bathed in a nutrient-rich aqueous solution. Over the years, hydroponic farming has been optimized and mastered, in part, by clandestine marijuana growers who often grow their crops indoors, hidden from law enforcement. Today, hydroponic farming is being used to grow crops in indoor plantations.

At Infinite Harvest in Lakewood, Colorado, hydroponic herbs are grown

THEgreenBEAT *continued*

in a 5,400 square foot facility under artificial lights using a fraction of the water that is used in traditional farming. Since there is no soil, there are no weeds or insects, eliminating herbicides and insecticides. Furthermore, because many genetically-modified (GMO) plants are engineered to keep insects away and to guarantee weed-free crops, there is no need to use GMO seeds. For this reason, most soil-free farms are naturally GMO-free. What's more, because there is no dirt, there is no need to wash your hydroponic lettuce three times to remove it! Perhaps the biggest advantage to this approach is that these farmers can ignore the changing of the seasons. They grow any crop any time, anywhere, regardless of the temperature outside.

Some modern farmers have taken their farms a step further. *Aeroponics* is the same as hydroponics except that a fine mist of nutrient-rich water is sprayed on the plant roots (see **Figure 14.29**). According to aeroponics advocates, this reduces the amount of water needed for a given crop by as much as 40% over hydroponics.

You may soon see aeroponic-based "grow trucks" at your neighborhood farmer's market (see **Figure 4.30**). The trailer part of these trucks are modified to become "aerofarms" with plants grown in vertical towers or in stackable beds to maximize space. Lighting can be powered with rooftop solar panels, and seeds can be germinated in cloth made from recycled plastic water bottles. These trucks can achieve up to 25 times the productivity of a traditional farm, and they are mobile. A grow truck can pull up to your local farmer's market and deliver plants that were harvested only minutes before. If you want the freshest, cleanest food available, skip your grocery store's produce aisle and find a local grow truck.

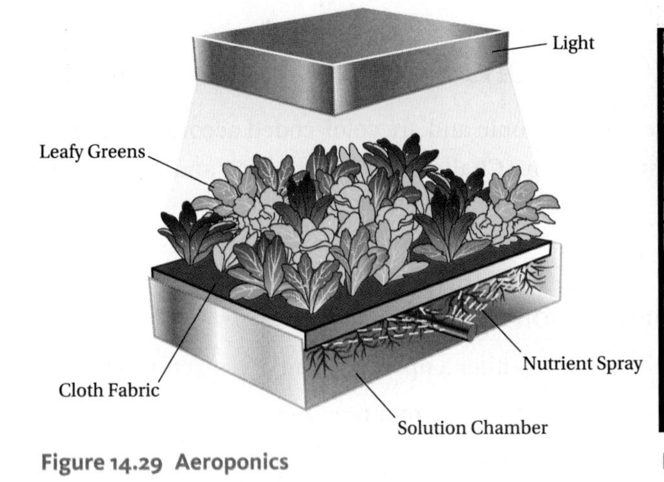

Figure 14.29 Aeroponics

Courtesy of Art Garden Growing Systems, LLC

Figure 4.30 A Grow Truck

Chapter 14 Themes and Questions

recurring themes in chemistry chapter 14

recurring theme 8 The Earth has a finite supply of raw materials that humans can use. (p. 418)

recurring theme 6 The structure of carbon-containing substances is predictable because each carbon atom almost always makes four bonds to other atoms. (p. 420)

recurring theme 7 Individual bonds between atoms dictate the properties and characteristics of the substances that contain them. (pp. 422, 442)

A list of the eight recurring themes in this book can be found on p. xv.

ANSWERS TO FOLLOW-UP QUESTIONS

Follow-Up Question 14.1: True. All three major macronutrients are required for a healthy diet. *Follow-Up Question 14.2:* Insects are low in fat and high in protein. They also require fewer natural resources than traditional protein sources do. *Follow-Up Question 14.3:* False. Vegans can get a variety of amino acid from plant-based food sources. *Follow-Up Question 14.4:* The groups on each neighboring chain interdigitate and hold the zigzagging chains together to strengthen the silk. *Follow-Up Question 14.5:* The complement to 3′-CGCGTTCGCGAATAG-5′ is 5′-GCGCAAGCGCTTATC-3′. *Follow-Up Question 14.6:* This RNA sequence codes for the amino acids Phe (full name: phenylalanine) and Lys (full name: lysine). *Follow-Up Question 14.7:* (b) They are the same age. *Follow-Up Question 14.8:* (a) hamburger meat. *Follow-Up Question 14.9:* Before the structure of carbohydrates was known, we considered them to be carbon plus water. Today, we know that their structures are like those shown in Figure 14.23 and they do not contain water molecules. *Follow-Up Question 14.10:* False. A polyunsaturated fatty acid has more than one double bond and a carboxylic acid functional group. *Follow-Up Question 14.11:* All but arachidonic acid will be solid. Arachidonic acid is a liquid at that temperature.

End-of-Chapter Questions

Questions are grouped according to topic and are color-coded according to difficulty level: **The Basic Questions***, **More Challenging Questions****, and **The Toughest Questions*****.

The Fundamentals of Nutrition

***1.** Label each of these substances as either a micronutrient or a macronutrient.

(a) vitamin B_{12} (b) zinc (c) lipid (d) protein

***2.** True or false? Because the United States is one of the wealthiest countries on Earth, few of its citizens are malnourished.

***3.** Which of the following is not a recommended part of a healthy plate of food, according to MyPlate?

(a) whole grains (b) protein (c) vegetables (d) butter

****4.** The three major macronutrients are carbohydrate, fat, and protein. For each of the following, indicate which of these macronutrients is a primary component of each food:

(a) bread (b) olive oil (c) chicken (d) tofu (e) eggs (f) pastry

****5.** The three major macronutrients are carbohydrate, fat, and protein. For each of the following, indicate which of these macronutrients is a primary component of each food:

(a) butter (b) beef (c) pretzels (d) oatmeal (e) whole milk (f) donuts

****6.** In your own words, explain what is meant by the term *whole grain*. Besides whole grains, what other type of grain is there? Which type of grain is healthier, and why?

**7. The MyPlate eating guidelines include the recommendation to "make half of your grains whole." What does this phrase mean? What does this recommendation tell us about the type of grain we should include in our diets?

***8. Critique the following meal using MyPlate and other nutritional information in this chapter as a guide: 50% of the plate contains beef, 25% of the plate contains white bread, 10% of the plate contains strawberries, and 15% of the plate contains green beans.

***9. Critique the following meal using MyPlate and other nutritional information in this chapter as a guide: 65% of the plate contains broccoli, 25% of the plate contains brown rice, and 10% of the plate contains blueberries.

Protein Structure

*10. True or false? Enzymes are a subclass of proteins that perform chemical reactions.

*11. Which of the following is NOT a role that proteins play in the human body?

(a) hormone (b) enzyme (c) oxygen carrier

(d) storage for genetic information (e) antibody

*12. Which of the following is NOT a place where you might find structural proteins?

(a) skin (b) hair (c) insect exoskeletons (d) oxygen carrier (e) horns

13. True or false? Nonessential amino acids are not needed by the human body.

14. What feature of the molecular structure of silk is responsible for its strength?

15. In your own words, explain why structural proteins are referred to as "nature's scaffolding."

***16. The amino acid sequence shown here contains several cysteine amino acids. Sketch structures that include each cysteine participating in one disulfide bond. How many structures can you draw? (Hint: You can change the chain shape as needed to link cysteine molecules together.)

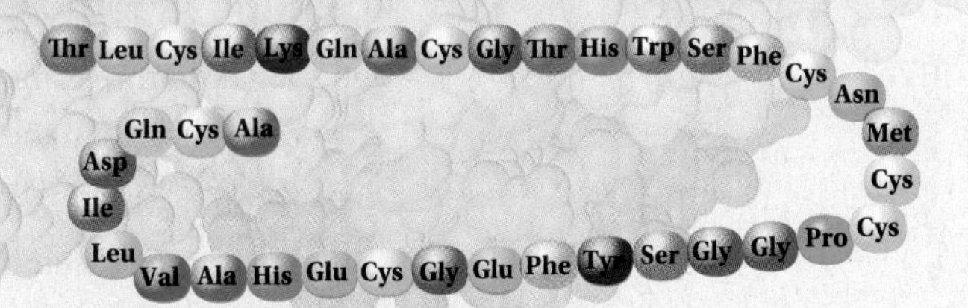

***17. The structural protein keratin, the protein that makes up human hair, has a special coiled structure. This same structure is found in other proteins. The coiled coil twists to the left, but the individual coils twist to the right. Explain why this arrangement makes a strong fiber.

How Proteins Are Made

*18. Is the nucleotide sequence 5′-CTCTAGCA-3′ found in a strand of DNA or RNA? How do you know?

*19. Write the RNA sequence complementary to the following strands:

(a) 5′-CGTAATTCGG-3′ (b) 3′-CGGTATTTACGG-5′

(c) 3′-CCTAATCGTTAA-5′

***20.** True or false? The instructions for making proteins are stored over the long term within the RNA in a cell.

****21.** In your own words, explain how proteins are made in the cell. What do the terms *transcription* and *translation* mean?

***22.** The 1997 film *GATTACA,* starring Ethan Hawke and Uma Thurman, addresses what might happen if genetic modification techniques were applied to human beings.

(a) In an RNA molecule, what sequence is complementary to GATTACA?

(b) Which dipeptide is produced by the first six letters of the film title, reading it left to right?

*****23.** One strand of a double-stranded piece of DNA has the sequence 5′-CAAAAAAATCCTTATCGGAAC-3′.

(a) Write the DNA sequence that is complementary to this sequence.

(b) Write the RNA sequence that is complementary to this sequence.

(c) Write the amino acid coded for by the RNA sequence of (b), starting with the 5′ end.

*****24.** We say that the genetic code is redundant because, for most amino acids, more than one triplet that codes for it. (a) Given this, write two RNA sequences that would result in the following peptide:

Gln-Trp-His-Gly-Pro-Gly-Pro-Arg-Ala-Thr

(b) Which amino acid(s) has/have only one triplet that codes for it/them?

AP Images/David Parry

*****25.** What point is this cartoon making about our biochemical makeup? Why are the men behind bars? Is this cartoon more or less relevant since the advent of genetic modification?

Genetic Modification and GMOs

***26.** What do the terms *GMO* and *transgenic organism* mean? Are there any differences between the meanings of these two terms?

***27.** If you take a gene from a pig and introduce it into the DNA of an orange, will the orange resemble the pig in every way?

***28.** What is glyphosate? What concerns have arisen over its use on crops?

****29.** Do GMOs pose a direct risk to human health? Do they pose an indirect risk to human health? If you answered yes to either of these questions, provide one example for each.

****30.** How is it possible for crops to withstand the application of Roundup? Why does Roundup kill only weeds (and not crop plants) in some cases?

****31.** What are two reasons that some people support the use of GMOs? What are two reasons that some people oppose the use of GMOs?

****32.** Name two GMO organisms that you could find in your grocery store.

*****33.** We discussed two specific genetic modifications that are used in modern agriculture. Outline each of them, and explain what purpose they serve. Which one is used in more major crops in the United States?

***34. In your own words, describe the reasons that we may someday see GMO orange juice in our grocery stores. According to the text, what are the pros and cons of developing GMO orange juice?

Carbohydrates

*35. Carbohydrate terminology can be confusing. In your own words, define the following terms: simple sugar, monosaccharide, polysaccharide, disaccharide, complex carbohydrate, dietary fiber.

*36. Describe the two major ways that complex carbohydrates are broken down in your digestive tract.

**37. What type of carbohydrate is not broken down by the digestive tract? What role does this type of carbohydrate play in a healthy diet?

**38. Each of the following foods is made up primarily of one type of biomolecule. Identify the biomolecule as a complex carbohydrate, simple sugar, or protein: (a) potato (b) peppermint candy (c) chicken breast (d) white bread

**39. Turn back to Figure 14.23. Do you think that sugar molecules are polar or nonpolar? Explain your answer.

***40. If you heat a carbohydrate—glucose, for example—to a very high temperature for a long time, you will notice steam escaping, and you will be left with a pile of black soot. Based on what you know about the origin of the word *carbohydrate,* account for these observations.

***41. In this chapter, we discuss the anatomy of a grain. Which parts of the grain are dietary fiber?

***42. If you chew on a piece of bread long enough, you will begin to notice a sweet taste in your mouth. In your own words, describe the source of that sweet taste and explain why it develops as you chew.

Lipids, Fatty Acids, and Trans Fat

*43. Identify the following fatty acids as saturated or unsaturated based on their shapes in the solid phase:

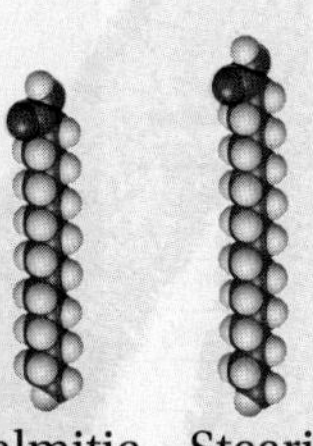
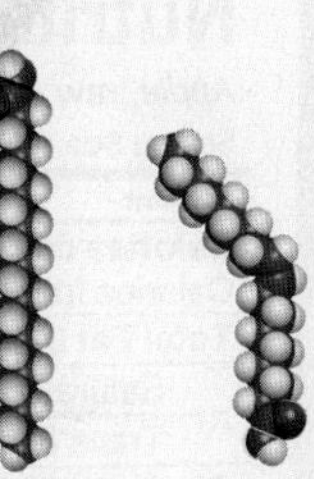
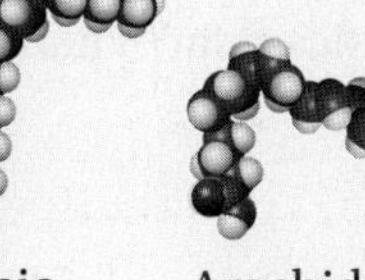
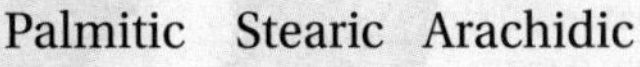
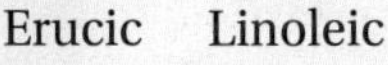

Palmitic Stearic Arachidic Oleic Erucic Linoleic Arachidonic Linolenic

*44. Of the four major classes of biomolecules—nucleic acids, protein, carbohydrate, and lipid—what do the first three have in common that is not true of lipids?

*45. Why are fats described as being hydrophobic? What happens when you mix a fat with water?

**46. True or false? Oils are composed mainly of monounsaturated and polyunsaturated fatty acids. Thus, at room temperature, they are usually in the liquid phase. Fats, however, contain a higher proportion of saturated fatty acids, so, at room temperature, they are usually in the solid phase.

**47. Fatty acids have polar and nonpolar characteristics. Explain which parts of a fatty acid have each of these traits. Why are fatty acids considered to be lipids even though they have some polar characteristics?

**48. In your own words, explain why some fats are solid and others are liquids. What dictates the phase of a fat?

**49. True or false? Trans fats are a by-product of the hydrogenation of unsaturated oils, which is a process intended to produce completely saturated hydrocarbon chains.

***50. Three fatty acids—labeled I, II, and III—have the following freezing points: 2°C, 12°C, and 37°C, respectively. State the phase for each at room temperature (about 25°C), at 0°C, and at 40°C. If the three fatty acids are mixed together at a temperature where they are all liquids, which one will solidify first? If the mixture is then cooled down to the point where they are all solids, which fatty acid will liquefy first as the mixture is warmed up again?

INTEGRATIVE QUESTIONS

*51. The following diagram shows a part of a biological polymer's sequence, and the same biomolecular sequence is compared among different animal species such as salamander, chicken, and human. Is this a sequence for protein, DNA, or RNA? How do you know?

```
Salanander    --ESFGSTSLGGLHSLNSVPPSPVVFLQTAPQLSPFIHHHG------QQVPYYLENEQGTYGAREAAPPAFYRPSSDLRRQTGR
Xenopus       --ETFGSSSLTGLHTLNNVPPSPVVFLAKLPQLSPFIHHHG------QQVPYYLESEQGTFAVREAAPPTFYRSSSDNRRQSGR
Chicken       --ESFGSSSLAGFHSLNNVPPSPVVFLQTAPQLSPFIHHHS------QQVPYYLENEQGSFGMREAAPPAFYRPSSDNRRHSIR
Alligator     --ESYGSSSLGGFHSLNNVPPSPVVFLQTAPQLSPFIHHHS------QQVPYYLENDQSGFGMREAAPSTFYRPGADSRRQSGR
Human         EAAAFGSNGLGGFPPLNSVSPSPLMLLHPPPQLSPFLQPHG------QQVPYYLENEPSGYTVREAGPPAFYRPNSDNRRQGGR
Zebrafish     YYPAPPDPHEEHLQTLGGGSSSPLMFAPSSPQLSPYLSHHGGHHTTPHQVSYYLDSSSSTVYRSSVVSSQQAAVGLCEELCSAT
```

*52. The apple shown below would be a healthy addition to a balanced diet. Based on the "food label" shown, what nutritional benefits does this apple have to offer? Is this apple a good source of protein? Does it contain significant fat? Vitamins? Fiber?

Nutrition Facts

Apple, raw

Serving Size 100g/3.5oz

Amount	% Daily Value
Calories 55	
Calories from Fat 1	
Total Fat 0.3 g	1%
Saturated Fat 0 g	0%
Trans Fat 0 g	
Cholesterol 0 mg	0%
Sodium 0 mg	0%
Carbohydrate 15 g	6%
Fiber 3 g	11%
Sugars 10 g	
Protein 0.2 g	
Vitamin A	1%
Vitamin C	8%
Calcium	1%
Iron	1%

Ekaterina_Minaeva/Shutterstock

***53. Consider the label, shown right, for flour made from crickets.

(a) Based on this label, what nutritional benefits does cricket flour have to offer?

(b) What mass of cricket flour, in grams, should you eat to get 100% of the recommended daily allowance of protein?

(c) What vitamin is present in high quantity in cricket flour? What mass, in grams, of cricket flour should you eat to get the 100% recommended daily allowance of that vitamin?

Nutrition Facts

Serving Size 2 Tbsp (17g)
Servings Per Container 26.8

Amount Per Serving

Calories 80.16

	% Daily Values*
Total Fat 4.08g	**6%**
Saturated Fat 0.04g	**0%**
Trans Fat 0.04g	
Cholesterol 38.76mg	**13%**
Potassium 187mg	**5%**
Sodium 52.7mg	**2%**
Total Carbohydrate 1.43g	**0%**
Dietary Fiber 1.02g	**4%**
Sugars 0.09g	
Protein 10g	**20%**

Calcium 1.9% • Iron 1.9%

Vitamin B12 68%

*Percent Daily Values are based on a 2,000 calorie diet. Your Daily Values may be higher or lower depending on your calorie needs.

	Calories	2,000	2,500
Total Fat	Less than	65g	80g
Sat Fat	Less than	20g	25g
Cholesterol	Less than	300mg	300mg
Sodium	Less than	2400mg	2400mg
Total Carbohydrate		300g	375g
Dietary Fiber		25g	30g

INGREDIENTS:
CRICKET POWDER (GRYLLODES SIGILLATUS)

***54. *Glypromate* is a drug that helps to protect brain tissue in patients with brain injuries and neurological diseases such as Alzheimer's disease and Parkinson's disease. Its line structure is shown below.

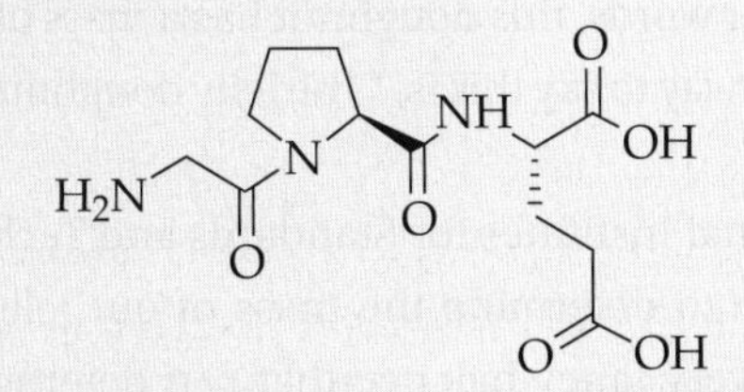

This drug is a very short protein, also known as a *peptide*. It includes three amino acids linked together by two peptide bonds.

(a) Locate the two peptide bonds in this structure.

(b) Refer back to our discussion of Kevlar in Section 13.3. What organic functional group does a peptide bond contain?

(c) Now refer back to Figure 14.4. Which three amino acids does this drug contain?

55. INTERNET RESEARCH QUESTION

Aeroponic farms are growing in number nationwide. Open your search engine and type in "aeroponic farm near me" or "aeroponic farms in [enter your state here]". Some of your hits may be stores that carry supplies for hydroponic and aeroponic farmers, and some of your hits should be aeroponic farms. Follow a link to one of the aeroponic farms in your area and try to answer the following questions.

(a) What crops do they grow? Are these crops grown out of season?

(b) What type of farming is done at this farm? Is it vertical, in towers, or perhaps stacked beds? Can you find a photograph of their growing facility?

(c) Does this farm use artificial lighting, or is it a greenhouse?

(d) How often are the crops at the farm harvested?

(e) Where can you buy their products? Do they sell their food at farmers' markets, restaurants, or grocery stores?

Appendix A

Working with Measured Numbers: Significant Figures

Measurements can be performed using very primitive or very sophisticated equipment. For this reason, it is important to be able to distinguish one type of measurement from the other by looking at the value measured. For example, let's say we want to determine the mass of a jelly doughnut. We use a kitchen scale that we bought at our local department store for $14.00 to perform this measurement. The balance tells us that the doughnut has a mass of 86 grams (g). By convention, the last number in a measurement—in this case, the number 6—is the one that is least reliable. This number could vary, perhaps by one unit in either direction, depending on the particular scale. In other words, this doughnut has a mass of 86 g, which could also be 87 g or 85 g. Another way to say this is, "The jelly doughnut has a mass of 86 g plus or minus 1 g ($\pm$1 g)."

If, instead, we go to the National Institutes for Standards and Technology and ask the scientists who work there to determine the mass of our jelly doughnut, they may be willing to use their very fancy balance that can determine mass to $\pm$0.00001 g. After a very careful measurement, the mass of our jelly doughnut is officially reported to be 85.32055 g. If we report the mass of our doughnut to others, we now have the right to report it with several decimal places, in contrast to the measurement we obtained from our kitchen balance, which only allows us to report mass with zero decimal places. We would not, for example, round off our very exact measurement to zero decimal places, because we earned those five decimal places by using an extremely sophisticated balance. In short, *the way we express a measured value tells us something about the equipment or method used to make that measurement.*

Because the last number of any measurement is estimated, we refer to it as *uncertain* and the other numbers preceding it as *certain*. Scientists usually keep track of certain and uncertain numbers by using *significant figures,* the total number of both in a measurement. To determine the number of significant figures in a measured value with a decimal point, we begin at the left end of the number and move to the right. When we reach the first nonzero digit (that is, the first digit that is not a zero), we stop. That nonzero digit and every digit to its right is considered to be a significant figure. Consider this example:

0.00009800600 seconds

Beginning at the left end of the number and moving right, we reach the first nonzero digit at the red number 9. This is the first significant figure. Each of the remaining numbers to its right is also a significant figure. Therefore, this number has seven significant figures. We also know that this measured value was measured using a more exact timing device than the value 0.00009800 seconds, which has only four significant figures.

Let's consider another set of measured values:

1.45 g has three significant figures

1.450 g has four significant figures

1.4500 g has five significant figures

1.45000 g has six significant figures

Clearly, the topmost mass measurement is the least exact, and the measurements become more exact as we move down the list. From this example, we can see that the zeroes at the ends of these measurements are an indicator of exactness. But what about numbers without a decimal point? What do we do with the zeroes in such numbers? This is an ambiguous situation, and we can count different numbers of significant figures. For example:

45,000 mg has 2, 3, 4, or 5 significant figures

300 mg has 1, 2, or 3 significant figures

BUT

300.00 mg definitely has five significant figures

We could report the number of significant figures in the measured number 45,000 mg as 2, 3, 4, or 5. However, in practice, we usually report the smallest number of significant figure because it is the most conservative number and it presumes the least exact value.

We have seen that significant figures are an indicator of exactness. The following example gives another reason that significant figures are important. In this equation, measured values are multiplied and divided:

$$\frac{281.00 \text{ g} \times 0.000\,66}{351 \text{ g}} = ?$$

If we punch this equation into our calculator, we will get 0.0005283761 g (different calculators may show more or fewer digits depending on how they are programmed). How do we know how many of these numbers we can include in the answer? The answer comes from the number of significant figures in each value in our calculation:

281.00 g has five significant figures

0.00066 g has two significant figures

351 g has three significant figures

Whenever we are performing a calculation where measured values are multiplied and divided, as in this example, *we limit the number of significant figures in the answer to the fewest found in the measured values in the calculation.* The smallest number of significant figures in this case is two (in 0.00066 g), and so our answer has two significant figures as well: 0.00053 g.

QUESTION A.1 How many significant figures are contained in each measured value? (a) 0.0900009 (b) 800 (c) 760.000 (d) 30,004

ANSWER A.1 (a) 6 (b) 1, 2, or 3 (c) 6 (d) 5

QUESTION A.2 Perform the following calculations:

(a) $\dfrac{5.600 \text{ kg} \times 0.000\,8}{7.9223 \text{ kg}} =$

(b) $\dfrac{4.555 \text{ L} \times 0.000\,551 \text{ L} \times 102.8 \text{ L}}{0.900\,9 \text{ L} \times 300.0 \text{ L}} =$

ANSWER A.2 (a) 0.0006 kg (b) 0.000955 L

QUESTION A.3 In the 1940s, the United States minted both steel pennies and copper pennies. You can distinguish the two types of pennies by their masses: the steel penny has a mass of 2.70 g, and the copper penny has a mass of 3.11 g. If you are going to buy a balance that you will use to distinguish one type of penny from the other, what is the minimum number of decimal places that your balance should display?

ANSWER A.3 If you round off both masses to one number only, both masses will be 3 g. If you round off to one decimal point, the masses will be 2.7 g and 3.1 g, respectively. Thus, a balance that determines mass to one decimal place should be adequate to distinguish one type of penny from another.

In cases where measured values are added or subtracted, there is a different rule for determining the number of significant figures in the answer. Consider this example:

$$2.335 \text{ s} + 33.20 \text{ s} + 6.1 \text{ s} = ?$$

When adding and subtracting, we rely on the number of decimal places that each number has, rather than the number of significant figures. In this calculation, the measured values have 3, 2, and 1 decimal place, respectively. Therefore the decimal places in the answer are limited to the fewest of these, which is 1. If we punch this calculation into our calculator, we will get the number 41.635 s. We must round this off so it has only one decimal place. Therefore, the correct answer is 41.6 s.

QUESTION A.4 Perform the following calculations:
(a) $3206.56 \text{ s} - 289.6554 \text{ s} =$
(b) $32.78900 \text{ g} + 25.544 \text{ g} + 6.32200 \text{ g} =$

ANSWER A.4 (a) Two decimal places are permitted in this answer: 2916.90 s.
(b) Three decimal places are permitted in this answer: 64.655 g.

When we perform calculations, we often use numbers that are not measured values. For example, let's say we measured the length of a car *very* exactly and found it to be 4670.15 millimeters. If we would like to express this measurement in meters, we must convert millimeters to meters by using the following conversion factor:

$$1000 \text{ mm} = 1 \text{ m}$$

If we set up a dimensional analysis, it looks like this:

$$(4670.15 \cancel{\text{mm}})\left(\frac{1 \text{ m}}{1000 \cancel{\text{mm}}}\right) = 4.670\,15 \text{ m}$$

It might seem like we just violated one of the rules we learned earlier. If the value 1 m has only one significant figure, then why isn't our answer limited to one significant figure? The answer is that the values 1 m and 1000 mm are not *measured* numbers. Rather, they are *exact numbers,* which are values that are either counted or defined by convention. In this case, the SI system of measurement defines 1000 mm as equal to exactly 1 m. Exact numbers such as these are considered to have an infinite number of significant figures. Therefore, in the calculation we just completed, the conversion factor does not limit the number of significant figures, but the measured value 4670.15 mm does.

The other instance where we see exact numbers is when things are counted. For example, if we use the conversion factor 12 eggs = 1 dozen, this is an exact number because the number 12 is counted. Because counted values increase in whole-number increments—we cannot count fractions of an egg—they are also considered to have an infinite number of significant figures and will not limit the number of significant figures in a calculation.

QUESTION A.5 Which of the following conversion factors limits the number of significant figures in a calculation?
(a) 1,000,000,000 μm = 1 km
(b) 58.93 g = the mass of 1 mole of Fe atoms
(c) 24,500 seats = capacity of the local football stadium
(d) 0.08350 nm = the average diameter of a human hair

ANSWER A.5 Choices (b) and (d) must be measured values. Therefore, the number of significant figures in these values must be considered when they are used in a calculation. Choice (a) is a metric unit definition, and choice (c) is a counted number.

Appendix B

Answers to Odd-Numbered End-of-Chapter Questions

Chapter 1

1. As stated in Section 1.1, a theory is based on the most recent experiments that have been performed. When new information comes to light and new experiments are performed, a new theory must be presented. In other words, a theory is always based on what is currently known, but it may change at any time in the future.

3. (a) While this may be a hypothesis, or a best guess, it is not a scientific hypothesis because it cannot be tested by experiment.

   (b) This is a scientific hypothesis. Experiments can be designed to test this hypothesis.

   (c) While this may be a hypothesis, or a best guess, it is not a scientific hypothesis because it cannot be tested by experiment.

   (d) This is a scientific hypothesis. Experiments can be designed to test this hypothesis.

5. Putting a red dot on the bread or the toast helps to identify the side of the bread or toast that is facing up before it is pushed off the table. The red dot is used only when the bread or toast is not buttered.

7. NASA reported that they had discovered a bacterium that was able to live in water laden with arsenic; they also reported that the organisms could actually live on arsenic. Furthermore, NASA suggested that the organisms could swap arsenic for phosphorus, a vital part of the genetic material in all living things. This story was big news because arsenic is very toxic to many living things and because the possibility had never before been raised that phosphorus in genetic material could be replaced with anything else. Scientists disproved the theory put forth by NASA by attempting to repeat the experiments that had been performed by NASA. Because other scientists were unable to repeat the results that NASA obtained, the theory was disproven.

9. (a) This should be 4.45×10^5. The number part must be a value between 1 and 10.

   (b) This one is correct.

   (c) This should be 6.700×10^{13}. The number part must be a value between 1 and 10.

   (d) This one is correct.

11. (a) 4.5×10^{10}

    (b) 4.4412×10^{-13}

    (c) 1.3445×10^{1}

    (d) 1.30000×10^{2}

13. In a glass about the size shown in Figure 1.6, there are about 7.9×10^{24} water molecules. When this value is not written in scientific notation, it becomes 7 900 000 000 000 000 000 000 000.

15. (a) 51

    (b) 0.000 000 001

    (c) 2 200 000 000 000 000 000

    (d) 9 100

17. $18 \text{ mountains} \times \frac{8800 \text{ trees}}{1 \text{ mountain}} \times \frac{125{,}000 \text{ leaves}}{1 \text{ tree}} = 2.0 \times 10^{10} \text{ leaves}$

19. (a) The letter μ represents the metric prefix *micro–*, and *mol* represents the unit mole; μmol is the abbreviation for micromole.

    (b) The letter *f* represents the metric prefix *femto–*, and *L* represents the unit liter; fL is the abbreviation for femtoliter.

    (c) The letter *M* represents the metric prefix *mega–*, and *m* represents the unit meter; Mm is the abbreviation for megameter.

    (d) The letter *d* represents the metric prefix *deci–*, and *s* represents the unit second; ds is the abbreviation for decisecond.

21. (a) An ounce is not a metric system base unit.

    (b) A gram is a base unit in the metric system.

    (c) A second is a base unit in the metric system.

    (d) A mile is not a metric system base unit.

    (e) A gallon is not a metric system base unit.

    (f) A minute is not a metric system base unit.

23. (a) According to Table 1.2, the metric prefix *kilo–* is 1000 times larger than the base unit. (The joule is a base unit for energy, although it is not listed in Table 1.2.) There are 1000 joules in 1 kilojoule.

    (b) According to Table 1.2, the metric prefix *mega–* is 10^{6} times larger than the base unit. Thus, there are 10^{6} bytes in 1 megabyte.

    (c) According to Table 1.2, the metric prefix *giga–* is 10^{9} times larger than the base unit. Because the meter is a base unit, this means there are 10^{9} meters in 1 gigameter.

25. (a) According to Table 1.2, the metric prefix *femto–* is 10^{15} times smaller than the base unit while the metric prefix *milli–* is 1000 times smaller than the base unit, so a millisecond is larger than a femtosecond.

    (b) According to Table 1.2, the metric prefix *kilo–* is 1000 times larger than the base unit while the metric prefix *nano–* is 10^{9} times smaller than the base unit, so a kiloliter is larger than a nanoliter.

(c) According to Table 1.2, the metric prefix *mega*– is 10^6 times larger than the base unit while the metric prefix *giga*– is 10^9 times larger than the base unit, so a gigabyte is larger than a megabyte.

27. (a) According to Table 1.2, the base unit for mass is the gram. A gram—or milligram or kilogram, for example—could be used to express the numerator of a density unit.

    (b) According to Table 1.2, the base unit for volume is the liter. The liter—or milliliter or microliter, for example—could be used to express the denominator.

    (c) If you know the density and you also know the mass, you know what volume the sample fills.

29. $140 \text{ kg} \times \dfrac{2.2 \text{ pounds}}{1 \text{ kg}} = 310 \text{ pounds}$

    Note that this answer was rounded using the rules for significant figures.

31. We can write a conversion factor and solve the problem using dimensional analysis.

$$1 \text{ drop} \times \frac{1 \text{ mL}}{20 \text{ drops}} = 0.05 \text{ mL}$$

33. Referring to Table 1.2, you can write conversion factors and convert each unit into grams. Then you can compare them to one another.

$$0.000\,045 \text{ kg} \times \frac{1000 \text{ g}}{1 \text{ kg}} = 0.045 \text{ g}$$

$$450{,}000 \text{ µg} \times \frac{1 \text{ g}}{10^6 \text{ µg}} = 0.45 \text{ g}$$

Based on these results, the first number listed, 4.5 g, is the largest of the three numbers.

35. (a) $20{,}000{,}000{,}000{,}000{,}000{,}000 \text{ fs} \times \dfrac{1 \text{ s}}{1{,}000{,}000{,}000{,}000{,}000 \text{ fs}} = 20{,}000 \text{ s}$

    (b) $20{,}000 \text{ s} \times \dfrac{1 \text{ min}}{60 \text{ s}} = 300 \text{ min}$ This is a very long time to wait on hold!

    (c) $300 \text{ min} \times \dfrac{1 \text{ hour}}{60 \text{ min}} \times \dfrac{1 \text{ day}}{24 \text{ hours}} \times \dfrac{1 \text{ year}}{365 \text{ days}} = 6 \times 10^{-4} \text{ years}$

    Note that the answers to (b) and (c) are rounded off according to the rules for significant figures.

37. Usually, kilometers are used to measure race distances. Using the information from Table 1.2 and dimensional analysis, you can convert this to a more reasonable metric unit.

$$42{,}195{,}000{,}000{,}000{,}000{,}000 \text{ fm} \times \frac{1 \text{ m}}{10^{15} \text{ fm}} \times \frac{1 \text{ km}}{1000 \text{ m}} = 42.195 \text{ km}$$

39. Using the conversion formula for Fahrenheit to Celsius, convert this temperature to degrees Celsius.

$$°C = \frac{5}{9}(°F - 32)$$

$$°C = \frac{5}{9}(375°F - 32)$$

$$°C = 191$$

41. By assigning two different temperatures in °F, you can use the conversion formula for Fahrenheit to Celsius to find the difference between the two temperatures. For example, you can assume the °F temperatures are 42°F and 72°F.

Recall that: $°C = \frac{5}{9}(°F - 32)$

First, find the Celsius temperature at 42°F.

$°C = \frac{5}{9}(42°F - 32)$

$°C = 5.6$

Next, find the Celsius temperature at 72°F.

$°C = \frac{5}{9}(72°F - 32)$

$°C = 22$

The difference between these temperatures is $22 - 5.6 = 16°C$.

Note that this answer is rounded using the rules of significant figures.

43. First, recall that volume can be found by multiplying length times width times height (or in this case, depth). The value 27,000 cm^3 is the volume of the bucket. (Remember from Section 1.4 that 1 cm^3 = 1 mL.) The product of 30 cm (length) × 30 cm (width) × D (which represents depth) is equal to the volume of the box.

$$\text{length} \times \text{width} \times \text{depth} = \text{volume}$$

$$30 \text{ cm} \times 30 \text{ cm} \times D = 27{,}000 \text{ cm}^3$$

Solving for D gives you 30 cm for the depth of the box.

45. $10 \text{ s} \times \frac{1{,}000{,}000{,}000{,}000{,}000 \text{ fs}}{1 \text{ s}} \times \frac{1 \text{ chemical process}}{200 \text{ fs}} = 5 \times 10^{13} \text{ chemical processes}$

47. If Joe's scale fluctuates by 7 pounds over a span of just 10 minutes, you could say that his scale is not very precise. A precise scale, by contrast, might change by 0.5 pound or 1 pound over 10 minutes.

49. (a) A balance should always be both accurate and precise. It should display values that are grouped closely together for the same measurement (precise) and it should display values that are as close as possible to the true value (accurate).

(b) If a balance is not calibrated properly, it can be inaccurate. A balance can be imprecise if, for example, someone leaves trace amounts of substance on the balance after use. Yes, a balance can be both inaccurate and imprecise.

(c) It can be precise but not accurate if it is working properly but has not been calibrated and so it reads each measurement as either too high or too low.

(d) It can be accurate but imprecise if it gives readings that are not tightly grouped around the true value.

51. The smallest volume is (a), 5.34×10^4 mL.

53. (a) Average temperatures are: 45.30°C (Rory); 51.43°C (Joaquin); 48.03°C (Michael).

(b) The equivalent Fahrenheit temperatures are: 113.5°F (Rory); 124.6°F (Joaquin); 118.4°F (Michael).

(c) Michael's data is most precise.

(d) Rory's data is most accurate.

55. *This has no answer because every student will create a unique answer.*

Chapter 2

1. (a) An element is defined by its atomic number—the number of protons its atoms contain. An atom containing 33 protons must be the element whose atomic number is 33, the element arsenic.

(b) In a neutral atom, the number of protons is the same as the number of electrons. So, each atom of arsenic contains 33 electrons.

3. An atom is the smallest unit of matter that cannot be broken down by chemical or physical means. The particles contained in the atom are the proton, the electron, and the neutron.

5. The atomic number of curium is 96, so an atom of curium will contain 96 protons and 96 electrons.

7. $(9.11 \times 10^{-31}\text{ kg}) \times \frac{1000\text{ g}}{1\text{ kg}} = 9.11 \times 10^{-28}\text{ g}$

$$(9.11 \times 10^{-31}\text{ kg}) \times \frac{1000\text{ g}}{1\text{ kg}} \times \frac{10^{15}\text{ fg}}{1\text{ g}} = 9.11 \times 10^{-13}\text{ fg}$$

9. This statement is false. The lanthanides and actinides are found at the bottom of the periodic table because they do not fit nicely into the periodic table. To fit them in where they belong, the periodic table would have to be much wider.

11. Element D is Ne (neon, its atomic number is 10, and it has 10 protons), and element C is Nd (neodymium, its atomic number is 60, and it has 60 protons).

13. (a) To determine this value, divide the mass of the proton by the mass of the electron.

$$\frac{1.67 \times 10^{-27}\text{ kg}}{9.11 \times 10^{-31}\text{ kg}} = 1830$$

(b) The neutron has the same mass as the proton, so you can use the same calculation as in (a) If you divide the mass of a neutron by the mass of an electron, you find that an electron is 1830 times lighter than a neutron. So, 1830 electrons have the same mass as 1 neutron.

(c) To convert from kilograms to picograms, use the following conversions.

$$(1.67 \times 10^{-27}\ \text{kg}) \times \frac{1000\ \text{g}}{1\ \text{kg}} \times \frac{10^{12}\ \text{pg}}{1\ \text{g}} = 1.67 \times 10^{-12}\ \text{pg}$$

$$(9.11 \times 10^{-31}\ \text{kg}) \times \frac{1000\ \text{g}}{1\ \text{kg}} \times \frac{10^{12}\ \text{pg}}{1\ \text{g}} = 9.11 \times 10^{-16}\ \text{pg}$$

15. This statement is true. In any atom, the number of protons must equal the number of electrons.

17. Gold, Au, has 79 protons. The noble gas with more protons than this is radon (Rn) with 86 protons.

19. If you look closely at the periodic table, you will note that after element 57, La, the next element is element 72, Hf. The elements from 58–71 are located in one of the two rows that are placed below the periodic table. If all of these elements were included in this row, the periodic table would be quite wide! Since tungsten is element 74, it comes after elements 58–71, and thus occupies the 60th box, if you don't take into account boxes 58–71.

21. Praseodymium has 59 protons and tin has 50 protons. The sum of these is 109, which is the number of protons found in meitnerium, named for Lise Meitner.

23. The number of protons plus the number of neutrons is the mass number, so if you subtract the number of neutrons from the mass number, you can find the number of protons (which is the same as the atomic number). Because $188 - 112 = 76$ and the atomic number of osmium (Os) is 76, the element must be osmium.

25. (a) The atomic number for tungsten is 74, which indicates that an atom of tungsten contains 74 protons and 74 electrons. Since the mass number is 180, and is the sum of the number of protons and neutrons, there are 106 neutrons ($180 - 74 = 106$).

    (b) The atomic number for chromium is 24, which indicates that an atom of chromium contains 24 protons and 24 electrons. Since the mass number is 54, and is the sum of the number of protons and neutrons, there are 30 neutrons ($54 - 24 = 30$).

    (c) The atomic number for hydrogen is 1, which indicates that an atom of hydrogen contains 1 proton and 1 electron. Since the mass number is 2, and is the sum of the number of protons and neutrons, there is 1 neutron ($2 - 1 = 1$).

    (d) The atomic number for palladium is 46, which indicates that an atom of palladium contains 46 protons and 46 electrons. Since the mass number is 104, and is the sum of the number of protons and neutrons, there are 58 neutrons ($104 - 46 = 58$).

27. (a) The mass number is the sum of the protons and neutrons, so the mass number for this element is 138. You can identify the element through the atomic number, which is equal to the number of protons: Ba.

    (b) Since the atomic number is the same as the atom in (a), it is the same element, but a different isotope of that element. In part (b), the isotope is barium-134 instead of barium-138.

29. This statement is false because elements can have different isotopes, which means that they have varying number of neutrons. While the number of protons for all the atoms in a sample of an element stays the same, the number of neutrons does not.

31. In a neutral atom, the number of electrons equals the number of protons. And, the number of protons is the same as the atomic number of that element. The element with atomic number 84 is polonium. So the element that is formed from radon is Po-218.

33. (a) The element with 82 electrons is lead, Pb.
    (b) The element with 79 electrons is gold, Au.
    (c) The element with 29 electrons is copper, Cu.
    (d) The element with 53 electrons is iodine, I.
    (e) The element with 17 electrons is chlorine, Cl.

35. Electrons can be described as elusive because, although it is possible to describe the general area where you might find an electron in an atom, it is not possible to say exactly where any one electron is at any particular moment in time.

37. Energy level 1 has 2 electrons. Energy level 2 has 8 electrons. Energy level 3 has 7 electrons.

39. An atom of chlorine has 17 electrons, an atom of nickel has 28 electrons, and an atom of silver has 47 electrons. The sum of these is 92, which means that the unknown atom is uranium.

41. (a) 0.000 000 000 000 000 000 000 000 000 000 911 kg

    (b) $1.0\text{ g} \times \frac{1\text{ kg}}{1000\text{ g}} \times \frac{1\text{ electron}}{9.11 \times 10^{-31}\text{ kg}} = 1.1 \times 10^{27}\text{ electrons}$

    (c) $\frac{9.11 \times 10^{-31}\text{ kg}}{1\text{ electron}} \times \frac{1000\text{ g}}{1\text{ kg}} = 9.11 \times 10^{-28}\text{ g per one electron}$

43. Ranked from highest energy to lowest energy: violet > blue > green > yellow > orange > red.

45. The term wavelength refers to the distance from the peak of one wave to the peak of the next wave in a wave of light.

47. This statement is false. When an atom absorbs energy, an electron in the atom moves from a lower energy level to a higher energy level.

49. $0.0015\text{ nm} \times \frac{1\text{ m}}{1{,}000{,}000{,}000\text{ nm}} \times \frac{1{,}000{,}000{,}000{,}000\text{ pm}}{1\text{ m}} = 1.5\text{ pm}$

51. (a) 62 protons
    (b) 91 neutrons
    (c) A wavelength of 0.001 nm corresponds to gamma radiation.

53. (a) 502.5 nm is dark purple, 546.1 nm is blue-purple, 577 nm and 579 nm are green, and 615.2 nm and 623.4 nm are green-yellow.
    (b) 502.5 nm is the highest-energy line. 623.4 nm is the lowest-energy line.
    (c) Mercury-200 possesses 80 protons and electrons as well as 120 neutrons.

55. (a) Hydrogen, fluorine, cesium, xenon, and mercury have 2, 1, 8, 9 and 7 naturally-occurring isotopes, respectively, as well as additional artificially-made isotopes. Xenon has the greatest number of naturally-occurring isotopes.

    (b) Fluorine has the fewest naturally-occurring isotopes.

    (c) Organesson has one known isotope with mass number 294 (Organnesson-294).

Chapter 3

1. The following are metals: (c) titanium, (d) cerium, and (e) erbium.
3. Carbon is typically found in organic compounds.
5. Nonmetals are described as (a) dull and (c) brittle.
7. From least abundant to most abundant: chlorine < sulfur < manganese < potassium.
9. $(85 \text{ kg}) \times \left(\frac{0.006}{100}\right) \times \left(\frac{\$0.13}{1 \text{ kg}}\right) = \$0.00066$

   Note that this answer was rounded according to the rules for significant figures.
11. Elements in the same group or family share common properties. Since (a) lithium, (d) rubidium, and (e) potassium are in the same family as sodium, you would expect them to share common properties with sodium.
13. There are more groups than periods on the periodic table.
15. A period on the periodic table refers to a horizontal row of the periodic table. A group and a family both refer to a column on the periodic table.
17. The following do not make up a significant part of the air we breathe: (b) fluorine, (c) xenon, and (e) helium.
19. (a) The noble gas with more protons than osmium is radon (Rn).

    (b) The halogen with the fewest number of electrons is fluorine (F).

    (c) The alkaline earth metal found in electronic devices is beryllium (Be).
21. (a) Glucose is a pure substance.

    (b) Filtered apple juice is a homogeneous mixture; a homogeneous mixture is completely uniform in appearance.

    (c) Unfiltered apple juice is a heterogeneous mixture; heterogeneous mixtures have boundaries where one component stops and another begins.

    (d) Chocolate chip ice cream is a heterogeneous mixture; heterogeneous mixtures have boundaries where one component stops and another begins.
23. (a) Gravel is a heterogeneous mixture; heterogeneous mixtures have boundaries where one component stops and another begins.

    (b) Granite is a heterogeneous mixture; heterogeneous mixtures have boundaries where one component stops and another begins.

    (c) Glass is a homogeneous mixture; a homogeneous mixture is completely uniform in appearance.

(d) Plywood is a heterogeneous mixture; heterogeneous mixtures have boundaries where one component stops and another begins.

(e) White glue is a homogeneous mixture; a homogeneous mixture is completely uniform in appearance.

25. Because compounds are substances made up of atoms of two or more elements, only (a) BF_3 and (d) C_6H_{12} are compounds. Answers (b) H_2 and (c) S_8 are elements because they are made up of atoms of the same element.

27. This statement is false. The element neon is made up of individual neon atoms unattached to one another, and does agree with this definition. However, the element gold is a solid that is made up of many, many gold atoms joined together. Another example is the element fluorine, which exists in nature in pairs of fluorine atoms that are connected together. Thus, the statement is only true for certain elements, not all elements.

29. (a) Aspirin is a compound—a substance made up of atoms of two or more elements.

    (b) Aspirin is a pure substance.

    (c) Vitamin C is a compound—a substance made up of atoms of two or more elements.

    (d) Vitamin C is a pure substance.

31. The most common form of phosphorus is white phosphorus and has the chemical formula P_4.

33. This substance cannot be absolutely pure, especially if the bottle is unlabeled and sitting around a chemistry laboratory. Any substance, even a very pure substance, that is in an unopened bottle contains impurities.

35. An element is the simplest unit of matter. Elements are listed on the periodic table. A compound is two or more elements combined in a simple, whole number ratio. A mixture is one or more elements or compounds combined together, but not in a whole number ratio. A pure substance is a single element or a single compound, not mixed with other elements or compounds. An element can be a pure substance. A compound can be a pure substance. A mixture cannot be a pure substance. A mixture cannot be an element. A mixture cannot be a compound. An element cannot be a compound. If two elements are combined chemically, they form a compound. When more than one compound is mixed with another compound, they form a mixture.

37. (a) In pentane, C_5H_{12}, there are 5 carbon atoms and 12 hydrogen atoms.

    (b) In hydrogen, H_2, there are 2 hydrogen atoms.

    (c) In hydrogen sulfide, H_2S, there are 3 atoms in total including 2 hydrogen atoms and 1 sulfur atom, for a total of 3 atoms.

39. In levonorgestrel, there are 21 carbon atoms, 28 hydrogen atoms, and 2 oxygen atoms, for a total of 51 atoms.

41. There are 10 carbon atoms and 10 hydrogen atoms. There are 2 oxygen atoms. Therefore, there are 22 atoms in total.

43. C_8H_6BrN

45. (a) Since a chemical change occurs when one substance is converted to another, gasoline burning in a car engine is a chemical change because the gasoline is converted into other substances, such as carbon dioxide and water.

    (b) Since snow melting only involves the conversion of water in its solid form to water in its liquid form, this is not a chemical change. It is instead a physical change.

    (c) The boiling of alcohol converts liquid alcohol into alcohol in the gas phase. Alcohol does not change into another substance, so this is a physical change, not a chemical change.

    (d) The explosion of dynamite converts the dynamite into other substances, so this is a chemical change.

47. Only answers (a), (c), and (d) are phase changes. The water in the lake (a) evaporates, which is a phase change from liquid to gas. (c) The candle undergoes a change of phase as the wax melts and changes from a solid to a liquid. The gasoline in (d) experiences a change of phase from liquid to gas.

49. When a substance undergoes a phase change, it is only changing from one phase to another. When water boils, the water molecules do not break apart into hydrogen and oxygen. They are still H_2O molecules.

51. (a) $C_{10}H_{14}N_2$

    (b) (i) freezing; (ii) deposition; (iii) sublimation

    (c) Liquid

53. (a) Nonmetals

    (b) 793 atoms

    (c) 32.3% are carbon atoms, 48.8% are hydrogen atoms, 8.2% are nitrogen atoms, 10.0% are oxygen atoms, and 0.7% are sulfur atoms. These sum to 100.0%, but this value will vary slightly depending on rounding.

55. *This has no answer because every student will create a unique answer.*

Chapter 4

1. An atom with an octet of electrons has 8 valence electrons. (b) Radon and (d) neon each have 8 valence electrons and therefore have an octet of electrons.

3. The total number of electrons is equal to the number of protons, which is the same as the atomic number. Valence electrons are the outermost electrons, and core electrons are the electrons between the valence electrons and the nucleus.

    (a) Sulfur has a total of 16 electrons. It has 6 valence electrons and 10 core electrons.

    (b) Fluorine has a total of 9 electrons. It has 7 valence electrons and 2 core electrons.

    (c) Lithium has a total of 3 electrons. It has 1 valence electron and 2 core electrons.

(d) Aluminum has a total of 13 electrons. It has 3 valence electrons and 10 core electrons.

5. A duet of electrons means that atom has 2 valence electrons. All except

(a) H^+ has 1 fewer electron than the neutral atom, so it has 0 valence electrons, not a duet of electrons.

(b) He has 2 valence electrons.

(c) Li^+ has 1 fewer electron than the neutral atom, and so it has 2 valence electrons (a duet).

(d) H^- has had 1 electron added to the neutral atom, it has a duet of electrons.

(e) Be^{2+} has 2 fewer electrons than the neutral atom, and so it has 2 valence electrons (a duet).

7. Each bromine atom has 7 valence electrons. When these 14 electrons (7 from each of the 2 bromine atoms) are distributed between the 2 atoms in the molecule, a total of 2 electrons are shared between the bromine atoms to form a single bond. The rest of the electrons exist as lone pairs, 3 pairs on each bromine atom. This arrangement provides each bromine atom with an octet of electrons in the molecule.

:Br:Br: or :Br—Br: or Br_2

9. Each dot in a Lewis dot diagram represents a valence electron.

:Ar: ·S· Na Ca ·N·

11. (a) Phosphorus has 5 valence electrons and 10 core electrons.

(b) Lithium has 1 valence electron and 2 core electrons.

(c) Helium has 2 valence electrons and 0 core electrons.

(d) Fluorine has 7 valence electrons and 2 core electrons.

13. (a) The sulfide ion has gained 2 electrons. It has 8 valence electrons and 10 core electrons. Its symbol is S^{2-}.

(b) The magnesium ion has lost 2 electrons. It has 0 valence electrons and 10 core electrons. Its symbol is Mg^{2+}.

(c) The potassium ion has lost 1 electron. It has 0 valence electrons and 18 core electrons. Its symbol is K^+.

(d) The chloride ion has gained 1 electron. It has 8 valence electrons and 10 core electrons. Its symbol is Cl^-.

15. If the core electrons from Cl (10), F (2), and Na (10) are added together, the sum is 22. And 22 × 3 equals 66, which is the number of protons in an atom of Dy, dysprosium.

17. Draw a Lewis dot diagram for ions of the following elements: astatine, oxygen, magnesium, lithium.

$:\ddot{\underset{..}{At}}:^-$ $:\ddot{\underset{..}{O}}:^{2-}$ Mg^{2+} Li^+

19. The Mg^+ ion is least likely to exist because alkaline earth metals form ions with a +2 charge.

21. (a) BaO (b) Rb_3N (c) $MgCl_2$

23. (a) The repeating unit is ABBB.

(b) The repeating unit is CD.

(c) The repeating unit is EEF.

(d) The repeating unit is GHH.

25. (a) Each of these ions has a specific number of negative charges in its anionic form because each has gained the number of electrons needed to have the same number as the nearest noble gas, which has an "ideal" number of electrons.

(b) The noble gases already have an ideal number of electrons, which makes each of the noble gases very stable. Neither Ne nor Ar would be likely to gain electrons.

27. True. As shown in Section 4.4, it is proper to use a line to connect atoms in a covalent bond.

29. Separately, the atoms could each gain one electron to form an iodide ion, I^-, which has a noble gas configuration of electrons. Alternatively, each iodine atom could share one of its 7 valence electrons can be shared in a single covalent bond to give each iodine atom an octet of electrons.

31. Each of the 3 hydrogen atoms has 1 valence electron. The nitrogen atom has 5 valence electrons. The 8 total electrons are distributed around the molecule so that two electrons are shared between each hydrogen and the nitrogen atom, each one a single bond. The two remaining electrons stay on the nitrogen atom.

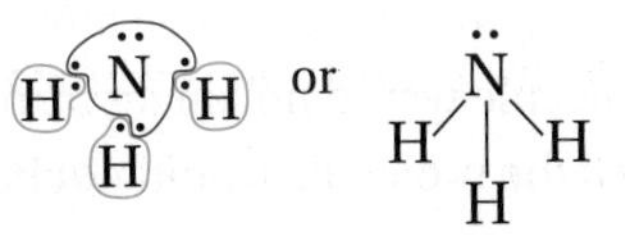

H:N:H (with H below) or H–N–H (with H below) or NH_3

33. (a) Argon is a noble gas and does not have any vacancies in its Lewis dot diagram.

(b) Nitrogen has three vacancies in its Lewis dot diagram.

(c) Carbon has four vacancies in its Lewis dot diagram.

(d) Phosphorus has three vacancies in its Lewis dot diagram.

35. (a) Xenon is a nonmetal.

(b) Copper is a metal.

(c) Manganese is a metal.

(d) Carbon is a nonmetal.

37. In metallic bonding, the electrons exist in a sea of electrons. They are free to move about the array of metal atoms and are not localized near a particular atom. The ability of electrons to move freely makes metals excellent conductors of heat and electricity.

39. (a) Na^+

(b) Ra^{2+}

(c) Li^+

(d) Ba^{2+}

41. Mo^{5+}: Based on the number of protons, the element must be molybdenum. Since it has 37 electrons, it must have lost 5 electrons, giving it a 5+ charge.

43. This statement is false. Atoms with the highest electronegativity tend to be located in the upper right corner of the periodic table.

45. The polarity of the bond between Cs and Cl is similar to that of the bond between Mg and O. These are both more polar than the bond between Pb and O. The Cl–Cl bond is the least polar, because there is no difference in electronegativity.

    (a) Cs and Cl is an ionic bond, because Cs is a metal and Cl is a nonmetal.

    (b) Cl and Cl form a nonpolar covalent bond.

    (c) Mg and O form an ionic bond, because Mg is a metal and O is a nonmetal.

    (d) O and Pb form an ionic bond, because Pb is a metal and O is a nonmetal.

47. The difference in electronegativity between K and F is 3.2, between O and N is 0.5 and between C and Co is 0.6.

    (a) The K–F bond is the most polar and is ionic.

    (b) The C–Co bond is less polar than K–F and more polar than O–N. This bond is covalent.

    (c) The O–N bond is the least polar and is covalent.

49. A molecule such as carbon dioxide can contain polar bonds and still be a nonpolar molecule as long as the lopsidedness that exists within one bond is canceled out by an equal amount of lopsidedness in another bond on the opposite side of the molecule. The carbon dioxide molecule can be thought of as a well-balanced seesaw. If one of the oxygen atoms were replaced by a different atom, then the molecule would no longer be nonpolar because the lopsidedness on each side of the molecule would not be equally balanced.

51. An ionic bond is an attraction between oppositely charged ions. A covalent bond results when electrons are shared in a molecule. Metallic bonds result from the attraction between metal nuclei and a delocalized sea of electrons. Covalent and metallic bonds involve sharing, but ionic bonds do not.

53. Protons have a +1 charge and are located within the nucleus. Neutrons have zero charge and are located within the nucleus. Electrons have a −1 charge and are located outside the nucleus. Electrons are added to, shared by, and taken away from atoms during chemical reactions. Protons and neutrons remain in the nucleus during chemical reactions.

55. In general, the melting points of column 2 metals are much greater than those in column 1 metals because column 2 metals make +2 ions, whereas column 1 metals make +1 ions. There is a stronger interaction between +2 metal ions and the sea of electrons compared to +1 metal ions.

Chapter 5

1. (a) An atom of fluorine is smaller than an atom of iodine. Recall from Section 5.1 that as you move down and to the left on the periodic table, the size of elements increases.

(b) An atom of tungsten is larger because it is below chromium on the periodic table.

(c) An atom of beryllium is smaller than an atom of calcium because it is higher up on the periodic table. Recall from Section 5.1 that as you move down the periodic table, the size of the elements increases.

3. The trend you expect to see is that oxygen is the smallest atom in this family. Since oxygen is a member of Period 2, just like carbon is, you expect it to share properties with other members of its group, but you also expect it to have unique characteristics based on the uniqueness principle and the fact that it is much smaller than other members of Group 16.

5. Looking at Figure 5.1, note that carbon measures 130 picometers. This is equivalent to

$$130 \text{ pm} \times \frac{1 \text{ m}}{10^{12} \text{ pm}} \times \frac{10^{10} \text{ angstrom}}{1 \text{ m}} = 1.3 \text{ angstroms}$$

$$130 \text{ pm} \times \frac{1 \text{ m}}{10^{12} \text{ pm}} \times \frac{10^{9} \text{ nm}}{1 \text{ m}} = 0.13 \text{ nm}$$

7. We learned from looking at carbon that atoms of an element tend to be smaller than expected when the atom contains few electrons. Because helium contains only two electrons, we would expect that a helium atom would have a radius that is much smaller than neon. It contains even fewer electrons than the elements in Period 2, indicating that it will be even smaller than we would expect in comparison to the elements of Period 2. A reasonable prediction for its radius is 25–35 pm. The measured value is 31 pm.

9. Here is one example that meets the criteria, shown as a Lewis dot diagram, a full structure, and a chemical formula.

```
  H   H  H  H  H                 H   H   H   H   H
  ..  .. .. .. ..                |   |   |   |   |
H:C ::C :C :C :C :H           H—C═C—C—C—C—H            C5H10
         .. .. ..                        |   |   |
         H  H  H                         H   H   H
```

C_5H_{10}

11. (a) The weakest carbon–carbon bonds are single bonds. According to Table 5.1, the bond energy of a carbon–carbon single bond is 0.624 attojoules. The single bond is the longest bond, with a length of 154 pm. The lower the energy of a bond, the weaker the bond is, and the further the distance between atoms in that bond.

(b) The strongest bond is the carbon–carbon triple bond, which has a bond energy of 1.39 attojoules. The length of this bond is 120 pm, which is much shorter than a single bond. The high bond energy for a triple bond indicates that the bond will be short and strong.

13. $750 \text{ kcal} \times \frac{4.184 \text{ kJ}}{1 \text{ kcal}} = 3100 \text{ kJ}$

Note that this answer is rounded according to the rules for significant figures.

15. 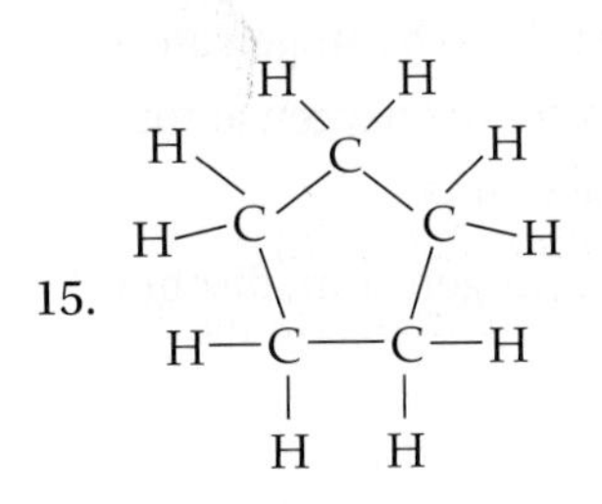

17. Graphite consists of layers of carbon atoms arranged in stacks of flat planes. Each carbon atom forms 4 bonds: 2 single bonds and 1 double bond. Graphene is a single layer of graphite.

19. You would expect the bonds between the carbon atoms within a sheet of graphite to be shorter than those in diamond. The carbon atoms in graphite form single and double bonds, while the carbon atoms in diamond form single bonds. The text states that double bonds are shorter than single bonds, so some of the bonds between carbon atoms in graphite are shorter than the bonds between carbon atoms in diamond.

21. In buckyballs, the carbon atoms are arranged in a set of 60 carbon atoms, and each set of 60 atoms is not strongly bonded to other carbon atoms. While the 60 atoms within the buckyball form a strong and stable structure, they do not form strong bonds to atoms outside of the buckyball. The lack of strong bonds between individual buckyballs, which is similar to the lack of strong bonds between layers in graphite, makes both of these substances good lubricants.

23. (a) This shows several sheets of graphite.

    (b) This is a layer of graphene.

    (c) This is a carbon nanotube.

    (d) This is a buckyball.

25. Bromine is a Period 4 element that has a tendency to make one covalent bond to other atoms.

27. The octet rule is the tendency of certain atoms to combine in order to obtain 8 electrons. The duet rule is the tendency of certain atoms to combine in order to obtain 2 electrons. The duet rule usually applies to hydrogen and to atoms of other small elements at the top of the periodic table. Hydrogen is too small to hold 8 electrons, but it combines with other atoms to have 2 electrons. Larger atoms, such as carbon, combine to obtain 8 electrons.

29. (a) $\cdot\underset{\cdot}{\ddot{\mathrm{P}}}\cdot$

    (b) Phosphorus usually forms 3 covalent bonds.

    (c) Nitrogen, a nonmetal, forms the same number of bonds in organic molecules.

31. :N:::N:

33. (a) This molecule disobeys the octet rule since the top carbons share only 6 electrons instead of 8. The molecule obeys the duet rule, since each hydrogen atom shares 2 electrons.

    (b) This molecule obeys both the octet and duet rules. Each hydrogen atom shares 2 electrons, each carbon atom shares 8 electrons, and the oxygen atom shares 8 electrons.

    (c) This molecule obeys both the octet and duet rules. Each hydrogen atom shares 2 electrons, each carbon atom shares 8 electrons, the oxygen atom shares 8 electrons, and the nitrogen atom shares 8 electrons.

    (d) This molecule obeys the duet rule. However, the nitrogen atom disobeys the octet rule because it shares only 6 electrons.

35. The chemical formula for cinnamaldehyde is C_9H_8O. The chemical formula for anethole is $C_{10}H_{12}O$.

37. The chemical formula for Prozac is $C_{17}H_{18}F_3NO$. The chemical formula for cocaine is $C_{17}H_{21}NO_4$.

39. Organic molecules can be very complex and contain many atoms, most of which are carbon and hydrogen. Line structures provide a simpler way to draw these molecules without having to draw a C or an H for each carbon or hydrogen atom, respectively.

41.

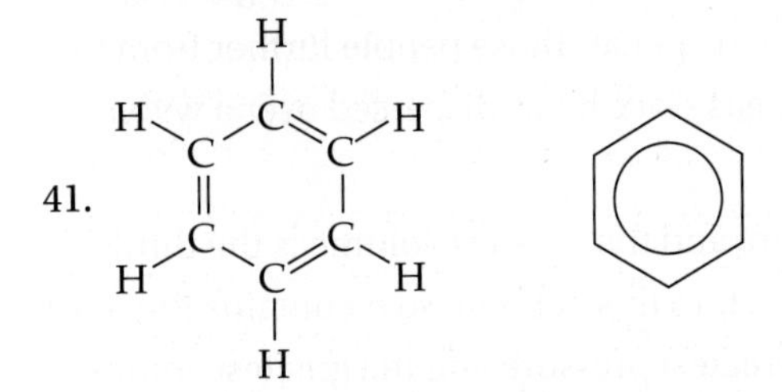

43. The ring inside the hexagon represents alternating single and double bonds.

45. The functional group in this organic molecule is a sulfide.

47. The term *functional group* refers to the part of an organic molecule that gives it its specific characteristics. In general, organic molecules are composed mostly of a hydrocarbon framework.

49. The hydrocarbon frameworks within most organic molecules contain carbon atoms that make four bonds to other atoms. Within this framework, the presence of a nitrogen atom with its lone pair of electrons and three bonds to other atoms causes any molecule containing an amine group to take on a unique shape.

51. (a) It contains two amine groups.

(b) It may have a foul odor.

(c) $C_6H_{16}N_2$

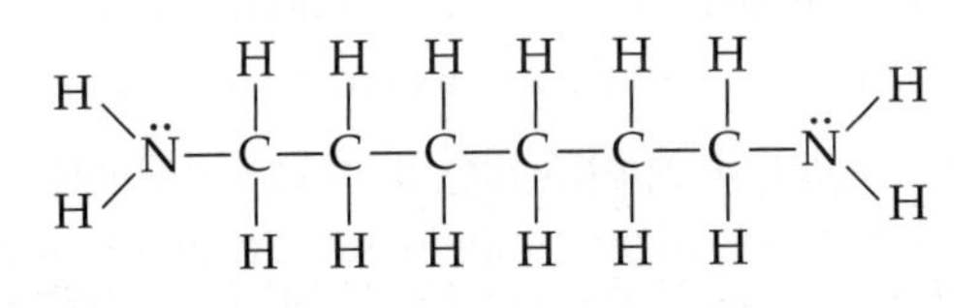

53. (a)

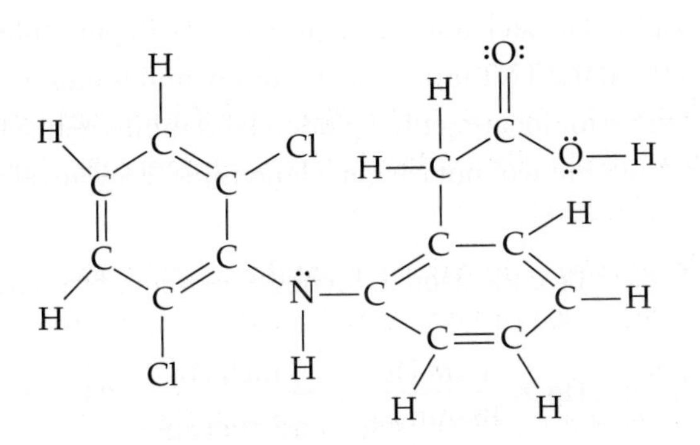

(b) $C_{14}H_{11}Cl_2NO_2$

(c) Yes, they obey the rules.

55. *This has no answer because every student will create a unique answer.*

Chapter 6

1. At higher temperatures, the particles of a gas move more quickly. Thus on a hot day the odor brought into a room by a very smelly person will travel more quickly than on a cold day.

3. This statement is true. All gases will mix completely under normal circumstances.

5. Gases disperse through the air from the source outward. Those people closest to the source are exposed to the highest concentration of noxious gases since the gases will have had less chance to disperse. Those people further from the gas source are exposed to less of the gas since it has dispersed over a wider area by the time it reaches them.

7. The balloon with the greatest pressure and the greatest volume is the third balloon. The most air has been blown into this balloon, so it contains the most gas particles and therefore has the greatest pressure and the greatest volume.

9. (a) A substance with a melting point of 234°C is a solid at room temperature. Since room temperature is about 25°C, the substance will not have achieved a high enough temperature to melt.

   (b) A substance that adopts the shape of the container but does not fill it is a liquid. A gas adopts the shape of its container but also completely fills it.

   (c) A substance that maintains its shape when removed from the box that held it would be a solid. If a liquid or gas were removed from its container, neither would maintain their original shape.

   (d) A substance that disappears immediately when you uncork the bottle that contains it is a gas.

11. (a) Mercury gas and (d) krypton gas are not found in significant quantities in air.

13. The types of matter you would find in air would be (a), (b), and (d). Diamond and granite are both dense solids and would not exist in a sample of air.

15. The tube will be smaller in Death Valley than in San Diego. The atmospheric pressure in Death Valley is higher than in San Diego, so the pressure from the outside air would press in on the tube and make it smaller.

17. Bees can be trained to seek out certain molecules. In particular, they can be trained to sniff out the TNT molecule in buried land mines, making them an excellent resource for locating unexploded land mines. Since bees are much smaller than dogs and will not trigger a land mine, they are also a safer choice for this task.

19. (a) 730 mm Hg can be converted to inches of mercury by converting from mm to inches:

$$730 \text{ mm Hg} \times \frac{1 \text{ cm Hg}}{10 \text{ mm Hg}} \times \frac{1 \text{ inch Hg}}{2.54 \text{ cm Hg}} = 29 \text{ inches Hg}$$

   (b) $730 \text{ mm Hg} \times \frac{1 \text{ atmosphere}}{760 \text{ mm Hg}} = 0.96 \text{ atmospheres}$

21. It is possible to express atmospheric pressure with a distance rather than a pressure unit because distance is used to determine how far the mercury in a barometer tube travels.

23. 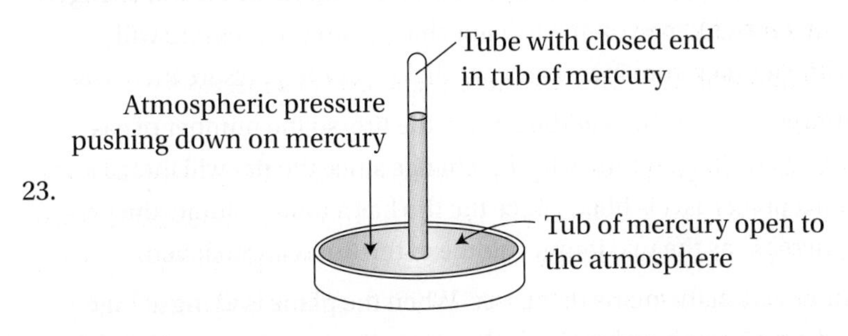

A barometer measures atmospheric pressure as follows: Pressure is the force exerted on a given surface area, so the higher the atmospheric pressure, or the pressure on the surface of the mercury in the open dish of the barometer, the higher up the tube the mercury moves. Therefore, the height of the mercury column in the tube is a way to measure atmospheric pressure.

25. Neither (b) nor (c) would be pressure units. Answer (b) does not use a square surface, and answer (c) does not use an area measurement, or square surface. Answer (a) is the only one that would be a unit of pressure. A unit of pounds per square centimeter is similar to pounds per square inch—both of these show a mass pushing on a square surface.

27. Water would not be a good choice for use in a barometer because the density of water is much, much lower than that of mercury. Mercury barometers are relatively short because the density of mercury is very high. A water barometer would need to be about 13 times taller than a mercury barometer because it is about 13 times less dense.

29. The molar volume of a gas is 22.4 L. The number of moles of gas is 1. The pressure of the gas is 1 atmosphere. The temperature of the gas is 273 K.

31. One-carat diamonds would likely not be counted using moles, because 1 mole of objects large enough to be seen by the naked eye is a huge quantity.

33. The variables that are the same are volume, since the volume of the box is fixed, and temperature, since both boxes are at standard temperature. The pressure inside the box will be different. It will be lower at high altitude. The number of moles of gas will be different, too. At high altitude, there will be fewer moles of gas in the box.

35. (a) For an experiment in which n and V change, you would need to use an insulated container (to keep the temperature constant) that could expand to keep the pressure constant.

    (b) For an experiment in which V and T change, you would need to use a container that was impermeable to keep the number of moles of gas the same, and a container that could expand to keep the pressure constant.

37. For the potatoes, the relationship between the temperature of the lake and the price of potatoes is inversely proportional. For the cherries, the relationship between the temperature of the lake and the price of cherries is directly proportional.

39. (a) *P* and *V* are inversely proportional to each other. The rest are directly proportional to one another.

41. If you add gas to a rigid, sealed box, the number of moles of gas will change. Since the box ix rigid, the volume will not change but the pressure will increase. The temperature increases very slightly as the pressure increases.

43. If a kid pumps up a tire, he is adding gas to the tire, so the number of gas particles changes. The volume will also change since the tire will increase in volume (to a point) as it is filled. Since the tire has a finite volume, the pressure will increase as the maximum volume of the tire is approached.

45. The pressure within the plane decreases. When the plane is flying at high altitudes, the cabin air is pressurized. The atmospheric pressure at 32,000 feet is much lower than at ground level, so the pressure must be increased inside the plane for passengers to have enough oxygen to breathe at this high altitude.

47. (a) The 2 billion particles have a longer mean free path than the 4 billion particles of gas, because there are fewer particles and they can move longer distances before colliding with one another.

    (b) The 4 billion particles in a 4-L container have the longer mean free path because this container is larger and the particles have more volume in which to move before colliding with one another.

49. The volume of 1 mole of gas at STP is 22.4 L. If 0.5 mole fits into half of the volume of STP (11.2 liters), then the gas must be at STP. So the pressure must be 760 mm Hg.

51. (a) Pressure will be greater than atmospheric pressure.

    (b) Pressure will be less than atmospheric pressure.

    (c) Pressure will be less than atmospheric pressure.

    (d) Pressure will be greater than atmospheric pressure.

53. 580 psi

55. *This has no answer because every student will create a unique answer.*

Chapter 7

1. A chemical equation is a notation chemists use to represent a chemical reaction. In a chemical reaction, a substance undergoes some type of rearrangement of its atoms and the electrons within those atoms. The chemical equation shows what these rearrangements are.

3. H_2O_2 is the reactant and exists as a liquid. H_2O and O_2 are the products. H_2O exists as a liquid, and O_2 exists as a gas.

5. There are 47 atoms in one molecule of ATP.

7. This statement is false. You do not need to have the same number of reactants and products in any balanced chemical equation. The law of conservation of mass states that any balanced chemical equation must have the same number of each type of atom on both the reactant side and the product side.

9. The value of *x* is 2.

11. The value of z is 6.

13. (a) The coefficient in front of HCl is not correct.

    (b) The coefficient in front of HCl should be changed to a 6.

    (c) Once this coefficient is changed, the equation is balanced because both sides of the equation now contain 2 Fe atoms, 3 O atoms, 6 H atoms, and 6 Cl atoms.

15. $4KClO_3\,(s) \rightarrow 3KClO_4\,(s) + KCl\,(s)$

17. (a) Yes, this equation is balanced.

    (b) There are 4 atoms of hydrogen on the reactant side. There are 2 atoms of sulfur on the reactant side. There are 6 atoms of oxygen on the reactant side.

    (c) There are 4 atoms of hydrogen on the product side. There are 2 atoms of sulfur on the product side. There are 6 atoms of oxygen on the product side.

    (d) In this case, 6 molecules of water are produced.

19. If you view chemical reactions from the atomic perspective, you think about what's happening in a chemical reaction at the atomic level. For instance, how are atoms and molecules rearranging themselves during a chemical reaction? If you think about the reaction from a laboratory perspective, you view the chemical reaction from a macroscopic viewpoint and record changes that are visible to the naked eye. For instance, can you see bubbles forming, the color changing, a solid forming, and so on?

21. (a) The perspective is at the atomic scale.

    (b) The perspective is at the laboratory scale.

23. Since 1 mole contains 6.02×10^{23} atoms, there are $5 \times 6.02 \times 10^{23}$ atoms in 5 moles. This is equal to 3.01×10^{24} atoms.

25. If you have 6 moles of acetic acid, then you have $6 \times 4 = 24$ moles of hydrogen atoms.

27. If 1 mole contains 6.02×10^{23} atoms, then 1.81×10^{24} atoms is equal to 3 moles of carbon.

$$(1.81 \times 10^{24} \text{ atoms}) \times \frac{1 \text{ mole}}{6.02 \times 10^{23} \text{ atoms}} = 3.01 \text{ moles}$$

29. The mass of 1 mole of galena is (Pb) 207.2 g/mol + (S) 32.01 g/mol = (PbS) 239.2 g/mol.

31. The molar mass of $Cu_2(CO_3)(OH)_2$ is (Cu) 2×63.55 g/mol + (C) 1×12.01 g/mol + (O) 5×16.00 g/mol + (H) 2×1.01 g/mol = 221.13 g/mol. The molar mass of $Cu_3(CO_3)_2(OH)_2$ is: (Cu) 3×63.55 g/mol + (C) 2×12.01 g/mol + (O) 8×16.00 g/mol + (H) 2×1.01 g/mol = 344.69 g/mol.

33. (a) The molar mass is (C) 33×12.01 g/mol + (F) 19.00 g/mol + (H) 35×1.01 g/mol + (N) 2×14.00 g/mol + (O) 5×16.00 g/mol = 558.68 g/mol

    (b) If the mass of 1 mole of Lipitor is 558.68 g, then the mass of 1 molecule is 9.28×10^{-22} g.

$$1 \text{ molecule} \times \frac{1 \text{ mole}}{6.02 \times 10^{23} \text{ molecules}} \times \frac{558.68 \text{ g}}{1 \text{ mole}} = 9.28 \times 10^{-22} \text{ g}$$

35. (a) Beryl most likely gets its name from beryllium, Be.

(b) The molar mass is (Be) 3×9.01 g/mol + (Al) 2×26.98 g/mol + (Si) 6×28.09 g/mol + (O) $18 \times 16.00 = 537.53$ g/mol.

37. This statement is true. A balanced chemical equation can be used to determine the amounts of substances needed in a chemical reaction.

39. Since 3 moles of oxygen are needed to react with 2 moles of hydrogen sulfide, then 9 moles of oxygen are needed to react with 6 moles of hydrogen sulfide.

41. To run a chemical reaction in the laboratory, you need information such as the amount of each reactant required and the best method to measure out those quantities. You may also need information such as the temperature and heating times required, if appropriate.

43. (a) $4NH_3 + 5O_2 \rightarrow 6H_2O + 4NO$

(b) Since you start with 4 moles of NH_3 and 5 moles of oxygen gas, you will form 4 moles of NO, based on the balanced chemical equation.

$$68.12 \text{ g } NH_3 \times \frac{1 \text{ mol } NH_3}{17.03 \text{ g } NH_3} = 4.000 \text{ moles } NH_3$$

$$160 \text{ g } O_2 \times \frac{1 \text{ mol } O_2}{32.0 \text{ g } O_2} = 5.0 \text{ moles } O_2$$

45. This reaction is endothermic. Since heat is included as a reactant, it means that heat was added to the reaction and was not released during the course of a reaction, as is the case for exothermic reactions.

47. The reaction shown here is exothermic. Based on the energy diagram, the products have less energy than the reactants, meaning energy has been released during the course of the reaction, which is indicative of an exothermic reaction.

49. (a) The diagram on the top shows the faster chemical reaction. The energy hill is lower for this reaction, so the reaction will proceed more quickly.

(b) Both reactions here are exothermic.

51. (a) Ca^{2+} (*aq*) + CO_3^{2-} (*aq*) $\rightarrow CaCO_3$ (*s*)

(b) Molar mass = 100.09 g/mole

53. (a) $x = 5$

(b) 316.8 g/mole

55. (a) $2H_2$ (*g*) + O_2 (*g*) $\rightarrow 2H_2O$ (*l*)

(b) *This has no answer because every student will create a unique answer.*

Chapter 8

1. Three things that contribute to anyone's water footprint include showering, flushing the toilet, and watering the lawn. But a large contribution to an individual's water footprint also comes from the food he or she eats—namely, the water to irrigate crops and the water needed to raise livestock.

3. Water in the gas phase is steam; water in the liquid phase is liquid water; and water in the solid phase is ice or snow.

5. About 3% of natural water is freshwater, and about 30% of that is groundwater.

7. Sublimation is the phase change in which a solid is converted directly to a gas.

9. Since ice is less dense than water, 100 milliliters of water is heavier than 100 milliliters of ice.

11. Songbirds directly or indirectly take up water from the locations through which they pass. Thanks to this, we can track their migration routes. The water in different locations contains different amounts of deuterium, so analyzing the isotopic ratio of the water in the birds' feathers should enable us to determine where they have been.

13. Liquid water is denser than ice. In ice, the H_2O molecules are held in a hexagonal lattice due to hydrogen bonds. The ordered structure of the lattice creates empty spaces between the water molecules that leads to the lower density of ice compared to water. The molecules in liquid water are closer together than the molecules in ice.

15. The most abundant is H_2O, followed by HDO, HTO, D_2O, and T_2O.

17. The two types of intermolecular forces mentioned in this chapter are hydrogen bonds and dipole–dipole interactions. Hydrogen bonds exist between molecules. For hydrogen bonding to exist, a hydrogen atom must be bonded to an N, O, or F atom within the molecule. In the hydrogen bonds between molecules, a hydrogen from one molecule interacts with an N, O, or F in another molecule. A dipole exists if the molecule is polar, so a dipole–dipole interaction will exist between polar molecules. A dipole–dipole interaction is weaker than a hydrogen bond interaction.

19. Hydrogen bond

21. Polar molecules interact with one another through dipole–dipole forces.

23. This statement is false. Both hydrogen bonds and dipole–dipole interactions are affected when liquid water changes to water vapor.

25. The container with the acetone is empty. Water molecules hydrogen bond to one another, so water will evaporate more slowly than acetone because its molecules are held together more strongly than those of acetone.

27.

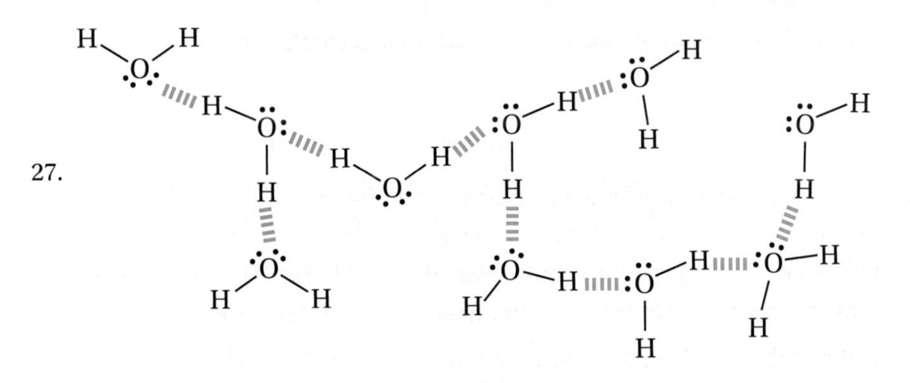

29. The bonds between C and O are polar, but only CO is a polar molecule. The CO_2 molecule has two C–O bonds, but the uneven electron density of one C–O bond is canceled out by the other C–O bond, so the molecule as a whole is not polar.

31. The melting of ice requires heat.

33. Evaporation is the phase transition from liquid to gas.

35. The boiling point of water decreases as altitude increases. It takes longer to cook foods at the lower temperature at which water boils at high altitude.

37. One phase transition mentioned is evaporation. Others are condensation, melting, freezing, and sublimation.

39. Cooking at high altitude will require more time and more cooking fuel due to the low temperatures at which water boils at high altitude. It would be best to pack foods that require little or no cooking.

41. Room temperature is 25°C; this substance is a gas at room temperature, since it boils at any temperature above -150°C.

43. Sublimation is not included on the heating curve in Figure 8.15.

45. If a substance has a boiling point of 30°C, then it is a gas at 300°C.

47. While the ice is being heated, the added heat goes to raising the temperature. Once the ice reaches 0°C, energy is used to break the bonds in the ice in order to make the transition to the liquid phase. Once the ice has made the transition to liquid water, heat that is then added causes the water to heat up.

49. If the line in the liquid phase for substance Z is half as steep as that for water, this means that substance Z absorbs less heat for each degree change in temperature it undergoes. Thus, substance Z is a less-efficient heat-storing fluid than water, and replacing water with Substance Z is *not* a good idea.

51. 920 grams

53. Mass of baby $=$ 131 oz $=$ 8.19 lb $=$ 3714 grams, and 0.78×3714 grams $=$ 2897 grams of water

55. *This has no answer because every student will create a unique answer.*

Chapter 9

1. (a) Mg^{2+} has the same number of electrons as neon.

   (b) Li^{+} has the same number of electrons as helium.

   (c) Br^{-} has the same number of electrons as krypton.

   (d) O^{2-} has the same number of electrons as neon.

3. (a) NaBr (b) Li_2S (c) $MgCl_2$ (d) CaO

5. Salts, or ionic solids, exist in very ordered lattice structures. The interactions between oppositely charged ions within lattice structures are very strong, which causes the melting points of ionic solids to be very high.

7. Rubidium iodide has the weakest bond strength because rubidium has a $+1$ charge and iodide has a -1 charge, and both are very large ions. Next is barium sulfide, since barium has a $+2$ charge and sulfide has a -2 charge; but both barium and sulfur are larger than magnesium and oxygen, so a bond between barium and sulfur is weaker than one between magnesium and

oxygen. Magnesium oxide has the highest bond strength because both ions have two charges and both ions are small.

9. (a) This is a molecule. (b) This is a simple salt.

   (c) This is a not a simple salt, because it contains the cyanide ion, CN^-, a polyatomic ion.

   (d) This is a simple salt.

   (e) This is a not a simple salt, because it contains the carbonate ion, CO_3^{2-}, a polyatomic ion.

11. Parentheses are used if there is more than one unit of a polyatomic ion in a salt. Since $Ca(NO_3)_2$ has two units of nitrate, parentheses are used here.

13. (a) K_3PO_4 (b) $KMnO_4$ (c) $Ca(HCO_3)_2$

    (d) NH_4Cl (e) $Ba(OH)_2$ (f) $RbClO_4$

15. (a) calcium nitrate; 164.10 g/mol (b) calcium sulfate; 136.15 g/mol

    (c) sodium phosphate; 163.94 g/mol (d) barium perchlorate; 336.20 g/mol

17. The pairs in (a) and (c) interact by ion–dipole interactions.

19. On a humid day, there will be water molecules in the air that can act to hydrate the ions in the salt. So, instead of just the salt in the salt shaker, there are salt molecules surrounded by water molecules. This makes your salt clumpy.

21. Natron was used in mummification because the goal was to remove water from the body and dehydrate it so that the body did not decompose. If water is removed from the body, bacteria will not grow.

23. Water is unusual for a number of reasons. First, it has a bent shape and is a very polar molecule. It also forms hydrogen bonds, which are a very strong type of intermolecular force.

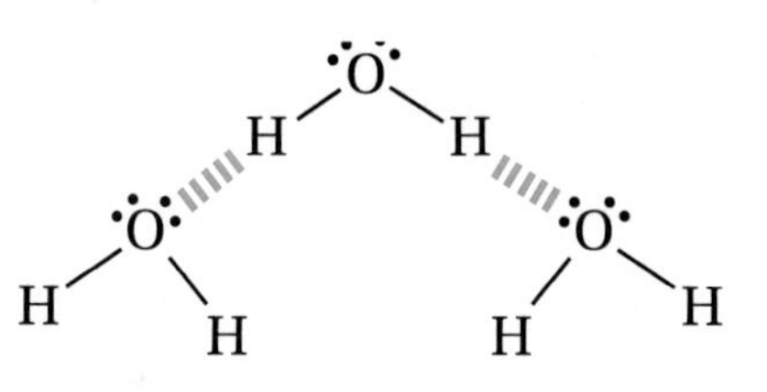

25. Salts are included in sports drinks so that an athlete can replace electrolytes lost during exercise—the salt composition of many sports drinks is very similar to the salt composition of human sweat. Sugar is mainly added to mask the taste of the electrolytes that have been added. Enough sugar is added to mask the taste, but not so much that the sugar causes the athlete to have stomach cramps.

27. (a) There is a higher concentration of magnesium ions in coconut milk than in human sweat.

    (b) The concentration of chloride ions is higher in human sweat than in Gatorade.

    (c) The concentration of sodium ions is the same in human sweat as it is in Gatorade.

    (d) The concentration of sugar in orange juice is higher than it is in human sweat.

    (e) The concentration of magnesium ions is lower in carrot juice than in human sweat.

29. (a) $MgCl_2 \longrightarrow Mg^{2+} + 2Cl^-$ (b) $LiI \longrightarrow Li^+ + I^-$

    (c) $Na_3N \longrightarrow 3\ Na^+ + N^{3-}$ (d) $KOH \longrightarrow K^+ + OH^-$

31. (a) Tomato juice contains lower sodium than Gatorade and no chloride ion, but it is high in potassium and low in sugar. It would work as an electrolyte replacement drink, but not as effectively as Gatorade. The no-salt tomato juice would be a better electrolyte replacement drink since the tomato juice containing salt is too high in sodium to be effective.

(b) The no-salt version of tomato juice is lower in sodium than Gatorade (tomato juice: 4.4 mM vs. Gatorade: 20 mM). Gatorade contains no magnesium, but tomato juice contains 4.5 mM and sweat contains 8 mM. Tomato juice contains no chloride, and Gatorade contains 27 mM. Tomato juice contains 56 mM potassium, while Gatorade contains 6.6 mM potassium. Tomato juice is 4.4% sugar, while Gatorade is 6%.

(c) The best choice for electrolyte replacement apart from Gatorade would be tomato juice because it is not too high in sugar and the electrolyte levels are the closest to Gatorade.

33. Begin by finding the number of moles of magnesium chloride from the number of moles and the molar mass. Divide this by the volume, in liters, to find molarity. The number of moles of chloride ion will be twice the number of moles of magnesium ion. To find the molarity of chloride ions, divide that number of moles by the volume, in liters.

$$87.5 \text{ g } MgCl_2 \times \frac{1 \text{ mol } MgCl_2}{95.2 \text{ g } MgCl_2} = 0.919 \text{ mol } MgCl_2$$

$$\frac{0.919 \text{ mol } Mg^{2+}}{5.0 \text{ L}} = 0.18 \text{ M } Mg^{2+}$$

$$2 \times 0.919 \text{ mol} = 1.84 \text{ mol } Cl^- \text{ and } \frac{1.84 \text{ mol } Cl^-}{5.0 \text{ L}} = 0.37 \text{ M } Cl^-$$

35. $\frac{6.50 \text{ g}}{3500 \text{ g}} \times 100\% = 0.186\%$

37. From least to most concentrated: 0.50 mM, 50 mM, 500 mM, 5 M, 50 M.

39. First, calculate the moles of silver nitrate from the mass of silver nitrate and the molar mass. Then divide the number of moles by the volume, in liters.

$$65.00 \text{ g } AgNO_3 \times \frac{1 \text{ mol } AgNO_3}{169.9 \text{ g } AgNO_3} = 0.3826 \text{ moles } AgNO_3$$

$$\frac{0.3826 \text{ moles } AgNO_3}{3.0 \text{ L}} = 0.13 \text{ M } AgNO_3$$

41. (a) $0.00650 \text{ M} \times \frac{1000 \text{ mM}}{1 \text{ M}} = 6.50 \text{ mM}$

(b) $0.044 \text{ M} \times \frac{1000 \text{ mM}}{1 \text{ M}} = 44 \text{ mM}$

(c) $0.887 \text{ M} \times \frac{1000 \text{ mM}}{1 \text{ M}} = 887 \text{ mM}$

(d) $4{,}000{,}000 \text{ nM} \times \frac{1 \text{ M}}{1 \times 10^9 \text{ nM}} \times \frac{1000 \text{ mM}}{1 \text{ M}} = 4 \text{ mM}$

43. To find the answer, solve for the mass of the solution, as shown below.

$$6.60\% = \frac{2.00 \text{ kg NaCl}}{? \text{ g solution}} \times 100\%$$

mass of solution = 30.3 kg

$$4.0 \text{ pints} \times \frac{0.5 \text{ quarts}}{1 \text{ pint}} \times \frac{1.00 \text{ L}}{1.0567 \text{ quarts}} = 1.9 \text{ L}$$

45. (a) $5.5 \text{ mM} \times \frac{1 \text{ M}}{1000 \text{ mM}} = 0.0055 \text{ M}$

$$\frac{0.0055 \text{ moles } Mg^{2+}}{1 \text{ L}} \times 1.9 \text{ L} = 0.010 \text{ moles } Mg^{2+}$$

$0.0055 \text{ M} \times 1.9 \text{ L} = 0.010 \text{ moles } Mg^{2+}$

(b) If magnesium were the only cation in the blood, then the molar concentration of chloride ions needed to balance the charge would be twice the concentration of magnesium ions, or 11.0 mM.

47. Osmosis is a process in which water moves across a semipermeable membrane from a solution that has a lower ion concentration to a solution that has a higher ion concentration. A concentration gradient is the gradual difference in concentration from one region of a solution to another. A semipermeable membrane is a membrane that allows certain substances to pass through it while blocking the passage of others.

49. As long as ions can flow from one arm of the tube into the other until the concentration is the same in both tubes, then the system will equilibrate to the same state as that shown in Figure 9.14.

51. (a) $CrCl_2$ (b) PbI_2 (c) $CoCl_2$

53. (a) 12% (b) 3.5% (c) 0.0018% (d) 0.053%

(e) 100% − (fat + cholesterol + sodium + carbohydrate + protein)
= 100% − (0% + 0.0018% + 0.053% + 12% + 3.5%) = 84% water

55. *This has no answer because every student will create a unique answer.*

Chapter 10

1. An acid is a substance that produces protons when dissolved in water. A base is a substance that produces hydroxide ions when dissolved in water. When looking at a chemical formula, you can recognize an acid because it contains a hydrogen atom, usually at the beginning of the formula (for example, HCl, H_2SO_4). You can recognize a base because it contains the hydroxide ion (OH^-) at the end of the formula [for example, NaOH, $Ca(OH)_2$].

3. At 25°C the concentration of hydrogen ion equals the concentration of hydroxide ions, so the pH is always equal to 7.00.

5. The bases are (a) and (b), and the acids are (c) and (d).

7. In 1 liter of pure water, there are 1.0×10^{-7} moles of H^+ ions. So in 8.0 liters of pure water, there are 8.0×10^{-7} moles of H^+ ions.

9. (a) This is a salt: $KClO_4(aq) \rightarrow K^+(aq) + ClO_4^-(aq)$

(b) This is a base: $RbOH(aq) \rightarrow Rb^+(aq) + OH^-(aq)$

(c) This is an acid: $H_2CrO_4(aq) \rightarrow 2H^+(aq) + CrO_4^{2-}(aq)$

(d) This is a salt: $Ca(CN)_2(aq) \rightarrow Ca^{2+}(aq) + 2CN^-(aq)$

(e) This is an acid: $H_3PO_4(aq) \rightarrow 3H^+(aq) + PO_4^{3-}(aq)$

(f) This is a salt: $Li_2CO_3(aq) \rightarrow 2Li^+(aq) + CO_3^{2-}(aq)$

11. (a) $HNO_3(aq) \rightarrow H^+(aq) + NO_3^-(aq)$

(b) The pH is the negative logarithm of the molar concentration of protons. Thus, the pH of a 4.5×10^{-4} M solution of nitric acid is 3.35.

13. (a) If you dissolve some NaOH in water, you would expect the pOH to be less than 7.

(b) If you dissolve some NaOH in water, you would expect the pH to be greater than 7.

15. (a) pH = 6.000 (b) pH = 2.000

17. (a) pOH = 10.0; acidic (b) pOH = 5.9; basic

(c) pOH = 3.0; basic (d) pOH = 12.7; acidic

19. Fruit Juice A will be more acidic. The difference in pH will be 2.0 pH units.

21. To find the pH, enter the concentration in your calculator and then find the logarithm and change the sign. The pH = 1.903.

23. To adjust the pH on the acidic side of the lake, acid could be carefully added to the lake. This would gradually reduce the pH in the lake.

25. $$\frac{9.0 \times 10^{19}\ H^+\ \text{ions}}{2.0\ L} \times \frac{1\ \text{mole}\ H^+\ \text{ions}}{6.02 \times 10^{23}\ H^+\ \text{ions}} = 7.5 \times 10^{-5}\ \text{mole/L}$$

The pH of the sample is the negative logarithm of this concentration. pH = 4.13.

27. The $HClO_4$ solution is red and the KOH solution is yellow.

29. Only solution (b) is acidic. It will turn blue litmus paper red.

31. The Clean Air Act of 1970 charged the EPA with setting limits on the amounts of gaseous pollutants in the United States. Passage of the Clean Air Act resulted in a 48% reduction in atmospheric NO_2 levels from 1980 to 2009. The event that prompted the passage of the Clean Air Act was the first Earth Day, held in 1970.

33. You would expect to find 75 molecules of SO_2 in 1 billion gas molecules in the air, so 2 billion would contain 150 molecules.

35. After the Clean Air Act, cars were required to have catalytic converters. Catalytic converters reduced the production of NO_x and carbon monoxide.

37. If you begin with 4 moles of NO_2 gas, then 4 moles of protons will result since every mole of nitric acid forms 1 mole of protons.

39. The type of coal used most in U.S. coal-fired plants is sub-bituminous. The type of coal that contains the most sulfur is bituminous.

41. The factors that contribute to the formation of dead zones are runoff from fertilizer, acid rain, chemicals found in sewage and detergents, and excesses of nitrogen.

43. In the Trout Lake study, scientists gradually acidified half of the lake while leaving the other half of the lake alone. As the lake was being slowly acidified, the scientists monitored the organisms in the lake and the characteristics of the lake water. As the pH of the water dropped, the offspring of the fish that were in the water were not able to survive. The water also became clear, allowing ultraviolet light from the sun to penetrate deeper into the water. This

caused algae to grow on the lake bottom. The lower pH also caused mercury from the lake floor to leach into the water, killing off the existing plankton. After acidification was complete, the scientists continued to study the lake. They found that the pH levels returned to normal after a few years. The fish and plants recovered, but this recovery took longer.

45. Acid rain is a problem in New England because the wind blows pollutants from the coal plants in Pennsylvania and other areas of the Ohio Valley into New England, where they cause damage. The pollutant gas that contributes to acid rain from these plants is SO_2.

47. The scientists were trying to protect the cathedral from the effects of pollution, in particular acid rain. Limestone was used to build the cathedral. Limestone is made of calcium carbonate, $CaCO_3$. Carbonate ions, CO_3^{2-}, are basic and react with the acids in acid rain. This causes the limestone to be slowly worn away.

49. The pH before acidification was 6.1. This represents a hydrogen ion concentration of 8×10^{-7} moles/L. If the lake contains 1.00×10^9 L, then this means there were 800 moles of hydrogen ions in the lake.

51. To reduce pH, add muriatic acid.

53. (a) $C_{18}H_{30}O_2$

    (b) Molar mass = 278.4 g/mol

    (c) Yes, it is because there is a double bond between the 6th and 7th carbons from the end of the chain.

55. *This has no answer because every student will create a unique answer.*

Chapter 11

1. Nuclear reactions involve changes to the nucleus of an atom, while chemical reactions only involve rearrangements of the electrons of an atom. Chemical reactions involve the making and breaking of chemical bonds. Chemical reactions do not involve changes in the number of protons and neutrons in the nucleus, but nuclear reactions do. Nuclear reactions produce much larger quantities of energy than chemical reactions.

3. (a) 5 neutrons, 5 protons, and 5 electrons

   (b) 52 neutrons, 38 protons, and 38 electrons

   (c) 92 neutrons, 78 protons, and 78 electrons

5. (a) $^{138}_{56}Ba$ (b) $^{76}_{35}Br$

7. (a) $^{100}_{48}Cd$ (b) $^{203}_{98}Cf$ (c) $^{154}_{70}Yb$

9. (a) Radon forms when uranium-238 decays through a series of nuclear reactions.

   (b) Six nuclear reactions are needed to convert uranium to radon.

   (c) There are four alpha emissions in this process.

   (d) There are two beta emissions in this process.

11. This is possible because the number of protons and or neutrons can change each time. Lead appears in this decay series as several different isotopes.

13. (a) $^{260}_{115}Sp$

    (b) It will be placed under Bi, bismuth.

15. (a) This is a chemical equation.

    (b) This is a nuclear equation.

    (c) This is a chemical equation.

17. It is not always possible to tell whether the reaction produces gamma radiation. First, gamma radiation is not a particle like alpha and beta radiation. Instead, it is a form of light. Since gamma radiation does not affect the balance of the nuclear equation, nuclear chemists sometimes write a nuclear equation without including it; gamma radiation often is understood to be present even if it is not included in the equation.

19. ${}^{238}_{92}\text{U} \longrightarrow {}^{4}_{2}\alpha + {}^{234}_{90}\text{Th}$

21. ${}^{218}_{84}\text{Po} \longrightarrow {}^{4}_{2}\alpha + {}^{214}_{82}\text{Pb}$

23. (a) This is not balanced. The number of neutrons and the number of protons is not the same on both sides.

    (b) This is not balanced. The number of neutrons and the number of protons and the number of protons are not the same on both sides.

25. (a) This is an alpha particle. (b) This is an alpha particle.

27. The design flaw in this bomb is that each portion contains more than the critical mass of 4 kg. If the total mass exceeds the critical mass, the bomb may explode prematurely and unexpectedly.

29. A reaction that amplifies itself by initiating more reactions is considered a chain reaction. The decay of uranium-235 creates a chain reaction because a neutron is used to initiate the reaction, but three neutrons are created when uranium-235 splits. These three neutrons can bombard other uranium-235 atoms, causing a chain reaction.

31. All reactors contain control rods, which can be inserted into the reaction chamber to slow down or stop the nuclear reaction. Reactors also have cooling systems to remove heat from the reactor core. Also, the core is kept in some type of containment system, usually made of concrete or other sturdy material. Used fuel is kept underwater in pools that block radiation and keep the used fuel cool. Finally, reactors have backup generators that are used in case of power failure.

33. Nuclear power plants need complex cooling systems because the nuclear reactions generate large amounts of heat. In Fukushima, the loss of cooling systems caused the water to boil away, which meant the spent fuel rods were exposed and the temperature at the reactor core skyrocketed. This caused a number of explosions and radiation leakage from the plant.

35. (a) ${}^{238}_{92}\text{U} + {}^{1}_{0}\text{n} \longrightarrow 2\,{}^{118}_{46}\text{Pd} + 3\,{}^{1}_{0}\text{n}$

    (b) This would promote a chain reaction because of the production of three neutrons are produced in the reaction.

37. The half-life of a radioactive substance is how long it takes for half of the substance to decay. If you know its half-life, you know how long the substance will give off radiation.

39. The parent material refers to the original radioactive substance. The daughter material refers to the substance that the parent changes into as it decays. For instance, uranium-238 (the parent material) decays through a series of reactions to radon-222 (the daughter material).

41. (a) 1.28×10^9 yr

(b) If 100% of the potassium-40 exists at the start, then 50% exists after 1.28×10^9 years, 25% remains after 2.56×10^9 years, and 12.5% remains after 3.84×10^9 years.

If you convert this into minutes, you have:

$$3.84 \times 10^9 \text{ yr} \times \frac{365 \text{ days}}{1 \text{ yr}} \times \frac{24 \text{ hr}}{1 \text{ day}} \times \frac{60 \text{ min}}{1 \text{ hr}} = 2.02 \times 10^{15} \text{ min}$$

43. If you start with 3.00 g, then after 27.8 days, $3.00/2 = 1.5$ grams will remain. After 55.6 days, $1.5/2 = 0.75$ grams will remain. After 83.4 days, $0.75/2 = 0.38$ grams will remain.

45. Alpha, beta, and gamma radiation are blocked by a thick piece of lead.

47. Gamma radiation most easily penetrates human skin. Since gamma radiation can penetrate the skin, it can be used to treat tumors in the body. However, while gamma radiation is effective in killing cancerous cells, it is also effective at killing healthy cells. This leads to the side effects seen in radiation treatment, such as nausea, fatigue, and hair loss.

49. Certain parts of the body are affected by radiation before other parts because of the rate at which these parts of the body grow or regenerate. The areas of the body most affected by radiation, such as bone marrow, have cells that multiply rapidly.

51. The time elapsed is about one half-life, 138 days. Therefore, one-half, or 10.00 mg, will remain.

53. The mass of Pu-239 will be halved, and the mass of Pu-240 will remain the same. After 87.74 years, the mass of the mixture will be 36.56 grams.

55. *This has no answer because every student will create a unique answer.*

Chapter 12

1. Power is the amount of energy used or produced in a certain amount of time. One unit for power is the watt. Energy is the ability to apply a force over a distance. One unit of energy is the joule.

3. The energy of the moving water hits the paddles of the water wheel and converts energy from the water's motion into the energy of moving paddles.

5. Another unit of power is horsepower.

7. The first law of thermodynamics states that energy cannot be created or destroyed. A patent for a machine that moves perpetually can be written and even granted by a patent office, but such a machine cannot exist because a machine that moves requires an input of energy.

9. To push the box five meters would require 20 joules. Since 1 watt = 1 joule/second, then

$$\frac{20 \text{ joules}}{3 \text{ s}} = 7 \text{ watts}$$

Note that this answer was rounded using the rules for significant figures.

11. Radioactive material (b) is not a type of fossil fuel.

13. (b) Gasoline has the lowest average number of carbon atoms.

15. Separation of the fractions of crude oil in a fractionating column occurs as a result of the differences in boiling points. Since boiling points of compounds of similar mass are similar, it is difficult to separate individual hydrocarbons. In addition, hydrocarbons that are isomers of each other but have the same molar mass are difficult to separate in the fractionating column.
17. The fraction taken from the top has the lower boiling point, meaning it boils off more easily than the fractions that are lower in the column.
19. (a) C_8H_{18} boils at about 120°C. (b) $C_{18}H_{38}$ boils at about 310°C.
21. To answer this question, you must add together all of the sources of energy for a given year. In 1915, roughly 25 EJ of energy were used in the U.S. In 2012 that value increased to roughly 100 EJ. Therefore, energy consumption increased by a factor of four from 1915 to 2012.
23. (d) 180 ppm
25. Ice-core measurements are used to study atmospheric conditions on Earth over hundreds of thousands of years. Scientists who work in this field are called *paleoclimatologists*.
27. In 1975, the U.S. Congress passed the Corporate Average Fuel Economy (CAFE) Law that required a doubling of fuel efficiency within 10 years to a value of 27.5 mpg. Within 10 years, the auto industry persevered and the new standard was met.
29. Before the 1990s, fewer trucks and SUVs were on the road, so the line labeled "both" was closer to the cars until that time.
31. The oxidation half-reaction occurs at the anode. The reduction half-reaction occurs at the cathode.
33. One bond connects the two hydrogen atoms in a hydrogen molecule. This is because each hydrogen atom has one valence electron. Since it takes 2 electrons to form one bond and there are a total of just 2 electrons available to bond. The result is a single bond.
35. In this reaction, hydrogen, H_2, is oxidized and oxygen, O_2, is reduced.
37. In some regions of the United States, the energy needed to produce the hydrogen may come from a source that produces larger amounts of greenhouse gases than a gasoline-powered car.
39. This statement is false. A hydrogen-fueled-car owner may obtain her or his hydrogen indirectly from nuclear fuel or fossil fuel, making it a potentially environmentally unfriendly choice.
41. $3Cu^{2+} + 2Al \longrightarrow 3Cu + 2Al^{3+}$
43. The wires and contacts that lie on top of the silicon in a solar cell carry the electric current generated within the cell.
45. There are regions of Norway, Finland, and Sweden that only experience 1 to 2 peak sun hours per day.
47. The redox reaction is: $Fe^{2+} + Zn \longrightarrow Zn^{2+} + Fe$
49. (a) 5,000,000 joules are contributed to the electrical grid for that day.

    (b) The homeowner earns $5 \times \$0.33 = \$1.65$ that day.

51. About 13 to 14 miles per one gallon. The 2010 truck gives you about 20 miles (or about 32 kilometers) more for one gallon of gas.

53. The refrigerator uses 3 kWh per day. *The second part of this question has no answer because every student will create a unique answer.*

55. *This has no answer because every student will create a unique answer.*

Chapter 13

1. One example of a cradle-to-grave process is the process of making and using a Styrofoam cup. Once the cup has been used, it is generally thrown away (the grave). One example of a cradle-to-cradle process begins with a new Nike shoe (cradle) and ends with the use of the old ground-up Nike shoes to make playground surfaces (a second cradle).

3. In a cradle-to-grave process, a substance is produced, used, and then discarded. In a cradle-to-cradle process, a substance is produced and then reformed into another substance to be used again.

5. Bakelite is a thermosetting polymer. It does not change shape when heated. It is not easily reshaped or reused, which is the reason manufacturers don't make items out of Bakelite anymore.

7. Some of the issues at stake include what resources are used in the production of each type of cup, the price of each type, the carbon and water footprints resulting from the production and use of each type, the biodegradability of each type, and the contribution each type makes to pollution. A plastic cup has a larger carbon footprint. Paper cups contribute more to water pollution.

9. BPA is a chemical called bisphenol A. It affects the human body because its structure is similar to that of estradiol, a hormone. BPA is able to bind to the same estrogen receptors in the body that estradiol does. Due to its ability to bind to estrogen receptor sites, BPA can disrupt normal development and hormonal balance in the body. It has been linked to genital defects, breast and prostate cancers, childhood obesity, and the early onset of puberty. It also contaminates recycled paper because it makes its way into the recycling stream and gets mixed into the recycled paper.

11. (a) The cup made in the nuclear-powered factory that requires several transportation steps would produce more pollution in the form of nitrogen oxides, since these pollutants are produced primarily by cars and trucks.

    (b) The cup made in the factory that gets its energy from a coal-fired power plant that is transported with electric vehicles is likely to produce more pollution in the form of sulfur oxides, since the coal-fired power plant is a major source of this type of pollution.

13. 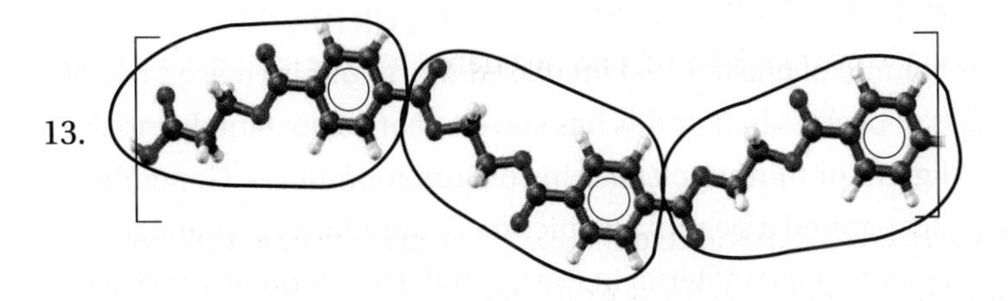

15. The resin ID code is 1, and the three-letter abbreviation is PET.

17. Yes, it is possible for the same polymer to exist in various lengths. By definition, polymers always contain more than one monomer unit. Usually, there is no set requirement for the length of a polymer.

19. 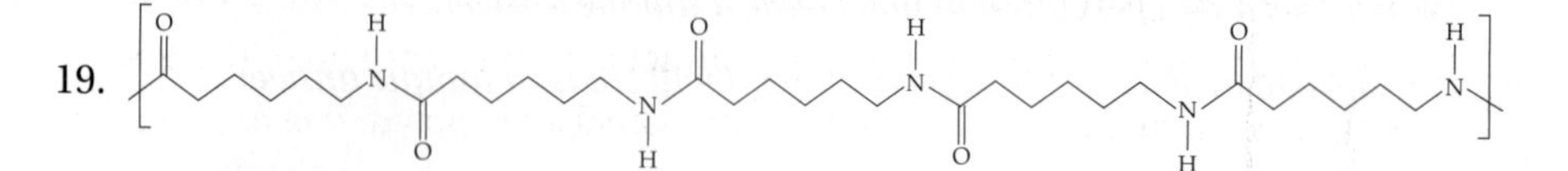

21. (a) The negative portion of molecule A will interact with the positive portion of molecule B.

    (b) Two molecules of A are not likely to react with each other.

    (c) Two molecules of B might react with each other, since the positive portion of one might interact with the nitrogen atom on the end of the other.

23. The molecule on the right (c) is the most polar because it contains electronegative oxygen atoms.

25. A sustainable polymer is a polymer made from sustainable materials. Sustainability is a broad term that refers to the capacity to endure; a process is *unsustainable* if it cannot be maintained. One example of a polymer that comes from a sustainable resource is polylactic acid, which can be made from corn, tapioca, or sugarcane. These substances are not in short supply and are easy to replace.

27. (a) Thermoplastic polymers are more easily recycled because they can be melted down and reformed into a new shape.

    (b) Bakelite is an example of a thermosetting polymer.

    (c) The Bakelite polymer does not have a resin ID code, because it is not possible to recycle it.

29. Items (b), (c), and (d) are probably not made from a thermosetting polymer.

31. (a) nonpolar (b) polar (c) nonpolar

33. (a)

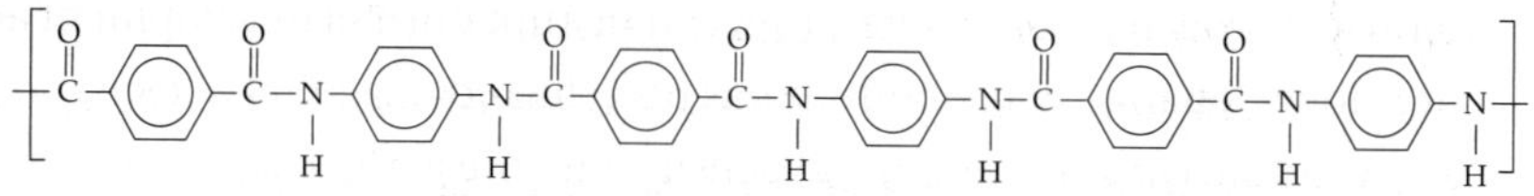

    (b) The molar mass of the smallest unit that makes up Kevlar is 238.3 g/mol.

    (c) Kevlar is so much stronger than most other polymers because the amide bond and the benzene ring make the polymer rigid and straight, so that the polymer strands can pack tightly together and form a very strong crystalline matrix. Also, the strands of the polymer interact through hydrogen bonding, cross-linking the strands and contributing to the strength of Kevlar.

35. One myth dispelled by the Garbage Project was the idea that organic materials such as paper decompose quickly. In fact, most of the volume of a landfill is made up of paper, and the organic matter in a landfill can stick around for a long time—sometimes for 50 years. Another myth dispelled by this project was the idea that the volume of plastics had increased due to our increased use of plastics. The mass of plastics in landfills has stayed nearly constant due to the increased use of lighter or thinner plastics in consumer products. Generally, the Garbage Project painted a pessimistic picture of our efforts to manage landfills. It found that organic material in the landfill doesn't break down as people often assumed it would, and that we continue to add more and more mass to the landfills in the form of plastics, especially single-use water bottles.

37. One or more heteroatoms (c) can make a polymer biodegradable.

39. To reshape a plastic, it must first be melted. After it is melted, the plastic is still composed of long polymer chains. When cooled, the polymer reverts to the solid phase and takes on any new shape it has been given. The end product of this process is the same as the starting product in terms of polymer identity. Biodegradation polymer, on the other hand, breaks the polymer chains, so the end product may be smaller pieces of the polymer chain, the original monomers, or even smaller molecules.

41. According to Figure 13.12, household recycling is taken to a facility where it is sorted according to the type of plastic it contains, which is determined by its resin ID code. The used plastic is then made into flakes and washed to make pellets of recycled plastic.

43. In order of easiest to recycle first: HDPE > PVC > polypropylene.

45. Recyced plastic containing only one type of plastic is made by separating that type of plastic from other types of plastic. That makes it more time and labor intensive to produce than a plastic made from a mixture of plastics, so the plastic made from the mixture will cost less.

47. They have been sorted according to resin ID code, but not yet flaked or washed.

49. Plastics that are not biodegradable usually come from petroleum. When these plastics are incinerated in a landfill, carbon dioxide is produced, leading to global warming and climate change. Incineration also can produce compounds such as dioxin, which are very toxic and can have detrimental effects on human health.

51. Parts (a) and (b) are student-dependent.

    (c) The contact lens polymer contains hydrophilic functional groups which make the lens suitable for use in the eye.

53. (a) $C_{12}H_{10}O_4S$; molar mass = 250.3 g/mol

    (b) Yes, because the molecule bonds by attaching at the ends, and BPS has the same groups as BPA in those locations.

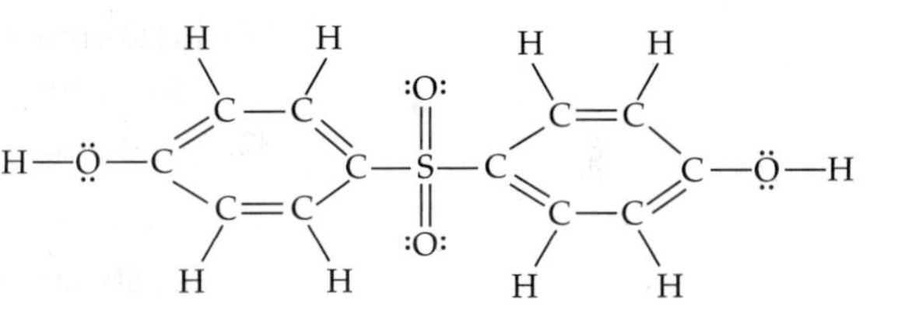

55. *This has no answer because every student will create a unique answer.*

Chapter 14

1. Answers (a) and (b) are micronutrients, and (c) and (d) are macronutrients.

3. Butter (d) is not a recommended part of a healthy plate of food.

5. (a) fat (b) protein (c) carbohydrate (d) carbohydrate

   (e) protein (and fat) (f) fat and carbohydrate

7. This phrase means that half of the grains you eat should be whole grains, not refined grains. This recommendation means that we should all eat more whole grains each day and fewer refined grains.

9. Since 75% of the plate is fruits and vegetables, it is too heavily weighted in fruits and vegetables at the expense of protein. There is no protein on this plate at all, even though a little less than 25% of the plate should be protein. Since 25% of the plate is brown rice, which is a whole grain, the plate fulfills the recommendation for the amount and type of grain you should consume.

11. Proteins do not store genetic information (d).

13. This statement is false. Your body needs nonessential amino acids but is able to produce them so that you don't have to get them from your diet.
15. Structural proteins have structures that are analogous to objects that we associate with strength or mechanical stability such as scaffolds. They are often fibrous in nature, and they are found in biological structures where strength or resilience is required.
17. A rope is made by taking a coiled strand and twisting it with other coiled strands in the reverse direction. This gives the rope strength. This matches the description of keratin, which has a ropelike structure.
19. (a) 3'-GCAUUAAGCC-5' (b) 5'-GCCAUAAAUGCC-3' (c) 5'-GGAUUAGCAAUU-3'
21. The first step in protein synthesis involves using DNA as a template to make RNA. Groups of three RNA bases, called triplets, then code for specific amino acids through the process known as translation. Amino acids are added to each other and connected using peptide bonds. Transcription is the process of making RNA from a DNA template. Translation is the process of translating the RNA nucleotides into a sequence of amino acids.
23. (a) 3'-GTT TTT TTA GGA ATA GCC TTG-5'
    (b) 5'-CAA AAA AAU CCU UAU CGG AAC-3'
    (c) Gln-Lys-Asn-Pro-Tyr-Arg-Asn
25. We are bound by our DNA. DNA puts boundaries around what we can change about ourselves. The men are behind bars because our DNA determines our biochemical makeup. We can't yet change our DNA. The cartoon is still relevant since we can modify genes in a lab but cannot easily modify genes in a human being.
27. The orange will not resemble the pig in every way. If only one gene is taken from the pig, then only the characteristic associated with that gene will show up in the orange.
29. GMOs may pose indirect risks to human health. For example, reports that study the use of Roundup suggest that it may cause birth defects and infertility in certain animals and that those who are routinely exposed to it may be at increased risk for certain serious diseases, such as multiple myeloma.
31. One reason that people might support the use of GMOs is that genetic modification can help provide disease resistance—this is the case in the development of citrus fruits that are resistant to the citrus greening bacterium. Another reason people might support the use of GMOs would be because genetic modification can increase plant yield. This could help provide larger amounts of food to feed a growing population. For instance, as fish supplies dwindle due to overfishing, GMO fish that mature faster might supplant the dwindling fish supply. One reason that people might not support the use of GMOs could be due to religious beliefs. Some religions preach that genetic modification is morally wrong. Another reason that people may not support GMOs is that they are unsure of the environmental impacts of genetically modified plants on the ecosystem and are concerned that there is insufficient monitoring of GMOs.
33. Two genetic modifications that are frequently employed in modern agriculture are modification of plants to increase resistance to insects and resistance

to herbicides. The former involves splicing a gene from the bacterium *Bacillus thuringiensis* into the plants DNA. This bacterium produces a protein that is toxic to certain insects. The latter modification helps to make plants more tolerant of herbicides. Thanks to this modification, herbicides can be applied to kill off weeds and leave the genetically modified plants unharmed. The one modification used in more major crops in the United States is the one that confers resistance to herbicides. It is used in corn, and at least 75% of our corn crops are transgenic corn.

35. Many types of carbohydrates are polymers. A saccharide is the monomer that makes up carbohydrates. A monosaccharide is made up of one monomer saccharide unit. A disaccharide is made up of two monomer saccharide units, though the two monomer units in a disaccharide can be the same or different monomers. A polysaccharide is made up of many monomer saccharide units. A simple sugar contains one or two monomer units. A complex carbohydrate contains hundreds or thousands of monomer units. Dietary fiber includes those carbohydrates that cannot be broken down by the body.

37. Fiber is not broken down in the digestive tract. Fiber helps maintain the health of the digestive tract. It does this by keeping moisture in the gut and helping to move waste through to its final destination.

39. Sugar molecules are polar due to the presence of multiple electronegative oxygen atoms. There are many locations on a sugar molecule where water can interact with it.

41. The bran and the germ are the parts of the grain that are dietary fiber.

43. Arachidic, stearic, and palmitic acid have very straight chains, which suggests that they are saturated. The rest are unsaturated.

45. Fats are hydrophobic because they are nonpolar. When you mix a fat with water, the fat will not dissolve, because polar (water) and nonpolar (fat) molecules do not interact.

47. The part of the fatty acid containing the carboxylic acid group is polar, since it contains numerous electronegative atoms and thus an uneven electron density. The long-chain hydrocarbons that make up the rest of the fatty acid are nonpolar. Fatty acids are considered to be lipids because the portion of the fatty acid that is nonpolar is much larger and tends to dominate the polar carboxylic acid portion of the fatty acid.

49. This statement is false. While it is true that trans fats are a by-product of the hydrogenation of unsaturated oils, the hydrogenation process is meant to *partially* saturate hydrocarbon chains, not to *completely* saturate them.

51. It is a protein sequence because the letters match the symbols for the 20 amino acids that compose proteins.

53. (a) It is high in protein and low in fat and carbohydrate.

    (b) One serving gives 20% of the RDA, so five servings would give 100% of the RDA. Five servings have a mass of 85 grams.

    (c) It is very high in vitamin B12. If 17 grams of cricket flour give 68% of the vitamin B12 RDA, then 25 grams will give 100% of the RDA for this vitamin.

55. *This has no answer because every student will create a unique answer.*

Glossary

accuracy How close a number is to a known and accepted standard value.

acid Any substance that increases the concentration of H^+ in a watery environment.

acid rain Rainfall that has a lower pH than natural, unpolluted rain.

acidic Describes an aqueous solution having a pH less than 7.00 at 25°C.

acidic solution An aqueous solution that contains an acid.

activation energy The energy required for a chemical reaction to scale the energy hill.

allotrope One of two or more different forms of an element.

alloy A mixture of metals.

alpha decay A type of radioactive decay in which alpha particles are emitted.

amide A functional group that contains nitrogen, carbon, and oxygen.

amine A functional group containing a single nitrogen atom.

amino acid One of 20 molecules that are building blocks of proteins.

Amontons' law The gas law that states that pressure is directly proportional to temperature.

anion A negatively charged ion.

anode The electrode in an electrochemical cell where oxidation takes place.

aqueous solution A homogeneous mixture of water molecules and another substance such as a salt.

atmosphere (atm) A common unit of pressure.

atmospheric pressure The pressure exerted by gases in the atmosphere.

atom The smallest possible unit of matter that cannot be broken down by chemical or physical means.

atomic number The number of protons in the nucleus of an atom.

atomic scale The perspective from the level of atoms and molecules.

atomic symbol A unique, one- or two-letter designation for an element.

autoionization A chemical reaction in which water splits to make H^+ and OH^-.

Avogadro's law The gas law that states that volume is proportional to moles of gas particles.

Avogadro's number The number of items in one mole; 6.02×10^{23}.

barometer A device used to measure atmospheric pressure.

base Any substance that adds hydroxide ions to an aqueous solution.

base unit The fundamental unit to which prefixes can be added for each type of metric measurement.

basic Describes an aqueous solution having a pH greater than 7.00 at 25°C.

battery An electrochemical cell based on two half-reactions.

beta decay A type of radioactive decay in which beta particles are emitted.

biodegrade To break down when exposed to natural forces in the environment.

biodiesel A fuel made from non-fossil-fuel sources.

biomolecule A molecule that is involved in the processes that support life.

bioplastic A plant-based polymer.

bisphenol A (BPA) A molecule found in some plastics that acts as an endocrine disruptor.

body mass index (BMI) An indicator that combines body weight and height in humans.

boiling point The temperature at which the phase transition from liquid to gas occurs.

bond energy The energy required to break a specific chemical bond.

bond length The distance between two atoms involved in a covalent bond.

Boyle's law The gas law that states that volume and pressure are indirectly proportional to one another.

buckyball An allotrope of carbon that contains 60 carbon atoms and has a spherical structure.

buffer Substances that offset acidity or basicity.

CAFE *See* Corporate Average Fuel Economy laws.

carbohydrate A macronutrient that is a polymer of saccharide molecules.

carbon footprint A tally of the amount of greenhouse gas emission that is associated with a person, business, production process, country, or the world.

carboxylic acid A functional group containing one carbon atom, two oxygen atoms, and one hydrogen atom.

catalyst A substance that increases the rate of a chemical reaction.

catalytic converter A device used in automobiles to remove undesirable gases from exhaust.

cathode The electrode in an electrochemical cell where reduction takes place.

cation A positively charged ion.

Celsius temperature scale A metric temperature scale based on the freezing and boiling points of water.

chain reaction A nuclear reaction that amplifies itself by initiating more reactions.

charge A property of matter that attracts it to other matter.

Charles's law The gas law that states that volume is directly proportional to temperature.

chemical bond A force that holds two atoms or ions together.

chemical change A change that alters the identity of a substance.

chemical equation A representation of a chemical process that uses chemical formulas to depict reactants and products.

chemical formula A way of representing the elemental composition of an element or compound.

chemical reaction An interaction between substances whereby electrons are rearranged, transferred, or shared.

Clean Air Act A law passed in 1970 that set limits on the amounts of gaseous pollutants that can be released into the air in the United States.

clean coal plant Coal plants that have installed equipment to remove some harmful pollutants.

climate change The effects of global warming, such as rising sea levels.

coefficient A number that precedes the chemical formula for a reactant and/or product in a balanced chemical equation.

combustion An exothermic reaction between a fuel and a source of oxygen.

complementary base pair One of the pairs of nitrogenous bases that exist on complementary strands of nucleic acids.

complex carbohydrate A carbohydrate composed of a long chain of saccharides.

compound A pure substance that contains more than one element.

concentration The amount of solute per unit volume.

condensation The phase transition from gas to liquid.

control experiment An experiment in which one or more variables are tightly controlled so that the effect of changing one variable can be determined.

control rods Solid rods that can be lowered into a nuclear reaction taking place in the core of a nuclear reactor to prevent overheating.

conversion factor Fractions that relate equal amounts expressed with different units; used to convert between units.

core electron An electron in an atom that is not a valence electron.

Corporate Average Fuel Economy (CAFE) laws U.S. laws that set minimum levels on fuel economy for cars and trucks.

corrosion A chemical reaction whereby a metal slowly deteriorates when it is exposed to the environment.

covalent bond Shared electron pairs between atoms.

cradle-to-cradle Describes a product made of a material that can be reused in a different product at the end of the original product's lifetime.

critical mass The minimum amount of fissionable material necessary to sustain a chain reaction.

cross-links Pieces of molecules that link together regions of larger molecules.

crystal When used to describe a salt, a solid with a repeating pattern of positive and negative ions.

crystallite Any highly ordered, crystalline region in a polymer.

dead zone An area where life cannot thrive.

deforestation Removal of forests so that land can be used for other purposes.

density The mass of a substance contained within a specified volume.

deoxyribonucleic acid (DNA) A nucleic acid that is the repository of an organism's genetic information.

diatomic molecule A molecule that is made up of two of the same atom.

dietary fiber Carbohydrates that are not digestible.

diffusion The mixing of one gas or liquid into another.

dimensional analysis A method for converting between units based on conversion factors and cancellation.

dipole A separation of positive and negative charge within a molecule.

dipole–dipole interaction A type of intermolecular force that exists between properly aligned dipoles in neighboring molecules.

dissociation The process whereby a salt or an acid separates into its ions, often in water.

dissolve To be incorporated into a liquid and become part of a solution.

disulfide bond A linkage, composed of two atoms of sulfur, that connects protein chains.

DNA (deoxyribonucleic acid) A nucleic acid that is the repository of an organism's genetic information.

double helix The shape formed by two complementary strands of DNA.

duet rule The equivalent of the octet rule for very small atoms.

electric current A flow of electrons.

electrochemical cell A device that interconverts chemical and electrical energy.

electrolysis The splitting of water into hydrogen and oxygen using electric current.

electrolyte A charged substance dissolved in water.

electromagnetic radiation Another word for light.

electromagnetic spectrum A continuum of radiant energies.

electron A negatively charged particle within an atom.

electron density The blur of fast-moving electrons around the nucleus of an atom.

electronegativity The tendency of an atom in a covalent bond to pull electron density to itself.

element A pure chemical substance, which cannot be broken down into simpler substances, consisting of a single type of atom.

endothermic Absorbing heat energy.

energy The ability to apply a force over a distance.

energy level An imaginary location that an electron can occupy in an atom.

enhanced greenhouse effect The amplified greenhouse effect that is attributable to fossil fuel use.

Environmental Protection Agency (EPA) The federal agency in the United States charged with protecting the environment.

enzyme A protein that catalyzes a chemical reaction.

equilibrium The condition in which there is no net change in a system, and reactions proceed in both directions at the same rate.

eutrophication An overgrowth of an organism, caused by a pollutant, that results in the depletion of oxygen in natural water.

evaporation One way that the phase transition from liquid to gas can take place.

excited state For an electron, any energy level higher in energy than its ground state.

exothermic Producing heat energy.

family A vertical column of elements on the periodic table. Also called a *group*.

fatty acid A type of lipid that contains a long hydrocarbon chain and a carboxylic acid functional group.

first law of thermodynamics The scientific law that states that energy cannot be created or destroyed.

fixation The process whereby a gaseous substance is incorporated into a larger, nongaseous molecule.

flocculant A substance that is used to remove particulates from water during the water treatment process.

flue gas desulfurization A method used in clean coal plants to remove sulfur-containing substances.

force A push or pull on an object.

formula unit When used to describe a salt, the repeating motif in a crystalline solid.

fossil fuel Carbon-based compounds derived from organic matter, including plant and animal fossils that have decayed over eons.

fractionation The separation of molecules in a mixture on the basis of their sizes.

freezing point The temperature at which the phase transition from liquid to solid occurs.

fresh water Water that is not salt water.

fuel cell A device that uses a redox reaction involving hydrogen and oxygen to make water and energy.

fuel economy The amount of fuel used by a vehicle, usually expressed in miles per gallon (mpg).

full structure A representation of an organic molecule that includes all the atoms and bonds.

functional group A specific motif within an organic molecule. Often, a heteroatom or group of heteroatoms.

gamma radiation A very high-energy form of type of electromagnetic radiation.

gas One of the three common phases of matter; a vaporous substance that completely fills its container.

gas law A relationship between changing gas variables.

gasoline A mixture of hydrocarbon molecules used as a fuel for combustion reactions.

gene The segment of a nucleic acid that codes for a specific protein.

genetic code The system by which the four types of nitrogenous base in RNA are translated into the 20 different amino acids.

genetic engineering Technology that allows humans to manipulate the hereditary material contained in the cells of living things.

genetically modified organism (GMO) An organism with modified genetic material.

global warming The increasing temperatures on Earth that result from human activities.

globular protein One of the two types of protein; formed by folded chains of amino acids.

graphene One isolated layer of graphite.

graphite An allotrope of carbon that consists of layers of two-dimensional carbon-containing sheets.

greenhouse effect The increase in Earth's temperature that results from the absorption of heat by molecules in the atmosphere.

greenhouse gas A gas that contributes to the warming of Earth's atmosphere by trapping heat in the atmosphere.

ground state The lowest possible energy level that an electron can occupy in an atom.

group A vertical column of elements on the periodic table. Also called a *family*.

half-life The time it takes for one-half of a radioactive sample to decay.

heating curve A graph illustrating phase changes for a substance.

heteroatom Any atom other than carbon or hydrogen in an organic molecule.

hydration The process whereby an ion is surrounded by water molecules.

hydrocarbon An organic molecule that contains only carbon atoms and hydrogen atoms.

hydrogen bond A type of intermolecular force that can exist among certain hydrogen-containing molecules.

hydrogenation The process whereby double bonds in fats are converted to single bonds via the addition of hydrogen molecules.

hydronium ion The H_3O^+ ion.

hydrophilic Describes a substance that attracts/mixes easily with water; water-loving.

hydrophobic Describes a substance that repels/does not mix easily with water; water-fearing.

hydroxide ion The OH^- ion.

hypothesis An initial best guess about some question concerning nature.

indicator A brightly colored molecule that changes color as the pH of a solution is changed.

inorganic Describes substances that are not organic and usually do not contain carbon.

intermolecular force An attractive force between molecules.

ion An atom with a net positive or a net negative charge.

ion–dipole interaction An attractive force between the dipole of a molecule and an ion.

ionic bond An interaction between oppositely charged ions.

ionic compound A compound composed of ions of opposite charge.

isotope Atoms of the same element that have different numbers of neutrons.

laboratory scale The perspective from the level of real quantities used in a lab.

law of conservation of mass The law that states that matter may not be created or destroyed.

law of constant composition The law that states that the number of atoms of each element in a compound must always be the same.

Lewis dot diagram A dot-based notation that shows valence electrons in an atom, ion, or molecule.

life-cycle assessment (LCA) An assessment of all the energy and materials that go into making a product and disposing of it.

light Radiant energy.

lightweighting The process of reducing the mass of plastic used to make a container.

liming The addition of lime, a base, to natural waters to increase the pH.

line spectrum Specific wavelengths of emitted light that are characteristic of one element.

line structure A drawing of an organic molecule that is made up of lines only rather than element symbols and bonds.

lipid An alternate word for fat; a type of biomolecule and a macronutrient.

liquid One of three common phases of matter; a fluid substance that fills the bottom of a container and has a surface.

liter The base unit for volume used in this book.

lone pair In a Lewis dot diagram, a pair of electrons on placed on the side or sides of an atom's symbol.

macronutrient A substance required in large amounts in the human diet.

malnutrition Lack of proper nutrition.

mass The quantity of matter contained in an object.

mass number The number of protons and neutrons in an isotope.

mass percent The percentage of the total mass of a mixture that is composed of a substance.

matter Anything that has mass and occupies space.

mean free path The average distance that a gas particle travels between collisions.

melting point The temperature at which the phase transition from solid to liquid occurs.

metal An element that tends to be shiny and malleable and to lose electrons; metals lie to the left of the stepped line on the periodic table.

metallic bond A chemical bond within a solid metal that is described as a "sea of electrons."

metalloid An element with properties intermediate between those of metals and nonmetals.

meter The base unit for length used in this book.

methane A simple carbon-containing molecule that is the major component of natural gas; its chemical formula is CH_4.

metric system An agreed-upon system of scientific measurement and units used throughout most of the world.

micronutrient A substance required in small amounts in the human diet.

millimeters of mercury (mm Hg) A unit used to measure the column of mercury in a barometer.

mixture A combination of two or more pure substances.

model A representation of reality that helps us to understand nature.

molar mass The mass of one mole of a pure substance.

molar volume The volume filled by one mole of gas at STP; 22.4 liters.

molarity The number of moles of solute in one liter of solution; a concentration unit.

mole A counting unit equal to 6.02×10^{23}.

molecular structure A representation of a molecule in which lines represent bonds and letters represent atoms.

molecule A group of atoms held together by covalent bonds.

monatomic ion An ion comprised of just one atom.

monomer A molecule that is repeated over and over to create a polymeric structure.

multiple bond For organic molecules, a double or a triple covalent bond.

net metering The process whereby power generated in a home with solar panels is given back to the power company.

network solid A substance that is made up of a continuous network of covalent bonds.

neutral Describes a solution that is neither acidic nor basic.

neutralization The chemical reaction between an acid and a base that creates a neutral solution.

neutron A neutral particle within an atom's nucleus.

nitrogenous base A component of a nucleotide that makes up the code required to make proteins.

noble gas A nonreactive gas that doesn't interact with anything because it has an octet of valence electrons; helium, neon, argon, krypton, xenon, and radon.

noble gas configuration An octet of valence electrons.

nonmetal An element that does not have metallic properties and tends to accept electrons; one of a small subset of elements that fall to the right of the stepped line on the periodic table.

nonpolar Describes a bond that has evenly distributed electron density.

nuclear fission The splitting of the atom's nucleus.

nuclear reaction A reaction in which the number of protons and/or neutrons changes in some way.

nucleic acid A biomolecule that is made up of nucleotides.

nucleotide One building block of a nucleic acid polymer.

nucleus The dense, positively charged region within an atom that contains protons and neutrons.

nutrient A substance that provides nourishment essential for growth and the maintenance of life.

octet rule Atoms often attain a total of eight valence electrons as a result of gaining or losing or sharing electrons.

organic Describes substances that contain a carbon framework.

osmosis Movement of water molecules across a semipermeable membrane from a region of lower ion concentration to a region of higher ion concentration.

oxidation A loss of electrons.

parts per billion (ppb) A concentration unit; one part in one billion parts.

parts per million (ppm) A concentration unit; one part in one million parts.

passive nuclear safety A safety backup system, which requires no human and no electronic signal, which initiates shutdown of a nuclear reactor.

passive solar A form of solar energy capture in homes or other buildings.

peptide bond The linkage between two amino acids in a protein.

period A horizontal row of elements on the periodic table.

periodic table The central organizing scheme for the elements.

pH The negative logarithm of the molar concentration of protons; a measure of acidity and basicity.

phase The state of existence of a sample of matter: solid, liquid, or gas.

photosynthesis The process plants use to convert sunlight, carbon dioxide, and water into oxygen and sugar.

photovoltaic cell A device that converts sunlight energy into electrical energy.

physical change A change that does not alter the identity of a substance.

plastic A polymeric organic material that can be shaped.

polar Describes a bond that has a lopsidedness of electron density.

polyatomic ion An ion comprised of two or more atoms.

polymer A large molecule that is made up of repeating monomer units.

potable Describes water that is safe to drink.

power The amount of energy used or generated in a specific amount of time.

precipitation (1) As used in Chapter 8: When water in the atmosphere returns to Earth in the form of snow, sleet, hail, or rain. (2) As used in Chapter 9: The return of an ion in solution to the solid crystalline lattice.

precision The closeness of measured numbers to one another.

pressure Force exerted over an area.

product A substance formed during a chemical reaction.

protein A macronutrient that is a polymer of amino acids.

proton A positively charged particle within an atom; one of the terms we use for H^+, because the H^+ ion contains one proton and no electrons.

pure substance A single element or compound that may have trace impurities.

radiation The emission of energy.

radioactive A term that refers to materials that release energy or particles when their nuclei undergo reactions.

radioactive decay The process by which certain isotopes give off energetic products.

radioactive decay series A sequential string of radioactive reactions that continues until a stable isotope is reached.

radioactive tracer A radioactive isotope used to visualize abnormalities in the human body.

radon A radioactive noble gas.

reactant A substance that undergoes a chemical reaction.

reaction energy diagram A depiction of energy changes as a function of the progress of a chemical reaction.

recyclable Describes something that contains reusable material (e.g., thermoplastic polymers that can be melted and reshaped into a new plastic product).

redox reaction A combined oxidation-reduction reaction.

reduction A gain of electrons.

rem A unit of measure for radiation exposure; 100 rem = 1 Sv.

resin ID code A number that classifies plastics according to their recyclability.

RNA (ribonucleic acid) A type of nucleic acid that includes the sugar ribose.

saccharide A sugar molecule; a building block for more complex sugars.

Safe Drinking Water Act U.S. law passed in 1974 to protect the quality of public water.

salt An ionic compound.

saturated solution A solution that contains as much dissolved solute as possible.

scientific method The general approach to scientific experimentation used worldwide; involves observation, experiment, measurement, hypothesis formulation, and rigorous, repeated testing of theories.

scientific notation A system for expressing a number using an exponential term.

semiconductor A substance that conducts an electric current only under specific conditions.

semipermeable membrane A thin, pliable sheet or barrier that allows only certain substances to pass through it.

side product A product of a chemical reaction that does not appear in the balanced chemical equation.

sievert (Sv) A unit used to measure the amount of radiation exposure in human beings.

simple salt An ionic compound made up of a monatomic cation and a monatomic anion.

solar energy Energy derived from the sun.

solid One of the three common phases of matter; a firm nonliquid, nongaseous substance does not adopt the shape of its container.

solubility The ability of a substance to dissolve in another substance.

solute A substance dissolved in a solvent.

solvent A substance in which one or more solutes are dissolved.

specific heat The amount of heat absorbed per gram per degree Celsius temperature change.

standard A precisely known quantity that others are measured against.

stoichiometry The use of balanced chemical equations to quantify reactants and products.

STP Standard temperature (0°C) and pressure (1 atm).

strong acid An acid that dissociates completely in water.

sublimation The phase transition directly from solid to gas.

sugar An alternative word for saccharide.

sulfide A functional group containing a single sulfur atom.

Superfund site An uncontrolled or abandoned place where hazardous waste is located.

sustainability The capacity to endure.

tetrahedron A four-sided shape that has four vertices; the shape carbon takes when it forms single bonds to four other atoms or groups of atoms.

theory A detailed explanation of data collected from scientific experiments.

thermoplastic polymer A polymer that can be melted and reshaped to make a new product.

thermosetting polymer Extremely tough polymers that do not change shape when heated.

trans fat A lipid that contains a double bond with a trans orientation.

transcription The process that creates a strand of RNA from a DNA template.

transgenic organism An organism that is modified through genetic engineering to contain hereditary material from another organism.

translation The process whereby the code within a strand of RNA bases dictates a sequence of amino acids.

triglyceride A form of stored fat that is composed of a glycerol molecule combined with three fatty acid molecules.

triplet In protein synthesis, the group of three RNA bases that is equated with a specific amino acid according to the genetic code.

uniqueness principle The principle that states that Period 2 elements are uniquely small and have special properties resulting from their small size.

valence electron An electron in the outermost energy level of an atom.

variable A condition that may change.

visible light The region of the electromagnetic spectrum that can be detected by human eyes.

VOC A volatile organic compound; an organic compound that easily enters the gas phase.

water cycle The way water is distributed among different locations and reservoirs as it circulates through the environment.

water footprint A tally of the amount of water used by a person, business, production process, country, or the world.

water vapor Water in the gas phase.

watt The number of joules (J) of energy transferred per second.

wavelength The distance between the peak of one wave and the peak of the next wave.

weak acid An acid that dissociates partially in water.

whole grain A grain that has not had the bran, the germ, and the endosperm removed by processing.

Index

Entries in bold are key terms in the text. They are defined in the glossary beginning on page 494.